T0176015

THE PRINCIPLES OF STATISTICAL MECHANICS

THE PRINCIPLES OF STATISTICAL MECHANICS

by
RICHARD C. TOLMAN
Late Professor of Physical Chemistry and
Mathematical Physics at the California
Institute of Technology

Dover Publications
Garden City, New York

This Dover edition, first published in 1979, is an unabridged and unaltered republication of the work originally published in 1938 by Oxford University Press. The Dover edition is published by special arrangement with Oxford University Press, Inc., 200 Madison Ave., New York, N.Y. 10016.

International Standard Book Number

ISBN-13: 978-0-486-63896-6
ISBN-10: 0-486-63896-0

Manufactured in the United States of America
63896024 2023
www.doverpublications.com

TO

MY FRIEND AND COLLEAGUE

J. ROBERT OPPENHEIMER

WITH WHOM ALL PARTS OF

THIS BOOK HAVE BEEN

DISCUSSED

PREFACE

In the preparation of this book it has been my hope and purpose to provide an exposition of the underlying principles of statistical mechanics which would give a reasonably clear and complete picture of the additional hypotheses that must be introduced and of the further methods that must be developed, both in the case of the classical and of the quantum mechanics, when it is desired to give appropriate treatment to the behaviour of mechanical systems in states that are less precisely specified than would be theoretically possible. It is even hoped that the present book may in some measure meet, from a modern point of view, the same needs for a fundamental exposition as were originally so successfully filled, for the classical mechanics, by the treatment given by Gibbs in his *Elementary Principles in Statistical Mechanics*. Throughout the book, although the work of earlier investigators will not be neglected, the deeper point of view and the more powerful methods of Gibbs will be taken as ultimately providing the most satisfactory foundation for the development of a modern statistical mechanics.

The present work is in no sense a revision or second edition of my earlier book, *Statistical Mechanics with Applications to Physics and Chemistry*, a book which was written at a time when the older quantum theory was being replaced by modern quantum mechanics, and a book which is now both out of date and out of print. Applications to actual physical-chemical systems are included in the present book only in so far as is desirable for illustrating the nature of the fundamental principles of statistical mechanics and the methods to be employed in applying them. The needs for a complete account of the manifold applications of statistical mechanics to physics and chemistry should be met for a long time by the exhaustive treatment given by Fowler in the second edition of his *Statistical Mechanics*.

In writing the book I have received helpful criticisms and suggestions from many physicists and chemists. In particular I must mention my gratitude in this connexion to my colleagues, Professors W. V. Houston, D. M. Yost, L. C. Pauling, R. M. Badger, R. G. Dickinson, J. H. Sturdivant, and E. T. Bell, of the California Institute of Technology, as well as to Professors R. H. Fowler of Cambridge, W. Pauli of Zürich, H. P. Robertson of Princeton, E. Bright Wilson of Harvard, E. H. Kennard of Cornell, and J. T. Hildebrand of California, with all of whom I have discussed one or another or many questions.

Above all, however, I must express my indebtedness to my friend and colleague, Professor J. R. Oppenheimer, of the University of California and of the California Institute, with whom I have completely discussed both specific details and general points of view as the manuscript was being prepared, and from whom I have received both minor suggestions and major enlightenment. This indebtedness to Professor Oppenheimer is so great that it can only be partially repaid by the dedication of the book which he has been willing to accept.

PASADENA, CALIFORNIA

R. C. T.

CONTENTS

Part One. THE CLASSICAL STATISTICAL MECHANICS

I. INTRODUCTION

II. THE ELEMENTS OF CLASSICAL MECHANICS

III. STATISTICAL ENSEMBLES IN THE CLASSICAL MECHANICS

IV. THE MAXWELL-BOLTZMANN DISTRIBUTION LAW

PART TWO. THE QUANTUM STATISTICAL MECHANICS

VII. THE ELEMENTS OF QUANTUM MECHANICS

A. HISTORICAL REMARKS

B. THE POSTULATES

C. THEOREMS ILLUSTRATING THE NATURE OF QUANTUM MECHANICS

D. Further Development of Quantum Mechanical Methods. Transformation Theory

VII. SOME SIMPLE APPLICATIONS OF QUANTUM MECHANICS

IX. STATISTICAL ENSEMBLES IN THE QUANTUM MECHANICS

X. THE MAXWELL-BOLTZMANN, EINSTEIN-BOSE, AND FERMI-DIRAC DISTRIBUTIONS

C. Specific Examples of Equilibrium

Part Three. STATISTICAL MECHANICS AND THERMODYNAMICS

XIII. STATISTICAL EXPLANATION OF THE PRINCIPLES OF THERMODYNAMICS

XIV. FURTHER APPLICATIONS TO THERMODYNAMICS

I
INTRODUCTION

1. The nature of statistical mechanics

It is the purpose of this book to expound the fundamental principles of statistical mechanics. Only incidental discussion of the applications of statistical methods to the problems of usual physical-chemical interest will be given. It is hoped, however, that the exposition of underlying assumptions and general methods will be sufficient to give a real insight into the physical nature and mathematical apparatus of statistical mechanics.

The science of statistical mechanics has the special function of providing reasonable methods for treating the behaviour of mechanical systems under circumstances such that our knowledge of the condition of the system is less than the maximal knowledge which would be theoretically possible. The principles of ordinary mechanics may be regarded as allowing us to make precise predictions as to the future state of a mechanical system from a precise knowledge of its initial state.† On the other hand, the principles of statistical mechanics are to be regarded as permitting us to make reasonable predictions as to the future condition of a system, which may be expected to hold on the average, starting from an incomplete knowledge of its initial state.

Since our actual contacts with the physical world are such that we never do have the maximal knowledge of systems regarded as theoretically allowable, the idea of the precise state of a system is in any case an abstract limiting concept. Hence the methods of ordinary mechanics really apply to somewhat highly idealized situations, and the methods of statistical mechanics provide a significant supplement in the direction of decreased abstraction and closer correspondence between theoretical methods and actual experience. Even in the case of simple systems of only a few degrees of freedom, where our lack of maximal knowledge is not due to difficulties arising from the complexity of the system, the methods of statistical mechanics may be applied to a system whose initial state is not completely specified.‡

† This is true both for the classical mechanics and for the quantum mechanics. In the latter case, however, a knowledge of the precise state of the system is not sufficient to determine precise values for all kinds of quantities. See Chapter VII.

‡ In the case of the classical mechanics, the possibility of applying statistical methods to a system of a small number of degrees of freedom was emphasized by Gibbs himself as well as by E. B. Wilson, *Annals of Math.* **10**, 129, 149 (1909). In the case of the quantum mechanics, the description of the change in time of atomic systems in terms

In its historical development, however, the science of statistical mechanics was specially devised for the treatment of complicated systems, composed of such an enormous number of individual molecules that it would be too difficult to try to calculate the precise behaviour of the system as a function of time by the methods of ordinary mechanics. For example, in the case of a gas, consisting, say, of a large number of simple classical particles, even if we were given at some initial time the positions and velocities of all the particles so that we could foresee the collisions that were about to take place, it is evident that we should be quickly lost in the complexities of our computations if we tried to follow the results of such collisions through any extended length of time. Nevertheless, a system such as a gas composed of many molecules is actually found to exhibit perfectly definite regularities in its behaviour, which we feel must be ultimately traceable to the laws of mechanics even though the detailed application of these laws defies our powers. For the treatment of such regularities in the behaviour of complicated systems of many degrees of freedom, the methods of statistical mechanics are adequate and especially appropriate.

The general nature of the statistical mechanical procedure for the treatment of complicated systems consists in abandoning the attempt to follow the precise changes in state that would take place in a particular system, and in studying instead the behaviour of a collection or *ensemble of systems* of similar structure to the system of actual interest, distributed over a range of different precise states. From a knowledge of the *average behaviour* of the systems in a *representative ensemble*, appropriately chosen so as to correspond to the partial knowledge that we do have as to the initial state of the system of interest, we can then make predictions as to what may be expected on the average for the particular system which concerns us.

The averages that we take for the properties of the systems in a representative ensemble may be mean values, or most probable values, or other average values as seems of interest.† The different kinds of average tend to coincide in the case of systems of large numbers of degrees of freedom. Furthermore, as we go to large numbers of degrees

of the language of transition probabilities often actually involves statistical mechanical methods even in the case of a system having only a few degrees of freedom, as will be seen in § 99.

† The term *average* is used throughout the book, in the usual sense of statisticians, as signifying a single number which is chosen as giving a good representation of a collection of numbers. The averages that will usually interest us will be the arithmetical mean and the mode or most probable. The mean square and the root mean square, however, will also occasionally appear.

of freedom, the average behaviour in an appropriately chosen representative ensemble is actually found to give a completely satisfactory account of the regularities of behaviour that are empirically known for individual systems.

It may seem surprising that the methods of statistical mechanics should be so simple and effective as they actually prove to be. The substitution of a whole ensemble of systems of similar structure, in place of a single system that was too complicated to treat by ordinary mechanical methods, would not of itself lead to increased simplicity of treatment. The simplification is actually introduced by the circumstance that the average behaviour of a suitably chosen collection of systems may be much easier to treat than the precise behaviour of a single complicated system.

Similar situations are typical of biological phenomena. A given man of interest, 40 years old, is altogether too complicated a system, surrounded by too complicated an environment, to permit a precise computation of the date of his death. Nevertheless, we have the possibility of preparing very reliable tables which will give the average number of men, of any selected age, in any geographical region, who can be expected to die within a given length of time. Both in the physical and biological sciences an appeal to statistical methods may be taken when the complexity of the situation of interest demands it.

The introduction of statistical methods in order to treat the behaviour of complicated systems may be carried out in two different ways which may be described by the terms *deductive* and *inductive*. A deductive introduction of statistical methods may be obtained by making a preliminary study of the nature of the system of interest and then introducing what seem to be reasonable postulates which will permit the calculation of average behaviour. An inductive introduction of statistical methods can be obtained by studying the actual behaviour in many examples of the situation of interest, and then codifying the average results obtained in order to make them available for use in further predictions.

The two modes of procedure may be illustrated in connexion with the results to be expected on flipping a coin. For a deductive approach we could make a preliminary study of the nature of the coin, thus assuring ourselves, for example, that the coin was not loaded, and then introduce the postulate of equal *a priori* probabilities for coming up heads or tails; this would then permit us to calculate the average number of heads and average fluctuations therefrom that might be expected in a thousand throws. For an inductive approach we could make empirical determinations of the average results obtained in groups of trials, and then use these for predictive purposes. Both methods can, of course, be applied to the same problem. As is familiar in other branches of scientific theory, the validity of the postulates introduced in a deductive exposition can be subjected to some measure of observational check.

The actual development of statistical mechanics has taken place along deductive lines. On the basis of the laws of ordinary mechanics, reasonable postulates are introduced which are sufficient to permit computations of the average results which may be expected on actual experimentation. For the development of statistical mechanics in its final quantum form it has been found necessary to introduce a pair of assumptions, which may be designated together as the hypothesis of *equal* a priori *probabilities and of random* a priori *phases* for the quantum states of a mechanical system. As is so often the case in theoretical developments, this hypothesis is not itself suitable for direct empirical test. It has, however, the indirect check that is provided by the extensive and almost astounding agreement, which is actually found between the deductive results obtained with its help and the observational results obtained on experiment.

2. The classical statistical mechanics

In the first part of this book, Chapters II to VI, we shall give a brief exposition of the classical statistical mechanics. This will be of advantage because of the desirability of first grasping the concepts and methods of statistical mechanics in their historically original and simple form, where the new kinds of principle involved can be most clearly seen. It will also be of advantage since the results of the classical development are still valid—in accordance with the correspondence principle—under limiting conditions where quantum deviations from classical findings can be neglected. There are, moreover, many problems where such limiting conditions prevail, and the classical point of view and treatment prove most satisfactory.

In Chapter II, 'The Elements of Classical Mechanics', we give a brief treatment of the classical mechanics itself, to the extent necessary to provide a basis for the development of the classical statistical mechanics, and also a basis for understanding the later change from the classical to the quantum mechanical point of view. We choose Hamilton's principle as the primary postulate for the development, and then obtain the equations of motion in the Lagrangian and canonical Hamiltonian forms. After treating the integrals of energy and momentum, we give a brief discussion of canonical transformations and the general theory of integration.

In Chapter III, 'Statistical Ensembles in the Classical Mechanics', we turn our attention to statistical methods by giving general consideration to the properties and behaviour of collections or ensembles of systems, each having a structure similar to that of some actual system of interest, and each independently obeying the laws of classical mechanics, but distributed over a range of different states. Early in the chapter we derive the *fundamental theorem of Liouville* which controls the temporal behaviour of ensembles. This permits us to discuss the conditions for statistical equilibrium, and to show the reason for introducing as the basic assumption of classical statistics the *hypothesis of equal* a priori *probabilities* for the different *classical states* defined by equal volumes in the phase space corresponding to the system of interest. We take this principle as a reasonable postulate to introduce, but one to be ultimately justified, however, only by the correspondence between deduced and observational results. We conclude the chapter, nevertheless, with a discussion of the historical attempt to obtain a more direct justification of statistical methods with the help of the so-called ergodic hypothesis.

In Chapter IV, 'The Maxwell-Boltzmann Distribution Law', we apply the methods of classical statistics to systems in equilibrium. To do this we regard a system in equilibrium with a specified energy as appropriately represented by an ensemble of systems, each having substantially that energy, distributed in accordance with the principle of equal *a priori* probabilities, the whole in statistical equilibrium. As a convenient average, we then consider the most probable condition in such an ensemble, and are led in the case of systems composed of weakly interacting molecules to the Maxwell-Boltzmann distribution for the expected numbers of molecules in their different individual states. This is one of the most important consequences of the classical statistics, and the chapter will include some discussion of its significance and use.

In Chapter V, 'Collisions as a Mechanism of Change with Time', we prepare the way for our later deduction of Boltzmann's H-theorem as to the temporal behaviour of complicated systems, by studying the nature of molecular collisions as a primary agent leading actual physical chemical systems to change with time. We begin with a discussion of the general *principle of dynamical reversibility* in time, not only because of its special bearing on the possibility of reverse collisions, but also because of its general bearing on the problem of the actual irreversibility found in the behaviour of physical systems when they are looked at from a gross, macroscopic point of view. We next classify the different kinds of *molecular states*, *constellations*, and *collisions* that can occur, and discuss the arrangement of corresponding collisions in a cyclical series. The consequences of Liouville's theorem when applied to a single collision are now investigated, and—making use of statistical considerations—the relations between the probability coefficients for any specified collision and the collisions reverse and corresponding thereto are then obtained.

In Chapter VI, 'Boltzmann's H-theorem', we are ready to complete the discussion that is needed to give a good insight into the classical statistics. With the help of the relations between the probability coefficients for collisions, obtained in the preceding chapter, we derive Boltzmann's famous H-theorem for the probable decrease in H with time, and the corresponding approach of systems towards a condition of equilibrium when they are left undisturbed. In the light of this H-theorem, discussion is then given to the problem of the actual irreversibility in the behaviour of physical-chemical systems when looked at from a gross macroscopic point of view, and to the problem of the continued fluctuations that must be expected around the ultimate condition of equilibrium. We then use the H-theorem to give a new discussion of the Maxwell-Boltzmann distribution as the equilibrium condition which will be reached in course of time, and give a discussion of the useful *principle of detailed balance* which holds under equilibrium conditions. We conclude the chapter with a discussion of the more general formulation and more powerful treatment of the problem of approach towards equilibrium which was discovered by Gibbs.

3. The quantum statistical mechanics

In the second part of the book, Chapters VII to XII, we undertake the exposition of quantum statistical mechanics. This part of the book is much longer than the first part dealing with the classical statistics. In the first place, greater detail is needed since the ideas of quantum mechanics, although intrinsically simple, must be expressed in less familiar and more intricate mathematical language than classical ones. In the second place, the introduction of quantum mechanics leads, as is well known, in addition to the Maxwell-Boltzmann type of statistical result, to the further types of possible result associated with the names Einstein-Bose and Fermi-Dirac. Furthermore, the treatment of quantum statistics is appropriately made more complete than that of classical statistics, since quantum mechanics is the fundamental discipline which includes all the results of classical mechanics as limiting cases, valid when specific quantum effects can be neglected.

In Chapter VII, 'The Elements of Quantum Mechanics', we give a fairly complete treatment of the fundamental principles of *non-relativistic* quantum mechanics. After discussing the necessity for modifying classical ideas, a rather full statement of the postulatory basis for quantum mechanics is presented, with the Schroedinger equation for the change in probability amplitudes with time playing a fundamental role. This is followed by the derivation of theorems which illustrate the general nature of quantum mechanical considerations. We then conclude the chapter by discussing the further development of quantum mechanical methods, including the ideas of the transformation theory.

In Chapter VIII, 'Some Simple Applications of Quantum Mechanics', we continue the consideration of quantum mechanics itself in order to obtain specific results which will either be advantageous in throwing light on the nature of quantum theory or be necessary for our later statistical considerations. The applications include treatments of simple particles in various fields of force, of interacting particles, of particles with spin, and of systems containing particles having intrinsically similar properties. This last application is of special importance when we use statistical mechanics to treat the behaviour of actual physical chemical systems containing many similar molecules.

In Chapter IX, 'Statistical Ensembles in the Quantum Mechanics', we can now turn our attention to statistical matters by giving general consideration to the properties and behaviour of ensembles of systems, each having a structure similar to that of some actual system of interest, and each independently obeying the laws of quantum mechanics, but distributed over a range of different quantum mechanical states. After discussing the apparatus involving the *density matrix* which is needed for the treatment of such quantum mechanical ensembles, the *quantum analogue of Liouville's theorem* is derived. This permits us to discuss the conditions for statistical equilibrium, and to show the reasons for introducing as the basic assumptions of quantum statistics the *hypothesis of equal* a priori *probabilities and of random* a priori *phases* for the quantum mechanical states of a system. At the correspondence principle limit these assumptions taken together agree with the classical hypothesis of equal *a priori* probabilities for classically defined states.

We shall regard these assumptions as reasonable postulates to introduce, but to be ultimately justified, however, by the agreement between deduced and empirical results. We conclude the chapter with a general discussion of the validity of statistical quantum mechanics together with remarks on the quantum mechanical status of the ergodic hypothesis.

In Chapter X, 'The Maxwell-Boltzmann, Einstein-Bose, and Fermi-Dirac Distributions', we apply the methods of quantum statistics to systems in equilibrium. To do this we regard a system in equilibrium with an approximately known value of its energy, as appropriately represented by an ensemble of systems having that approximate energy, distributed in accordance with the hypothesis of equal *a priori* probabilities and random *a priori* phases, and itself in statistical equilibrium. We then consider the most probable condition in such an ensemble, and are led in the case of systems composed of many similar weakly interacting elements either to the Maxwell-Boltzmann, Einstein-Bose, or Fermi-Dirac distributions, according to the type of symmetry restrictions—none, symmetric, or antisymmetric—to be satisfied by the solutions of Schroedinger's equation in the cases of the different kinds of particles or other elements that occur in nature. These results are of great practical importance, and we give some indication of their application to gases, crystals, conduction electrons, and radiation.

In Chapter XI, 'The Change in Quantum Mechanical Systems with Time', we now prepare the way for our later derivation of the quantum mechanical form of Boltzmann's H-theorem by studying the nature of the changes with time that take place in quantum mechanical systems. On account of the greater power of quantum as compared with classical methods, these considerations are fortunately much more general than our analogous classical considerations, which were mainly concerned with molecular collision as the mechanism of change. It is first shown that the *principle of dynamical reversibility* holds also in the quantum mechanics in appropriate form, indicating that quantum theory supplies no new kind of element for understanding the actual irreversibility in the macroscopic behaviour of physical systems. The next section of the chapter is then devoted to the mathematical problem of the *integration of Schroedinger's equation* for the change in state of a system with time, by the method of *variation of constants* and by the more general method of treating change with time as a *unitary transformation*. And a further section is devoted to the mathematical problem of treating the effect of *changes with time in the external para-*

meters of a system. We then turn to more physical matters by considering the character of the actual observations and specifications of state that could be appropriately made for a quantum mechanical system which is changing with time. We next consider the theory of transitions that take place in a system with a probability proportional to the time, and treat the special case of transitions that take place in a gas as a consequence of collisions. We conclude the chapter with a brief general treatment of changes with time in ensembles of systems.

In Chapter XII, 'The Quantum Mechanical H-theorem', we complete our study of the essential features of the quantum statistical mechanics. The chapter is divided into three parts. In Part A we derive the quantum mechanical H-theorem first in a form similar to that of Boltzmann, where H is regarded as a quantity directly characterizing the system of interest, and then in a form similar to that of Gibbs, where $\bar{\bar{H}}$ is introduced as a quantity directly characterizing the representative ensemble for the system of interest. In Part B we then discuss the relation of the decrease in $\bar{\bar{H}}$ with time to the ultimate *equilibrium condition* for a system of interest, and show the reasons for taking a *microcanonical ensemble* as representing equilibrium for a perfectly isolated system, and for taking a *canonical ensemble* as representing equilibrium for a system in contact with a heat bath or in what we shall call essential rather than perfect isolation. In Part C we make use of the canonical ensemble in studying some specific examples of equilibrium and conclude with remarks on the quantum form of the principle of detailed balance.

4. Statistical mechanics and thermodynamics

In the third and final part of this book, Chapters XIII and XIV, we discuss the application of statistical mechanics to the problem of obtaining a mechanical explanation for the phenomena of thermodynamic behaviour. We cannot refrain from doing this, even though our primary purpose has been to expound rather than to apply the principles of statistical mechanics. The explanation of the complete science of thermodynamics in terms of the more abstract science of statistical mechanics is one of the greatest achievements of physics. In addition, the more fundamental character of statistical mechanical considerations makes it possible to supplement the ordinary principles of thermodynamics to an important extent.

In Chapter XIII, 'Statistical Explanation of the Principles of Thermodynamics', we present the desired mechanical interpretation and explanation of the first and second laws of thermodynamics. To do this the thermodynamic variables ordinarily used to describe a thermodynamic system of interest, such as volume, pressure, energy, entropy, and temperature, are correlated with mechanical quantities belonging to a representative ensemble that corresponds in an appropriate manner to the system in question; and the thermodynamic behaviour of the system is then correlated with the mean behaviour in this representative ensemble. The proper statistical correlates for essentially mechanical quantities such as volume, pressure, and energy are readily obtained. The correlates for the quantities entropy and temperature, which thermodynamics has introduced on its own level of abstraction, need more discussion, however, and the greater part of the chapter is devoted to showing that the correlates chosen do have the correct properties. The ultimate outcome is the derivation of two statistical mechanical relations which are the same in form and equivalent in implications to the thermodynamic relations which express the first and second laws of thermodynamics.

In Chapter XIV, 'Further Applications to Thermodynamics', we now apply the consequences of the preceding chapter to a number of illustrative cases. Without such considerations, the increased power of statistical mechanics in obtaining actual expressions for the quantities used by thermodynamics would not be appreciated. As the statistical apparatus of calculation we use the sum-over-states for the thermodynamic system of interest, and with its help we treat the properties of gases, crystals, and solutions. The problems of vapour pressure and chemical equilibria are also treated, and the present status of the so-called third law of thermodynamics is discussed. This is followed by a discussion of the use of *grand canonical ensembles* as representing equilibrium for 'open' thermodynamic systems. The chapter ends with a discussion of fluctuations in statistical ensembles, the limitations thereby imposed on thermodynamic concepts, and the observational detection of fluctuations in systems at thermodynamic equilibrium.

5. Points of view and methods of presentation

Before undertaking the programme outlined above, a few remarks may be made concerning the points of view that will be adopted as fundamental, and methods of presentation that will be selected as convenient.

It is first to be emphasized, as already indicated above, that the treatments given in this book, to the classical and to the quantum mechanics, and to the two corresponding forms of statistics, are to be regarded as deductive expositions based on postulates. These postulates will be chosen in ways which are familiar or plausible, but their ultimate validity will be regarded as resting on the correspondence between deduced results and empirical findings. It is hoped that the postulatory expositions of classical and of quantum mechanics that are given will seem compact and useful treatments without reference to their further employment for statistical purposes.

In building up our deductive systems we shall not try to present a complete set of indefinables, definitions, and postulates, nor to achieve the rigour of demonstration that would be demanded for the perfect construction of a logical universe of discourse. We shall, however, try to give expositions that will provide a real insight into the physical nature of the situations to be treated. For the classical mechanics we take Hamilton's principle, with the associated ideas as to generalized coordinates and velocities and the Lagrangian function, as a starting-point which provides an appropriate generalization of Newton's laws of motion. For the quantum mechanics we take Schroedinger's equation including the time, with the associated ideas as to probability amplitudes and the Hamiltonian and other operators, as fundamental. For the extensions to statistical mechanics we take the assumptions of equal *a priori* probabilities for classical states, or of equal *a priori* probabilities and random *a priori* phases for quantum mechanical states, as the necessary additional hypotheses.

In developing the quantum mechanics itself we shall only strive for a non-relativistic theory. This will be sufficient for illustrating all the essential features of the quantum statistical mechanics and for treating all the commoner problems to which statistical methods are applied. The development will be carried out making use for the most part of coordinate representations for the quantum mechanical states of a system; thus it will start from Schroedinger's equation expressed in coordinate or q-language. The ideas of the transformation theory will be introduced later. The quantum mechanical notation used will be similar to that of Pauli in his *Handbuch* article, and our own development will lean heavily on his exposition.†

As the fundamental idea involved in statistical treatments, we take the procedure of correlating any actual mechanical system of interest, in an incompletely specified state, with an appropriately chosen representative ensemble of such systems, distributed over a range of states,

† Pauli, *Handbuch der Physik,* second edition, volume 24/1, Berlin, 1933.

followed by the procedure of then using average values for the members of this ensemble as furnishing good estimates as to what we can expect for the actual system. In this connexion different possibilities are open as to what representative ensemble we select as appropriate, and as to what averages we take as convenient.

In the treatment of equilibrium we shall first adopt the familiar method of representing a system at equilibrium by a microcanonical ensemble—composed of members all of which have substantially the same energy as the system of interest—and of then taking the most probable values of quantities in the ensemble as convenient averages. We shall use this method in the derivation of the classical Maxwell-Boltzmann distribution law which we give in Chapter IV, and in the derivations of the quantum Maxwell-Boltzmann, Einstein-Bose, and Fermi-Dirac distributions which we give in Chapter X. Such a method of procedure is theoretically quite sound provided the system of interest can be regarded as having a practically definite energy at equilibrium. This is, of course, true for perfectly isolated systems which have been allowed to come to equilibrium, and is nearly enough true for systems which have been allowed to come to thermal equilibrium with their surroundings, provided the number of degrees of freedom for the system is sufficiently large.

At a later stage of the development we shall then adopt the somewhat less common method of representing a system at equilibrium by a canonical ensemble—composed of members distributed to correspond to the temperature of the system of interest—and of then taking the mean values of quantities in the ensemble as convenient averages. We shall use this method both for a new and more satisfactory treatment of the three molecular distribution laws mentioned above, and shall make it fundamental in our treatment of thermodynamic equilibrium. Such a method of procedure, as first appreciated by Gibbs, is theoretically a sound one when the system of interest can be regarded as having a definitely specified temperature.

It will also be of interest in this connexion to note a further alternative method of treatment employed by Fowler in his extensive and important treatise on statistical mechanics.† There, a system in a condition of equilibrium is represented by a microcanonical ensemble of systems all having substantially the same energy, but mean rather than most probable values are taken as the averages to be considered. These means are calculated by an approximate method involving saddle-point

† Fowler, *Statistical Mechanics*, second edition, Cambridge, 1936.

integration. For systems of many degrees of freedom, all three of the methods mentioned lead to results which are in substantial agreement. The finding that somewhat altered treatments can lead to practically the same consequences has long been familiar in statistical mechanics.†

In carrying out our development of the quantum form of statistical mechanics we shall make special efforts to distinguish clearly between those uncertainties in an observational situation, which are inherent in the quantum mechanics itself, and those additional uncertainties, due to lack of maximal knowledge, which are the primary concern of statistical mechanics. In the quantum mechanics, even if we had a maximal observation, so that the system under consideration could be definitely assigned to a precise state, say k corresponding to an eigenvalue F_k of some observable quantity F, this would only be sufficient to permit assertions as to the probability W_m for finding the system in a state m corresponding to an eigenvalue G_m of some other kind of observable quantity G, except in the special case when the operators for the two quantities F and G commute. Hence, even the quantum mechanics itself has a statistical character that is foreign to the classical mechanics. In developing what we specifically designate as statistical mechanics, however, we are interested both classically and quantum mechanically in cases where our observations have not been sufficient to determine the precise state of a system, so that we have to consider the statistical properties of some suitable representative ensemble. We shall employ an appropriate notation to assist in keeping this distinction clear. Thus the symbol W_m would denote the probability that a given precisely specified quantum mechanical system will be found in a state m, and the symbol P_m would denote the probability that a system picked at random from a given ensemble will be found in state m. Furthermore, we shall use a single bar $^-$ to denote a mean which is obtained by averaging over different quantum mechanical possibilities, and a double bar $^=$ to denote a mean which includes the additional

† A word as to the relative advantages of the above three methods of treating systems in equilibrium may be of interest. The method of taking most probable values in a microcanonical ensemble is historically the earliest and often leads most expeditiously to the desired results. It runs into analytical difficulties when the most probable values are desired for integers which are too small to be regarded as corresponding to a continuous variable; for large numbers it gives averages which agree in ordinary cases with those obtained by the other two methods. The method of taking mean values in a canonical ensemble seems conceptually the most satisfactory and the computations are relatively simple and straightforward. The method of taking mean values in a microcanonical ensemble has been developed by Fowler into a very general and powerful scheme, but the computational apparatus is complicated. The means obtained agree precisely in ordinary cases with the means for a canonical ensemble.

process of averaging over the members of an ensemble of systems. Thus $\bar{G}$ would denote the mean or expectation value for the observable G in the case of a given precisely defined quantum system, and $\bar{\bar{G}}$ would denote the mean value for all the systems in a given ensemble.

We shall base our development of the relations between statistical mechanics and thermodynamics very definitely on the point of view of Gibbs,† that a thermodynamic system, in a condition specified by a limited number of macroscopic thermodynamic variables, can be naturally regarded as a mechanical system in an incompletely specified microscopic state, and hence correlated for purposes of treatment with a representative ensemble of similar systems appropriately distributed over a range of different precise states. Proceeding with this line of attack, we shall not only be able to obtain mechanical correlates for the variables—such as entropy and temperature—which thermodynamics has specially introduced on its own level of abstraction, but shall also be able to show that the laws of thermodynamics are satisfactorily correlated with the statistical mechanical laws for the behaviour of the appropriate representative ensembles. We shall not carry out this programme until after we have obtained the principles of statistical mechanics in their full quantum mechanical form. Nevertheless, we shall find that the classical methods of Gibbs need but slight modification to be specially suited for a quantum mechanical development of the relations between statistics and thermodynamics. One can only feel that the work of this great and modest scientist was guided by such fundamental considerations that he based his development on those essential features of the relations between mechanics and thermodynamics which have not been altered by the change from classical to quantum mechanics.

In addition to the transcription from classical to quantum language, there will be several less radical differences between the point of view and methods adopted by Gibbs and those employed in this book. To these we may devote a few words.

Impressed by the real failure of the classical statistical mechanics to give a valid account of the phenomena of radiant heat and of the specific heat of diatomic gases,‡ Gibbs was very cautious in making assertions as to the relation of the systems actually encountered in nature to the statistical ensembles which he discussed,|| and seemed to regard the properties of these ensembles as providing analogies rather than explanations for the principles of thermodynamics. However, as

† Gibbs, *Elementary Principles in Statistical Mechanics*, Yale University Press, 1902.

‡ Gibbs, loc. cit., p. 167.

|| Gibbs, loc. cit., p. x of the preface.

these apparent failures of statistical mechanics have since been solved by the quantum theory, we can now afford to adopt a more positive attitude. This we do by clearly introducing and distinguishing between the two concepts of system of interest and corresponding representative ensemble, and by then asserting that an appropriately chosen ensemble would really permit valid predictions as to the behaviour to be expected on the average for a system of interest started off in successive trials in the same partially specified condition under consideration. This procedure then involves a more explicit consideration of the hypothesis as to *a priori* probabilities and phases, used in constructing appropriate ensembles, that was needed in the work of Gibbs.

A difference of emphasis, as compared with Gibbs, will lie in a more complete discussion of the justification that can be given for the use of the canonical distribution as representing a system which has arrived in a condition of thermodynamic equilibrium (see §§ 111 and 112). We shall regard this justification as sufficiently sound so that it will no longer seem necessary to consider the so-called second and third analogues for thermodynamic quantities which were based by Gibbs on the use of the microcanonical instead of the canonical distribution as representing thermodynamic equilibrium.

In the application of statistical mechanics to thermodynamic problems, Gibbs† has employed the abstract concept of 'perfectly isolated' systems, in which there would be absolutely no interaction between a system and the walls of its container or its other immediate surroundings, and has used this as the appropriate idealization for the treatment of systems which would be regarded as isolated from the point of view of thermodynamics. In this connexion, however, we shall introduce (§ 111) a new concept of 'essentially isolated' systems, in which interaction between a system and its surroundings without change in mean energy would be allowed, and shall take this as the appropriate idealization to use in the treatment of systems regarded as isolated from the thermodynamic point of view. This will make it possible (§ 124 (*a*)) to give a better treatment to thermodynamic, reversible, adiabatic processes than was possible for Gibbs.

In conclusion it may be mentioned that in carrying out our treatment we shall try to be reasonably scrupulous in restating the significance of any symbol when it reappears in the text after a long interval. In addition, a list of symbols employed for different purposes will be found in Appendix I.

† Gibbs, loc. cit., p. 164.

II

THE ELEMENTS OF CLASSICAL MECHANICS

6. Introduction

It is the purpose of the present chapter to give a brief outline of the principles of the classical mechanics, which will be sufficient for the development of the classical statistical mechanics, and also sufficient for an understanding of the later change to the point of view of the quantum theory.

The fundamental ideas of the classical mechanics are, firstly, that the state of a mechanical system may be observationally determined at any time of interest; secondly, that the specification of this state combined with a knowledge of the nature of the system will then be sufficient to determine the values of all the mechanical properties of the system at that time; and, finally, that the specification of the state when further combined with the laws of mechanics will also be sufficient for a complete determination and prediction of the future behaviour of the system. For example, in the case of a system of interacting particles, the state at any time could be specified by the positions and velocities of the component particles; combined with a knowledge of the masses of the particles and the forces they exert this would then be sufficient to determine the values of mechanical quantities such as the kinetic and potential energy of the system, its components of momentum, etc.; and further combined with a knowledge of the laws of mechanics, say Newton's laws of motion, it would also be sufficient to determine the complete future behaviour of the system.

This general idea that the observation of instantaneous state should be sufficient for a complete determination of the condition and future behaviour of a system has found increasing application ever since the time of Newton. Combined originally with the laws of Newtonian mechanics, it provided treatments for the dynamics of systems of particles, of rigid bodies, and of continuous fluids and solids. Combined at a later time with the more precise laws of relativistic mechanics, it successfully provided a refined treatment for those same problems. And combined with laws as to the behaviour of electricity and of heat, it controlled the developments of electrodynamics and thermodynamics. By the end of the nineteenth century it so dominated the whole of scientific thought as to seem a necessary part of the axiomatic structure of any kind of scientific procedure.

From the point of view of modern quantum theory, however, we now know that this idea of the complete determination of a system by an observation of its state was based on too superficial view as to the effect of observation on the system itself; and with the development of a consistent and empirically verified quantum mechanics, we now realize that such an idea is not a necessary axiom of science. From our present viewpoint the classical mechanics then becomes a special case of the quantum mechanics, valid at the so-called correspondence principle limit where special quantum effects can be neglected. Nevertheless, this special case includes, of course, all the phenomena that could be successfully treated by classical methods, that is, all those phenomena involving sufficiently massive objects so that the unknown disturbing effects of observation are negligible. This makes the range of validity so wide that the classical mechanics may still be developed with profit as a separate discipline. To such a development we now turn.

7. State of a system. Generalized coordinates and velocities

In the classical mechanics we may regard the state of a system as specified by the momentary positions and velocities of its parts. For example, in the case of a system of particles, we could specify the state of the system at any moment of interest by the Cartesian coordinates x, y, z for the positions of the different particles, together with the corresponding components of velocity $\dot{x}$, $\dot{y}$, $\dot{z}$, which give the rates of change, $\dot{x} = dx/dt$, etc., taking place in these coordinates with respect to time.

In specifying the state of a mechanical system we need not regard ourselves as limited, however, to the use of Cartesian coordinates. Rather, we shall find it advantageous to introduce from the very start the idea of a set of *generalized coordinates*

$$q_1, q_2, q_3, \ldots, q_n \tag{7.1}$$

which can be used for specifying the instantaneous configuration of a system of interest, together with a corresponding set of *generalized velocities*

$$\dot{q}_1, \dot{q}_2, \dot{q}_3, \ldots, \dot{q}_n, \tag{7.2}$$

where the dots indicate rate of change with respect to time, which can be used for specifying the instantaneous motion of the system. For example, in the case of an oscillating pendulum we might find it convenient to take as the single necessary generalized coordinate the angle θ made by the pendulum with the vertical; in the case of a rigid body we might use the Cartesian coordinates for its centre of gravity together with the three angles necessary for describing the orientation of the body; and in the case of two interacting particles it might be convenient to choose the three Cartesian coordinates for the centre of gravity of the pair together with their distance apart and the two angles giving the orientation of the line joining them. The values of such coordinates together with their rates of change with time would then be sufficient for specifying the state of the system.

It will be possible to choose the generalized coordinates for a system in a great variety of ways; and transformations between the languages corresponding to different choices can be made with the help of the equations connecting the values of one set of coordinates with those of another. In the absence of specific statement to the contrary, it may be assumed that so-called *stationary coordinates* are being used for specifying the positions of the various parts of a system. These are coordinates which can be derived by direct transformations, not involving the time, from the coordinates corresponding to an unaccelerated set of Cartesian axes. More general coordinates can be treated by the methods of transformation to be discussed in § 14.

The number of coordinates introduced must, of course, always be sufficient for a complete specification of the position and configuration of the system. It will be advantageous, however, to choose coordinates in such a way as to take the minimum number of independent variables necessary for the description. In the absence of specific statement to the contrary it is to be assumed that this has been done.

Having chosen the minimum number of coordinates needed to specify the position and configuration, a system is called *holonomic* if each of these coordinates could be independently varied without violating any constraints inherent in the nature of the system. A system is called *non-holonomic*, on the other hand, when the structure of the system is such that the values of these coordinates are connected by relations which prevent their independent variation. These relations restricting the variations in the case of a non-holonomic system must be non-integrable, since otherwise they could be used to reduce the number of independent variables chosen, and we have already assumed that the minimum possible number has been taken. In the case of either kind of system, the number of actually *independent variations* that can be made in the coordinates is spoken of as the *number of degrees of freedom* of the system.

The systems ordinarily encountered in applications are holonomic in character, and the method of treatment applicable to them can be readily adapted to the non-holonomic case by introducing the restrictions on independent variations at the appropriate point in the development (see § 9). Hence we shall only give a treatment to holonomic systems.

Two remarks may be made in concluding this section on the specification of the state of a system, with the help of a set of generalized coordinates and corresponding generalized velocities.

In the first place, it may be emphasized that the procedure tacitly assumes the possibility of making a simultaneous observational determination of all the coordinates and velocities needed for the specification. When we come to the development of the quantum mechanics we shall find that this can be regarded as true only as a limiting case.

In the second place, it may be noted that the formulation assumes that a specification of coordinates and velocities alone is sufficient for the complete determination of instantaneous state, and hence also sufficient for the prediction of future behaviour when combined with a knowledge of the nature of the system and of the laws of mechanics. For example, it has been tacitly assumed that accelerations and higher derivatives do not have to be included in the specification. The development of classical methods has shown this to be actually realizable although it may be necessary—as in treating the interactions of electrically charged particles—to introduce the idea of a *field* with what amounts to three coordinates and velocities at each point of space, thus achieving the desired formulation only by making the number of degrees of freedom of the system infinite.

8. Hamilton's principle and the Lagrangian function

Having discussed what is meant by the instantaneous state of a system, we must now consider a formulation of the laws of mechanics which will determine the changes in state which take place with time. As a very general and satisfactory postulatory statement of these laws we shall choose *Hamilton's principle*, which can be expressed by the variational equation

$$\delta \int_{t_1}^{t_2} L\, dt = 0, \tag{8.1}$$

where the so-called Lagrangian function L is a quantity which depends on the coordinates and velocities, and in some cases also explicitly on the time in a manner to be discussed later.

This variational equation is to be understood in the following sense.

We consider the actual path by which the system passes from an initial configuration at time t_1 to a later configuration at time t_2, and compare the value of the integral $\int L\,dt$ over this actual path with the values which it would have on neighbouring paths by which the system could be carried *in the same time-interval from the same initial to the same later configuration*, without violating the constraints of the system. Hamilton's principle then states that the value of the integral $\int L\,dt$ for the actual path has in any case a *stationary value* as compared with such neighbouring paths.

The above compact formulation of the laws of mechanics has, of course, no actual content until we specify the form of the *Lagrangian function L*. Indeed from a certain point of view we can regard the empirical determination of the form of L, for the various kinds of systems —mechanical or otherwise—amenable to Hamilton's principle, as one of the major tasks of the classical physics. It should be remarked that Newtonian systems can be devised, for which the equations of motion cannot be derived from a variation principle; but for our purposes Hamilton's principle provides an adequately general basis.

In considering the form of the Lagrangian function we must first make a distinction between *conservative* and *non-conservative systems*. In the case of conservative systems the Lagrangian function depends explicitly only on the coordinates and velocities which specify the state of the system. Thus for a holonomic system of f degrees of freedom we have

$$L = L(q_1 \dots q_f, \dot{q}_1 \dots \dot{q}_f). \tag{8.2}$$

On the other hand, in the case of non-conservative systems the Lagrangian function depends also explicitly on the time, so that we have

$$L = L(q_1 \dots q_f, \dot{q}_1 \dots \dot{q}_f, t). \tag{8.3}$$

Resorting for the purposes of a quick description to language that will already be familiar, conservative systems will prove to be those in which the sum of the kinetic and potential energies remains constant in time. They are thus isolated systems, unacted on by external forces, and having no internal forces of dissipative—for example, frictional—character. Non-conservative systems, on the other hand, will be those in which the sum of the kinetic and potential energies is not constant because of the action of either external or dissipative forces.

It is evident in this connexion that any system may be treated as isolated, and hence unacted on by external forces, provided we include a large enough region inside the boundary that we take for the system. Furthermore, in view of the conservation of energy for fundamental particles, it is evident that dissipative forces may be eliminated from consideration if we introduce a sufficient number of coordinates, corresponding to the fundamental particles of which the system is constructed, so that thermal phenomena will themselves receive a mechanical treatment, which is one of the goals of statistical mechanics. Hence it becomes possible in principle to treat any situation by the mechanical principles applicable to conservative systems. Nevertheless, since it is desirable in the situations of interest for statistical mechanics to preserve the distinction between system and external surroundings which can act thereon, it is most useful to develop the principles of mechanics in their general form where the Lagrangian function L contains the time t explicitly and then consider the simplifications that result for conservative systems.

Having made the general distinction between the two possibilities, that the Lagrangian function L may or may not depend explicitly on the time, we must now give consideration to the specific form of L, since the actual content of Hamilton's principle, as already remarked, will be dependent on the form assigned to L as a function of the coordinates q_i, velocities $\dot{q}_i$, and time t.

For *simple mechanical systems*, where the forces acting on the different parts are derivable from a potential and the velocities are small compared with that of light, the Lagrangian function L can be taken equal to the difference between the kinetic energy of the system T and its potential energy V,

$$L = T-V. \tag{8.4}$$

The kinetic energy T in this expression is to be regarded as determined from the masses m and velocities v of the different parts of the system by a sum or integral over the kinetic energies $mv^2/2$ of these different parts. And the potential energy V is to be regarded as the work necessary to bring the system of interest from some standard internal configuration and location with respect to external bodies to its actual configuration and location at the instant being considered. In making any application, these quantities are, of course, to be expressed as functions of the particular set of generalized coordinates and corresponding velocities that have been chosen for use. Assuming the use of stationary coordinates, as defined in § 7, it will be seen that the kinetic energy T will necessarily be a homogeneous quadratic function of the generalized velocities $\dot{q}_i$, with coefficients which can in general depend on the coordinates q_i. It will also be seen that the potential energy V will be a function, of the coordinates q_i alone in the case of conservative systems, and of the coordinates q_i and time t in the case of non-conservative systems.

For more complicated mechanical systems, or even non-mechanical ones, suitable choices of the Lagrangian may be possible which will make Hamilton's principle still applicable—sometimes in generalized form. In the case of a system of particles having velocities v sufficiently high compared with that of light c so that the special theory of relativity must be applied, the quantity T appearing in (8.4) has only to be replaced by $\sum m_0 c^2\{1-\sqrt{(1-v^2/c^2)}\}$, where m_0 denotes the rest mass of a particle and the summation is over all the particles of the system. In the case of systems having additional forces beyond those derivable from a potential, the expression for the Lagrangian function given by (8.4) may sometimes be modified by making an appropriate addition of a further function. (See, for example, § 56 (*b*) for the added term corresponding to the magnetic force acting on a moving charged particle.) In the case of combined mechanical and electromagnetic systems, the Lagrangian function may be suitably set up by including terms which correspond to the conditions at all points of the electromagnetic field. In the case of gravitational effects so great that the general theory of relativity has to be applied, a variational equation can be employed which can be regarded as a generalization of Hamilton's principle. Thus, with a correct choice for the Lagrangian function L, Hamilton's principle can be regarded as having a wide range of applicability. We shall ourselves be mainly interested in the simple case of ordinary mechanical systems as given by (8.4).

In concluding this section on Hamilton's principle, it may be remarked that an important advantage of taking this particular principle as the starting-point for mechanics lies in the consideration that the form of expression, as given by (8.1), is independent of the coordinates which we may desire to use. The principle is valid without reference to the particular choice of generalized coordinates and velocities in terms of which the Lagrangian function is expressed.

It may be further remarked that the feasibility of applying Hamilton's principle to such a wide variety of different kinds of systems is certainly dependent to a considerable extent on the flexibility open to us in picking a form for the Lagrangian function L which will actually lead to a correspondence with empirical findings. Also connected with this is the possibility already mentioned in § 7 of introducing the notion of a field in order to make an infinite number of generalized coordinates and velocities available for inclusion in the Lagrangian function.

9. The equations of motion in the Lagrangian form

Having taken Hamilton's principle as a convenient, compact formulation of the laws of mechanics, we must now obtain its consequences in forms suitable for computational applications. For this purpose we may first derive the equations of motion of a system in their so-called Lagrangian form.

Considering a holonomic system of f degrees of freedom, we can write Hamilton's principle in the form

$$\delta \int_{t_1}^{t_2} L(q_1 \dots q_f, \dot{q}_1 \dots \dot{q}_f, t)\, dt = 0, \tag{9.1}$$

where L is a function of the f generalized coordinates $q_1,\dots, q_f$ and the corresponding velocities $\dot{q}_1,\dots,\dot{q}_f$, together with an explicit dependence on the time t in the case of non-conservative systems. Since the variations in path are to be taken without changing the time of transition t_1 to t_2 from the initial to the later configuration, the above may be rewritten in the form

$$\begin{aligned} 0 &= \int_{t_1}^{t_2} \delta L(q_1 \dots q_f, \dot{q}_1 \dots \dot{q}_f, t)\, dt \\ &= \int_{t_1}^{t_2} \sum_{i=1}^{f} \left(\frac{\partial L}{\partial q_i}\delta q_i + \frac{\partial L}{\partial \dot{q}_i}\delta \dot{q}_i\right) dt, \end{aligned} \tag{9.2}$$

where δq_i and $\delta \dot{q}_i$ denote the variations in the ith coordinate and velocity on passing from the actual to a displaced path, and we take a sum over all f degrees of freedom.

In order to simplify this expression, we note that we may put

$$\dot{q}_i + \delta \dot{q}_i = \frac{d}{dt}(q_i + \delta q_i), \tag{9.3}$$

since this merely asserts an equality between two different modes of expressing a generalized velocity on the displaced path. This then gives us

$$\delta \dot{q}_i = \frac{d}{dt}\delta q_i. \tag{9.4}$$

Returning to (9.2), and considering the terms in the integrand involving $\delta \dot{q}_i$, we can then write

$$\begin{aligned} \int_{t_1}^{t_2} \frac{\partial L}{\partial \dot{q}_i}\delta \dot{q}_i\, dt &= \int_{t_1}^{t_2} \frac{\partial L}{\partial \dot{q}_i}\frac{d}{dt}\delta q_i\, dt \\ &= \left|\frac{\partial L}{\partial \dot{q}_i}\delta q_i\right|_{t_1}^{t_2} - \int_{t_1}^{t_2} \frac{d}{dt}\frac{\partial L}{\partial \dot{q}_i}\delta q_i\, dt = -\int_{t_1}^{t_2} \frac{d}{dt}\frac{\partial L}{\partial \dot{q}_i}\delta q_i\, dt, \end{aligned} \tag{9.5}$$

where the term between limits has been taken as equal to zero, since by hypothesis the system is to have the same configuration at times t_1 and t_2 on the actual and displaced paths, thus making δq_i equal to zero at those times.

Substituting (9.5) into (9.2), and changing signs, we can then write

$$\int_{t_1}^{t_2} \sum_{i=1}^{f} \left(\frac{d}{dt}\frac{\partial L}{\partial \dot{q}_i} - \frac{\partial L}{\partial q_i}\right) \delta q_i \, dt = 0. \tag{9.6}$$

For a holonomic system, however, the variations δq_i would be arbitrary.† Hence the above equation can hold in general only if the f individual equations

$$\frac{d}{dt}\frac{\partial L}{\partial \dot{q}_i} - \frac{\partial L}{\partial q_i} = 0 \quad (i = 1,\ldots,f), \tag{9.7}$$

one for each degree of freedom, are themselves valid. These are the desired *equations of motion in the Lagrangian form* for a holonomic system of f degrees of freedom.

To illustrate the significance of these equations we may consider the simple case of a system of particles, moving with velocities small compared with that of light, and acted on by forces—depending on their positions relative to each other or relative to outside bodies—which are derivable from a potential. For the Lagrangian function we can then take

$$L = T - V, \tag{9.8}$$

where T is the kinetic energy and V is the potential energy of the system. Denoting the mass of the kth particle by m_k, and choosing as our generalized coordinates the coordinates x_k, y_k, z_k which would give the positions of the various particles k with respect to an unaccelerated set of Cartesian axes, the kinetic energy would be given by

$$T = \sum_k \tfrac{1}{2} m_k (\dot{x}_k^2 + \dot{y}_k^2 + \dot{z}_k^2), \tag{9.9}$$

and the potential energy would be given by

$$V = V(x_1\, y_1\, z_1 \ldots x_k\, y_k\, z_k \ldots, t), \tag{9.10}$$

where the explicit dependence on time t would be correlated with the behaviour of the external bodies exerting forces on the particles. Substituting these expressions for T and V into (9.8), and using the

† The development can be modified at this point for the treatment of non-holonomic systems by introducing the equations—arising from the structure of the system—which limit the independent variations of the coordinates q_i.

resulting form of L in the Lagrangian equations (9.7), we are led to results of the form

$$\frac{d}{dt}(m_k \dot{x}_k)+\frac{\partial V}{\partial x_k} = 0,$$
$$\frac{d}{dt}(m_k \dot{y}_k)+\frac{\partial V}{\partial y_k} = 0, \tag{9.11}$$
$$\frac{d}{dt}(m_k \dot{z}_k)+\frac{\partial V}{\partial z_k} = 0,$$

for each of the particles k. Furthermore, from the definition of the potential energy in terms of the work done by displacing the particles against the forces acting on them, these equations could also be written in the Newtonian form

$$F_{xk} = m_k \ddot{x}_k, \qquad F_{yk} = m_k \ddot{y}_k, \qquad F_{zk} = m_k \ddot{z}_k, \tag{9.12}$$

where F_{xk}, F_{yk}, and F_{zk} are the components of force acting on the kth particle in the x-, y-, and z-directions.

We thus see, for this simple case, that the starting-point for the development of mechanics provided by Hamilton's principle, combined with an appropriate choice for the Lagrangian function, leads to results in agreement with the original starting-point of Newton. The above simple example also indicates the possibility of supplementing the treatment by introducing an appropriate 'force function' in order to secure agreement with the Newtonian starting-point in more complicated cases where the forces acting on the particles cannot be derived from a potential. Indeed we may conclude the present section by emphasizing that Newton's laws of motion could themselves have been taken, in the special case of systems of Newtonian 'bodies' or particles, as the starting-point for a derivation of Hamilton's principle. By starting with Hamilton's principle, however, we have chosen a postulate which applies to systems in some respects more general than those of Newton, and which is conveniently expressed in language independent of the coordinate system.

10. Generalized momenta, the Hamiltonian function, and the canonical equations of motion

The equations of motion in the Lagrangian form

$$\frac{d}{dt}\frac{\partial L}{\partial \dot{q}_i}-\frac{\partial L}{\partial q_i} = 0 \quad (i = 1,...,f) \tag{10.1}$$

for a holonomic system of f degrees of freedom are seen to be a set of

f second-order equations. For many purposes it is desirable to replace them by an equivalent set of $2f$ first-order equations.

To secure this we may begin by defining a new set of quantities, $p_1,\ldots,p_f$, by the equations

$$p_i = \frac{\partial L}{\partial \dot{q}_i} \quad (i = 1,\ldots,f). \tag{10.2}$$

These we call the *generalized momenta* corresponding to the generalized coordinates $q_1,\ldots,q_f$. Using Cartesian coordinates x, y, z for the case of a single particle of mass m, it will be seen that the corresponding momenta would be equal to the ordinary components of momentum $m\dot{x}$, $m\dot{y}$, and $m\dot{z}$ for the particle.

With the help of these momenta we may now also define another new quantity H by the equation

$$H = p_1\dot{q}_1+\ldots+p_f\dot{q}_f-L, \tag{10.3}$$

where $\dot{q}_1,\ldots,\dot{q}_f$ are the generalized velocities and L is the Lagrangian function for the system. Differentiating this expression we obtain

$$dH = p_1 d\dot{q}_1+\ldots+p_f d\dot{q}_f+\dot{q}_1 dp_1+\ldots+\dot{q}_f dp_f- \\ -\frac{\partial L}{\partial q_1}dq_1-\ldots-\frac{\partial L}{\partial q_f}dq_f-\frac{\partial L}{\partial \dot{q}_1}d\dot{q}_1-\ldots-\frac{\partial L}{\partial \dot{q}_f}d\dot{q}_f-\frac{\partial L}{\partial t}dt,$$

or, making use of (10.2),

$$dH = \dot{q}_1 dp_1+\ldots+\dot{q}_f dp_f-\frac{\partial L}{\partial q_1}dq_1-\ldots-\frac{\partial L}{\partial q_f}dq_f-\frac{\partial L}{\partial t}dt \tag{10.4}$$

since two sets of terms in the previous form of expression are seen to cancel. This final form of expression shows that the quantity H could itself be regarded as a function of the coordinates $q_1,\ldots,q_f$, the momenta $p_1,\ldots,p_f$, and the time t, with the partial derivatives

$$\frac{\partial H}{\partial p_i} = \dot{q}_i, \qquad \frac{\partial H}{\partial q_i} = -\frac{\partial L}{\partial q_i}, \qquad \frac{\partial H}{\partial t} = -\frac{\partial L}{\partial t}. \tag{10.5}$$

When H actually is expressed in terms of the coordinates, momenta, and time

$$H = H(q_1 \ldots q_f, p_1 \ldots p_f, t) \tag{10.6}$$

it is called the *Hamiltonian function* for the system. It will be noted from the form of the original equation of definition (10.3) that the distinction between conservative and non-conservative systems will now be correlated respectively with the absence or presence of an explicit dependence of the Hamiltonian on time.

With the help of the foregoing we can now express the equations of motion in the so-called canonical form. Combining the definition of momenta given by (10.2) with the Lagrangian equations given by (10.1), we can evidently write

$$\frac{dp_i}{dt} = \frac{d}{dt}\left(\frac{\partial L}{\partial \dot{q}_i}\right) = \frac{\partial L}{\partial q_i} \tag{10.7}$$

for the rate of change in the ith component of momentum with time. Using this, the first two expressions in (10.5) can now be rewritten in the form

$$\frac{dq_i}{dt} = \frac{\partial H}{\partial p_i} \quad \text{and} \quad \frac{dp_i}{dt} = -\frac{\partial H}{\partial q_i}. \tag{10.8}$$

These are the desired *equations of motion in the Hamiltonian or canonical form.* With a known form for the Hamiltonian H as a function of the coordinates q_i, momenta p_i, and time t, their solution would evidently allow us to calculate the values of the coordinates q_i in their dependence on time. Since they are first-order equations, with a remarkably symmetrical form, they are often more convenient for fundamental investigations than the corresponding Lagrangian equations. The corresponding coordinates and momenta q_i and p_i which they contain are spoken of as being *conjugate variables.*

11. The change in quantities with time. Poisson brackets

With the help of the equations of motion in the canonical form we can now obtain a useful expression for the rate of change with time in any quantity F, which can be expressed as a function of the coordinates q_i and momenta p_i of the system of interest. From the assumed dependence of F on q_i and p_i, and not explicitly on the time, we can evidently write

$$\frac{dF}{dt} = \sum_i \left(\frac{\partial F}{\partial q_i}\frac{dq_i}{dt} + \frac{\partial F}{\partial p_i}\frac{dp_i}{dt}\right), \tag{11.1}$$

where we take a summation over all degrees of freedom i. And substituting from the canonical equations (10.8), for the rates of change with time in the coordinates and momenta, this assumes the convenient form

$$\frac{dF}{dt} = \sum_i \left(\frac{\partial F}{\partial q_i}\frac{\partial H}{\partial p_i} - \frac{\partial H}{\partial q_i}\frac{\partial F}{\partial p_i}\right). \tag{11.2}$$

Combinations of quantities such as that appearing on the right-hand side of (11.2) are quite important in the classical mechanics and have received the special name of Poisson brackets. For any two functions

M and N of the coordinates and momenta of a system, the *Poisson bracket* $\{M,N\}$ is defined by

$$\{M,N\} = \sum_i \left(\frac{\partial M}{\partial q_i}\frac{\partial N}{\partial p_i} - \frac{\partial N}{\partial q_i}\frac{\partial M}{\partial p_i}\right), \tag{11.3}$$

where a summation is taken over all the degrees of freedom of the system i.

Hence the expression for the rate of change with time of any function F of the coordinates and momenta of a system can also be written in the abbreviated form

$$\frac{dF}{dt} = \{F,H\}. \tag{11.4}$$

This mode of writing will be of interest when comparison is later made with the analogous quantum mechanical expression for the rate of change with time in the *expectation values* of quantities. See § 63 (*a*).

12. The integral of energy and the interpretation of the Hamiltonian function

With the help of our general expression for the rate of change of quantities with time, we can now investigate the rate of change of the Hamiltonian function itself. We shall be interested in the case of *conservative systems*, where H would not be an explicit function of the time t. Substituting H in place of F in the fundamental equation (11.2), for the rate of change in time of quantities which are solely functions of the coordinates and momenta, we then obtain

$$\frac{dH}{dt} = \sum_i \left(\frac{\partial H}{\partial q_i}\frac{\partial H}{\partial p_i} - \frac{\partial H}{\partial q_i}\frac{\partial H}{\partial p_i}\right) = 0. \tag{12.1}$$

We now see that the Hamiltonian function for a conservative system would be a quantity which does not change with time. We have thus obtained one integral of the equations of motion for a conservative system

$$H(q_i, p_i) = E, \tag{12.2}$$

where E is a constant.

The validity of this result depends on the circumstance that we have confined our considerations to conservative systems, and in case we employ stationary coordinates as defined in § 7 the constant E turns out to be equal to the quantity ordinarily defined as the energy of the system. Hence the Hamiltonian function H may now be interpreted as an expression for the energy of the system in terms of its coordinates and momenta.

In the case of simple mechanical systems the interpretation of the

Hamiltonian as the energy of the system agrees with familiar usage. For such systems the Lagrangian function is given by

$$L = T - V, \tag{12.3}$$

where T and V are the so-called kinetic and potential energies of the system, as already defined in connexion with (8.4). Substituting this into the original expression (10.3) used for defining the Hamiltonian, we have

$$\begin{aligned} H &= \sum_i p_i \dot{q}_i - T + V \\ &= \sum_i \dot{q}_i \frac{\partial L}{\partial \dot{q}_i} - T + V \\ &= \sum_i \dot{q}_i \frac{\partial T}{\partial \dot{q}_i} - T + V \\ &= 2T - T + V \\ &= T + V, \end{aligned} \tag{12.4}$$

where the second form of writing comes from the definition of the momenta p_i given by (10.2), the third form of writing comes from the fact that T is the only part of the Lagrangian containing the velocities $\dot{q}_i$, and the next to the last form of writing comes from the consideration that our definition of T was such as to make it a homogeneous quadratic function of the velocities $\dot{q}_i$, in any system of stationary coordinates. The final result, combined with our original definitions of T and V, shows that the Hamiltonian is equal to the usual expression for the energy of a simple mechanical system in terms of the spatial location, masses, and velocities of its parts.

In the case of more complicated conservative systems we can regard the constant quantity H as defining what we shall call the energy of the system. Thus, for example, in the case of a system of particles moving with high enough velocities so that the special theory of relativity has to be applied, the Lagrangian function may be taken as

$$L = \sum m_0 c^2\{1 - \sqrt{(1 - v^2/c^2)}\} - V \tag{12.5}$$

and the Hamiltonian is then found to be equal to the relativistic expression for energy

$$H = \sum \frac{m_0 c^2}{\sqrt{(1 - v^2/c^2)}} + V. \tag{12.6}$$

And similar treatments can be given in non-mechanical cases where Hamilton's principle applies.

13. The integrals of linear and of angular momentum

Our general expression for the rate of change in quantities with time can also be used to investigate the components of linear and of angular momentum for a system as a whole.† We may confine our considerations to systems of ordinary particles acting under the influence of forces which can be derived from a potential. For such systems, using Cartesian coordinates x_k, y_k, z_k for the different particles of masses m_k that compose the system, it will readily be seen that the Hamiltonian can be expressed in the form

$$H = \sum_k \frac{1}{2m_k}(p_{xk}^2+p_{yk}^2+p_{zk}^2)+V(x_k, y_k, z_k), \tag{13.1}$$

where the generalized momenta p_{xk}, p_{yk}, etc., corresponding to the coordinates used, are the ordinary components of momentum for the individual particles

$$p_{xk} = m_k \dot{x}_k, \qquad p_{yk} = m_k \dot{y}_k, \qquad p_{zk} = m_k \dot{z}_k. \tag{13.2}$$

The *linear momentum* in the x-direction for such a system may then be defined by

$$P_x = \sum_k p_{xk}, \tag{13.3}$$

which takes a sum over the individual components of momentum in the x-direction for the different particles k, and similar definitions may be given for the other directions. And the *angular momentum* around the z-axis for such a system may be defined by

$$M_z = \sum_k (x_k p_{yk} - y_k p_{xk}), \tag{13.4}$$

which takes a sum over the individual components of angular momentum for the different particles k, and similar definitions may be given for the other axes. We now wish to investigate the rate of change of these quantities in time, with the help of our general equation (11.2), for the rate of change of any function $F(q,p)$ of the coordinates and momenta

$$\frac{dF}{dt} = \sum_i \left(\frac{\partial F}{\partial q_i}\frac{\partial H}{\partial p_i} - \frac{\partial H}{\partial q_i}\frac{\partial F}{\partial p_i}\right), \tag{13.5}$$

where the summation is over all the degrees of freedom i.

For the rate of change in linear momentum with time, we immediately see from the foregoing that we shall obtain the general result

$$\frac{dP_x}{dt} = -\sum_k \frac{\partial H}{\partial x_k} = -\sum_k \frac{\partial V}{\partial x_k}, \tag{13.6}$$

† For a more general treatment of the integrals of linear and angular momentum and their connexion with the invariance of the Hamiltonian under infinitesimal transformation, see, for example, Whittaker, *Analytical Dynamics*, second edition, § 144, Cambridge, 1917.

where the summation is over all particles k. For an isolated system of particles, moreover, it is evident that the total potential energy V would depend only on the relative positions of the particles and would not be altered if they were all displaced by the same amount in the x-direction. Hence, for the case of an isolated conservative system, the result reduces to

$$\frac{dP_x}{dt} = 0. \tag{13.7}$$

Making use of the similar results for the y- and z-directions, we are thus provided with three further integrals for the equations of motion,

$$P_x = \text{const.}, \qquad P_y = \text{const.}, \qquad P_z = \text{const.}, \tag{13.8}$$

for the case of an isolated system of particles.

Turning now to angular momentum, the previous equations (13.1), (13.4), and (13.5) are seen to lead to the result

$$\frac{dM_z}{dt} = \sum_k \frac{1}{m_k}(p_{yk}p_{xk} - p_{xk}p_{yk}) + \sum_k \left(\frac{\partial V}{\partial x_k}y_k - \frac{\partial V}{\partial y_k}x_k\right). \tag{13.9}$$

The first term in this expression immediately cancels out, and by changing to polar coordinates $x_k = r_k \cos\phi_k$ and $y_k = r_k \sin\phi_k$ around the z-axis, the second term is seen to reduce to

$$\frac{dM_z}{dt} = -\sum_k \frac{\partial V}{\partial \phi_k}. \tag{13.10}$$

For an isolated conservative system in which the potential energy would be unaltered by a mere rotation of the system about the z-axis, this then gives

$$\frac{dM_z}{dt} = 0. \tag{13.11}$$

Making use of the similar results for the x- and y-axes, we are thus provided with three still further integrals for the equations of motion,

$$M_x = \text{const.}, \qquad M_y = \text{const.}, \qquad M_z = \text{const.}, \tag{13.12}$$

for the case of an isolated system of particles.

In this and the preceding section we have thus obtained in all seven integrals of the equations of motion, and similar results hold not only for systems of particles but, in general, for isolated conservative systems.† These results are often spoken of as the *conservation laws* for the energy, for the three components of linear momentum, and for the three components of angular momentum of an isolated system as a

† In the case of systems which have to be treated by the methods of *general relativity*, the analogue of the principle of the conservation of angular momentum has not been obtained, so far as the writer is aware.

whole. In the case of systems of a few degrees of freedom these integrals alone may be sufficient to provide a complete solution of the equations of motion.

14. Canonical transformations

In order to complete this brief account of the classical mechanics we must now give some consideration to the possibility of making more general transformations of variables than those which we have hitherto had in mind. This will then put us in a position to consider general methods of undertaking the integration of the equations of motion, which are due to Hamilton and to Jacobi.

In the foregoing we have already considered the possibility of using different sets of generalized coordinates in treating the behaviour of a given mechanical system. We have regarded any such set of coordinates, however, as directly connected by transformation equations with the coordinates provided by some unaccelerated set of Cartesian axes, and indeed have found it convenient to consider only so-called stationary coordinates in which the new variables $\bar{q}_i$ would be given in terms of the old variables q_i by transformation equations

$$\bar{q}_i = \bar{q}_i(q_1 \cdots q_f) \tag{14.1}$$

which do not depend explicitly on the time.

We now wish to consider the much more general possibility of transforming from an original set of coordinates q_i and momenta p_i to a new set $\bar{q}_i$ and $\bar{p}_i$, where the equations of transformation

$$\begin{aligned} \bar{q}_i &= \bar{q}_i(q_1 \cdots q_f, p_1 \cdots p_f, t), \\ \bar{p}_i &= \bar{p}_i(q_1 \cdots q_f, p_1 \cdots p_f, t) \end{aligned} \tag{14.2}$$

are such that the new coordinates and momenta can be functions both of the old coordinates and of the old momenta, and also depend explicitly on the time if so desired. In carrying out such transformations, nevertheless, we shall wish to limit ourselves to changes of a character which would make the new variables, whatever the original form of H, still satisfy equations of motion

$$\frac{d\bar{q}_i}{dt} = \frac{\partial \bar{H}}{\partial \bar{p}_i} \quad \text{and} \quad \frac{d\bar{p}_i}{dt} = -\frac{\partial \bar{H}}{\partial \bar{q}_i}, \tag{14.3}$$

of the same form as the canonical equations (10.8) applying to the original variables. Under such circumstances we shall then speak of the change of variables as being a *canonical transformation*.

Transformations of variables, having this desired property, can be carried out for a system of f degrees of freedom with the help of an arbitrary function F depending on f of the old and f of the new variables, together with the time t if desired; and such a function may be called the *generating function* for the transformation. Such generating functions are ordinarily chosen in one of the four forms

$$F(q_i, \bar{q}_i, t), \quad F(q_i, \bar{p}_i, t), \quad F(p_i, \bar{q}_i, t) \quad \text{or} \quad F(p_i, \bar{p}_i, t),$$

and the transformation equations obtained will then have a form which depends on this choice. We may illustrate the procedure for obtaining such equations of transformation, making use of the first of these four forms, and then give a statement as to the results which can be obtained starting with the other forms.

Choosing the generating function in the form $F(q_i, \bar{q}_i, t)$, where the symbols q_i and $\bar{q}_i$ are taken as representing the sets of old and new coordinates, we then assert that a canonical transformation of variables can be obtained by requiring the validity of the relation

$$\sum_i p_i \dot{q}_i - H(q_i, p_i, t) = \sum_i \bar{p}_i \dot{\bar{q}}_i - \bar{H}(\bar{q}_i, \bar{p}_i, t) + \frac{dF(q_i, \bar{q}_i, t)}{dt}, \tag{14.4}$$

where the summations $\sum$ are to be taken over all f degrees of freedom i, and the last term is to be taken as signifying the total rate of change with time in the generating function F.

We may first show that a relation of the form (14.4) would be sufficient to determine the desired equations of transformation. Here we encounter two cases, according as the transformation is complete enough so that there are no identical relations connecting the old and new coordinates q_i and $\bar{q}_i$, or is more limited so that the new coordinates are not completely independent of the old. In the first case, assuming the new coordinates independent of the old so that there will be no necessary connexion between the values of the $\dot{\bar{q}}_i$'s and the $\dot{q}_i$'s, and noting that in any case these generalized velocities are in each set independent among themselves, we then see that the general validity of the above equation can only be secured provided we satisfy the $2f+1$ individual relations

$$\begin{aligned} p_i &= \frac{\partial F(q_i, \bar{q}_i, t)}{\partial q_i}, \\ \bar{p}_i &= -\frac{\partial F(q_i, \bar{q}_i, t)}{\partial \bar{q}_i}, \\ H(q_i, p_i, t) &= \bar{H}(\bar{q}_i, \bar{p}_i, t) - \frac{\partial F(q_i, \bar{q}_i, t)}{\partial t}. \end{aligned} \tag{14.5}$$

In the second case, assuming the existence of identical relations between the old and new coordinates of the form†

$$Q_r(q_i, \bar{q}_i) = 0, \tag{14.6}$$

where the subscript r denotes a particular one of these relations, it is evident that the validity of the equations (14.4) and (14.6) taken together is to be secured by the individual relations

$$\begin{aligned} p_i &= \frac{\partial F(q_i, \bar{q}_i, t)}{\partial q_i} - \sum_r \lambda_r \frac{\partial Q_r(q_i, \bar{q}_i)}{\partial q_i}, \\ \bar{p}_i &= -\frac{\partial F(q_i, \bar{q}_i, t)}{\partial \bar{q}_i} + \sum_r \lambda_r \frac{\partial Q_r(q_i, \bar{q}_i)}{\partial \bar{q}_i}, \\ H(q_i, p_i, t) &= \bar{H}(\bar{q}_i, \bar{p}_i, t) - \frac{\partial F}{\partial t}(q_i, \bar{q}_i, t), \end{aligned} \tag{14.7}$$

where the quantities λ_r are undetermined constant multipliers and the summations $\sum$ are taken over all the identical relations r which exist. In the first case it will be seen that the equations given by the first two lines of (14.5) can be regarded as the desired equations of transformation, since they could be solved at will either for the $\bar{q}_i$ and $\bar{p}_i$ in terms of the q_i and p_i or vice versa. In the second case the first two lines of (14.7) together with (14.6) are enough to determine the $\bar{q}_i$, $\bar{p}_i$, and λ_r in terms of the q_i and p_i.

We must now show that a relation of the form (14.4) is not only sufficient to provide transformation equations, but will also guarantee that the transformation actually is a canonical one. To do this let us consider the result of multiplying both sides of equation (14.4) by dt and integrating over a given time-interval t_1 to t_2. We thus obtain

$$\int_{t_1}^{t_2} \Big[\sum_i p_i \dot{q}_i - H(q_i, p_i, t)\Big]\, dt = \int_{t_1}^{t_2} \Big[\sum_i \bar{p}_i \dot{\bar{q}}_i - \bar{H}(\bar{q}_i, \bar{p}_i, t)\Big]\, dt + F(q_i, \bar{q}_i, t)\Big|_{t_1}^{t_2} \tag{14.8}$$

Let us further now consider the result of performing a variation on both sides of this equation corresponding to an altered path for the system but to the same time-interval t_1 to t_2. In carrying out this variation we shall replace increments in the velocities such as $\delta\dot{q}_i$ by $d(\delta q_i)/dt$ and then treat by the familiar method of partial integration; we shall not impose any restrictions, however, on the values of δq_i and

† The treatment can also be extended to the case where the Q_r depend explicitly on the time.

$\delta\bar{q}_i$ at the limits corresponding to t_1 and t_2. We then obtain, as will be readily verified,

$$\int_{t_1}^{t_2} \sum_i \left[\left(\frac{dq_i}{dt}-\frac{\partial H}{\partial p_i}\right)\delta p_i-\left(\frac{dp_i}{dt}+\frac{\partial H}{\partial q_i}\right)\delta q_i\right]dt+\sum_i p_i\,\delta q_i\Big|_{t_1}^{t_2}$$

$$=\int_{t_1}^{t_2} \sum_i \left[\left(\frac{d\bar{q}_i}{dt}-\frac{\partial \bar{H}}{\partial \bar{p}_i}\right)\delta \bar{p}_i-\left(\frac{d\bar{p}_i}{dt}+\frac{\partial \bar{H}}{\partial \bar{q}_i}\right)\delta \bar{q}_i\right]dt+$$

$$+\sum_i \bar{p}_i\,\delta\bar{q}_i\Big|_{t_1}^{t_2}+\sum_i\left[\frac{\partial F}{\partial q_i}\delta q_i+\frac{\partial F}{\partial \bar{q}_i}\delta\bar{q}_i\right]\Big|_{t_1}^{t_2}. \qquad (14.9)$$

And since the first term on the left-hand side of this equation is equal to zero on account of the assumed validity of the canonical equations in the original variables, and the last term on the left-hand side taken together with the last two terms on the right-hand side will all cancel out, in accordance with (14.5) or in accordance with (14.6) and (14.7) in the two cases treated, we now obtain the result†

$$\int_{t_1}^{t_2} \sum_i \left[\left(\frac{d\bar{q}_i}{dt}-\frac{\partial \bar{H}}{\partial \bar{p}_i}\right)\delta \bar{p}_i-\left(\frac{d\bar{p}_i}{dt}+\frac{\partial \bar{H}}{\partial \bar{q}_i}\right)\delta \bar{q}_i\right]dt=0. \qquad (14.10)$$

Finally, since the $\delta\bar{q}_i$ and $\delta\bar{p}_i$ would be arbitrary, we now see that this can only be satisfied if we also have the equations of motion in the canonical form,

$$\frac{d\bar{q}_i}{dt}=\frac{\partial \bar{H}}{\partial \bar{p}_i} \quad \text{and} \quad \frac{d\bar{p}_i}{dt}=-\frac{\partial \bar{H}}{\partial \bar{q}_i}, \qquad (14.11)$$

holding in the new variables. Thus indeed we also find that the relation, expressed by (14.4), does determine a canonical transformation.

It may be remarked that the necessary and sufficient condition that a transformation $q,p \rightarrow \bar{q},\bar{p}$ be canonical may be put in the often convenient form

$$\{\bar{q}_i,\bar{p}_j\}=\delta_{ij}, \qquad \{\bar{q}_i,\bar{q}_j\}=0, \qquad \{\bar{p}_i,\bar{p}_j\}=0, \qquad (14.12)$$

† The first term in (14.9) which we took as equal to zero on account of the validity of the canonical equations of motion can also be written in the form

$$\delta\int_{t_1}^{t_2}\left[\sum_i p_i\,\dot{q}_i-H(q_i,p_i,t)\right]dt=0,$$

provided we now take the variation δ as one in which the time of transition is not to be altered and the configuration at times t_1 and t_2 are *now* to be regarded as the same on the actual and varied paths. Since the integrand in this equation is equal by (10.3) to the Lagrangian L for the system, although expressed in a different manner, the equation is often spoken of as the modified Hamilton's principle. Hence, comparing with (14.10), we can also regard a canonical transformation as defined by the property of leaving the modified Hamilton's principle invariant to the change in variables.

where the $\bar{q}$ and $\bar{p}$ are regarded as functions of the q and p in evaluating the above Poisson brackets as defined by (11.3).

As already mentioned, we could also obtain other forms for the equations of transformation corresponding to a canonical change of variables by choosing other forms for the generating function F. The methods of proof are entirely similar to those given above.† For convenience we give the following table, which lists, under each of the four principal forms for the generating function F, the corresponding forms which will be assumed by the equations of transformation, in the absence of identical relations between the variables on which F depends.

$$\begin{array}{llll} F = F(q_i, \bar{q}_i, t) & F = F(q_i, \bar{p}_i, t) & F = F(p_i, \bar{q}_i, t) & F = F(p_i, \bar{p}_i, t) \\ p_i = \dfrac{\partial F}{\partial q_i} & p_i = \dfrac{\partial F}{\partial q_i} & q_i = -\dfrac{\partial F}{\partial p_i} & q_i = -\dfrac{\partial F}{\partial p_i} \\ \bar{p}_i = -\dfrac{\partial F}{\partial \bar{q}_i} & \bar{q}_i = \dfrac{\partial F}{\partial \bar{p}_i} & \bar{p}_i = -\dfrac{\partial F}{\partial \bar{q}_i} & \bar{q}_i = \dfrac{\partial F}{\partial \bar{p}_i} \\ H = \bar{H} - \dfrac{\partial F}{\partial t} & H = \bar{H} - \dfrac{\partial F}{\partial t} & H = \bar{H} - \dfrac{\partial F}{\partial t} & H = \bar{H} - \dfrac{\partial F}{\partial t} \end{array} \tag{14.13}$$

In using these equations to secure canonical transformations, it is to be remembered that the generating function F can be taken as an arbitrary function of its arguments, so that an infinite variety of canonical transformations can be obtained. It is convenient to have the equations of transformation in their different forms, since a given problem may be more easily treated with one form than another. For example, if we are interested in a transformation that involves identical relations between the old and new coordinates q_i and $\bar{q}_i$, instead of using the first form of F and allowing for the identical relations by using the more elaborate equations (14.7), we may be able to avoid complications by changing to the second form of F provided that there are no identical relations between the q_i and $\bar{p}_i$. The equations prove useful for a considerable variety of different problems.‡ We shall ourselves

† For example, in the case of the second form of generating function $F(q_i, \bar{p}_i, t)$, a canonical transformation can be obtained from the equation

$$\sum_i p_i \dot{q}_i - H(q_i, p_i, t) = -\sum_i \bar{q}_i \dot{\bar{p}}_i - \bar{H}(\bar{q}_i, \bar{p}_i, t) + \frac{dF(q_i, \bar{p}_i, t)}{dt}.$$

By similar methods to those used above, it can readily be shown that this equation leads to the transformation equations given in the second column of (14.13), and that the transformation does have the desired canonical character.

‡ See, for example, Born, *The Mechanics of the Atom*, translation by Fisher and Hartree, London, 1927.

be specially interested in using the second form of the equations given by (14.13) for discussing two different general methods of integrating the equations of motion for a system. We now turn to a consideration of the first of these methods.

15. Integration of the equations of motion by transformation to cyclic coordinates. Hamilton's characteristic function

In treating the behaviour of mechanical systems with the help of the canonical equations of motion, we may encounter cases where the expression for the Hamiltonian of the system does not contain one or more of the coordinates q_i which are being used. For example, the Hamiltonian may be of the form

$$H = H(q_2 \dots q_f, p_1 \dots p_f, t), \tag{15.1}$$

where the coordinate q_1 is missing. In accordance with the equations of motion in the canonical form, we should then have for the rate of change with time of the corresponding momentum p_1

$$\frac{dp_1}{dt} = -\frac{\partial H}{\partial q_1} = 0,$$

and hence at once obtain $\quad p_1 = \text{const.}$

as one integral of the equations of motion.

The angular orientation around the axis of rotation of a freely rotating body provides a simple example of such a coordinate, the Hamiltonian expression for the energy of the body being independent of the value of this angle, and the corresponding (angular) momentum being constant. Coordinates which do not enter into the expression for the Hamiltonian are hence often called *cyclic coordinates*.

The above specially simple character of cyclic coordinates suggests a method of treating the equations of motion for a system by making a canonical transformation to new variables such that none of the new coordinates will appear in the altered expression for the Hamiltonian and hence will all have the so-called cyclic character. In the case of conservative systems this is then found to provide, at least in principle, a complete solution of the equations of motion.

Considering a conservative system, and making use of the second form of generating function given in (14.13), we can write equations of transformation in the form

$$p_i = \frac{\partial F(q_i, \bar{p}_i)}{\partial q_i}, \qquad \bar{q}_i = \frac{\partial F(q_i, \bar{p}_i)}{\partial \bar{p}_i}, \qquad H(q_i, p_i) = \bar{H}(\bar{q}_i, \bar{p}_i). \tag{15.2}$$

Substituting the first of these equations into the third, and requiring the altered Hamiltonian to be independent of the new coordinates, we can then write

$$H\left(q_1 \dots q_f, \frac{\partial F}{\partial q_1} \dots \frac{\partial F}{\partial q_f}\right) = \bar{H}(\bar{p}_1 \dots \bar{p}_f) \tag{15.3}$$

as an equation whose solution would provide a generating function $F(q_i, \bar{p}_i)$ which would give a canonical transformation to new variables such that none of the new coordinates $\bar{q}_i$ would appear in the altered Hamiltonian.

Since equations of motion in the canonical form would still apply in the transformed language, we can now write

$$\frac{d\bar{p}_1}{dt} = -\frac{\partial \bar{H}}{\partial \bar{q}_1} = 0 \quad \dots \quad \frac{d\bar{p}_f}{dt} = -\frac{\partial \bar{H}}{\partial \bar{q}_f} = 0 \tag{15.4}$$

for the rate of change of the new momenta with time. And thus write

$$\bar{p}_i = \alpha_1 \quad \dots \quad \bar{p}_f = \alpha_f \tag{15.5}$$

for the values of the new momenta where the quantities $\alpha_1,\dots,\alpha_f$ are constants. This then makes the altered Hamiltonian a function of the constants $\alpha_1,\dots,\alpha_f$,

$$\bar{H} = \bar{H}(\alpha_1 \dots \alpha_f). \tag{15.6}$$

Hence, again making use of the canonical equations of motion, we can write

$$\frac{d\bar{q}_1}{dt} = \frac{\partial \bar{H}}{\partial \alpha_1} = \omega_1 \quad \dots \quad \frac{d\bar{q}_f}{dt} = \frac{\partial \bar{H}}{\partial \alpha_f} = \omega_f \tag{15.7}$$

for the rate of change of the new coordinates with time, where $\omega_1,\dots,\omega_f$ would themselves be constants. Or, on integrating, we have

$$\bar{q}_1 = \omega_1 t + \beta_1 \quad \dots \quad \bar{q}_f = \omega_f t + \beta_f, \tag{15.8}$$

where $\beta_1,\dots,\beta_f$ are constants of integration. We are thus provided with a solution of the equations of motion for a conservative system by a transformation to new coordinates which depend linearly on the time.

The above treatment has made no specification as to the form of the generating function F except that it shall be such as to make the transformed Hamiltonian $\bar{H}$ depend in some way solely on the new momenta. A specially simple case, which makes the transformed Hamiltonian itself equal to one of the new momenta, arises when we choose Hamilton's so-called *characteristic function* S for the system as the generating function. This function S is defined as the general solution of the so-called Hamilton-Jacobi partial differential equation

$$H\left(q_1 \dots q_f, \frac{\partial S}{\partial q_1} \dots \frac{\partial S}{\partial q_f}\right) = E \tag{15.9}$$

and will be of the form

$$S = S(q_1 \dots q_f, E, \alpha_2 \dots \alpha_f) + \text{const.}, \tag{15.10}$$

where the f constants of integration are given by $\alpha_2, \dots, \alpha_f$ together with the arbitrary additive constant. Using this expression as giving the generating function F in equations (15.2) with $E, \alpha_2, \dots, \alpha_f$ as the new momenta, we then have the transformed Hamiltonian,

$$\bar{H} = E, \tag{15.11}$$

directly equal to one of the new momenta which is itself the energy E of the system. And since the canonical equations of motion still hold in the new variables, our previous solutions (15.5) and (15.8) now reduce to the form

$$\bar{p}_1 = E \qquad \bar{p}_2 = \alpha_2 \quad \dots \quad \bar{p}_f = \alpha_f \tag{15.12}$$

and

$$\bar{q}_1 = t + \beta_1 \quad \bar{q}_2 = \beta_2 \quad \dots \quad \bar{q}_f = \beta_f, \tag{15.13}$$

since $\omega_1 = \partial\bar{H}/\partial E$ becomes unity and $\omega_2, \dots, \omega_f$ become zero.

From the present point of view the problem of integrating the equations of motion for a conservative system, then, reduces to the problem of obtaining a general solution for the Hamilton-Jacobi partial differential equation (15.9). In principle we are thus provided with a method of completely solving for the motion of any conservative system. In practice, it is to be noted, however, that the transformed coordinates and momenta $\bar{q}_1 \dots \bar{q}_f, E, \alpha_2 \dots \alpha_f$ may have a very complicated relation to any actual variables which we should wish to use in describing the behaviour of the system. It is also to be noted that the solution of the Hamilton-Jacobi equation may be impracticable unless it is possible to apply the so-called method of separation of variables.

16. Integration of the equations of motion by transformation to constant coordinates and momenta. Hamilton's principal function

In concluding the chapter we may also consider another method of integrating the equations of motion by transforming to new variables, using Hamilton's so-called principal function as the generating function for the transformation. The method is applicable both to conservative and to non-conservative systems, and leads at least formally to new coordinates and momenta *all* of which are constants independent of the time.

Taking $H(q_1 \dots q_f, p_1 \dots p_f, t)$ as an expression for the Hamiltonian of the system of interest in terms of the original coordinates and momenta and the time, Hamilton's *principal function* W for the system may be defined as the general solution of the partial differential equation

$$H\left(q_1 \dots q_f, \frac{\partial W}{\partial q_1} \dots \frac{\partial W}{\partial q_f}, t\right)+\frac{\partial W}{\partial t} = 0. \tag{16.1}$$

The desired solution will then be of the form

$$W = W(q_1 \dots q_f, \alpha_1 \dots \alpha_f, t)+\text{const.} \tag{16.2}$$

with $f+1$ arbitrary constants.

For the special case of a conservative system, the equation defining the principal function will reduce to

$$H\left(q_1 \dots q_f, \frac{\partial W}{\partial q_1} \dots \frac{\partial W}{\partial q_f}\right)+\frac{\partial W}{\partial t} = 0, \tag{16.3}$$

and it is then evident that we can take the solution to be of the form

$$W = S-Et, \tag{16.4}$$

where S again denotes Hamilton's characteristic function for the system, as defined by the Hamilton-Jacobi equation

$$H\left(q_1 \dots q_f, \frac{\partial S}{\partial q_1} \dots \frac{\partial S}{\partial q_f}\right)-E = 0. \tag{16.5}$$

Let us now regard Hamilton's principal function as the generating function for a canonical transformation of the second type, listed in (14.13), with the new momenta equal to the constants $\alpha_1,\dots,\alpha_f$. In accordance with (14.13) we then have the *original momenta* given by

$$p_i = \frac{\partial W}{\partial q_i}, \tag{16.6}$$

the *new coordinates* given by

$$\bar{q}_i = \frac{\partial W}{\partial \alpha_i}, \tag{16.7}$$

and the *new Hamiltonian* given by

$$\bar{H} = 0. \tag{16.8}$$

Hence, since the canonical equations of motion would still hold in the new variables, and the new Hamiltonian has the value zero independent of the new coordinates and momenta, the equations of motion in the new variables would assume the specially simple form

$$\bar{q}_i = \beta_i \quad \text{and} \quad \bar{p}_i = \alpha_i, \tag{16.9}$$

where the α's and β's are arbitrary constants. Hence, in principle at least, we are thus provided with $2f$ integrals of the equations of motion.

It is of interest in connexion with the foregoing to consider the behaviour of a moving surface which can be defined, in the space corresponding to the coordinates $q_1,..., q_f$, by setting the principal function W equal to a constant, in accordance with

$$W(q_1 ... q_f, \alpha_1 ... \alpha_f, t) = W_0, \tag{16.10}$$

where $\alpha_1,..., \alpha_f$ and W_0 are constants which determine a class of motions of the system. The surface defined by (16.10) would evidently assume successively changing positions in the $q_1,..., q_f$ space as time proceeds.

For example, in the case of conservative systems, where we can use $W = S-Et$ as given by (16.4) to express the principal function W in terms of the characteristic function S, the position of the surface could be taken at any time t as agreeing with the surface defined by

$$S(q_1 ... q_f, E, \alpha_2 ... \alpha_f) = W_0+Et, \tag{16.11}$$

where the energy E together with $\alpha_2,..., \alpha_f$ and W_0 are $f+1$ constants determining the class of motions under consideration.

Returning to the general equation for our moving surface (16.10), let us now pick any point $q_1,..., q_f$ lying on the surface at time t, and consider the vector that could be erected at this point with the components

$$\frac{\partial W}{\partial q_1}, \frac{\partial W}{\partial q_2}, ..., \frac{\partial W}{\partial q_f}. \tag{16.12}$$

This vector has two important properties. In the first place, it will be seen, in accordance with (16.6), that its components are equal to the momenta $p_1,..., p_f$, which systems of the class under consideration would have when their coordinates $q_1,..., q_f$ are those for the point taken. In the second place, it will be seen from the values for its components given by (16.12) that the vector would be normal to the instantaneous surface under consideration. Hence the surface defined by (16.10) has the useful property of assuming successive configurations and positions such that a normal erected at any point on the surface would have components proportional to the momenta of a system arriving at that point with a motion of the class specified by the $f+1$ constants $\alpha_1,..., \alpha_f, W_0$.

This result becomes specially simple and interesting if we take a single particle of mass m as the system to be treated, and use its Cartesian coordinates x, y, z as determining the space in which the surface defined by $W =$ const. moves. The normal at any point on this surface then has components proportional to the ordinary components of momenta

$$\frac{\partial W}{\partial x} = p_x = m\dot{x}, \quad \frac{\partial W}{\partial y} = p_y = m\dot{y}, \quad \frac{\partial W}{\partial z} = p_z = m\dot{z} \tag{16.13}$$

for the particle and hence lies in the same direction as the velocity with which a particle of the class considered would pass through the point taken. Hence the moving surface and the system of normals which it determines might now be considered as having the nature of a moving *wave-front* which determines a system of *rays* along which a class of particle motions can take place. It should be noted in this connexion that the velocity of the wave-front and that of the associated particle motion along the rays would not in general be the same.

In concluding this section on the properties of Hamilton's principal function for a mechanical system, it may be well to emphasize the distinction between the two different but related roles which we can regard this function as playing in the explanation of mechanical phenomena. On the one hand, we can think of the principal function as making possible the integration of the equations of motion, when it is used as a function for transforming to new coordinates and momenta which are all of them constants. This was the point of view of Jacobi. On the other hand, we can think of the principal function as making it possible to obtain a description of mechanical motions in terms similar to those of wave-front and associated rays as used in geometrical optics. This was the point of view of special interest to Hamilton in the original development of the ideas we have been discussing. It may also be remarked that this latter point of view was specially suggestive to Schroedinger in his original method of developing a wave mechanics for treating quantum phenomena, his idea being that a complete wave mechanics should be related to ordinary mechanics in the same way that a complete wave theory of optics is related to geometrical optics.

III

STATISTICAL ENSEMBLES IN THE CLASSICAL MECHANICS

17. Ensemble and phase space

In the preceding chapter we have developed the principles of the classical mechanics, and are thus provided with methods for treating the behaviour of any given mechanical *system* of interest as it changes with time from one precisely defined state to another. We are now ready to begin our consideration of the statistical methods that can be employed when we need to treat the behaviour of a system concerning whose condition we have some knowledge but not enough for a complete specification of precise state. For this purpose we shall wish to consider the average behaviour of a collection of systems of the same structure as the one of actual interest but distributed over a range of different possible states. Using the original terminology of Gibbs we may speak of such a collection as an *ensemble of systems*.†

In order to investigate the behaviour of such ensembles of systems it is convenient to have a quasi-geometrical language which can be used in specifying the state of each system in the ensemble, and in describing the condition of the ensemble as a whole. For this purpose, corresponding to any system of f degrees of freedom, we can construct a conceptual Euclidean space of $2f$ dimensions, with $2f$ rectangular axes, one for each of the coordinates $q_1,\ldots,q_f$ and one for each of the momenta $p_1,\ldots,p_f$, whose values would determine the state of the system. Again using the terminology of Gibbs, who designated the changes in state of a system as being changes in phase, we can speak of such a conceptual space as a *phase space* for the kind of system under consideration.

The instantaneous state of any system in the ensemble can then be regarded as specified by the position of a *representative point* in the phase space, and the condition of the ensemble as a whole can be

† The terminology in this book agrees in general with the original terminology of Gibbs (loc. cit.). The word *system* is used, as in mechanics and as in thermodynamics, to denote the physical object of interest. Such a system may, of course, be composed of a large number of *elements* such as particles, atoms, molecules, electrons, modes of vibration, or what not, but this is not involved in the fundamentals of the theory. A collection of such systems which is studied from a statistical point of view is called an *ensemble of systems.*

The terminology of Fowler (loc. cit.) differs in that the word system is applied to the molecules or other elements composing the object of interest, and the whole is called an assembly of systems. This terminology would be clumsy for our purposes, however, since we should then have to speak of ensembles of assemblies of systems.

regarded as described by a 'cloud' of such representative points, one for each system in the ensemble. The behaviour of the ensemble as time proceeds can then be associated with the 'streaming' motion of this cloud of points as they describe trajectories in the phase space in accordance with the laws of mechanics. The representative points for the different systems are often spoken of as *phase points*. Since the position of a phase point completely determines the state of the corresponding system, it will be noted that the motion of any phase point is an unambiguous function of its instantaneous position.

A number of remarks as to the nature of such ensembles and phase spaces may now be made to assure complete understanding.

It is to be emphasized that the various systems which compose an ensemble are not to be regarded as interacting with each other, but each as carrying out its own independent behaviour in accordance with the laws of mechanics. To be sure, at a later stage of development, Chapters XIII and XIV, when we study the nature of heat transfer to or from a system, we shall wish to consider ensembles which can be regarded as constructed by placing members of one ensemble representing a given system of interest in thermal contact with the members of another ensemble representing a heat reservoir; but in the absence of specific statement to the contrary, each system in an ensemble is to be regarded as carrying out its own independent motion. In particular it should be emphasized that a system composed of a large number of similar interacting elements or subsystems, such as a gas composed of a large number of similar molecules colliding with each other, is not to be confused with an ensemble of independent systems. Indeed, in using statistical mechanics to study the behaviour of such a gas, it will be necessary to consider an ensemble of independent systems each of which will itself be a separate sample of the whole gas of interest.

In the case of systems composed of a large number of individual elements, such as molecules, it is evident that the coordinates and momenta, which could be assigned separately to the individual molecules, can, if desired, be taken as furnishing the total collection of coordinates and momenta to be used for the system as a whole. Furthermore, since we could construct a phase space for each individual molecule with axes corresponding to its own coordinates and momenta, it is evident that we could then also regard the collection of phase spaces for the individual molecules as providing the total phase space for the system as a whole.

In this connexion a terminology which is sometimes convenient has been introduced by Ehrenfest.† The phase space for the system as a whole (gas) is called a *γ-space*, and the phase space for any individual kind of element (molecule) contained in the system is called a *μ-space* for that species of element. Using an appropriate γ-space, the state of any system can then be completely specified by the position therein of a single representative point. Or using a μ-space for each kind of molecule involved, the condition of the system may also be described by giving the numbers of representative points for the component molecules which lie in the different regions of the μ-space for each species of molecule. The latter description of condition is, of course, less complete than the former, since it makes no distinction between different molecules of the same kind. By interchanging such-like molecules different precise states could be obtained, all of which would agree with the condition described by merely specifying the numbers of molecules whose representative points fall in different regions of the appropriate μ-space.

Returning to the phase space for a system as a whole, it is also sometimes convenient to regard it as constructed out of the *configuration space* that corresponds to the set of coordinates chosen for the system taken together with the *momentum space* that corresponds to the set of momenta conjugate to those coordinates. These two spaces regarded in combination then give the complete phase space which is conveniently used in specifying the state of the system.

It would, of course, also be possible to consider a conceptual space constructed out of a configuration space for the system taken together with the corresponding *velocity space*. This procedure would also furnish a combined space such that the state of the system could be completely specified by the position of a representative point. Nevertheless, when we come to the derivation of Liouville's theorem, we shall find that the phase space, with rectangular axes corresponding to each coordinate and momentum, has such specially simple properties that it is the convenient one to choose.

18. Density of distribution in the phase space. Averages for the ensemble

In the case of an ensemble of systems it is evident, in accordance with the preceding section, that the state of each individual system in the ensemble could be precisely specified at any instant by giving the

† P. and T. Ehrenfest, 'Mechanik der aus sehr zahlreichen diskreten Teilen bestehenden Systeme', *Encykl. d. math. Wiss.* iv. 2, ii, Heft 6, Leipzig.

position of the corresponding representative point in a suitable phase space. In using ensembles for statistical purposes, however, it is to be noted that there is no need to maintain distinctions between individual systems, since we shall be interested merely in the numbers of systems at any time which would be found in the different states that correspond to different regions in the phase space. Moreover, it is also to be noted for statistical purposes that we shall wish to use ensembles containing a large enough population of separate members so that the numbers of systems in such different states can be regarded as changing continuously as we pass from the states lying in one region of the phase space to those in another. Hence, for the purposes in view, it is evident that the condition of an ensemble at any time can be regarded as appropriately specified by the *density* ρ with which representative points are distributed over the phase space.

This density of distribution

$$\rho = \rho(q_1 \dots q_f, p_1 \dots p_f, t) \tag{18.1}$$

is to be taken, in the case of systems of f degrees of freedom, as a function of the $2f$ coordinates and momenta $q_1,\dots, q_f, p_1,\dots, p_f$, which correspond to the different coordinate axes in the phase space. It is also to be taken as a function of the time t, since the density would, in general, change with time at any point in phase space owing to the motion of the phase points as the coordinates and momenta for the corresponding systems change their values in the manner prescribed by the principles of mechanics. As a convenient abbreviation we may use

$$\rho = \rho(q, p, t) \tag{18.2}$$

as an expression to indicate this dependence on coordinates, momenta, and time.

The quantity ρ is then to be understood as determining the number of systems δN, which would be found at time t to have coordinates and momenta lying in any selected infinitesimal range $\delta q_1 \dots \delta q_f\, \delta p_1 \dots \delta p_f$, in accordance with the equation

$$\delta N = \rho(q, p, t)\, \delta q_1 \dots \delta q_f\, \delta p_1 \dots \delta p_f. \tag{18.3}$$

We assume a large enough total population of systems so that ρ and δN can be regarded with sufficient approximation as changing continuously as we go from one region in the phase space to another.

By integrating over the whole of phase space, we can write

$$N = \int \dots \int \rho(q, p, t)\, dq_1 \dots dp_f \tag{18.4}$$

as an expression for the *total number of systems* N in the ensemble or phase points in the phase space. For the physical interpretation of this formalism it is essential that the integral (18.4) for N converge; we shall in general be concerned only with densities ρ which have this property. Equation (18.4) then gives

$$\frac{\rho(q,p,t)}{N} = \frac{\rho(q,p,t)}{\int \dots \int \rho(q,p,t)\, dq_1 \dots dp_f} \tag{18.5}$$

as the *probability per unit extension in the phase space* that the phase point for a system chosen at random from the ensemble would be found at time t to have the specified values of the q's and p's. With the help of this expression and a knowledge of the actual dependence of ρ on the q's and p's, this then makes it possible to calculate any desired kind of average, over all the systems in the ensemble, of any quantity which depends on the coordinates and momenta of the systems. For example, if we consider a mechanical quantity $F(q, p)$, which characterizes systems of the kind under consideration, its *mean value* for all the systems in the ensemble would be given at time t by

$$\begin{aligned}\overline{\overline{F}}(q,p) &= \frac{1}{N}\int \dots \int F(q,p)\rho(q,p,t)\, dq_1 \dots dp_f \\ &= \frac{\int \dots \int F(q,p)\rho(q,p,t)\, dq_1 \dots dp_f}{\int \dots \int \rho(q,p,t)\, dq_1 \dots dp_f}.\end{aligned} \tag{18.6}$$

We use the double bar $=$ to indicate a mean over all the systems of the ensemble, since we shall wish to reserve the single bar $^-$ to denote the mean or expectation value for a quantum mechanical quantity in the case of a single system.

Instead of taking the quantity ρ precisely as above, it is sometimes convenient to regard it as *normalized to unity*, in accordance with the equation

$$1 = \int \dots \int \rho(q,p,t)\, dq_1 \dots dp_f. \tag{18.7}$$

The quantity ρ itself then gives directly the probability per unit volume of finding the phase point for a system picked at random from the ensemble in different regions of the phase space, and the expression for the mean value of any function $F(q, p)$ of the coordinates and momenta reduces to the simpler form

$$\overline{\overline{F}}(q,p) = \int \dots \int F(q,p)\rho(q,p,t)\, dq_1 \dots dq_f. \tag{18.8}$$

In the original development of Gibbs these two senses in which ρ may be used were distinguished by the separate designations, *density in phase D* and *coefficient of probability of phase P*. Nowadays, however, it is usual to regard the two possibilities as merely corresponding to two different possible modes of normalizing ρ.

In order to avoid confusion we shall adopt the practice of always using the density ρ in the obvious sense first given above, unless a statement is made to the contrary. It seems desirable to point out the two possibilities, however: in the first place since there is no universally accepted convention in the matter, and in the second place since the corresponding quantum mechanical quantity—the density matrix ρ_{nm} to be introduced later—is most frequently taken as normalized to unity.

19. Liouville's theorem for the change in density with time

Having seen in the foregoing section that the condition of an ensemble can be appropriately described at any instant by the density of distribution ρ of representative points in the phase space, we may now inquire into the changes, which take place in this quantity with time, as the representative points describe trajectories in the phase space in accordance with the principles of mechanics. We may first consider the rate of change of ρ with time at any given point in the qp-phase space on which we fix our attention.

To treat this, let us consider, at any point $q_1,\ldots,q_f,p_1,\ldots,p_f$ in the phase space, the differential element of extension that would be defined at that point by $\delta q_1\ldots\delta q_f\,\delta p_1\ldots\delta p_f$. For the number of phase points inside this element at any instant, we can write

$$\delta N = \rho\,\delta q_1\ldots\delta q_f\,\delta p_1\ldots\delta p_f, \tag{19.1}$$

where ρ is the density at the point and instant under consideration. This number will, in general, be changing with the time, however, since the number of phase points entering the element of hyper-volume through any 'face' will, in general, be different from the number which are leaving through the opposite 'face'. Thus, if we consider the two faces perpendicular to the q_1-axis, which are located at q_1 and $q_1+\delta q_1$, we could write for the number of phase points entering the first of these surfaces per unit time the expression

$$\rho\dot{q}_1\,\delta q_2\ldots\delta q_f\,\delta p_1\ldots\delta p_f, \tag{19.2}$$

where ρ and $\dot{q}_1$ are the density and indicated component of velocity for representative points at $q_1\ldots q_f\,p_1\ldots p_f$, and should then have as the corresponding expression for the number of phase points leaving the opposite face

$$\left(\rho+\frac{\partial\rho}{\partial q_1}\delta q_1\right)\left(\dot{q}_1+\frac{\partial\dot{q}_1}{\partial q_1}\delta q_1\right)\delta q_2\ldots\delta q_f\,\delta p_1\ldots\delta p_f, \tag{19.3}$$

where in both expressions higher order differentials have been neglected. Combining these two expressions, again with neglect of higher order differentials, and summing up over all such resulting terms i for the f coordinates and for the f momenta, we then obtain

$$\frac{d(\delta N)}{dt} = -\sum_{i=1}^{f}\left[\rho\left(\frac{\partial \dot{q}_i}{\partial q_i}+\frac{\partial \dot{p}_i}{\partial p_i}\right)+\left(\frac{\partial \rho}{\partial q_i}\dot{q}_i+\frac{\partial \rho}{\partial p_i}\dot{p}_i\right)\right]\delta q_1 \ldots \delta q_f\, \delta p_1 \ldots \delta p_f \tag{19.4}$$

as an expression for the rate of change with time in the number of representative points δN lying in the specified element of phase space.

This result, however, can be immediately simplified. In accordance with the equations of motion in the canonical form (10.8), we can write

$$\dot{q}_i = \frac{\partial H}{\partial p_i} \quad \text{and} \quad \dot{p}_i = -\frac{\partial H}{\partial q_i}, \tag{19.5}$$

as expressions for the rate of change with time in the coordinates and momenta of a system of the kind under consideration. The quantity H appearing in these expressions is the Hamiltonian for the system as a function of the coordinates and momenta, and hence, since the order of differentiation is immaterial, we obtain

$$\frac{\partial \dot{q}_i}{\partial q_i} = -\frac{\partial \dot{p}_i}{\partial p_i} \quad \text{or} \quad \sum_i \left(\frac{\partial \dot{q}_i}{\partial q_i}+\frac{\partial \dot{p}_i}{\partial p_i}\right) = 0, \tag{19.6}$$

which leads to a cancellation of the first terms on the right-hand side of (19.4). Furthermore, dividing (19.4) through by the element of extension $\delta q_1 \ldots \delta q_f\, \delta p_1 \ldots \delta p_f$, we evidently obtain the rate of change in the density itself at the point of interest, so that we can now write our result in the desired simple form

$$\left(\frac{\partial \rho}{\partial t}\right)_{q,p} = -\sum_i \left(\frac{\partial \rho}{\partial q_i}\dot{q}_i+\frac{\partial \rho}{\partial p_i}\dot{p}_i\right). \tag{19.7}$$

We use the symbol of partial differentiation with respect to time to indicate that we are fixing our attention on a given stationary point in the phase space.

The result given by (19.7) proves to be of fundamental importance for statistical mechanics. It is often spoken of as *Liouville's theorem*, since the so-called equation of incompressibility (19.6) on which it is based is of a form considered by that investigator.† The result can be expressed in a variety of different forms which prove convenient under different circumstances.

† Liouville, *Journ. de Math.* **3**, 349 (1838).

Substituting the expressions for $\dot{q}_i$ and $\dot{p}_i$ given by (19.5) into (19.7), we can write Liouville's theorem in the form

$$\left(\frac{\partial\rho}{\partial t}\right)_{q,p} = -\sum_i \left(\frac{\partial\rho}{\partial q_i}\frac{\partial H}{\partial p_i} - \frac{\partial\rho}{\partial p_i}\frac{\partial H}{\partial q_i}\right). \tag{19.8}$$

Or, noting the definition given by (11.3) for a Poisson bracket, we can also write this in the form

$$\left(\frac{\partial\rho}{\partial t}\right)_{q,p} = -\{\rho, H\}. \tag{19.9}$$

The rate of change with time, at a given point of phase space, in the density of distribution $\rho(q,p,t)$ for an ensemble of systems is thus given by an expression of the same form—but with opposite sign—as that for the rate of change with time in a function of the coordinates and momenta $F(q,p)$ for a single system. This will be of interest when we later consider the analogous quantum mechanical expressions. See § 81 (*c*).

Returning once more to (19.7), and transposing the summation of terms to the other side of the equation, we can also write Liouville's theorem in the form

$$\left(\frac{\partial\rho}{\partial t}\right)_{q,p} + \sum_i \left(\frac{\partial\rho}{\partial q_i}\dot{q}_i + \frac{\partial\rho}{\partial p_i}\dot{p}_i\right) = 0. \tag{19.10}$$

Hence, taking cognizance of the full dependence of the density $\rho(q,p,t)$ on the coordinates, momenta, and time, and noting that $\dot{q}_i$ and $\dot{p}_i$ are themselves expressions for the components of the velocity with which a representative point would move through the phase space, we now obtain the simple result

$$\frac{d\rho}{dt} = 0, \tag{19.11}$$

when we consider the rate of change of density in the neighbourhood of any selected moving phase point instead of in the neighbourhood of a fixed point in the phase space. This form of expression may be called, in accordance with Gibbs, the principle of the *conservation of density in phase.*

Finally, with the help of (19.11), we may obtain one further form of expression for Liouville's theorem which will prove useful. For this purpose let us consider a region in the phase space

$$\delta v = \int ... \int dq_1 ... dp_f, \tag{19.12}$$

which is taken small enough so that the density ρ can be regarded as uniform over its extension. For the number of representative points inside this region we can then write

$$\delta N = \rho\, \delta v = \rho \int \dots \int dq_1 \dots dp_f. \tag{19.13}$$

Let us now follow the motion of this little region through the phase space, allowing its boundaries to be permanently determined by the representative points originally lying thereon. Then, since no representative points can be created or destroyed on account of their correlation with mechanical systems, and since no representative points can cross the boundaries on account of the unambiguous determination of mechanical motions, we must have the result

$$\frac{d(\delta N)}{dt} = \frac{d\rho}{dt}\, \delta v + \rho \frac{d}{dt}(\delta v) = 0. \tag{19.14}$$

In accordance with (19.11), however, the term containing $d\rho/dt$ is equal to zero, since we are following a natural motion in the phase space. Hence, we then obtain the desired result

$$\frac{d}{dt}(\delta v) = \frac{d}{dt} \int \dots \int dq_1 \dots dp_f = 0. \tag{19.15}$$

It will be noted from the possibility of combining small elements that this final form of expression would evidently also apply to an extension of any magnitude in the phase space provided its boundaries are permanently determined by the same selection of representative points.

Again adopting the terminology of Gibbs, this last form of Liouville's theorem may be called the principle of the *conservation of extension in phase*. In accordance with this theorem any extension in the phase space, with moving boundaries specified as above, would retain a constant 'volume' as time proceeds. Its 'shape', however, would, in general, change with time, and after a sufficient interval this shape might become so 'filamentous' as to extend into many different parts of the phase space.

The simplicity of the results obtained in the above derivations depends on the choice of a phase space—constructed by the combination of configuration space with momentum space—as the apparatus for describing the condition of an ensemble. Principles so simple in form as those of the conservation of density in phase and the conservation of extension in phase would not appear if we used the language provided by a combination of configuration space with velocity space. It is in large measure to Boltzmann† that we owe the recognition of the importance of using momenta instead of velocities in statistical mechanical considerations.

20. Invariance of density and of extension to canonical transformations

In making use of the foregoing ideas as to phase spaces for the treatment of systems and ensembles, we shall usually regard the coordinates and momenta, for the system of interest as a whole, as provided by the total collection of coordinates and momenta which would be naturally assigned to the individual molecules composing the system in simple and familiar ways. It is evident, however, that there is nothing involved in the conceptual introduction of a qp-phase space to represent the state of a system, or of a density of phase points ρ for the description of an ensemble, which implies the use of any particular set of coordinates and momenta. Furthermore, it is evident that the validity of the various forms of Liouville's theorem derived in the last section depends solely on the assumption of the equations of motion in the canonical form for the systems treated. Hence the foregoing methods and results would apply using any set of canonical variables for the kind of system of interest.

The circumstance, that Liouville's theorem would continue to hold after any canonical transformation of variables, now makes it easy to show that the density of distribution at any selected point in the phase space would have a numerical value invariant to such canonical transformation. To investigate this let us consider two different sets of canonical variables, say $q_i p_i$ and $\bar{q}_i \bar{p}_i$, and at any selected time t_1 let us use the symbols ρ_1 and $\bar{\rho}_1$ to denote the density of distribution, in the neighbourhood of any selected point in the phase space, in the two languages respectively. Our problem, then, is to prove the equality of ρ_1 and $\bar{\rho}_1$.

To show this let us now consider a representative phase point carrying out its natural motion and so chosen as to coincide in position at time t_1 with the arbitrary point which we have selected in the phase space. At time t_1 the density in the neighbourhood of this phase point would then be ρ_1 and $\bar{\rho}_1$ in the two languages respectively. At a later time t_2 we may denote the density in its neighbourhood by ρ_2 and $\bar{\rho}_2$ in the

† See Boltzmann, *Vorlesungen über Gastheorie*, II. Teil, zweiter, unveränderter Abdruck, Leipzig, 1912.

two languages, and in accordance with Liouville's theorem in the form of the conservation of density in phase we must then have

$$\rho_1 = \rho_2 \quad \text{and} \quad \bar{\rho}_1 = \bar{\rho}_2. \tag{20.1}$$

However, we may now evidently introduce† a third system of canonical variables $\bar{\bar{q}}_i \bar{\bar{p}}_i$ which would coincide with the set $q_i p_i$ at time t_1 and with the set $\bar{q}_i \bar{p}_i$ at time t_2. Using similar notation to that above, this then gives us the relations

$$\rho_1 = \bar{\bar{\rho}}_1, \quad \bar{\rho}_2 = \bar{\bar{\rho}}_2, \quad \text{and} \quad \bar{\bar{\rho}}_1 = \bar{\bar{\rho}}_2. \tag{20.2}$$

Combining with the second of the two equations (20.1), this then leads to the required result

$$\rho_1 = \bar{\rho}_1. \tag{20.3}$$

Furthermore, combining with the first of equations (20.1), it also gives us the analogous result, for the later time t_2,

$$\rho_2 = \bar{\rho}_2. \tag{20.4}$$

We have thus shown, as desired, the invariance of the density ρ at any arbitrary point and time to a canonical transformation of variables.

The finding that the density in phase would be invariant to canonical transformations now makes it evident that the magnitude of any extension in phase, with boundaries defined by a selection of definite phase points, would also have to be invariant to canonical transformation. To see this let us first consider an extension in phase small enough so that we could regard the density ρ as not varying appreciably within its boundaries. Again taking $q_i p_i$ and $\bar{q}_i \bar{p}_i$ as two different sets of canonical variables, we could then write as expressions for the number of phase points in the extension

$$\delta N = \rho \int \dots \int dq_1 \dots dp_f$$

and

$$\delta \bar{N} = \bar{\rho} \int \dots \int d\bar{q}_1 \dots d\bar{p}_f \tag{20.5}$$

in the two languages respectively. Since we have proved $\rho = \bar{\rho}$, however, and since we must in any case have $\delta N = \delta \bar{N}$ as exactly the same phase points are involved in the two cases, we obtain the result

$$\int \dots \int dq_1 \dots dp_f = \int \dots \int d\bar{q}_1 \dots d\bar{p}_f. \tag{20.6}$$

† Let $F(q_i, \bar{p}_i, t)$ be the generating function for the transformation from the set $q_i p_i$ to the set $\bar{q}_i \bar{p}_i$. Then the function

$$\Phi(q_i, \bar{\bar{p}}_i, t) = \sum_i q_i \bar{\bar{p}}_i \frac{t-t_2}{t_1-t_2} + F(q_i, \bar{\bar{p}}_i, t) \frac{t-t_1}{t_2-t_1}$$

would be a possible generating function for the transformation from $q_i p_i$ to $\bar{\bar{q}}_i \bar{\bar{p}}_i$.

Moreover, from the possibility of combining one element of extension with another, we see that this invariance to canonical transformation would have to hold for an extension in phase of any magnitude provided its boundaries are objectively defined.

By making use of Liouville's theorem in the form of the conservation of extension in phase, the above result could also have been proved by a derivation similar to that which we gave for the invariance of the density ρ to canonical transformation. In addition, the result (20.6) could also be deduced by a direct consideration of the consequences of making a canonical transformation of variables.†

Making use of the above invariance of any extension in phase space $\int ... \int dq_1 ... dp_f$, we now always have the possibility, without resulting effect on the numerical magnitude, of changing to a system of canonical variables such that the q's and p's would be coordinates and momenta for the various parts of the mechanical system in the ordinary sense of the words. We then see that any extension in phase space would have the *dimensions*

$$\left(\frac{LML}{T}\right)^f = \left(\frac{ML^2}{T}\right)^f, \tag{20.7}$$

that is, of *action* raised to a power equal to the number of degrees of freedom f of the system.‡

This will be of interest in connexion with our later introduction of the quantum mechanical point of view. We shall then find that the exact specification of a quantum mechanical state could furnish information as to the values of a coordinate q and its conjugated momentum p only within limits corresponding to the Heisenberg uncertainty relation

$$\Delta q \Delta p \approx h, \tag{20.8}$$

where h is Planck's fundamental quantum of action. Hence, in the case of a system of f degrees of freedom, an extension in phase space of the magnitude

$$\int ... \int dq_1 ... dp_f = h^f \tag{20.9}$$

may be regarded as an approximate classical analogue for a precisely defined quantum mechanical state.

† For a general treatment of integrals invariant to canonical transformation, see Poincaré, *Méthodes nouvelles de la mécanique céleste*, vol. iii, chaps. 22–4, Paris, 1899. See also Brody, *Zeits. f. Phys.* **6**, 224 (1921). Starting with the invariance of extension in the phase space it is possible to proceed in the reverse direction and derive Liouville's theorem. The order of treatment given above is that of Gibbs.

‡ It will be noted that the numerical magnitude of an extension in phase, although invariant to canonical transformation, would depend, of course, on the units chosen for mass, length, and time.

21. Conditions for statistical equilibrium

Leaving this digression as to the consequences of transforming from one set of variables to another, we may now return to our main line of development for the ideas of the classical statistical mechanics. In previous sections we have seen that the condition of an ensemble of systems at any instant can be satisfactorily described by the density ρ with which the corresponding representative points are distributed in the appropriate phase space; we have found that a knowledge of this density of distribution would be sufficient to determine average values for different mechanical properties of the systems composing the ensemble; and we have obtained (19.7) the simple expression

$$\left(\frac{\partial\rho}{\partial t}\right)_{q,p} = -\sum_i \left(\frac{\partial\rho}{\partial q_i}\dot{q}_i + \frac{\partial\rho}{\partial p_i}\dot{p}_i\right), \tag{21.1}$$

for the rate at which this density would be changing with time at any selected fixed point in the phase space.

We may now inquire into the nature of the distributions of ρ as a function of the q's and p's which would make ρ permanently independent of time at all points in the phase space. Under such circumstances the probabilities of finding phase points in the various regions of the phase space and the average values for the properties of the systems in the ensemble would be independent of time, and we could describe the ensemble as being in *statistical equilibrium*.

Examining (21.1), we immediately see that a very simple method of obtaining statistical equilibrium would be to set up an ensemble with ρ originally distributed uniformly over the whole of phase space,

$$\rho(q,p) = \text{const.} \tag{21.2}$$

This would then make

$$\frac{\partial\rho}{\partial q_i} = \frac{\partial\rho}{\partial p_i} = 0, \tag{21.3}$$

and hence, in accordance with (21.1), the uniform distribution would be permanently maintained.

As a more general condition for statistical equilibrium, we could take ρ originally distributed as any function

$$\rho = \rho(\alpha) \tag{21.4}$$

of some constant of the motion α for the systems considered. Since α would itself be a function of the coordinates and momenta with a value for any given system which would not change with time, we should then have

$$\frac{d\alpha}{dt} = \sum_i \left(\frac{\partial\alpha}{\partial q_i}\dot{q}_i + \frac{\partial\alpha}{\partial p_i}\dot{p}_i\right) = 0, \tag{21.5}$$

provided $\dot{q}_i$ and $\dot{p}_i$ are rates of change with time for any actual system of the kind considered. Hence, since the quantities $\dot{q}_i$ and $\dot{p}_i$ appearing in (21.1) are such rates of change for the natural motion of a system whose phase point is located at the position in phase space under consideration, we now obtain

$$\left(\frac{\partial\rho}{\partial t}\right)_{q,p} = -\frac{d\rho}{d\alpha}\sum_i\left(\frac{\partial\alpha}{\partial q_i}\dot{q}_i+\frac{\partial\alpha}{\partial p_i}\dot{p}_i\right) = 0, \tag{21.6}$$

and see also under these circumstances that the original distribution would be permanently maintained. In the case of a conservative system, the energy of the system is the most natural constant of the motion to use in setting up ensembles which are in permanent statistical equilibrium.

Ensembles which have the property of being in statistical equilibrium prove to be important both for the representation of systems which are themselves in a condition of macroscopic equilibrium, and also for obtaining an insight into the fundamental hypothesis of statistical mechanics which we shall introduce later. We now turn in the next section to a more detailed consideration of certain examples of ensembles in statistical equilibrium.

22. The uniform, microcanonical, and canonical ensembles

(*a*) **The uniform ensemble.** We have already seen in the preceding section that an ensemble with its representative phase points distributed uniformly over the phase space with

$$\rho = \text{const.} \tag{22.1}$$

would retain this distribution permanently. Such an ensemble may be called a *uniform ensemble.*

As a representative ensemble for making predictions as to the behaviour of a single system in a condition corresponding to a partial specification of precise state, the uniform ensemble is not of interest. At the most we could only say that such an ensemble represented an individual system concerning whose state we had no information whatever. The uniform ensemble is of interest, however, in showing that there is no inherent tendency for the representative points to 'crowd' into any particular region of the phase space, as we shall emphasize in the next section.

An important property of the uniform ensemble arises from the invariance to canonical transformations which we have found in § 20 for the density of distribution ρ in the phase space. In accordance with this invariance, we see that an ensemble which is set up with uniform distribution with respect to one set of coordinates and momenta would also be uniformly distributed, and indeed with a value for ρ of the same numerical magnitude, using any other set of coordinates and momenta. Regarding a uniform ensemble as representing a system concerning whose state we have no knowledge, the above conclusion may be suggestively described by saying that we should be equally completely ignorant as to the state in all languages.

(*b*) **The microcanonical ensemble.** In the case of conservative systems, with which we shall for the most part be concerned, the energy $H(q,p) = E$ would be a constant of the motion. Hence for conservative systems, in accordance with the preceding section, an ensemble with the density ρ distributed as any function of the energy E would be in statistical equilibrium.

A very useful ensemble of this kind can be obtained by taking the density as equal to zero for all values of the energy except in a selected narrow range E to $E+\delta E$. Using the terminology of Gibbs, such an ensemble, specified by

$$\begin{aligned} \rho &= \text{const.} \quad (E \text{ to } E+\delta E), \\ \rho &= 0 \qquad\quad (\text{outside the above range}), \end{aligned} \tag{22.2}$$

may be called a *microcanonical ensemble.*

An ensemble of this kind can be regarded as obtained from an originally uniform ensemble by discarding all systems having phase points with positions that do not fall within the limits in the phase space that correspond to the energy range E to $E+\delta E$. Since no phase point could in any case cross over a surface of constant energy in the phase space, we thus obtain an immediate insight into the reasons why such an ensemble would remain in statistical equilibrium.

The microcanonical distribution is often employed as giving a representative ensemble for predicting the properties of a system which is itself known to be in a state of macroscopic equilibrium with an energy in the range E to $E+\delta E$. Indeed before the work of Gibbs it was the only ensemble extensively considered.

If we consider the possibility of making the selected range of energies δE narrower and narrower, we can regard ourselves as approaching the limit of a surface distribution of phase points all of which would correspond to systems having precisely the same energy. The resulting distribution might be called a *surface ensemble*, and in the older classical statistical mechanics it was often felt that such a surface distribution was specially appropriate for treating conservative systems, since classically a system of interest would be regarded as having some perfectly definite value of its energy even if we did not know precisely what that value was. For such a surface ensemble the surface density of distribution σ would be given by

$$\sigma = \frac{\text{const.}}{\left[\left(\frac{\partial E}{\partial q_1}\right)^2 + \dots + \left(\frac{\partial E}{\partial q_f}\right)^2 + \left(\frac{\partial E}{\partial p_1}\right)^2 + \dots + \left(\frac{\partial E}{\partial p_f}\right)^2\right]^{\frac{1}{2}}} = \frac{\text{const.}}{v_p}, \tag{22.3}$$

where v_p is the magnitude of the velocity with which a representative point moves through phase space. There was seldom, if ever, any real advantage, however, in going to this more complicated formulation.

(*c*) **The canonical ensemble.** Another kind of ensemble of importance for conservative systems may be defined by taking the density of distribution ρ as given by

$$\rho = Ne^{\frac{\psi - E}{\theta}}, \tag{22.4}$$

where N is the total number of systems in the ensemble, E is the energy as a function of the coordinates and momenta, and ψ and θ are parameters, independent of the q's and p's, whose values complete the description of the distribution. Such distributions were first introduced by Gibbs under the name *canonical ensemble.* Since the energy E would be a constant of the motion for conservative systems, these distributions as defined by (22.4) would then satisfy the conditions for statistical equilibrium given in the preceding section.

The *distribution parameters* ψ and θ appearing in (22.4) are constants having the dimensions of energy. Their values are evidently not independent since by integrating over the whole of phase space we obtain

$$N = \int \dots \int \rho \, dq_1 \dots dp_f = N \int \dots \int e^{\frac{\psi - E}{\theta}} \, dq_1 \dots dp_f \tag{22.5}$$

for the total number of systems in the ensemble. And this gives

$$e^{-\psi/\theta} = \int \dots \int e^{-E/\theta} \, dq_1 \cdots dp_f \tag{22.6}$$

as a necessary relation which must be satisfied by ψ and θ. By different choices of values for these parameters, that do satisfy the above relation, we can then obtain different canonical distributions for a given kind of system which will prove useful for representing the same system of interest in different macroscopic conditions.

Under the circumstances ordinarily encountered in statistical mechanical applications, the canonical distribution as defined by (22.4) is such that nearly all the systems in the ensemble have energies which are very close to the average energy for the ensemble. This arises from the combined effect of two factors. On the one hand, in the case of systems of many degrees of freedom, the volume in the phase space increases very rapidly as we include regions corresponding to higher and higher energies, and this tends to depress the numbers of systems in the ensemble with less than the average energy. On the other hand, the form of the distribution law itself, with the negative of the energy appearing in the exponent, tends to depress the numbers of systems with energies much greater than the average. The combined effect of the two factors usually leads to a high concentration in the neighbourhood of the average, as we shall see in more detail later (§ 141 (*b*)). This makes it possible to employ canonical ensembles for the representation of systems of interest which themselves have an energy which is precisely or at least nearly precisely defined. In particular, as first shown by Gibbs, canonical ensembles prove to be specially appropriate for the representation of systems in a condition of macroscopic thermodynamic equilibrium. The justification for this use of canonical ensembles will be presented very completely from a modern quantum mechanical point of view in Chapters XII and XIII.

The uniform, microcanonical, and canonical ensembles considered in this section all have distributions which are in statistical equilibrium. It will be appreciated, however, that ensembles having distributions which change with time can also be important and will indeed be essential for the representation of systems of interest which are not themselves in equilibrium.

23. The fundamental hypothesis of equal *a priori* probabilities in the phase space

We must now undertake a consideration of the fundamental hypothesis, as to equal *a priori* probabilities for equal regions in the phase space, which has to be introduced into the classical statistical mechanics in order to make applications to situations of actual interest. Although we shall endeavour to show the reasonable character of this hypothesis, it must nevertheless be regarded as a postulate which can be ultimately justified only by the correspondence between the conclusions which it permits and the regularities in the behaviour of actual systems which are empirically found.

The introduction of some kind of postulate as to *a priori* probabilities is always involved in the use of statistical methods for predictive purposes. In the case of statistical mechanics the typical situation, in which statistical predictions are desired, arises when our knowledge of the condition of some *system of interest* is not sufficient for a specification of its precise state. Under such circumstances we take recourse to a *representative ensemble* of systems of similar structure to the one of actual interest, appropriately distributed over a variety of different precise states, and then take the average properties and average behaviour of the systems in this ensemble as providing our best estimates as to the properties and behaviour of the actual system of interest. In setting up such a representative ensemble we shall, of course, assign the systems of the ensemble to different states in such a way as to agree with our partial knowledge of the state of the actual system of interest. We need some postulate as to *a priori* probabilities, however, to guide us in making assignments of systems to states that agree equally well with our knowledge as to the actual condition of the system. This necessity for an additional postulate arises not from any incompleteness nor inexactness in the principles of the classical mechanics, when applied to the conceptual situations for which this discipline was immediately devised, but from the real extension in theory which is needed when we are confronted by incompleteness and inexactness in the specification of states for the systems we wish to treat.

For the above purpose we now introduce the hypothesis of *equal* a priori *probabilities for different regions in the phase space* that correspond to extensions of the same magnitude. By this we mean that the phase point for a given system is just as likely to be in one region of the phase space as in any other region of the same extent which corresponds equally well with what knowledge we do have as to the condition of the system. Thus, if we consider groups of neighbouring states that correspond to differently located regions in the phase space R_1, R_2, R_3, etc., having the extensions in phase

$$v_1 = \int_{R_1} \dots \int dq_1 \dots dp_f, \qquad v_2 = \int_{R_2} \dots \int dq_1 \dots dp_f, \text{ etc.,} \tag{23.1}$$

we shall take the probabilities of finding the system in these different groups of states as proportional to the extensions v_1, v_2, v_3, etc., provided our actual knowledge as to the condition of the system is *equally well* represented by the states in any of the groups considered. For example, if all we know about the state of a system is that its energy lies in the range E to $E+\delta E$, and we consider different extensions in the phase space v_1, v_2, v_3, etc., which themselves lie entirely in the range E to $E+\delta E$, we shall take these extensions as giving the relative probabilities that the system itself would be found on examination to be in the corresponding different conditions.

As already emphasized, this principle must be regarded as a postulate. Nevertheless, we can see the reasonable character of the principle if we consider the behaviour of the uniform ensemble which was studied in the last section. It was there shown, if we set up an ensemble with the phase points for its member systems uniformly distributed, with the same number in all equal regions of the phase space without reference to location, that this uniform density of distribution would be permanently maintained as time proceeds. We thus find that the principles of mechanics do not themselves include any tendency for phase points to concentrate in particular regions of the phase space. (This may be contrasted, for example, with the tendency to concentrate which would be found in general if we used a combined configuration and velocity space for plotting the changing states of our systems.) Under the circumstances we then have no justification for proceeding in any manner other than that of assigning equal probabilities for a system to be in different equal regions of the phase space that correspond, to the same degree, with what knowledge we do have as to the actual state of the system. And, as already intimated, we shall, of course, find that the results which can then be calculated as to the properties and behaviour of systems do agree with empirical findings.

In concluding this section on the hypothesis of equal *a priori* probabilities for equal extensions $\int ... \int dq_1 ... dp_f$ in the phase space, attention may be called to our previous finding, see § 20, that the magnitudes of extensions in phase space are invariant to canonical transformation. It will hence be appreciated that our hypothesis as to equal *a priori* probabilities holds equally well in the language of any set of canonical variables.

Attention may also be called to our previous remark—see end of § 20—that an extension in phase space of the magnitude

$$\int ... \int dq_1 ... dp_f = h^f \tag{23.2}$$

might be regarded as an approximate classical picture for a precisely defined quantum mechanical state. At the correspondence principle limit, where conditions prevail such that either classical or quantum mechanical methods can be applied, we may then expect the principle of equal *a priori* probabilities in the phase space to imply—at least for those conditions—a principle of equal *a priori* probabilities for different quantum mechanical states. And we shall indeed find when we come to the development of quantum statistics that our postulatory basis will contain an hypothesis which will assure this agreement.

24. System of interest and representative ensemble

Accepting the postulate as to equal *a priori* probabilities for equal regions in the phase space, we are now provided with principles for selecting an appropriate ensemble to represent an individual system in a partially specified state, or vice versa of determining what knowledge as to the condition of an individual system is represented by a given ensemble. The essence of the relationship between representative ensemble and system of interest is that the distribution of the members of the ensemble over different states agrees with what is known as to the actual state of the system of interest but is otherwise uniform in the phase space in accordance with the hypothesis as to equal *a priori* probabilities.

With the help of this connexion we can then use the appropriate representative ensemble for making estimates as to the properties and behaviour of the corresponding system of interest itself. To do this we take the fraction of all the systems in the representative ensemble in any range of precise states as equal to the probability of finding the system itself, on examination, to be in that range of states; and we take the average value for any property for all the systems of the ensemble as equal to the average value of that property for the system itself. By the *probability* of finding the system in any given condition, we mean the fractional number of times it would actually be found in that condition on repeated trials of the same experiment. And by the *average* value of any property of the system we mean the average that would actually be found on such repeated trials. As averages we may, of course, take means, most probable values, or other forms of average, as seems convenient or appropriate. We may then use such averages as furnishing estimates for the properties of the system itself.

The above may also be expressed in more specific mathematical form. If we consider a condition of the system that corresponds to a group of states lying in a given region of the phase space R, we can take the probability P of finding the system in that condition as given by

$$P = \frac{1}{N} \int_R \cdots \int \rho(q,p)\, dq_1 \ldots dp_f, \tag{24.1}$$

where N is the total number of systems in the representative ensemble and $\rho(q,p)$ is their density of distribution. If we consider two such conditions, corresponding to regions R_1 and R_2, we can take the ratio of probabilities of finding the system in these conditions as given by

$$\frac{P_1}{P_2} = \frac{\int\limits_{R_1} \dots \int \rho(q,p)\, dq_1 \dots dp_f}{\int\limits_{R_2} \dots \int \rho(q,p)\, dq_1 \dots dp_f}. \tag{24.2}$$

And if, for example, we are interested in the mean as the average for any property of the system, which is a function $F(q,p)$ of the coordinates and momenta, we can use therefor our previous formula (18.6) for the mean over all the members of the ensemble

$$\bar{\bar{F}}(q,p) = \frac{1}{N} \int \dots \int F(q,p)\rho(q,p)\, dq_1 \dots dp_f. \tag{24.3}$$

In concluding this section on the relation between system of interest and corresponding representative ensemble, it may be noted that systems which are themselves in a steady condition of macroscopic equilibrium will be correlated with ensembles, which are in statistical equilibrium, and which hence exhibit constant average properties. On the other hand, systems whose macroscopic properties are changing with time will be correlated with ensembles, which are not in statistical equilibrium, and which hence give predictions as to average properties that change with time. In the next chapter we shall consider systems and ensembles in steady conditions, and in following chapters systems and ensembles that change with time.

25. Validity of statistical mechanics

It has been made clear by the foregoing that statistical methods can be employed in a very natural manner for predicting the properties and behaviour of any given system of interest in a partially specified state, by the procedure of taking averages in an appropriately chosen ensemble of similar systems as giving reasonable estimates for quantities pertaining to the actual system. We may now conclude the present chapter with some remarks concerning the point of view as to the validity of these methods which is adopted for the purposes of the present book.

In the first place, it is to be emphasized, in accordance with the viewpoint here chosen, that the proposed methods are to be regarded as *really statistical* in character, and that the results which they provide are to be regarded as true *on the average* for the systems in an appropriately chosen ensemble, rather than as necessarily precisely true in any individual case. In the second place, it is to be emphasized that the representative ensembles chosen as appropriate are to be constructed with the help of an hypothesis, as to equal *a priori* probabilities, which is introduced at the start, *without proof*, as a necessary postulate.

Concerning the first of these apparent limitations, it is to be remarked that we have, of course, no just grounds for objecting to the fact that our methods provide us with average rather than precise results. This is merely an inevitable consequence of the statistical nature of our attack, and we have committed ourselves to statistical rather than precise methods, either because we are forced thereto by lack of precise initial knowledge or because the practical problems which we have in mind are otherwise too complicated for treatment. Moreover, it is to be noted that the proposed methods make it possible to compute not only the average values of quantities but also the average *fluctuations* around those values. This, then, makes it possible to draw conclusions also as to the frequency with which we may expect to find systems with properties differing from the average to any specified extent. In the case of typical applications the computed fluctuations are extremely small. In the special cases where they are large enough they may be compared with what is found experimentally.

Concerning the second of the above-mentioned limitations on the character of the proposed methods, two remarks already made in the preceding section may again be emphasized. In the first place, it is to be appreciated that *some* postulate as to the *a priori* probabilities for different regions in the phase space has in any case to be chosen. This again is merely a consequence of our commitment to statistical methods. It is analogous to the necessity of making some preliminary assumption as to the probabilities for heads or tails in order to predict the results to be expected on flipping a coin. In the second place, it is to be emphasized that the actual assumption, of equal *a priori* probabilities for different regions of equal extent in the phase space, is the only general hypothesis that can reasonably be chosen. With the help of Liouville's theorem, it has been shown that the principles of mechanics do not themselves contain any tendency for phase points to crowd into one region in the phase space rather than another; and hence, in the absence of any knowledge except that our systems do obey the laws of mechanics, it would be arbitrary to make any assumption other than that of equal *a priori* probabilities for different regions of equal extent in the phase space. The procedure may be regarded as roughly analogous to the assumption of equal probabilities for heads and tails, after a preliminary investigation has shown that the coin has not been 'loaded'.

In further support of the validity of the proposed methods it may, of course, again be emphasized that they have the *a posteriori* justification of leading to conclusions which do agree with empirical facts. This includes agreement with conclusions not only as to average values but also as to fluctuations.

Hence the present point of view as to the validity of the methods of statistical mechanics may be summarized as follows. The methods are essentially statistical in character and only purport to give results that may be expected on the average rather than precisely expected for any particular system. The methods lead to calculated fluctuations around the averages which are exceedingly small in the case of the usual typical applications, and in other cases can be compared with empirical findings. The methods being statistical in character have to be based on some hypothesis as to *a priori* probabilities, and the hypothesis chosen is the only postulate that can be introduced without proceeding in an arbitrary manner. The methods lead to results which do agree with empirical findings.†

In the course of the historical development of statistical mechanics, the above point of view as to the validity of its methods was not the one ultimately adopted by Maxwell and Boltzmann. With the help of a different assumption, rather than our own bald postulate as to equal *a priori* probabilities, it was hoped to justify the methods of statistical mechanics by showing that the *time average* of any quantity pertaining to any single system of interest would actually agree with the *ensemble average* for that quantity calculated by the methods of statistical mechanics for all the members of the corresponding representative ensemble. The postulate leading to this conclusion was called by Boltzmann the *ergodic hypothesis*, and by Maxwell the assumption of *continuity of path.* It states that the phase point for any isolated system would pass in succession through every point compatible with the energy of the system before finally returning to its original position in the phase space.

† The above point of view does not conflict with that of Gibbs. He was much less explicit, however, concerning the hypothesis of equal *a priori* probabilities for equal regions in the phase space, and much less confident about the validity of statistical mechanics, presumably because of its failure to account for phenomena—such as specific heats—which we now know must be treated by the methods of quantum rather than classical statistical mechanics. The point of view appears in several ways sympathetic th that adopted by Fowler in his *Statistical Mechanics.*

We may first show that this ergodic hypothesis *would* lead to the above-stated simple relation between time averages and ensemble averages. To see this let us consider the behaviour of an isolated system of energy E and the behaviour of the corresponding microcanonical ensemble of systems with phase points uniformly distributed in the shell between E and $E+\delta E$, where δE is regarded as going to zero. In accordance with the ergodic hypothesis all the systems in this ensemble, and the system of interest itself, would have to exhibit exactly the same long time behaviour—as we let δE go to zero—since each would have to pass in succession through all states compatible with the energy E. Hence, if we divide our phase space up into different *elementary regions* $k, l, m,\ldots$, we could regard all systems in the ensemble, and the system of interest itself, as spending the same fractions, w_k, w_l, w_m,... of any very long time interval 0 to T in those different regions.

Taking N as the total number of systems in the ensemble, we could then write $Nw_k T$, $Nw_l T$, $Nw_m T$,... as expressions for the total time, during the interval 0 to T, which would be spent by members of the ensemble in the regions specified. On the other hand, if we denote by N_k, N_l, N_m,... the numbers of systems at any time t in these regions, we could also write $\int N_k\,dt$, $\int N_l\,dt$, $\int N_m\,dt$,... for these same quantities, where we integrate over the time interval 0 to T. Equating, we then have

$$\begin{aligned} Nw_k T &= \int_0^T N_k\,dt, \\ Nw_l T &= \int_0^T N_l\,dt, \\ Nw_m T &= \int_0^T N_m\,dt, \\ &\quad \cdots\cdots \end{aligned} \tag{25.1}$$

The quantities N_k, N_l, N_m,... are, however, independent of time since the microcanonical ensemble has been shown to be one that remains in statistical equilibrium. This then gives us

$$w_k = \frac{N_k}{N}, \qquad w_l = \frac{N_l}{N}, \qquad w_m = \frac{N_m}{N}, \quad \ldots. \tag{25.2}$$

This result then states that the fraction of the time during a long interval 0 to T which a given conservative system of interest would spend in any region of the phase space, would be equal to the fraction of all the systems in the corresponding representative ensemble, which would be in that same region. The probability of finding any given system of interest at a random instant of time in any specified state would then be equal to the probability of finding a system picked at random from the corresponding representative ensemble in that state, and the time average of any quantity pertaining to a single system of interest would be given by the corresponding ensemble average.

Hence the older point of view, based on the ergodic hypothesis, might seem at first sight to furnish a more satisfactory justification for the methods of statistical mechanics than is furnished by the point of view which has been adopted in this book. In the first place, the older development was regarded as primarily based on an hypothesis which might be an actual consequence of the *exact* laws of mechanics, while our development is based on a definite postulate as to *a priori* probabilities, which can only be justified if its consequences do correspond with *statistical* findings. In the second place, the older development claims that every individual system will exhibit time averages that agree with ensemble averages, while our development only assumes that the averages obtained on successive trials of the same experiment will agree with ensemble averages, thus permitting any particular individual system to exhibit a behaviour in time very different from the average. Nevertheless, it now seems clear, firstly, that the ergodic hypothesis cannot be strictly maintained in its original form, and, secondly, that its employment even as a working hypothesis would deny to statistical ensembles their full function of representing the relative probabilities for all the different kinds of states, including the unusual ones, which might be exhibited by a given system of interest.

With regard to the possible truth of the ergodic hypothesis itself, it is evident—except for the trivial one-dimensional case—that a mechanical system could not possibly obey this hypothesis in the original strict form which would require the phase point to pass precisely through every point in the phase space compatible with the system's energy. For a system of f degrees of freedom there would be $2f-1$ quantities besides the energy which would be *arbitrary* constants of the motion, and in effect only one of these would be 'used up' in specifying the initial position of the phase point along any particular trajectory. Hence the phase point for a system with one assignment of constant values for the remaining $2f-2$ quantities could certainly not pass through points in the phase space corresponding to any other assignment even though these points did correspond to the right energy.

In the case of 'simple' quantities, which are constants of the motion, for example the components of linear and angular momentum for the system as a whole, we should presumably regard ourselves as knowing the values of these quantities for the actual system of interest, and hence the above limitation would have no effect on our considerations since we should then only use systems all having the same (e.g. zero) components of total momenta in constructing the appropriate representative ensemble. And in the case of constants of the motion, having a very complicated dependence on the coordinates and momenta, the limitation to a particular choice of values might usually be such as to permit the phase point to visit substantially all regions of the ergodic surface even though it could not pass precisely through every point. Nevertheless, it is easy to think of cases where the departure from the consequences of the ergodic hypothesis would be very great. Thus, to take a system of the kind usual in classical statistical applications, let us consider a gas, contained in an exactly rectangular box, and composed of rigid particles moving back and forth perpendicular to a pair of parallel faces in such a way as to avoid collisions. Since this to-and-fro motion would be permanently maintained, it is evident that the phase point for this system would certainly not visit all significant regions in the phase space compatible with the energy, and that the time averages for quantities pertaining to the system (e.g. pressure on the walls) would be very different from the ensemble averages over the corresponding microcanonical distribution.

The foregoing now also makes it evident why the use of the ergodic hypothesis, even merely as a heuristic principle, would prevent the methods of statistical mechanics from exhibiting their full statistical character and usefulness. Since it is conceptually possible to have a system of interest permanently remain in a condition very different from the average condition for all the members of the corresponding representative ensemble, it would be quite inappropriate to employ an hypothesis which would regard all the systems of the representative ensemble as being in a condition to give the same time averages for their properties. Rather we must wish our representative ensembles to be capable of furnishing predictions as to the frequency with which we may expect to find a system of interest in any kind of condition, including even such rare kinds as that given by the example of a gas composed of particles moving permanently back and forth between a pair of parallel faces. Moreover, from the point of view adopted in the present exposition, this complete function of our statistical methods *is* achieved by our employment of the hypothesis of equal *a priori* probabilities for equal regions in the phase space in the construction of representative ensembles. This procedure, which, as we have seen, is the only possible non-arbitrary one, does provide ensembles which would give an exceedingly small representation to conditions that we recognize as rare, and indeed in general provides ensembles which give predictions that do accord with actual phenomena.

A further unsatisfactory aspect of the original introduction of the ergodic hypothesis may be emphasized. Although the ergodic hypothesis, if true in its original form, would have secured an equality between the time averages for any individual system and the ensemble averages over the corresponding microcanonical distribution, the certainty of this result was only demonstrated for exceedingly long time intervals, and no assurance as to the pertinence of the result with respect to the short time intervals involved in actual experimentation was provided.† Thus even the ergodic hypothesis leads to the desired kind of prediction for a system of interest only when supplemented by the further essentially statistical assumption that the properties of a system at any one time could be successfully taken as their long time average.

It must hence be felt that the point of view of the early founders did not give due recognition to the truly statistical character of the problems to be attacked.‡ Impressed by the exact character of the principles of classical mechanics, and also by the actual regularities in the macroscopic behaviour of systems composed of many molecules, they apparently hoped to secure really precise results for such systems by the temporary introduction of an hypothesis which might itself be ultimately validated from the principles of mechanics proper, and did not sufficiently appreciate that a further essentially statistical assumption would be needed even if their hypothesis were valid. As emphasized in § 23, however, the treatment of systems which are not in precisely specified states *must involve* an actual *extension of theory*, which requires the introduction of an additional postulate not derivable from those included in the mechanics of precise states.

Returning to the course of historical development, after it was

† Compare with the similar view expressed by Fowler, *Statistical Mechanics*, second edition, Cambridge, 1936, § 1.4.

‡ This failure to adopt a truly statistical viewpoint is not to be ascribed to Gibbs.

appreciated that the ergodic hypothesis in its original form could not be strictly true, attention was turned to the study of the so-called *quasi-ergodic hypothesis* in accordance with which the phase point would in time approach as near as desired to any selected point on the ergodic surface, and also to the study of the additional conditions which would be sufficient to secure desirable relations between time and ensemble averages.† These studies have recently been advanced by the work of von Neumann, Birkhoff, and others‡ showing that there would be no inconsistency in the existence of a class of mechanical systems of a character such that there would be a great 'tendency' for individual paths to exhibit time averages which would be 'nearly' the same as the ensemble averages over the corresponding microcanonical ensemble. Nevertheless, it is evident that such studies can neither contradict the mechanical possibility for some paths, which would exhibit time averages quite different from the ensemble averages, nor eliminate the ultimate necessity for recourse to a postulate as to *a priori* probabilities.

From the point of view adopted in the present book, the hypothesis, of equal *a priori* probabilities for equal regions in the phase space, must in any case be regarded as an essential element of statistical mechanics, which has to be introduced by postulation and which is then sufficient to provide the statistical methods used. *Without* this postulate there would be nothing to correspond to the circumstance that nature does not have any tendency to present us with systems in conditions which we regard as mechanically entirely possible but statistically improbable; and *with* the postulate the use of statistical mechanics for the determination of averages and fluctuations then becomes merely a matter for computation.

† The possibility for a quasi-ergodic hypothesis and the necessity for additional assumptions, to secure the desired relation between time and ensemble averages, were first appreciated by P. and T. Ehrenfest, *Encykl. d. Math. Wiss.* IV. 2, ii, Heft 6, Leipzig. See footnotes 89a and 90.

‡ See von Neumann, *Proc. Nat. Acad.* **18,** 70 (1932); ibid. **18,** 263 (1932); Birkhoff, ibid. **17,** 656 (1931); Birkhoff and Koopman, ibid. **18,** 279 (1932).

IV

THE MAXWELL-BOLTZMANN DISTRIBUTION LAW

26. The microcanonical ensemble as representing a system in equilibrium

We are now ready to commence the application of the classical statistical mechanics to the study of systems composed of large numbers of individual molecules. This is the kind of system for whose treatment the methods of statistical mechanics were originally especially devised.

In the present chapter we shall be interested in the properties of such systems when they can be regarded, from a gross point of view which neglects the behaviour of individual molecules, as being in a steady condition of macroscopic equilibrium. These properties will be determined by the manner in which the molecules composing the system are then distributed among their own individual states, and our main task in the present chapter will be a derivation of the so-called Maxwell-Boltzmann distribution law, which gives the distribution over different molecular states that can be expected to prevail when the system has come to macroscopic equilibrium.

Owing to the complicated character and, in practical situations, incomplete specification of state of a system composed of an enormous number of individual molecules, we shall wish to employ the methods of statistical mechanics in solving the above problem. We must hence first determine an appropriate statistical ensemble of systems, similar in structure to the one of actual interest, which can be used to represent the average properties to be expected for the system itself. In the case of a system which is given a definite fixed energy and which is enclosed in a fixed container in order to avoid the dissipation of matter, we shall find that a microcanonical distribution will give a suitable representative ensemble.

Consider a system composed of a large number of individual molecules, having the total energy E, and enclosed in a stationary container of fixed volume v. Let the walls of the container be of an idealized non-conducting character so that there will be no direct flow of energy through them as a result of molecular collisions. And let the container itself be sufficiently massive so that it can be regarded as approximately stationary with no appreciable transfer of kinetic energy between the system and the container when collisions take place. Furthermore, if any external fields of force act on the molecules, let them be conservative in character and unchanging in time.

The total energy of the system would then have a constant value independent of the time. On the other hand, the components of total linear momentum for the system would be able to adjust themselves by interaction with the walls of the approximately stationary container, and the components of angular momentum would also be able to make such an adjustment, except for the special case of a container with walls which are ideally smooth and have the shape of a surface of revolution.

Let us now assume that all we know about the mechanical state of the system of molecules is the constant value of its total energy E. In view of the partial character of this specification of precise state, we must then apply the methods of statistical mechanics in order to obtain estimates as to the expected properties and behaviour of the system. This we do by resorting to an appropriate representative ensemble of similar systems.

In accordance with the relations between system of interest and representative ensemble as discussed in § 24, the phase points for the members of the representative ensemble are to be taken—in the first place—as distributed in a manner to agree with our partial knowledge of the actual state, and—in the second place—as otherwise distributed in as uniform a manner as possible in agreement with the principle of equal *a priori* probabilities for equal regions in the phase space. In the present case this can evidently be secured by taking a *microcanonical ensemble* of systems, all having the structure and volume of interest, and having their representative points uniformly distributed over that portion of the phase space which lies in a narrow shell between the surfaces of constant energy E to $E+\delta E$, that is, by taking

$$\begin{aligned} \rho &= \text{const.} \quad (E \text{ to } E+\delta E), \\ \rho &= 0 \qquad\quad (\text{outside the above range}). \end{aligned} \tag{26.1}$$

As we let δE go to zero, we then fulfil the requirements that the distribution should be left uniform except in so far as is needed to represent our partial knowledge of the state of the system, namely in this case its energy E.

The particular characteristic of such a microcanonical ensemble, which we now wish to emphasize, is the fact that it has a distribution which is known (§ 22) to remain in statistical equilibrium. Thus the most probable or other kind of average values which it predicts for the properties of the system of interest will themselves not be changing with the time. Hence a microcanonical ensemble may be regarded as representing a system which is itself—as far as we know—in a steady macroscopic condition when looked at from a point of view that neglects the behaviour of individual molecules.

This important character of the microcanonical ensemble arises from the fact that it represents a system concerning whose state nothing but the energy is known, and a mere specification of energy does not lead to the expectation of any particular direction of change for the macroscopic properties of the system. Fluctuations of those properties around their most probable values may be expected, and would indeed be predicted from the microcanonical ensemble but, nevertheless, as taking place with equal frequency in any given direction and its reverse. The situation is usually quite different when we have a system for which other things besides the energy are known. For example, if we consider a gas known to have a given total energy, and also known at some given instant to be completely situated in one-half of the total container available to it, the corresponding representative ensemble would be such as to give a practically certain prediction that the gas would start at once in the direction of distributing itself more uniformly throughout the container.

Instead of using the microcanonical ensemble, as we shall in the present chapter, for studying the properties of a system of molecules in a condition of equilibrium, it would also be possible to use the canonical ensemble. Such an ensemble also represents a system which is in a steady condition of macroscopic equilibrium, but one for which the energy instead of being precisely specified has a most probable value. In the classical statistics, as developed by Gibbs, such canonical ensembles were already found especially appropriate for studying systems having a precisely specified temperature rather than a precisely specified energy. In the quantum statistics, as developed later in this book, we shall also find reasons for such use of the canonical ensemble. For our present purposes, however, we shall find the microcanonical ensemble adequate.

In concluding this section on the classical use of the microcanonical ensemble to represent a system of known energy in a macroscopically steady state, it is to be noted that the ensemble is such as to give equal probabilities for finding the representative point for the actual system of interest in all equal elements δv_γ of the corresponding phase space or γ-space which are located within the volume available for the molecules and within the infinitesimal energy range E to $E+\delta E$. This is the result which we shall need in following sections in calculating the probabilities for the different conditions of a system of molecules that will interest us.

27. Specification of condition for a system of many similar molecules

We may now turn to the problem of specifying different conditions of interest for a system composed of many molecules. We shall wish to specify these conditions in terms of the distribution of the molecules composing the system among their own possible states, and may confine our attention at the start to systems containing molecules of a single kind, since the later extension to molecules of more than one kind (see § 30) will involve no additional principles. We may take n as the total number of similar molecules composing the system, and r as the number of degrees of freedom for a single molecule of the kind in question. The total number of degrees of freedom for the system as a whole will then be $f = nr$.

We may begin by considering the specification of state for a single molecule in the system. For this purpose we may use any desired set of r conjugate coordinates and momenta

$$q_1 \cdots q_r,\ p_1 \cdots p_r \tag{27.1}$$

whose values would describe the state of the molecule, considered itself as a minute mechanical system. It will often be convenient to regard the first three of the above coordinates, q_1, q_2, q_3, as giving the position of the centre of gravity of the molecule with respect to a set of Cartesian axes, which is taken stationary relative to the container for the system, and regard the remaining coordinates as giving the internal configuration of the molecule. Our set of coordinates and momenta for the individual molecule then becomes

$$x,\ y,\ z,\ q_4 \cdots q_r,\ p_x,\ p_y,\ p_z,\ p_4 \cdots p_r, \tag{27.2}$$

where x, y, z are the Cartesian coordinates for the centre of gravity of the molecule, p_x, p_y, p_z are the corresponding components of linear momentum for the molecule as a unit, and the remaining coordinates and momenta refer to the internal configuration and motion of the molecule.

Having chosen the kind of coordinates and momenta to be used for the individual molecules, it will then be convenient to introduce the idea of a *μ-space*, provided with $2r$ rectilinear axes, one for each of the r coordinates and r momenta. The exact position of a representative point in such a μ-space would then give an exact specification of the state of the molecule under consideration.

For our later statistical applications, however, we shall not be interested in the precise states of molecules but rather in the various small ranges within which the values of their coordinates and momenta might fall. For this purpose we may now consider the μ-space for any molecule of interest as divided up into a collection of *equal elementary regions*, corresponding to different ranges

$$\delta v_\mu = \delta q_1 \dots \delta q_r \, \delta p_1 \dots \delta p_r, \tag{27.3}$$

all having the same magnitude of extension. Such elementary regions are often spoken of as *cells* in the μ-space. Thinking of these different cells as labelled by the integers, 1, 2, 3,..., i,..., we can then specify the state of any molecule, within the range desired, by giving the particular cell in the μ-space inside of which the representative point for the molecule is taken as situated.

We may next turn to the specification of the state of the system as a whole. For this purpose we shall regard the $2r$ coordinates and momenta, as assigned above to each of the individual molecules, as now furnishing a total collection of $2nr$ coordinates and momenta suitable for the system itself.† The state of the whole system can then be regarded as determined by the individual states of its n component molecules. Furthermore, by taking a μ-space of the kind discussed above for each of the different molecules and combining these individual spaces together, we can now construct a γ-space or phase space suitable for the whole system. The exact position of a single representative point in this γ-space will then give the exact location of each particular molecule in its own μ-space, and hence give an exact specification of the state of the whole system.

However, as already remarked above, we shall be interested for our later statistical applications in specifications of state which are not precise, but which correspond to ranges within which the coordinates and momenta for individual molecules might lie. For this purpose we may now regard the equal elementary regions, into which we have

† This implies that we can regard the coordinates and momenta of the molecules themselves as giving a sufficiently precise specification of the intermolecular field, and can neglect the circumstance that, in general, a knowledge of additional variables would really be necessary to give an exact specification to the complicated electromagnetic field existing between the molecules. This approximation is valid when the velocities of the electric charges involved are small compared with that of light.

divided the μ-space for a single molecule, as also providing a division of the whole γ-space into *equal elementary regions*, all having the same magnitude of extension, but corresponding to different ranges of the form

$$\delta v_\gamma = (\delta q_1 \dots \delta p_r)_1 (\delta q_1 \dots \delta p_r)_2 \dots (\delta q_1 \dots \delta p_r)_k \dots (\delta q_1 \dots \delta p_r)_n, \quad (27.4)$$

where the subscripts 1, 2,..., k,..., n denote the particular molecule to which the indicated range applies. Within the desired ranges of values for the coordinates and momenta of the component molecules we could then specify the state of the system as a whole by giving the particular region in the γ-space, of the form (27.4), inside of which the representative point for the whole system is taken as situated.

Having obtained this method of using elementary regions δv_γ of the form (27.4) for specifying the (approximate) state of our system, it is now to be specially noted that a number of such regions would, in general, correspond to the same condition of the system from the point of view of its macroscopic properties and behaviour, since from that viewpoint it makes no difference which particular molecules of the n similar ones are taken as lying in specified cells of the μ-space. For example, if our system is specified by an elementary region δv_γ having such a character that each of the n molecules is assigned to a different cell in the μ-space, it is then evident by considering the possibilities for permuting the molecules among the different occupied cells that there would be a total of $n!$ different regions in the γ-space all corresponding to the same gross observational properties.

For this reason we shall now finally be specially interested in specifying what may be called *the condition* of our system by stating merely the numbers of molecules

$$n_1, n_2, n_3, \dots, n_i, \dots \quad (27.5)$$

which are assigned to different cells i in the μ-space, without specifying just which molecules are used to provide the quotas for those different cells. Taking any such condition of the system, it is then evident that there would be a total of

$$G = \frac{n!}{n_1!\, n_2!\, n_3! \dots n_i! \dots} \quad (27.6)$$

different elementary regions δv_γ of the form given by (27.4), which would all correspond to this same specified condition. By changing from one such elementary region in the γ-space to another we change the *state* of the system—as specified within a range of δv_γ of the form (27.4)—but leave what we have called the *condition* of the system unaltered.† Such changes from one δv_γ to another could be brought about by merely interchanging pairs of entirely similar molecules which were originally assigned to different cells of the μ-space, and hence these changes would have no effect on the gross observational properties of the system.

In this connexion it may be noted in passing that our procedure in regarding the interchange of two similar molecules as corresponding to a significant change in the mechanical state of a system, even though not in its condition, evidently implies the possibility of keeping a continuous observation on the system which would let us know whether two similar molecules do change roles or not. This, however, is in entire agreement with the point of view of the *classical mechanics*, which would permit such a continuous observation, at least in principle, without any disturbing effect on the behaviour of the molecules. On the other hand, when we come to the *quantum mechanics* we must be prepared to find limitations on the idea of maintaining a precise observational control on the behaviour of molecules without introducing disturbances. Indeed we shall actually find in the quantum mechanics that we should lose all knowledge as to the distinction between two similar molecules, if they pass through an encounter sufficiently intimate so that it could not be described in terms of classical trajectories on account of the complementarity expressed by the Heisenberg uncertainty principle. This feature of the quantum mechanics proves essentially to be involved in the origin of certain striking differences between the classical and quantum statistics of systems containing *colliding* molecules, which can become very important under some circumstances, as we shall see later.

† A remark may be desirable as to the terminology adopted in the present book. The word *state*—when without qualification—is used as corresponding to the most precise specification of instantaneous condition theoretically possible. In the classical mechanics this means precise values for all the $2f$ coordinates and momenta for the system; in the quantum mechanics it means precise values for f instead of $2f$ variables, as we shall see later. The word *state*—when qualified—for example, *state specified within a range* δv_γ, can be used to correspond to a less precise specification of condition. The word *condition* will be used in general corresponding to specifications of different degrees of precision.

This is not the only terminology that might have been adopted. Some authors use the phrases 'microscopic state' and 'macroscopic state' to denote respectively what we have described above by 'the state of the system as specified within a range δv_γ' and 'the condition as specified by the numbers of molecules n_i in the various cells i of the μ-space'. Such a terminology is not fortunate, however, since the word 'microscopic' would suggest the greatest possible precision, and the word 'macroscopic' would suggest primarily a lesser degree of precision rather than the neglect of distinctions between different similar molecules.

28. The probabilities for different conditions of the system

We may now combine the results of the two preceding long sections. In the first of these sections, § 26, it has been shown that a system in macroscopic equilibrium with a specified total energy can be appropriately represented by a microcanonical ensemble, and that this ensemble gives equal probabilities for all equal regions δv_γ in the γ-space which correspond to the range of energies E to $E+\delta E$ considered. In the second of the foregoing sections, § 27, it has been shown for a system of n similar molecules that there would be a total of $n!/(n_1!\, n_2!\, n_3!\, \dots\, n_i!\, \dots)$ different equal regions δv_γ in the γ-space which would all correspond to the condition of the system specified by taking $n_1, n_2, n_3,\dots, n_i,\dots$ as the numbers of molecules assigned to different equal cells in the μ-space for the kind of molecule involved. Combining these two findings, we can now write

$$P = \frac{n!}{n_1!\, n_2!\, n_3!\, \dots\, n_i!\, \dots} \times \text{const.} \tag{28.1}$$

for the probability P of finding the system of interest in a condition specified by the numbers of molecules n_i in the different cells i of the μ-space, where the factor denoted by 'const.' would have the same value irrespective of the condition considered.

For our further purposes it will be more convenient to take the logarithm of this probability, which can be expressed with the help of a summation over all cells i, in the form

$$\log P = \log n! - \sum_i \log n_i! + \text{const.} \tag{28.2}$$

In problems of ordinary interest, moreover, not only will the total number of molecules n in the system be exceedingly large, but the numbers of molecules n_i in most of the occupied cells i can also be taken as large compared with unity. It then becomes possible to simplify (28.2) by introducing Stirling's approximation for the factorials of large numbers:

$$\begin{aligned} n! &\approx \sqrt{(2\pi n)}\left(\frac{n}{e}\right)^n, \\ \log n! &\approx n\log n - n + \tfrac{1}{2}\log 2\pi n. \end{aligned} \tag{28.3}$$

Doing so, noting the cancellation of the two terms $-n$ and $+\sum_i n_i$, and dropping terms of order $\log n$, the result (28.2) then leads to

$$\log P = n\log n - \sum_i n_i \log n_i + \text{const.} \tag{28.4}$$

as the finally desired expression for the probability P of a condition of the system specified by the numbers of molecules n_i in different equal cells i of the μ-space for the kind of molecule involved.

29. Condition of maximum probability. Maxwell-Boltzmann distribution law

With the help of the foregoing expression for the probabilities of different conditions of a system, composed of n similar molecules in macroscopic equilibrium with a specified energy E, we shall now be interested in determining which condition of the kind specified will be the most probable one. To carry out the analysis, let us take the numbers of molecules n_i in most of the occupied cells i as being large enough, not only to permit the foregoing use of Stirling's approximation for their factorials, but also to justify us in treating the numbers themselves as continuous variables in applying the calculus of variations. We can then use (28.4) as an appropriate expression for the probability P of any specified condition, and can take the variational equation

$$\delta \log P = -\sum_i (\log n_i + 1)\,\delta n_i = 0 \tag{29.1}$$

as a necessary condition for a maximum value of P.

The above variations δn_i in the numbers of molecules n_i in the various states i cannot be carried out, however, in a completely arbitrary manner. In the first place, the total number of molecules n cannot be altered, so that we must also have the subsidiary equation

$$\delta n = \sum_i \delta n_i = 0. \tag{29.2}$$

In the second place, the total energy of the system E has to remain within the *infinitesimal* range which we have specified for our micro-canonical ensemble, so that we shall also have the subsidiary equation

$$\delta E = \sum_i \epsilon_i\,\delta n_i = 0, \tag{29.3}$$

where ϵ_i is the rate of increase in the energy of the system per molecule introduced into the ith region of the μ-space for equilibrium values of the n_i, as will be further discussed below. (The variation δE in (29.3) should, of course, not be confused with the infinitesimal range in energy for the ensemble.)

Equations (29.1), (29.2), and (29.3) are thus three simultaneous expressions which must be satisfied in order to secure the desired conditional maximum of $\log P$ or of P itself. Combining them by Lagrange's method of undetermined multipliers, we obtain a single equation which can be written in the form

$$\sum_i (\log n_i + \alpha + \beta\epsilon_i)\,\delta n_i = 0, \tag{29.4}$$

where α and β are undetermined constants. Since the variations δn_i

can now be treated as arbitrary, this then leads for each region in the μ-space to the equation

$$\log n_i+\alpha+\beta\epsilon_i = 0. \tag{29.5}$$

Solving for n_i, we then finally obtain

$$n_i = e^{-\alpha-\beta\epsilon_i} \tag{29.6}$$

as the desired expression for the *most probable* number of molecules n_i, in the ith one of the equal ranges $\delta q_1 ... \delta p_f$, which we have set up for the coordinates and momenta of the kind of molecules that we are considering. This result, which gives the most probable distribution of molecules among their own individual states for a system in a macroscopically steady condition, may be called the *Maxwell-Boltzmann distribution law*.

The quantities α and β occurring in (29.6) are constants whose values are related to the total number of molecules and to the energy (or temperature) of the system in a manner which we shall examine more closely in a later section. The quantity ϵ_i will be seen from the manner of its introduction in (29.3) to be the rate of increase in the energy of the system,

$$\epsilon_i = \frac{\partial E}{\partial n_i}, \tag{29.7}$$

per molecule introduced into the ith region of the μ-space, when we have the most probable distribution of molecules over the different regions. In the case of a sufficiently dilute gas, where the energy of interaction between the molecules can be taken as negligible, ϵ_i will simply be the energy that would be assigned to a molecule having the position, internal configuration, external and internal momenta corresponding to the ith range of coordinates and momenta $\delta q_1 ... \delta p_r$. In the case of more concentrated systems, where the energy of a molecule would also depend appreciably on interaction with its neighbours, the interpretation of ϵ_i would be less simple. We shall be primarily interested in the case of dilute gases and shall usually treat ϵ_i simply as equal to the actual energy of a single molecule in the ith cell $\delta q_1 ... \delta p_r$ of the μ-space.†

The above derivation of the Maxwell-Boltzmann distribution law was obtained with the help of the Stirling approximation for the factorials $n_i!$ and with the added approximation involved in treating

† In the expositions usually given, it is tacitly assumed from the start that the system is a highly dilute gas, and that ϵ_i can be treated in this manner. The more general treatment of ϵ_i, suggested above and previously by the writer, might prove useful in studying more concentrated systems.

the numbers n_i themselves as continuous variables. This was done in order to employ the usual methods of the calculus of variations as applied to continuous variables. The results obtained are practically exact, however, for all regions i where the numbers n_i *are* large compared with unity. Hence the introduction of the approximations has no serious consequences in the treatment of systems of ordinary interest composed of many molecules.

Other methods of studying the equilibrium distribution of molecules are open to us, depending on different possibilities of choice as to representative ensemble and computed average. Two methods which have received considerable attention may be briefly mentioned here.

In one of these methods the *microcanonical distribution* is retained as the representative ensemble, but instead of the most probable numbers the *mean numbers* of molecules in the various cells i are calculated for the systems composing the ensemble. This procedure has the advantage that it no longer proves necessary to take the number of molecules in a given cell as large in order to perform the computation; nevertheless the actual calculation is found to involve approximations of its own, and the method is less expeditious than the historically familiar one presented above. The new averages—mean numbers instead of most probable numbers—are also found† to obey the Maxwell-Boltzmann distribution law (29.6).

In the other of the alternative methods to be mentioned, the *canonical distribution* instead of the microcanonical is taken as furnishing the representative ensemble, and the *mean numbers* of molecules in the various cells i are the averages calculated for the systems composing the ensemble. As already noted in § 26, the change from a microcanonical to a canonical ensemble corresponds to a change from a system of interest for which the energy is precisely specified to one in which there is a spread in the values which might be found for that quantity. We may regard this last method of attack as the most satisfactory of all. Using this method we shall be able to show (see § 113) that calculations made without introducing any approximations can lead precisely to a result of the form (29.6).

As a final remark in connexion with the derivation, which we have given for the Maxwell-Boltzmann distribution law, it may be emphasized for the case of large numbers of molecules that the Maxwell-Boltzmann can be shown to be not only the most probable distribution but also to be much more probable than other distributions which differ

† See Fowler, *Statistical Mechanics*, second edition, Cambridge, 1936.

appreciably therefrom. We shall leave the calculation of fluctuations, however, until we have developed the full apparatus of the quantum statistics.

30. Maxwell-Boltzmann distribution for molecules of more than a single kind

We must next consider the form of the equilibrium distribution law for a system composed of interacting but not chemically reacting molecules of more than a single kind. For the total numbers of molecules of the different kinds present, we may now use the symbols n, n', n'', etc., and may specify a given condition of the system by taking n_i, n'_j, n''_k, etc., as being the numbers of molecules of these different kinds which are assigned to the various equal cells i, j, k, etc., into which we divide the μ-spaces for these different kinds of molecules.

In accordance with methods of treatment similar to those previously developed, it is now evident that we can take the probability P for any such condition of our system, when it has come to macroscopic equilibrium with a definite energy E, as given by

$$\log P = n\log n - \sum_i n_i \log n_i + n'\log n' - \sum_j n'_j \log n'_j + \\ + n''\log n'' - \sum_k n''_k \log n''_k + \ldots + \text{const.} \quad (30.1)$$

As the equations determining the most probable condition of the system under the imposed requirements as to total numbers of molecules and total energy, we shall then have

$$\delta \log P = -\sum_i (\log n_i + 1)\,\delta n_i - \sum_j (\log n'_j + 1)\,\delta n'_j - \sum_k (\log n''_k + 1)\,\delta n''_k - \ldots \\ = 0, \quad (30.2)$$

$$\delta n = \sum_i \delta n_i = 0,$$

$$\delta n' = \sum_j \delta n'_j = 0, \quad (30.3)$$

$$\delta n'' = \sum_k \delta n''_k = 0,$$

$$\ldots\ldots\ldots,$$

and

$$\delta E = \sum_i \epsilon_i\,\delta n_i + \sum_j \epsilon'_j\,\delta n'_j + \sum_k \epsilon''_k\,\delta n''_k + \ldots = 0. \quad (30.4)$$

Combining once more by the method of undetermined multipliers, this gives us as the single equation which must be satisfied

$$\sum_i (\log n_i + \alpha + \beta\epsilon_i)\,\delta n_i + \sum_j (\log n'_j + \alpha' + \beta\epsilon'_j)\,\delta n'_j + \\ + \sum_k (\log n''_k + \alpha'' + \beta\epsilon''_k)\,\delta n''_k + \ldots = 0, \quad (30.5)$$

where α, α', α'',... and β are undetermined constants. Since the variations can now be treated as independent, we then have

$$\begin{aligned} n_i &= e^{-\alpha-\beta\epsilon_i}, \\ n'_j &= e^{-\alpha'-\beta\epsilon'_j}, \\ n''_k &= e^{-\alpha''-\beta\epsilon''_k}, \\ &\cdots\cdots \end{aligned} \tag{30.6}$$

as the desired expressions for the most probable numbers of molecules of the different kinds in the different cells of their μ-spaces, when macroscopic equilibrium prevails.

It will be noted that each kind of molecule will be distributed in accordance with an expression having the form of the original Maxwell-Boltzmann distribution law (29.6). The constants α, α', α'',... will, in general, be different for the different kinds of molecules, and will have values determined by the total number of molecules of the kind in question, as we shall examine in more detail presently. The constant β will be *the same* for the different kinds of molecules, and will be related to the temperature of the system, as we shall also see presently. For sufficiently dilute systems the quantities ϵ_i, ϵ'_j, ϵ''_k,... will be the energies to be ascribed to a single molecule in the indicated cells i, j, k,... into which we have divided the μ-spaces for the different kinds of molecule involved.

31. Re-expression of Maxwell-Boltzmann law in differential form

The expression for the Maxwell-Boltzmann distribution law

$$n_i = e^{-\alpha-\beta\epsilon_i} \tag{31.1}$$

gives the equilibrium number of molecules n_i in the ith equal region $\delta q_1 \ldots \delta p_f$ into which we have divided the μ-space for the kind of molecule under consideration. It will now be natural to re-express this in a form which recognizes the dependence of the number of molecules in any region of the μ-space on the extension of that region and on the total number of molecules available.

To do this we may introduce a new constant C defined in terms of our previous constant α by the expression

$$e^{-\alpha} = nC\,\delta q_1 \ldots \delta p_r, \tag{31.2}$$

where n is the total number of molecules and $\delta q_1 \ldots \delta p_r$ is the extension in the μ-space of the different equal cells into which we have divided that space. Substituting (31.2) into (31.1), and using the symbol δn to denote the number of molecules in the elementary cell, we can then rewrite the Maxwell-Boltzmann distribution law in the more familiar form

$$\delta n = nCe^{-\beta\epsilon}\,\delta q_1 \ldots \delta p_r. \tag{31.3}$$

This may now be interpreted as a differential form of expression for the Maxwell-Boltzmann distribution law. It makes the number of molecules δn in any region of the μ-space proportional to the extension $\delta q_1 \ldots \delta p_r$ of that region, and also indicates the dependence of δn on the energy ϵ of a molecule in that region, where this quantity is now to be treated as a function

$$\epsilon = \epsilon(q_1 \ldots p_r) \tag{31.4}$$

of the coordinates and momenta of the molecule.

32. Evaluation of constants in the Maxwell-Boltzmann distribution law

(*a*) **Value of constant α (or C).** With the help of the foregoing differential form of expression for the Maxwell-Boltzmann distribution law we can now investigate the constants which it contains. The consideration of the constant α, or more conveniently the constant C with which we have replaced it, is very simple.

Starting with the Maxwell-Boltzmann law

$$\delta n = nCe^{-\beta\epsilon}\,\delta q_1 \ldots \delta p_r, \tag{32.1}$$

and considering an integration over all possible values of the coordinates and momenta for a molecule, we must evidently obtain the total number of molecules n. This then gives us

$$n = nC\int\ldots\int e^{-\beta\epsilon}\,dq_1 \ldots dp_r. \tag{32.2}$$

And we hence obtain the result

$$C = \frac{1}{\int\ldots\int e^{-\beta\epsilon}\,dq_1 \ldots dp_r} \tag{32.3}$$

for the constant C itself.

The integration in this expression is to be carried out over all possible values of the coordinates and momenta $q_1 \ldots p_r$. In principle this can, of course, always be done when we know the functional dependence of the energy ϵ of a molecule on its coordinates and momenta. In the case of systems composed of more than a single kind of molecule an expression of form similar to (32.3) will hold for each of the constants $C, C', C'', \ldots$ which can be correlated with the constants $\alpha, \alpha', \alpha'', \ldots$ occurring in the original Maxwell-Boltzmann expressions (30.6) for each kind of molecule.

(*b*) **Introduction of the idea of a perfect gas thermometer.** In order to proceed to an understanding of the other constant β, which occurs in the Maxwell-Boltzmann expression, it will now be convenient to regard our system of interest as including as one of its parts a vessel of volume v, containing a dilute monatomic gas, composed of n simple particles of mass m. Such a gas, if sufficiently dilute, would obey the known laws for a perfect gas, and the adjunct to our system that has thus been introduced may itself be spoken of as a perfect gas thermometer. We shall now show that it is possible to relate the constant β to the temperature T of this thermometer.

To obtain the desired relation we must first consider the Maxwell-Boltzmann distribution for the n particles of mass m that compose the contents of the thermometer. Introducing Cartesian coordinates x, y, z for these particles, together with the corresponding momenta $m\dot{x}$, $m\dot{y}$, $m\dot{z}$, the distribution law (32.1) may be written in the form

$$\delta n = nCm^3e^{-\beta\epsilon}\,\delta x\delta y\delta z\delta\dot{x}\delta\dot{y}\delta\dot{z}, \tag{32.4}$$

where δn gives the number of particles at equilibrium which may be expected to have coordinates in the range $\delta x\delta y\delta z$ and components of velocity in the range $\delta\dot{x}\delta\dot{y}\delta\dot{z}$. Integrating the coordinates over the total volume v of the container, we may then write

$$\delta n = nCvm^3e^{-\beta\epsilon}\,\delta\dot{x}\delta\dot{y}\delta\dot{z},$$

or, denoting the product of the constant factors Cvm^3 by the single letter A,

$$\delta n = nAe^{-\beta\epsilon}\,\delta\dot{x}\delta\dot{y}\delta\dot{z}, \tag{32.5}$$

for the number of particles having components of velocity in the indicated range.

We may now use this result to obtain an expression for the pressure p of the gas. Consider an element δS on the surface of the container, taken for convenience as lying perpendicular to the x-axis in such a way that it is bombarded by particles travelling towards it in the positive x-direction; and consider now the number of those particles, having a component of velocity between $\dot{x}$ and $\dot{x}+\delta\dot{x}$, which would collide with this element of surface δS in unit time. Making use of (32.5), and noting that $\dot{x}\,\delta S$ would give the volume from which such particles could come in unit time, we can evidently take

$$\frac{\dot{x}\,\delta S}{v}\int\limits_{\infty}^{+\infty}\int\limits_{-\infty}^{+\infty} nAe^{-\beta\epsilon}\,\delta\dot{x}d\dot{y}d\dot{z}$$

as an expression for the number of such collisions, where the integrations are taken over all possible values from minus to plus infinity for the unspecified components of velocity $\dot{y}$ and $\dot{z}$. Furthermore, we can also evidently take $2m\dot{x}$ as an expression for the momentum which would be transferred to the surface by the impact of each such particle. Hence for the total force acting on the element of surface we can now write

$$p\,\delta S = 2\frac{\delta S}{v}\int_{-\infty}^{+\infty}\int_{-\infty}^{+\infty}\int_{0}^{+\infty} nAe^{-\beta\epsilon}m\dot{x}^2\,d\dot{x}d\dot{y}d\dot{z},$$

which, by changing the limits of integration for $\dot{x}$, gives us

$$p = \frac{1}{v}\int_{-\infty}^{+\infty}\int_{-\infty}^{+\infty}\int_{-\infty}^{+\infty} nAe^{-\beta\epsilon}m\dot{x}^2\,d\dot{x}d\dot{y}d\dot{z} \tag{32.6}$$

as a preliminary expression for the pressure p of the gas.

To evaluate the integral occurring in (32.6) let us now return to our original expression (32.5) for the number of particles lying in the velocity range $\delta\dot{x}\delta\dot{y}\delta\dot{z}$. By integrating this over all possible values of the components of velocity $\dot{x}$, $\dot{y}$, $\dot{z}$ we shall obtain an expression for the total number of particles in the container, which can evidently be written in the successive forms

$$\begin{aligned} n &= \int_{-\infty}^{+\infty}\int_{-\infty}^{+\infty}\int_{-\infty}^{+\infty} nAe^{-\beta\epsilon}\,d\dot{x}d\dot{y}d\dot{z} \\ &= \int_{-\infty}^{+\infty}\int_{-\infty}^{+\infty}\int_{-\infty}^{+\infty} nAe^{-\beta\frac{1}{2}m(\dot{x}^2+\dot{y}^2+\dot{z}^2)}\,d\dot{x}d\dot{y}d\dot{z} \\ &= \int_{-\infty}^{+\infty}\int_{-\infty}^{+\infty} \Big| nAe^{-\beta\frac{1}{2}m(\dot{x}^2+\dot{y}^2+\dot{z}^2)}\dot{x}\Big|_{\dot{x}=-\infty}^{\dot{x}=+\infty} d\dot{y}d\dot{z} + \\ &\qquad + \int_{-\infty}^{+\infty}\int_{-\infty}^{+\infty}\int_{-\infty}^{+\infty} nAe^{-\beta\frac{1}{2}m(\dot{x}^2+\dot{y}^2+\dot{z}^2)}\beta m\dot{x}^2\,d\dot{x}d\dot{y}d\dot{z} \\ &= \beta\int_{-\infty}^{+\infty}\int_{-\infty}^{+\infty}\int_{-\infty}^{+\infty} nAe^{-\beta\epsilon}m\dot{x}^2\,d\dot{x}d\dot{y}d\dot{z}, \end{aligned} \tag{32.7}$$

where the second form of writing is obtained by substituting for the energy of a particle of mass m its known form of dependence on velocity, the third form of writing comes from performing a partial integration with respect to $\dot{x}$, and the last form comes from the consideration that the term between the limits $\dot{x} = +\infty$ and $\dot{x} = -\infty$ is seen to go to zero since β must in any case be a positive quantity in order that the total energy of the system be finite. The final form of expression in (32.7) then gives the desired evaluation for the integral occurring in (32.6), showing indeed that it has the simple value n/β.

By combining (32.7) with (32.6) we now obtain

$$p = \frac{n}{v\beta} \tag{32.8}$$

as the desired *statistical mechanical* expression for the pressure of the gas in terms of the number of particles n, total volume v, and the constant of interest β. On the other hand, since the gas by hypothesis is sufficiently dilute to behave as perfect, we can also write

$$p = \frac{nkT}{v}, \tag{32.9}$$

where k is a constant, as a *phenomenological expression* for the pressure of the gas in terms of its temperature T. Equating the two expressions for pressure given by (32.8) and (32.9), we now obtain the desired relation

$$\beta = \frac{1}{kT} \tag{32.10}$$

between the constant β appearing in the Maxwell-Boltzmann expression for the particles in the perfect gas thermometer and the temperature T of that gas.

(c) **Value of constant β.** The relation given by (32.10) applies in the first instance to the n particles which form the contents of the perfect gas thermometer which we have regarded as in contact with or as a part of our actual system of interest. Nevertheless, assuming appropriate thermal interaction so that energy transfer can readily take place, it is evident from the fact that we are considering equilibrium conditions that the temperature T would be the same in the rest of the system as in the thermometer, and it is evident from the considerations of § 30 that the same value of β would appear in the Maxwell-Boltzmann expressions for any class of molecules in the system as in the expression for the particles in the thermometer. Hence we may now take the relation

$$\beta = \frac{1}{kT} \tag{32.11}$$

as applying in general to any system obeying the Maxwell-Boltzmann distribution law.

Equation (32.11) may be regarded as providing a connexion between the purely statistical mechanical quantity β and the empirically defined temperature T. Since temperature is a quantity with which we already have familiarity, the use of (32.11) to replace the quantity β in statistical mechanical equations by $1/kT$ is of real advantage in giving an understanding of statistical mechanical results. At a later stage of our considerations (see §§ 122 and 131) we shall present what may seem to be a more fundamental method of introducing the thermodynamic notion of temperature into statistical mechanical considerations.

The quantity k occurring in the relation between β and T will be seen from (32.9) to be the ordinary gas constant R per mol of gas divided by Avogadro's number N_A, that is, by the number of molecules in a mol. This quantity is ordinarily called Boltzmann's constant, and has the known value†

$$k = 1{\cdot}380\times 10^{-16} \text{ ergs per degree centigrade.} \qquad (32.12)$$

33. Useful forms of expression for the distribution law

In the present section we may now consider the expression of the Maxwell-Boltzmann distribution law in several different forms which prove useful when applications are to be made. Detailed consideration of these applications will nevertheless have to be omitted in view of the character of the present book.

Substituting the expression obtained in the last section for the constant β into our previous expression (31.3) for the Maxwell-Boltzmann law, we may begin by writing

$$\delta n = nCe^{-\epsilon/kT}\,\delta q_1 \ldots \delta p_r \qquad (33.1)$$

as a general expression for the number of molecules at equilibrium in any infinitesimal range of coordinates and momenta $\delta q_1 \ldots \delta p_r$, where n is the total number of molecules and ϵ is the energy contributed by a single molecule in the above range. In agreement with (32.3), the constant C in the above expression will then have the value

$$C = \frac{1}{\int \ldots \int e^{-\epsilon/kT}\,dq_1 \ldots dp_r}, \qquad (33.2)$$

where the integrations are carried out over all possible values of the r coordinates and r momenta for a molecule, $q_1 \ldots q_r, p_1 \ldots p_r$. The actual evaluation of C will thus involve a knowledge of the energy per molecule ϵ as a function of the coordinates and momenta, and the result obtained will depend on the temperature T of the system.

For many purposes it is possible and convenient to take the first

† Birge, *Phys. Rev. Supplement*, **1**, 1 (1929); modified to correspond to the new value $4{\cdot}803\times 10^{-10}$ e.s.u. for e.

three coordinates q_1, q_2, q_3 appearing above as being the Cartesian coordinates x, y, z for the centre of gravity of a molecule with respect to axes fixed relative to the container for the system, and then take the momenta corresponding to these coordinates as equal to $m\dot{x}$, $m\dot{y}$, $m\dot{z}$, where m is the mass of a molecule and $\dot{x}$, $\dot{y}$, $\dot{z}$ are its components of velocity. We may also assume the possibility of expressing the energy per molecule ϵ as a sum of the *potential energy* $\epsilon(x,y,z)$ associated with its position in any (constant) external field of force present, of the *kinetic energy* $\frac{1}{2}m(\dot{x}^2+\dot{y}^2+\dot{z}^2)$ due to its velocity as a whole, and of the *internal energy* $\epsilon(q_4 \dots q_r, p_4 \dots p_r)$ associated with the remaining coordinates and momenta

$$\epsilon = \epsilon(x,y,z)+\tfrac{1}{2}m(\dot{x}^2+\dot{y}^2+\dot{z}^2)+\epsilon(q_4 \dots q_r, p_4 \dots p_r). \tag{33.3}$$

Substituting into (33.1), we may then express the Maxwell-Boltzmann distribution law in the more explicit form

$$\delta n = nCm^3e^{-\frac{\epsilon(x,y,z)+\frac{1}{2}m(\dot{x}^2+\dot{y}^2+\dot{z}^2)+\epsilon(q_4\dots q_r,p_4\dots p_r)}{kT}}\times$$
$$\times\,\delta x\delta y\delta z\delta q_4 \dots \delta q_r\,\delta\dot{x}\delta\dot{y}\delta\dot{z}\delta p_4 \dots \delta p_r. \tag{33.4}$$

With the help of this expression we may then investigate the distribution of our molecules as a function of position, velocity, and internal condition. When ϵ is separable as assumed, it will be noted that these three aspects of the distribution can be regarded as independent. Specifying other classes of variables in any desired way, the distribution of molecules will be seen to decrease as we go to positions of higher potential energy, *or* to velocities corresponding to higher kinetic energy, *or* to internal variables corresponding to higher internal energy.

The distribution over the different possible *ranges of velocity* is often of special interest. To investigate this by itself we may regard the expression given by (33.4) as integrated over all possible values of the positional coordinates x, y, z and of the internal coordinates and momenta $q_4 \dots q_r, p_4 \dots p_r$. We shall then evidently obtain an expression of the form

$$\delta n = nAe^{-m(\dot{x}^2+\dot{y}^2+\dot{z}^2)/2kT}\,\delta\dot{x}\delta\dot{y}\delta\dot{z}, \tag{33.5}$$

where A is now a new constant containing the results of the integrations which have been performed. This result, which gives the number of molecules having components of velocity in the indicated range $\delta\dot{x}\delta\dot{y}\delta\dot{z}$, may be called the *Maxwell distribution law for velocities*, in accordance with Maxwell's original discovery of this specialized form of the more general Maxwell-Boltzmann relation.

By integrating (33.5) over all possible velocities we readily obtain for the constant A the value

$$A = \frac{1}{\int\limits_{-\infty}^{+\infty}\int\limits_{-\infty}^{+\infty}\int\limits_{-\infty}^{+\infty} e^{-m(\dot{x}^2+\dot{y}^2+\dot{z}^2)/2kT}\, d\dot{x}d\dot{y}d\dot{z}}$$

$$= \left(\frac{m}{2\pi kT}\right)^{\frac{3}{2}} \tag{33.6}$$

(see formulae of integration in Appendix II). This then makes it possible to write the Maxwell distribution law in the following different forms which prove useful under different circumstances.

For the number of molecules, having *components of velocity* $\dot{x}$, $\dot{y}$, $\dot{z}$ in a range $\delta\dot{x}\delta\dot{y}\delta\dot{z}$, we have

$$\delta n = n\left(\frac{m}{2\pi kT}\right)^{\frac{3}{2}} e^{-m(\dot{x}^2+\dot{y}^2+\dot{z}^2)/2kT}\, \delta\dot{x}\delta\dot{y}\delta\dot{z}. \tag{33.7}$$

For the number, having a *speed* c and a *direction of motion*—given by the polar angles θ and ϕ—lying in the range $\delta\theta\delta\phi\delta c$, we have

$$\delta n = n\left(\frac{m}{2\pi kT}\right)^{\frac{3}{2}} e^{-mc^2/2kT}\, c^2 \sin\theta\, \delta\theta\delta\phi\delta c. \tag{33.8}$$

For the number, having a *speed* c in the range δc without reference to direction, we have

$$\delta n = 4\pi n\left(\frac{m}{2\pi kT}\right)^{\frac{3}{2}} e^{-mc^2/2kT} c^2\, \delta c. \tag{33.9}$$

And for the number, having a *kinetic energy* ϵ lying in the range $\delta\epsilon$, we have

$$\delta n = 2\pi n\left(\frac{1}{\pi kT}\right)^{\frac{3}{2}} e^{-\epsilon/kT} \epsilon^{\frac{1}{2}}\, \delta\epsilon. \tag{33.10}$$

The foregoing specific results often prove very useful. They are particularly important in investigating the frequency and character of the collisions that can be expected to occur between the molecules of a gas. The results obtained find application in studying such physical questions as the viscosity, thermal conductivity, and diffusion of gases, and such chemical questions as the activation by collision of the molecules of a chemically reacting gas.

34. Mean values obtained from the distribution law

Since the Maxwell-Boltzmann distribution law gives information as to the numbers of molecules under equilibrium conditions which would have coordinates and momenta falling in any range of interest, we can use the distribution law to calculate the mean values of any functions of those variables. Thus if $f(q_1 \dots q_r, p_1 \dots p_r)$ is a function of the coordinates and momenta of a single molecule, it is evident from the general form of the Maxwell-Boltzmann distribution law as given by (33.1) that the mean value of this quantity for the n molecules of the system would be given by

$$\bar{f} = \int \dots \int Cf(q_1 \dots p_r) e^{-\epsilon/kT}\, dq_1 \dots dp_r$$
$$= \frac{\int \dots \int f(q_1 \dots p_r) e^{-\epsilon/kT}\, dq_1 \dots dp_r}{\int \dots \int e^{-\epsilon/kT}\, dq_1 \dots dp_r}, \tag{34.1}$$

where the integrations are over all values of the coordinates and momenta $q_1 \dots p_r$, and where in this section and the next we shall use a single bar to indicate a mean value taken over a distribution of molecules. A few illustrations of the application of this equation may be of interest.

We have already seen (33.2) that the parameter C occurring in the Maxwell-Boltzmann distribution law has a value which could be computed—at least in principle—from the equation

$$C = \frac{1}{\int \dots \int e^{-\epsilon/kT}\, dq_1 \dots dp_r}. \tag{34.2}$$

The actual evaluation of the indicated integral over all values of the coordinates and momenta may nevertheless be difficult to carry out. Hence it is of interest to note that we can in any case obtain a simple expression for the temperature coefficient of the parameter C. We achieve this by taking a logarithmic differentiation of the quantity C with respect to temperature T. This gives us

$$\frac{d \log C}{dT} = \frac{\int \dots \int e^{-\epsilon/kT}(\epsilon/kT^2)\, dq_1 \dots dp_r}{\int \dots \int e^{-\epsilon/kT}\, dq_1 \dots dp_r},$$

since the only occurrence of T is the *explicit* one in the exponent. With the help of (34.1), we then see that our temperature coefficient has the simple value

$$\frac{d \log C}{dT} = \frac{\bar{\epsilon}}{kT^2}, \tag{34.3}$$

where $\bar{\epsilon}$ is the mean energy of the molecules under consideration. This result at times proves useful.

It is also of interest to apply the method of calculating mean values given by (34.1) to a determination of the *average speed* c of the molecules of a gas. For this purpose we can make use of Maxwell's equation for the distribution of speeds (33.9), where the needed integrations have already been performed over all the variables except the one of interest c. For the *mean speed* of the molecules in a gas under equilibrium conditions we then easily obtain

$$\bar{c} = \sqrt{\left(\frac{8kT}{\pi m}\right)} = 14{,}500\sqrt{\frac{T}{M}}\ \frac{\text{cm.}}{\text{sec.}} \tag{34.4}$$

For the *root-mean-square speed* we can similarly obtain

$$\sqrt{(\bar{c^2})} = \sqrt{\left(\frac{3kT}{m}\right)} = 15{,}800\sqrt{\frac{T}{M}}\ \frac{\text{cm.}}{\text{sec.}} \tag{34.5}$$

And by considering the conditions for a maximum, instead of taking a mean, we can obtain for the *most probable speed*

$$\tilde{c} = \sqrt{\left(\frac{2kT}{m}\right)} = 12{,}900\sqrt{\frac{T}{M}}\ \frac{\text{cm.}}{\text{sec.}} \tag{34.6}$$

The somewhat close agreement between these different kinds of average for molecular speed results from a fairly sharp convergence around the most probable value. The numerical values† are in centimetres per second, with T in degrees centigrade absolute, and M the molecular weight of the gas.

The *mean energy* of the molecules in a system is an average of great importance. For this we have in accordance with our general equation (34.1) the formal expression

$$\bar{\epsilon} = \frac{\int \dots \int e^{-\epsilon/kT}\epsilon\, dq_1 \dots dp_r}{\int \dots \int e^{-\epsilon/kT}\, dq_1 \dots dp_r}, \tag{34.7}$$

which can be evaluated in principle for any known dependence of ϵ on the q's and p's. Multiplying by the total number of molecules present we can then also obtain an expression for the *total energy* of the system, and by differentiating with respect to the temperature T we can obtain an expression for its *heat capacity at constant volume*.

The treatment of mean energies is specially simplified when we can express the energy of a molecule ϵ as a sum of terms, one or more of which depend only on a *single variable* out of the total set of coordinates and momenta $q_1 \dots q_r, p_1 \dots p_r$. For example, it may be possible to express the energy ϵ in the form

$$\epsilon = \epsilon(p_1) + \epsilon(q_1 \dots q_r, p_2 \dots p_r), \tag{34.8}$$

† Dushman, *High Vacuum*, Schenectady, 1922.

where the momentum p_1 occurs only in a single term. Using our general formula for mean values (34.1), it is then evident that the mean energy associated with this particular variable could be calculated from the formula

$$\overline{\epsilon(p_1)} = \frac{\int \epsilon(p_1) e^{-\epsilon(p_1)/kT}\, dp_1}{\int e^{-\epsilon(p_1)/kT}\, dp_1}, \tag{34.9}$$

where the variables other than p_1 have been integrated out.

It frequently happens that the part of the energy under consideration has a *quadratic dependence* on the variable involved,

$$\epsilon(p_1) = bp_1^2, \tag{34.10}$$

where b is a constant, as, for example, in the case of the kinetic energy $p_x^2/2m = m\dot{x}^2/2$ associated with the x-component of the velocity of a molecule of mass m. Substituting (34.10) into (34.9) and performing the indicated integrations, we then readily obtain†—with the limits of integration from minus to plus infinity—

$$\overline{\epsilon(p_1)} = \tfrac{1}{2}kT. \tag{34.11}$$

In accordance with this finding the mean energy associated with each variable, which contributes a quadratic term to the total energy of the molecule, has the same value $\frac{1}{2}kT$. This result is ordinarily spoken of as the *principle of the equipartition of energy*. It should be emphasized that the principle is a consequence of the assumed quadratic form of dependence on the variables involved rather than a general consequence of statistical mechanics.

With the help of the principle of the equipartition of energy we can now easily obtain expressions for the *total energy* E and *heat capacity at constant volume* C_v in certain simple cases where the principle would apply at least in first approximation. For a dilute *monatomic gas* composed of n simple particles we obtain

$$E = \tfrac{3}{2}nkT \quad \text{and} \quad C_v = \tfrac{3}{2}nk, \tag{34.12}$$

corresponding to the three component momenta that contribute quadratic terms to the kinetic energy of the molecule. Similarly, for the case of a *diatomic gas* composed of n molecules which are thought of as rigid rotators, we obtain

$$E = \tfrac{5}{2}nkT \quad \text{and} \quad C_v = \tfrac{5}{2}nk, \tag{34.13}$$

corresponding to the two additional rotational energy terms that are now present. And, as a final example, for a *crystal* composed of n atoms, which are thought of as point particles, oscillating around their

† See Appendix II for the necessary formulae of integration.

equilibrium positions under the restraint of forces obeying Hooke's law, we obtain

$$E = 3nkT \quad \text{and} \quad C_v = 3nk, \tag{34.14}$$

corresponding to the three quadratic terms for the kinetic energy and the three further quadratic terms for the potential energy of a particle.

The foregoing simple consequences of the principle of equipartition of energy are approximately correct under appropriate circumstances, and are among the results of the classical statistical mechanics which have proved most illuminating for the development of physical-chemical ideas. The deviations from these simple results on account of the necessity of using relativistic mechanics in the case of high-velocity particles are of interest,† and deviations which arise in problems where we must use quantum instead of classical mechanics are very important as we shall see later in the present book.

To conclude this section on mean values, we may also consider, in addition to the energy of a system, its components of linear and of angular momenta. Using Cartesian coordinates x, y, z corresponding to axes stationary with respect to the container for our system, and assuming the separation of the energy into the potential, kinetic, and internal energy of a molecule as in (33.3) and (33.4), we readily obtain, by integrating out over the internal coordinates and momenta,

$$\overline{p_x} = \overline{m\dot{x}} = \frac{\int \dots \int m\dot{x}\, e^{-\frac{\epsilon(x,y,z,)+\frac{1}{2}m(\dot{x}^2+\dot{y}^2+\dot{z}^2)}{kT}}\, dxdydzd\dot{x}d\dot{y}d\dot{z}}{\int \dots \int e^{-\frac{\epsilon(x,y,z)+\frac{1}{2}m(\dot{x}^2+\dot{y}^2+\dot{z}^2)}{kT}}\, dxdydzd\dot{x}d\dot{y}d\dot{z}} \tag{34.15}$$

for the *mean linear momentum* of the molecules parallel to the x-axis, and

$$\overline{m_z} = \overline{m(x\dot{y}-y\dot{x})}$$

$$= \frac{\int \dots \int m(x\dot{y}-y\dot{x})e^{-\frac{\epsilon(x,y,z,)+\frac{1}{2}m(\dot{x}^2+\dot{y}^2+\dot{z}^2)}{kT}}\, dxdydzd\dot{x}d\dot{y}d\dot{z}}{\int \dots \int e^{-\frac{\epsilon(x,y,z)+\frac{1}{2}m(\dot{x}^2+\dot{y}^2+\dot{z}^2)}{kT}}\, dxdydzd\dot{x}d\dot{y}d\dot{z}} \tag{34.16}$$

for the *mean angular momentum* of the molecules around the z-axis. Examining the above expressions, however, we note that in each little spatial region $dxdydz$ the components of velocity would be distributed symmetrically around the value nought. Hence we obtain mean values per molecule and thus also total values of linear and angular momentum for the system as a whole which are zero:

$$P_x = n\bar{p}_x = 0, \qquad M_z = n\overline{m}_z = 0. \tag{34.17}$$

† See Jüttner, *Ann. d. Physik*, **34**, 856 (1911); Tolman, *Phil. Mag.* **28**, 583 (1914).

This result that the most probable values of the linear and angular momentum of the system will be zero, when referred to axes stationary in the container, is fundamentally a consequence of our original starting-point in which we set a definite value for the energy of the system but permitted exchange of linear and angular momenta through molecular impacts with the walls of our assumed massive container.

35. The general principle of equipartition

It has been emphasized in the preceding section that the principle of the equipartition of energy, which leads to the mean value $\frac{1}{2}kT$ for the energy associated with a single coordinate or momentum $q_1 \ldots p_r$, only holds in the case of a quadratic dependence of energy on the variable under consideration. For this reason it is of interest to show that a considerably more general form of equipartition principle can be derived,† which applies under wider circumstances and reduces to the equipartition of energy in the case of an actual quadratic dependence on the variable involved.

To obtain the desired generalized principle, let us start with the Maxwell-Boltzmann distribution law in the general form

$$\delta n = nCe^{-\epsilon/kT}\,\delta q_1 \ldots \delta p_r, \tag{35.1}$$

where n is the total number of molecules and δn is the number in the indicated range of coordinates and momenta $\delta q_1 \ldots \delta p_r$. Integrating over all possible values of the coordinates and momenta, and dividing through by n, we must then obtain the result

$$\int \ldots \int Ce^{-\epsilon/kT}\,dq_1 \ldots dp_r = 1. \tag{35.2}$$

In this expression let us then carry out a partial integration with respect to any one of the coordinates and momenta $q_1 \ldots p_r$, denoted for simplicity by q_1 whether in actuality a coordinate or a momentum. We then obtain a result which can be written in the form

$$\int \ldots \int \left| Ce^{-\epsilon/kT} q_1 \right|_{q_1=a}^{q_1=b} dq_2 \ldots dp_r + \int \ldots \int Ce^{-\epsilon/kT}\frac{q_1}{kT}\frac{\partial \epsilon}{\partial q_1}\,dq_1 \ldots dp_r = 1, \tag{35.3}$$

where a and b are respectively the lower and upper limits to the possible range of values for q_1.

In connexion with the above equation we usually encounter one or the other of two different types of situation.

In the first type of situation the energy of a molecule ϵ is independent

† Tolman, *Phys. Rev.* **11**, 261 (1918).

of the variable q_1 under consideration. For example, the variable q_1 is a cyclic positional coordinate in the absence of any corresponding field of force. Under such circumstances the second term in (35.3) is equal to zero, since $\partial\epsilon/\partial q_1$ is itself zero, and the equation reduces to the obvious but unimportant result

$$\int \dots \int \left| Ce^{-\epsilon/kT} q_1 \right|_{q_1=a}^{q_1=b} dq_2 \dots dp_r = 1. \tag{35.4}$$

In the second type of situation, which is frequently encountered and is the one of actual interest, the first term in equation (35.3) proves to be equal to zero. This occurs very often because the energy ϵ of the molecule goes to infinity at both limits a and b, thus leading to the value zero at both those limits owing to the occurrence of ϵ as a negative exponent. This is the case when q_1 is a positional coordinate having associated potential energy which increases without limit for positive and negative values of q_1, as in the case of a bound particle, or when q_1 is a momentum having associated kinetic energy that goes to infinity as q_1 goes to plus and minus infinity. The first term in equation (35.3) can also turn out to be zero because q_1 is zero at one limit and ϵ infinity at the other. Under such circumstances the first term in (35.3) disappears and the equation leads to the result

$$\int \dots \int C\left(q_1 \frac{\partial \epsilon}{\partial q_1}\right) e^{-\epsilon/kT} dq_1 \dots dp_r = kT. \tag{35.5}$$

Comparing this result with our general expression (34.1) for the calculation of mean values of functions of the coordinates and momenta of a molecule, we now obtain a *general equipartition principle* which can be written in the form

$$\overline{\left(q_i \frac{\partial \epsilon}{\partial q_i}\right)} = \overline{\left(p_j \frac{\partial \epsilon}{\partial p_j}\right)} = kT, \tag{35.6}$$

and which applies to the indicated mean values for each coordinate q_i or momentum p_j that behaves at the limits to its range of values in the frequently encountered manner described above.

When the energy of a molecule does depend quadratically on the variable in question, so that the portion of the energy associated with the variable in question has the form

$$\epsilon_i = aq_i^2 \quad \text{or} \quad \epsilon_j = bp_j^2, \tag{35.7}$$

the general principle of equipartition as given by (35.6) is seen to reduce to the equipartition of energy

$$\overline{\epsilon_i} = \overline{\epsilon_j} = \tfrac{1}{2}kT. \tag{35.8}$$

On the other hand, the general principle is valid even when we do not have this simple quadratic form of dependence. Thus, for example, in the case of a gas composed of particles moving with velocities $(\dot{x}, \dot{y}, \dot{z})$ high enough compared with the velocity of light c so that *relativistic effects* must be considered, we have

$$\epsilon = c(p_x^2+p_y^2+p_z^2+m^2c^2)^{\frac{1}{2}} \tag{35.9}$$

for the dependence of the energy of a particle ϵ on its components of momentum

$$p_x = \frac{m\dot{x}}{\sqrt{(1-u^2/c^2)}}, \quad p_y = \frac{m\dot{y}}{\sqrt{(1-u^2/c^2)}}, \quad p_z = \frac{m\dot{z}}{\sqrt{(1-u^2/c^2)}}, \tag{35.10}$$

where m is the rest mass of the particle and $u = \sqrt{(\dot{x}^2+\dot{y}^2+\dot{z}^2)}$ its total velocity. Applying the general equipartition principle (35.6), we then obtain as a relation between mean values

$$\overline{\left(\frac{c^2p_x^2}{\epsilon}\right)} = \overline{\left(\frac{c^2p_y^2}{\epsilon}\right)} = \overline{\left(\frac{c^2p_z^2}{\epsilon}\right)} = kT,$$

or, by substituting from (35.9) and (35.10),

$$\overline{\left(\frac{m\dot{x}^2}{\sqrt{(1-u^2/c^2)}}\right)} = \overline{\left(\frac{m\dot{y}^2}{\sqrt{(1-u^2/c^2)}}\right)} = \overline{\left(\frac{m\dot{z}^2}{\sqrt{(1-u^2/c^2)}}\right)} = kT. \tag{35.11}$$

This consequence of the general equipartition principle is of considerable interest.†

In the first place, we note that it gives immediate information as to the mean rate of transport of momentum by the molecules of the gas which leads at once to the familiar result

$$p = \frac{nkT}{v} \tag{35.12}$$

for the pressure of the gas even when the particles do have high velocities.

In the second place, by adding together the individual equations given by (35.11), we obtain the result

$$\overline{\left(\frac{mu^2}{\sqrt{(1-u^2/c^2)}}\right)} = 3kT. \tag{35.13}$$

We may compare this with the expression for mean kinetic energy

$$\bar{\epsilon} = \overline{\left(\frac{mc^2}{\sqrt{(1-u^2/c^2)}}-mc^2\right)}. \tag{35.14}$$

† Compare Tolman, *Phil. Mag.* **28**, 583 (1914).

Expanding both expressions in series form and combining, we easily obtain the result

$$\bar{\epsilon} = \tfrac{3}{2}kT + mc^2\left(\frac{1}{8}\frac{\overline{u^4}}{c^4} + \frac{1}{8}\frac{\overline{u^6}}{c^6} + \frac{15}{128}\frac{\overline{u^8}}{c^8} + \ldots\right). \tag{35.15}$$

We thus find on the one hand from (35.12) that our gas containing high-velocity particles would obey the perfect gas laws, but on the other hand from (35.15) that its kinetic energy would be greater than for an ordinary perfect gas.

This must now conclude our chapter on the pre-quantum status of the Maxwell-Boltzmann distribution law for the molecules of a system which has come to a steady condition of macroscopic equilibrium. We have tried to give a real insight into the statistical principles involved in the derivation of this law, but have only endeavoured to give a rough indication of the nature of its applications.

V

COLLISIONS AS A MECHANISM OF CHANGE WITH TIME

36. Introduction

In the preceding chapter we have considered the application of statistical mechanics to systems which have come to a steady condition of equilibrium when looked at from a macroscopic point of view that neglects the behaviour of individual molecules. We now wish to commence the study of systems that are in a condition which is changing with time. In the present chapter we shall be concerned with the consideration of molecular collisions as an important mechanism involved in the changes with time that do take place. And in the following chapter we shall then complete the classical part of our study by deriving and considering Boltzmann's famous H-theorem, which gives a fundamental kind of information as to the direction in which changes in condition with time may be expected to occur.

Except in the specially simple case of collisions between spherically symmetrical molecules, a treatment of the classical theory of collisions, adequate for a derivation of the H-theorem, proves to be a somewhat complicated matter. This necessary complexity was appreciated by Boltzmann himself, but has not always been emphasized by later expositors who have tended to confine their demonstrations of the tendency for H to decrease with time to the case of systems composed of rigid, elastic spheres. In view of the elaboration actually needed, it will be desirable to give a brief outline of the treatment to be presented in this chapter.

We begin the development in the next section, § 37, by first giving a preliminary discussion of two useful principles pertaining to the temporal behaviour of mechanical systems, which may be called the principles of dynamical reversibility and of dynamical reflectability. In accordance with these principles, both the reverse and the mirror image of any possible motion for a dynamical system would also itself be a possible motion. We shall find these principles of use in our discussion of the different kinds of molecular collisions that can take place. We shall also be interested later in reconciling the reversibility possible for any mechanical behaviour with the irreversibilities actually observed in the behaviour of physical chemical systems.

We then commence our detailed treatment of molecular collisions, in § 38, by considering an appropriate method of specifying the different possible states i of a molecule when the effect of collisions is to be studied. This we do by leaving the coordinates q_1, q_2, q_3 for the centre of gravity of the molecule undefined, and specifying the values for the remaining coordinates and the momenta $q_4 \ldots q_r, p_1 \ldots p_r$ within an infinitesimal range $\delta\omega_i$. This is followed by a useful classification correlating any state i of a molecule with possible identical states i, congruent states i', reverse states $-i$, and enantiomorphous states $\underline{i}$.

In §§ 39 and 40 we can then classify the possible constellations (j, i) for a pair of colliding molecules, and consider the different kinds of collisions $\begin{pmatrix} j, i \\ k, l \end{pmatrix}$ that can take place in which a pair of molecules enter the encounter in states i and j and leave in states k and l. For any chosen collision $\begin{pmatrix} j, i \\ k, l \end{pmatrix}$ it is possible to construct congruent collisions $\begin{pmatrix} j', i' \\ k', l' \end{pmatrix}$, enantiomorphous collisions $\begin{pmatrix} \underline{j}, \underline{i} \\ \underline{k}, \underline{l} \end{pmatrix}$, and the reverse collision $\begin{pmatrix} -k, -l \\ -j, -i \end{pmatrix}$. It is also always possible to construct what is called the corresponding collision $\begin{pmatrix} l, k \\ x, y \end{pmatrix}$ in which the molecules enter the new collision in the states with which they left the originally chosen collision. It is not in general possible, however, to construct an inverse collision $\begin{pmatrix} l, k \\ i, j \end{pmatrix}$ such that the molecules would be returned from their final back to their initial states.

In §§ 41 and 42 we then demonstrate the general possibility of arranging all possible collisions for a given pair of molecules in closed cycles containing a succession of collisions each of which corresponds to the preceding one in the list until we finally arrive at the first member with which we started. We also show for the case of spherically symmetrical molecules that such a closed cycle of corresponding collisions would reduce to only two members, each of which in this very special case would then be the inverse collision to the other.

After completing this rather long account of formal consequences of the definitions selected for molecular states and collisions, we are then ready to apply the principles of mechanics to molecular collisions. In § 43 we first call attention in passing to the results of applying the conservation laws for energy and for the components of linear and angular momentum to a collision between two molecules. In § 44, returning to the line of argument needed to prepare for our later derivation of the H-theorem, we then apply Liouville's theorem, for the conservation of extension in phase, to the behaviour of two molecules when they pass through a collision. This leads, after quite a long and complicated calculation, to an important relation (44.19) which connects the velocity of approach and range of variables defining the initiation of any collision with the velocity of recession and range of variables defining the completion of that collision.

With the help of this consequence of the conservation of extension in phase, it then finally becomes possible in § 45 to apply the methods of statistical mechanics to the study of the probable frequencies with which different kinds of collisions would take place in a gas having a specified distribution over different molecular states. Considering any given collision $\begin{pmatrix} j, i \\ k, l \end{pmatrix}$ together with its reverse collision $\begin{pmatrix} -k, -l \\ -j, -i \end{pmatrix}$ and its corresponding collision $\begin{pmatrix} l, k \\ x, y \end{pmatrix}$, it is found that the probable numbers of such collisions, taking place in unit time, would be proportional to the densities of molecular distribution for the pairs of initial states involved in the collision, with a proportionality constant or probability coefficient having just the same value for all three of the related collisions.

The chapter then closes in § 46 with some remarks on the necessity of supplementing the study of classical bimolecular collisions by further considerations in order to obtain a complete picture of the mechanisms leading to the changes that take place with time in the condition of actual physical chemical systems.

The two findings, first that all possible kinds of collision between two molecules can be arranged into closed cycles in which each collision corresponds to the preceding collision in the list, and second that the probability coefficients for the frequencies of any given collision and the collision corresponding thereto have the same value, are the two final consequences of the present chapter which will be needed for the proof of Boltzmann's H-theorem. The derivation of these findings is much simplified when attention is restricted to collisions between rigid, elastic spheres. The closed cycles of corresponding collisions then reduce to pairs of collisions which are the inverse of each other as shown in § 42, and the elaborate discussion of the application of the principle of the conservation of extension in phase as given in § 44 becomes shorter and easier to follow. A consideration of this simplified case can be readily carried through if desired in order to obtain a valid idea of the general nature of the argument.

37. The principles of dynamical reversibility and reflectability

Before proceeding to the detailed treatment of molecular collisions as a mechanism of change with time, we shall devote the present section to two general principles pertaining to the temporal behaviour of mechanical systems. Taking for the moment the point of view of the special theory of relativity and considering different sets of Galilean coordinates which might be employed, it will be appreciated that the equations of mechanics would be covariant to the translation or rotation of spatial axes, to the Lorentz transformation involving both spatial and temporal axes, and to inversions of the time axis or of any chosen spatial axis. The two possibilities of inverting the time axis or a chosen spatial axis may be called the principles of dynamical reversibility and reflectability. These principles will be sufficiently important for our work so that we can afford to give them detailed treatment. For our purposes it will be sufficient to take the point of view of Newtonian rather than of relativistic mechanics.

(*a*) **The principle of dynamical reversibility.** We may first consider a derivation of the principle of dynamical reversibility. For this purpose, let us consider an isolated, conservative mechanical system of f degrees of freedom, corresponding to the coordinates $q_1 \dots q_f$, and let us take its behaviour as governed by Lagrange's equations of motion (9.7) in the form

$$\frac{d}{dt}\frac{\partial L}{\partial \dot{q}_i}-\frac{\partial L}{\partial q_i}=0 \quad (i=1,2,3,\dots,f). \tag{37.1}$$

The system is taken for simplicity as isolated and conservative since we regard it as possible to look at any system from a point of view which would make this true, and the system is taken for simplicity as purely mechanical. The result obtained will apply under somewhat more general circumstances to systems, which may be electromagnetic as well as mechanical in nature, and which may be acted on by external forces provided these are derivable from a potential or, if magnetic, are taken as produced by external currents which are themselves reversed in direction when the reversed motion is considered.

On account of the conservative character of the system the Lagrangian function L will not contain the time t explicitly. In addition we can make use of the circumstance that the Lagrangian function for mechanical systems is known to be a quadratic or in any case an even function of the generalized velocities $\dot{q}$. With these restrictions on the form of L, it is then evident that equations (37.1) could also be written in the form

$$\frac{d}{d(-t)}\frac{\partial L}{\partial(-\dot{q}_i)}-\frac{\partial L}{\partial q_i}=0 \quad (i=1,2,3,\ldots,f). \tag{37.2}$$

Hence either of the sets of equations (37.1) or (37.2) can be taken as giving a valid description of the possible motions of the system.

The fact that the equations of motion for our system may be written in either of the forms (37.1) or (37.2) now makes it easy to draw the desired conclusion. Corresponding to any particular solution of equations (37.1) which gives the coordinates q_i as a function of time

$$q_i=\phi_i(t) \quad (i=1,2,3,\ldots,f), \tag{37.3}$$

it is evident that there will be a solution of equations (37.2) which can be written in the form

$$q_i'=\phi_i(-t) \quad (i=1,2,3,\ldots,f). \tag{37.4}$$

Both of these solutions, however, agree with possible motions of the system since we appreciate that they are both possible solutions of the Lagrangian equations for the system.

To make the nature of the above result evident, let us—for the sake of specificity—designate the possible behaviour described by (37.3) as a *forward motion* of the system and the behaviour described by (37.4) as the corresponding *reverse motion*; and let us consider two systems of the kind under consideration, S and S', one carrying out the forward motion and the other the reverse motion. At time $t=0$ the configurations of the two systems S and S', as expressed by the values of their coordinates

$$q_i'(0)=q_i(0), \tag{37.5}$$

would be in agreement, in accordance with (37.3) and (37.4), and their motions, as expressed by the values of their generalized velocities

$$\dot{q}_i'(0)=-\dot{q}_i(0), \tag{37.6}$$

would be in reverse directions. At any later time t the configuration for the system S' would agree with that for system S at an earlier time $(-t)$,

$$q_i'(t)=q_i(-t), \tag{37.7}$$

and the motion would be in the reverse direction to that which prevailed for S at that earlier time,

$$\dot{q}_i'(t)=-\dot{q}_i(-t). \tag{37.8}$$

Thus we see, corresponding to any possible motion of a system of the kind mentioned, that there would be a possible reverse motion in which the same values of the coordinates would be reached in the reverse order with reversed values for the velocities. This is the content of the principle of *dynamical reversibility*.

This principle, which is a direct consequence of the classical mechanics, shows that any motion of an isolated mechanical system and its reverse are equally possible. We shall find this result of importance in understanding the different kinds of molecular collision which can take place. We shall also, however, be interested in the principle from a broader point of view, since we shall have to reconcile the reversibility in the microscopic behaviour of physical-chemical systems, which would be predicted by the above principle of mechanics, with the actual irreversibilities, which can be observed in their macroscopic behaviour. It will perhaps already be evident that the reconciliation will depend on recognizing that the predicted reversibility is an attribute of changes from one precisely defined state to another, which must be treated by the methods of exact mechanics, while the observed irreversibilities are an attribute of changes from one less precisely specified condition to another, which must be treated by the methods of statistical mechanics.

(*b*) **The principle of dynamical reflectability.** In connexion with the above finding that any solution of the equations of motion for an isolated conservative mechanical system remains a valid description of possible behaviour after a change in the sign of the time, it is also of interest to show a similar possibility for changes in the sign of an appropriate group of the coordinates. For this purpose it will be necessary to distinguish between different spatial directions, and Lagrange's equations for our isolated conservative system may now be written in the form

$$\begin{aligned}\frac{d}{dt}\frac{\partial L}{\partial \dot{x}_i}-\frac{\partial L}{\partial x_i} &= 0,\\ \frac{d}{dt}\frac{\partial L}{\partial \dot{y}_i}-\frac{\partial L}{\partial y_i} &= 0,\\ \frac{d}{dt}\frac{\partial L}{\partial \dot{z}_i}-\frac{\partial L}{\partial z_i} &= 0,\end{aligned} \tag{37.9}$$

where the subscript i now refers to the different component parts of the system, and the coordinates x_i, y_i, z_i are with respect to some set of Cartesian axes. If we now take the Lagrangian function $L = T-V$ as an even function of each of the velocities $\dot{x}_i$, $\dot{y}_i$, $\dot{z}_i$, since the kinetic energy T will now have this dependence, and also take L as an even function of the coordinate differences between parts, of the type (x_i-x_k), (y_i-y_k), (z_i-z_k), since the potential energy V will depend on the distances between parts, it is evident that the above equations could also be written in the form

$$
\begin{aligned}
\frac{d}{dt}\frac{\partial L}{\partial \dot{x}_i}-\frac{\partial L}{\partial x_i} &= 0,\\
\frac{d}{dt}\frac{\partial L}{\partial \dot{y}_i}-\frac{\partial L}{\partial y_i} &= 0,\\
\frac{d}{dt}\frac{\partial L}{\partial(-\dot{z}_i)}-\frac{\partial L}{\partial(-z_i)} &= 0,
\end{aligned}
\tag{37.10}
$$

where all the coordinates z_i and velocities $\dot{z}_i$ parallel to one of the axes, in this case the z-axis, have been changed in sign.

With the help of these two forms in which the equations of motion for our system could be taken, we now see, writing

$$x_i = f_i(t), \qquad y_i = g_i(t), \qquad z_i = h_i(t) \tag{37.11}$$

as a possible particular solution of the equations of motion, that

$$x_i' = f_i(t), \qquad y_i' = g_i(t), \qquad z_i' = -h_i(t) \tag{37.12}$$

would also be a possible solution, where z_i' is the negative of z_i' for all the component parts i at all times t. Since a system S carrying out the motion (37.11) and a system S' carrying out the motion (37.12) would bear to each other the relation of mirror images as reflected at the xy-plane, this result might be called *the principle of dynamical reflectability*. It will now be appreciated, of course, that the position and orientation of such a reflecting plane might be chosen arbitrarily, and we may now say that any motion of a system and any mirror image or *enantiomorph* of that motion would be equally possible. There will be a place in what follows (§ 42) where the principle of dynamical reflectability can be employed to advantage.

38. Molecular states

(*a*) **Specification of molecular states appropriate for the consideration of collisions.** We are now ready to prepare the way for the treatment of molecular collisions by considering an appropriate method for specifying the different states into which a molecule may be thrown as a result of collision with other molecules.

In accordance with our previous procedure, we may regard the *precise state* of a molecule of r degrees of freedom as *completely specified* at any time of interest by the exact values of r coordinates and r momenta, $q_1 \ldots p_r$. It is often convenient to choose the first three of these variables as Cartesian coordinates x, y, z for the centre of gravity of the molecule, so that the total collection will become

$$x, y, z, q_4 \ldots q_r, p_x, p_y, p_z, p_4 \ldots p_r, \tag{38.1}$$

where p_x, p_y, and p_z are the components of linear momentum for the molecule as a whole.

However, for the purpose of treating the changes in molecular condition that occur on collision, such a complete and precise specification of state would usually be more than we desire. In the first place, we shall for the most part not be interested in the spatial location x, y, z of a molecule when it is changed by collision from one condition to another. In the second place, in agreement with the circumstance that we shall ultimately wish to employ the methods of statistical rather than of exact mechanics, we shall be interested in the specification of the other variables $q_4 \ldots q_r$ only to within a small range of values. For these reasons we shall now often wish to specify the state of a molecule only to the degree corresponding to an *infinitesimal range* $\delta\omega$ of the form

$$\delta\omega = \int \ldots \int dq_4 \ldots dp_r, \tag{38.2}$$

where we regard the limits of integration for the variables $q_4 \ldots q_r$, p_x, p_y, p_z, $p_4 \ldots p_r$ as lying close together in correspondence with the infinitesimal character of $\delta\omega$. By assigning indices to designate different possible such regions, $\delta\omega_1$, $\delta\omega_2$, $\delta\omega_3, \ldots$, $\delta\omega_i, \ldots$, which we may wish to consider, we can then take a molecular state as sufficiently specified for many of our purposes by the index i which denotes some particular selected range $\delta\omega_i$ out of the various ones that interest us.

(*b*) **Classification of molecular states.** For the treatment of molecular collisions, it will also be useful to have a classification of molecular states into what we may call identical, congruent, reverse, and enantiomorphous states. In order to introduce this classification it is convenient to note that any particular specification of the internal coordinates and the external and internal momenta of a molecule, $q_4 \ldots p_r$, may be regarded as corresponding to a geometrical figure which gives the configuration of the constituent parts of the molecule and also carries appropriately attached vectors to represent the magnitudes and directions of the velocities of those parts. The desired classification can then be made to depend on the possibilities for the conceptual superposition of such geometrical figures, regarding them as rigid but movable structures, as in procedures familiar in elementary geometry.

Identical states for a pair of similar molecules may then be defined as those where it would be possible to secure superposition, in the above sense, with the help of a simple translation of one of the molecules, leaving the configuration of its constituent parts and the directions of the vectors representing velocities unchanged, and not changing the relation of the molecules to any external field of force that might be acting. This definition of identity is thus chosen to agree with the circumstance that we shall not be specially interested in the mere spatial location when molecules change by collision from one state to another. The possibility for the existence of identical states is obvious.

Congruent states for a pair of molecules may be defined as those where a rotation as well as a translation might be needed to secure superposition. Congruent states thus become identical after a suitable rotation. In making such rotations the configuration, the relative directions of the vectors representing velocities, and the relation to external fields of force are all to be kept unaltered. Congruence for molecular physics is thus rather similar to congruence for ordinary geometry, provided we consider the additional circumstances that arise from the presence of vectors representing velocities and fields of force. The possible existence of congruent states is obvious from the invariance of the laws of mechanics to the translation and rotation of axes.

Reverse states for a pair of molecules may be defined as those which become identical, i.e. superposable by translation alone, after there has been a reversal in direction without change of magnitude in all the velocities for one of the molecules.† The idea of reverse states is thus a physical one for which there is no very immediate geometrical analogue. We shall find the idea of reverse states quite important in our consideration of molecular collisions. The possible existence of reverse states is guaranteed by the principle of dynamical reversibility.

Enantiomorphous states for a pair of molecules may now finally be defined as those where one state would be identical with the mirror image of the other. The idea of enantiomorphous states is already familiar in organic chemistry, but is here extended to include the reflection of velocities and fields of force as well as of configuration. It is interesting to note that this provides possibilities for truly enantiomorphic forms even in the absence of any asymmetric atom. There is really only one place in what follows where this idea as to enantio-

† The definition of reverse states here adopted differs from that in the author's earlier book, *Statistical Mechanics with Applications to Physics and Chemistry*, where reverse states were defined as becoming merely congruent rather than identical after the reversal of velocities for one of the pair of structures under consideration. The present definition is less likely to lead to confusion. Similar remarks apply to the following case of enantiomorphous states.

morphous states will be of importance to us. The possible existence of such states is guaranteed by the principle of dynamical reflectability.

The foregoing definitions as to what we shall mean by identical, congruent, reverse, and enantiomorphous states have been chosen in such an obviously natural manner as perhaps to make the detailed statements that we have given unnecessary. Nevertheless, we shall find these terms quite useful. They can evidently be applied not only to states of single molecules, but also to systems consisting of more than one molecule, and may be used to characterize the nature of processes as well as of states. If we use the symbol i to designate a particular state of a molecule, the symbol i' may be used to designate a congruent state, the symbol $-i$ to designate the reverse state, and the symbol $\underline{i}$ in the rare cases when necessary to designate an enantiomorphous state. Such symbols can be taken when necessary as referring to states which are precisely defined except as to spatial location, but will ordinarily be used as referring to states defined merely to the extent of correspondence with an infinitesimal range $\delta\omega$ as discussed in the foregoing.

39. Molecular constellations

We are now ready to turn to the consideration of pairs of molecules which are undergoing collision. Introducing the nomenclature of Boltzmann,† such a pair of molecules may be called a *constellation.*

In classifying the different kinds of constellation that can occur we may evidently use the phrases *identical*, *congruent*, *reverse*, or *enantiomorphous constellation* as descriptive terms having the same significance as applied above to molecular states. In addition, however, we shall need some further terms which we must now consider.

As a convenient abstraction, originally employed by Boltzmann, we shall regard two molecules as having a negligible effect on each other until their centres of gravity approach within a critical distance b. In our consideration of collisions we shall then regard the collision as commencing when the centres of gravity arrive at this critical distance and as being completed when they again reach the distance b on their way apart. In accordance with this procedure we shall now introduce the term *critical constellation* to denote a situation where the two molecules are at the critical distance apart, and shall introduce the terms *initial constellation* and *final constellation* to distinguish between critical constellations where a collision is just beginning or just ending.

† See Boltzmann, *Vorlesungen über Gastheorie*, zweiter, unveränderter Abdruck, II. Teil, Leipzig, 1912.

One further descriptive term will also be needed. If we consider any given critical constellation it is evident that another critical constellation could be constructed by interchanging the positions of the centres of gravity for the two molecules, or, in other words, by translating one of the molecules relative to the other a distance $2b$ along the line connecting their centres of gravity, so as again to arrive at a situation where they would be separated by the critical distance b. If such an interchange or translation is carried out, without changing the states of the two molecules either as to internal configuration or as to the directions of velocities, we shall define the new constellation as the *corresponding constellation* to the one originally taken.

In the case of two corresponding constellations, it is evident, if one of the two is an initial constellation, that the other could only be a final constellation, since if the two molecules were originally approaching each other it is obvious that they would have to be receding from each other after an interchange of positions without disturbance of velocities, and vice versa if they were originally receding. Furthermore, it may be remarked, as will be made evident in the next section, that for any arbitrary initial constellation the corresponding constellation would be a final constellation with which a conceivable collision actually could end, and that for any arbitrary final constellation the corresponding constellation would be an initial collision with which a conceivable collision actually could begin.

In order to specify a given critical constellation it is evident, in the first place, that we should have to specify the states of the two molecules involved, to the extent of describing the external motions of the molecules as a whole and the internal configurations and motions of their parts. And it is evident, in the second place, that we should also have to specify the direction of the line joining their centres of gravity, thus describing the angle of attack in the case of an initial constellation, or angle of departure in the case of a final constellation.

To specify the states of the molecules, in the above sense, we could, if desired, give the precise values of the necessary variables $q_4 \cdots q_r$, p_x, p_y, p_z, $p_4 \cdots p_r$ in the case of a molecule of r degrees of freedom. But for our purposes it will usually be sufficient to regard such states as specified by the indices i, j, k, etc., which denote infinitesimal ranges for such values $\delta\omega_i$, $\delta\omega_j$, $\delta\omega_k$, etc., as described in the preceding section. Since the critical distance b has been defined as sufficient to make the influence of the molecules on each other negligible, it is evident that such possible states would be those for a free molecule.

To specify the direction of the line of centres for the critical constellation under consideration, we could give the precise values for the polar and equatorial angles θ and ϕ which together with the critical distance b as a radius would give the position of the centre of gravity, say of the second of the two molecules, with respect to the centre of gravity of the first as an origin. But here, too, it will usually be sufficient to regard the specification as given only to the extent of an infinitesimal solid angle $\delta\lambda$, or corresponding infinitesimal surface $b^2\,\delta\lambda$ on the sphere of radius b surrounding the first molecule.

To denote the different critical constellations that could be specified in the above manner we shall use Boltzmann's symbols of the form (j, i), where the letters i and j can be thought of as denoting the states of the two molecules involved. Such symbols are evidently not complete since they contain no reference to the angle of attack or departure, but they will be sufficient for our purposes. If a given critical constellation is denoted by (j, i), a possible congruent constellation will be denoted by (j', i'), a possible enantiomorphous constellation by $(\underline{j}, \underline{i})$, the reverse constellation by $(-j, -i)$, and the corresponding constellation by (i, j). If (j, i) is an initial constellation, it will be noted that (j', i') and $(\underline{j}, \underline{i})$ will also be initial constellations, while $(-j, -i)$ and (i, j) will be final constellations, and vice versa if (j, i) is a final constellation.

40. Molecular collisions

Having completed what may seem to be an unduly elaborate development of preliminary apparatus, we are now ready to consider molecular collisions themselves. To specify any such collision it is evident that we should only have to specify the initial constellation with which it begins, since the nature of the collision itself would then be determined by the laws of mechanics. Thus a specification of the initial constellation, say (j, i), with which a collision begins would also amount to a specification of the final constellation, say (k, l), with which it ends.

In accordance with this determinate relation between initial and final constellations, we may now introduce Boltzmann's symbols of the form $\begin{pmatrix} j, i \\ k, l \end{pmatrix}$ to denote different possible collisions, where the upper letters i and j designate, in our previous sense, the states of the two molecules in the initial constellation, and the lower letters k and l designate the states in the final constellation. It will be appreciated that such a symbol provides a more complete description of the collision intended than would be given by the mere symbol (j, i) for the initial constellation, since by including a designation of the final states k and l we lay necessary requirements on the angle of attack, when the collision starts, which, as was noted above, have been omitted from our incomplete symbols for constellations.

In using symbols of the form $\begin{pmatrix} j, i \\ k, l \end{pmatrix}$ to designate collisions, it will be convenient to introduce the convention that the right-hand letter in the upper row and the left-hand letter in the lower row refer to states of one of the molecules, while the other two letters refer to states of the other molecule. Thus in the above case i and k are the initial and final states of one of the molecules and j and l of the other. This convention corresponds with the idea that we could regard the two molecules in states i and j as first approaching each other in an initial constellation which could be more completely symbolized than previously by $(j \rightarrow, \leftarrow i)$, as then passing through each other's field of influence, and as later separating from each other in states k and l in a final constellation which could be symbolized by $(\leftarrow k, l \rightarrow)$. It is to be noted that there is nothing in the idea of a collision to make it necessary for the two colliding molecules to be of the same kind, and that our symbolism is equally adapted both for cases of like and of unlike molecules.

Considering some *given collision* symbolized by $\begin{pmatrix} j, i \\ k, l \end{pmatrix}$, we shall regard a symbol of the form $\begin{pmatrix} j', i' \\ k', l' \end{pmatrix}$ as designating a *congruent collision*, which would agree with the original collision after translation and rotation in the sense discussed in § 38. It will be appreciated that an infinite variety of collisions congruent to any given collision would be possible, and it will be noted that a mere congruence for the initial states with which two molecules enter a collision without any specification as to angle of attack would not in general be sufficient to secure a completely congruent collision.

Again considering some given collision $\begin{pmatrix} j, i \\ k, l \end{pmatrix}$, we shall regard a symbol of the form $\begin{pmatrix} \underline{j}, \underline{i} \\ \underline{k}, \underline{l} \end{pmatrix}$ as designating an *enantiomorphous collision*. An infinite variety of such enantiomorphous collisions, using different planes of reflection, would be possible in accordance with the principle of dynamical reflectability.

Again considering some given collision $\begin{pmatrix} j, i \\ k, l \end{pmatrix}$, we shall regard the symbol $\begin{pmatrix} -k, -l \\ -j, -i \end{pmatrix}$ as designating the *reverse collision* thereto. This is to be understood as the collision that would occur if we took the final constellation (k, l) of the original collision and then reversed all the velocities involved, both of the molecules as a whole and of their internal parts, so as to secure an initial constellation $(-k, -l)$. In accordance with the principle of dynamical reversibility the molecules would then retrace their previous configurations to arrive in the final constellation $(-j, -i)$, with the two molecules now in states which are the reverse of those with which the original collision started. It may be specially noted that the principle of dynamical reversibility guarantees for any given collision the possible existence of a reverse collision, that for any precisely defined collision there is only a single such reverse collision, and that the reverse collision does *not* end with the molecules in the initial states which they had in the original given collision, but in the *reverse* of those states.

Again considering some given collision $\begin{pmatrix} j, i \\ k, l \end{pmatrix}$, we shall now define the *corresponding collision* thereto as that which would occur if we took the final constellation (k, l), constructed therefrom the corresponding initial constellation (l, k), and then allowed that collision to take place which would be demanded in accordance with the laws of mechanics. To denote such a corresponding collision we may use the symbol $\begin{pmatrix} l, k \\ x, y \end{pmatrix}$, where x and y denote whatever final states do arise from the collision that we have thus prescribed. From the procedure given it is evident for any given precisely defined collision that there always would be one and only one corresponding collision.

Finally, again considering some given collision $\begin{pmatrix} j, i \\ k, l \end{pmatrix}$, we shall often be specially interested in the possibility of returning the two molecules from their final states k and l to their initial states i and j. If this can be done in a single collision we shall call it an *inverse collision* to the original one. Such a collision can be denoted by the symbol $\begin{pmatrix} l, k \\ i, j \end{pmatrix}$ provided we now allow the molecules in states l and k to come together with any angle of attack that would assure the desired final result that the molecules are to emerge from the collision in states i and j. As we shall emphasize later, it is in general *not possible* to find an inverse collision for any given collision. Inverse collisions only exist in special simple cases, and are thus unlike reverse and corresponding collisions which must exist in any case.

With the help of the foregoing formalism for the treatment of collisions we may now consider some interesting properties of collisions and constellations.

As already mentioned in the preceding section, for any final constellation there will be a corresponding initial constellation with which a collision might actually begin, and for any initial constellation there will be a corresponding final constellation with which a collision might actually end. The validity of the first half of the above statement is immediately evident, since for any final constellation (k, l) we can obviously construct the corresponding constellation (l, k), and this will then be an initial constellation for whatever collision does occur, as we have already remarked in defining corresponding collisions. The validity of the second half of the statement is less immediately evident, since considering any initial constellation (y, x) it now becomes necessary to prove that the corresponding constellation (x, y) which we obtain by an interchange of centres of gravity really is one that could occur at the end of an actual collision. To demonstrate this let us consider the constellation (x, y), which can in any case be set up, and construct the reverse constellation $(-x, -y)$. Since the two molecules were approaching in (y, x), they would be receding in (x, y), and thus again approaching in $(-x, -y)$. Hence $(-x, -y)$ is an initial constellation that will lead to some collision which we may designate by $\begin{pmatrix} -x, -y \\ -v, -u \end{pmatrix}$. By the principle of dynamical reversibility, however, the reverse of this collision $\begin{pmatrix} v, u \\ x, y \end{pmatrix}$ would also be a possible one. This ends, moreover, with the constellation of interest (x, y), thus giving the desired demonstration that this is a possible final constellation for an actually possible collision.

As a consequence of the above finding we now see that by starting out with the totality of all possible initial constellations with which a collision could begin and constructing the constellations that correspond thereto we should obtain the totality of all final constellations, and, vice versa, starting with the totality of all final constellations the corresponding constellations would give the totality of all initial ones. It will also be noted that similar relations would hold if we started with the totality of all initial constellations or of all final constellations and constructed the reverse constellations thereto.

41. The closed cycle of corresponding collisions

We must now consider an important property of corresponding collisions which was first discovered by Boltzmann. The result which he obtained may be called *the theorem of the closed cycle of corresponding collisions*, and this theorem proves to be fundamental for the derivation of Boltzmann's H-theorem in a general enough form to be valid for molecules of any degree of complexity.

Let us write down a series of conceptually possible collisions

$$\begin{pmatrix}2,1\\3,4\end{pmatrix}\begin{pmatrix}4,3\\5,6\end{pmatrix}\begin{pmatrix}6,5\\7,8\end{pmatrix}\begin{pmatrix}8,7\\9,10\end{pmatrix}\cdots \qquad (41.1)$$

which is constructed by starting with some arbitrary collision $\begin{pmatrix}2,1\\3,4\end{pmatrix}$, taking next the collision $\begin{pmatrix}4,3\\5,6\end{pmatrix}$ that corresponds thereto, and so on in succession setting down each time the collision that corresponds to the preceding one in the list. We now wish to prove that such a list would have the nature of a closed cycle which would ultimately start to repeat itself.

To show this, let us regard the initial constellation of all (2, 1), with which we commence the series, as specified in the manner already described in § 39, by giving the infinitesimal ranges $\delta\omega_1$ and $\delta\omega_2$ within which we take the initial values of the variables $q_4 \dots q_r$, p_x, p_y, p_z, $p_4 \dots p_r$ that specify the internal condition and external momenta for the two molecules, and by giving the infinitesimal solid angle $\delta\lambda$, or the corresponding infinitesimal area $b^2\,\delta\lambda$, within which the centre of gravity of the second of the two molecules lies at the start of the collision with respect to the centre of gravity of the first molecule as an origin. With such a specification of the initial constellation it will then be noted that the final constellation for the first collision would itself be specified only to a similar set of infinitesimal ranges $\delta\omega_3$, $\delta\omega_4$, and $\delta\lambda'$. So also for the initial and final constellations for all the following collisions that we have written down, the specifications will only be to within a set of infinitesimal ranges.

In view of the foregoing, it will now be appreciated, with any choice of size for our original small ranges $\delta\omega_1$, $\delta\omega_2$, and $\delta\lambda$, that we shall ultimately have to come to a collision in the series with a final constellation that corresponds to the initial constellation for some earlier collision that has already appeared in the list, since, with a fixed value for the energy of the two molecules and with small but nevertheless non-vanishing sizes for ranges of the kind $\delta\omega$ and $\delta\lambda$, an unlimited number of completely separate possibilities for the final constellations is not available.† We now wish to show that the first time when this occurs we shall have the final constellation (1, 2) which corresponds to the first initial constellation of all (2, 1) and thus gives us a closed cycle.

To prove this, let us write down our series of collisions (41.1) in the somewhat more general form

$$\begin{pmatrix}2,1\\3,4\end{pmatrix}\begin{pmatrix}4,3\\5,6\end{pmatrix}\begin{pmatrix}6,5\\7,8\end{pmatrix}\cdots\begin{pmatrix}k, & k-1\\k+1, & k+2\end{pmatrix}\begin{pmatrix}k+2,k+1\\k+3,k+4\end{pmatrix}\begin{pmatrix}k+4,k+3\\k+5,k+6\end{pmatrix}\cdots \tag{41.2}$$

and let us consider any collision in the series that does have a final constellation which corresponds to some initial constellation that has already appeared earlier in the series. For example, let us assume that the final constellation $(k+5, k+6)$ corresponds to the earlier initial constellation (6, 5). Making a formal use of the equality sign to denote the identity of constellations or of collisions that are expressed in different form but are really the same, we can then write

$$(k+5, k+6) = (5, 6),$$

and hence
$$\begin{pmatrix}k+4,k+3\\k+5,k+6\end{pmatrix} = \begin{pmatrix}k+4, & k+3\\5, & 6\end{pmatrix}.$$

But we already have an expression $\begin{pmatrix}4,3\\5,6\end{pmatrix}$ for the collision which ends with the constellation (5, 6), and hence now obtain

$$(k+4, k+3) = (4, 3) \quad \text{or} \quad (k+3, k+4) = (3, 4),$$

which gives us for the next to last collision written in the list (41.2)

$$\begin{pmatrix}k+2,k+1\\k+3,k+4\end{pmatrix} = \begin{pmatrix}k+2, & k+1\\3, & 4\end{pmatrix}.$$

Repeating a similar argument, based on the consideration that we already have the expression $\begin{pmatrix}2,1\\3,4\end{pmatrix}$ for a collision ending with the constellation (3, 4), we then obtain

$$(k+2, k+1) = (2, 1) \quad \text{or} \quad (k+1, k+2) = (1, 2),$$

† See the end of the present section for a discussion of the degree to which this final constellation could be regarded as corresponding to the earlier initial constellation.

which finally gives us the desired relation

$$\begin{pmatrix} k, & k-1 \\ k+1, & k+2 \end{pmatrix} = \begin{pmatrix} k, & k-1 \\ 1, & 2 \end{pmatrix}. \tag{41.3}$$

This result then shows, starting with any particular place of occurrence in the series (41.2) of a final constellation which corresponds to some previous initial constellation, that we could always trace back to the first occurrence of such a place, where the series would be closed by the appearance of a final constellation which corresponds to the initial constellation for the first collision of all with which the list was commenced.

This is the desired theorem of the closed cycle of corresponding collisions. The theorem will be needed in the next chapter in deriving Boltzmann's H-theorem.

There is a point in connexion with the above deduction of the closed cycle of corresponding collisions that needs further elaboration. We have based the deduction on the idea that there must at least be some place in the series of collisions (41.2) where we encounter a final constellation which corresponds to an initial constellation that has already appeared in the list, and have justified this idea by the consideration that an unlimited number of completely separate possibilities for final constellations would not be available, since they are specified by ranges of the kind $\delta\omega$ and $\delta\lambda$ which, though small, would nevertheless ultimately exhaust the total range of possibilities. Two remarks, however, must now be made in this connexion.

In the first place, we shall actually wish to regard the ranges $\delta\omega_1$, $\delta\omega_2$, and $\delta\lambda$, needed for starting off the list of collisions (41.2), as differential quantities that can be made to approach the limit zero as we go to treatments of higher and higher accuracy. As we make these ranges smaller and smaller, however, it is then evident that we may encounter cases where we have to take a longer and longer list of collisions before we get to a final constellation with ranges that overlap those of some earlier initial constellation. Hence we shall have to allow the possibility that the list may become infinitely long before the desired closure is established.

In the second place, it is evident, as we look along the list of collisions given by (41.2) for a final constellation that does correspond to some earlier initial constellation, that we may expect in general to find a final constellation that corresponds only incompletely with an earlier initial one by a partial overlap of ranges. Thus, returning to the line of proof given in the foregoing, it might be that only a part of the precisely defined constellations in the range denoted by $(k+5, k+6)$ would correspond to constellations in the range denoted by $(6, 5)$. Hence, when we trace back to the constellation $(k+1, k+2)$, we could only conclude that there was partial correspondence with the original constellation $(2, 1)$. Nevertheless, if this degree of correspondence is not sufficiently exact to satisfy us, it is evident that we can always continue to add members to the list of collisions given by (41.2) until we ultimately get a final constellation which corresponds just as exactly as we please with an earlier initial constellation.† Thus, if we again recognize the possibility that closure may only result when the list becomes infinitely long, our theorem of the closed cycle of corresponding collisions still remains valid.

We thus find, for two different reasons, that the length, which we have to assign to a list of corresponding collisions before we get closure, may depend at least in some cases on the degree of accuracy which we demand of our treatment. This complicating circumstance does not appear, however, to invalidate the proof of Boltzmann's H-theorem which we shall give in the next chapter.

42. The closed cycle of two members in the case of spherical molecules

In the case of molecules having states that can be defined by specifying the external motion of the centre of gravity together with an internal condition which is static and exhibits spherical symmetry around the centre of gravity,‡ the foregoing theorem reduces to a very simple form, since it can be shown that the whole closed cycle of corresponding collisions reduces to a series of only two members. The simplicity of the result depends on the circumstance that the various possible states i, j, k, etc., for a molecule can then be taken as depending only on the magnitudes and directions of the velocities $\mathbf{v}_i$, $\mathbf{v}_j$, $\mathbf{v}_k$, etc., for the centre of gravity of the molecule in those states, and on an internal condition which will not be affected by reversal of velocities or by reflection parallel to any plane.

† Our later application of Liouville's theorem to collisions will show that there is nothing in the nature of a progressive change in the size of the ranges of the kind $\delta\omega$ and $\delta\lambda$, which we use in describing constellations, that would still further prolong the arrival at a point in the list where the desired degree of precision for the correspondence is attained.

‡ We thus exclude from our present consideration cases where the state of the *free* molecule would also depend on such factors as rotation or vibration.

To prove the desired result, let us start with any desired *arbitrary collision* between the two molecules

$$\begin{pmatrix} j, i \\ k, l \end{pmatrix} \tag{42.1}$$

and then construct the *reverse collision* thereto

$$\begin{pmatrix} -k, -l \\ -j, -i \end{pmatrix}. \tag{42.2}$$

This will necessarily have to exist in accordance with the principle of dynamical reversibility.

Let us next construct a *congruent collision* to (42.2)

$$\begin{pmatrix} -k', -l' \\ -j', -i' \end{pmatrix}, \tag{42.3}$$

by considering a rotation through the angle π parallel to any plane P. This will be a conceptually realizable collision since all congruent collisions are equally possible.

It will now be appreciated, however, that the reversal of velocities followed by the process of rotation through the angle π has changed the original vectors determining the states of the molecules in (42.1) into their mirror images with respect to the plane P. And it will also be appreciated that the rotation through the angle π has changed the positions of the two molecules into the mirror images—with respect to the plane P—of the positions that would have been brought about by interchanging the original positions of the molecules in the constellations involved in (42.1). Hence the collision (42.3) will be seen to have a character which can be represented by

$$\begin{pmatrix} \underline{l}, \underline{k} \\ \underline{i}, \underline{j} \end{pmatrix}, \tag{42.4}$$

where the constellations $(\underline{l}, \underline{k})$ and $(\underline{i}, \underline{j})$ are seen to be mirror images with respect to the plane P of those that would correspond to the original constellations appearing in (42.1).

Let us now construct the *enantiomorphous collision* to (42.4)

$$\begin{pmatrix} l, k \\ i, j \end{pmatrix}, \tag{42.5}$$

which can be obtained by reflection with respect to the plane P. This will be a possible collision in accordance with the principle of dynamical reflectability. It will now be seen, in accordance with the foregoing discussion, that this final collision (42.5) is not only the corresponding one to the original arbitrary collision (42.1) with which we started, but in addition that it ends with a final constellation which corresponds to the initial constellation of (42.1). Hence, in the case of molecules having such spherical symmetry, we find that the closed cycle of corresponding collisions would always reduce to two members, which could be written in the form

$$\begin{pmatrix} j, i \\ k, l \end{pmatrix} \begin{pmatrix} l, k \\ i, j \end{pmatrix}. \tag{42.6}$$

It is of interest to note in the present case that the final constellation (i, j) would correspond exactly to the initial constellation (j, i), without any necessity for considering a partial overlap of ranges of specification which might occur in some cases as discussed at the end of the preceding section. It is also of interest to note for this simple case that each of the above collisions would be the *inverse collision* to the other, as already defined in § 40, since each starts with the two molecules in the states which they would occupy at the end of the other collision, and returns them to the states with which that collision would start.

In contrast to this simple result in the case of molecules having spherical symmetry, it may now be emphasized in more general cases that we cannot expect to find any inverse collision at all for an arbitrarily chosen collision. In these more general cases the above simple proof for the existence of the inverse collision fails, since the first step of reversing velocities does not lead in general for a non-spherical molecule to a state congruent with the original one. And hence the following steps of rotation in the plane P through the angle π and reflection through that plane do not lead to constellations that correspond to those of the original collision.

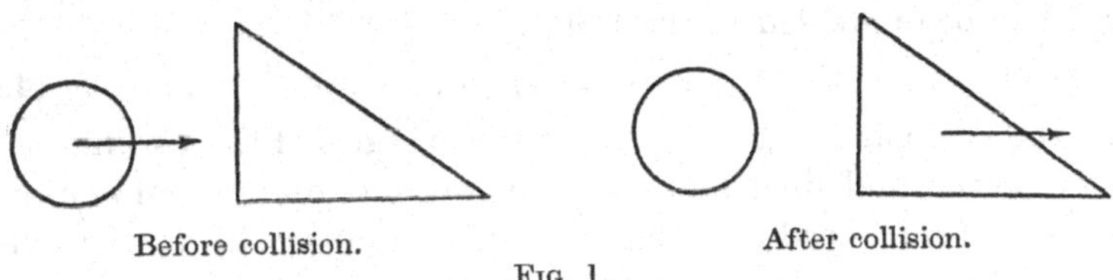

FIG. 1.

It will be of interest to illustrate the general case of the failure for inverse collisions to exist, with the help of a very simple if somewhat artificial example. For this purpose let us consider a collision between a sphere and a wedge-shaped body as shown in Fig. 1. We take the masses of the two bodies as equal, and the point of collisional contact as lying on the line connecting the centres of gravity of the two bodies, in order to have the specially simple situations before and after collision shown in the figure. After such a collision it is then evident that no possible inverse collision could be constructed which would bring the wedge to rest without rotational motion and return the sphere to its original motion from left to right. Similar results are evidently to be expected whenever the colliding bodies do not have spherical symmetry, and a somewhat complete treatment for the case of ellipsoidal bodies has been given by Lorentz.†

The specially simple possibility in the case of spherically symmetrical molecules, of immediately correlating with any arbitrary collision an inverse collision that would return the molecules to their original states, provides the reason why the derivations usually presented for Boltzmann's H-theorem confine themselves to the treatment of collisions between rigid, spherical molecules. It appears to have been clearly appreciated by the original discoverer of the theorem, however, that this is an artificially simplified case and that it is necessary to consider a complete cycle of corresponding collisions in order to obtain a general treatment.

43. Application of conservation laws to collisions

We are now ready to commence the application of the laws of mechanics to collisions. In the present brief section we shall merely point out that the ordinary conservation laws of mechanics for energy and for the components of linear and angular momentum would, of course, hold for any collision.

If we consider a collision $\binom{j,i}{k,l}$, the principle of the conservation of energy can be expressed in the form

$$\epsilon_i+\epsilon_j = \epsilon_k+\epsilon_l, \tag{43.1}$$

where the symbols ϵ_i, ϵ_j; ϵ_k, ϵ_l denote the energies of the two molecules in their initial and final states, no expressions for mutual energy of interaction being needed owing to our assumption as to the extent of separation of the two molecules by the critical distance b at the times which we take as marking the beginning and ending of a collision.

Similarly, as an expression of the principle of the conservation of linear momentum for this collision, we can write

$$\begin{aligned} p_{ix}+p_{jx} &= p_{kx}+p_{lx}, \\ p_{iy}+p_{jy} &= p_{ky}+p_{ly}, \\ p_{iz}+p_{jz} &= p_{kz}+p_{lz}, \end{aligned} \tag{43.2}$$

† Lorentz, *Sitzungsber. d. Akad. d. Wiss. zu Wien*, 2 Abt., **95**, 115 (1887).

where the symbols denote the initial and final components of momentum, parallel to Cartesian axes, calculated for the molecules as a whole from their masses and the velocities of their centres of gravity. If of interest we could also write expressions for the conservation of angular momentum.

44. Application of Liouville's theorem to a collision

We now come to a more complicated application of the classical mechanics to collisions which will be very important for our further work. Since a pair of colliding molecules is itself a mechanical system, obeying the laws of mechanics in the Hamiltonian form, it is evident that Liouville's theorem, which was derived from those laws in Chapter III, should be applicable to collisional processes. We shall now make such an application, taking Liouville's theorem in the form of the principle of the conservation of extension in phase.

Let us consider any collision between two molecules $\binom{j,i}{k,l}$ which could be specified by giving the ranges $\delta\omega_i$ and $\delta\omega_j$ for the states with which the first and second molecules enter the collision, and by giving the range $\delta\lambda_1$ for the direction of the line connecting the centres of gravity of the two molecules—i.e. the angle of attack—when the collision is commencing at time t_1. This will give us a range of neighbouring processes to which the principle of the conservation of extension in phase can be applied. It will first be desirable, however, to have an exact description of one of these collisional processes.

For the purposes of exact description, a fairly complicated notation will be necessary. We shall introduce single and double accents, $'$ and $''$, to denote the values of variables which apply to the first and second molecules respectively at the start of the collision, when they are in the states partially specified by i and j, and introduce triple and quadruple accents, $'''$ and $''''$, to denote similar values for the two molecules at the end of the collision when they are in states k and l. In addition we shall regard the collision under consideration as starting at time t_1 when the centres of gravity of the two molecules first reach the critical distance b, and as ending at time t_2 when they again reach that distance, and shall use the subscripts, $_1$ and $_2$, in general as denoting the values of variables that apply respectively at the beginning and end of the collision.

With the help of this notation we may now describe a precise collision, of the group $\binom{j,i}{k,l}$ that interests us, by taking

$$x', y', z', q'_4 \dots q'_r, p'_x, p'_y, p'_z, p'_4 \dots p'_r \tag{44.1}$$

as the values for the coordinates and momenta of the *first molecule* at the beginning of the collision, and by taking

$$x'', y'', z'', q_4'' \dots q_r'', p_x'', p_y'', p_z'', p_4'' \dots p_r'' \tag{44.2}$$

as the values for the coordinates and momenta of the *second molecule* at the beginning of the collision. Furthermore, since the centres of gravity of the two molecules will be separated by the distance b at time t_1, when the collision starts, we can write

$$\begin{aligned} x'' &= x' + b\sin\theta_1\cos\phi_1, \\ y'' &= y' + b\sin\theta_1\sin\phi_1, \\ z'' &= z' + b\cos\theta_1 \end{aligned} \tag{44.3}$$

as a connexion between the positional coordinates for the two molecules, where θ_1 and ϕ_1 are the polar and equatorial angles which locate the centre of gravity of the second molecule, with respect to that of the first as an origin, at time t_1.

Similarly, at time t_2, when the collision ends, we can take

$$x''', y''', z''', q_4''' \dots q_r''', p_x''', p_y''', p_z''', p_4''' \dots p_r''' \tag{44.4}$$

as the values of the coordinates and momenta of the *first molecule*,

$$x'''', y'''', z'''', q_4'''' \dots q_r'''', p_x'''', p_y'''', p_z'''', p_4'''' \dots p_r'''' \tag{44.5}$$

as the values for the coordinates and momenta of the *second molecule*, and

$$\begin{aligned} x'''' &= x''' + b\sin\theta_2\cos\phi_2, \\ y'''' &= y''' + b\sin\theta_2\sin\phi_2, \\ z'''' &= z''' + b\cos\theta_2 \end{aligned} \tag{44.6}$$

as a connecting relation between positional coordinates, where θ_2 and ϕ_2 are now the polar and equatorial angles for the centre of gravity of the second molecule with respect to that of the first at the later time t_2.

We are now ready to investigate the application of the principle of the conservation of extension in phase. To do this, instead of considering merely the above precisely defined collisional process, let us now consider a group of such processes of neighbouring character that would correspond to the ranges $\delta\omega_i$, $\delta\omega_j$, and $\delta\lambda_1$ by which we specify what we have called the collision $\begin{pmatrix} j,i \\ k,l \end{pmatrix}$. By equating the extensions in phase for such a group of processes at times t_1 and t_2, we can then obtain the desired application of Liouville's theorem. We must first examine suitable ranges to take for the initial values of the coordinates and momenta of the two molecules in order to obtain such a group of processes.

Let us begin by considering the initial values of the variables for the first of the two molecules, as given by (44.1). Since we shall not be interested in the particular position where a process of the kind $\binom{j,i}{k,l}$ occurs, we may take the coordinates x', y', z' that locate the centre of gravity of the first molecule at time t_1 as having values anywhere within an arbitrary volume

$$v_1 = \iiint dx'dy'dz'. \tag{44.7}$$

We may then complete the specification of initial conditions for the first molecule by taking the remaining variables as lying within an infinitesimal range

$$\delta\omega_i = \int \dots \int dq_4' \dots dp_r', \tag{44.8}$$

of the kind we have adopted for use in the specification of collisions.

Let us now consider the initial values of the variables for the second of the two molecules, as given by (44.2). Since we shall be interested in processes of the kind $\binom{j,i}{k,l}$, we must first pick out a range of values for the initial coordinates x'', y'', z'' that will place the centre of gravity of the second molecule in a position to secure results of this character. For this purpose, let us return to the relation between the positional coordinates for the two molecules given by (44.3), and for each set of values x', y', z', that locate the centre of gravity of the first molecule, choose a range for x'', y'', z'' such that the centre of gravity of the second molecule would enter the critical sphere surrounding the first molecule within a range of polar angles θ_1 to $\theta_1+\delta\theta_1$, within a range of equatorial angles ϕ_1 to $\phi_1+\delta\phi_1$, and within an interval of time t_1 to $t_1+\delta t_1$. Using the symbol $\delta\lambda_1$ to denote the *infinitesimal solid angle* $\sin\theta_1\,\delta\theta_1\,\delta\phi_1$, and the symbol k_1 to denote the *radial component of the relative velocity* of approach of the two molecules, we see for each point x', y', z' that the above procedure would then place x'', y'', z'' within an infinitesimal volume, which—neglecting higher order differentials—has the magnitude

$$b^2\,\delta\lambda_1\,k_1\,\delta t_1 = \iiint dx''dy''dz''. \tag{44.9}$$

The nature and location of this infinitesimal volume can be understood with the help of the *rough diagrammatic* representation provided by Fig. 2, where x', y', z' and x'', y'', z'' give a pair of precise initial positions for the centres of gravity of the two molecules, which would lead to a collision of the exact kind that was discussed above, and the cross-hatching illustrates the infinitesimal volume $b^2\,\delta\lambda_1\,k_1\,\delta t_1$ within which the centre of gravity of the second molecule would lie in order to obtain our group of closely similar collisions. To complete the initial specification of conditions for the second molecule, we may now take the

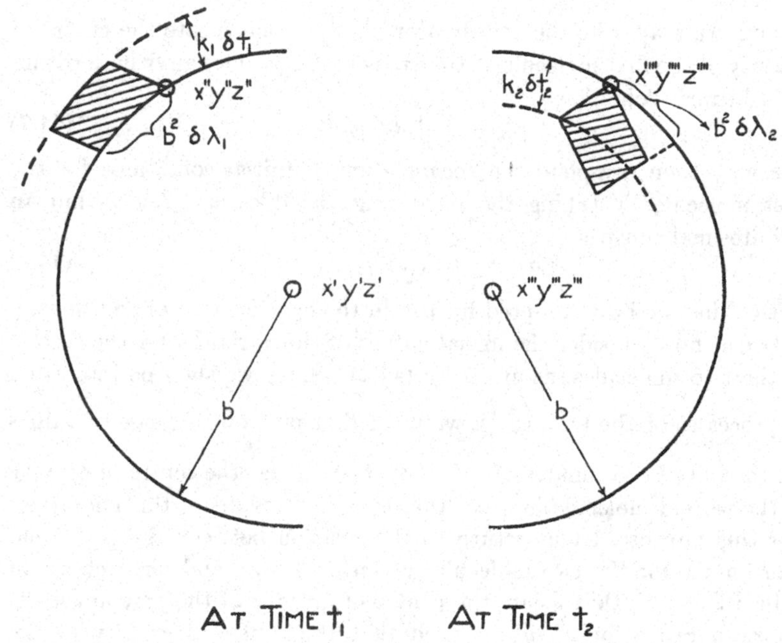

Fig. 2.

remaining variables for that molecule as lying within an infinitesimal range

$$\delta\omega_j = \int \dots \int dq_4'' \dots dp_r''. \tag{44.10}$$

With the help of the foregoing specifications of the ranges for the initial values of the coordinates and momenta for our system of two molecules we can now write, by combining (44.7), (44.8), (44.9), and (44.10),

$$V_1 = \int \dots \int dx'dy'dz'dq_4' \dots dp_r'\,dx''dy''dz''dq_4'' \dots dp_r''$$

$$= v_1\,b^2\,\delta\lambda_1\,k_1\,\delta t_1\,\delta\omega_i\,\delta\omega_j \tag{44.11}$$

as an expression for the *initial value of the extension in phase* for this system.

We must now turn to a consideration of the ranges of final values for the coordinates and momenta.

Let us begin by considering the final values for the positional coordinates x''', y''', z''' for the first molecule. Since all the collisions that we are considering are substantially identical in character, the final range of values for these variables—neglecting second-order differentials—will be the same as their initial range, thus giving us

$$v_2 = \iiint dx'''dy'''dz''' \tag{44.12}$$

with

$$v_2 = v_1 \tag{44.13}$$

as the volume within which the centre of gravity of the first molecule would lie at the final time t_2 when collisions of the kind discussed above would end. To complete the specification of conditions for the first molecule at the end of the collision we may now take

$$\delta\omega_k = \int \dots \int dq_4''' \dots dp_r''' \tag{44.14}$$

as an expression for the infinitesimal range of values, for the other variables describing that molecule, which actually would apply at time t_2 to the processes that we are considering.

Let us now consider the final values for the positional coordinates x'''', y'''', z'''' for the second molecule. To study the range in values for these coordinates at time t_2 we note, for any given final position x''', y''', z''' of the first molecule, that a collision of the exact kind previously discussed would put the centre of gravity of the second molecule at a precise point x'''', y'''', z'''' located on the critical sphere with x''', y''', z''' as a centre, as shown in the second half of Fig. 2. The centres of gravity of the second molecule in the case of any one of the other slightly different collisions which we are considering would then lie at time t_2 in the neighbourhood of this point, as illustrated by the cross-hatching in the diagram. Let us now denote by $b^2\,\delta\lambda_2$ the *infinitesimal area*, at the above point on the surface of the sphere, through which the centres of gravity of the second molecules would pass, denote by k_2 the *radial component of the relative velocity* of separation of the two molecules at the end of the collisions under consideration, and denote by δt_2 the *time interval* between the passage of the earliest and latest second molecule through the point x'''', y'''', z''''. It will then be seen from the diagram, neglecting second-order differentials, that we can write

$$b^2\,\delta\lambda_2\,k_2\,\delta t_2 = \iiint dx''''dy''''dz'''' \tag{44.15}$$

as an expression for the infinitesimal volume within which the centre of gravity of the second molecule would lie at time t_2 for each position of the centre of gravity of the first molecule. Furthermore, since δt_1 is the time interval between the passage of the earliest and latest second molecule through the point x'', y'', z'', and δt_2 is such an interval for passage through the point x'''', y'''', z'''', we can also write within second-order terms

$$\delta t_1 = \delta t_2. \tag{44.16}$$

To complete the specification of conditions for the second molecule at the end of the collision we may now take

$$\delta\omega_l = \int ... \int dq_4'''' ... dp_r'''', \tag{44.17}$$

as an expression for the infinitesimal range of values, for the other variables describing the second molecule, that actually would apply at time t_2 to the processes that we are considering.

We have now completed the specification of the ranges for the final values of the coordinates and momenta for our system of two molecules, and, by combining (44.12), (44.14), (44.15), and (44.17), can write

$$V_2 = \int ... \int dx'''dy'''dz'''dq_4''' ... dp_r''' \, dx''''dy''''dz''''dq_4'''' ... dp_r''''$$
$$= v_2 b^2 \delta\lambda_2 k_2 \delta t_2 \delta\omega_k \delta\omega_l \tag{44.18}$$

as an expression for the *final value of the extension in phase* for this system.

In accordance with Liouville's theorem, however, we may now equate this final value of the extension in phase V_2 to the original value V_1 for the extension given by (44.11). Doing so, cancelling b^2 from both sides together with the equal factors given by (44.13) and (44.16), we then obtain as the finally desired result

$$k_1 \delta\lambda_1 \delta\omega_i \delta\omega_j = k_2 \delta\lambda_2 \delta\omega_k \delta\omega_l. \tag{44.19}$$

This important consequence of the principle of the conservation of extension in phase applies in general to any collision $\begin{pmatrix} j, i \\ k, l \end{pmatrix}$ between two molecules, where k_1 and k_2 are the radial components of the relative velocities of approach and of recession of the centres of gravity of the two molecules at the beginning and at the end of the collision respectively, and the quantities $\delta\lambda_1$, $\delta\omega_i$, $\delta\omega_j$ and $\delta\lambda_2$, $\delta\omega_k$, $\delta\omega_l$ specify respectively the initial and final constellations for the collision, in the manner described in § 39.

45. The probability coefficients for collisions

Having considered the properties of the different kinds of collisions which can take place between a pair of molecules, we now turn to a consideration of the frequency with which such collisions could be expected to occur in an actual system. Thus far our treatment of collisions has not involved statistical considerations. To treat our present problem, however, we shall now have to introduce the ideas of *statistical mechanics*, at least in a simple form.

Let us consider a gas, taken for simplicity sufficiently dilute so that we may confine our attention to bimolecular collisions, and let us inquire into the number of collisions of the kind specified by the symbol $\binom{j,\,i}{k,\,l}$ which would take place within some time interval of interest t to $t+\delta t$. If we were provided with an exact and complete specification, at time t, of the coordinates and momenta for each of the molecules composing the system, it is then evident—with the help of the exact laws of the classical mechanics—that we could make in principle a calculation of the precise number of collisions of the above kind that would take place in the selected time interval δt. If, however, we are actually furnished with only an incomplete specification of state, we shall then instead have to make a calculation—with the help of the principles of statistical mechanics—of the number of collisions of the above kind that can be expected on the average for samples of gas in the condition that is specified.

Let us take the condition of the gas at time t as specified merely by its total volume v, total energy E, and by the numbers of molecules n_i and n_j having internal coordinates and external and internal momenta $(q_4 \ldots p_r)$ lying in the ranges $\delta\omega_i$ and $\delta\omega_j$ involved in a collision of the kind $\binom{j,\,i}{k,\,l}$ that interests us. We assume that no specification is given for the positions of any of the molecules. Furthermore, for simplicity, we take the gas as highly dilute and as enclosed in a fixed container located in a field-free region, in order to avoid restrictions on the positions of the molecules that would otherwise result from the necessity of allowing for their mutual energy of interaction and their potential energy in the field in meeting the requirement as to total energy.

To apply the principles of statistical mechanics, by the methods already discussed in Chapter III, we must first take an appropriate representative ensemble of systems, each a gas of the kind under consideration, with their individual phase points so located in the qp-phase space as to agree with the above partial specification of state, but otherwise distributed uniformly in accordance with the hypothesis of equal *a priori* probabilities for equal regions in the phase space. We can then take the average behaviour of the systems in such an ensemble as giving a good estimate for the behaviour of the actual system of interest; or, more particularly in the present case, we can take the average number of collisions of the kind $\binom{j,i}{k,l}$ that would occur in the systems of the ensemble in the time δt as an estimate for the number that would occur in the actual sample of gas of interest. If we choose the *most probable* number of collisions of the kind $\binom{j,i}{k,l}$ in the time δt as the average of interest, the application of this procedure can be made in a very simple manner.

We must begin by considering the character of the appropriate representative ensemble. In accordance with our methods, the phase points for the members of this ensemble are to be taken as uniformly distributed in so far as this does agree with the partial specification of state provided. In the present case, since the positions of the molecules are completely unspecified, and also unrestricted except for their location in the volume v, this means a uniform distribution—within the limits allowed by v—with respect to those axes in the phase space which correspond to the positional coordinates $x_1 y_1 z_1$, $x_2 y_2 z_2$,..., $x_s y_s z_s$,... for the centres of gravity of the various molecules s. As a consequence of this uniformity we can then conclude for samples selected at random from the ensemble that the probability P of finding the centre of gravity of a particular molecule of interest inside a specified volume δv will be given by the ratio

$$P = \frac{\delta v}{v} \tag{45.1}$$

which the specified volume bears to the total volume v of the container.

With the help of this simple and perhaps obvious consequence of the methods of statistical mechanics we can now determine for the systems of the ensemble the most probable number of pairs of molecules which are so located that a collision of the kind $\binom{j,i}{k,l}$ would be initiated within the time δt. In order for such a collision to occur, the critical sphere of radius b, surrounding a molecule with coordinates and momenta $q_4 \ldots p_r$ lying in the range $\delta\omega_i$, must be entered through a specified area $b^2\,\delta\lambda_1$ on the critical sphere by the centre of gravity of a second molecule with variables lying in the range $\delta\omega_j$. Thus, for any given molecule in the range $\delta\omega_i$, a collision of the kind $\begin{pmatrix} j, i \\ k, l \end{pmatrix}$ will be initiated in the time δt, if the centre of gravity of a second molecule in the range $\delta\omega_j$ is situated within a specified volume $b^2\,\delta\lambda_1\,k_1\,\delta t$, where k_1 is the radial component of the relative velocity of approach of the two centres of gravity. Hence, making use of the above result (45.1) for the probability of finding the centre of gravity of a given molecule in a specified volume, and making use of obvious approximations for the case of high dilution, we can now write

$$P_{ij} = \frac{b^2k_1\,\delta\lambda_1\,\delta t}{v}n_i\,n_j \tag{45.2}$$

for the most probable number of pairs of molecules which are so situated as to lead to a collision of the above kind in time δt, where n_i and n_j are the specified numbers of molecules in the conditions denoted by $\delta\omega_i$ and $\delta\omega_j$. Or, dividing through by δt, we obtain for the most probable number of collisions $\begin{pmatrix} j, i \\ k, l \end{pmatrix}$ occurring per unit of time†

$$Z_{k,l}^{j,i} = \frac{b^2k_1\,\delta\lambda_1}{v}n_i\,n_j. \tag{45.3}$$

It will be noted that the number of such collisions will be dependent on the numbers of molecules in states i and j, but quite independent of the states of other molecules.

In order to put this result into a more useful form it will now be desirable to introduce expressions, for the numbers of molecules n_i and n_j in the ranges $\delta\omega_i$ and $\delta\omega_j$, which show the explicit dependence of such numbers on the size of the ranges selected, and their explicit dependence on the time in cases where a steady state of equilibrium has not been reached. For this purpose, let us take the expression

$$\delta n = f(q_1 \ldots p_r, t)\,\delta q_1 \ldots \delta p_r \tag{45.4}$$

as providing a general method of specifying the number of molecules, of any kind contained in the system, which have coordinates and momenta $q_1 \ldots p_r$ lying in the indicated range $\delta q_1 \ldots \delta p_r$, at time t. We take the function f which specifies the distribution as depending in

† For simplicity we shall regard the time necessary for the completion of a collision as sufficiently short so that it will not be necessary for us to introduce a distinction between the number of collisions 'initiated' and 'occurring' per unit time.

general, at a selected initial time t, in any arbitrary way desired on the coordinates and momenta $q_1 \dots p_r$. For the special case of a steady condition of equilibrium the function reduces to the constant form $f = nCe^{-\epsilon/kT}$ independent of the time but depending on the energy $\epsilon(q_1 \dots p_r)$, and the general expression reduces to the Maxwell-Boltzmann distribution law

$$\delta n = nCe^{-\epsilon/kT}\,\delta q_1 \dots \delta p_r. \tag{45.5}$$

In the cases now under consideration, where the positions of the molecules are left unspecified, the general expression (45.4) can be rewritten in the form

$$\delta n = f(q_4 \dots p_r, t)\,\delta x \delta y \delta z \delta q_4 \dots \delta p_r, \tag{45.6}$$

where x, y, and z are the Cartesian coordinates for position. Integrating over the total volume v, and over a selected small range

$$\delta\omega_i = \int \dots \int dq_4 \dots dp_r$$

for the momenta and internal coordinates of a molecule, we then obtain expressions of the desired form

$$n_i = f_i v\,\delta\omega_i \quad \text{and} \quad n_j = f_j v\,\delta\omega_j \tag{45.7}$$

for the total numbers of molecules n_i and n_j in the ranges $\delta\omega_i$ and $\delta\omega_j$ that interest us, where f_i and f_j are the values of the function f at the time t, and at the locations of the regions $\delta\omega_i$ and $\delta\omega_j$.

Substituting (45.7) into our previous equation (45.3), we can now write

$$Z_{k,l}^{j,i} = vb^2 k_1\,\delta\lambda_1\,\delta\omega_i\,\delta\omega_j\,f_i f_j \tag{45.8}$$

as the finally desired expression for the rate of the collisions that we have been considering. This expression applies in general in the case of a dilute, homogeneous gas to a collision denoted by $\begin{pmatrix} j, i \\ k, l \end{pmatrix}$. The quantity v is the volume of the container, b is the critical radius defining the beginning and end of the collision, k_1 is the radial component of the relative velocity of approach of the centres of gravity of the two molecules at the start of the collision, $\delta\lambda_1$ is the solid angle with respect to the centre of gravity of the first molecule as an origin within which the centre of gravity of the second molecule lies at the start of the collision, and the quantities $vf_i\,\delta\omega_i$ and $vf_j\,\delta\omega_j$ give the total numbers of molecules n_i and n_j which have momenta and internal coordinates lying in the ranges $\delta\omega_i$ and $\delta\omega_j$, these quantities together with $\delta\lambda_1$ being those that specify the character of the collision. With this understanding of the quantities involved, the expression then gives the *most probable rate*, with which collisions of the kind $\binom{j,\, i}{k,\, l}$ will be initiated, in the systems of an appropriate representative ensemble all of which have the above numbers of molecules n_i and n_j in the indicated conditions.

We shall now wish to compare this expression for the rate of collisions of the kind $\binom{j,\, i}{k,\, l}$, with the similar expressions for the rates of collisions of the reverse kind $\binom{-k,\, -l}{-j,\, -i}$, and of the corresponding kind $\binom{l,\, k}{m,\, n}$. Since expressions similar to (45.8) can evidently be used for any collision, the appropriate expression for the rate of the reverse collision will be of the form

$$Z^{-k,-l}_{-j,-i} = vb^2 k_r\, \delta\lambda_r\, \delta\omega_{-k}\, \delta\omega_{-l}\, f_{-k}\, f_{-l}, \tag{45.9}$$

and for the corresponding collision of the form

$$Z^{l,k}_{m,n} = vb^2 k_c\, \delta\lambda_c\, \delta\omega_k\, \delta\omega_l\, f_k\, f_l, \tag{45.10}$$

where we use $k_r\,\delta\lambda_r$ and $k_c\,\delta\lambda_c$ to designate for these new collisions the quantities analogous to the $k_1\,\delta\lambda_1$ appearing in (45.8), and the rest of the notation will be obvious.

In order to compare the above three expressions for the rates of these related collisions, it will now be necessary to recall the previous equation (44.19), derived with the help of Liouville's theorem, which gives a connexion between quantities pertaining to the beginning and to the end of any collision $\binom{j,\, i}{k,\, l}$. This connecting relation has the form

$$k_1\,\delta\lambda_1\,\delta\omega_i\,\delta\omega_j = k_2\,\delta\lambda_2\,\delta\omega_k\,\delta\omega_l, \tag{45.11}$$

where, as quantities pertaining to the end of the collision, k_2 is the radial component of the relative velocity of separation of the centres of gravity of the two molecules, $\delta\lambda_2$ is the infinitesimal solid angle specifying the range of positions for the second molecule on the critical sphere surrounding the first, and $\delta\omega_k$ and $\delta\omega_l$ are the ranges that actually apply to the momenta and internal coordinates of the molecules as a consequence of the kind of collision considered.

With the help of this relation the remainder of the argument is now quite simple. Since the reverse collision $\binom{-k,\, -l}{-j,\, -i}$ can be obtained by merely reversing all velocities at the end of the originally specified collision $\binom{j, i}{k, l}$, we must evidently have the relations

$$k_r = k_2, \quad \delta\lambda_r = \delta\lambda_2, \quad \delta\omega_{-k} = \delta\omega_k, \quad \text{and} \quad \delta\omega_{-l} = \delta\omega_l, \tag{45.12}$$

and since the corresponding collision can be obtained by merely interchanging the centres of gravity of the two molecules at the end of the originally specified collision $\binom{j, i}{k, l}$, we must also have the relations

$$k_c = k_2 \quad \text{and} \quad \delta\lambda_c = \delta\lambda_2. \tag{45.13}$$

Combining with the consequences of Liouville's theorem given by (45.11), and examining our original expressions (45.8), (45.9), and (45.10) for the rates of the three related collisions, we then see that we can write in general for the case of a dilute homogeneous gas

$$\begin{aligned} Z^{j,i}_{k,l} &= Cf_i f_j, \\ Z^{-k,-l}_{-j,-i} &= Cf_{-k} f_{-l}, \\ Z^{l,k}_{m,n} &= Cf_k f_l, \end{aligned} \tag{45.14}$$

as respective expressions for the *most probable rates* of a given collision $\binom{j, i}{k, l}$, the reverse collision thereto $\binom{-k, -l}{-j, -i}$, and the corresponding collision thereto $\binom{l, k}{m, n}$, where the *probability coefficient* C has the *same value,*

$$C = vb^2k_1\,\delta\lambda_1\,\delta\omega_i\,\delta\omega_j, \tag{45.15}$$

for all three collisions, and the instantaneous values for the *numbers of molecules* n_i, in different ranges of the type $\delta\omega_i$, used in specifying the nature of the collisions, are given by expressions of the form

$$n_i = vf_i\,\delta\omega_i. \tag{45.16}$$

This final conclusion as to the equal probability coefficients for these related collisions will be very important in the treatment of Boltzmann's H-theorem in the next chapter.

46. Concluding remarks on molecular collisions

This now concludes our somewhat elaborate and perhaps tiresome treatment of molecular collisions as a mechanism leading to changes that take place with time in the condition of actual physical-chemical systems of interest. The complicated character of the discussion was made necessary by our desire to give a general treatment applying to molecules of any kind, rather than to restrict ourselves to the specially simple case of collisions between rigid elastic spheres, as is so often done in developments that are given for the classical statistical mechanics.

It is evident, however, that we cannot regard this general treatment of bimolecular collisions in a dilute homogeneous gas as giving a complete account of all the processes by which actual physical-chemical systems change with time from one condition to another. In addition to collisions between pairs of molecules there will be collisions of higher order and interactions with radiation. Furthermore, although in the absence of complete homogeneity we can often apply our expressions for the rates of collisions to small regions of the gas where the numbers of molecules in different conditions are specified, it is evident that processes such as pressure equalization and diffusion must also be considered—using known but supplementary methods—in order to obtain a complete picture of the time behaviour of physical systems. Nevertheless, bimolecular collisions provide a sufficiently typical and important mechanism of change, so that their use in connexion with the derivation of the H-theorem in the next chapter will be appropriate for giving a fundamental understanding of the nature—if not a complete picture—of the changes that take place with time in the condition of actual physical-chemical systems.

There is, of course, also another serious limitation on the general applicability of the methods of the present chapter in that we have based them on the classical rather than on the quantum mechanics. Particularly in the case of the internal coordinates and momenta of a molecule we must expect many cases where the classical mechanics gives a very poor approximation for the more strictly valid results of the quantum mechanics. This difficulty will be remedied when we come to the analogous quantum mechanical part of our development of statistical mechanics.

In this connexion it is also of interest to note that the quantum mechanical treatment, which we shall give to the changes in systems with time in Chapter XI, will be much more complete in character than the considerations of the present chapter. We shall, to be sure, also there give a quantum mechanical treatment to collisions as a mechanism of change, but on account of the greater power and simplicity of quantum mechanical as compared with classical methods, it will be found possible to supplement this with somewhat formal but nevertheless quite general treatments applicable to any kind of processes that take place with time in actual physical-chemical systems.

VI

BOLTZMANN'S *H*-THEOREM

47. Definition of the quantity *H*

In Chapter IV we have studied the Maxwell-Boltzmann expression which describes the distribution of molecules in different states that can be expected to prevail under conditions of macroscopic equilibrium; and in Chapter V we have made a detailed study of collisions as an important mechanism which leads to the change of molecules from one state to another. In the present chapter we are now ready to study Boltzmann's famous H-theorem, demonstrating the actual tendency for the molecules of a system to approach their equilibrium distribution when started off in any desired arbitrary manner. The derivation of this theorem and the appreciation of its significance may be regarded as among the greatest achievements of physical science.

In this section, as a preliminary task, we must consider the definition of a quantity, called H by Boltzmann, which, as we shall see, can be regarded as providing an appropriate measure of the extent to which the condition of a system deviates from that corresponding to equilibrium. The later proof of the H-theorem will then consist in showing the actual tendency for this quantity to decrease with time to a minimum, and thus for the system to approach its equilibrium condition.

To obtain the desired definition we must first have a general method of specifying molecular distributions, suitable for use whether equilibrium has been attained or not. For this purpose we may again introduce our previous expression (45.4) for the number of molecules,

$$\delta n = f(q_1 \dots p_r, t)\, \delta q_1 \dots \delta p_r, \tag{47.1}$$

in any small region of the μ-space denoted by $\delta q_1 \dots \delta p_r$, where the function f depends on the coordinates and momenta $q_1 \dots p_r$ for the molecules under consideration, and on the time t, in whatever manner that is needed to describe the distribution. In the special case of equilibrium, the function f would then assume the form $nCe^{-\epsilon/kT}$, and (47.1) would reduce to the Maxwell-Boltzmann distribution law,

$$\delta n = nCe^{-\epsilon/kT}\, \delta q_1 \dots \delta p_r, \tag{47.2}$$

where the energy ϵ depends on the coordinates and momenta $q_1 \dots p_r$ but not on the time t.

To use such specifications of molecular condition for the definition of the quantity H, let us now regard the total μ-space, for each kind of molecule present in the system, as divided up into a collection of small finite regions all having equal magnitudes of extension

$$\delta v_\mu = \delta q_1 \ldots \delta p_r \tag{47.3}$$

in the μ-space. In this way we are provided, as in § 27, with a collection of equal 'cells' in the μ-space, which we can think of as labelled by the integers 1, 2, 3,..., i,.... In agreement with (47.1) we can then write

$$n_i = f_i(t)\, \delta q_1 \ldots \delta p_r \tag{47.4}$$

as a general expression for the number of molecules n_i in the ith such region at time t.

With the help of these quantities f_i we can then define the quantity H for our system by

$$H = \sum_i f_i \log f_i\, \delta q_1 \ldots \delta p_r, \tag{47.5}$$

where we take a sum, for each kind of molecule present, over all the regions i, of equal magnitudes $\delta q_1 \ldots \delta p_r$, into which we regard the μ-spaces for the different kinds of molecules as divided. Or, replacing summation by integration over the whole of the μ-spaces, we can also write our defining expression in the form

$$H = \int \ldots \int f \log f\, dq_1 \ldots dp_r, \tag{47.6}$$

provided we realize that we must treat the differential ranges $dq_1 \ldots dp_r$ as of high enough order so that no trouble arises from regarding the number of molecules $f\, dq_1 \ldots dp_r$ in such ranges as a continuous function of position in the μ-space. Or, in accordance with (47.3–5), it will be seen that we can also express H in terms of the numbers of molecules present in the different cells i in the form

$$\begin{aligned} H &= \sum_i (n_i \log n_i - n_i \log \delta v_\mu) \\ &= \sum_i n_i \log n_i + \text{const.}, \end{aligned} \tag{47.7}$$

where we take a summation over all such numbers of molecules n_i.

This last form of expression is useful in showing, for any specified condition of an isolated system, the close relation between our new quantity H and our earlier quantity P, giving the probability of finding a system chosen at random from the corresponding microcanonical ensemble in the specified condition. Comparing (47.7) with our previous expression (28.4), we can write

$$H = -\log P + \text{const.}, \tag{47.8}$$

where the constant term only contains quantities of a kind that do not depend on the different conditions of any given isolated system. In accordance with this relation we now see that the quantity H has been defined in such a way as to give a measure of the extent to which the condition of an isolated system deviates from equilibrium, since decreases in this new quantity H would correspond to increases in the quantity $\log P$, which assumes its maximum possible value at equilibrium. It is also to be noted, for a given isolated system—with specified energy content, volume, and total number of molecules—that there would be a minimum possible value of H corresponding to our previously investigated maximum possible value of $\log P$. Hence the approach of a system towards the condition of equilibrium may be regarded as corresponding to a decrease in H towards its minimum possible value.

One further form of expression for H will also prove useful in a later chapter—see § 102—when we wish to set up the quantum analogue of the classical quantity H. In accordance with § 28, the probability P for any condition of interest is to be taken proportional to the number of regions of equal volume δv_γ in the phase space for the system which correspond to that condition. Hence, using G to denote the number of such regions, we can also write our expression for H in the form

$$H = -\log G + \text{const.}, \tag{47.9}$$

where G may be conveniently spoken of as the number of classical states—defined by the equal volumes δv_γ—which correspond to the condition of interest.

48. Derivation of the H-theorem

(*a*) **Rate of change of H with time.** We are now ready to consider the verification of the H-theorem, which asserts an actual tendency for H to decrease with time, and hence for the molecules of a system to approach their equilibrium distribution. To do this we shall first obtain a general formal expression for the rate of change in H with time, and then show the effect of actual molecular processes in contributing to a negative value for this rate of change.

Let us start with our definition for the quantity H in the form

$$H = \int \dots \int f \log f \, dq_1 \dots dp_r, \tag{48.1}$$

where the function $f(q_1 \dots p_r, t)$ determines at time t the distribution in the μ-space for each kind of molecule present, in accordance with the expression

$$\delta n = f(q_1 \dots p_r, t) \, \delta q_1 \dots \delta p_r. \tag{48.2}$$

If we now differentiate (48.1) with respect to t, we can write for the rate of change in H with time

$$\frac{dH}{dt} = \int \dots \int \left(\frac{df}{dt}\log f + \frac{df}{dt}\right) dq_1 \dots dp_r, \tag{48.3}$$

since f is the only factor that depends on the time. Noting, however, the significance of the second term df/dt that appears on the right-hand side of (48.3), we appreciate in accordance with (48.2) that the integration of this term over all values of the coordinates and momenta would give an expression for the rate of change with time in the total number of molecules in the system. And this can be taken equal to zero, since we shall only consider systems composed of a fixed number of molecules. Hence we can now write

$$\frac{dH}{dt} = \int \dots \int \frac{df}{dt}\log f \, dq_1 \dots dp_r \tag{48.4}$$

as a general, formal expression for the rate of change of H with time, in terms of the function f, which specifies the distribution of molecules among their own different states, and in terms of its time rate of change df/dt.

Starting with this formal expression for the time rate of change in H, a general derivation of the H-theorem would now consist in showing, once and for all, for every possible kind of molecular process, and for any possible specification of the distribution function f, that we could expect values of df/dt which would make H decrease with time to its minimum equilibrium value. With the present method of attack, however, no such general derivation appears feasible, since each kind of molecular process, that can lead to changes in the distribution f, needs what amounts to a separate investigation. Hence we now turn to a special consideration of the effect of collisions as an important and typical mechanism determining the change of H with time, thereby specializing to sufficiently dilute systems so that bimolecular collisions can be considered as a well-defined source of change. We follow this with some qualitative remarks as to other mechanisms. And we shall later consider more general methods of attacking the approach of a system to equilibrium, in § 51 of this chapter for the case of the classical mechanics, and in §§ 105 and 106 of Chapter XII for the case of the quantum mechanics.

(*b*) **Effect of collisions on change of H with time.** To study the effect of collisions in leading to changes in H with time, let us consider a dilute gas composed of one or more kinds of molecules, enclosed in a fixed container of volume v in a region where no external fields of force are acting, having a total energy E, and having for each kind of molecule a distribution over molecular states which is specified at time t by an expression of the form

$$\delta n = f(q_4 \dots p_r, t)\, \delta x \delta y \delta z\, \delta q_4 \dots \delta p_r, \tag{48.5}$$

consonant with the specified value E of the total energy. We purposely take the distribution function f independent of the positional variables x, y, z in order that we may concentrate our attention solely on the effect of collisions, using the methods of Chapter V, and leave for later mention other processes such as pressure equalization and interdiffusion which could also contribute to changes in H in the case of a non-uniform spatial distribution. In accordance with the circumstance that the function f does not depend on the coordinates x, y, z for the positions of the molecules, we may then rewrite (48.5) in the simplified form

$$\delta n = f(\omega, t)\, \delta v \delta \omega, \tag{48.6}$$

where δv represents an infinitesimal range in position, and $\delta\omega$ represents an infinitesimal range in the internal coordinates and external and internal momenta $q_4 \dots p_r$. Using this simple specification of the distribution, our general formula (48.4) for the rate of change in H with time now assumes the form

$$\begin{aligned} \frac{dH}{dt} &= \int \dots \int \frac{df}{dt} \log f\, dq_1 \dots dp_r \\ &= \iint \frac{df(\omega, t)}{dt} \log f(\omega, t)\, dv d\omega \\ &= v \int \frac{df(\omega, t)}{dt} \log f(\omega, t)\, d\omega, \end{aligned} \tag{48.7}$$

where the integration may be regarded, if necessary, as including terms for more than a single kind of molecule.

Returning to the methods of treating collisions developed in the preceding chapter, we are now ready to consider the effect of any particular collision—say $\begin{pmatrix} j, i \\ k, l \end{pmatrix}$—in contributing to the above expression for the rate of change of H. In the first place, in agreement with (45.7), we can write

$$n_i = vf_i\, \delta\omega_i, \quad n_j = vf_j\, \delta\omega_j, \quad n_k = vf_k\, \delta\omega_k, \quad n_l = vf_l\, \delta\omega_l \tag{48.8}$$

as expressions for the numbers of molecules in those ranges—$\delta\omega_i$, $\delta\omega_j$, $\delta\omega_k$, and $\delta\omega_l$—from and to which such a collision leads, where f_i, f_j, f_k, and f_l are the values of $f(\omega, t)$ at the location of these ranges and at the time of interest t. In the second place, in accordance with (45.14), we can write

$$Z_{k,l}^{j,i} = Cf_i f_j \tag{48.9}$$

as an expression for the most probable number of such collisions per unit time, where the value of the coefficient C will be considered later. Hence, since each such collision would result in the transfer of a pair of molecules from the ranges $\delta\omega_i$ and $\delta\omega_j$ to the ranges $\delta\omega_k$ and $\delta\omega_l$, we can now write

$$\begin{aligned} Cf_i f_j &= -\frac{dn_i}{dt} = -v\frac{df_i}{dt}\delta\omega_i \\ &= -\frac{dn_j}{dt} = -v\frac{df_j}{dt}\delta\omega_j \\ &= +\frac{dn_k}{dt} = +v\frac{df_k}{dt}\delta\omega_k \\ &= +\frac{dn_l}{dt} = +v\frac{df_l}{dt}\delta\omega_l \end{aligned} \tag{48.10}$$

as an expression for the probable rates of change in the quantities listed at the right. Comparing these expressions with the integral (48.7), which gives the total rate of change in H, we see that we can then take

$$\begin{aligned} \left(\frac{dH}{dt}\right)_{\substack{j,i\\k,l}} &= Cf_i f_j(-\log f_i - \log f_j + \log f_k + \log f_l) \\ &= Cf_i f_j \log\frac{f_k f_l}{f_i f_j} \end{aligned} \tag{48.11}$$

as an expression for the probable contribution to this total rate of change that would be made by collisions of the kind $\begin{pmatrix} j, i \\ k, l \end{pmatrix}$.

This contribution to dH/dt could, of course, be either positive or negative, according as $f_k f_l$ is greater or less than $f_i f_j$. Hence no immediate conclusion as to the sign of dH/dt can be drawn until we see the result of combining the effect of this particular collision with that of others which would also be taking place in the gas. In this connexion we may now give separate treatment, first to the specially simple case of collisions between spherical molecules, and then to the more complicated case of collisions between molecules in general.

In the case of *spherical molecules*, it has been shown in § 42 that the existence of any collision $\begin{pmatrix} j, i \\ k, l \end{pmatrix}$ implies the existence of another collision $\begin{pmatrix} l, k \\ i, j \end{pmatrix}$, which is both the corresponding and the inverse collision thereto.

Furthermore, it has been shown in § 45 that the probability coefficient C would have the same value for the rates of occurrence of any collision and the collision that corresponds thereto. Hence, in analogy to (48.11), we may now also write

$$\left(\frac{dH}{dt}\right)_{\substack{l,k\\ i,j}} = Cf_k f_l \log\frac{f_i f_j}{f_k f_l}, \tag{48.12}$$

with the same coefficient C, as an expression for the probable contribution to the total rate of change in H, which would be made by the inverse collision to that first considered. Adding (48.11) and (48.12) together, we then obtain as an expression for the combined effect of the pair of inverse collisions $\begin{pmatrix} j, i \\ k, l \end{pmatrix}$ and $\begin{pmatrix} l, k \\ i, j \end{pmatrix}$

$$\frac{dH}{dt} = C(f_i f_j - f_k f_l)\log\frac{f_k f_l}{f_i f_j}. \tag{48.13}$$

This result, however, is of the form

$$C(x-y)\log\frac{y}{x},$$

where C, x, and y are all essentially positive quantities. Hence the total product itself can only be equal to or less than zero, since with x greater than y the factor $\log(y/x)$ would be negative, and with x less than y the factor $(x-y)$ would be negative. Thus, since every possible collision in the case of spherical molecules could be considered along with its inverse, we now obtain the final result

$$\frac{dH}{dt} \leqslant 0, \tag{48.14}$$

as a description of the combined effect of all collisions, the equality sign applying in accordance with (48.13) only when we have a distribution which makes

$$f_i f_j = f_k f_l \tag{48.15}$$

for all possible collisions $\begin{pmatrix} j, i \\ k, l \end{pmatrix}$.

In the more complicated case of collisions between *non-spherical molecules* the above treatment has to be modified, since we have seen in § 42 that an arbitrary collision will in general have no inverse collision at all. In this general case it then becomes necessary to consider the combined effect of the whole closed cycle of corresponding collisions

$$\begin{pmatrix} 2, 1 \\ 3, 4 \end{pmatrix} \begin{pmatrix} 4, 3 \\ 5, 6 \end{pmatrix} \begin{pmatrix} 6, 5 \\ 7, 8 \end{pmatrix} \cdots \begin{pmatrix} k, k-1 \\ 1, 2 \end{pmatrix},$$

which, as shown in § 41, can always be set down as a closed list taking any desired arbitrary collision $\binom{2,1}{3,4}$ as a starting-point. Making use of the equality of values found in § 45 for the probability coefficients C that appear in the expressions for the rates of corresponding collisions, and making use of the general form of expression (48.11) for the contribution of any particular collision to the probable rate of change in H, we can now take

$$\frac{dH}{dt} = C\left(f_1 f_2 \log\frac{f_3 f_4}{f_1 f_2} + f_3 f_4 \log\frac{f_5 f_6}{f_3 f_4} + \ldots + f_k f_{k-1} \log\frac{f_1 f_2}{f_k f_{k-1}}\right) \tag{48.16}$$

as an expression for the combined effect on the quantity H of all the collisions that would appear in such a closed list.

This result is of the form

$$C \log\left(\frac{\beta}{\alpha}\right)^{\alpha}\left(\frac{\gamma}{\beta}\right)^{\beta}\left(\frac{\delta}{\gamma}\right)^{\gamma}\left(\frac{\epsilon}{\delta}\right)^{\delta} \cdots \left(\frac{\alpha}{\mu}\right)^{\mu},$$

where the quantities C, α, β, γ, δ, ϵ,..., μ are all essentially positive, and can itself readily be shown to have a value that can only be equal to or less than zero. To see this we begin by noting that somewhere in the series of quantities α, β, γ, δ, ϵ,..., μ there will have to be some quantity which is not greater than either of its two neighbours. Let us assume γ to be such a quantity, and rewrite the above result in the form

$$C \log\left(\frac{\gamma}{\delta}\right)^{\beta-\gamma} + C \log\left(\frac{\beta}{\alpha}\right)^{\alpha}\left(\frac{\delta}{\beta}\right)^{\beta}\left(\frac{\epsilon}{\delta}\right)^{\delta} \cdots \left(\frac{\alpha}{\mu}\right)^{\mu}.$$

Since we have picked out a quantity γ such that

$$\beta \geqslant \gamma \leqslant \delta,$$

we now see that we have replaced our original result by the sum of a term which cannot be greater than zero and a term of exactly the same form as before but containing one less factor in the argument of the logarithm. Proceeding in this fashion, it is then evident that we should finally obtain a sum of terms none of which could be greater than zero, and have thus obtained the desired proof that the total result can only be equal to or less than zero.

Hence, since any collision whatever could be treated as a member of a similar closed cycle of collisions, we can now conclude—as in the special case of spherical molecules—that the combined effect of all collisions would be to give

$$\frac{dH}{dt} \leqslant 0 \tag{48.17}$$

as the most probable rate of change of H with time. It will also be seen from the foregoing that the equality sign would again apply only in the case of a distribution giving

$$f_i f_j = f_k f_l \tag{48.18}$$

for all possible collisions $\begin{pmatrix} j, i \\ k, l \end{pmatrix}$. This, moreover, is a relation, as we shall see in detail in § 50 (b), that would characterize the condition of the system when H has reached its minimum possible value and the molecules of the system have attained their equilibrium Maxwell-Boltzmann distribution. Furthermore, it will also be seen from the foregoing that the negative values, predicted for dH/dt by (48.16), would be greater and greater the larger the deviations from the equilibrium relation (48.18). Hence we can say in a qualitative way that the probability of finding a negative value of dH/dt becomes more and more nearly certain for values of H farther and farther above the minimum.

We thus complete the desired demonstration of the H-theorem for the general case of bimolecular collisions between any kinds of molecules, and may regard the results obtained as typical for other kinds of processes as well.†

(c) **Effect of other processes on change of H with time.** In addition to bimolecular collisions as a mechanism that leads to changes in the condition of physical-chemical systems, by changing the states of the pairs of molecules involved, it is evident that other processes can also contribute to changes in condition. Hence, as already remarked above, a complete verification of the H-theorem would necessitate an investigation of the effect of every such process. In this connexion, nevertheless, we shall content ourselves with a few qualitative remarks as to other processes. We need feel no hesitation on the score of this incompleteness, however, since we shall later give more general methods of attacking the approach to equilibrium, both for the classical and for the quantum mechanics, as we have already remarked above.

† In the original development of the H-theorem by Boltzmann, interest was primarily centred on the effect of collisions in changing the internal conditions and external momenta of molecules, and the treatment was thrown into such a form that H was not regarded as changing as a result of molecular motions not involving collisions. See Boltzmann, *Vorlesungen über Gastheorie*, Leipzig, 1910 and 1912, Part I, § 18, footnote 2, and Part II, § 75. This procedure of Boltzmann is not appropriate, however, if we take the condition of a system as specified by the numbers of molecules in small but finite ranges $\delta x \delta y \delta z\, \delta q_4 \ldots \delta p_r$. See P. and T. Ehrenfest, *Encykl. d. Math. Wiss.* IV. 2, ii, Heft 6, p. 69, footnote 195.

Collisions of higher order than the second, where more than two molecules would simultaneously come close enough to influence each other, immediately suggest themselves as furnishing a kind of process which cannot be neglected. The treatment of such higher order collisions makes it necessary to introduce additional complications in defining critical constellations, in discussing the closed cycle of corresponding collisions, in applying Liouville's theorem, and in proving the equality of coefficients for the probability of occurrence of corresponding collisions. Nevertheless, it appears possible to carry out deductions quite similar to that given above for bimolecular collisions. For example, in the case of trimolecular collisions we can expect an expression of the form

$$\frac{dH}{dt} = C\left(f_1 f_2 f_3 \log\frac{f_4 f_5 f_6}{f_1 f_2 f_3} + f_4 f_5 f_6 \log\frac{f_7 f_8 f_9}{f_4 f_5 f_6} + \dots + f_k f_{k+1} f_{k+2} \log\frac{f_1 f_2 f_3}{f_k f_{k+1} f_{k+2}}\right) \tag{48.19}$$

for the contribution to the probable value of dH/dt, which would now be made by an appropriately selected closed cycle of corresponding collisions. And this can be shown to be essentially a negative quantity, with the value zero only when the distribution is such as to give equalities of the form

$$f_i f_j f_k = f_l f_m f_n \tag{48.20}$$

for all possible trimolecular collisions. In this manner we can obtain the desired verification of the H-theorem for collisional processes of higher order than the second.

The two familiar processes, of *pressure equalization* for a gas as a whole and of *interdiffusion* of its constituents in the case of a mixture, also need mention. Starting with any non-uniform distribution of pressure or concentration, the end result of these processes, neglecting for simplicity the effect of external fields of force, is well known to be the establishment of a uniformly distributed, uniform mixture of the constituents. Two remarks may now be made in connexion with such processes.

In the first place, it will be appreciated that the principles of statistical mechanics would themselves indeed lead us to expect a general tendency for gases to distribute themselves uniformly. The application of these principles would consist in examining the average behaviour of a representative ensemble of systems so chosen that each member of the ensemble would exhibit any lack of uniformity of distribution specified for the actual gas of interest. With any non-uniform spatial distribution it is evident, however, that highly artificial restrictions on precise molecular positions and velocities would be necessary to prevent a net transport of molecules from regions of high concentration to those of low. Hence, since no such special restrictions on positions and velocities would be warranted by any actual knowledge that we should have as to the gas of interest, substantially every system in the representative ensemble would exhibit a tendency to change in the direction of more uniform distribution, and this same tendency would then be predicted as probable for the sample of interest itself.

As a second remark, it is to be emphasized that the occurrence of such spontaneous changes, in the direction of increased uniformity of spatial distribution, would be accompanied by decreases with time in the quantity H. To show this let us return to our previous general expression (48.4) for the rate of change of H with time,

$$\frac{dH}{dt} = \int \dots \int \frac{df}{dt} \log f \, dxdydz \, dq_4 \dots dp_r. \tag{48.21}$$

Examining this equation, we then see at least qualitatively that dH/dt would be negative, when changes are taking place that make the distribution function f depend in a more uniform manner on the spatial variables x, y, z, since, by and large, the occurrence of such changes would imply negative values of df/dt in spatial regions where f is large and positive values where f is small. To take a specific example, let us consider the transport of molecules between two equal elementary cells in the μ-space, $\delta v_1 \, \delta\omega$ and $\delta v_2 \, \delta\omega$, which correspond to equal but differently located ranges δv_1 and δv_2 for the positional coordinates x, y, z of the molecules, and to the same range $\delta\omega$ for the remaining variables $q_4 \dots p_r$ that determine the condition of the molecules. Using f_1 and f_2 to denote the values of the distribution function corresponding to these two cells, we could then write

$$\frac{dH}{dt} = \delta v_1 \, \delta\omega \frac{df_1}{dt} \log f_1 + \delta v_2 \, \delta\omega \frac{df_2}{dt} \log f_2 = \text{const.} \times \frac{df_1}{dt} \log \frac{f_1}{f_2} \tag{48.22}$$

as an expression for the contribution to the total rate of change in H, which would correspond to the transport of molecules between the two cells, where the second form of writing is made possible by the consideration that any increase in f_1 would be at the expense of f_2. We immediately see, however, that this expression would indeed be negative, since with f_1 greater than f_2 the direction of spontaneous transport would make df_1/dt negative, and with f_1 less than f_2 the factor $\log(f_1/f_2)$ would itself be negative. Hence we can thus obtain also for the processes of pressure equalization and interdiffusion the desired verification of the H-theorem.

There are, of course, many other important, spontaneous processes such as fusion, vaporization, chemical reaction, and the interactions between matter and radiation which might be considered in order to extend our verifications of the H-theorem. Further investigations along present lines would hardly be justified, however, in view of the limited range of validity for classical methods, and in view of the more general treatment of the approach towards equilibrium which will be given later.

In conclusion, it may be noted that the decrease of H to its final, lowest possible value would depend in general on the co-operation of more than a single process. Hence a complete treatment of the rate of change of H with time would involve a combination of the treatments given to the different individual kinds of process. Thus, for example, our treatment of the effect of bimolecular collisions in leading to a redistribution of molecular states in the case of a homogeneous gas can be readily extended to give an account of the effects of collisions in a non-homogeneous gas by regarding the gas as divided up into elementary volumes taken small enough to be treated as homogeneous, but this treatment then has to be further supplemented by including an account of those processes of pressure equalization and of interdiffusion, which in the absence of external forces would finally lead to a uniform mixture for different kinds of molecules and of molecular states. Similarly, the treatment of the effect of collisions in changing molecular states has to be supplemented by a treatment of the effects resulting from the absorption and emission of radiation. For this purpose it proves possible, guided by ideas from the quantum theory of radiation, to define a quantity H which depends not only in the previous manner on the numbers of molecules in different states, but also contains further terms depending on the condition of the radiation field.† It then becomes possible to show that this combined H would decrease to a minimum value in such a manner as to require the simultaneous satisfaction of the Maxwell-Boltzmann distribution law for molecules and of the Planck distribution law for radiation, thus providing a combined derivation of these two laws.

† See Tolman, *Journ. of the Franklin Inst.* **203**, 814 (1927); and *Statistical Mechanics*, New York, 1927, § 245, p. 198.

49. Discussion of the H-theorem

(*a*) **Statistical character of the H-theorem.** We must now undertake a somewhat lengthy and detailed discussion of the foregoing H-theorem in order to have a clear appreciation of its significance. For this purpose it is especially important to recognize the character of Boltzmann's result as a theorem of *statistical mechanics* which makes a reasonable prediction as to the future behaviour that may be expected for a system in an incompletely specified state, rather than as a theorem of *exact mechanics* which could furnish an exact prediction as to the precise behaviour of a system starting from a precisely specified state. The lack of appreciation of this distinction has often led in the past to seeming paradoxes and difficulties.

As we have just suggested, the statistical character of the H-theorem depends in the first place on the circumstance that we treat systems in states that are only incompletely specified, and in the second place on the consideration that we are thus forced to employ the methods of statistical rather than of exact mechanics.

The incomplete nature of the specifications is inherent in the kind of descriptions which we use for the different conditions of a system that interest us. These descriptions are given by expressions (47.1) of the form

$$\delta n = f(q_1 \dots p_r, t)\, \delta q_1 \dots \delta p_r, \tag{49.1}$$

where δn is the number of molecules with coordinates and momenta in the range $\delta q_1 \dots \delta p_r$, and f is taken at the time of interest as a function of the coordinates and momenta $q_1 \dots p_r$. Such an expression, however, does not give an exact specification of the state of a system, since we regard the various ranges $\delta q_1 \dots \delta p_r$ as having a small but nevertheless *finite extension*, corresponding to an *approximate observation* of the condition of the system which might actually be made in practice. Indeed we treat the ranges $\delta q_1 \dots \delta p_r$ as infinitesimals of an order large enough, so that δn can in general be taken as a large number and f as a continuous function of the variables $q_1 \dots p_r$. Hence no exact specification of the positions, velocities, and internal conditions of the molecules is provided.

In view of this incomplete specification of state we are then forced to employ the methods of statistical mechanics, when we come to compute the temporal behaviour of the quantity of interest

$$H = \int \dots \int f \log f \, dq_1 \dots dp_r, \tag{49.2}$$

where it may be emphasized that this integral is itself to be regarded as the limit of a sum of terms of the form $f \log f\, \delta q_1 \ldots \delta p_f$ in which the ranges $\delta q_1 \ldots \delta p_r$ are treated as infinitesimals of the order mentioned above. As a consequence, our final theorem as to the decrease of H with time has the statistical character of being a statement only of what can be expected on the average for a system in the condition that has been described.

It will be of interest to follow the statistical character of the considerations in detail in the case of the treatment which we have given for the effect of collisions in leading to changes of H with time in a homogeneous gas. The condition of the gas at the time of interest t was there described by an expression (48.5) of the form

$$\delta n = f(q_4 \ldots p_r, t)\, \delta x \delta y \delta z\, \delta q_4 \ldots \delta p_r, \tag{49.3}$$

where the function f was taken as independent of the positional coordinates x, y, z. This description of condition is not sufficient for a precise specification of state, since it does not prescribe precise values for coordinates and momenta but only gives the numbers of molecules which would have values of those quantities falling within infinitesimal ranges of large enough order so that f can be treated as a continuous function. In particular it is to be noted that the lack of dependence of f on x, y, and z only means a *macroscopic* homogeneity of spatial distribution which puts no restrictions on the number of pairs of molecules that would have mutual positions such as to initiate a collision of any kind of interest. In view of the lack of precise specification, we then have to employ the methods of statistical mechanics to predict the expected changes with time in the condition of the system. This we do by using our previous expressions for the average (most probable) rates of collision in a representative ensemble of systems, all of which are in states that agree with a description of condition of the kind (49.3), but are otherwise distributed in accordance with the hypothesis of equal *a priori* probabilities for equal regions in the phase space. Hence the final use of these expressions in calculating the value of dH/dt will only give us a result which can be expected to hold on the average under the circumstances so far as they have been specified.

As the result of the foregoing discussion, we now recognize in general the statistical character of the H-theorem. In accordance with this theorem, a system in a condition that does not correspond to the lowest attainable value of H will exhibit on the average a negative value of dH/dt, but this does not have to be true in each individual case. The recognition of this possibility for exceptional behaviour is important in clarifying difficulties that have sometimes arisen in understanding the H-theorem, as will be seen in what follows.

In passing it may be remarked that the specific calculations, which were made in § 48 (*b*) for the effect of collisions, were concerned with the most probable value of dH/dt, as the average to be computed for the members of the representative ensemble. The choice of this particular average is based, however, merely on considerations as to mathematical simplicity and historical familiarity. It would also be possible to consider other averages, for example to study the mean value of dH/dt and the fluctuations around this mean, in the case of different kinds of systems and processes.

(*b*) **Observations on the continued decrease of H with time.** In accordance with the foregoing, if we have an isolated system at a given initial time t_1 in a condition which does not correspond to the lowest possible value of H, we can expect the value of H to be decreasing with the time. Hence it seems natural to conclude that the condition of such a system could be expected to continue to change with time in the direction of lower and lower values of H until the minimum possible is attained. Nevertheless, it will be necessary to examine the grounds for this conclusion somewhat more in detail, since our actual finding that the most probable value of dH/dt would be negative at the initial time t_1 cannot be regarded as immediately equivalent to an assertion as to the long time behaviour of H, for which some appropriate form of integration over the time would be needed. Indeed, since we realize that occasional positive values of dH/dt are not eliminated by statistical considerations as to the most probable value of that quantity, the possibility has to be entertained that the very systems which exhibit a negative value of dH/dt at some initial time t_1 would be the actual ones most liable to exhibit positive values of that quantity at a later time.

A justification for the above conclusion as to the long time behaviour of H, which is at least partially satisfactory, may be obtained by considering the nature of the succession of actual observations which would be needed in order to follow the behaviour of H as dependent on time in the case of any given system of interest. To carry out such a succession of observations, let us take some suitable isolated system, and, assuming no prior knowledge as to its history or condition, let us make an appropriate observation at an initial time t_1 on the distribution of its component molecules, from which we can then compute the corresponding value of $H(t_1)$. Let us assume that this initial value of H is much larger than the lowest possible value.

On the basis of this information let us next consider what can be expected as to the future behaviour of the system. Since the above initial determination of molecular distribution would have—in accordance with our previous discussion—only an approximate character, it will not be sufficient for an exact specification of state, and we shall have to employ the methods of statistical rather than of ordinary mechanics in order to predict the future behaviour. This we do by considering a representative ensemble of systems set up so that all of its members are in states that agree with the initially observed condition, but are otherwise distributed in accordance with our fundamental postulate as to equal *a priori* probabilities. By considering the behaviour in this ensemble we can then calculate the average (most probable) value of dH/dt for the members of the ensemble. Since we find that this average value is a negative quantity, we then predict that the value of dH/dt will also presumably be negative for the actual system of interest, and indeed in cases where H is very much larger than its minimum value the prediction that dH/dt will be negative can be taken as almost certain.

So far our considerations have not gone beyond the original arguments leading to the expectation of a negative value of dH/dt when H has a value greater than its minimum. Let us now, however, proceed a step farther by using our average value of dH/dt to make a prediction as to the value of H for the system at a later time t_2. For this purpose we may take

$$H(t_2) = H(t_1)+\left(\frac{dH}{dt}\right)_{\text{av}}(t_2-t_1) \tag{49.4}$$

as giving a reasonable prediction, provided the time interval (t_2-t_1) is short enough to justify the neglect of higher order terms. As the average value of dH/dt is negative, we then predict that $H(t_2)$ will be less than $H(t_1)$, provided the difference in time (t_2-t_1) is not too great.

At the later time t_2, let us now consider that we actually make a new observation of molecular distribution, which will permit us to determine the actual value of $H(t_2)$ for the system of interest at time t_2. In accordance with the fact that the value of dH/dt inserted in (49.4) was only an estimate of what could be expected on the average, the actual value $H(t_2)$ will presumably not agree precisely with the value which we have predicted. Nevertheless, it will almost certainly be less than the original value $H(t_1)$, since the negative character of dH/dt is very highly probable.

Starting with the new observation of molecular distribution, which has given us the actual value of $H(t_2)$ at time t_2, let us then again make a prediction as to the value of dH/dt. Let us do this by now setting up a *new representative ensemble*, constructed so that all its members are in states that agree with our *new observation* of condition at time t_2, but are otherwise again distributed in accordance with our postulate as to equal *a priori* probabilities for equal regions in the phase space. Since this new ensemble would evidently again lead to the prediction of a negative value for dH/dt, assuming that H has not yet reached its minimum, we could then proceed as before to the conclusion that $H(t_3)$ at a still later neighbouring time t_3 would almost certainly be less than $H(t_2)$ at time t_2. Continuing in this fashion, we are thus provided with arguments for regarding it as highly probable that successive observations on the condition of the system of interest would be found to correspond to lower and lower values of H.

Nevertheless, as already intimated above, this treatment may not seem entirely satisfactory. From the point of view of present interest, legitimate reasons for dissatisfaction could, of course, only lie in the possibility that the above application of the methods of statistical mechanics has not really been carried out rigorously in the manner demanded by our postulatory basis. In accordance with that basis we are to draw conclusions as to the expected behaviour of a given system of interest, in a partially specified state, by considering the average behaviour in an appropriate representative ensemble of similar systems. To obtain such an appropriate ensemble we distribute the representative points for its members throughout the phase space, in the first place in such a manner as to agree with what partial knowledge we do have as to the state of the system of interest, and in the second place—in accordance with our fundamental postulate as to equal *a priori* probabilities—in as uniform a manner as is consistent with that partial knowledge of state. We then take the average behaviour in this ensemble as giving a good prediction for the behaviour of the actual system.

Looking back now on the foregoing attempt to apply these methods, it will be agreed that our first application, at the initial time t_1, was correctly made, since we then considered the behaviour of a representative ensemble which was constructed to agree with all that we knew about the state of the system, namely the results given by our initial observation of molecular distribution at time t_1. It may be contended, however, that our second application of the prescribed methods at the later time t_2 was not so correctly made, since we then considered a representative ensemble which was constructed to agree with the results of our new observation at that time, but not in such a manner as to make any allowance for the additional circumstance that the system was known to have passed in the time interval t_1 to t_2 from the actually observed initial to the actually observed later condition. Similar remarks would, of course, also apply at times later than t_2.

In principle it would be possible to overcome this difficulty, encountered in applying the methods of statistical mechanics at a time when our knowledge of the instantaneous condition of a system is supplemented by knowledge as to previous behaviour, by constructing a representative ensemble so distributed as to agree both with the instantaneous condition and with the past behaviour of the system of actual interest, and otherwise left uniform in agreement with the fundamental postulate as to equal *a priori* probabilities. Thus, returning to the above example, our second application of the methods of statistical mechanics at time t_2 would seem thrown into a justifiable form if we should eliminate from the ensemble representing the known condition of the system at time t_2 all those members which could not have come in the interval t_1 to t_2 from the condition known to prevail at the earlier time, and should then use this appropriately modified ensemble to draw conclusions as to the expected value of dH/dt for the system of interest. The rigorous application of this corrected procedure would not be simple, however, nor perhaps indeed fall within the range of treatments for which statistical methods are specially suited, since the elimination of systems which could not pass in the time t_1 to t_2 from states corresponding to the original to those corresponding to the later condition could only be carried out with the help of an elaborate investigation of the precise mechanical behaviour of the systems composing the original representative ensemble at time t_1. Hence we shall now content ourselves with the statement that it seems reasonable to believe, in cases where the initial and later observations of the conditions at times t_1 and t_2 are not sufficiently exact to lead to a close specification of precise state, that the states to be eliminated from the second representative ensemble—as not agreeing with the known time of passage—could usually be regarded as scattered among the states corresponding to the condition at time t_2 in such a random manner as not to disturb the previous conclusion that the average value of dH/dt for the members of the ensemble would be a negative quantity.† If this be accepted, we could then maintain the conclusion that any actual isolated system of interest could usually be expected to exhibit lower and lower values of H on its way towards the lowest possible value, at least until the successive observations of conditions and times of passage could be used in principle for such a precise determination of state that the use of a representative ensemble corresponding merely to the latest observation of condition became clearly inappropriate.

This demonstration of the probability for a continued decrease of H with time is therefore now clearly seen to lack rigour and to apply to what may be called typical cases. Indeed one can readily invent examples where a knowledge of the past behaviour of a system would be highly relevant to the future changes to be expected in H. We shall later supplement the above considerations as to the long time behaviour of H by a different method of treatment based on a generalization of the H-theorem. Note in particular the remarks at the end of § 51.

(c) **H-theorem and the principle of dynamical reversibility.** We now turn to the treatment of an apparent inconsistency between exact and statistical mechanics which originally seemed very puzzling. On the one hand, in accordance with the H-theorem, as discussed in the present chapter, it is evident that the methods of statistical mechanics make it necessary to expect a preferential tendency for isolated systems to change in the direction of smaller values of H, provided the lowest possible value of that quantity has not been attained. On the other hand, in accordance with the principle of dynamical reversibility, as discussed in the preceding chapter, it is well known that the laws of exact mechanics make it equally possible for any internal behaviour of an isolated system and the reverse of that behaviour to take place. Hence at first sight, as originally suggested by Loschmidt,‡ some possibility for conflict between the conclusions drawn from statistical and from exact mechanics might seem to be possible, since for any behaviour of a system such that H was decreasing with time it would be equally possible to have another behaviour in which H was increasing with time.

† Compare the similar point of view expressed by P. and T. Ehrenfest, *Encykl. d. Math. Wiss.* IV. 2, ii, Heft 6, p. 50, § 18c. It may be mentioned in this connexion, nevertheless, that Burbury (*Kinetic Theory of Gases*, Cambridge, 1899) was always of the opinion that the continued decrease of Boltzmann's H as a result of collisions could not be maintained, since the collisions would themselves result in a correlation of velocities for neighbouring molecules which would invalidate the use of the hypothesis of molecular chaos—i.e. of the postulate as to equal *a priori* probabilities—in calculating the effect of further collisions.

‡ Loschmidt, *Wien. Ber.* **73**, 139 (1876), and **75**, 67 (1877).

To have a sharp example of this apparent possibility for conflict, let us consider a gas enclosed in an isolated container, which is itself divided into two equal compartments by a partition perforated with a small hole, and let us start out with the gas distributed quite unequally in amount between the two compartments. In accordance with the H-theorem, we can then expect the gas to flow through the connecting hole from the high- to the low-pressure side of the partition, since the process of equalizing the distribution would lead—as we have seen in § 48 (*c*)—to lower and lower values of H. In accordance with the principle of dynamical reversibility, nevertheless, it would be conceptually possible, by reversing all molecular velocities at any time during the outward flow, to obtain a natural mechanical motion of the system with the molecules of the gas retracing their paths from the low-pressure side back through the hole in the partition into the high-pressure side. Furthermore, it may be emphasized that the possibility for such reverse behaviour would exist for any possible precisely specified motion by which the flow from the high- to the low-pressure side might be taking place. We may then state the apparent possibility for conflict, with the help of anthropomorphic language, in the form: Why does the gas prefer to obey the H-theorem and flow from the high- to the low-pressure side, when the principle of dynamical reversibility provides it in every case with an equally valid mechanical behaviour in the opposite direction?

The resolution of the apparent difficulty lies in maintaining a clear distinction between the use of statistical mechanics to treat the behaviour of systems in conditions that correspond to a range of different possible precise states, and the use of exact mechanics to treat the behaviour of systems in precisely defined mechanical states. With this distinction in mind we then see that there is really no opportunity for conflict between the H-theorem and the idea of dynamical reversibility since the two principles apply to different observational situations. The H-theorem is a principle of statistical mechanics which applies to a system which is only known to be in a *condition* that will correspond in typical cases to many different precise states; when this condition is very different from that of equilibrium the principle asserts that most of these states will be such that there is a high probability for the system to move towards conditions with a lower value of H. The idea of dynamical reversibility is a principle of exact mechanics which applies to a system known to be in some particular *precise state* and hence to be carrying out some particular precise motion; the principle then asserts that it is entirely possible to specify another (reverse) precise state for the system such that the system would carry out the reverse of the first motion. The H-theorem predicts a given direction as probable for spontaneous changes in the condition of a system known to be in one or another of a collection of different states. The principle of dynamical reversibility says that it would be conceivably possible, with an exact knowledge of state and adequate apparatus for the reversal of velocities, to secure a reversal for any precise motion of the system. The two principles thus apply to different situations and there is no occasion for conflict.

The above discussion may be illustrated with the help of our previous example of the gas in the container having two connected compartments. The initial specification that the gas is unevenly divided between the two equal compartments gives nowhere nearly enough information to define a precise state of the system, and by examining the states that could correspond to the condition described, making use of the principle of equal *a priori* probabilities for different equal extensions in the phase space, we see that the gas has an overwhelming probability of being in a state such that flow would take place from the high- to the low-pressure side. This conclusion is in no way contradicted by the fact, that for any precisely defined succession of mechanical states by which the equalizing flow might take place, it would be theoretically possible to set up a behaviour in the opposite direction. The reason for the flow from the high- to the low-pressure side may be said to lie in the circumstance that it is very easy to put the gas unequally in the two compartments in such a way that flow would take place in a direction to equalize the pressure, but that it would be very difficult to adjust the positions and velocities of the molecules so as to obtain a net transfer of molecules through the hole in the opposite direction from the low- to the high-pressure side.

It may be regarded as a great achievement of Boltzmann that we are now able to comprehend the compatibility between the reversibility, implied by the laws of exact mechanics, and those actual irreversibilities observed in mechanical behaviour, which must be treated by the methods of statistical mechanics. The compatibility depends on the different degrees of observational precision involved in mechanical situations which are appropriately treated by the methods of exact or of statistical mechanics. The reconciliation between reversibility and irreversibility, which has thus been obtained, may even be thought of as providing the appropriate reconciliation between the divergent points of view of Newtonian and of Aristotelian mechanics as to the natural state of a material body being one of uniform motion or of rest.

(*d*) **H-theorem and the occurrence of continued fluctuations.** We may now complete this long discussion of Boltzmann's H-theorem by considering another apparent difficulty which was first pointed out by Zermelo.† As we have seen in the foregoing, we can expect to obtain an immediate verification of the tendency for H to decrease with time whenever we definitely set up an isolated system in a condition having a value of H much greater than its minimum, since there will then be an overwhelming probability that the system will actually be in a precise state from which changes in the direction of lower values of H must ensue. The situation is altered, however, if we desire to obtain a verification of the H-theorem by observing the long time behaviour of an isolated system, since it then turns out that we must regard this continued behaviour as a succession of fluctuations in which the value of H will increase as often as it decreases. At first sight such a behaviour might seem to contradict the previous conclusion as to the tendency for H to decrease.

The reason for the conclusion that the long time behaviour of an isolated system must be such as to include increases as well as decreases in H may be based on a theorem of Poincaré.‡ In accordance with this theorem, an enclosed isolated mechanical system if left to itself would in general carry out a quasi-periodic sort of motion, so that the system would always return after a finite interval of time to pass through states at least in the very close neighbourhood of those through which it has previously passed. The derivation of the theorem depends, on the one hand, on the finite total extension of phase space that would be available to the phase points representing spatially confined and isolated systems, and depends, on the other hand—in accordance with Liouville's theorem—on the constant extension which would be permanently needed for any group of phase points representing systems originally started off in neighbouring states. As a consequence of the theorem, we must conclude that any decreases in H which take place during the continued behaviour of such a system must be balanced in the course of time by corresponding increases, and vice versa.

In this connexion it is also to be noted that our fundamental principle as to equal *a priori* probabilities has a bearing on the conclusion that

† Zermelo, *Wied. Ann.* **57**, 485 (1896).
‡ Poincaré, *Acta Math.* **13**, 67 (1890).

increases as well as decreases in H must take place in isolated systems. In accordance with this principle, if we select a system which is merely known to be isolated with an energy content, say, between E and $E+\delta E$, there will then be equal probabilities for finding the system, on precise examination, in states corresponding to any one of the equal regions into which we might divide the shell in the phase space lying between E and $E+\delta E$. But this means that an isolated system, chosen at random to use for an investigation of long time behaviour, is equally likely to be found in any state and in the reverse of that state, since such pairs of states are equally represented in the phase space. Hence, in making tests on the behaviour of isolated systems, the samples that we use are equally likely to be carrying out any particular kind of motion and the reverse of that motion. This, too, makes it necessary to conclude that the long time behaviour of isolated systems can exhibit increases as well as decreases in H, in agreement with the above conclusion that any one isolated system would exhibit a succession of balancing increases and decreases of H as time proceeds.

We must now show that there really is no necessary conflict between this new conclusion, that an isolated system left to itself would exhibit increases and decreases of H with equal frequency, and our previous conclusion given by the H-theorem that the most probable behaviour for an isolated system with a value of H greater than the minimum would be for H then to decrease with time. The possibility of combining these two conclusions without contradiction depends on two factors. In the first place it is to be noted that the H-theorem only makes an assertion as to the probable or average behaviour of H, and hence does not rule out the possibility for actual increases as well as decreases in that quantity to occur. In the second place it is then to be noted that a succession of increases and decreases of H with time is not incompatible with a large probability for that quantity to decrease with time from any specified high value, provided that value of H usually stands at the peak of a trajectory from which changes back to lower values do take place.

The combination of the H-theorem with the conclusion that H must rise and fall with equal frequency thus provides us with a good qualitative picture of the long time behaviour of an isolated system, as consisting of a succession of fluctuations in the value of H, up from the neighbourhood of the minimum and back again, with an almost certain immediate return downward from any high value of H which is attained. For example, if we take H_0 as the minimum possible value and $H_1, H_2, H_3,...$ as values which are considerably higher in the order given, we may regard the succession of steps

$$\begin{matrix} & H_2 & \\ H_1 & & H_1 \end{matrix} \tag{49.5}$$

as taking place much more frequently than the successions

$$\begin{matrix} & & H_3 \\ & H_2 & \\ H_1 & & \end{matrix} \quad \text{or} \quad \begin{matrix} H_3 & & \\ & H_2 & \\ & & H_1, \end{matrix} \tag{49.6}$$

and these again much more frequently than the succession

$$\begin{matrix} H_3 & & H_3 \\ & H_2. & \end{matrix} \tag{49.7}$$

Under these circumstances a high value of H, such as H_2 in the present illustration, would almost certainly be attained as the top point of a fluctuation, and immediate return to lower values would result. We then see that increases and decreases of H could take place with equal frequency, and yet the conclusion still be maintained that the system would almost certainly move towards lower values when known to have a high value of H, such as H_2 in the above illustration.

This finally satisfactory elucidation of the problem raised by Zermelo is particularly due to P. and T. Ehrenfest,† who appreciated the possibility of an ultimate analysis of the changes in H with time into a succession of steps brought about by the transfer of individual molecules from one cell $\delta v_\mu = \delta q_1 \ldots \delta p_f$ in the μ-space to another. It is this analysis of the behaviour of H into discontinuous changes that makes it possible to understand the sudden reversals that can occur at the peak of an upward fluctuation.

Several remarks of interest may be made in connexion with the fluctuations that we must now regard as taking place in enclosed isolated mechanical systems.

In the first place it will be noticed that the successions of steps (49.5–7), which we have considered above, are entirely symmetrical with respect to the forward and backward directions of time, and hence that the immediate return to lower values of H from the high point of a fluctuation can be regarded as taking place with great probability equally well for any particular motion of a system and for the reverse of that motion. Again we see that the conclusions of statistical

† P. and T. Ehrenfest, *Encykl. d. Math. Wiss.* IV. 2, ii, Heft 6, pp. 41 ff.

mechanics stand in no kind of conflict with the reversibility of time implied by the fundamental laws of exact mechanics.

It will be further noticed that the greater probability of occurrence which we have assigned to the downward succession of steps given by (49.6), as compared with the reversal of steps given by (49.7), is in agreement with the conclusion discussed in § 49 (*b*) that the decrease in H from an originally high value, in this case H_3, can be regarded as continuing on with great probability down through lower values towards the minimum.

It is also of interest to consider the relative probabilities for fluctuations of different extents away from the minimum value of H. In accordance with the circumstance that the probable value of dH/dt approaches zero as we approach the minimum of H, we can regard small fluctuations away from the minimum as occurring with great frequency. In the case of large upward fluctuations, however, not only does the probable value of dH/dt become more and more negative as we go to higher values of H, but also the succession of steps in which H must increase rather than decrease becomes longer and longer. Hence we may regard large fluctuations of H away from the minimum as occurring very infrequently in the case of an isolated system. In this connexion it is of interest to note the estimate of Boltzmann† that times enormously great compared with $10^{10^{10}}$ years would be needed before an appreciable separation would occur spontaneously in a 100 cubic centimetre sample of two mixed gases.

We thus bring to an end our long discussion of this famous H-theorem, which has outlived so many attacks and misunderstandings. We can only conclude with words of admiration for the genius of Boltzmann. His fruitful discovery of a suitable function for measuring the displacement of a whole system of molecules from equilibrium, and his elegant mastery of the complicated effect of collisions in making such displacements decrease with time, alike compel attention. His reconciliation of phenomenological irreversibility with the reversible character of the laws of exact mechanics, and his understanding of the compatibility of continued fluctuations with a tendency towards equilibrium, are among the great achievements of theoretical physics. And his penetrating remarks on the great role that fluctuations might play in the long time behaviour of the universe as a whole, show Boltzmann's preoccupation with the deepest problems of physics.

† Boltzmann, *Vorlesungen über Gastheorie*, Leipzig, 1912, Part II, p. 254.

50. H-theorem and the condition of equilibrium

(*a*) **Maxwell-Boltzmann distribution when H is a minimum.** We see from the foregoing that an enclosed isolated mechanical system, definitely started off with a large value of H, can be expected first to change in the direction of lower and lower values of H, and then to end up by carrying out a succession of fluctuations which will rarely carry it to values of H much above the minimum for that quantity. We must now investigate the equilibrium condition of a system when H is at or near its minimum value.

To study the approach towards equilibrium, let us consider an isolated system composed of n similar molecules, enclosed in a fixed container of volume v, and with a specified energy content lying in the narrow range E to $E+\delta E$. And let us express H for this system, as in (47.7), in the form of the summation

$$H = \sum_i n_i \log n_i + \text{const.}, \tag{50.1}$$

where n_i gives the number of molecules in the ith equal cell into which we have divided the μ-space for the kind of molecules under consideration, and we take a sum over all possible cells i. The approach to equilibrium will now consist in the decrease of H towards the minimum possible value allowed by the specified constitution, volume, and energy of the system.

The equilibrium distribution of the molecules will hence be characterized by a minimum value of H, as determined by the variational equation

$$\delta H = \sum_i (\log n_i + 1)\,\delta n_i = 0, \tag{50.2}$$

under the subsidiary conditions

$$\delta n = \sum_i \delta n_i = 0, \tag{50.3}$$

imposed by the fixed number of molecules, and

$$\delta E = \sum_i \epsilon_i\,\delta n_i = 0, \tag{50.4}$$

imposed by the fact that the mechanisms responsible for the change of H cannot alter the total energy of the system. We immediately see, however, that these equations have exactly the same consequences as our earlier equations of § 29, requiring a conditional maximum for $\log P$ instead of a conditional minimum for H. Hence we shall thus find a complete agreement between our present treatment of equilibrium, from the point of view of the behaviour of H with time, and our previous treatment, from the point of view of the most probable condition for the systems in an appropriate microcanonical ensemble.

In accordance with this finding, our previous solution (29.6) for the distribution of molecules at equilibrium,

$$n_i = e^{-\alpha-\beta\epsilon_i}, \tag{50.5}$$

where α and β are undetermined constants, will also satisfy our new equations (50.2), (50.3), and (50.4). Furthermore, this solution may also be rewritten as before in the more familiar form

$$\delta n = nCe^{-\epsilon/kT}\,\delta q_1 \ldots \delta p_r, \tag{50.6}$$

where δn is the number of molecules with coordinates and momenta in the range $\delta q_1 \ldots \delta p_r$, ϵ is the energy per molecule introduced into that range, C and k are constants, and T is the temperature of the system. We are thus once more led to the Maxwell-Boltzmann distribution law for the molecules of a system that has come to equilibrium.

(*b*) **Steady condition when H is a minimum.** It is also of interest to show that this Maxwell-Boltzmann distribution, prevailing when H is a minimum, satisfies the conditions, previously considered in § 48, for a probable cessation of the changes in H with time. This will then agree with our ideas that an isolated system, which once arrives at the condition of minimum H, will thereafter tend to remain in the neighbourhood of that condition.

Comparing the Maxwell-Boltzmann distribution (50.6) with our previous general expression (47.1) for the number of molecules

$$\delta n = f(q_1 \ldots p_r, t)\,\delta q_1 \ldots \delta p_r \tag{50.7}$$

in the specified range $\delta q_1 \ldots \delta p_r$, we see that our general distribution function f will assume the special form

$$f(q_1 \ldots p_r) = nCe^{-\epsilon(q_1 \ldots p_r)/kT} \tag{50.8}$$

when the Maxwell-Boltzmann distribution, prevailing with H a minimum, has been attained. We may now consider several examples showing that this value of the distribution function f does satisfy our previous conditions for the probable value of dH/dt to go to zero.

In the case of *bimolecular collisions* $\begin{pmatrix} j, i \\ k, l \end{pmatrix}$, in accordance with (50.8), we can write

$$\begin{aligned} f_i &= nCe^{-\epsilon_i/kT}, & f_j &= nCe^{-\epsilon_j/kT}, \\ f_k &= nCe^{-\epsilon_k/kT}, & f_l &= nCe^{-\epsilon_l/kT} \end{aligned} \tag{50.9}$$

as the values of the distribution function f that would apply to molecules in the states i, j and k, l from and to which such a collision leads; and, in accordance with the principle of the conservation of energy, we can write

$$\epsilon_i+\epsilon_j = \epsilon_k+\epsilon_l \tag{50.10}$$

as a connexion between the energies of the molecules before and after collision. By combining (50.9) with (50.10), we are then led for any collision of the form $\begin{pmatrix} j, i \\ k, l \end{pmatrix}$ to the result

$$f_i f_j = f_k f_l. \tag{50.11}$$

This, however, is our previous condition (48.18) for the probable value of dH/dt to go to zero, in so far as it depends on the effect of bimolecular collisions. Similarly, in the case of *trimolecular collisions* we shall be led from (50.8) to results of the form

$$f_i f_j f_k = f_l f_m f_n \tag{50.12}$$

for a collision initiated and completed by molecules in the states i, j, k and l, m, n; and this is our previous condition (48.20) for the probable value of dH/dt to go to zero, in so far as it depends on trimolecular collisions. Again, in the case of a *transport of molecules* between cells in the μ-space, say $\delta v_1 \delta\omega$ and $\delta v_2 \delta\omega$, which correspond to the same energy but to different spatial locations, we are led by (50.8) to results of the form

$$f_1 = f_2, \tag{50.13}$$

and this, as discussed in § 48 (*c*), is the condition for the probable rate of transport and corresponding rate of change in H to go to zero. We thus appreciate in general that we are justified in regarding the minimum of H as corresponding to a steady condition of an isolated system around which, however, fluctuations can take place.

(*c*) **Detailed balance when H is a minimum.** In accordance with the above, the Maxwell-Boltzmann distribution of molecules, which is attained when the quantity H for an isolated system reaches its minimum value, has a character such that the further action of collisions or of other internal molecular processes can be expected to leave the overall condition of the system unchanged. In order to maintain such a steady condition it is evident that the rates of different molecular processes, by which we can regard the molecules as being transferred between different regions in the μ-space, must be related to each other in such a manner that the resulting deficits and excesses in the molecular populations of the different regions involved are kept balanced out. Thus, for example, it is evident from the foregoing discussion, in the case of bimolecular collisions in a homogeneous gas, that the probable rates for the different collisions

$$\begin{pmatrix}2,1\\3,4\end{pmatrix}\begin{pmatrix}4,3\\5,6\end{pmatrix}\begin{pmatrix}6,5\\7,8\end{pmatrix}\cdots\begin{pmatrix}k,k-1\\1,2\end{pmatrix} \tag{50.14}$$

in any closed cycle of *corresponding collisions*, assume such values, when the Maxwell-Boltzmann distribution is attained, that there will be a balance between the rates at which these collisions are adding and subtracting molecules from the pairs of regions involved. Furthermore, in the specially simple case of *spherical molecules*, where the cycle of corresponding collisions reduces to a pair of *inverse collisions* of the form

$$\begin{pmatrix}j,i\\k,l\end{pmatrix}\begin{pmatrix}l,k\\i,j\end{pmatrix} \tag{50.15}$$

it is evident that we can regard the effect of each such collision as directly compensated by the effect of the inverse collision.

It will now be of interest to consider a general method, when H is a minimum, of setting up a specially simple kind of balance, between the effects of different molecular processes, which is often important in giving an insight into the treatment of physical chemical problems. To obtain such a balance we cannot, of course, in general correlate each molecular process with an inverse process, and thus obtain a direct compensation of effects as in the above-mentioned simple case of collisions between spherical molecules, since in general, as we have seen in § 42, processes which are the inverse of each other do not exist. Nevertheless, it does prove possible to set up a useful balance by correlating any given molecular process with the reverse process thereto, which, as we know from the principle of dynamical reversibility, must necessarily be capable of existence.

To investigate this we note, in accordance with the Maxwell-Boltzmann distribution law (50.6), that the numbers of molecules, assigned as equilibrium quotas to different equal infinitesimal regions in the μ-space, depend solely on the energies ϵ corresponding to those regions. Thus, since the energy of a molecule is unaltered by a mere reversal of velocities, it is evident that the probability of finding a molecule at equilibrium in any specified range i of states in the μ-space is equal to the probability of finding a molecule in the range $-i$ composed of the reverse states thereto. As a consequence we can also assign equal values at equilibrium to the probabilities of finding any constellation of interacting molecules in specified ranges of the μ-space $i, j, \ldots$ and of finding the reverse constellation of molecules in the ranges $-i, -j, \ldots$.

Hence we can now conclude, under equilibrium conditions, that any molecular process and the reverse of that process will be taking place on the average at the same rate, provided, of course, that we use equal ranges in the μ-space in defining the two processes. This result may be called the *principle of microscopic reversibility at equilibrium.*†

As an illustration of this principle we may consider, in the case of a homogeneous gas which has come to equilibrium, the two reverse processes which are provided by the *reverse collisions* $\begin{pmatrix} j, i \\ k, l \end{pmatrix}$ and $\begin{pmatrix} -k, -l \\ -j, -i \end{pmatrix}$ as defined in Chapter V. It is then immediately evident, in agreement with the above principle, that we can equate the probable rates of these two collisions,

$$Z^{j,i}_{k,l} = Z^{-k,-l}_{-j,-i}, \tag{50.16}$$

since we at once appreciate that we can expect, at any time under equilibrium conditions, to find the same number of pairs of molecules present in the gas in the constellation (j, i), with which the first of these collisions is beginning, and present in the constellation $(-j, -i)$, with which the second of the collisions is ending. It may further be noted that the above equality could also be obtained from our previous general expressions (45.14) for the rate of these two collisions, provided we substitute therein the consequences of the Maxwell-Boltzmann distribution and of the conservation of energy for any collision.

As a more general formulation of the principle of microscopic reversibility at equilibrium, it will be convenient to employ the equation

$$Z_{i,j,\ldots\to k,l,\ldots} = Z_{-k,-l,\ldots\to -i,-j,\ldots}, \tag{50.17}$$

where $Z_{i,j,\ldots\to k,l,\ldots}$ is the average rate of occurrence under equilibrium conditions of any specified process by which molecules are transferred between regions in the μ-space denoted by $i, j, \ldots$ and by $k, l, \ldots$, and $Z_{-k,-l,\ldots\to -i,-j,\ldots}$ is the average rate of the reverse process, where the regions denoted by $-i, -j, \ldots$ and $-k, -l, \ldots$ consist of the reverse states to those previously included. Using appropriately specified regions in the μ-space, this equation can be regarded as holding at equilibrium for molecular processes of any degree of complexity.

With the help of this general expression of the principle of microscopic reversibility at equilibrium we may now also obtain a useful

† The phrase, 'principle of microscopic reversibility', was first suggested by the writer, *Phys. Rev.* **23**, 699 (1924). See also Tolman, *Proc. Nat. Acad.* **11**, 436 (1925), and *Statistical Mechanics*, New York, 1927, § 204. The phrase is not very appropriately descriptive. It might be better to replace it by the phrase, 'principle of equal frequency for reverse molecular processes at equilibrium'. The principle should in any case be distinguished from any statement as to equal frequencies for *inverse molecular* processes, since these, in general, do not exist.

expression for the total rates at which molecules would be transferred, by the combined action of all different processes, between any one pair of regions in the μ-space and between the pair of regions containing the corresponding reverse states. For the total rates at which molecules would be transferred at equilibrium, say from regions i to k in the μ-space and in the reverse direction from $-k$ to $-i$, we may evidently write

$$N_{i\to k} = \sum_{j,l,\ldots} Z_{i,j,\ldots\to k,l,\ldots}$$

and

$$N_{-k\to -i} = \sum_{-j,-l,\ldots} Z_{-k,-l,\ldots\to -i,-j,\ldots}, \tag{50.18}$$

where the summations are taken over all possible processes which result in the transfer of a molecule from i to k and from $-k$ to $-i$ in the two cases respectively. It will be noted, however, that the two summations are made up of rates applying to individual processes that are reverse to each other. And hence, in accordance with (50.17), we can now write the equality,

$$N_{i\to k} = N_{-k\to -i}, \tag{50.19}$$

between the total average rates with which molecules would be transferred at equilibrium from regions i to k in the μ-space and from regions $-k$ to $-i$.

The relation given by (50.19) can now be used to show the possibility mentioned above of setting up a simple kind of balance between the effects of different molecular processes. In addition to processes by which molecules are transferred from i to k and from $-k$ to $-i$, processes may also be possible in which molecules are transferred in other ways between those four conditions. The total rates of transfer will in any case, however, be subject to relations of the general form (50.19), so that we can now evidently write

$$\begin{aligned} N_{i\to k} &= N_{-k\to -i}, \\ N_{i\to -k} &= N_{k\to -i}, \\ N_{-i\to k} &= N_{-k\to i}, \\ N_{-i\to -k} &= N_{k\to i} \end{aligned} \tag{50.20}$$

as relations connecting the total rates for all possible kinds of transfer between the four specified regions in the μ-space. Adding these four equations together we then obtain the desired result, which can be written in the symbolic form

$$N_{\pm i\to \pm k} = N_{\pm k\to \pm i}, \tag{50.21}$$

where the left-hand side denotes the total rate of transfer of molecules at equilibrium from either of the regions i or $-i$ in the μ-space to either of the regions k or $-k$, and the right-hand side denotes the rate from regions k or $-k$ back to i or $-i$.

The result expressed by this equation may be called the *principle of detailed balance at equilibrium*.† In accordance with this principle we can now regard the steady condition, corresponding to the minimum value of H for an isolated system, as maintained in such a way that the number of molecules which pass per unit time from any pair of regions $\pm i$ in the μ-space—containing states that are reverse to each other—to any other similar pair $\pm k$ can be balanced against the number which pass per unit time from $\pm k$ to $\pm i$. Since we are usually not interested in maintaining any distinction between reverse molecular states, this result often proves very useful when the direct calculation turns out to be much simpler for one of the two related rates of transition than for the other. This is illustrated in the well-known method of drawing conclusions as to rates of chemical activation from the much more simply calculated rates of chemical deactivation.

51. The generalized H-theorem

(*a*) **Definition of the quantity $\bar{\bar{H}}$ for an ensemble.** In the preceding parts of this chapter we have given consideration to a method for treating the approach of an isolated system of molecules towards its final equilibrium distribution which was originally discovered by Boltzmann. We shall now bring the chapter to a close by turning our attention to a more general method for treating the same problem, which was subsequently devised by Gibbs.

To carry out the Boltzmann method of treatment we first defined a quantity H, which directly characterizes the condition of any system of interest by having a value that depends at any instant on the distribution of the molecules of that system among their different possible states. We next introduced the methods of statistical mechanics by showing, for an ensemble of systems, all members having the same condition as that of the actual system of interest, that the average (most probable) value of dH/dt would be negative for the members of the ensemble. We then drew the conclusion that the value of H, for the

† The phrase 'principle of detailed balance' is due to the frequent consideration given to the 'detailed balancing' of elementary processes by Fowler; see, for example, his *Statistical Mechanics*, Cambridge, 1929. The phrase is an excellent one. The difficulty mentioned by Fowler, *Phil. Mag.* **47**, 264 (1924), as to the necessity of ruling out cyclical molecular processes as a mechanism of balance in order to maintain the validity of the principle, is met above and in *Statistical Mechanics*, New York, 1927, § 204, by the device of grouping each molecular state with its reverse in considering the rates of transfer between different molecular conditions.

system of interest itself, could be expected to decrease with time towards its minimum equilibrium value.

To carry out the Gibbs method of treatment we shall introduce the ideas of statistical mechanics at the very start by defining a new quantity $\bar{\bar{H}}$, which characterizes the condition of the ensemble of systems which we use as appropriate for representing the continued behaviour of the actual system of interest. This new quantity will have a value that depends at any instant on the distribution of the members of the ensemble among their different states. We shall show that the quantity $\bar{\bar{H}}$ for such a representative ensemble can be expected to decrease with time towards a final minimum value, requiring a uniform distribution in the phase space for the members of the ensemble lying in any energy range E to $E+\delta E$. We shall then draw the conclusion that the system of interest itself—if originally started off in a specified energy range E to $E+\delta E$—could be expected to approach a final equilibrium condition that would be represented by such a final microcanonical distribution. As we shall see in what follows, especially when we come to the quantum mechanics, the Gibbs point of view is really more appropriate and powerful than that of Boltzmann.

Before proceeding to the definition of our new quantity $\bar{\bar{H}}$ for an ensemble of systems, it will be necessary to introduce a distinction between two different quantities for an ensemble, which may be called its fine-grained density and its coarse-grained density of distribution in the phase space.†

In accordance with the methods of Chapter III, we can regard the precise state of an ensemble as specified, at any instant, by the density $\rho(q_1 \ldots p_f)$ with which the phase points for its members are distributed in the phase space that corresponds to the $2f$ coordinates and momenta $q_1 \ldots p_f$ for the kind of system under consideration. Taking N as the total constant number of systems in the ensemble, and assuming *normalization to unity*, as now proves convenient, we can then write

$$\frac{dN}{N} = \rho(q_1 \ldots p_f)\, dq_1 \ldots dp_f \tag{51.1}$$

for the probability of finding a member of the ensemble in the infinitesimal range $dq_1 \ldots dp_f$ at the point $q_1 \ldots p_f$ in the phase space, and can write

$$\int \ldots \int \rho(q_1 \ldots p_f)\, dq_1 \ldots dp_f = 1 \tag{51.2}$$

† The clear-cut distinction between these two quantities is due to P. and T. Ehrenfest, *Encykl. d. Math. Wiss.* IV. 2, ii, Heft 6, p. 60, § 23.

as the result of integration over all possible values of the q's and p's. The quantity $\rho(q_1 \dots p_f)$ appearing in these expressions may be spoken of as the *fine-grained density* of distribution in the phase space, since—in agreement with the assumptions of Chapter III as to a large enough total population N for the ensemble—we can regard (51.1) as giving a precise expression for the probability of finding a member in the specified range as we let $dq_1 \dots dp_f$ approach zero as nearly as we desire.

In making any actual measurement of the coordinates and momenta of a system, however, it is evident that we ordinarily do not achieve the precise knowledge of their values theoretically permitted by the classical mechanics. For this reason we shall also be interested in the probability of finding members of an ensemble within small but finite regions having extensions $\delta q_1 \dots \delta p_f$, which correspond to the limits of accuracy actually available to us. For this purpose we may then define another kind of density P (this can be read capital rho if desired) at any point of the phase space by the equation

$$P = \frac{\int \dots \int \rho \, dq_1 \dots dp_f}{\delta q_1 \dots \delta p_f}, \tag{51.3}$$

where the integration is taken over such a small finite range $\delta q_1 \dots \delta p_f$. The new quantity $P(q_1 \dots p_f)$ is thus a mean of the fine-grained densities in the neighbourhood of the point $q_1 \dots p_f$, and may itself be spoken of as the *coarse-grained density* at that point.

In accordance with this definition of coarse-grained density, we can now write

$$\frac{\delta N}{N} = P(q_1 \dots p_f) \, \delta q_1 \dots \delta p_f \tag{51.4}$$

for the probability of finding a member of the ensemble in the specified small but finite region denoted by $\delta q_1 \dots \delta p_f$. Furthermore, by regarding the whole phase space as divided up into a collection of such regions, all of the same extent $\delta q_1 \dots \delta p_f$, we can evidently write the equation

$$\sum_m P_m \, \delta q_1 \dots \delta p_f = 1, \tag{51.5}$$

where P_m is the coarse-grained density for the mth such region, and we take a summation over all such regions m. Or, replacing the summation by an integration over the whole of phase space, we can also write this relation in a form similar to (51.2)

$$\int \dots \int P(q_1 \dots p_f) \, dq_1 \dots dp_f = 1, \tag{51.6}$$

where we still regard the coarse-grained density P at any point as an appropriate mean of the fine-grained densities ρ in the neighbourhood of that point.

With the help of the coarse-grained density P we may now define the new quantity $\bar{\bar{H}}$ for any ensemble under consideration by the summation

$$\bar{\bar{H}} = \sum_m P_m \log P_m \, \delta q_1 \dots \delta p_f \tag{51.7}$$

taken over all the regions m of equal extent into which we regard the phase space as divided. Or, again replacing summation by integration over the whole of phase space, we can also write the defining expression in the form†

$$\bar{\bar{H}} = \int \dots \int P \log P \, dq_1 \dots dp_f, \tag{51.8}$$

provided we still maintain our understanding of the coarse-grained density P at any point as being a mean of the fine-grained densities ρ taken over a small but finite extension $\delta q_1 \dots \delta p_f$ in the phase space in the neighbourhood of that point. It is clear that the $\bar{\bar{H}}$ so defined depends in an essential manner on the size and character of the regions $\delta q_1 \dots \delta p_f$ which we introduce as corresponding to the observations contemplated.

In connexion with this definition we may emphasize the distinction between the two different quantities

$$\int \dots \int \rho \log \rho \, dq_1 \dots dp_f \quad \text{and} \quad \int \dots \int P \log P \, dq_1 \dots dp_f, \tag{51.9}$$

the latter being the one by which $\bar{\bar{H}}$ is actually defined. It may be noted, however, that we could also write

$$\bar{\bar{H}} = \int \dots \int \rho \log P \, dq_1 \dots dp_f \tag{51.10}$$

since $\log P$ would be constant over each one of the small regions $\delta q_1 \dots \delta p_f$ that were introduced above, and the integration of ρ over such a region would give us $P \, \delta q_1 \dots \delta p_f$. In accordance with our definition we see that our new quantity $\bar{\bar{H}}$ may be regarded as the mean value of $\log P$ for the ensemble as a whole. The symbolism that we have adopted for the new quantity thus agrees with our general convention of using the double bar $=$ to indicate a mean value for an ensemble as a whole.

(*b*) **Two necessary lemmas.** Two lemmas may now be presented which will be needed for the desired investigation of the rate of change in $\bar{\bar{H}}$ with time.

† This is the $\bar{\eta}$ of Gibbs, *Elementary Principles in Statistical Mechanics*, Yale University Press, 1902, except for added emphasis on the necessity of using the coarse-grained density P rather than the fine-grained density ρ in order to obtain a suitable quantity.

As the first of these lemmas we have the fact that the quantity $\int ... \int \rho \log \rho \, dq_1 ... dp_f$, defined for an ensemble in terms of the fine-grained density ρ, must actually have a value which remains constant independent of the time. The validity of this lemma depends on Liouville's theorem, in the form of the principle of the conservation of density in phase (19.11), which gives us

$$\frac{d\rho}{dt} = 0 \tag{51.11}$$

for the fine-grained density in the neighbourhood of any moving phase point. This lets us write

$$\frac{d}{dt}\int ... \int \rho \log \rho \, dq_1 ... dp_f = 0$$

for the rate of change with time of the whole integral, since this is taken so as to include all phase points. We hence obtain an equality,

$$\int ... \int \rho_1 \log \rho_1 \, dq_1 ... dp_f = \int ... \int \rho_2 \log \rho_2 \, dq_1 ... dp_f, \tag{51.12}$$

between the values of the integral at any two successive times t_1 and t_2. The result given by (51.12), however, does not, of course, necessitate a constant value for the integral of primary interest,

$$\int ... \int P \log P \, dq_1 ... dp_f,$$

since the two quantities ρ and P will not in general be equal.

As the second lemma to be presented we have the fact that the two quantities ρ and P can be combined in such a way as to give a quantity Q which must itself be essentially positive in character without reference to the values of ρ and of P. This lemma has the form

$$Q = \rho \log \rho - \rho \log P - \rho + P \geqslant 0, \tag{51.13}$$

and holds for any pair of quantities ρ and P which cannot themselves assume negative values, as is true in the present case on account of the interpretation of ρ and P as probabilities. To see the validity of (51.13) we note for any fixed value of P that we shall have

$$Q = 0 \quad \text{when} \quad \rho = P$$

and

$$\frac{\partial Q}{\partial \rho} = \log \frac{\rho}{P} \quad \begin{cases} > 0 & \text{when} \quad \rho > P \\ < 0 & \text{when} \quad \rho < P. \end{cases}$$

Hence the equality sign in (51.13) will hold only when ρ equals P, and the quantity Q will become increasingly positive as ρ becomes either greater or less than P.

(c) Change in $\overline{\overline{H}}$ with time. We are now ready to study the change in $\overline{\overline{H}}$ with time for an ensemble representing some actual system whose temporal behaviour we wish to predict. For this purpose let us consider that we make an appropriate approximate observation on the state of the system at some initial time t_1, and then set up an ensemble to represent the initial condition that we have found. In accordance with the methods of statistical mechanics and our fundamental postulate as to equal *a priori* probabilities, this initial state of the ensemble can be obtained by taking uniform distributions of phase points throughout those regions in the phase space that do correspond to our partial knowledge of the state of the actual system. These regions will have small but, nevertheless, finite extensions $\delta q_1 \ldots \delta p_f$ in the phase space on account of the limited accuracy of our observation. Since the fine-grained density ρ will have a constant value inside each such region, it will also be equal to the coarse-grained density P for that same region. Hence, at the *initial time* t_1, we shall have

$$\rho_1 = P_1 \tag{51.14}$$

at all points in the phase space. And, in accordance with our definition (51.8), we can then write for the initial value of the quantity $\overline{\overline{H}}$ that interests us

$$\overline{\overline{H}}_1 = \int \ldots \int \rho_1 \log \rho_1 \, dq_1 \ldots dp_f \tag{51.15}$$

owing to the equality of ρ and P at the initial time.

In the special case, in which the initial condition of the system is found to be such as to be represented by a uniform distribution throughout the whole of each shell in the phase space that lies within any specified energy range E to $E+\delta E$, for example by a canonical or microcanonical ensemble, the distribution in the phase space will remain unaltered as time proceeds (see § 21), and ρ and P will continue to remain permanently equal to each other at all points in the phase space. This is the case when the system is already in its equilibrium condition at the initial time t_1, and its properties may then be investigated by the methods previously employed in Chapter IV. In the general case that now interests us, however, the initial condition of the system will not be such as to correspond to a uniform density ρ throughout the whole of each shell E to $E+\delta E$, and the distribution of the ensemble will change as time proceeds. Under these circumstances we can then no longer expect the initial equality between the two densities ρ and P at time t_1 to persist at all points in the phase space as we go to later times.

To investigate this we note, in accordance with Liouville's theorem, in the form of the conservation of extension in phase (19.15), that the phase points initially present with the density ρ in any one of the small regions $\delta q_1 \ldots \delta p_f$ into which we have regarded the phase space as divided, will in the course of time continue to occupy a total extension of unchanged magnitude $\delta q_1 \ldots \delta p_f$, provided we fix our attention on the precise boundaries of that extension. Nevertheless, this extension cannot be expected at later times to retain its original configuration or 'shape', since phase points in different parts of the original small but finite region $\delta q_1 \ldots \delta p_f$ will correspond to systems that carry out quite different motions. Indeed, as time continues, we must expect the boundaries of such an extension to assume a very complicated 'shape', so that the phase points contained therein will ultimately be distributed in 'filaments', of the original density ρ, throughout many regions $\delta q_1 \ldots \delta p_f$ of the kind into which we have divided the whole phase space. Thus, at any time t_2 later than the initial time t_1, we must expect to find regions $\delta q_1 \ldots \delta p_f$ inside of which the fine-grained density ρ will have a variety of different values, corresponding to the densities of the original regions that have contributed phase points to the region now in question, including, of course, the value zero when the original density at time t_1 was zero in certain regions that might otherwise have made a contribution.

Hence, at the *later time* t_2, we shall in general have points in the phase space where the fine-grained density ρ and the coarse-grained density P will not be equal,

$$\rho_2 \neq P_2, \tag{51.16}$$

since the latter will be a mean of a variety of different values of the former. And we shall now have to write, at this later time,

$$\bar{\bar{H}}_2 = \int \ldots \int P_2 \log P_2 \, dq_1 \ldots dp_f, \tag{51.17}$$

for the value of the quantity $\bar{\bar{H}}$ that interests us, since we can no longer substitute ρ in place of P as at the initial time.

We are now ready to compare the values of $\bar{\bar{H}}$ at the initial time t_1 and at the later time t_2. Subtracting (51.17) from (51.15), we may begin by writing

$$\bar{\bar{H}}_1 - \bar{\bar{H}}_2 = \int \ldots \int (\rho_1 \log \rho_1 - P_2 \log P_2) \, dq_1 \ldots dp_f. \tag{51.18}$$

This equation may be changed, however, so as to give a more informing expression. In accordance with the first of our previous lemmas as given by (51.12), we can replace $\rho_1 \log \rho_1$ in the integrand by $\rho_2 \log \rho_2$, since the integral of $\rho \log \rho$ over the whole of the phase space has been shown to have a value that does not change with time. Furthermore, in accordance with the possibility of expressing the value of $\bar{\bar{H}}$ in either of the forms (51.8) or (51.10), we see that we can replace $P_2 \log P_2$ in the integrand by $\rho_2 \log P_2$. Finally, in accordance with (51.2) and (51.6), it is evident that we can add the two terms $-\rho_2$ and $+P_2$ to the integrand, since the effects of the two additions will cancel out on integration. Making these changes, we can then rewrite (51.18) in the form

$$\bar{\bar{H}}_1 - \bar{\bar{H}}_2 = \int \dots \int (\rho_2 \log \rho_2 - \rho_2 \log P_2 - \rho_2 + P_2)\, dq_1 \dots dp_f. \quad (51.19)$$

In accordance with the second of our previous lemmas, however, as given by (51.13), we see that our integrand now consists of a quantity which will be zero at any point in the phase space, where the fine-grained and coarse-grained densities ρ_2 and P_2 are equal, but will be greater than zero at all those points discussed above, where the two densities are no longer equal. Hence we now obtain the desired result

$$\bar{\bar{H}}_2 < \bar{\bar{H}}_1, \quad (51.20)$$

showing that the value of $\bar{\bar{H}}$ for the ensemble will be less at the later time t_2 than at the initial time t_1. This result may be called the *generalized H-theorem*. It is clear from the method of derivation that the asymmetry between the earlier and later values $\bar{\bar{H}}_1$ and $\bar{\bar{H}}_2$ is essentially dependent on the asymmetry between the expressions $\rho_1 = P_1$ and $\rho_2 \neq P_2$, which corresponds to a decrease with time in the definite character of our information as to the condition of the system of interest.

We thus find that we have indeed defined for our representative ensemble of systems a quantity

$$\bar{\bar{H}} = \int \dots \int P \log P \, dq_1 \dots dp_f, \quad (51.21)$$

which as time proceeds will change to values lower than that which it has initially when the ensemble is set up to represent some system of interest in a known observationally determined condition. Furthermore, since this change with time depends on a development of inequalities, at different points in the phase space, between the originally equal values of ρ and P, we can expect a *continued decrease* in $\bar{\bar{H}}$ as time proceeds and more and more such inequalities develop.

Moreover, realizing that the constancy of energy will keep the total number of phase points in each energy range E to $E+\delta E$ constant, and taking the ensemble as in any case only containing members which have the same values (e.g. zero) as the system of interest itself for such simple constants of the motion as the components of linear and angular momentum, it seems reasonable to assume in typical cases that the changes in distribution considered above would continue until each small but finite region $\delta q_1 \ldots \delta p_f$ in any energy range E to $E+\delta E$ would contain approximately the same proportionate contribution of phase points from each of the regions in that range which was originally populated. In drawing this conclusion we regard the restrictions imposed by less simple constants of the motion as leading to a scattered exclusion of phase points from precise positions within the various small but finite regions $\delta q_1 \ldots \delta p_f$ in a manner which does not interest us. Hence, in typical cases, we may take the distribution of the ensemble as continuing to change and the value of $\bar{\bar{H}}$ as continuing to decrease, until we arrive at an *ultimate approximately uniform distribution* of coarse-grained probability

$$P = \text{const.} \quad (E \text{ to } E+\delta E) \tag{51.22}$$

within each energy range E to $E+\delta E$ which originally contained representative points.

When this uniform distribution has been reached, further decrease in $\bar{\bar{H}}$ would no longer be possible, since $\bar{\bar{H}}$ will have then reached its minimum possible value. This is seen from the consideration that equation (51.22) is the solution of the condition for a minimum value of $\bar{\bar{H}}$,

$$\delta\bar{\bar{H}} = \int \ldots \int (\log P+1)\, \delta P\, dq_1 \ldots dp_f = 0, \tag{51.23}$$

under the necessary subsidiary restriction

$$\int_E^{E+\delta E} \ldots \int \delta P\, dq_1 \ldots dp_f = 0 \tag{51.24}$$

applying in each energy range. On once arriving at such a distribution, we can expect the approximate uniformity to persist over exceedingly long intervals of time.

In case our original knowledge of the condition of a system of interest is such as to confine it to a single energy range E to $E+\delta E$, we see that the ultimate condition of its representative ensemble would be an approximately uniform density of distribution within that energy range. Hence, in accordance with the methods of statistical mechanics, we may now conclude that any actual system of interest when left to itself with a specified value of energy can be expected to arrive in an ultimate condition of equilibrium, such that its properties can then be correlated, as in Chapter IV, with the average properties in a corresponding uniform microcanonical ensemble. This gives us a very satisfactory justification for our previous treatment of equilibrium.

In case our knowledge of the system of interest is such as to be represented by a distribution over various energy ranges E to $E+\delta E$, we see that the ultimate condition of the ensemble would be such as still to preserve the same relative probabilities for different values of the energy.

The above conclusions apply, of course, to the case of perfect isolation where there is no opportunity for the system of interest to interchange energy with its surroundings. In a later place, §§ 111 and 112, we shall discuss the long time behaviour of ensembles representing systems in contact with their surroundings, and show under appropriate circumstances that the condition of equilibrium can then be best represented by a canonical distribution with respect to energy.

(*d*) **Relation between the two forms of H-theorem.** We may next consider the relation of the original form of the H-theorem, which gives the temporal behaviour of a quantity H defined so as to depend on the condition of a single system, to the above generalized form of the H-theorem, which gives the temporal behaviour of a quantity $\bar{\bar{H}}$ defined so as to depend on the condition of the corresponding representative ensemble of systems.

For this purpose, let us consider a system composed of n similar molecules, and let us regard the μ-space, corresponding to the $2r$ coordinates and momenta $q_1 \dots p_r$ for the kind of molecule involved, as divided into small but finite cells, all having equal extensions

$$\delta v_\mu = \delta q_1 \dots \delta p_r \tag{51.25}$$

in the μ-space. The condition of such a system can then be specified by giving the numbers of molecules $n_1, n_2, n_3, \dots, n_i, \dots$ having values of their coordinates and momenta lying in the different cells i; and the value of the quantity H, corresponding to this condition, will be given, in accordance with (47.7), by

$$H = \sum_i n_i \log n_i + \text{const.}, \tag{51.26}$$

where we take a summation over all cells i in the μ-space.

Let us next consider an ensemble of such systems, and let us regard the γ-space, corresponding to the $2f = (2r)^n$ coordinates and momenta $q_1 \dots p_f$ for the n molecules of a system, as divided into small but finite regions, all having equal extensions

$$\delta v_\gamma = (\delta q_1 \dots \delta p_r)_1 (\delta q_1 \dots \delta p_r)_2 \dots (\delta q_1 \dots \delta p_r)_n = (\delta v_\mu)^n \tag{51.27}$$

in the γ-space, which can be regarded as obtained—in the manner indicated—by combining cells from the μ-spaces for the n different molecules. The condition of the ensemble can then be described by giving the probabilities $P_k\,\delta v_\gamma$ for finding a member of the ensemble in the different regions k into which we have thus divided the γ-space. And the quantity $\overline{\overline{H}}$, corresponding to this condition of the ensemble, will be given in accordance with (51.7) by

$$\overline{\overline{H}} = \sum_k P_k \log P_k\,\delta v_\gamma, \tag{51.28}$$

where we take a summation over all regions k in the γ-space.

We now wish to obtain an interrelation between the two different kinds of quantities H and $\overline{\overline{H}}$ given by (51.26) and (51.28). In agreement with the considerations of § 27, we note that there would be a total of

$$G = \frac{n!}{\prod_i n_i!} \tag{51.29}$$

individual equivalent elementary regions, of extension δv_γ in the phase space, between which we do not distinguish when we determine the condition of a system by the numbers of molecules n_i assigned to the different cells i of the μ-space. Hence, using the symbol G_κ to designate the number of regions k in the γ-space, which correspond to a particular such condition indicated by the subscript κ, we may now write as an expression for the total probability P_κ of finding a member of our ensemble in the condition κ

$$P_\kappa = G_\kappa P_k\,\delta v_\gamma, \tag{51.30}$$

where $P_k\,\delta v_\gamma$ is the probability of finding a member in each elementary region k that corresponds to the condition κ. Solving (51.30) for P_k and substituting in (51.28), we then see that we can write our expression for $\overline{\overline{H}}$ in the form

$$\begin{aligned}\overline{\overline{H}} &= \sum_\kappa G_\kappa \frac{P_\kappa}{G_\kappa\,\delta v_\gamma} \log \frac{P_\kappa}{G_\kappa\,\delta v_\gamma}\,\delta v_\gamma \\ &= \sum_\kappa P_\kappa \log \frac{P_\kappa}{G_\kappa\,\delta v_\gamma},\end{aligned} \tag{51.31}$$

where we now have a summation over the different kinds of condition κ for a single system.

This expression will now give us the desired interrelation between the two kinds of quantities H and $\overline{\overline{H}}$. For this purpose we note that the expressions for H and G, given by (51.26) and (51.29), together with the use of Stirling's approximation for factorials, will make it possible for any particular condition κ to substitute

$$\log \frac{1}{G_\kappa \delta v_\gamma} = H_\kappa + \text{const.}, \tag{51.32}$$

where H_κ is the value of our original quantity H for a single system in the condition κ. Hence (51.31) may now be rewritten in the form

$$\begin{aligned} \overline{\overline{H}} &= \sum_\kappa P_\kappa H_\kappa + \sum_\kappa P_\kappa \log P_\kappa + \text{const.} \\ &= \overline{\overline{H}}_{\text{syst.}} + \sum_\kappa P_\kappa \log P_\kappa + \text{const.}, \end{aligned} \tag{51.33}$$

where we use the symbol $\overline{\overline{H}}_{\text{syst.}}$ to denote the mean value, for all the members of the ensemble, of the quantity H applying in the original form of the H-theorem to a single system. This expression now makes it easy to obtain a clear idea as to the relation between the new and old forms of H-theorem.

In the first place, we see that (51.33) expresses the present quantity $\overline{\overline{H}}$ by which we characterize the condition of the representative ensemble for a system of interest as the sum of two terms, the first of which, $\overline{\overline{H}}_{\text{syst.}}$, is the mean value for the members of the ensemble of the previous quantity H by which we directly characterized the system itself, and the second of which, $\sum P_\kappa \log P_\kappa$, describes the distribution of the members of the ensemble over different conditions κ. For the special case $P_\kappa = 1$ corresponding to an ensemble which has just been set up to represent a system which has been observed to be in a particular condition κ, we note that the above connexion would reduce to the simple form

$$\overline{\overline{H}} = H + \text{const.}, \tag{51.34}$$

where $\overline{\overline{H}}$ applies to the ensemble and H applies directly to the system of interest.

In the second place, differentiating (51.33) with respect to the time and making use of the circumstance that the total probability for all conditions κ has to remain constant, we can write

$$-\frac{d\overline{\overline{H}}}{dt} = -\frac{d\overline{\overline{H}}_{\text{syst.}}}{dt} - \sum_\kappa \log P_\kappa \frac{dP_\kappa}{dt}. \tag{51.35}$$

We then see that the rate of decrease in $\overline{\overline{H}}$ with time, which we now take as characterizing the behaviour of the representative ensemble, is the sum of two terms, the first of which expresses a decrease with time in the mean value for the members of the ensemble of the quantity H

which we previously took as probably decreasing with time for the system of interest itself, and the second of which expresses a change in the direction of more uniform occupation of the different possible conditions κ. Since it will be seen from (51.33) that a minimum for $\bar{\bar{H}}$ would require the distribution

$$P_\kappa = \text{const. } e^{-H_\kappa} \tag{51.36}$$

in each energy range E to $E+\delta E$, it will be appreciated that the generalized theorem really provides the more direct understanding of the tendency for a system to proceed towards conditions κ where low rather than the absolutely smallest possible value of H may be expected.

(*e*) **Concluding remarks on the generalized H-theorem.** We may now bring this long section to a close with a few further remarks as to the character and validity of the generalized H-theorem.

In the first place, it will be of interest to make some qualitative remarks as to the rapidity with which we can expect the decrease in $\bar{\bar{H}}$ with time to commence, when we set up an ensemble to represent a system in a known initial condition. In accordance with the discussion of § 51 (*c*), we see that this decrease results from the failure of representative points of the ensemble, originally present in any particular small but finite region $\delta q_1 \ldots \delta p_f$ with the uniform density ρ, to move together as a compact whole without change in 'shape' of the extension which they occupy. We see, however, that we can usually expect such changes in 'shape' to start occurring at once since the precise behaviours of representative points in different parts of the *finite* region $\delta q_1 \ldots \delta p_f$ will be very different. For example, in the case of molecular collisions, since the precise states of the different members of the ensemble can correspond to widely different spatial positions for molecules within finite volumes $\delta x \delta y \delta z$, it is evident that such differences in behaviour will occur as soon as collisions take place. We also see from such considerations that we can usually expect a continuance of such differences in behaviour and a continued decrease in $\bar{\bar{H}}$ as further collisions occur.

In the second place, it will be of interest to make some further remarks as to the validity of our conclusion that the final condition of our representative ensemble would be given in typical cases by an approximate uniformity of the coarse-grained density P throughout any energy range E to $E+\delta E$. Here it is evident that the arguments leading to such a conclusion can only be of a qualitative and plausible character, unless we are willing and able to undertake the more precise kind of mechanical investigation which it is the function of statistical mechanics to try to avoid. In this connexion the analogy of Gibbs† as to the effect of stirring on a mixture of water and ink is appropriate.

If we consider a vessel containing portions of water and *non-diffusible* inks of different degrees of blackness, originally in an unmixed condition, it is evident that nearly *any kind of stirring* can be expected to result in an ultimate uniform grey, when looked at from a rough-grained point of view that neglects the fine-grained densities of the individual filaments of water and ink that have been drawn out by the stirring. The analogy is sufficiently good to provide sound reasons for believing, in the case of the ensembles discussed, that the different fine-grained densities of distribution ρ, within a shell of the phase space corresponding to an energy range E to $E+\delta E$, would ultimately be so mixed up as to give us an approximately uniform coarse-grained density P within that range.

Some further remarks with respect to the approach of an ensemble towards its ultimate condition will also be of interest. It will be noted that the behaviour of ensembles that we have described is evidently what we can expect in general, but that singular behaviours of a different kind might occasionally occur. For example, if we consider the behaviour of an ensemble over an exceedingly long time, it is evident that we might encounter a reconcentration of distribution in the phase space, which would be analogous in the above illustration to an 'unmixing' of the water and ink that could occur during an infinite period of stirring. This we regard as improbable rather than impossible. As another example, if the possible motions of the system are of an exceedingly simple type, it is evident that our original determination of condition might be sufficient to exclude the distribution permanently from large regions of the phase space of the correct energy, as would be analogous in the above illustration to the possibility that no amount of stirring of the kind employed could lead to a really complete mixing. Nevertheless, in the case of systems composed of many molecules, it is evident that the highly singular behaviours, which a single system could carry out if originally started off with a very special arrangement of molecules, would usually be unimportant owing to the finite size of the regions $\delta q_1 \ldots \delta p_f$ which we originally populate with representative points.

† Gibbs, *Elementary Principles in Statistical Mechanics*, Yale University Press, 1902, chapter xii.

It is also of interest to inquire into the rapidity with which the final equilibrium distribution of the ensemble would be approached. It is evident that quantitative answers to such questions would depend in a specific manner on the properties of the kind of system under consideration. It is important to note, nevertheless, if we divide the phase space into regions $\delta q_1 \ldots \delta p_f$ of a magnitude large enough to correspond to the kind of observational accuracy usually attainable, that we can often expect a reasonably uniform distribution of representative points into different regions of the right energy to take place quite rapidly. There is no reason to assume, with observational control of ordinary accuracy, that the intervals of time, necessary for a practically uniform distribution of coarse-grained density P, would be of the order of the Poincaré periods necessary for a single system to return to the immediate neighbourhood of its original state.†

In concluding this chapter on the H-theorem it is evident that we must now regard the original discovery of that theorem by Boltzmann as supplemented in a fundamental and important manner by the deeper and more powerful methods of Gibbs.‡

† Note the contrary opinion expressed by P. and T. Ehrenfest, *Encykl. d. Math. Wiss.* IV. 2, ii, Hoft 6, p. 61.

‡ For a modern appreciation of the contributions to statistical mechanics of Gibbs, see the article by Epstein, 'Critical Appreciation of Gibbs' Statistical Mechanics', in *Commentary on the Scientific Writings of J. Willard Gibbs*, vol. ii, Yale University Press, 1936.

VII
THE ELEMENTS OF QUANTUM MECHANICS

A. HISTORICAL REMARKS

52. The necessity for modifying classical ideas

The development of physical theory from the time of Galileo and Newton to the end of the nineteenth century took place as a continuous process validated at each step by an increased understanding and mastery of the physical world. To be sure, further concepts beyond those of Newton were added, more powerful mathematical methods were introduced, and the considerations of mechanics were increasingly supplemented by those of electrodynamics, thermodynamics, and chemistry. At no time, however, did it appear obligatory to re-examine the fundamental ideas as to the nature of space and time and as to the character of observation and measurement, which were implicit in Newtonian procedure. Nevertheless, already before the end of the nineteenth century, physical theory had encountered two difficulties whose later treatments were to involve the re-examination of those fundamental ideas.

The first of these difficulties was that of reconciling the accepted wave-like propagation of electromagnetic disturbances with the failure of all attempts—in particular that of Michelson and Morley—to detect the earth's motion through an ether suitable for such a propagation. The solution of this difficulty, at the hands of Einstein, was made possible by a deeper criticism of the nature of the processes by which spatial and temporal determinations can be made. The new ideas thus obtained as to the nature of space-time specifications, as now incorporated in the special and general theories of relativity, have affected the whole of physical thought. The scope and character of the effects in the realm of macroscopic physics, including the gravitational behaviour of astronomical bodies, are now well understood; and the character of many of the effects in the realm of microscopic, atomic physics are also clear.

The other of the two difficulties for nineteenth-century physics was that of explaining the failure of electromagnetic energy to distribute itself uniformly over all the possible modes of vibration in an enclosure containing radiation which has come to thermal equilibrium. A reasonably satisfactory solution of this problem, and of others which proved to be associated with it, has only been made possible by a criticism of the very nature of physical observation, with a resulting appreciation of the uncontrolled character of the effects that measurement itself must produce on systems—particularly microscopic ones—when under observation. Our present system of quantum mechanics may be regarded as the ultimate outcome of such criticism. The actual process, however, by which physical theory was led from classical mechanics to quantum mechanics involved several steps which we may first briefly describe.

(*a*) **Discrete energy levels.** In the year 1900 a formula was derived by Planck† which did agree with the observed distribution of radiation as a function of frequency inside a hollow enclosure which has come to thermal equilibrium. To obtain the derivation the radiation was treated as being in interaction with a set of electric oscillators, of all different frequencies, which were themselves assumed to be capable of absorbing or emitting energy only in discrete *quanta*. For each given frequency ν the magnitude of these definite amounts of energy was found to be $h\nu$, where h is the new physical constant, having the dimensions of action, to which Planck's name is now attached.

Thus was introduced into physics the new idea of atomic systems—i.e. the electric oscillators—having a set of *discrete energy levels* separated by differences in energy which are related to the frequency of the radiation absorbed or emitted in passing from one level to another by the expression

$$E_2-E_1 = h\nu. \tag{52.1}$$

The further usefulness of this idea of discrete energy levels was soon made evident by the work of Einstein,‡ Debye,‖ and others†† on the specific heat of solids. By assigning discrete energy levels—$h\nu$ apart—to the different modes of vibration of a crystalline solid, it was shown possible to give a good account of the actual energy content of crystals as a function of temperature, including in particular the low temperature range where the content is much less than would be predicted in accordance with the classical law of Dulong and Petit.

The theoretical fertility of the idea of discrete energy levels was perhaps most importantly extended, however, by Bohr's‡‡ work of 1913 on the hydrogen spectrum. The atom of hydrogen was treated, on the basis of the nuclear model of Rutherford, as consisting of a single

† Planck, *Verh. deutsch. Phys. Ges.* **2**, 202, 237 (1900).
‡ Einstein, *Ann. der Phys.* **22**, 180 (1907).
‖ Debye, *Ann. der Phys.* **39**, 789 (1912).
†† Born and Kármán, *Phys. Zeits.* **13**, 297 (1912); **14**, 15 (1913).
‡‡ Bohr, *Phil. Mag.* **26**, 1, 476, 857 (1913).

electron rotating around a central proton with centrifugal force balanced by the electric attraction, and a simple rule was given for selecting discrete quantized orbits from the manifold of possible classical orbits. The energies of these quantized orbits then provided a set of discrete energy levels, whose use in connexion with equation (52.1) would account for the actual frequencies of the many lines of the spectrum of monatomic hydrogen. And natural extensions of this procedure, with the help of somewhat generalized rules for the selection of quantized states for any kind of periodic motion,† soon brought nearly the whole of the complicated field of spectroscopic facts into a considerable measure of order.

Furthermore, on the experimental side the reality of atomic energy levels was clearly demonstrated in 1914 by the experiments of Franck and Hertz,‡ in which gaseous atoms could be bombarded by electrons of known and controllable velocity. When the kinetic energy of the electrons was kept less than that necessary to raise the bombarded atoms from their normal state to that of the next highest energy level, only elastic collisions were found to occur, with no appreciable loss of energy on the part of the electrons as shown by their continued ability to pass through an appropriate opposing electric field. As soon, however, as the energy of the electrons was increased to that necessary for raising the atom to the higher level, inelastic collisions set in and the energy absorbed by the atoms was found to be re-emitted in the form of radiation of the expected frequency.

This new idea, that atoms are characterized by sets of discrete energy levels so that radiation can be absorbed and emitted in definite quanta, is the feature of the new developments which has led to the *name* quantum mechanics. Its introduction marks a considerable step away from classical ideas since there was nothing in the classical picture of an electric oscillator or of a planetary atom which would lead us to expect that unique properties should be assigned to any particular energy levels chosen out of all the possible ones. As we shall see, however, it is not the feature of the new developments which is really most characteristic or significant for the new mechanics.

(*b*) **Wave-particle duality.** As a consequence of the idea that atoms absorb and emit radiation in definite quanta, the further idea was introduced in 1905 by Einstein‖ that radiation itself—even in spite of

† Sommerfeld, *Ann. der Phys.* **51,** 1 (1916); Wilson, *Phil. Mag.* **29,** 795 (1915); Ishiwara, *Tokyo Math. Phys. Proc.* **8,** 106 (1915).

‡ Franck and Hertz, *Verh. deutsch. phys. Ges.* **16,** 457, 512 (1914).

‖ Einstein, *Ann. der Phys.* **17,** 132 (1905); **20,** 199 (1906).

the successes of the wave theory—might be regarded at least from some points of view as having a corpuscular character. The energy of each corpuscle or *photon* associated with radiation of frequency ν would be given by

$$E = h\nu \tag{52.2}$$

in agreement with our previous relation between energy levels and frequency; and the momentum of the photon in the direction of propagation would be given by

$$p = \frac{h\nu}{c} = \frac{h}{\lambda} \tag{52.3}$$

in agreement with the relativistic relations between mass, energy, and momentum, where the wave-length $\lambda = c/\nu$ is substituted in the second form of writing.

This idea of the corpuscular character of radiation was immediately useful in explaining the photo-electric emission of electrons from an irradiated metal, since these are found to come off with energies which do not depend on the intensity of illumination but do increase with the frequency of the radiation in the way to be expected if due to the action of corpuscles of energy of amount $h\nu$. And Einstein† was also able to derive the Planck law for the distribution of radiation at thermal equilibrium by studying the steady state which would be reached if atoms can absorb and emit such photons, introducing appropriate expressions for the probabilities of these elementary processes.

Further support for the corpuscular character was furnished by the discovery of Compton‡ that the frequency of X-radiation scattered by free electrons was reduced by the amount to be expected as a result of elastic encounters of photons with electrons. While the later work of Bothe and Geiger|| and of Compton and Simon†† gave direct evidence that the scattered photon and electron do indeed come off at the same time and in the directions to be expected if energy and momentum are conserved in the collision. The justification for assigning a dual wave-like and particle-like character to radiation thus became evident.

The idea of such duality was then extended also to matter by L. de Broglie‡‡ in 1924. As the inverse of Einstein's association of particles with the well-known waves of electromagnetic radiation, it was now proposed to associate some kind of waves with the well-known particles

† Einstein, *Verh. deutsch. phys. Ges.* **18**, 318 (1916).
‡ Compton, *Phys. Rev.* **21**, 483 (1923).
|| Bothe and Geiger, *Zeits. f. Phys.* **32**, 639 (1925).
†† Compton and Simon, *Phys. Rev.* **26**, 289 (1925).
‡‡ L. de Broglie, *Phil. Mag.* **47**, 446 (1924).

of matter. This can be done by assigning to a particle of energy E and momentum p waves of the frequency ν and wave-length λ given by the previous equations (52.2) and (52.3) for the case of the photon. Using relativistic expressions for the energy and momentum of the particle, and the usual dependence of phase velocity and group velocity on frequency and wave-length, this is then found to provide waves whose phase velocity is greater than that of light but whose group velocity is that of the particle itself, in agreement with the idea, as we shall see more clearly later, that the motion of a particle can be related to the behaviour of a wave packet.

The importance of this idea of associating some kind of waves with material particles was made very evident in 1927 by the electron diffraction experiments of Davisson and Germer.† A nickel crystal was bombarded with a stream of slow electrons of controlled velocity, and the angles investigated at which a maximum reflection of electrons occurred. Regarding the surface of the nickel as a grating with a spacing determined by the known crystal structure, it was then indeed found that selective reflection did occur at the velocities and angles which would be predicted as a result of the interference and reinforcement of waves with the de Broglie wave-length (see 52.3)

$$\lambda = \frac{h}{p} = \frac{h}{mv}, \tag{52.4}$$

where m and v are the mass and velocity of the electron. Many further examples of such interference phenomena with other methods and other particles are now known.

We are thus led, both in the case of radiation and of matter, to some kind of *wave-particle duality* which seems at first sight to involve the assignment of contradictory properties to the same entity. As we shall later see, one of the main tasks and successes of the quantum mechanics has been to resolve the apparent conflict between opposing points of view as to structure which seemed classically irreconcilable.

(*c*) **Uncertainty, complementarity, and indetermination.** We must now turn to the fundamental considerations of Heisenberg‡ and of Bohr‖ as to the conceptual possibilities of observation and measurement in the case of atomic systems. In classical thought it is tacitly assumed that the changes that take place in a mechanical system with time can be followed by making observations and measurements which

† Davisson and Germer, *Nature*, **119**, 558 (1927).

‡ Heisenberg, *Zeits. f. Phys.* **43**, 172 (1927).

‖ Bohr, *Nature*, **121**, 580 (1928); *Naturwissenschaften*, **16**, 245 (1928).

do not themselves disturb that behaviour in an uncontrolled manner. As soon as we think about the possibilities of observing the behaviour of atomic systems, however, we immediately see that effects resulting from the very act of measurement, which would be negligible in the case of the macroscopic systems of classical mechanics, may now become very important.

It is characteristic of the new kind of considerations that they incorporate the two types of phenomena, described by particle language and by wave language, as will be strikingly illustrated by the example of the γ-ray microscope which we are about to discuss, where we shall have to use both particle concepts and wave concepts in different parts of the discussion, in order to give a complete treatment of the situation. The consistency of such dual descriptions is ultimately saved by a careful limitation of the extent to which the two modes of description and the regularities to which they correspond do in fact preclude each other; in this connexion a recognition of the disturbances produced by observation becomes essential (cf. remarks at the beginning of § 61).

The nature of the disturbances produced by observation may be illustrated by Heisenberg's treatment of the effects that would result from a measurement of the position of a free electron, let us say its positional coordinate $q_x = x$ along the x-axis. To carry out such a measurement, we may think of the electron as illuminated with radiation of short enough wave-length to give an accurate determination of position—say γ-rays—and as then observed in a 'γ-ray microscope', pointed at right angles to the x-axis, and having a large enough aperture to give the desired resolution. It is, of course, evident that the scattering of the γ-rays by Compton collisions with the electron will produce a change in the momentum of the electron, and to make this as small as possible we may allow a single photon to be scattered into the microscope and recorded on a sensitive plate. We immediately perceive that the situation presents a competition between the accuracy with which we can observe the position of the electron q_x and the uncontrollable change in the corresponding momentum p_x, which will be introduced by the Compton collision. On the one hand, treating the γ-rays as waves obeying the principles of optics, we see that the accuracy of the determination of position will be improved the higher we make the frequency of the radiation and the larger we take the aperture and hence resolving power of the microscope. On the other hand, treating the γ-rays as photons and regarding only a single photon as scattered, we see that the extent of the uncontrolled change in the momentum of the electron, as a consequence of Compton scattering, will become greater the higher the frequency and hence momentum of the photon, and the greater the aperture and hence uncertainty as to the direction of scattering and consequent fraction of the momentum transferred to the electron in a direction at right angles to the microscope.

Calling Δq_x the uncertainty in our knowledge of the position of the electron and Δp_x the uncertainty introduced by the collision into the corresponding component of momentum, simple calculation shows that the product of the two uncertainties will at least be of the order of Planck's constant h. Since similar relations will evidently hold for the y and z directions, we may now write

$$\begin{aligned}\Delta p_x \Delta q_x &\approx h,\\ \Delta p_y \Delta q_y &\approx h,\\ \Delta p_z \Delta q_z &\approx h\end{aligned} \tag{52.5}$$

as general expressions for the products in question.

It is to be noted that the uncertainties introduced into the components of momenta by the type of collision considered above would destroy any previous more exact knowledge as to momentum which we might have obtained, for example by examining the Doppler effect in radiation reflected by the moving electron. Hence the above equations prescribe limits to the knowledge of the simultaneous values of the variables p and q which can be obtained by this method of measurement, and it is found that these limits cannot be lessened by other methods of measurement, for example by observing the position first and the momentum afterwards. It should be noted that the limitation applies only to the conjugated pairs of variables consisting of a coordinate and its corresponding momentum, since the same difficulties would not be encountered in the simultaneous determination of two coordinates, or of one coordinate and of a component of momentum at right angles to the coordinate axis. Later, in § 62, we shall also find that the conjugated quantities energy and time are connected by a similar restrictive relation

$$\Delta E \Delta t \approx h, \tag{52.6}$$

having a physical content which will be investigated. The above equations (52.5) and (52.6) are particular examples of the general relation known as Heisenberg's *uncertainty principle*.

The relationship between two classical variables, whose values are subject to interconnected uncertainties as in the above, has been called *complementarity* by Bohr.† It is to be particularly emphasized in the case of such complementary variables that increased accuracy in our knowledge of one of the variables can be obtained only at the expense of decreased accuracy in our knowledge of the other. This may be regarded as an illustration of a general type of complementarity in which one kind of knowledge or one method of treatment is connected in a mutually exclusive fashion with another supplementary kind of knowledge or method of treatment; for example, the complementarity between the treatment of radiation from the wave and from the particle points of view.

The most striking consequence of Heisenberg's uncertainty principle is the *indeterminacy* which it introduces into the possibilities of physical prediction. In the classical mechanics we grew accustomed to the idea that an exact knowledge of the coordinates and momenta of a system at a given initial time would then make it possible, with the help of the equations of motion, to make an exact prediction as to the future behaviour of the system. We now see, however, that such exact knowledge of the initial values of both the coordinates and momenta is not possible, and hence must give up our older ideas of the possibility of exact prediction and of a complete causal dependence of the later on the earlier behaviour of a mechanical system.

Such a conclusion produces a drastic change in the ideology of science, the progress of which has hitherto been dominated by the concepts of strict causal determinism and predictability; and, to be sure, any *entire* elimination of the ideas of causation and prediction would surely be fatal to science itself. We note, however, that our foregoing considerations have not necessarily eliminated the possibility for a sufficient degree of causal dependence to permit predictions as to the probable behaviour of mechanical systems from the actually allowable knowledge of their initial states, even though exact predictions are no longer possible. Indeed we shall regard as the most characteristic feature of the quantum mechanics the methods which it does provide for making predictions as to the expected or *average* behaviour of systems.

(*d*) **The correspondence principle.** These new ideas as to the possible functions of the science of mechanics are so different from classical ones that some anxiety might be felt as to the preservation of an appropriate status for classical mechanics in the new scheme of thought, since the great achievements of the older mechanics certainly have a wide range of valid application. This range of validity, however, is satisfactorily

† Bohr, *Nature*, **121**, 580 (1928); *Naturwissenschaften*, **16**, 245 (1928).

preserved in the case of macroscopic systems of the kind whose study led to the formulation of classical mechanics, since we shall find that the probability predictions of the quantum mechanics are then affected by such a narrow range of uncertainty as to explain and justify the classical treatment, in agreement with our intuitive appreciation of the fact that disturbing effects connected with methods of observation must be negligible in the case of macroscopic systems. Furthermore, the whole development of quantum theory—also when concerned with microscopic systems as well—has been carried out in the light of Bohr's *correspondence principle* which recognizes the fairly general existence of limiting conditions where the corresponding classical and quantum treatment of a problem lead to converging results. We shall expect these limiting conditions to be approached when the degree of uncertainty, arising from the appearance of the elementary quantum of action h in the uncertainty relations (52.5) and (52.6), becomes negligible for the problem in hand. Hence the correspondence principle may be applied to the conclusions of quantum mechanics by seeing if they approach the corresponding classical ones as h goes to zero.

(*e*) **Plan of treatment.** We have now completed a meagre account of the new ideas which seem most important for the quantum theory; and we might hence try to trace the steps by which these new ideas could be made to lead to a new theory of mechanics. Such a plan of treatment would not be very satisfactory, however, since it is always difficult to show that a given set of specific facts and considerations are sufficient to give a unique determination of a general theoretical structure; and since the actual path of development was, of course, affected by historical accidents which no longer seem important. Hence we shall actually adopt the preferable plan of a deductive rather than an inductive form of exposition, in which we shall lay down a postulatory basis for the quantum mechanics and then show that our postulates do imply conclusions in agreement with the new ideas as to energy levels, wave-particle duality, indetermination, and correspondence.

In the next four sections we shall first present and discuss the postulatory basis necessary for the new mechanics, which will then be summarized in § 57. This will be followed in §§ 58 and 59 by the derivation of two fundamental theorems which are necessary for the whole of quantum mechanics, and then in §§ 60 to 63 we shall be able to show that our new system does incorporate the changed ideas discussed above. The remainder of the chapter will be devoted to the further development of quantum mechanical methods, and the following chapter to a derivation of needed consequences of the theory. We shall then be ready to discuss the statistical methods which are to be introduced for the treatment of systems which are not in a sufficiently well defined state for treatment by the direct methods of quantum mechanics.

The postulatory treatment which we shall give will not pretend to complete logical elegance and rigour, since we shall not try to obtain the smallest possible number of mutually independent and compatible postulates, nor to make sure that all the postulates are actually explicitly stated. Furthermore, the postulatory treatment will actually be put in terms especially appropriate for application to the non-relativistic mechanics of particles so important for atomic physics. By a suitable generalization—in the sense of the classical Hamiltonian theory of fields —of the meaning of the coordinates and momenta in terms of which our postulates are formulated, we should obtain a suitable basis for a relativistic quantum mechanics of particles and a quantum treatment of radiation. In spite of inadequacies it is hoped in any case that the exposition will be sufficient to assure a correct feeling for the essential nature of quantum theory.

B. THE POSTULATES

53. The existence of probability densities and amplitudes

Our first postulate for quantum mechanics gives explicit recognition to the idea that the statements and predictions of the new theory must for the most part be confined to assertions as to the probability of finding particular values for the coordinates and momenta of a mechanical system at a given time of interest, rather than assertions as to their exact values.

Consider a mechanical system of f degrees of freedom, which would be described classically by specifying the exact values of its $2f$ coordinates and momenta $q_1 \ldots q_f$ and $p_1 \ldots p_f$, and let us now denote by

$$W(q_1 \ldots q_f, t)\, dq_1 \ldots dq_f \quad \text{and} \quad W(p_1 \ldots p_f, t)\, dp_1 \ldots dp_f \qquad (53.1)$$

the probabilities at time t that the system would be found on measurement to have values for its coordinates or momenta lying in the differential ranges indicated. The quantities $W(q_1 \ldots q_f, t)$ and $W(p_1 \ldots p_f, t)$ may appropriately be called *probability densities*. They will evidently have to be real positive quantities on account of the physical nature which we have assigned them.

Let us now express the real positive character of these two kinds of probability density by equating them to the squares of the absolute magnitudes of certain corresponding quantities

$$\psi(q_1 \dots q_f, t) \quad \text{and} \quad \phi(p_1 \dots p_f, t),$$

writing

$$W(q_1 \dots q_f, t) = \psi^*(q_1 \dots q_f, t)\psi(q_1 \dots q_f, t)$$

and

$$W(p_1 \dots p_f, t) = \phi^*(p_1 \dots p_f, t)\phi(p_1 \dots p_f, t), \tag{53.2}$$

where complex conjugates are denoted by the use of asterisks. We may give the name *probability amplitudes* to the newly introduced quantities ψ and ϕ for reasons which will appear later (§§ 59, 61). In general they will actually turn out to be complex quantities as indicated by our notation.†

Our first fundamental postulate may now be regarded as definitely asserting the existence of such probability densities $W(q,t)$ and $W(p,t)$ with the physical significance which we have attached to them, and as carrying the suggestion that an appropriate apparatus for their calculation will be found by relating them, as we have done in equations (53.2), to the corresponding probability amplitudes $\psi(q,t)$ and $\phi(p,t)$.

We must now examine some of the explicit or tacit implications of the postulatory material which we have thus introduced. We may begin by considering the expressions given by (53.1) for the probabilities of finding values for the coordinates and momenta of a system that lie in specified ranges.

Here it is to be emphasized first of all that the probability densities $W(q,t)$ and $W(p,t)$ are to be thought of as quantities whose values could actually be empirically determined by setting up the same physical situation over and over again and finding the frequency with which the coordinates *or* momenta do fall at the time of interest t in different ranges $dq_1 \dots dq_f$ and $dp_1 \dots dp_f$. This implies in the first place that we are going to retain the general classical idea of the existence of coordinates and momenta whose values can be measured by recognized classical methods; and implies in the second place that the results can be regarded as applying at a perfectly definite time t. The first of these implications means that at least *some* classical methods of measurement are going to remain appropriate in the quantum mechanics; and the

† Two remarks as to notation may be made at this point. For the purpose of physics it is convenient to denote complex conjugates by an asterisk *, rather than by a bar $^-$ as is done in pure mathematics, since the bar can then still be used to denote mean values as is customary in physics. In expressing the functional dependence of probability densities and amplitudes on the coordinates and momenta, it will often be convenient to denote the collections of quantities $q_1 \dots q_f$ and $p_1 \dots p_f$ by the single symbols q and p, as is done in the next paragraph.

second implication means that probabilities can be specified in our present non-relativistic quantum mechanics *without* introducing any range dt in the time similar to our ranges $dq_1 \ldots dq_f$ and $dp_1 \ldots dp_f$ for the coordinates and momenta.

The existence of the probability densities $W(q,t)$ and $W(p,t)$ can also be regarded as implying that there is no theoretical limitation—which we need now consider—on the accuracy with which all of the coordinates *or* all of the momenta of a system could be determined. On the other hand, the introduction of separate probability densities for the coordinates and momenta does suggest that the question of limits of accuracy in the simultaneous determination of coordinates *and* of momenta is to be left open for later examination. The future development of physical theory may also make it necessary to consider the limits which would be imposed—even in the separate measurement of coordinates or momenta for themselves—by the necessary atomic nature of our measuring instruments. But in the present theory we have to neglect the possible effect of such considerations.†

The fact that the introduction of the probability densities $W(q,t)$ and $W(p,t)$ has implied the general classical idea of coordinates and momenta as observable quantities should not be interpreted as limiting the quantum mechanics solely to the treatment of observables which have a classical analogue. Indeed, when we come to a consideration of the spin of the electron or of other fundamental particles, we shall be led to introduce observables which have no classical analogue, and shall then make use of probability densities and amplitudes which are functions of non-classical spin variables as well as of the classical coordinates and momenta of the particle. Until we come to that extension, however, our statements will not be worded so as to give explicit recognition to the possibility of observables having no classical analogue.

As a necessary consequence of the significance ascribed to the probability densities, it is evident, since the coordinates and momenta must have unit chance of falling somewhere, that the following equations must be true at all times:

$$\int \ldots \int W(q,t)\, dq_1 \ldots dq_f = \int \ldots \int \psi^*(q,t)\psi(q,t)\, dq_1 \ldots dq_f = 1$$

$$\text{and} \quad \int \ldots \int W(p,t)\, dp_1 \ldots dp_f = \int \ldots \int \phi^*(p,t)\phi(p,t)\, dp_1 \ldots dp_f = 1, \qquad (53.3)$$

† See Ruark, *Proc. Nat. Acad.* **14**, 322 (1928); and Landau and Peierls, *Zeits. f. Phys.* **69**, 56 (1931), concerning limitations on the measurement of position alone. Limitations on the possibility of locating an electron within a region of the order of the Compton wave-length have since been found related to the production of pairs of positive and negative electrons.

where the integrations are to be taken over the whole range of possible values for the coordinates or momenta, and where (53.3) implies that the integrals must converge.

Turning now to the quantities on the right-hand side of equations (53.2), the probability amplitudes ψ and ϕ will in general actually turn out to be complex numbers consisting of a real and imaginary part, and are not themselves measurable but are to be regarded as summarizing the directly observable properties of the system. For example, the squares of their absolute magnitudes are real quantities equal to the probability densities which can be empirically observed. This is in agreement with the idea that the equations of mathematical physics are to provide a formalism for computation which leads to results capable of empirical determination even though the formalism itself contains symbols which have no direct reference to observable physical quantities.†

The justification for the introduction of these complex quantities, the probability amplitudes, whose squares are to give the observable probability densities, will, of course, only appear in the sequel as the usefulness and validity of the whole formalism becomes apparent. We shall actually find that these probability amplitudes play a primary role in the quantum mechanics, since we shall see later that all we can know about a system at any time of interest will be determined by giving the instantaneous dependence of such a quantity on the variables of which it is a function; for example, by giving the dependence of ψ on $q_1 \ldots q_f$ at the time of interest t. For this reason we shall speak of the probability amplitudes for a system—such as ψ, ϕ, or others which prove possible—as being quantities by which we can specify the *quantum mechanical state of a system*.

It may be remarked in passing that the attempt to specify the state of a system by using a single *real* quantity as a probability amplitude would not be sufficient to determine the further behaviour of the probability field. Just as a dual specification of coordinates and velocities is necessary to determine the state of a system in the classical mechanics, so, too, a dual specification corresponding to the two numbers specifying our *complex* probability amplitudes is necessary to determine the quantum mechanical state of a system.

† This is contrary to the apparently pleasing but somewhat unfortunate statement that the equations of mathematical physics should contain only quantities which are susceptible of direct measurement.

54. The interrelation of probability amplitudes

In the preceding section we have introduced both the probability amplitude $\psi(q_1 \dots q_f, t)$, which determines the chance of finding the coordinates in a given range, and the analogous probability amplitude $\phi(p_1 \dots p_f, t)$ for the momenta. These quantities, however, are not independent but are related to each other, so that either can be calculated from a knowledge of the other. As our second postulate for the quantum mechanics, we must now give a statement of the mutual interdependence of these quantities.

Since the explicit form of the relation connecting the values of $\phi(p,t)$ with those of $\psi(q,t)$ is not entirely independent of the kind of coordinates and momenta which are being employed, we shall state the relation in the form for so-called *canonical quantum mechanical coordinates and momenta*. In the case of a mechanical system composed of particles, the ordinary Cartesian coordinates and corresponding momenta of the individual particles furnish such a canonical set, and, starting with these, a transformation to other kinds of coordinates and momenta can be undertaken whenever desirable. Coordinates and momenta, which are canonical in the quantum mechanical sense, always have the important property of being physically interpretable over the whole range of possible values from minus to plus infinity.

Using such canonical coordinates and momenta, the relation of the probability amplitudes ϕ and ψ can now be simply expressed with the help of a Fourier integral by the equation

$$\phi(p_1 \dots p_f, t) = h^{-\frac{1}{2}f} \int \dots \int \psi(q_1 \dots q_f, t) e^{-\frac{2\pi i}{h}(p_1 q_1 + \dots + p_f q_f)}\, dq_1 \dots dq_f, \tag{54.1}$$

where h is Planck's constant, i is the symbol for the imaginary root $\sqrt{(-1)}$, and the integrations are to be taken over the total range from $-\infty$ to $+\infty$ for which the coordinates q have significance.

By the ordinary rules for the inversion of Fourier integrals (see Appendix II c) this equation can be solved in a form to permit the calculation of ψ as a function of the q's for any given distribution of ϕ as a function of the p's. Furthermore, we can readily write down the analogous relations for the complex conjugate quantities ψ^* and ϕ^*. Introducing some obvious abbreviations, we can then rewrite equation (54.1), together with the three further equations which it implies, in the following forms which will be useful for future reference:

$$
\begin{aligned}
\phi(p,t) &= h^{-\frac{1}{2}f} \int \psi(q,t) e^{-\frac{2\pi i}{h}(p_1q_1+\dots+p_fq_f)}\,dq, \\
\psi(q,t) &= h^{-\frac{1}{2}f} \int \phi(p,t) e^{+\frac{2\pi i}{h}(p_1q_1+\dots+p_fq_f)}\,dp, \\
\phi^*(p,t) &= h^{-\frac{1}{2}f} \int \psi^*(q,t) e^{+\frac{2\pi i}{h}(p_1q_1+\dots+p_fq_f)}\,dq, \\
\psi^*(q,t) &= h^{-\frac{1}{2}f} \int \phi^*(p,t) e^{-\frac{2\pi i}{h}(p_1q_1+\dots+p_fq_f)}\,dp.
\end{aligned}
\tag{54.2}
$$

This part of our postulatory basis is of great significance for quantum mechanics. It shows that the specification at a given time t *either* of the probability amplitude ψ as a function of the coordinates q, *or* of the probability amplitude ϕ as a function of the momenta p, is alone sufficient to determine *both* the probability density $\psi^*\psi$ which permits us to calculate the chances for finding given values of the coordinates, *and* the probability density $\phi^*\phi$ which permits us to calculate the chances for finding given values of the momenta, and thus also the chances for finding given values of various other dynamical quantities which are functions of the coordinates and momenta. Indeed, in the quantum mechanics we regard the instantaneous *state of a system* as fully given when we have specified the dependence of a single suitable probability amplitude—such as ψ, ϕ, or others which prove possible—on the variables of which it is a function.

This possibility of describing the state of a system in alternative ways by using one or another of various possible probability amplitudes is a very characteristic feature of the quantum mechanics. The different kinds of probability amplitudes are themselves functions of different sets of variables, ψ and ϕ, for example, being functions of the q's and p's respectively. Hence it is sometimes convenient to say that we are using a particular kind of language, for example the q-language or the p-language, to describe the state of our system; or we can also say that we are using a q-representation or a p-representation of that state. Equations (54.2) give us the necessary apparatus for transforming from the q-language to the p-language and vice versa, and similar equations become available for transforming to other kinds of language. Factors such as the quantities

$$
h^{-\frac{1}{2}f} e^{-\frac{2\pi i}{h}(p_1q_1+\dots+p_fq_f)} \quad \text{and} \quad h^{-\frac{1}{2}f} e^{+\frac{2\pi i}{h}(p_1q_1+\dots+p_fq_f)}
$$

occurring in equations (54.2), which are used in transforming from one language to another, are known as *transformation functions* for the sets of variables involved.

Our postulatory basis for transforming between the q-language and p-language as given by equations (54.2) was set up for the case of canonical coordinates and momenta whose range of significant values was assumed to go from minus infinity to plus infinity. If we desire to use coordinates and momenta for which this is not true, for example to use polar coordinates and their corresponding momenta, it is best to set the particular problem up in Cartesian coordinates and carry out a later translation to the final coordinates and momenta desired. The consideration of transformations to other than q- and p-representations and of transformations when non-classical variables are involved will have to be postponed.

55. The operators corresponding to observable quantities and their use in calculating expectation values

(*a*) **Preliminary discussion.** In accordance with the foregoing, the state of a system at any time of interest can be specified by giving the instantaneous dependence of a suitable probability amplitude on the variables of which it is a function. For this purpose we may use either the probability amplitude $\psi(q_1 \ldots q_f, t)$, which is a function of the f coordinates of the system, or the probability amplitude $\phi(p_1 \ldots p_f, t)$, which is a function of the corresponding f momenta; and making use of equations (54.2) we can transform the specification of state from one of these sets of variables to the other; i.e. translate between the q- and p-languages. We already appreciate that such specifications of state, with the help of the relation between probability amplitudes and probability densities given by (53.2), make it possible to calculate the probabilities of finding different values of the coordinates and momenta when a system in some specified state of interest is examined. In the present section we now wish to discuss the possibility of using such specifications of state to calculate the *mean value*, or so-called *expectation value*, which would be found for any function $F(q,p)$ of the coordinates and momenta, as the average result of a series of measurements made on that quantity when the system is in a specified state of interest.

In the case of quantities which are functions solely of the coordinates q or solely of the momenta p of the system, the calculations of such mean values can be carried out without further addition to our postulatory basis. In the case of a quantity $F(q_1 \ldots q_f)$ which depends solely on the coordinates of the system, the calculation can be made most simply with the help of a specification of state in the q-language. The probability of finding the coordinates of the system in any specified range $dq_1 \dots dq_f$ will then be given, in accordance with (53.2), by $\psi^*(q,t)\psi(q,t)\,dq_1 \dots dq_f$, and the mean value of $F(q)$ at the time t will hence be given by

$$\bar{F}(q,t) = \int \dots \int \psi^*(q,t)F(q)\psi(q,t)\,dq_1 \dots dq_f, \tag{55.1}$$

where the integration is over all possible values of the coordinates $q_1 \dots q_f$, and the reason for choosing the order of writing with $F(q)$ placed between ψ^* and ψ will be seen later. Similarly, in the case of a quantity $F(p)$, which depends solely on the momenta, the mean value can be simply expressed in the p-language by the equation

$$\bar{F}(p,t) = \int \dots \int \phi^*(p,t)F(p)\phi(p,t)\,dp_1 \dots dp_f, \tag{55.2}$$

where we now take an integration over all possible values of the momenta.

We thus see that the calculation of the expectation value for a function $F(q)$ of the coordinates will be straightforward in the q-language, and the similar calculation for a function $F(p)$ of the momenta will be straightforward in the p-language. Nevertheless, it will also be possible, in the case of functions which can be expressed as polynomials in the q's or the p's, to carry out the calculations with the other choice of language. This we must now investigate, since it will lead us to the important idea of the operators which correspond to observable quantities in the quantum mechanics and which make it possible to express the expectation values for such quantities in any desired quantum mechanical language.

As a simple illustration, let us investigate the possibility of expressing the mean value of one of the momenta for the system, say p_k, in terms of the q-language. For this purpose we can start out by using the p-language and write

$$\bar{p}_k = \int \dots \int \phi^* p_k \phi\,dp_1 \dots dp_f \tag{55.3}$$

as an evident expression for the mean value of p_k in terms of the momentum probability amplitudes $\phi^*(p_1 \dots p_f,t)$ and $\phi(p_1 \dots p_f,t)$, which correspond to the state of the system at any time t. This, however, can be re-expressed in the q-language with the help of our fundamental postulated relation

$$\phi(p_1 \dots p_f,t) = h^{-\frac{1}{2}f} \int \dots \int \psi(q_1 \dots q_f,t)e^{-\frac{2\pi i}{h}(p_1q_1+\dots+p_fq_f)}\,dq_1 \dots dq_f, \tag{55.4}$$

which connects the two kinds of probability amplitudes. Substituting (55.4) in (55.3), we obtain

$$\bar{p}_k = h^{-\frac{1}{2}f} \int \dots \int \int \dots \int \phi^* p_k \psi e^{-\frac{2\pi i}{h}(p_1 q_1 + \dots + p_f q_f)} \, dq_1 \dots dq_f \, dp_1 \dots dp_f.$$

Introducing a differentiation with respect to the coordinate q_k that corresponds to the momentum p_k of interest, this can be rewritten in the form

$$\bar{p}_k = h^{-\frac{1}{2}f} \int \dots \int \int \dots \int \phi^* \psi \left(\frac{-h}{2\pi i}\right) \frac{\partial}{\partial q_k} e^{-\frac{2\pi i}{h}(p_1 q_1 + \dots + p_f q_f)} \, dq_1 \dots dq_f \, dp_1 \dots dp_f.$$

And carrying out a partial integration with respect to q_k, this then gives us

$$\bar{p}_k = h^{-\frac{1}{2}f} \int \dots \int \int \dots \int \phi^* \frac{h}{2\pi i} \frac{\partial \psi}{\partial q_k} e^{-\frac{2\pi i}{h}(p_1 q_1 + \dots + p_f q_f)} \, dq_1 \dots dq_f \, dp_1 \dots dp_f, \tag{55.5}$$

since we take $\psi(q, t)$ as equal to zero at the limits of integration $q_k = \pm\infty$ owing to our present interest in systems where the probability of finding infinite values of the coordinates is vanishingly small. As our final step we may now again introduce the connecting relation between the two languages, this time in the form

$$\psi^*(q_1 \dots q_f, t) = h^{-\frac{1}{2}f} \int \dots \int \phi^*(p_1 \dots p_f, t) e^{-\frac{2\pi i}{h}(p_1 q_1 + \dots + p_f q_f)} \, dp_1 \dots dp_f, \tag{55.6}$$

which can be obtained from the original form (55.4), as we have already seen in the preceding section. Substituting (55.6) in (55.5), we then obtain the desired result

$$\bar{p}_k = \int \dots \int \psi^* \frac{h}{2\pi i} \frac{\partial \psi}{\partial q_k} dq_1 \dots dq_f, \tag{55.7}$$

which provides an expression for the mean value of the momentum p_k in terms of the coordinate probability amplitudes $\psi^*(q, t)$ and $\psi(q, t)$.

Comparing (55.3) with (55.7), we now see that the two expressions for the mean value of p_k,

$$\bar{p}_k = \int \dots \int \phi^* p_k \phi \, dp_1 \dots dp_f = \int \dots \int \psi^* \frac{h}{2\pi i} \frac{\partial}{\partial q_k} \psi \, dq_1 \dots dq_f, \tag{55.8}$$

are actually quite similar in form in the two different languages. In the p-language we insert the quantity p_k itself between the two functions $\phi^*(p, t)$ and $\phi(p, t)$, which describe the state of the system in that language, and then integrate over all values of their arguments. In the q-language we insert instead the differential operator $\frac{h}{2\pi i} \frac{\partial}{\partial q_k}$ between the two functions $\psi^*(q, t)$ and $\psi(q, t)$, which now describe the state of the system, and again integrate over all values of their arguments. Moreover, this result can be immediately generalized, since by a simple extension of the above calculation we readily find that the mean value of any rational product of the momenta of the form $p_1^\alpha p_2^\beta \dots p_f^\mu$ can be calculated in the q-language with the help of the corresponding operator

$$\left(\frac{h}{2\pi i}\frac{\partial}{\partial q_1}\right)^\alpha \left(\frac{h}{2\pi i}\frac{\partial}{\partial q_2}\right)^\beta \cdots \left(\frac{h}{2\pi i}\frac{\partial}{\partial q_f}\right)^\mu,$$

where the exponents indicate the number of times the differential operator is to be applied in succession to $\psi(q,t)$. Similarly, by the analogous calculation, we also find that the mean value of any rational product of the coordinates of the form $q_1^\alpha q_2^\beta \dots q_f^\mu$ can be calculated in the p-language with the help of the corresponding operator

$$\left(-\frac{h}{2\pi i}\frac{\partial}{\partial p_1}\right)^\alpha \left(-\frac{h}{2\pi i}\frac{\partial}{\partial p_2}\right)^\beta \cdots \left(-\frac{h}{2\pi i}\frac{\partial}{\partial p_f}\right)^\mu.$$

Hence it now proves possible to calculate the expectation values for any functions $F(q_1 \dots q_f)$ or $F(p_1 \dots p_f)$, which are *polynomials* in the q's and p's alone, with the help of expressions of the form

$$\begin{aligned} \bar{F}(q) &= \int \dots \int \psi^* F(q_1 \dots q_f)\psi \, dq_1 \dots dq_f \\ &= \int \dots \int \phi^* F\left(-\frac{h}{2\pi i}\frac{\partial}{\partial p_1} \dots -\frac{h}{2\pi i}\frac{\partial}{\partial p_f}\right)\phi \, dp_1 \dots dp_f \end{aligned}$$

and

$$\begin{aligned} \bar{F}(p) &= \int \dots \int \psi^* F\left(\frac{h}{2\pi i}\frac{\partial}{\partial q_1} \dots \frac{h}{2\pi i}\frac{\partial}{\partial q_f}\right)\psi \, dq_1 \dots dq_f \\ &= \int \dots \int \phi^* F(p_1 \dots p_f)\phi \, dp_1 \dots dp_f. \end{aligned} \quad (55.9)$$

In accordance with the above findings, it now becomes convenient in the quantum mechanics to regard any such *observable quantity* F as correlated in each language with a *corresponding operator* $\mathbf{F}$, which can then be used to calculate the *expectation value* $\bar{F}$ of that observable quantity by substitution in an expression of the general form

$$\bar{F} = \int \psi^*(x)\mathbf{F}\psi(x) \, dx, \quad (55.10)$$

where the probability amplitudes ψ^* and ψ and the operator $\mathbf{F}$ are all to be expressed in the same language, and the integration is to be taken over all values of the variables x which are characteristic of that language. As indicated above, the operator corresponding to any observable quantity can be conveniently denoted by the same letter printed in Clarendon or bold-face type, and there will be little danger of confusing this with an occasional use of bold-face type to designate vector quantities. The properties which we have already found for such operators may be tabulated in the form

q-language	p-language
$\mathbf{q}_k = q_k$	$\mathbf{p}_k = p_k$
$\mathbf{p}_k = \frac{h}{2\pi i}\frac{\partial}{\partial q_k}$	$\mathbf{q}_k = -\frac{h}{2\pi i}\frac{\partial}{\partial p_k}$
$\mathbf{F}(\mathbf{q}_1 \dots \mathbf{q}_f) = F(q_1 \dots q_f)$	$\mathbf{F}(\mathbf{p}_1 \dots \mathbf{p}_f) = F(p_1 \dots p_f)$
$\mathbf{F}(\mathbf{p}_1 \dots \mathbf{p}_f) = F\left(\frac{h}{2\pi i}\frac{\partial}{\partial q_1} \dots \frac{h}{2\pi i}\frac{\partial}{\partial q_f}\right)$	$\mathbf{F}(\mathbf{q}_1 \dots \mathbf{q}_f) = F\left(-\frac{h}{2\pi i}\frac{\partial}{\partial p_1} \dots -\frac{h}{2\pi i}\frac{\partial}{\partial p_f}\right)$

(55.11)

where in both columns of the table the last form can only be applied in the case of quantities which can be expressed as polynomials in the p's or q's respectively.

As was intimated earlier, it will be seen that we have been able to obtain the foregoing special results as to the expectation values and operators for quantities which are functions either of the q's or p's alone, without additions to our previous postulatory basis. It is evident, however, that some additions may well be necessary in the case of quantities containing both kinds of variables, since the foregoing treatments were all based on the possibility of starting out with expressions which contained only a single kind of variable. We must hence turn to a consideration of the problem of constructing operators corresponding to quantities that depend both on the coordinates and momenta, and shall find it desirable to preface the treatment with a somewhat general account of the properties of the kinds of operators that will interest us.

(*b*) **Operator manipulation.** When a function $u(x_1 \dots x_n)$ of certain variables $x_1 \dots x_n$ is changed into another function v of the same or other variables by the application of a definite rule denoted by $\mathbf{F}$, we can regard the process as an *operation*, symbolized by the equation

$$v = \mathbf{F}u, \tag{55.12}$$

where v may be called the *result* produced by the action of the *operator* $\mathbf{F}$ on the *operand* u.

For our present purposes we shall be interested in operations which leave the result v a function of the same variables $x_1 \dots x_n$ as appeared in the operand u. This limitation arises from the fact that operators corresponding to observable quantities, when a particular language is being employed, will contain no variables not already appearing in the probability amplitudes, such as $\psi(q,t)$, on which they operate. In general, however, the quantum mechanics can also be interested in operations which lead to functions of new variables, as illustrated by equations (54.2) for transforming from one language to another. The immediately following remarks will be valid for either kind of operation.

The *sum* of two operators

$$\mathbf{S} = \mathbf{F}+\mathbf{G} \tag{55.13}$$

may be defined with the help of the equation

$$\mathbf{S}u = (\mathbf{F}+\mathbf{G})u = \mathbf{F}u+\mathbf{G}u \tag{55.14}$$

as the operator which will give the sum of the results of the two individual operators.

The *product* of two operators

$$\mathbf{P} = \mathbf{FG} \tag{55.15}$$

may be defined with the help of the equation

$$\mathbf{P}u = \mathbf{FG}u = \mathbf{F}(\mathbf{G}u) \tag{55.16}$$

as the operator which will give the result obtained by operating with the first of the two operators on the result which is itself obtained from the operation of the second of the two operators. Since the order of application can make a difference in the final result, operators are in general *non-commutative*, i.e.

$$\mathbf{FG} \neq \mathbf{GF}. \tag{55.17}$$

However, many pairs of operators are, of course, *commutative*.

The successive application of the same operator can be indicated with the help of exponents obeying the *usual law of indices*

$$\mathbf{F}^{n+m} = \mathbf{F}^n\mathbf{F}^m. \tag{55.18}$$

If the successive application of the two different operators $\mathbf{F}$ and $\mathbf{F}^{-1}$ produces no change in the operand, they are called the *reciprocals* of each other. The situation may be symbolized by the equation

$$\mathbf{FF}^{-1} = \mathbf{F}^{-1}\mathbf{F} = \mathbf{I}, \tag{55.19}$$

where $\mathbf{I}$ may be called the *identity operator*.

As will be seen from the above, relations between operators can often be conveniently symbolized by equations from which the operands have been omitted. It will be realized, however, that a particular class of operands may have to be kept in mind when operator equations are stated without the insertion of a specific operand.

(*c*) **Linear operators.** For the purposes of the quantum mechanics we shall be specially interested in so-called *linear* or *distributive* operators. These have the property, when applied to the sum of more than one operand, of giving the sum of the results obtained by application to each operand separately, in accordance with the equation

$$\mathbf{F}(u_1+u_2) = \mathbf{F}u_1+\mathbf{F}u_2. \tag{55.20}$$

As an additional property of linear operators we shall also require them to satisfy the relation

$$\mathbf{F}(cu) = c\mathbf{F}u, \tag{55.21}$$

where c is any constant. For the case when c is an integer this is a simple consequence of (55.20), but for linear operators it is a required condition when c is any constant, real or complex.

If $\mathbf{F}$ and $\mathbf{G}$ are linear operators it is readily seen that sums and products of the form

$$\begin{aligned} \mathbf{S} &= c_1\,\mathbf{F}+c_2\,\mathbf{G}, \\ \mathbf{P} &= c_3\,\mathbf{FG} \end{aligned} \tag{55.22}$$

are also linear, where c_1, c_2, and c_3 are arbitrary constants.

A special significance of linear operators for the quantum mechanics arises, as will be seen in § 59, from an important additive property possessed by the solutions of equations involving such operators. Consider the equation

$$\mathbf{F}u(x_1 \ldots x_n) = 0, \tag{55.23}$$

where $\mathbf{F}$ is a linear operator involving the same variables $x_1 \ldots x_n$ as the operand u. And let

$$u_1 = u_1(x_1 \ldots x_n),$$
$$u_2 = u_2(x_1 \ldots x_n)$$

be any two solutions for the above equation, giving the explicit dependence of u_1 and u_2 on the variables $x_1 \ldots x_n$. Since these are both solutions, we shall have

$$\mathbf{F}u_1 = 0$$

and

$$\mathbf{F}u_2 = 0,$$

and, in accordance with (55.20) and (55.21), we shall then also have

$$\mathbf{F}(c_1\,u_1+c_2\,u_2) = 0. \tag{55.24}$$

Hence, in the case of linear operators, any *linear combination* of individual solutions for an equation such as (55.23) will also be a solution thereof.

The quantum mechanical operators corresponding to observable quantities will all be linear.

(*d*) **Hermitian operators.** In the quantum mechanics we shall be specially interested in operators which are not only linear but also

Hermitian in character as well. This further property of operators must be defined with reference to the class of functions, $u(x_1 \ldots x_n)$, on which they are to operate, and the range of values for the arguments, $x_1 \ldots x_n$, of these functions which is to be considered. If $u_1(x_1 \ldots x_n)$ and $u_2(x_1 \ldots x_n)$ are any two of the class of operands to be considered, a linear Hermitian operator **H** is one satisfying the relation

$$\int \ldots \int u_1^*(x_1 \ldots x_n)\mathbf{H}u_2(x_1 \ldots x_n)\, dx_1 \ldots dx_n$$
$$= \int \ldots \int u_2(x_1 \ldots x_n)[\mathbf{H}u_1(x_1 \ldots x_n)]^*\, dx_1 \ldots dx_n,$$

where the formation of a complex conjugate is indicated by an asterisk, and the integrations are to be taken over the total range of the variables $x_1 \ldots x_n$ to which significance is ascribed. If, for example, the functions u were probability amplitudes depending on the Cartesian coordinates or momenta of one or more particles, the range of integrations would be taken from minus to plus infinity, and the class of functions considered would be those which are free from non-integrable singularities and which vanish sufficiently rapidly as the limits of integration are approached. Introducing some obvious abbreviations, the condition for linear Hermitian operators given above can be written in the simplified form

$$\int u_1^*\mathbf{H}u_2\, dx = \int u_2(\mathbf{H}u_1)^*\, dx. \tag{55.25}$$

We must now investigate the possibility of combining linear Hermitian operators to give further operators which are themselves also linear and Hermitian. It is evident that the combination of linear Hermitian operators by addition will lead to operators which retain the desired character. For the case of multiplication, however, a special investigation is necessary.

Let **F** and **G** be two linear Hermitian operators suitable for use with a class of operands $u(x_1 \ldots x_n)$, and let their action on any member of the class lead—as is the customary case of interest—to results which are themselves members of the class. In accordance with (55.25) we can then write

$$\int (\mathbf{F}u_1)^*\mathbf{G}u_2\, dx = \int u_2(\mathbf{G}\mathbf{F}u_1)^*\, dx,$$

and

$$\int (\mathbf{G}u_2)^*\mathbf{F}u_1\, dx = \int u_1(\mathbf{F}\mathbf{G}u_2)^*\, dx.$$

It will be noticed, however, that the left-hand sides of these two equations have been chosen so as to be the complex conjugates of each other. We may then equate the right-hand side of one of these equations to the complex conjugate of the right-hand side of the other, which will give us the general relation

$$\int u_1^* \mathbf{FG} u_2 \, dx = \int u_2 (\mathbf{GF} u_1)^* \, dx. \tag{55.26}$$

As another example of this general relation we can also write

$$\int u_1^* \mathbf{GF} u_2 \, dx = \int u_2 (\mathbf{FG} u_1)^* \, dx. \tag{55.27}$$

We see at once from the above that the product of two Hermitian operators is not itself in general Hermitian unless the two operators commute.

Adding the two equations (55.26) and (55.27), however, we obtain

$$\int u_1^* (\mathbf{FG}+\mathbf{GF}) u_2 \, dx = \int u_2 [(\mathbf{FG}+\mathbf{GF}) u_1]^* \, dx. \tag{55.28}$$

Hence, comparing with equation (55.25), we now see that in general the *symmetrized product* of two linear Hermitian operators,

$$\mathbf{H} = \mathbf{FG}+\mathbf{GF}, \tag{55.29}$$

will itself be linear and Hermitian.

Furthermore, subtracting equation (55.27) from (55.26), we obtain

$$\int u_1^* (\mathbf{FG}-\mathbf{GF}) u_2 \, dx = \int u_2 [(\mathbf{GF}-\mathbf{FG}) u_1]^* \, dx,$$

and, multiplying by i, this can evidently also be written in the form

$$\int u_1^* i(\mathbf{FG}-\mathbf{GF}) u_2 \, dx = \int u_2 [i(\mathbf{FG}-\mathbf{GF}) u_1]^* \, dx, \tag{55.30}$$

so that we can also combine Hermitian operators to form another operator which is Hermitian by the rule

$$\mathbf{H} = i(\mathbf{FG}-\mathbf{GF}). \tag{55.31}$$

In case the two operators $\mathbf{F}$ and $\mathbf{G}$ *commute* with each other, we see from (55.29) that their simple product $\mathbf{FG}$ would be Hermitian without the necessity for symmetrization, and that the Hermitian operator given by (55.31) would reduce to zero. Since any operator commutes with itself, any polynomial constructed from an Hermitian operator will be Hermitian.

The expression $(\mathbf{FG}-\mathbf{GF})$ occurring in (55.31) is often called the *commutator* for the two operators $\mathbf{F}$ and $\mathbf{G}$. Similarly, the expression $(\mathbf{FG}+\mathbf{GF})$ is sometimes called the *anticommutator* for the two operators.

The commutator for two operators is so important that it will be convenient to introduce the abbreviation

$$[\mathbf{F}, \mathbf{G}] = \mathbf{FG}-\mathbf{GF}. \tag{55.32}$$

The following properties of the commutator of two operators can be readily verified. They apply to linear operators in general:

$$\begin{aligned} [\mathbf{F},\mathbf{G}] &= -[\mathbf{G},\mathbf{F}], \\ [\mathbf{F},(\mathbf{G}+\mathbf{H})] &= [\mathbf{F},\mathbf{G}]+[\mathbf{F},\mathbf{H}], \\ [\mathbf{F},\mathbf{GH}] &= [\mathbf{F},\mathbf{G}]\mathbf{H}+\mathbf{G}[\mathbf{F},\mathbf{H}], \\ [\mathbf{GH},\mathbf{F}] &= [\mathbf{G},\mathbf{F}]\mathbf{H}+\mathbf{G}[\mathbf{H},\mathbf{F}]. \end{aligned} \tag{55.33}$$

In the case of two Hermitian operators which do not commute, it is sometimes desirable to re-express their product as a sum containing the commutator and anticommutator as follows:

$$\begin{aligned} \mathbf{FG} &= \frac{\mathbf{FG}+\mathbf{GF}}{2}+\frac{\mathbf{FG}-\mathbf{GF}}{2} \\ &= \frac{\mathbf{FG}+\mathbf{GF}}{2}+\frac{i(\mathbf{FG}-\mathbf{GF})}{2i}. \end{aligned} \tag{55.34}$$

Noting (55.29) and (55.31), this re-expression may be regarded as a decomposition into a real and imaginary Hermitian part.

(*e*) **The operators q and p.** With the help of the foregoing we are now ready to discuss the nature of the operators to be associated in the quantum mechanics with observable quantities.

Making use of the q-language, let us first consider the fundamental operators corresponding to the canonical coordinates and momenta q_k and p_k. These will then be given, as we have already seen, by the expressions

$$\begin{aligned} \mathbf{q}_k &= q_k, \\ \mathbf{p}_k &= \frac{h}{2\pi i}\frac{\partial}{\partial q_k}, \end{aligned} \tag{55.35}$$

where the q_k will have a range of significant values from $-\infty$ to $+\infty$. They will be used to operate on probability amplitudes $\psi(q_1 \dots q_f, t)$ which are functions of the coordinates $q_1 \dots q_f$ and also of the time t. Nevertheless, since we shall actually be interested in the application of these operators at some particular time, say t_0, it will be more convenient to write

$$u(q_1 \dots q_f) = \psi(q_1 \dots q_f, t_0) \tag{55.36}$$

as a general expression for the kind of functions on which the above operators are to act, where t_0 is to be regarded as a parameter whose choice will determine the form of function to be considered.

First of all, it will be important to investigate the *commutation properties* of our fundamental operators. Making use of the explicit form given by (55.35) we can write

$$(\mathbf{p}_k\mathbf{q}_k-\mathbf{q}_k\mathbf{p}_k)u = \frac{h}{2\pi i}\frac{\partial}{\partial q_k}(q_k u)-\frac{h}{2\pi i}q_k\frac{\partial u}{\partial q_k}$$

$$=\frac{h}{2\pi i}u. \tag{55.37}$$

We hence see that the pair of operators corresponding to a coordinate and its conjugated momentum do not commute. On the other hand, we easily see by a similar treatment that the operators for two coordinates, or for two momenta, or for a coordinate and a momentum that are not conjugated do commute. Hence we may now summarize the commutation properties for our fundamental operators by the equations

$$\mathbf{p}_k\mathbf{q}_l-\mathbf{q}_l\mathbf{p}_k = \frac{h}{2\pi i}\delta_{kl},$$

$$\mathbf{q}_k\mathbf{q}_l-\mathbf{q}_l\mathbf{q}_k = 0, \tag{55.38}$$

$$\mathbf{p}_k\mathbf{p}_l-\mathbf{p}_l\mathbf{p}_k = 0,$$

where δ_{kl} is equal to unity or zero according as k and l are the same or different. These commutation properties, which can be shown to be preserved in all forms of representation as well as in the q-language, are fundamental for the quantum mechanics. The lack of commutability for the quantum mechanical operators $\mathbf{q}_k$ and $\mathbf{p}_k$ compared with the commutability of the classical variables q_k and p_k can be regarded as an expression of one of the fundamental differences between quantum mechanics and classical mechanics.

As a second important property of the fundamental operators for coordinates and momenta, it will immediately be seen from the form given by (55.35) that these operators are *linear* in character. Hence also, in accordance with (55.22), any polynomial constructed from these operators will also be linear.

As a third important property, it will also be seen that these operators are *Hermitian*. This is immediately evident in the q-language for the operator $\mathbf{q}_k$ since we shall obviously have

$$\int\dots\int u_1^*\mathbf{q}_k u_2\,dq_1\dots dq_f = \int\dots\int u_2(\mathbf{q}_k u_1)^*\,dq_1\dots dq_f, \tag{55.39}$$

owing to the fact that the operator $\mathbf{q}_k$ signifies multiplication by the *real* variable q_k and the order of multiplication makes no difference. To show the Hermitian character of $\mathbf{p}_k$ in the q-language is a little more complicated. We can see this, nevertheless, from the following equations:

$$\begin{aligned}\int \cdots \int u_1^* \mathrm{p}_k u_2 \, dq_1 \ldots dq_f \\ &= \int \cdots \int u_1^* \frac{h}{2\pi i} \frac{\partial u_2}{\partial q_k} dq_1 \ldots dq_f \\ &= \int \cdots \int \left| u_1^* \frac{h}{2\pi i} u_2 \right|_{q_k=-\infty}^{q_k=+\infty} dq_1 \ldots dq_f - \int \cdots \int u_2 \frac{h}{2\pi i} \frac{\partial u_1^*}{\partial q_k} dq_1 \ldots dq_f \\ &= \int \cdots \int u_2 \left(-\frac{h}{2\pi i} \frac{\partial u_1^*}{\partial q_k} \right) dq_1 \ldots dq_f \\ &= \int \cdots \int u_2 (\mathrm{p}_k u_1)^* \, dq_1 \ldots dq_f, \end{aligned} \tag{55.40}$$

where the first form of the equality is obtained by substituting the explicit expression for the operator in the q-language; the second form is obtained by partial integration with respect to q_k; the third form is obtained by dropping the integrated term, since in problems of interest probability amplitudes such as u_1 and u_2 are taken as going to zero at the limits $q_k = \pm\infty$ in order that there should be no probability of finding infinite values of the coordinates; and the fourth form of the equality is obtained with the help of the explicit form for the operator p_k and the rule for finding the complex conjugate of a function.

This investigation of the commutability, linearity, and Hermitian character of the fundamental operators q_k and p_k was actually carried out in the q-language, making use of the specific forms that these operators have in that particular language. It will easily be seen, however, from the forms which these operators assume in the p-language,

$$\begin{aligned}\mathrm{q}_k &= -\frac{h}{2\pi i} \frac{\partial}{\partial p_k}, \\ \mathrm{p}_k &= p_k,\end{aligned} \tag{55.41}$$

that all the same properties would still be found. It may also be mentioned that these properties are still retained when transformations are made to other possible modes of representation than those provided by the q- and p-languages.

(*f*) **The operators corresponding to observable quantities in general.** Having found the properties of the fundamental operators q_k and p_k, we may now state the *general rule* that the quantum mechanical operator corresponding to any classical function of the coordinates and momenta is to be obtained therefrom by replacing the variables q_k and

p_k by the operators $\mathbf{q}_k$ and $\mathbf{p}_k$, in such a way as to secure a *linear Hermitian operator*. This rule can be symbolized, for any classical function $F(q,p)$ of the coordinates and momenta, by the equations

$$\begin{array}{cc} q\text{-language} & p\text{-language} \\ \mathbf{F}(\mathbf{q},\mathbf{p}) = F\left(q, \frac{h}{2\pi i}\frac{\partial}{\partial q}\right), & \mathbf{F}(\mathbf{q},\mathbf{p}) = F\left(-\frac{h}{2\pi i}\frac{\partial}{\partial p}, p\right). \end{array} \quad (55.42)$$

We thus supplement the original, evidently appropriate, method of obtaining the operators that correspond to functions of the q's or p's alone, as given by (55.11), by a natural extension, as given by (55.42), which permits us to obtain operators that correspond to more complicated functions as well.

In case the classical expression $F(q,p)$ is a polynomial in the q's and p's, so that it can be written as a *symmetrized sum* of terms of the form

$$F = \sum \tfrac{1}{2}c(q_1^\alpha \dots q_f^\zeta\, p_1^\rho \dots p_f^\mu + p_1^\rho \dots p_f^\mu\, q_1^\alpha \dots p_f^\zeta), \quad (55.43)$$

there will be no difficulty in applying the above rules, after symmetrization has been introduced in accordance with (55.29) to secure the desired Hermitian character. It will be noticed, however, that the method of symmetrization which we have given is not necessarily unique for terms containing more than one pair of coordinates and momenta, since $\mathbf{q}_1\mathbf{q}_2\mathbf{p}_1\mathbf{p}_2+\mathbf{p}_1\mathbf{p}_2\mathbf{q}_1\mathbf{q}_2$ and $\mathbf{q}_1\mathbf{p}_2\mathbf{q}_2\mathbf{p}_1+\mathbf{q}_2\mathbf{p}_1\mathbf{q}_1\mathbf{p}_2$ would both be Hermitian operators. Nevertheless, difficulties due to such sources of uncertainty have not actually proved serious in the quantum mechanics.†

In case the classical expression is a polynomial in the p's but contains arbitrary functions Q of the coordinates, so that it can be written as a sum of terms of the form

$$F = \sum \tfrac{1}{2}(Qp_1^\rho \dots p_f^\mu + p_1^\rho \dots p_f^\mu\, Q), \quad (55.44)$$

it will be straightforward to obtain the form of the corresponding operator in the q-language; its form in the p-language could then be very complicated. Similar remarks hold for arbitrary functions in the momenta alone which would allow straightforward treatment in the p-language.

In the case of classical quantities, which have an *explicit dependence on the time t*, owing to the possibility that the form of the function may change with time, the above rule is readily extended by using at each

† The problem of a unique correlation of operators with classical functions has been studied by Weyl, *Zeits. f. Phys.* **46**, 1 (1927), and McCoy, *Proc. Nat. Acad.* **18**, 674 (1932).

instant the form of $F(q, p)$ which then applies. This can be symbolized by writing

$$\begin{array}{cc} q\text{-language} & p\text{-language} \\ \mathbf{F}(\mathrm{q}, \mathrm{p}, t) = F\left(q, \frac{h}{2\pi i}\frac{\partial}{\partial q}, t\right), & \mathbf{F}(\mathrm{q}, \mathrm{p}, t) = F\left(-\frac{h}{2\pi i}\frac{\partial}{\partial p}, p, t\right). \end{array} \tag{55.45}$$

The foregoing considerations assume that the coordinates and momenta are canonical in the quantum mechanical sense, for example Cartesian coordinates and momenta for the particles of a system, and the use of other coordinates can be most safely undertaken by transforming thereto after setting the problem up in Cartesian coordinates. In the case of functions dependent on observables such as the spin, which has no classical analogue, special treatment of the quantum mechanical operators will be necessary. See § 75.

(*g*) **The calculation of expectation values in general.** With the help of the above generalized definition for the operators that correspond to functions of the coordinates and momenta, we shall now postulate in general that the *expectation value* $\overline{F}(q, p)$, for any *measurable function* $F(q, p)$ of the coordinates and momenta of a system, can be calculated with the help of the *corresponding operator* $\mathbf{F}(\mathrm{q}, \mathrm{p})$ by substitution into equations of the general form

$$\overline{F}(q, p) = \int \psi^*(x)\mathbf{F}(\mathrm{q}, \mathrm{p})\psi(x)\, dx, \tag{55.46}$$

where the probability amplitudes ψ^* and ψ and the operator $\mathbf{F}$ are all to be expressed in the same language, and the integration is to be taken over all values of the variables x which are characteristic of that language. This makes a definite addition to our postulatory basis, since we have previously only been able to show that such a procedure would necessarily lead to correct expectation values in the case of functions of the coordinates or momenta alone.

By the expectation value for such a quantity $F(q, p)$ we are to understand the mean result, which would be obtained from a series of measurements made when the system is known to be in the specified state of interest. Hence we shall only maintain that operators and expectation values have a necessary significance in cases where some conceivable experimental method for measuring the quantity in question can be devised. The possibility of calculating expectation values is very important for the quantum mechanics, since we can now make assertions as to what values we can expect to find on the average for a system in a given state, even though we can no longer make all the assertions as to exact values which were thought possible in the classical mechanics.

It is to be specially emphasized that our formalism has been correctly devised so that the calculation of expectation values will lead to the same result, without dependence on our choice of the q- or p-language in which to make the computation. This can be seen in detail with the help of transformations similar to those by which we passed from the expression (55.3) for $\bar{p}_k$ in momentum language to its expression (55.7) in coordinate language. It is also to be emphasized that the formalism has been devised, so that expectation values will be *real numbers*, in agreement with their significance as a mean result of the measurement of real numbers. To see this we note, in accordance with (55.46), that we could write for any such expectation value $\bar{F}$, for example in the q-language

$$\bar{F} = \int \psi^*(q)\mathbf{F}(\mathbf{q}, \mathbf{p})\psi(q)\, dq,$$

and hence also for its complex conjugate

$$\bar{F}^* = \int \psi(q)[\mathbf{F}(\mathbf{q}, \mathbf{p})\psi(q)]^*\, dq.$$

In accordance with the Hermitian character of the operator **F**, however, the right-hand sides of these two equations are equal, so that we obtain the result

$$\bar{F}^* = \bar{F} \tag{55.47}$$

showing that $\bar{F}$ actually is a real number. This is, of course, one of the reasons for demanding that the operators **F** should have Hermitian character as was done above.

Finally, it may be noted, since we can compute not only the mean value of any measurable quantity F but also the mean value for any power thereof, that we can obtain a complete prognosis as to the probabilities of finding different values of F when a series of measurements is made on a system in a given state.

56. The Schroedinger equation for change in state with time

In accordance with preceding sections, the instantaneous state of a system can be described in the quantum mechanics by specifying the dependence of a suitable probability amplitude on the variables of which it is a function at any time of interest, and from such a specification it is possible to evaluate the probabilities at that time for finding different values of the coordinates, momenta, and other measurable quantities belonging to the system. This possibility of evaluation, however, applies in the first instance only at the particular time for which the state of the system has been specified, and we must now consider the fundamental question of the change in the state of a system with time.

We shall find just as in the classical mechanics that the state of an undisturbed system can be regarded as a definite function of the time, so that the state at any later time can be calculated from a knowledge of its state at a selected initial time. In this connexion, however, it is specially necessary to keep clearly in mind two important differences between classical and quantum mechanical states. In the first place we must always remain cognizant of the fact that a knowledge of the quantum mechanical state of a system allows us in general to make statements only as to the *probabilities* of finding given values for its various dynamical variables. In the second place we must not neglect the fact that the very process of observing a system will itself introduce disturbances, so that a new calculation of the *undisturbed* behaviour will, in general, have to be initiated after each new observation. Indeed it is characteristic of the quantum mechanics that we use the information obtained by observation at some initial time to set up an appropriate amplitude function to represent the system, with the help of which we can then make predictions as to the probabilities of obtaining various results at the later time of a second observation. The calculus of the quantum mechanics is thus suitable for connecting the behaviour of a system at a single earlier observation with that at a single later observation, rather than useful for a continuous description of the behaviour of the system as is essayed in the classical mechanics.

(*a*) **Postulated form of Schroedinger equation.** We are now ready to introduce the postulate by which changes in the state of an isolated undisturbed system can be calculated. This we do with the help of a simple partial differential equation with respect to the time t, which can be written in the form

$$\mathbf{H}\psi+\frac{h}{2\pi i}\frac{\partial\psi}{\partial t}=0, \tag{56.1}$$

where ψ is the *probability amplitude* which specifies the state of the system, and $\mathbf{H}$ is the so-called *Hamiltonian operator* for the system, which corresponds to the classical expression for its energy, as a function of coordinates and momenta, in the manner described in the preceding section. In accordance with its discovery by Schroedinger,† and

† See Schroedinger, *Wave Mechanics*, translated from second German edition, London, 1928.

in order to distinguish it from a related equation which applies to steady states and does not contain the time, the above expression may be spoken of as the *Schroedinger equation including the time*.

The above form of expression (56.1) for the Schroedinger equation may be regarded as a general one, applying in any language in which the amplitude ψ and the operator **H** are expressed. Nevertheless, since the classical Hamiltonian expression for the energy of familiar systems will usually be a simple polynomial in the momenta, even though depending in a complicated manner on the coordinates, it is often most convenient to regard the equation as expressed in *coordinate language* in the form

$$H\left(q, \frac{h}{2\pi i}\frac{\partial}{\partial q}\right)\psi(q,t)+\frac{h}{2\pi i}\frac{\partial\psi(q,t)}{\partial t} = 0, \tag{56.2}$$

since the Hamiltonian operator $H\left(q, \dfrac{h}{2\pi i}\dfrac{\partial}{\partial q}\right)$ can then be readily obtained from the classical Hamiltonian function $H(q,p)$ for the system by substituting a differential operator of the form $\dfrac{h}{2\pi i}\dfrac{\partial}{\partial q_k}$ in place of each of the momenta p_k for the system.

This construction of the appropriate Hamiltonian operator for a system is, of course, to be carried out in accordance with the general methods described in the preceding section for obtaining the *linear and Hermitian*† operators that correspond to observable quantities. As already made evident in that section, some difficulties and ambiguities‡ may be encountered in applying the methods, and extensions of method may be made necessary by the existence of variables—e.g. spin—which have no classical analogue. Nevertheless, for the most part the construction of the appropriate Hamiltonian operators for simple systems has turned out to be quite straightforward. Furthermore, in the case of difficulties or needs for extension of method the problem may be looked at from the broader point of view of discovering that Hamiltonian operator for the system which will give results that agree with experiment. This is, then, similar to the classical problem of picking out a Lagrangian function, which will apply when the simple rule

† For discussion of a further necessary character of the Hamiltonian operator, see Pauli, *Handbuch der Physik*, xxiv/1, second edition, Berlin, 1933, p. 142.

‡ In case an expression for the Hamiltonian operator is desired in polar or other coordinates which are not canonical in the quantum mechanical sense, it is best to begin by setting it up in canonical coordinates, e.g. Cartesian coordinates for the particles, and then transform to the desired coordinates. See Podolsky, *Phys. Rev.* **32**, 812 (1928).

of setting that function equal to the difference between kinetic and potential energy no longer holds as in the usual systems of Newtonian mechanics.

Having obtained an appropriate expression for the Hamiltonian operator, in coordinate language, for any system of interest, we can then make use of the Schroedinger equation in the form (56.2) to treat the problem of predicting the state of the system in its dependence on time. This we do by solving that equation for the state $\psi(q,t)$ as a function of t, with boundary conditions so chosen as to agree with our knowledge of the state $\psi(q,t_0)$ at the time of initial observation t_0. This solution can then be used to make predictions as to the expectation values for different quantities at any later time t when a second observation is undertaken. And, with a *new choice* of boundary conditions to agree with the results actually obtained in any such second observation, we can then make further predictions for yet later times.

The foregoing formalism applies in the treatment of isolated systems, which carry out their behaviour without any action from the outside, except for the disturbances introduced at the times when a new observation and a new selection of boundary conditions is made. In the case of systems which are acted on from the outside in a definite manner, so that the form of the Hamiltonian operator can be regarded as a specified function of the time, the formalism can be extended by taking

$$H\left(q, \frac{h}{2\pi i}\frac{\partial}{\partial q}, t\right)\psi(q,t)+\frac{h}{2\pi i}\frac{\partial\psi(q,t)}{\partial t} = 0, \tag{56.3}$$

where the explicit dependence of the operator $\mathbf{H}$ on the time t has been indicated.

(*b*) **Some specific examples of the Schroedinger equation.** The above remarks as to the general form of Schroedinger's equation may now be illustrated by some actual examples of the form of the equation expressed in coordinate language.

As our first example let us consider a single particle of mass m moving in a field of force corresponding to a potential V which is a definite function of position. In the classical mechanics we should write

$$H = \frac{1}{2m}(p_x^2+p_y^2+p_z^2)+V(x,y,z) \tag{56.4}$$

as the Hamiltonian expression for the energy of the particle, in terms of its Cartesian coordinates x, y, z and its corresponding components of momenta p_x, p_y, and p_z. In accordance with the rule given by (55.42), we then obtain as the Hamiltonian operator for the system

$$\mathbf{H} = \frac{1}{2m}\left\{\left(\frac{h}{2\pi i}\frac{\partial}{\partial x}\right)^2+\left(\frac{h}{2\pi i}\frac{\partial}{\partial y}\right)^2+\left(\frac{h}{2\pi i}\frac{\partial}{\partial z}\right)^2\right\}+V(x,y,z)$$

$$= -\frac{h^2}{8\pi^2 m}\left\{\frac{\partial^2}{\partial x^2}+\frac{\partial^2}{\partial y^2}+\frac{\partial^2}{\partial z^2}\right\}+V(x,y,z)$$

$$= -\frac{h^2}{8\pi^2 m}\nabla^2+V(x,y,z), \tag{56.5}$$

where the second form of expression is obtained in accordance with the convention that an operator is to be successively applied the number of times indicated by the power to which it is raised, and the third form of expression is obtained by introducing the familiar abbreviation ∇^2 for the Laplacian operator. Substituting this expression for the Hamiltonian operator into (56.2) and changing signs, we can now write, as the *Schroedinger equation for a single particle in a potential field*,

$$\frac{h^2}{8\pi^2 m}\left(\frac{\partial^2\psi}{\partial x^2}+\frac{\partial^2\psi}{\partial y^2}+\frac{\partial^2\psi}{\partial z^2}\right)-V\psi-\frac{h}{2\pi i}\frac{\partial\psi}{\partial t} = 0$$

or

$$\frac{h^2}{8\pi^2 m}\nabla^2\psi-V\psi-\frac{h}{2\pi i}\frac{\partial\psi}{\partial t} = 0, \tag{56.6}$$

where V is to be treated as a function of the Cartesian coordinates x, y, and z.

As a second simple example we may next consider a system composed of several particles having the different masses m_1, m_2, m_3, etc., moving in a field of force which can again be described by a potential V which will now be a function of the positions of all the particles. Taking our coordinates for the system as a whole as being the Cartesian coordinates for the particles of which it is composed, and applying the same methods as above, we easily obtain, as the *Schroedinger equation for a system of particles in a potential field*, the result

$$\sum_k \frac{h^2}{8\pi^2 m_k}\left(\frac{\partial^2\psi}{\partial x_k^2}+\frac{\partial^2\psi}{\partial y_k^2}+\frac{\partial^2\psi}{\partial z_k^2}\right)-V\psi-\frac{h}{2\pi i}\frac{\partial\psi}{\partial t} = 0,$$

$$\sum_k \frac{h^2}{8\pi^2 m_k}\nabla_k^2\psi-V\psi-\frac{h}{2\pi i}\frac{\partial\psi}{\partial t} = 0, \tag{56.7}$$

where m_k, x_k, y_k, and z_k are the mass and Cartesian coordinates corresponding to the kth particle and the indicated summations are to be taken over all the particles composing the system.

As a third example of the explicit form of Schroedinger's equation let us now consider a particle of mass m which carries an electric charge e, and moves in a combined electric and magnetic field. On account of the presence of magnetic forces, the action of such a field on a moving charge can no longer be described by a single potential function V, but can be treated with the help of the scalar potential $\Phi(x,y,z,t)$ together with the components A_x, A_y, and A_z of the vector potential $\mathbf{A}(x,y,z,t)$ used in electromagnetic theory. The intensities $\mathbf{E}$ and $\mathbf{H}$ of the electric and magnetic fields are then given by

$$\mathbf{E} = -\operatorname{grad}\Phi - \frac{1}{c}\frac{\partial \mathbf{A}}{dt} \tag{56.8}$$

and

$$\mathbf{H} = \operatorname{curl}\mathbf{A} \tag{56.9}$$

and the force acting on the particle by the Lorentz expression

$$\mathbf{F} = e\left(\mathbf{E} + \frac{\mathbf{u}}{c}\times\mathbf{H}\right), \tag{56.10}$$

where $\mathbf{u}$ is the velocity of the particle.

As first shown by Larmor,† the classical behaviour of such a particle under the action of this force can be described by the usual methods of mechanics with the help of the Lagrangian function

$$L = \tfrac{1}{2}m(\dot{x}^2+\dot{y}^2+\dot{z}^2) - e\Phi + \frac{e}{c}(A_x\dot{x}+A_y\dot{y}+A_z\dot{z}). \tag{56.11}$$

In accordance with (10.2) this leads to the components of momenta

$$p_x = m\dot{x} + \frac{e}{c}A_x, \qquad p_y = m\dot{y} + \frac{e}{c}A_y, \qquad p_z = m\dot{z} + \frac{e}{c}A_z, \tag{56.12}$$

and in accordance with (10.3) to the classical Hamiltonian

$$\begin{aligned} H &= p_x\dot{x}+p_y\dot{y}+p_z\dot{z}-L \\ &= \frac{1}{2m}\left[\left(p_x-\frac{e}{c}A_x\right)^2+\left(p_y-\frac{e}{c}A_y\right)^2+\left(p_z-\frac{e}{c}A_z\right)^2\right]+e\Phi \\ &= \frac{1}{2m}\Big[p_x^2-\frac{e}{c}p_xA_x-\frac{e}{c}A_xp_x+\frac{e^2}{c^2}A_x^2+p_y^2-\frac{e}{c}p_yA_y-\frac{e}{c}A_yp_y+\frac{e^2}{c^2}A_y^2+ \\ &\qquad +p_z^2-\frac{e}{c}p_zA_z-\frac{e}{c}A_zp_z+\frac{e^2}{c^2}A_z^2\Big]+e\Phi, \end{aligned} \tag{56.13}$$

where the second form is obtained from the first by substituting (56.11) and (56.12), and the third form has been *symmetrized* in accordance with (55.29) in order to secure an operator having Hermitian character.

Applying our rule (55.42) for changing this classical Hamiltonian into the corresponding quantum mechanical operator and substituting into

† Larmor, *Aether and Matter*, Cambridge, 1900.

(56.2), we can now write the *Schroedinger equation for a charged particle in an electromagnetic field* in the form

$$\frac{1}{2m}\sum_k\left[\left(\frac{h}{2\pi i}\frac{\partial}{\partial x_k}\right)^2-\frac{e}{c}\frac{h}{2\pi i}\frac{\partial}{\partial x_k}A_k-\frac{e}{c}\frac{h}{2\pi i}A_k\frac{\partial}{\partial x_k}+\frac{e^2}{c^2}A_k^2\right]\psi+ \\ +e\Phi\psi+\frac{h}{2\pi i}\frac{\partial\psi}{\partial t}=0,$$

where the subscripts $k = 1, 2, 3$ refer to the Cartesian coordinates x, y, z. Carrying out the application of the operators as indicated in the first term and changing signs, this can be re-expressed in the more convenient form

$$\frac{h^2}{8\pi^2 m}\nabla^2\psi+\frac{1}{2m}\sum_k\left[\frac{he}{2\pi ic}\frac{\partial}{\partial x_k}(A_k\psi)+\frac{he}{2\pi ic}A_k\frac{\partial\psi}{\partial x_k}-\frac{e^2}{c^2}A_k^2\psi\right]- \\ -e\Phi\psi-\frac{h}{2\pi i}\frac{\partial\psi}{\partial t}=0. \qquad (56.14)$$

It will be noticed that each of these three examples of the explicit form of the Schroedinger equation has provided us with a differential equation for the probability amplitude $\psi(q_1 \ldots q_f, t)$ of the system as a function of the coordinates and time. Since the quantum mechanical state of a system is determined at any time by the form of ψ, solutions of these equations—with boundary conditions chosen to correspond to any desired initial state—will then provide explicit solutions of the problem proposed at the beginning of the present section, of describing how the quantum mechanical state of a system depends on the time.

(*c*) **Transformation of Schroedinger equation from coordinate to momentum language.** As already remarked above, it is often most convenient to express the Schroedinger equation in coordinate language since the dependence on momenta will frequently be such as to lead at once to a simple expression for the Hamiltonian operator in that language. Nevertheless, we regard our original expression of the Schroedinger equation (56.1) as valid in any language provided we use the appropriate expression for the operator **H** in that language. In the case of a classical Hamiltonian which can be expressed as a polynomial in the coordinates as well as in the momenta, the transformation of the Schroedinger equation from coordinate to momentum language proves simple.

To investigate this, let us start with the Schroedinger equation (56.2), expressed in coordinate language in the form

$$H\left(q, \frac{h}{2\pi i}\frac{\partial}{\partial q}\right)\psi(q,t)+\frac{h}{2\pi i}\frac{\partial\psi(q,t)}{\partial t}=0, \qquad (56.15)$$

and assuming the coordinates to be canonical in the quantum mechanical sense, replace $\psi(q,t)$ by $\phi(p,t)$ with the help of the transformation equation (54.2)

$$\psi(q,t) = h^{-\frac{1}{2}f} \int \dots \int \phi(p,t) e^{\frac{2\pi i}{h}(p_1 q_1 + \dots + p_f q_f)} \, dp_1 \dots dp_f, \qquad (56.16)$$

where the integrations are over all values of the p's from $-\infty$ to $+\infty$. Substituting (56.16) into (56.15), and noting that the H does not now contain the p's, we can write

$$\int \dots \int \Big[\phi(p,t) H\Big(q, \frac{h}{2\pi i}\frac{\partial}{\partial q}\Big) e^{\frac{2\pi i}{h}(p_1 q_1 + \dots + p_f q_f)} + \\ + \frac{h}{2\pi i} \frac{\partial \phi(p,t)}{\partial t} e^{\frac{2\pi i}{h}(p_1 q_1 + \dots + p_f q_f)} \Big] \, dp_1 \dots dp_f = 0.$$

It will be seen, however, in case H is a polynomial in the coordinates q and in the differential operators $\frac{h}{2\pi i}\frac{\partial}{\partial q}$, as assumed, that this can be rewritten in the form

$$\int \dots \int \Big[\phi(p,t) H\Big(\frac{h}{2\pi i}\frac{\partial}{\partial p}, p\Big) e^{\frac{2\pi i}{h}(p_1 q_1 + \dots + p_f q_f)} + \\ + \frac{h}{2\pi i} \frac{\partial \phi(p,t)}{\partial t} e^{\frac{2\pi i}{h}(p_1 q_1 + \dots + p_f q_f)} \Big] \, dp_1 \dots dp_f = 0.$$

By carrying out partial integrations with respect to the momenta and making use of the circumstance that $\phi(p,t)$ will go to zero at the limits of integration, this can then be rewritten in the form

$$\int \dots \int \Big[H\Big(-\frac{h}{2\pi i}\frac{\partial}{\partial p}, p\Big) \phi(p,t) + \frac{h}{2\pi i} \frac{\partial \phi(p,t)}{\partial t} \Big] e^{\frac{2\pi i}{h}(p_1 q_1 + \dots + p_f q_f)} \, dp_1 \dots dp_f = 0.$$

Since this equation must be satisfied for arbitrary choices of the q's, this now leads to the desired result

$$H\Big(-\frac{h}{2\pi i}\frac{\partial}{\partial p}, p\Big) \phi(p,t) + \frac{h}{2\pi i} \frac{\partial \phi(p,t)}{\partial t} = 0, \qquad (56.17)$$

giving a simple expression of the Schroedinger equation in momentum language, with the operator **H** of the expected form. It should be borne in mind, unless the dependence of H on q is simple (polynomial) as assumed above, that the operator $H\Big(-\frac{h}{2\pi i}\frac{\partial}{\partial p}, p\Big)$ will have to be interpreted as an integral and not a differential operator.

For a special case, this then illustrates the validity of the assumed possibility of expressing the Schroedinger equation either in the q-language using the amplitude $\psi(q,t)$ or in the p-language using the amplitude $\phi(p,t)$. We shall usually find it most convenient, however, to use a coordinate representation in setting up our problems. The possibility of using a Schroedinger equation with probability amplitudes corresponding to still other observables will appear later. (See § 67 (c).)

57. Summary of postulatory basis

We have now completed our exposition of the postulatory basis for non-relativistic quantum mechanics. The account has been so long that a brief summary in the q- and p-languages, as given by the following equations, will not be out of place.

For the relation between probability densities and amplitudes we have

$$\begin{aligned} W(q,t)\,dq &= \psi^*(q,t)\psi(q,t)\,dq, \\ W(p,t)\,dp &= \phi^*(p,t)\phi(p,t)\,dp. \end{aligned} \tag{57.1}$$

For the transformation of probability amplitudes we have

$$\begin{aligned} \phi(p,t) &= h^{-\frac{1}{2}f} \int \psi(q,t)e^{-\frac{2\pi i}{h}(p_1 q_1+\ldots+p_f q_f)}\,dq, \\ \psi(q,t) &= h^{-\frac{1}{2}f} \int \phi(p,t)e^{+\frac{2\pi i}{h}(p_1 q_1+\ldots+p_f q_f)}\,dp. \end{aligned} \tag{57.2}$$

For the operators corresponding to observable quantities we have

$$\begin{aligned} \mathbf{F}(\mathbf{q},\mathbf{p}) &= F\left(q, \frac{h}{2\pi i}\frac{\partial}{\partial q}\right), \\ \mathbf{F}(\mathbf{q},\mathbf{p}) &= F\left(-\frac{h}{2\pi i}\frac{\partial}{\partial p}, p\right). \end{aligned} \tag{57.3}$$

For the mean or expectation value of an observable quantity we have

$$\begin{aligned} \overline{F}(q,p) &= \int \psi^*\mathbf{F}\psi\,dq, \\ \overline{F}(q,p) &= \int \phi^*\mathbf{F}\phi\,dp. \end{aligned} \tag{57.4}$$

For the change of state with time we have the Schroedinger equation

$$\begin{aligned} \mathbf{H}\psi+\frac{h}{2\pi i}\frac{\partial\psi}{\partial t} &= 0, \\ \mathbf{H}\phi+\frac{h}{2\pi i}\frac{\partial\phi}{\partial t} &= 0. \end{aligned} \tag{57.5}$$

The equations as written contain some obvious abbreviations and must be used, of course, in the light of the preceding discussion. They are sufficient for a non-relativistic quantum mechanics except for special assumptions that have to be introduced in connexion with observables having no classical analogue that can be expressed as a function of coordinates and momenta.

C. THEOREMS ILLUSTRATING THE NATURE OF QUANTUM MECHANICS

58. Probability density and probability current

We must now derive a number of theorems from the foregoing postulates which will illustrate the nature of quantum mechanics. These will include in §§ 60, 61, 62 those consequences of the new theory which are the analogues of the ideas as to energy levels, wave-particle duality, and uncertainty that have led us to the belief that the classical mechanics is not sufficient. For the most part we shall carry out the derivations in the q-language.

(*a*) **The conservation of total probability.** First of all, we must investigate an important question as to the consistency of the formalism which we have set up. In equations (57.1) we have related certain probability amplitudes to the actual probabilities of finding values for coordinates and momenta that lie in specified ranges, and in equations (57.5) we have then also stated how these amplitudes are to depend on the time. It is evident for consistency that this time dependence must at least be such as to preserve constant the total probability of finding values for our variables that lie somewhere within the ranges that are possible for them. This we can easily show to be the case.

For the rate of change with time in the total probability of finding values for the coordinates, that lie somewhere, we can write

$$\frac{d}{dt}\int \psi^*\psi\,dq = \int \left[\psi\frac{\partial\psi^*}{\partial t}+\psi^*\frac{\partial\psi}{\partial t}\right]dq,$$

where the integrations are to be taken over the total range of possible values for the coordinates. With the help of the Schroedinger equation we can then re-express this, however, in the form

$$\frac{d}{dt}\int \psi^*\psi\,dq = \frac{2\pi i}{h}\int [\psi(\mathbf{H}\psi)^*-\psi^*(\mathbf{H}\psi)]\,dq = 0, \qquad (58.1)$$

where the value zero arises in accordance with (55.25) from the Hermitian character of the Hamiltonian operator $\mathbf{H}$. We now see why it has been specially important to prescribe this character for the Hamiltonian operator.

Having found that the total probability does remain constant, it will of course be possible to *normalize* our probability amplitudes in such a way as to make the total probability of finding the system somewhere in configuration space permanently equal to unity:

$$\int \psi^*\psi \, dq = 1. \tag{58.2}$$

Similar treatment can be given to the total probability of finding the system at least somewhere in momentum space.

(*b*) **The concept of probability current.** It is also possible to investigate somewhat more in detail the nature of the processes by which the probability density can change at a given point while its integrated value over the whole of configuration space remains constant. It will be sufficient to illustrate this for the case of a simple particle of mass m moving in a potential field $V(x,y,z)$.

In agreement with (56.6) we can then write the Schroedinger equation in the form

$$\frac{\partial\psi}{\partial t}+\frac{h}{4\pi im}\left(\frac{\partial^2\psi}{\partial x^2}+\frac{\partial^2\psi}{\partial y^2}+\frac{\partial^2\psi}{\partial z^2}\right)+\frac{2\pi i}{h}V\psi = 0$$

and the corresponding conjugate equation in the form

$$\frac{\partial\psi^*}{\partial t}-\frac{h}{4\pi im}\left(\frac{\partial^2\psi^*}{\partial x^2}+\frac{\partial^2\psi^*}{\partial y^2}+\frac{\partial^2\psi^*}{\partial z^2}\right)-\frac{2\pi i}{h}V\psi^* - 0.$$

Multiplying the first of these equations by ψ^* and the second by ψ and then adding, we obtain

$$\psi^*\frac{\partial\psi}{\partial t}+\psi\frac{\partial\psi^*}{\partial t}+$$
$$+\frac{h}{4\pi im}\left(\psi^*\frac{\partial^2\psi}{\partial x^2}+\psi^*\frac{\partial^2\psi}{\partial y^2}+\psi^*\frac{\partial^2\psi}{\partial z^2}-\psi\frac{\partial^2\psi^*}{\partial x^2}-\psi\frac{\partial^2\psi^*}{\partial y^2}-\psi\frac{\partial^2\psi^*}{\partial z^2}\right) = 0,$$

and this can easily be shown equivalent to the form

$$\frac{\partial}{\partial t}(\psi^*\psi)+\frac{h}{4\pi im}\operatorname{div}(\psi^*\operatorname{grad}\psi-\psi\operatorname{grad}\psi^*) = 0. \tag{58.3}$$

Hence, if we now define the density of probability current as the vector

$$\mathsf{S} = \frac{h}{4\pi im}(\psi^*\operatorname{grad}\psi-\psi\operatorname{grad}\psi^*), \tag{58.4}$$

which will be seen to be a *real* quantity, we can connect *probability density*

$$W = \psi^*\psi \tag{58.5}$$

and *probability current density* S by the simple relation

$$\frac{\partial W}{\partial t}+\operatorname{div}\mathsf{S} = 0, \tag{58.6}$$

which shows, in agreement with the conservation of total probability, that the change in probability in a given region can be regarded as due to flow through its boundary.

In our present case the components of **S** have a simple pictorial interpretation, S_x for example being the probability in unit time that the particle would be found to pass through unit area perpendicular to the x-axis, moving in the positive direction, diminished by the similar probability for passage in the negative direction.

Analogous treatments of probability current can be given for more complicated cases. For a system of charged particles in a combined electric and magnetic field we can use a configuration space of $3n$ dimensions, where n is the number of particles, and denote the x, y, z coordinates for the successive particles by the general symbol q_k. For the kth component of the generalized probability current we then have†

$$S_k = \frac{h}{4\pi i m_k}\left(\psi^*\frac{\partial\psi}{\partial q_k}-\psi\frac{\partial\psi^*}{\partial q_k}\right)-\frac{e_k}{m_k c}A_k\psi^*\psi, \tag{58.7}$$

where m_k and e_k are the mass and charge of the particle having q_k as one of its Cartesian coordinates and A_k is the component of the vector potential in the direction of the q_k-axis.

In analogy to (58.6) we then have as the equation of continuity

$$\frac{\partial W}{\partial t}+\sum_k \frac{\partial S_k}{\partial q_k} = 0. \tag{58.8}$$

59. The principle of superposition

Since the two operators—for the Hamiltonian and for differentiation with respect to time—which occur in the Schroedinger equation

$$\mathbf{H}\psi+\frac{h}{2\pi i}\frac{\partial\psi}{\partial t} = 0 \tag{59.1}$$

are both linear, it is evident—in agreement with the discussion of § 55 (c)—that any linear combination of individual solutions of this equation will also be a solution thereof. Thus, if $\psi_1(q,t)$ and $\psi_2(q,t)$ are solutions, the combination

$$\psi(q,t) = c_1\psi_1(q,t)+c_2\psi_2(q,t) \tag{59.2}$$

will also be a solution, where c_1 and c_2 can be any constants real or complex. This result expresses an important character of the quantum mechanics which may be called the *principle of superposition*.

† See Pauli, *Handbuch der Physik*, xxiv/1, second edition, Berlin, 1933, p. 109.

The principle can also be expressed in the somewhat more general form

$$\psi = c_1\psi_1+c_2\psi_2, \tag{59.3}$$

where, with an appropriate choice of c_1 and c_2, we can regard ψ as expressed in one quantum mechanical language and ψ_1 and ψ_2 in another. As an important example of this possible way of superimposing solutions, we may take our fundamental equation (57.2)

$$\psi(q,t) = h^{-\frac{1}{2}f} \int \phi(p,t)e^{+\frac{2\pi i}{h}(p_1 q_1+...+p_f q_f)}\,dp, \tag{59.4}$$

which gives the possibility of expressing a given state of a system either directly in coordinate language or by a superposition of states in momentum language.

The possibility of superposing individual solutions of the Schroedinger equation is somewhat analogous to the possibility of superposing solutions corresponding to different wave motions in the classical mechanics; and the actual solutions of (59.1) often show an explicit trigonometric or exponential dependence of ψ on q and t which is reminiscent of familiar wave motions. This is in agreement with the frequent use of the term *wave mechanics* as an alternative for quantum mechanics, and the practice of speaking of the Schroedinger equation as a *wave equation* to be used for calculating the amplitude ψ of *probability waves*.

Nevertheless, the actual functions superimposed in equations such as (59.2) have a physical significance quite different from that of functions which can be superimposed in the older mechanics. Thus, at a given time t, ψ_1 and ψ_2 represent possible quantum mechanical states of the system and the result of their superposition ψ also represents a possible state. This will make it possible in the quantum mechanics to treat a system in such a state ψ also from the point of view of its being partly in state ψ_1 and partly in state ψ_2, and to make predictions as to the relative probability after a suitable experimental observation that it will actually be found to be in one or the other of these latter states. The quantum mechanics thus provides possibilities for the *superposition and decomposition of states*—which latter can be made in a variety of ways—for which the classical mechanics with its perfectly definite states provides no analogy.

Characteristic for the whole structure of the wave mechanics is the fact that the wave functions ψ which obey the superposition principle are not the physically observed things, and that the observable expectation values for dynamical variables depend bilinearly on ψ and ψ^*.

Thus the calculation of these expectation values will exhibit interference effects, in the sense that the expectation values corresponding to the state

$$\psi = c_1\psi_1 + c_2\psi_2$$

will not in general be the weighted average of the expectation values for the states ψ_1 and ψ_2. For example, considering (59.4), we see in predicting the behaviour of a system whose state in coordinate language is expressed as a superposition of various states in momentum language, that the interference between these states may be all-important when this form of expression is used. It is clear that such a system cannot be represented as a collection of systems with various but well-defined momenta. As especially emphasized by Dirac, this may be regarded as a fundamental expression of the complementary character of the physical description of a given situation in terms of the concepts of position and of momentum.

60. Energy levels for systems in steady states. Eigenvalues and eigenfunctions

We are now ready to treat the quantum mechanical explanation of the existence of discrete energy levels. The actual occurrence of such levels was the first of the considerations which we mentioned as leading to a necessity for modifying classical mechanics.

We shall define a quantum mechanical system—having a Hamiltonian not itself dependent on the time—as being in a *steady state*, when both the probability density (58.5)

$$W = \psi^*(q,t)\psi(q,t) \tag{60.1}$$

and the probability current (58.7)

$$S_k = \frac{h}{4\pi i m_k}\left(\psi^*\frac{\partial\psi}{\partial q_k} - \psi\frac{\partial\psi^*}{\partial q_k}\right) - \frac{e_k}{m_k c}A_k\psi^*\psi \tag{60.2}$$

are themselves independent of the time. For such a state, both the chance for finding a given particle at a specified place and for finding it moving through a specified area will be independent of the time. Hence the designation steady state is appropriate.

To satisfy the first of the above conditions we must evidently take our amplitudes of the form

$$\psi^* = u^*(q)e^{if(q,t)} \quad \text{and} \quad \psi = u(q)e^{-if(q,t)},$$

where $u(q)$ is a function of the coordinates alone, while $f(q,t)$ may for the present be any *real* function of the q's and t.

To satisfy the second of our conditions for a steady state—since A_k

being part of the Hamiltonian is not an explicit function of t—we shall then have to have

$$\psi^*\frac{\partial\psi}{\partial q_k}-\psi\frac{\partial\psi^*}{\partial q_k} = u^*\frac{\partial u}{\partial q_k}-u\frac{\partial u^*}{\partial q_k}-2iu^*u\frac{\partial f(q,t)}{\partial q}$$

independent of the time. And this will only be true if f is of the form

$$f(q,t) = g(q)+h(t).$$

Hence, absorbing $e^{-ig(q)}$ into $u(q)$, we can now write the probability amplitude for a steady state in the necessary form

$$\psi(q,t) = u(q)e^{-ih(t)}.$$

This expression for the amplitude, however, must of course be a solution of the Schroedinger equation (57.5), and we can obtain further information as to the form $h(t)$ by substituting therein. Doing so we have

$$\mathbf{H}u(q)e^{-ih(t)}-\frac{h}{2\pi i}u(q)e^{-ih(t)}\frac{dh(t)}{dt} = 0,$$

and evidently this can only be satisfied if $h(t)$ is a constant multiplied by the time,

$$h(t) = \text{const.}\times t.$$

Hence we may now write the *probability amplitude* for a system in a *steady state* in the following general form, which will prove convenient,

$$\psi(q,t) = u(q)e^{-\frac{2\pi i}{h}Et}, \tag{60.3}$$

where E must be a *real* constant, in order that $\psi^*\psi$ shall retain its independence of t. Furthermore, by substituting this result back into the general expression (57.5) for the Schroedinger equation we now obtain, as a special case applicable to steady states, the so-called Schroedinger *wave equation without the time*

$$\mathbf{H}u(q) = Eu(q). \tag{60.4}$$

With the help of these equations we can now derive some important general results for systems in steady states.

In the first place, substituting (60.3) into our general expression (57.4) for the mean value of an observable quantity, we can evidently write

$$\overline{H^n} = \int u^*(q)\mathbf{H}^n u(q)\,dq$$

as an expression for the mean value of the nth power of the energy of a system in a stationary state. In accordance with (60.4) this will give us

$$\overline{H^n} = E^n \tag{60.5}$$

valid for any integral power n; and this can only be satisfied if the energy of the system has a value exactly equal to our constant E. Hence a system in a given *steady state* is characterized by a *precise value* E which will be found for its energy.

In the second place it is to be specially noted that not all values of the energy may be physically possible, since we can obviously only permit such values for the constant E in equation (60.4) as will lead to solutions for $u(q)$ which allow a sensible interpretation of u^*u as a probability density. Values of E which make $u(q)$ single-valued, continuous, and finite throughout the range of the coordinates q will evidently be satisfactory.† These allowable values of E may be called *eigenvalues* for the energy of the system, and the associated solutions for $u(q)$ *eigenfunctions*.‡ Corresponding eigenvalues and eigenfunctions are conveniently designated by attaching the same suffix, thus E_n and $u_n(q)$.

Applying the indicated methods of treatment to the actual problems of atomic mechanics, cases both of *discrete* and of *continuous* spectra of energy eigenvalues are encountered. These make it possible for the quantum mechanics to explain both the *discrete energy levels* exhibited by atoms with energies less than their ionization limit and the *continuous changes* in energy exhibited above that limit. Furthermore, as first shown for the case of the hydrogen atom by Schroedinger,|| the quantum mechanics gives a correct prediction of the energy levels actually found.

Returning to a consideration of the general results which can be obtained for systems in steady states, let us take E_n and E_m as two possible eigenvalues of the energy, and u_n and u_m as the corresponding eigenfunctions. In accordance with (60.4) and the corresponding complex conjugate equation, we can then write

$$\mathbf{H}u_n = E_n u_n$$

and

$$(\mathbf{H}u_m)^* = E_m^* u_m^*.$$

Multiplying the first of these equations by u_m^* and the second by u_n, and integrating over the total range of coordinates q, we obtain

$$\int u_m^* \mathbf{H}u_n \, dq = E_n \int u_m^* u_n \, dq$$

and

$$\int u_n (\mathbf{H}u_m)^* \, dq = E_m^* \int u_m^* u_n \, dq,$$

† For a deeper discussion of the requirements for a satisfactory eigenfunction, see Pauli, *Handbuch der Physik*, xxiv/1, second edition, Berlin, 1933, p. 123.

‡ The earlier English designations, 'characteristic values' and 'characteristic functions', are not so commonly used in quantum mechanics.

|| Schroedinger, *Ann. der Phys.* **79**, 361 (1926).

where the constants E_n and E_m^* can evidently be taken outside the integral sign. Since the left-hand sides of these equations are equal, owing to the Hermitian character (55.25) of the operator $\mathbf{H}$, we can now write the useful result

$$(E_n - E_m^*) \int u_m^* u_n \, dq = 0. \tag{60.6}$$

This equation has some important implications.

In the first place let us consider the case $m = n$, where the two indices refer to the same eigenvalues and eigenfunctions. Since the integral $\int u_n^* u_n \, dq$ will not be zero, owing to the essentially positive character of $u_n^* u_n = |u_n|^2$, we must then conclude that

$$E_n = E_n^*. \tag{60.7}$$

Hence the allowable values for the energy will all be *real*, in agreement with our earlier conclusion noted when the constant E was first introduced, and also in agreement with a more general principle by which it can be shown that the eigenvalues for any observable quantity must be real. (See (64.9).)

As a second case let us consider the situation $m \neq n$ corresponding to the different eigenvalues $E_m \neq E_n$. To preserve the truth of equation (60.6) we shall then have to have

$$\int u_m^* u_n \, dq = 0. \tag{60.8}$$

We thus find that eigenfunctions corresponding to different eigenvalues of the energy are necessarily *orthogonal* to each other. This is a very important property of such eigenfunctions which is often made use of in the quantum mechanics.

Finally, let us consider the possible situation $m \neq n$ with $E_m = E_n$ which arises in so-called cases of *degeneracy*, where the Schroedinger equation (60.4) exhibits the possibility for quite different solutions u_m and u_n that correspond nevertheless to eigenvalues E_m and E_n which turn out to be equal. In such a case, if g is the number of *independent* solutions, we say that the energy level is g-fold degenerate. Such degeneracies can in general be removed by subjecting the system to perturbing fields which will differently affect the different eigensolutions and break up the energy level into g neighbouring levels.

In the case of degeneracy we can no longer conclude that two eigenfunctions u_m and u_n belonging to the same energy level are necessarily orthogonal, since the previous argument leading to equation (60.8) now breaks down. Nevertheless, if we have g independent solutions which are not orthogonal, we can always form new solutions by linear combination which will give us an equivalent set of g functions which are orthogonal. (See § 64 (*c*).)

We shall find that the quantum mechanics often makes use of these solutions of the Schroedinger equation (60.4), or energy eigenfunctions, which we have been discussing. They are customarily *normalized*, so that

$$\int u_n^* u_n \, dq = 1. \tag{60.9}$$

They are either necessarily orthogonal or may be *orthogonalized* by the process described above, so that

$$\int u_m^* u_n \, dq = 0. \tag{60.10}$$

And as a rule they are known or assumed to provide a *complete* set of functions in terms of which other functions $f(q)$ may be expanded in accordance with the expression

$$f(q) = \sum_k c_k u_k(q), \tag{60.11}$$

where the c_k are constant coefficients, where summation may be replaced by integration in the case of a continuous spectrum of eigenvalues, and where, as will be explained later, this equation is in general to be interpreted in terms of convergence in the mean.

61. Wave-particle duality. De Broglie waves for free particles

Let us now turn to a consideration of the quantum mechanical point of view with respect to wave-particle duality. The difficulty of understanding the nature of such a duality from a classical point of view provided the second of the considerations which we mentioned as necessitating a modification of classical mechanics.

In a general way it can be said that the quantum mechanics is able to incorporate the idea of an entity having dual particle-like and wave-like properties, by restricting the particle-like properties so that they do not permit the description or prediction of an exact space-time motion, and by introducing probability waves as the actually appropriate apparatus for the making of predictions. The restrictions on particle-like properties are such that we can still determine at will the position of the entity, for example, by catching it in a small receptacle, or its velocity, for example, by the Doppler effect in reflected light, but are such that they do not permit a simultaneous knowledge of position and velocity sufficient for an exact kinematic description. The introduction of superposable waves for calculating the probability that the entity will appear in a given place carries with it those possibilities for interference and reinforcement which have seemed experimentally necessary. The limitations on the kinematic description of the particle-like behaviour of the entity can be shown in detail just sufficient to avoid conflict with the complementary description of its wave-like behaviour.†

It will be instructive to consider an entity—such as an electron or proton—which we have grown accustomed to regard purely from the particle point of view, and examine the nature of the probability waves associated with it. If we take the motion as being in free space, where the potential energy $V(x,y,z)$ can be taken as zero, the classical Hamiltonian would have the form

$$H = \frac{1}{2m}(p_x^2+p_y^2+p_z^2), \tag{61.1}$$

where m is the mass of the classical particle. This Hamiltonian will permit us to write the Schroedinger equation (57.5) in a specially simple form if we use the p-language. We obtain

$$\frac{1}{2m}(p_x^2+p_y^2+p_z^2)\phi+\frac{h}{2\pi i}\frac{\partial\phi}{\partial t} = 0,$$

which has the simple solution

$$\phi(p_x,p_y,p_z,t) = a(p_x,p_y,p_z)e^{-\frac{2\pi i}{h}\frac{1}{2m}(p_x^2+p_y^2+p_z^2)t}, \tag{61.2}$$

where $a(p_x,p_y,p_z)$ is an arbitrary function of its arguments. This expression gives the probability amplitude for finding particular values of the momenta p_x, p_y, p_z. In this case of free space, however, these momenta determine the energy in accordance with the equation

$$\frac{1}{2m}(p_x^2+p_y^2+p_z^2) = E. \tag{61.3}$$

Hence there will be no change in significance if we rewrite our expression for this probability amplitude in the simpler form

$$\phi(p_x,p_y,p_z,t) = a(p_x,p_y,p_z)e^{-\frac{2\pi i}{h}Et}. \tag{61.4}$$

We have thus obtained a simple expression for the probability amplitude $\phi(p,t)$. We shall be specially interested, nevertheless, in the probability amplitude $\psi(q,t)$ expressed in the q-language. Substituting (61.4) into the appropriate transformation relation (57.2), this will be given by

$$\psi(x,y,z,t) = \frac{1}{h^{\frac{3}{2}}}\iiint\limits_{-\infty}^{+\infty} a(p_x,p_y,p_z)e^{\frac{2\pi i}{h}(p_x x+p_y y+p_z z-Et)}\,dp_x\,dp_y\,dp_z. \tag{61.5}$$

† See Pauli, *Handbuch der Physik*, xxiv/1, second edition, Berlin, 1933, p. 86.

It can readily be shown, however, that this expression can be regarded as resulting from the superposition of plane waves, travelling through the space x, y, z with directions, frequencies, and phase relations that can be chosen at will.

To see this let us first introduce the so-called *de Broglie relations*, connecting the components of momenta and energy with *wave numbers and frequency*. These can be written in the form

$$\frac{p_x}{h} = \sigma_x, \qquad \frac{p_y}{h} = \sigma_y, \qquad \frac{p_z}{h} = \sigma_z, \qquad \frac{E}{h} = \nu, \tag{61.6}$$

and may be regarded as *defining* the new quantities—wave numbers σ_x, σ_y, σ_z, and frequency ν. Substituting these quantities into (61.5), this expression can evidently be rewritten in the form

$$\psi(x,y,z,t) = \iiint\limits_{-\infty}^{+\infty} A(\sigma_x, \sigma_y, \sigma_z) e^{2\pi i(\sigma_x x+\sigma_y y+\sigma_z z-\nu t)}\, d\sigma_x\, d\sigma_y\, d\sigma_z, \tag{61.7}$$

where $A(\sigma_x, \sigma_y, \sigma_z)$ is now an arbitrary function of the new arguments. And, making use of (61.3) and (61.6), we can write

$$\nu = \frac{h}{2m}(\sigma_x^2+\sigma_y^2+\sigma_z^2) \tag{61.8}$$

as a necessary relation connecting frequency and wave numbers. Equation (61.7) is, however, a well-known form of expression for the result of superposing plane waves, and equation (61.8) can be spoken of as the *law of dispersion* for the waves under consideration.

In accordance with (61.7) the individual waves which are superposed to give the total probability amplitude $\psi(x,y,z,t)$ will be of the form

$$\psi = Ae^{2\pi i(\sigma_x x+\sigma_y y+\sigma_z z-\nu t)}. \tag{61.9}$$

(It should be noted that this expression for ψ does not go to zero as x, y, and z go to infinity, and must be regarded as a limiting case of the quadratically integrable ψ's that we have admitted.) They may be called the *de Broglie waves* associated with a free particle. By introducing a familiar relation between exponential and trigonometric language, this expression for the de Broglie waves may be rewritten in the form

$$\psi = A\cos 2\pi(\sigma_x x+\sigma_y y+\sigma_z z-\nu t)+iA\sin 2\pi(\sigma_x x+\sigma_y y+\sigma_z z-\nu t). \tag{61.10}$$

And, by noting the periodicity of the expression with respect to spatial and temporal displacement, it may also be written as

$$\psi = A\cos 2\pi\left(\frac{x\cos\alpha}{\lambda}+\frac{y\cos\beta}{\lambda}+\frac{z\cos\gamma}{\lambda}-\frac{t}{\tau}\right)+$$
$$+iA\sin 2\pi\left(\frac{x\cos\alpha}{\lambda}+\frac{y\cos\beta}{\lambda}+\frac{z\cos\gamma}{\lambda}-\frac{t}{\tau}\right), \qquad (61.11)$$

where $$\lambda = \frac{\cos\alpha}{\sigma_x} = \frac{\cos\beta}{\sigma_y} = \frac{\cos\gamma}{\sigma_z} \quad\text{and}\quad \tau = \frac{1}{\nu} \qquad (61.12)$$

are seen to be the wave-length and period of the wave, and α, β, and γ the angles which describe its direction of propagation with respect to the x, y, and z axes.

Some of the properties of these de Broglie waves may now be considered.

In the first place it will be noticed, for example from the form of expression (61.10), that the amplitude ψ which is propagated by these waves will be a complex quantity for all non-vanishing values of the arguments of the sine and cosine terms and of A, which latter can itself in general be a real, imaginary, or complex quantity. This is in agreement with our earlier mention of the fact that the probability amplitudes would in general turn out to be complex quantities.

In the second place, if we consider the Schroedinger equation for our present system as given in accordance with (56.6) in the q-language,

$$\frac{h^2}{8\pi^2 m}\left(\frac{\partial^2\psi}{\partial x^2}+\frac{\partial^2\psi}{\partial y^2}+\frac{\partial^2\psi}{\partial z^2}\right)-\frac{h}{2\pi i}\frac{\partial\psi}{\partial t} = 0,$$

and substitute the expression (61.9) for an individual de Broglie wave, we readily obtain, after cancelling common factors, the result

$$\frac{h}{2m}(\sigma_x^2+\sigma_y^2+\sigma_z^2)-\nu = 0$$

in agreement with the necessary relation (61.8) which we have called the law of dispersion for the waves in question. We thus see that each of the de Broglie waves of the form (61.9), which are superposed in the integrated expression (61.7), obeys the Schroedinger wave equation. This is, of course, a necessary consequence of our method of development. Furthermore, it will be seen that the full expression (61.7) is the most general solution of that wave equation.

In the third place it will be seen that these de Broglie waves can evidently be superposed in such a way as to form 'wave packets' which at any selected time will give a large probability for finding approximately such values for the position and momentum of the particle as we may desire to specify. The nature of such wave packets will be more thoroughly considered in the next section. It may be noted now, however, that a wave packet which gives a large probability for finding specified values of the momenta p_x, p_y, and p_z can only be obtained by assigning relatively large values to the coefficients $A(\sigma_x, \sigma_y, \sigma_z)$ in (61.7) in the neighbourhood of the values of the wave numbers σ_x, σ_y, and σ_z that correspond to the momenta which we wish to favour. Making use of the known approximate connexion between group velocity and dispersion,

$$v_x = \frac{\partial \nu}{\partial \sigma_x}, \qquad v_y = \frac{\partial \nu}{\partial \sigma_y}, \qquad v_z = \frac{\partial \nu}{\partial \sigma_z}, \tag{61.13}$$

and substituting from the dispersion equation (61.8), we then obtain

$$v_x = \frac{h\sigma_x}{m}, \qquad v_y = \frac{h\sigma_y}{m}, \qquad v_z = \frac{h\sigma_z}{m} \tag{61.14}$$

for the approximate velocity of such a wave packet, where σ_x, σ_y, and σ_z are the values of the wave numbers that have been favoured in the construction of the packet. Substituting the de Broglie relations (61.6), this now gives us the satisfactory result

$$v_x = \frac{p_x}{m}, \qquad v_y = \frac{p_y}{m}, \qquad v_z = \frac{p_z}{m} \tag{61.15}$$

as a connexion between the approximate velocity of the wave packet and the probable values which will be found for the components of momentum.

Most important of all, it will now be appreciated that the de Broglie waves provide the possibilities for interference and reinforcement which are needed to explain experiments on the diffraction of particles, such as those of Davisson and Germer on the preferred directions for the reflection of electrons from a crystal grating. Since these effects are obtained with electrons whose momentum is first fixed by dropping through an appropriate field, the de Broglie waves will be such as to give a large probability for finding that value of the momentum

$$p = \sqrt{(p_x^2+p_y^2+p_z^2)}.$$

Hence, in accordance with our previous considerations, noting (61.6) and (61.12) in particular, these will be waves whose wave-length will satisfy

$$p = \sqrt{\{h^2(\sigma_x^2+\sigma_y^2+\sigma_z^2)\}} = \frac{h}{\lambda}.$$

It is experimentally known, however, that the diffraction effects are such as would be accounted for by the interference and reinforcement of waves having the de Broglie wave-length

$$\lambda = \frac{h}{p} = \frac{h}{mv}. \tag{61.16}$$

The foregoing considerations are sufficient to give an idea of the quantum mechanical treatment of wave-particle duality in the case of entities which were customarily regarded solely from the particle point of view in the past. Somewhat similar considerations apply to the reverse case of electromagnetic radiation, which was in the past regarded solely from the wave point of view, but which may now be treated from the particle point of view with the help of the concept of photons. Nevertheless, as these photons travel with the velocity of light, a non-relativistic treatment would not be sufficient, and, as a consequence of this, one cannot assign a precise value for the position of a photon.

62. The Heisenberg uncertainty relation

(*a*) **Case of a free particle. Wave packets.** We now come to the quantum mechanical treatment of the uncertainty relations connecting coordinates and momenta. The necessity for some restriction on the possibility of exact knowledge as to the simultaneous values of conjugated coordinates and momenta was the third of the considerations which we discussed as necessitating a revision of classical mechanics.

To treat the problem we may first consider the simple case of a single free particle and investigate the extent to which its position and momentum can be specified by a properly chosen quantum mechanical state. Since the probability amplitude $\psi(q, t)$ representing any such state can be obtained, as was seen in the preceding section, by the superposition of de Broglie waves, this means that we must investigate the nature of the *wave packets* which can be built up from these waves in such a way as to give an approximate specification of position and momentum.

The individual de Broglie waves to be used in constructing the wave packet will be of the form (61.9), namely

$$\psi = A(\sigma_x, \sigma_y, \sigma_z)e^{2\pi i(\sigma_x x + \sigma_y y + \sigma_z z - \nu t)}.$$

If now we desire at some given time to build up a packet which gives a large probability for finding the particle within a specified range Δx, it is evident that we shall at least have to make use of waves with sufficient difference in their wave numbers σ_x and $\sigma_x+\Delta\sigma_x$ so that they can reinforce in the middle of the range Δx and interfere outside. Since the number of wave crests per unit distance along the x-axis for two such waves will be equal to σ_x and $\sigma_x+\Delta\sigma_x$, the condition for such reinforcement and interference will then evidently be

$$(\sigma_x+\Delta\sigma_x)\Delta x-\sigma_x\Delta x = \Delta\sigma_x\Delta x \approx 1,$$

as this is what is needed to give us one more wave crest in the distance Δx for one of the waves than the other.

Hence, considering all three dimensions, we may now write

$$\begin{aligned}\Delta\sigma_x\Delta x &\approx 1,\\ \Delta\sigma_y\Delta y &\approx 1,\\ \Delta\sigma_z\Delta z &\approx 1\end{aligned} \tag{62.1}$$

as expressions giving the ranges in wave numbers $\Delta\sigma_x$, $\Delta\sigma_y$, $\Delta\sigma_z$ which will be needed to construct a wave packet which locates our particle in the approximate range $\Delta x\Delta y\Delta z$. If, however, we have to assign appreciable amplitudes to waves having these differences in wave number, it is evident from the preceding section that we shall have to assign appreciable probability amplitudes $\phi(p,t)$ for finding momenta that differ in accordance with the de Broglie relations (61.6) by amounts of the order

$$\Delta p_x = h\Delta\sigma_x, \qquad \Delta p_y = h\Delta\sigma_y, \qquad \Delta p_z = h\Delta\sigma_z. \tag{62.2}$$

Combining (62.1) and (62.2), we then see that the quantum mechanical state, corresponding to a wave packet specially constructed for the purpose of specifying the position and momentum of a particle, will only do so subject to the *Heisenberg uncertainty relations*

$$\begin{aligned}\Delta p_x\Delta x &\approx h,\\ \Delta p_y\Delta y &\approx h,\\ \Delta p_z\Delta z &\approx h.\end{aligned} \tag{62.3}$$

The formalism of the quantum mechanics is thus devised to agree with limitations on classical measurability which are conceptually necessary as soon as we consider the disturbing effect that observation itself would have on small systems.

For particles of large mass the uncertainties corresponding to (62.3) would be far below the limits of practical precision. Thus for a particle having a mass of 1 gram we could write

$$\Delta v_x\Delta x \approx h = 6{\cdot}5\times 10^{-27},$$

where Δv_x is the uncertainty in the x-component of velocity, and hence could have a simultaneous specification of position and velocity with accuracies of the order of 8×10^{-14} centimetres and centimetres per second respectively. It is therefore not surprising that the classical mechanics, which was devised to describe the behaviour of large masses, was able to employ the idea of exact kinematic description.

The uncertainty relations (62.3) connecting coordinates and momenta also imply an uncertainty relation connecting energy and time. If we consider for simplicity the motion of a particle of mass m in the x-direction with the approximate velocity v in that direction, we can re-express (62.3) in the form

$$m\Delta v\Delta x = (mv\Delta v)\left(\frac{\Delta x}{v}\right) \approx h.$$

And this result can be written as

$$\Delta E\Delta t \approx h, \tag{62.4}$$

where ΔE denotes the uncertainty in the energy of the particle, and Δt the uncertainty as to the time at which this energy is transferred from one side to the other of a selected surface. This means, for example, if we arrange a shutter for letting particles out from a container, that the time Δt during which the shutter is kept open and the uncertainty ΔE in the energy of a particle which escapes would be connected by the above relation.

Before turning to a more general consideration of the nature of uncertainty relations it will be of interest to point out certain further properties of the wave packets which we have used for investigating the simple case of a free particle.

Let us denote by $\bar{x}$ and $\bar{p}_x$ the mean values or expectation values for the x-coordinate and x-component of momentum of the particle as obtained by integrating over the whole of the wave packet. We have as expressions for these quantities

$$\bar{x} = \iiint \psi^* x\psi\, dxdydz \quad \text{and} \quad \bar{p}_x = \iiint \psi^* \frac{h}{2\pi i}\frac{\partial\psi}{\partial x}\, dxdydz, \tag{62.5}$$

in accordance with the rules for obtaining expectation values given by (57.4). Differentiating these expressions with respect to the time, and using Schroedinger's equation (57.5) to re-express the values of $\partial\psi^*/\partial t$ and $\partial\psi/\partial t$ thus introduced, it can readily be shown that we obtain for the case of a *free particle*

$$\frac{d\bar{x}}{dt} = \overline{\left(\frac{\partial H}{\partial p_x}\right)} = \frac{\bar{p}_x}{m}, \qquad \frac{d\bar{p}_x}{dt} = -\overline{\left(\frac{\partial H}{\partial x}\right)} = 0, \tag{62.6}$$

these results being a special case of equations (63.9) which we shall derive in the next section. For sharply defined wave packets, such as can be obtained with particles of large mass, the results given by (62.6) approach the classical results which give the velocity of a particle in terms of its momentum and assert that for a free particle the momentum will be constant.

To mention a further property of wave packets for a free particle, it is of interest to consider the change in their dimensions with time. To do this let us make our ideas as to degree of uncertainty somewhat more precise by defining the mean square uncertainties in the position and momentum of the particle by

$$\overline{\Delta x^2} = \iiint (x-\bar{x})^2\psi^*\psi\, dxdydz$$

and

$$\overline{\Delta p_x^2} = \iiint (p_x-\bar{p}_x)^2\phi^*\phi\, dp_x dp_y dp_z \tag{62.7}$$

It can then be shown† that the time dependence of the first of these quantities is given by

$$\overline{\Delta x^2} = \overline{(\Delta x^2)}_{t=0}+2t\iiint [(x-\bar{x})S_x]_{t=0}\, dxdydz+\frac{t^2}{m^2}\overline{\Delta p_x^2}, \tag{62.8}$$

while the uncertainty in momentum will remain constant in the case of a free particle unacted on by forces.

At first sight it might seem that equation (62.8) contains nothing particularly characteristic of the quantum mechanics, since it would also hold for classical particles distributed with the current density S_x and mean-square uncertainty in momentum $\overline{\Delta p_x^2}$. It is to be appreciated, however, that in such a classical distribution the quantities $\overline{(\Delta x^2)}_{t=0}$ and $\overline{\Delta p_x^2}$ could be taken as small as desired, which would not be possible in the quantum mechanical interpretation.

In accordance with (62.8) it is seen that the dimensions of a wave packet for a free particle would be a quadratic function of the time, increasing without limit after passing through a possible minimum, thus making it increasingly inappropriate as time proceeds to attempt a classical kinematic description of behaviour.

(*b*) **General treatment of uncertainty relations.** We may now turn to a somewhat mathematical consideration of uncertainty relations applicable to more general kinds of systems and observable quantities.‡

† See, for example, Kemble, *The Fundamental Principles of Quantum Mechanics*, New York, 1937, equation (33.18), which has been re-expressed above in terms of current density.

‡ See Robertson, *Phys. Rev.* **34**, 163 (1929); **35**, 667 (1930); **46**, 794 (1934); Schroedinger, *Berl. Ber.* 1930, p. 296.

Let the quantities of interest correspond to the operators **F** and **G**. At a particular given time we can then write for the mean or expectation values of the two quantities

$$\bar{F} = \int \psi^* \mathbf{F} \psi \, dq \quad \text{and} \quad \bar{G} = \int \psi^* \mathbf{G} \psi \, dq, \tag{62.9}$$

where ψ is the probability amplitude for the state, at that time, expressed in the q-language. We may now define the uncertainties ΔF and ΔG in these quantities somewhat more definitely than previously as the root mean square of the deviations from the expectation value, in accordance with the equations

$$(\Delta F)^2 = \int \psi^* (\mathbf{F} - \bar{F})^2 \psi \, dq \quad \text{and} \quad (\Delta G)^2 = \int \psi^* (\mathbf{G} - \bar{G})^2 \psi \, dq. \tag{62.10}$$

As a simplification let us now use the symbols

$$\mathbf{f} = \mathbf{F} - \bar{F} \quad \text{and} \quad \mathbf{g} = \mathbf{G} - \bar{G} \tag{62.11}$$

as an abbreviation for those operators; and let us define two new wave functions by

$$\psi_1 = \mathbf{f}\psi \quad \text{and} \quad \psi_2 = \mathbf{g}\psi. \tag{62.12}$$

Substituting in (62.10), we obtain

$$(\Delta F)^2 = \int \psi^* \mathbf{f} \psi_1 \, dq = \int \psi_1 (\mathbf{f}\psi)^* \, dq = \int \psi_1 \psi_1^* \, dq$$

$$\text{and} \qquad (\Delta G)^2 = \int \psi^* \mathbf{g} \psi_2 \, dq = \int \psi_2 (\mathbf{g}\psi)^* \, dq = \int \psi_2 \psi_2^* \, dq, \tag{62.13}$$

where the second forms of expression come from the Hermitian character of the operators **f** and **g**, in accordance with (55.25).

We are now ready to employ the known Schwarzian inequality†

$$\int \psi_1 \psi_1^* \, dq \int \psi_2 \psi_2^* \, dq \geqslant \int \psi_1^* \psi_2 \, dq \int \psi_2^* \psi_1 \, dq. \tag{62.14}$$

Substituting from above, and again making use of (55.25), we obtain

$$\begin{aligned}(\Delta F)^2 (\Delta G)^2 &\geqslant \int (\mathbf{f}\psi)^* \mathbf{g}\psi \, dq \int (\mathbf{g}\psi)^* \mathbf{f}\psi \, dq \\ &\geqslant \int \psi (\mathbf{g}\mathbf{f}\psi)^* \, dq \int \psi (\mathbf{f}\mathbf{g}\psi)^* \, dq \\ &\geqslant \int \psi^* \mathbf{g}\mathbf{f}\psi \, dq \int \psi^* \mathbf{f}\mathbf{g}\psi \, dq.\end{aligned} \tag{62.15}$$

Since in general **f** and **g** will not commute, we shall find it useful to re-express the right-hand side of this inequality with the help of the

† See Weyl, *Gruppentheorie und Quantenmechanik*, tr. Robertson, London, 1931, p. 393.

method for decomposing products given by (55.34). We can then write

$$\int \psi^* \mathbf{fg}\psi \, dq \int \psi^* \mathbf{gf}\psi \, dq = \int \psi^* \left(\frac{\mathbf{fg}+\mathbf{gf}}{2} + i\frac{\mathbf{fg}-\mathbf{gf}}{2i} \right) \psi \, dq \times$$

$$\times \int \psi^* \left(\frac{\mathbf{fg}+\mathbf{gf}}{2} - i\frac{\mathbf{fg}-\mathbf{gf}}{2i} \right) \psi \, dq$$

$$= \left(\frac{\overline{fg+gf}}{2} \right)^2 + \left(\frac{\overline{fg-gf}}{2i} \right)^2, \tag{62.16}$$

where, in accordance with (55.29) and (55.31), we now have the mean values corresponding to Hermitian operators.

Introducing (62.16) into (62.15), and also using the relations (62.11) by which **f** and **g** were defined, we now easily find that our uncertainty relation for any two quantities can be expressed in the general form

$$(\Delta F)^2(\Delta G)^2 \geqslant \left(\frac{\overline{FG+GF}}{2} - \bar{F}\bar{G} \right)^2 + \left(\frac{\overline{FG-GF}}{2i} \right)^2. \tag{62.17}$$

Fixing our attention on the right-hand side of this expression, we see that the product of the two uncertainties would at least have to be large enough to correspond to the second of the two terms given, in the case of two operators which do not commute. It should be noted that the limit of the complementary uncertainty itself depends in general on the state considered, since expectation values appear on the right in (62.17).

As a special example of the applicability of (62.17), we may now consider the case of conjugated canonical coordinates and momenta having the commutator $h/2\pi i$ given by (55.38). We then obtain from (62.17)

$$\Delta p_k \Delta q_k \geqslant \frac{h}{4\pi} \tag{62.18}$$

as an expression of the Heisenberg uncertainty relation, which is now applicable to conjugate canonical coordinates and momenta in general, and is also somewhat more definite than our original expression (62.3) for the case of a free particle.

In addition to the uncertainty relations connecting coordinates and momenta, we also had for free particles an uncertainty relation between energy and time (62.4) which can now be written somewhat more definitely in the form

$$\Delta E \Delta t \geqslant \frac{h}{4\pi}. \tag{62.19}$$

For the case of a free particle, this relation connects the uncertainty ΔE in the energy of the particle with the uncertainty Δt in the time at which the particle passes through a given boundary. In the case of the more general systems which now interest us, since a change in their energy content could be brought about by the impingement of particles, it will be natural to retain this same relation, with ΔE now denoting the uncertainty in any amount of energy E which is transferred to the system, and Δt the uncertainty in the time t at which the transfer takes place.

This new form of the principle would also have a bearing on the *accuracy* with which the *total energy* of a system could be regarded as *determined*. To measure this total energy we should have to allow for the effects of interaction between the system and the apparatus of measurement, and with a time Δt available for the measurement we should at least have an uncertainty ΔE in the result of the order given by (62.19), since this would express our lack of knowledge as to the transfer between system and measuring apparatus. Conversely, in order to fix the time of occurrence of an event within a system with a precision Δt in terms of a clock external thereto, an uncontrollable exchange of energy ΔE with the system of the order given by (62.19) would be necessary.

63. Correspondence between classical and quantum mechanical results

(*a*) **Change in expectation values with time.** We have now completed our account of the quantum mechanical point of view with regard to those problems connected with energy levels, wave-particle duality, and uncertainty which seemed so difficult from a classical standpoint. It remains to study the very extensive correspondence which relates quantum mechanical with classical results. Among other things this will be helpful in explaining and justifying the use of the classical mechanics under those limiting conditions where it remains applicable.

To investigate this matter it will first of all be advantageous to obtain a general expression for the rate of change with time in the expectation value for any observable quantity. This will then make it possible to compare quantum mechanical expressions for the rate of change with time in the expectation values of observable quantities with classical expressions for the rate of change in their precise values.

In accordance with our fundamental expression (57.4), the average or expectation value for any function $F(q,p)$ of the coordinates and momenta can be calculated from a knowledge of the corresponding operator $\mathbf{F}(\mathbf{q},\mathbf{p})$ with the help of the equation

$$\bar{F} = \int \psi^* \mathbf{F} \psi \, dq. \tag{63.1}$$

Considering for simplicity cases where $\mathbf{F}$ is not itself an explicit function of the time, we then obtain by differentiation

$$\begin{aligned}\frac{d\bar{F}}{dt} &= \int \left[\frac{\partial\psi^*}{\partial t}\mathbf{F}\psi + \psi^*\mathbf{F}\frac{\partial\psi}{\partial t}\right] dq \\ &= \frac{2\pi i}{h}\int [(\mathbf{H}\psi)^*\mathbf{F}\psi - \psi^*\mathbf{F}(\mathbf{H}\psi)]\, dq \\ &= \frac{2\pi i}{h}\int [\psi(\mathbf{F}\mathbf{H}\psi)^* - \psi^*\mathbf{F}\mathbf{H}\psi]\, dq \\ &= \frac{2\pi i}{h}\int \psi^*(\mathbf{H}\mathbf{F} - \mathbf{F}\mathbf{H})\psi \, dq.\end{aligned} \tag{63.2}$$

The first form in which the equation is written is a direct result of differentiation; the second form containing the Hamiltonian operator $\mathbf{H}$ comes by substituting for $\partial\psi^*/\partial t$ and $\partial\psi/\partial t$ with the help of the Schroedinger equation (57.5); the third form results from the definition of Hermitian operators (55.25); and the last form results from the derived property of Hermitian operators (55.26).

Introducing our previous abbreviation (55.32) for the *commutator* for two operators, we can also write (63.2) in the forms

$$\begin{aligned}\frac{d\bar{F}}{dt} &= \frac{2\pi i}{h}\int \psi^*[\mathbf{H},\mathbf{F}]\psi \, dq = \frac{2\pi i}{h}\overline{[H,F]} \\ &= \frac{2\pi}{ih}\int \psi^*[\mathbf{F},\mathbf{H}]\psi \, dq = \frac{2\pi}{ih}\overline{[F,H]},\end{aligned} \tag{63.3}$$

where $\overline{[F,H]}$ is the expectation value for the quantity that corresponds to the operator $[\mathbf{F},\mathbf{H}]$. Since we have already seen (55.31) that i times the commutator for two Hermitian operators is itself Hermitian, it is evident that the rate of change of $\bar{F}$ with time will be a real quantity, in satisfactory agreement with the reality of $\bar{F}$ itself, as given by (55.47).

Comparing the quantum mechanical equation (63.3) for the rate of change in the *expectation value* of $F(q,p)$, with the classical equation (11.4) for the rate of change in the *exact value* of $F(q,p)$, we also see that the quantity $\frac{2\pi}{ih}\overline{[F,H]}$ may be regarded in the present connexion as the quantum mechanical analogue of the classical Poisson bracket $\{F, H\}$. This is the first of several examples—compare (67.27) and (81.21)—showing a close relation between the classical Poisson bracket $\{M, N\}$ for two quantities and the quantum mechanical operator $\frac{2\pi}{ih}[\mathbf{M}, \mathbf{N}]$ which depends on the commutator for the two operators that correspond to those quantities. This relationship is made to play a fundamental role in the Dirac method of developing the quantum mechanics.†

(*b*) **The analogue of the Hamiltonian equations of motion.** We shall first apply the result given by (63.3) to the calculation of the rates of change in the expectation values for the coordinates and momenta themselves.

To do this let us first consider the character of the commutator $[\mathbf{H}, \mathbf{q}_k]$. We shall assume that the energy operator is given in accordance with the discussion of § 55 (*f*) as a sum of symmetrized terms of the form

$$\mathbf{H} = \sum \tfrac{1}{2}C(\mathbf{QP}+\mathbf{PQ}), \tag{63.4}$$

where $\mathbf{Q}$ and $\mathbf{P}$ are operators which are dependent on the $\mathbf{q}$'s and $\mathbf{p}$'s separately. For the commutator in question we can then write

$$\begin{aligned}[\mathbf{H}, \mathbf{q}_k] &= \sum \tfrac{1}{2}C[(\mathbf{QP}+\mathbf{PQ}), \mathbf{q}_k] \\ &= \sum \tfrac{1}{2}C\{[\mathbf{QP}, \mathbf{q}_k]+[\mathbf{PQ}, \mathbf{q}_k]\} \\ &= \sum \tfrac{1}{2}C\{\mathbf{Q}[\mathbf{P}, \mathbf{q}_k]+[\mathbf{Q}, \mathbf{q}_k]\mathbf{P}+\mathbf{P}[\mathbf{Q}, \mathbf{q}_k]+[\mathbf{P}, \mathbf{q}_k]\mathbf{Q}\} \\ &= \sum \tfrac{1}{2}C\{\mathbf{Q}[\mathbf{P}, \mathbf{q}_k]+[\mathbf{P}, \mathbf{q}_k]\mathbf{Q}\},\end{aligned} \tag{63.5}$$

where the justification for the various forms of writing is either evident or provided by the fundamental properties of commutators previously given by equations (55.33).

For the commutator $[\mathbf{P}, \mathbf{q}_k]$, however, we can obtain, with the help of an intermediate expression in the p-language,

$$\begin{aligned}[\mathbf{P}, \mathbf{q}_k] &= \mathbf{P}\mathbf{q}_k - \mathbf{q}_k\mathbf{P} \\ &= P\left(-\frac{h}{2\pi i}\frac{\partial}{\partial p_k}\right)-\left(-\frac{h}{2\pi i}\frac{\partial}{\partial p_k}\right)P \\ &= \frac{h}{2\pi i}\frac{\partial \mathbf{P}}{\partial \mathbf{p}_k}.\end{aligned} \tag{63.6}$$

Substituting into (63.5), we then obtain, with the help of the original expression (63.4) for the Hamiltonian itself,

$$[\mathbf{H}, \mathbf{q}_k] = \frac{h}{2\pi i}\sum \tfrac{1}{2}C\left(\mathbf{Q}\frac{\partial \mathbf{P}}{\partial \mathbf{p}_k}+\frac{\partial \mathbf{P}}{\partial \mathbf{p}_k}\mathbf{Q}\right) = \frac{h}{2\pi i}\frac{\partial \mathbf{H}}{\partial \mathbf{p}_k}. \tag{63.7}$$

† Dirac, *The Principles of Quantum Mechanics*, second edition, Oxford, 1935.

Furthermore, it is readily seen that analogous expressions can be obtained for the momenta. The important results thus obtained are summarized by the equations

$$[\mathbf{H}, \mathbf{q}_k] = \frac{h}{2\pi i}\frac{\partial \mathbf{H}}{\partial \mathbf{p}_k}$$

and

$$[\mathbf{H}, \mathbf{p}_k] = -\frac{h}{2\pi i}\frac{\partial \mathbf{H}}{\partial \mathbf{q}_k}. \tag{63.8}$$

These equations may now be substituted in our general expression (63.3) for the rate of change in expectation values with time, to give the desired results

$$\frac{d\bar{q}_k}{dt} = \overline{\left(\frac{\partial H}{\partial p_k}\right)} \quad \text{and} \quad \frac{d\bar{p}_k}{dt} = -\overline{\left(\frac{\partial H}{\partial q_k}\right)}. \tag{63.9}$$

We thus obtain the quantum mechanical analogues of the *equations of motion in the Hamiltonian form*. They show that the expectation values of the coordinates and momenta of a quantum mechanical system depend on the expectation values of partial derivatives of the energy in the same way that the precise values of the coordinates and momenta of a classical system depend on the precise values of those partial derivatives.†

As a result of this finding we now see that the classical mechanics may be regarded as the limiting form assumed by the quantum mechanics when the dispersion around the expectation values for the coordinates and momenta can be neglected. Since this dispersion, in accordance with the Heisenberg uncertainty relations, is due to the finite size of h, we can also regard the classical mechanics as the limiting form approached by the quantum mechanics as h goes to zero. We shall investigate the nature of the process by which the approach to the limit takes place somewhat more thoroughly later in the present section.

(c) **The conservation of energy in quantum mechanics.** Returning to our fundamental equation for the rate of change in the expectation value of any function of the coordinates and momenta,

$$\frac{d\bar{F}}{dt} = \frac{2\pi i}{h}\int \psi^*[\mathbf{H}, \mathbf{F}]\psi\, dq = \frac{2\pi i}{h}\overline{(HF - FH)}, \tag{63.10}$$

it will also be profitable to apply this equation to the case of the *energy* itself, where we take $\mathbf{F} = \mathbf{H}$.

† See Ehrenfest, *Zeits. f. Phys.* **45**, 455 (1927).

Since any operator commutes with itself, we at once obtain the result

$$\frac{d\bar{H}}{dt} = 0. \tag{63.11}$$

Furthermore, since **H** will commute with any power of itself, we shall also have

$$\frac{d\overline{H^n}}{dt} = 0 \tag{63.12}$$

for the time dependence of the expectation value of any integral power n of the energy.

These results, which apply, of course, only to conservative systems where **H** is not an explicit function of the time, show that both the *mean value* of the energy and the *relative probabilities* of finding different particular values for the energy of an undisturbed system will not change with the time. These conclusions may be regarded as the quantum mechanical analogue of the *principle of the conservation of energy*.

For the special case when the state of the system at some initial time is such that there is unit probability of finding only a particular value of the energy and zero probability for finding others, it is evident from the above that unit probability for finding that particular value of the energy will be retained until the time of the next measurement. We can then use the idea of the conservation of energy in the simple sense of the classical theory.

(*d*) **The conservation of momentum in quantum mechanics.** We may also apply our present methods of investigation to the total *linear momentum* of a system. Let us consider a system of particles and let q_k and p_k be the Cartesian coordinates and momenta of the individual particles. For the total momentum, in the x-direction for example, we can then write

$$P_x = \sum_k p_{xk} \tag{63.13}$$

by summing the x-components of momenta for all the particles $k = 1,..., n$.

For the commutator of the operators corresponding to the energy and to this component of the total momentum we shall have

$$[\mathbf{H}, \mathbf{P}_x] = \sum_k [\mathbf{H}, \mathbf{p}_{xk}] = -\frac{h}{2\pi i}\sum_k \frac{\partial \mathbf{H}}{\partial \mathbf{x}_k}, \tag{63.14}$$

where the last form results from the expression (63.8) already obtained for the commutator of energy with any individual component of momentum. For an isolated system, however, in which the Hamiltonian is dependent only on the *relative position* of the particles, we shall evidently have

$$\sum_k \frac{\partial \mathbf{H}}{\partial \mathbf{x}_k} = 0, \tag{63.15}$$

just as in the classical mechanics. (See § 13.) Hence the commutator (63.14) will be equal to zero, and, in accordance with (63.10), we then obtain

$$\frac{d\overline{P_x}}{dt} = 0 \tag{63.16}$$

for the time dependence of the mean value of the total momentum of the system in the x-direction.

Furthermore, the commutator for any power n of the momentum could be expanded in the form

$$[\mathbf{H}, \mathbf{P}_x^n] = [\mathbf{H}, \mathbf{P}_x]\mathbf{P}_x^{n-1} + \mathbf{P}_x[\mathbf{H}, \mathbf{P}_x^{n-1}], \tag{63.17}$$

and hence by a continuation of such steps into a succession of terms each containing the commutator $[\mathbf{H}, \mathbf{P}_x]$ which we have seen to be equal to zero for an isolated system. Hence we also obtain

$$\frac{d\overline{P_x^n}}{dt} = 0, \tag{63.18}$$

for the time rate of change of the expectation value of any power of the total momentum in the x-direction of an isolated system.

Equations (63.16) and (63.18) provide the quantum mechanical analogue of the *principle of the conservation of momentum.* Just as in the case of energy, we see for an isolated system which is known at some initial time to have a particular value of the momentum, we shall have conservation of momentum in the simple classical sense up to the time of the next measurement.

We can also treat the total *angular momentum* of a system around a selected axis. Let us consider a system of particles, for simplicity in the absence of any magnetic field, and let us use Cartesian coordinates x_k, y_k, z_k and the corresponding momenta p_{xk}, p_{yk}, p_{zk} for the different particles $k = 1,..., n$ of the system. Taking the angular momentum of the system around the z-axis as an example, we shall have as the operator corresponding thereto

$$\mathbf{M}_z = \sum_k (\mathbf{x}_k \mathbf{p}_{yk} - \mathbf{y}_k \mathbf{p}_{xk}), \tag{63.19}$$

where the summation is taken over all the n particles of the system, and no symmetrization of the operator has been necessary owing to the commutability of non-conjugated coordinates and momenta. For the commutator of this operator with H, we have a sum of terms

$$[\mathbf{H}, \mathbf{M}_z] = \sum_k \{[\mathbf{H}, \mathbf{x}_k \mathbf{p}_{yk}] - [\mathbf{H}, \mathbf{y}_k \mathbf{p}_{xk}]\}. \tag{63.20}$$

And treating the individual terms by our general rules (55.33) for manipulating commutators, and substituting from the fundamental expressions (63.8) for the commutators of energy with individual coordinates and momenta, we obtain results of the form

$$[\mathbf{H}, \mathbf{x}\mathbf{p}_y] - [\mathbf{H}, \mathbf{y}\mathbf{p}_x] = [\mathbf{H}, \mathbf{x}]\mathbf{p}_y + \mathbf{x}[\mathbf{H}, \mathbf{p}_y] - [\mathbf{H}, \mathbf{y}]\mathbf{p}_x - \mathbf{y}[\mathbf{H}, \mathbf{p}_x]$$
$$= \frac{h}{2\pi i}\left\{\frac{\partial \mathbf{H}}{\partial \mathbf{p}_x}\mathbf{p}_y - \mathbf{x}\frac{\partial \mathbf{H}}{\partial \mathbf{y}} - \frac{\partial \mathbf{H}}{\partial \mathbf{p}_y}\mathbf{p}_x + \mathbf{y}\frac{\partial \mathbf{H}}{\partial \mathbf{x}}\right\}. \tag{63.21}$$

The first and third terms on the right-hand side of this expression will cancel, however, for the case of a particle where the momenta enter the Hamiltonian as a sum of squared terms $(p_x^2+p_y^2+p_z^2)$. Hence, by substitution into our general expression (63.10) for the rate of change in expectation values, we now obtain the desired résult

$$\frac{d\overline{M}_z}{dt} = -\sum_k \overline{\left(x_k \frac{\partial H}{\partial y_k} - y_k \frac{\partial H}{\partial x_k}\right)}. \tag{63.22}$$

Just as in the classical mechanics, however, this result will be equal to zero for an isolated system in which the Hamiltonian is invariant to a rotation of axes. (See § 13.) We can, of course, also treat the time rate of change in the expectation value for any power of the angular momentum. We thus obtain the quantum mechanical analogue of the *principle of the conservation of angular momentum.*

(*e*) **Approach of quantum mechanical behaviour to the classical limit.** We have already emphasized, in connexion with equations (63.9), that the classical mechanics could be regarded as a limiting case approached by the quantum mechanics as h goes to zero. We may now amplify our preceding treatment by considering somewhat more in detail the nature of the approach to the limit.† For purposes of illustration it will be sufficient to consider the behaviour of a single particle of mass m, with Cartesian coordinates denoted by q_k $(k = 1, 2, 3)$, moving in a potential field $V(q_k)$.

Making use of coordinate language, we shall find it convenient to express the probability amplitude for this system in the quite general possible form

$$\psi(q,t) = Ae^{2\pi iW/h}, \tag{63.23}$$

† The method is due to Wentzel, *Zeits. f. Phys.* **38**, 5 (1926), and Brillouin, *Comptes Rend.* **183**, 24 (1926).

where A and W are real functions of the coordinates and time. Substituting this expression into the Schroedinger equation (56.6) for such a particle

$$\sum_k \frac{h^2}{8\pi^2 m}\frac{\partial^2\psi}{\partial q_k^2} - V\psi - \frac{h}{2\pi i}\frac{\partial\psi}{\partial t} = 0,$$

and cancelling the factor $e^{2\pi iW/h}$, we obtain

$$\sum_k \frac{h^2}{8\pi^2 m}\left\{\frac{\partial^2 A}{\partial q_k^2} + \frac{4\pi i}{h}\frac{\partial A}{\partial q_k}\frac{\partial W}{\partial q_k} - \frac{4\pi^2}{h^2}A\left(\frac{\partial W}{\partial q_k}\right)^2 + \frac{2\pi i}{h}A\frac{\partial^2 W}{\partial q_k^2}\right\} - \\ -VA - \frac{h}{2\pi i}\frac{\partial A}{\partial t} - A\frac{\partial W}{\partial t} = 0. \quad (63.24)$$

Separating the real and imaginary parts, this then leads to two equations which can be written—after some simplification—in the forms

$$\sum_k \frac{1}{2m}\left\{\left(\frac{\partial W}{\partial q_k}\right)^2 - \frac{h^2}{4\pi^2}\frac{1}{A}\frac{\partial^2 A}{\partial q_k^2}\right\} + V + \frac{\partial W}{\partial t} = 0$$

and

$$\frac{\partial (A^2)}{\partial t} + \sum_k \frac{\partial}{\partial q_k}\left(\frac{A^2}{m}\frac{\partial W}{\partial q_k}\right) = 0. \quad (63.25)$$

These equations have been obtained without approximation and provide the means for an exact description of the quantum mechanical behaviour of the particle. The second of the two equations relates the rate of change in probability density (A^2) to the divergence of the appropriate expression for probability current.

We must now examine the effect on these equations of letting h go to zero. The first of the two equations (63.25) can then be written in the form

$$\sum_k \frac{1}{2m}\left(\frac{\partial W}{\partial q_k}\right)^2 + V + \frac{\partial W}{\partial t} = 0. \quad (63.26)$$

And comparing with our previous classical equation (16.1), as presented in Chapter II, we see at our present stage of approximation that the solution of the above equation

$$W = W(q_k, \alpha_k, t) + \text{const.}, \quad (63.27)$$

where the α_k are constants of integration, would be Hamilton's *principal function* W for the classical motion of a particle of mass m in the potential field V. Hence, if we now consider for particles in the potential field V the classical trajectories that would correspond to a particular selection of the constants, we can write

$$\frac{\partial W}{\partial q_k} = p_k = m\dot{q}_k, \quad (63.28)$$

in accordance with (16.13), where the $\dot{q}_k$'s, regarded as functions of the q_k's and t, now give the velocity with which a classical particle would be moving at any point q_k along such trajectories.

Returning now to the second of the two equations (63.25), and substituting (63.28), we can then write

$$\frac{\partial(A^2)}{\partial t}+\sum_k \frac{\partial}{\partial q_k}(A^2\dot{q}_k) = 0, \tag{63.29}$$

which shows that the probability density (A^2) would depend on position and time in the same manner as the density of a cloud of representative points moving along classical trajectories for particles of mass m in the potential field V. In the classical mechanics, however, it would be possible to consider the motion of a little region, containing such points, which could be chosen as small as desired. Hence, by letting h go to zero, we should approach the possibility of constructing a wave packet giving unit probability for finding a particle as time proceeds within as small a region as desired which would itself move along a classical trajectory.

To investigate the range of validity for such a method of treatment, we note, in accordance with the first of equations (63.25), that the treatment depends on an approximation which is justified only when

$$\left|\sum_k \frac{1}{A}\frac{\partial^2 A}{\partial q_k^2}\right| \ll \sum_k \frac{4\pi^2}{h^2}\left(\frac{\partial W}{\partial q_k}\right)^2. \tag{63.30}$$

Or noting that the original form of solution (63.23) actually has a periodic character corresponding to an approximate wave-length λ, and a corresponding approximate momentum p, which would be given by

$$\sum_k \frac{1}{h^2}\left(\frac{\partial W}{\partial q_k}\right)^2 = \frac{1}{\lambda^2} = \frac{p^2}{h^2}, \tag{63.31}$$

we see that the approximation would only be justified with

$$\left|\sum_k \frac{1}{A}\frac{\partial^2 A}{\partial q_k^2}\right| \ll \frac{4\pi^2}{\lambda^2} = \frac{4\pi^2 p^2}{h^2}. \tag{63.32}$$

We thus see that the construction of small wave packets with A fairly permanently concentrated in the neighbourhood of a classical trajectory will in any case only be possible for particles of high momentum and the corresponding small wave-lengths.

D. FURTHER DEVELOPMENT OF QUANTUM MECHANICAL METHODS. TRANSFORMATION THEORY

64. Characteristic states. Eigenvalues and eigenfunctions in general

The foregoing completes our account of the quantum theoretical explanation of those changes away from classical ideas which have seemed empirically or conceptually necessary, and also completes our discussion of the limiting relations still preserved between classical and quantum mechanics. We must now consider some further general developments of quantum mechanical methods that can be obtained from the postulatory basis already introduced. We shall then be ready in the next chapter to undertake applications of quantum mechanics to specific problems, which will be necessary for our later statistical considerations.

(*a*) **Equation determining a characteristic state.** In the present section we shall study the possibility of specifying the state of a quantum mechanical system, at any time of interest, in such a way that some selected observable dynamical quantity has a precise value. Such a state may be called a *characteristic state* for the observable quantity in question.

To carry out the investigation, let us take $F(q,p)$ as the observable quantity which is to exhibit a precise value (eigenvalue) F_e at the time of interest t_0; and let us make use of a coordinate representation so that the state of the system at that time can be specified in accordance with the foregoing by a function of the coordinates

$$u(q_1 \dots q_f) = \psi(q_1 \dots q_f, t_0) \tag{64.1}$$

which gives the probability amplitude $\psi(q,t)$ at the time t_0. Using this notation, it will then be found that the equation

$$\mathbf{F}u(q_1 \dots q_f) = F_e u(q_1 \dots q_f), \tag{64.2}$$

where $\mathbf{F}$ is the quantum mechanical operator corresponding to the observable $F(q,p)$, will determine the form of the probability amplitude $u(q)$ necessary to specify a state such that this observable will exhibit the precise value F_e.

To prove this principle we note in the first place that the mean value of the nth power of the quantity $F(q,p)$ would in any case be given in accordance with the quantum mechanics by an equation having the form of (57.4), namely

$$\overline{F^n} = \int u^*(q)\mathbf{F}^n u(q)\, dq, \tag{64.3}$$

where $u(q)$ is the probability amplitude for the system at the time of interest. And hence, if we actually specify the state of the system by a function $u(q)$ which is a solution of (64.2), we see that we shall then be able to write

$$\begin{aligned} \overline{F^n} &= \int u^*(q)F_e^n u(q)\,dq \\ &= F_e^n \int u^*(q)u(q)\,dq \\ &= F_e^n, \end{aligned} \tag{64.4}$$

where the second form of expression is justified by the fact that F_e is merely a number, and the third form by the consideration that the integrated probability for finding the coordinates somewhere in their possible range must be equal to unity. This result shows, however, that the mean value of any integral power of F is equal to the same power of the specified quantity F_e, which can only be realized if F itself has the precise value

$$F = F_e, \tag{64.5}$$

as was to be proved.

Equation (64.2) is hence a very important one, since it can be used to study the form of the probability amplitude when any chosen observable quantity has any one of the different precise values which it can assume. Indeed, starting from a different point of view, we have already discussed in § 60 the important equation (60.4)

$$\mathbf{H}u(q) = Eu(q), \tag{64.6}$$

which we now see to be the special form assumed by the general equation (64.2) for the case when the energy of the system is the quantity which has a definite specified value.

(*b*) **Eigenvalues and eigenfunctions corresponding to characteristic states.** Returning to the more general expression

$$\mathbf{F}u(q_1 \ldots q_f) = F_e u(q_1 \ldots q_f), \tag{64.7}$$

it will now be profitable to consider several implications of this equation for determining the forms of the probability amplitude $u(q)$ which give characteristic states for the observable quantity $F(q,p)$.

In the first place it is evident that any solutions of (64.7) which have physical interest must give expressions for the probability amplitude $u(q)$ such that $u^*(q)u(q)$ can actually be interpreted as a probability density. This will be the case if $u(q)$ is a single-valued, continuous, and finite function of the coordinates q over their range of significant values.† It is evident, however, that such allowable solutions for $u(q)$ may in general be possible only for very special values of F_e.

Hence equation (64.7) now permits us to determine those values of $F(q,p)$ which are physically possible. These may be called the *eigenvalues* of this quantity. Having determined what these possible eigenvalues are, we may then speak, as was previously done in the special case of energy, of the *spectrum of possible eigenvalues* for any observable quantity. Such spectra can, of course, turn out to be *discrete, continuous*, or *partly both* as the case may be.

Corresponding to each eigenvalue F_k, there will be one or more independent solutions of (64.7) for the form of $u(q)$, and these may be called the *eigenfunctions* corresponding to the different possible eigenvalues. In case there is only one such independent solution, we may call the state *non-degenerate*, and may label the solution with an appropriate subscript, e.g. $u_k(q)$, to indicate the eigenvalue to which it corresponds. In case we find it possible to obtain more than one independent allowable solution corresponding to a given eigenvalue F_k, we shall say that the state is *degenerate* in the same way that we have previously spoken of degenerate energy levels. Under these circumstances we may either label the different eigenfunctions with single subscripts which do not try to specify the different eigenvalues to which the function may correspond, or we may use pairs of subscripts, as, for example, $u_{kl}(q)$, which would denote the lth independent solution corresponding to the kth eigenvalue.

(c) **Properties of eigenvalues and eigenfunctions.** The properties of these eigenvalues and eigenfunctions may now be studied, using similar methods to those employed in the latter part of § 60 for studying the eigenvalues and eigenfunctions for the particular case of states characteristic of the energy. It will be profitable to carry out such a study, treating degenerate states somewhat more in detail than in § 60.

In accordance with our general equation (64.7) for a characteristic state, and the complex conjugate equation corresponding to it, we may write

$$\mathbf{F}u_k = F_k u_k$$

and

$$(\mathbf{F}u_j)^* = F_j^* u_j^*$$

as equations applying to the eigenvalues and eigenfunctions indicated. Multiplying the first of these equations by u_j^* and the second by u_k and

† For a discussion of the somewhat less restricted requirements actually necessary for a satisfactory solution, see Pauli, *Handbuch der Physik*, xxiv/1, second edition, Berlin, 1933, p. 123.

integrating over the whole range of the coordinates q, we obtain

$$\int u_j^* \mathbf{F} u_k \, dq = F_k \int u_j^* u_k \, dq$$

and
$$\int u_k (\mathbf{F} u_j)^* \, dq = F_j^* \int u_j^* u_k \, dq.$$

Since the operator $\mathbf{F}$ is Hermitian, this then leads, in accordance with the definition of Hermitian operators (55.25), to the useful result

$$(F_k - F_j^*) \int u_j^* u_k \, dq = 0. \tag{64.8}$$

With the help of this equation we may now obtain some important conclusions.

In the first place let us consider the case $j = k$, where the subscripts refer to the same eigenfunctions. On account of the essentially positive character of $u_k^* u_k = |u_k|^2$, we then obtain

$$F_k = F_k^* \tag{64.9}$$

and must conclude that all possible eigenvalues are real quantities, in agreement with our previous finding (55.47) that the expectation values of observable quantities are always *real*.

As a second possibility let us consider the case $j \neq k$, corresponding to the *different eigenvalues* $F_j \neq F_k$. To preserve the truth of equation (64.8) we must then have

$$\int u_j^* u_k \, dq = 0 \tag{64.10}$$

and are led to the important conclusion that the eigenfunctions corresponding to different eigenvalues are necessarily *orthogonal*.

Finally we must consider the possibility $j \neq k$ but $F_j = F_k$, which would arise in the case of a *degenerate state* having more than one independent eigensolution, i.e. both u_k and u_j corresponding to the same eigenvalue F_k. Under these circumstances it would no longer be possible to conclude from (64.8) that these different eigenfunctions were necessarily orthogonal. We must now give special consideration to the eigenfunctions corresponding to degenerate states.

Considering a particular eigenvalue F_k, we shall call the corresponding state g-fold degenerate in case we can find g essentially different solutions $u_{kl}(q)$ $(l = 1, 2,..., g)$ of equation (64.7) corresponding to this eigenvalue. As the criterion of essential difference or *linear independence* of the different solutions, $u_{k1}, u_{k2},..., u_{kg}$, we shall require that no linear relation can be found of the form

$$a_1 u_{k1} + a_2 u_{k2} + ... + a_l u_{kl} = 0, \tag{64.11}$$

where the a's are constant coefficients, which for arbitrary values of the independent variables q would permit us to solve for one of the functions u_{kn} in terms of the others. Having found such a set of independent functions, it will then be possible, however, to express any further possible solution of (64.7), corresponding to F_k, as a linear combination of the members of this set.

As already noted, the independent eigenfunctions composing the set would not necessarily be orthogonal since they correspond to the same eigenvalue F_k. Nevertheless, it will always be possible—and indeed in an infinite number of ways—to construct therefrom a satisfactorily equivalent set the members of which will be orthogonal. To do this we may, for example, take u_{k1} itself as the first member of the new set

$$u'_{k1} = u_{k1}.$$

As the second member of the set we may then take

$$u'_{k2} = u_{k2}+\alpha u_{k1},$$

where the constant α may evidently be chosen in such a way as to secure the desired orthogonality

$$\int u'^*_{k1} u'_{k2}\, dq = \int u^*_{k1} u_{k2}\, dq + \alpha \int u^*_{k1} u_{k1}\, dq = 0.$$

Continuing, we may then take the next member of the set as

$$u'_{k3} = u_{k3}+\beta u_{k1}+\gamma u_{k2},$$

where β and γ are chosen to secure orthogonality of u'_{k3} with both the previous eigenfunctions u'_{k1} and u'_{k2}. In this way we can evidently proceed until we obtain a new set of g independent eigenfunctions which *are* orthogonal and are equivalent to the original set for the purpose of forming linear combinations to express the different solutions of (64.7) that correspond to the eigenvalue F_k of the degenerate state in question.

As a result of the foregoing discussion we now see that our equation (64.7) for the states characteristic of any observable quantity F will provide a series of independent eigenfunctions $u_n(q)$ in terms of which we can express any solution characteristic of that observable. These eigenfunctions will either be necessarily *orthogonal*,

$$\int u^*_m(q)u_n(q)\, dq = 0, \tag{64.12}$$

or can be orthogonalized by the method described above. Furthermore, they can evidently be multiplied by suitable factors so as to be *normalized* in accordance with the equation

$$\int u^*_n(q)u_n(q)\, dq = 1. \tag{64.13}$$

In later sections we shall study the important use made of such eigenfunctions in the quantum mechanics in obtaining expansions for other more general functions of the coordinates.

(*d*) **States characteristic of more than one observable.** We must now inquire into the possibility of specifying states in such a way that more than one kind of observable quantity, corresponding, say, to the different operators **F** and **G**, should simultaneously exhibit precisely defined values.

In order to achieve this, we at once see from our previous general treatment of uncertainty relations given in § 62 (*b*) that it will in any case be *necessary* for the expectation value of the commutator of the two operators to vanish,

$$\overline{(FG-GF)} = 0,$$

since otherwise the two quantities would be subject to uncertainties ΔF and ΔG which, in accordance with (62.17), could not simultaneously go to zero. This necessary condition will, of course, always be satisfied if the two operators commute,

$$\mathbf{FG} = \mathbf{GF}. \tag{64.14}$$

We shall now be able to show, moreover, that such commutability will actually be a *sufficient* condition for obtaining states such that F and G will both be precisely defined.

To investigate this let us consider the equation

$$\mathbf{F}u_{kl}(q) = F_k u_{kl}(q) \qquad (l = 1, 2,..., g), \tag{64.15}$$

which would determine, in the case of g-fold degeneracy, the lth independent eigenfunction corresponding to the eigenvalue F_k. Applying the operator **G** to this equation and making use of the commutability (64.14), we then obtain

$$\mathbf{F}(\mathbf{G}u_{kl}) = F_k(\mathbf{G}u_{kl}). \tag{64.16}$$

From this result we see that $\mathbf{G}u_{kl}$ must itself express a state characteristic of the eigenvalue F_k, and hence can be written as a linear combination

$$\mathbf{G}u_{kl} = \sum_m G_{ml} u_{km} \qquad (l, m = 1, 2,..., g) \tag{64.17}$$

of the independent eigenfunctions u_{km} corresponding to that eigenvalue.

In the special case of no degeneracy, (64.17) then solves the problem, since it shows that the single u_{km} will be an eigenfunction of **G** as well as of **F**, and the single G_{ml} an eigenvalue of G.

In the general case, the g^2 quantities G_{ml} occurring in (64.17) will be constants whose values could evidently be determined, making use of the normalization and orthogonality of the eigenfunctions u_{km}, from expressions of the form

$$G_{nl} = \int u_{kn}^* \mathbf{G} u_{kl}\, dq. \tag{64.18}$$

As a result of (64.18) and the Hermitian character of the operator $\mathbf{G}$, we then see that the quantities G_{nl} are the components of an Hermitian matrix.

This now makes it possible to apply a known theorem concerning Hermitian forms.† In accordance with this theorem, if we have a set of g independent, normalized, orthogonal eigenfunctions $u_l(q)$, it will be possible to change to a new set of similar, independent, normalized, orthogonal eigenfunctions $v_n(q)$ such that a general Hermitian form in the original eigenfunctions will be connected with a simplified Hermitian form in the new eigenfunctions by the equation

$$\sum_{l,m} u_l^* G_{ml} u_m = \sum_n v_n^* G_n v_n, \tag{64.19}$$

where the g quantities G_n are real constants, and the equations connecting the old and new eigenfunctions are those for a unitary transformation (see § 67 (e))

$$v_n = \sum_l u_l S_{ln} \quad \text{and} \quad u_l = \sum_n v_n S_{nl}^{-1}, \tag{64.20}$$

with the components of the transformation matrices subject to the relations

$$\begin{gathered} S_{mn} = \int u_m^* v_n\, dq, \qquad S_{mn}^{-1} = S_{nm}^* = \int v_m^* u_n\, dq, \\ \sum_l S_{ln}^* S_{lm} = \sum_l S_{nl}^* S_{ml} = \delta_{nm}. \end{gathered} \tag{64.21}$$

To apply this theorem we return to (64.17) and for simplicity temporarily drop the index k which labels all the eigenfunctions involved. Multiplying (64.17) by u_l^* and summing over l, we then obtain

$$\sum_l u_l^* \mathbf{G} u_l = \sum_{l,m} u_l^* G_{ml} u_m,$$

which gives us, in accordance with (64.19),

$$\sum_l u_l^* \mathbf{G} u_l = \sum_n v_n^* G_n v_n,$$

and this in turn can be rewritten, in accordance with (64.20) and (64.21), in the forms

$$\begin{aligned} \sum_{l,n,r} v_n^* (S_{nl}^{-1})^* \mathbf{G} v_r S_{rl}^{-1} &= \sum_{l,n,r} S_{ln} S_{lr}^* v_n^* \mathbf{G} v_r = \sum_{n,r} \delta_{nr} v_n^* \mathbf{G} v_r \\ &= \sum_n v_n^* \mathbf{G} v_n = \sum_n v_n^* G_n v_n. \end{aligned}$$

† See Weyl, *Gruppentheorie und Quantenmechanik*, tr. Robertson, London, 1931, p. 21.

Taking cognizance of the independent character of the functions v_n, the last two members of this series of equations then lead to the desired final expression, in which the common index k has again been introduced,

$$\mathbf{G}v_{kn} = G_n v_{kn} \qquad (n = 1, 2,..., g). \tag{64.22}$$

In accordance with this result and our fundamental expression for characteristic states (64.7), we now see that our new functions v_{kn} describe states of the system such that the quantity G will exhibit one or another of the precise values G_n. Furthermore, since the v_{kn} are themselves expressible in accordance with (64.20) as linear combinations of the original u_{kl}, we see that the states v_{kn} will also be such that the system exhibits the precise value F_k. We have hence given the desired demonstration that states simultaneously characteristic of different observable quantities can be found, provided only that the operators corresponding to these observables commute.

The above possibility of obtaining characteristic states can, of course, be extended to more than two observables, provided all the corresponding operators commute. For example, a state such as v_{kn} considered above, which is characterized by the specific eigenvalues F_k and G_n, may itself still be a degenerate state, and a further application of the foregoing methods of treatment may make it possible to introduce new functions w_{knm} characterized by eigenvalues F_k, G_n, and H_m of essentially different observable quantities F, G, and H. In the case of a system of f degrees of freedom, f would be the maximum number of actually independent observables which could be so employed.

(*e*) **Eigenvalues and eigenfunctions for the coordinates and momenta.** Included among the general possibilities for states characteristic of different observable quantities we have the special cases of states characteristic of a coordinate or of a momentum. The eigenvalues and eigenfunctions then present some special properties which deserve consideration. It will be sufficient to illustrate these properties for a one-dimensional problem with q and p as the canonical coordinate and conjugate momentum.

If we are using coordinate language, the operator q will reduce to q itself and our general equation (64.7) for a state characteristic of this quantity will reduce to the simple form

$$qu(q) = q_e u(q), \tag{64.23}$$

where q_e is an eigenvalue, i.e. some particular observable value for the coordinate q. As the solution of this equation we may take the improper function

$$u(q) = \delta(q-q_e), \tag{64.24}$$

to which Dirac has given the name delta function, and which is usually normalized by the condition

$$\int \delta(q-q_e)\,dq = 1. \tag{64.25}$$

It has the important property that for any continuous function $f(q)$ we have

$$\int f(q)\delta(q-q_e)\,dq = f(q_e). \tag{64.26}$$

Turning now to solutions characteristic of the momentum, the operator $\mathbf{p}$ in coordinate language will have the form $\frac{h}{2\pi i}\frac{\partial}{\partial q}$, and our general equation (64.7) for a state characteristic of this quantity will have the form

$$\frac{h}{2\pi i}\frac{\partial u(q)}{\partial q} = p_e u(q), \tag{64.27}$$

where p_e is some eigenvalue for the momentum. As the solution of this equation we can take

$$u(q) = e^{(2\pi i/h)p_e q}. \tag{64.28}$$

In evaluating expectation values from (64.24) and (64.28) by the formula

$$\overline{F} = \int u(q)\mathbf{F}u(q)\,dq \tag{64.29}$$

one must approximate to the singular eigenfunctions, in the one case by normalized packets more and more concentrated about the point q_e, in the other by normalized packets corresponding to sharper and sharper momentum definition at the value p_e, and must carry out the integration (64.29) before passing to the limit.

Thus also in the somewhat special cases of coordinates and momenta, with their continuous ranges of eigenvalues, we can obtain formally appropriate characteristic state solutions.

65. Expansions in terms of eigenfunctions

The preceding section has discussed the methods of obtaining eigenfunctions $u(q)$, for any quantum mechanical system, which would describe states such that observable quantities exhibit precisely defined values. With the help of these methods we can construct complete sets of eigenfunctions which are very important in the quantum mechanics. Such sets may be composed of independent eigenfunctions selected to correspond to all the possible eigenvalues of some one particular observable quantity F, or, if desired, they may be composed—in the case of systems of more than one degree of freedom—of independent eigenfunctions selected to correspond simultaneously to the eigenvalues of more than one observable quantity F, G, etc.

In the case of such sets of eigenfunctions, different notations may be used to distinguish one member of the set from another. The simplest notation is to regard each member of the set as labelled with a specific index. Thus we may write

$$u_n(q)$$

to denote the nth eigenfunction of the set, or may also write this in the form

$$u(n,q)$$

to emphasize that the value of u depends both on the index n and on the coordinates q. In the case of a system of one degree of freedom, the index n could be regarded as a number assuming integral or continuous values according as the corresponding eigenvalues gave a discrete or continuous spectrum. In the case of f degrees of freedom the index n could be regarded as a collection of f such numbers. We shall for the most part use this simple notation, although sometimes it is profitable to use more complicated notations, such, for example, as $u_{kl}(q)$ in the case of a degenerate state to denote the lth eigenfunction of the set that corresponds to an eigenvalue F_k, or $u(F,G,\ldots,q)$ in the case of a state characteristic of more than one observable to indicate the dependence on the eigenvalues F, G,... .

In accordance with the discussion of the preceding section we shall always regard the eigenfunctions of such sets as normalized and orthogonal in agreement with the expression

$$\int u_m^*(q)u_n(q)\,dq = \delta_{mn}. \tag{65.1}$$

Such sets of normalized orthogonal functions are often used in the quantum mechanics for the purpose of expressing some selected function of the coordinates, say $f(q)$, as an expansion into a series of the form

$$f(q) = c_1 u_1(q)+c_2 u_2(q)+\ldots+c_n u_n(q)+\ldots = \sum_k c_k u_k(q), \tag{65.2}$$

where the c_k are constant coefficients which may be real, imaginary, or complex.

When such an expansion is possible, the value of any coefficient c_n can easily be expressed by multiplying (65.2) through by the function $u_n^*(q)$ and integrating over the full range of the coordinates q. We thus obtain

$$\int u_n^*(q)f(q)\,dq = \sum_k c_k \int u_n^*(q)u_k(q)\,dq,$$

which, in accordance with the normalization and orthogonality expressed by (65.1), gives us the simple result

$$c_n = \int u_n^*(q)f(q)\,dq. \tag{65.3}$$

This general method of obtaining the values of the coefficients is thus the same as that familiar in the special case when the orthogonal functions are the sine and cosine terms used in a Fourier expansion. Since the values of the coefficients are definitely determined by (65.3), an expansion in terms of a given set of orthogonal functions is unique.

In order that a successful expansion can be made in the form (65.2) we need some criterion for the *convergence* of the series and for the *completeness* of the set of orthogonal functions. The following remarks may be made in this connexion.

In the quantum mechanics, expansions such as the above are used for functions $f(q)$ such that the integral $\int |f(q)|^2\,dq$, over the whole range of coordinates, exists in the sense of having a definite non-infinite value. And it is sufficient to regard an expansion as satisfactory when it gives *convergence in the mean*, i.e. when

$$\lim_{n\to\infty} \int \left|f(q)-\sum_{k=1}^{n} c_k u_k\right|^2 dq = 0, \tag{65.4}$$

rather than to impose convergence of the series in the ordinary sense as a necessary requirement.

By evaluating the indicated square in (65.4) and making use of the normalization and orthogonality expressed by (65.1), the above equation can readily be shown equivalent to a relation, having the simpler form,

$$\int |f(q)|^2\,dq = \sum_{k=1}^{\infty} c_k^* c_k. \tag{65.5}$$

The application of this equation, using values of the c_k determined by (65.3), can be used as a test for *convergence in the mean*, and therefore also as a test for *completeness* since its fulfilment would show that no further eigenfunctions could be added to those already employed.

In making use of a series expansion (65.2) of the kind that we have suggested,

$$f(q) = \sum_k c_k u_k(q), \tag{65.6}$$

it will be appreciated that some of the coefficients c_k may turn out to be zero, in case $f(q)$ has special properties, such, for example, as being an odd or even function, and it is also apparent that cases may arise where only a few eigenfunctions may be necessary to give a satisfactorily approximate representation of a function $f(q)$.

From a more general point of view, nevertheless, it is evident that we must contemplate the appearance in our series of all possible eigenfunctions corresponding to the different eigenvalues of the observable or observables of interest. In case these eigenvalues present a *continuous spectrum*, this means that we must make use of the continuous range of corresponding eigenfunctions; and in the case of *degeneracy* it means that we must make use of all the independent eigenfunctions that may correspond to a single eigenvalue. If, however, we do include *all* possible eigenfunctions corresponding to the different eigenvalues of *any selected observable or set of observables*, it appears possible to obtain satisfactory expansions of the general type (65.6) for all the functions of $f(q)$, with $\int |f(q)|^2\,dq$ existing, which we need to treat by this method for the purposes of the quantum mechanics.†

In the practical treatment of actual problems the simple summation, expressed by (65.6) with each independent eigenfunction labelled by its own subscript k, is often replaced by formalisms which give explicit recognition to the occurrence of continuous ranges of eigenvalues, or to the nature of any degeneracy that may be present. This may involve the introduction of integration in place of summation, and the designation of eigenfunctions in ways already mentioned to indicate their correlation with particular eigenvalues. Nevertheless, since the process of integration must be regarded as the limit approached by processes of summation, and since we are regarding our subscripts k as labelling all possible eigenfunctions, the theoretical aspect of the problem is not fundamentally altered by such changes in formalism. Furthermore, a variety of such formalisms, which are often quite complicated in structure, are employed in different special cases. Hence in the remainder of the present chapter we shall demonstrate those uses of *eigenfunction expansions* which are of interest to us, with the help of the simple formalism (65.6), calling special attention when necessary to aspects of the situation resulting from continuous spectra or from degeneracy.

66. Expansion of the probability amplitude $\psi(q,t)$

(*a*) **Expansion at a given time of interest.** For the purposes of the quantum mechanics, a very important application of the foregoing possibility for series expansions arises when the function $f(q)$ is itself taken as the probability amplitude

$$u(q) = \psi(q,t_0) \tag{66.1}$$

† See von Neumann, *Mathematische Grundlagen der Quantenmechanik*, Berlin, 1932, chapter ii.

for the state of a system at some particular time of interest t_0. We can then write

$$u(q) = \sum_k c_k u_k(q) \tag{66.2}$$

as an expansion for the probability amplitude $u(q)$, which describes the actual state of the system, in terms of the eigenfunctions $u_k(q)$, which describe states characteristic of some selected observable or set of observables.

It will be noted that we might expect an *expansion which converges in the mean* to give a sufficiently satisfactory expression for probability amplitudes, since the failure of such expansions at individual points would not appear serious for the uses made of probability amplitudes in the quantum mechanics. It will also be noted that this method of expression is in agreement with the general principle of the superposition of states discussed in § 59.

The possibility of expressing the instantaneous state of a system in terms of the eigenfunctions $u_k(q)$ characteristic of some particular observable quantity F is specially valuable when we are interested in the probabilities that the system will exhibit one or another of the different possible eigenvalues of that quantity. To investigate this use we note, in accordance with the general expression (57.4) for obtaining expectation values, that the nth power of F in the state specified by (66.2) would exhibit the mean value

$$\begin{aligned}\overline{F^n} &= \int u^*(q)\mathbf{F}^n u(q)\,dq \\ &= \int \sum_l c_l^* u_l^* \Big(\sum_k c_k \mathbf{F}^n u_k\Big)\,dq \\ &= \sum_{k,l} c_l^* c_k \int u_l^* F_k^n u_k\,dq \\ &= \sum_k |c_k|^2 F_k^n,\end{aligned} \tag{66.3}$$

where the third form of writing depends on the association of the eigenvalue F_k with the eigenfunction u_k, and the last form of writing makes use of the normalization and orthogonality of eigenfunctions expressed by (65.1).

Since equation (66.3) is valid for any integral power n, we now see that the probabilities in the state of interest of finding different values F_k for our observable quantity F would be determined by the squares of the corresponding coefficients in the series expansion (66.2). If the eigenvalue F_k is that for a single non-degenerate state, the probability of finding this value would be equal to the square of the corresponding coefficient

$$W(F_k) = |c_k|^2, \tag{66.4}$$

and in the case of g-fold degeneracy, where the same eigenvalue F_k corresponds to the g different eigenfunctions $u_{k+1},\ldots, u_{k+g}$, the probability of finding this value would be given by the sum of the squares of the corresponding coefficients

$$W(F_k) = \sum_{l=k+1}^{l=k+g} |c_l|^2. \tag{66.5}$$

The above expressions give actual rather than merely relative probabilities, provided, of course, that the original probability amplitude was normalized to give

$$\int u^*(q)u(q)\,dq = \sum_k |c_k|^2 = 1. \tag{66.6}$$

(*b*) **Expansion as a function of the time.** By an obvious extension the above method of expansion can also be used to give a form of expression for the probability amplitude of the system not only at one time but as a function thereof. To do this, since the probability amplitude is assumed expressible at any given time by an expansion with constant coefficients c_k, we can evidently change to coefficients $a_k(t)$ which vary with time, and obtain a satisfactory expression for the probability amplitude as a function of the time by an expansion of the form

$$\psi(q,t) = \sum_k a_k(t)u_k(q). \tag{66.7}$$

The coefficients in this series are seen by the same considerations that led to (65.3) to be given by

$$a_n(t) = \int u_n^*(q)\psi(q,t)\,dq. \tag{66.8}$$

Furthermore, in case the eigenfunctions $u_k(q)$ are chosen so as to correspond to the eigenvalues F_k of some selected observable quantity, we see that the probabilities of finding the different possible values for that quantity would be given by expressions of the form

$$W(F_k,t) = |a_k(t)|^2 \tag{66.9}$$

in the absence of degeneracy, and

$$W(F_k,t) = \sum_{l=k+1}^{l=k+g} |a_l(t)|^2 \tag{66.10}$$

in the case of g-fold degeneracy.

(*c*) **Special case of expansion in energy eigenfunctions.** The eigenfunctions $u_k(q)$ used in making expansions of the kind we are discussing may be picked out to correspond to the eigenvalues F_k of any particular observable quantity that we may desire, or to correspond to the eigenvalues F_k, G_n, etc., of more than a single observable. A case of frequent interest proves to be that of expansion in terms of the solutions corresponding to the eigenvalues E_n of the energy. The coefficients $a_n(t)$ are then found in the case of an undisturbed system to have a particularly simple form, which agrees with the simple time dependence for stationary state solutions already discussed in § 60.

To see this we may substitute the expression for the probability amplitude (66.7) into the Schroedinger equation (57.5) and write

$$\sum_k \left[\mathbf{H}+\frac{h}{2\pi i}\frac{\partial}{\partial t}\right]a_k(t)u_k(q) = 0,$$

which is readily seen to lead to

$$\sum_k \left[a_k(t)E_k+\frac{h}{2\pi i}\frac{\partial a_k(t)}{\partial t}\right]u_k(q) = 0$$

when we remember that the operator $\mathbf{H}$ is not dependent on t for an isolated system, and that we have expanded in terms of energy eigenfunctions. Multiplying this last expression by $u_n^*(q)$, and integrating over the whole range of coordinates q, we then obtain, with the help of the normalization and orthogonality of the eigenfunctions, the result

$$a_n(t)E_n+\frac{h}{2\pi i}\frac{\partial a_n(t)}{\partial t} = 0$$

which has the solution

$$a_n(t) = c_n\, e^{-(2\pi i/h)E_n t}, \tag{66.11}$$

where the quantities c_n are constants.

Hence the expansion of an arbitrary probability amplitude in terms of steady state solutions has the specially simple form in the case of an isolated system

$$\psi(q,t) = \sum_k c_k\, u_k(q)e^{-(2\pi i/h)E_k t}. \tag{66.12}$$

The constant coefficients in this expression are seen by our previous methods to be given by

$$c_n = \int u_n^*(q)e^{+(2\pi i/h)E_n t}\,\psi(q,t)\,dq. \tag{66.13}$$

Furthermore, the probability of finding a particular value for the energy E_k will be given at any time by the *constant expressions*

$$W(E_k,t) = |c_k|^2 \tag{66.14}$$

or

$$W(E_k,t) = \sum_{l=k+1}^{l=k+g} |c_l|^2 \tag{66.15}$$

respectively in the absence or presence of a g-fold degeneracy. This is in agreement with our previous finding in § 63 (c) that the relative probabilities for the system to exhibit different values of its energy would remain constant for an isolated system so long as it remains undisturbed.

67. Transformation theory

In the development of quantum mechanical methods so far given we have for the most part used the q-language in setting up and solving our problems, although we have appreciated and sometimes used the possibility of taking a momentum instead of a coordinate representation. In the present section we shall give a brief treatment of the quantum mechanical transformation theory,† which investigates the possibility of using representations corresponding to any desired observable quantity or compatible combination of observables. This branch of the theory is to some extent the analogue of the theory of canonical transformations in the classical mechanics; and fairly general possibilities of using other than coordinate language would seem necessary if our new mechanics is to be regarded as a satisfactory extension of the old. Our development of the transformation theory will show that such possibilities do exist without essentially new additions to our postulatory basis, and will provide us with a very general kind of language which is useful in the treatment of fundamental quantum mechanical problems.

(*a*) **Probability amplitudes in general.** The possibility of a variety of different modes of representing the state of a quantum mechanical system is already implicit in our previous equations (66.7) and (66.8) for the expansion of probability amplitudes, which we may now rewrite in the forms

$$\psi(q,t) = \sum_k a(k,t)u(k,q) \tag{67.1}$$

and

$$a(n,t) = \int u^*(n,q)\psi(q,t)\,dq. \tag{67.2}$$

In accordance with these equations we see that the state of a quantum mechanical system which is specified at any time t by a knowledge of the probability amplitude $\psi(q,t)$ as a function of the coordinates q is also equally well specified by a knowledge of the quantity $a(n,t)$ as a function of the index n.

Either function can be equally well calculated from the other with the help of the eigenfunctions $u(k,q)$ and $u^*(n,q)$ which play the role of *transformation functions* between the two modes of representation. Furthermore, the quantity $a(n,t)$ may be appropriately designated as

† First developed by Dirac and by Jordan; see in particular *The Principles of Quantum Mechanics*, by Dirac, second edition, Oxford, 1935.

the *transformed probability amplitude,* corresponding to the observable or observables whose eigenvalues are denoted by the different values of the index n, since the probability of finding such eigenvalues will be given by

$$W(n,t) = a^*(n,t)a(n,t) \tag{67.3}$$

in agreement with (66.10).

To be sure, the functions $\psi(q,t)$ and $a(k,t)$ enter the two equations (67.1) and (67.2) in a somewhat unsymmetrical manner. This arises, however, because the formalism is devised, on the one hand, to give explicit recognition to the consideration that the coordinates q are known to exhibit a continuous spectrum of eigenvalues, but, on the other hand, to leave the possibility open for the eigenvalues corresponding to the index n to exhibit discontinuous or continuous spectra. When the eigenfunctions $u(k,q)$ are such as to correspond simultaneously to the eigenvalues of f independent observables $F_1 \ldots F_f$, one for each degree of freedom, and these observables do exhibit continuous spectra, summation over the index k can be replaced by an appropriate integration, and equations (67.1) and (67.2) can be replaced by the entirely symmetrical formulation

$$\psi(q_1 \ldots q_f, t) = \int \ldots \int a(F_1 \ldots F_f, t)u(F_1 \ldots F_f, q_1 \ldots q_f)\, dF_1 \ldots dF_f \tag{67.4}$$

and

$$a(F_1 \ldots F_f, t) = \int \ldots \int \psi(q_1 \ldots q_f, t)u(q_1 \ldots q_f, F_1 \ldots F_f)\, dq_1 \ldots dq_f, \tag{67.5}$$

where we introduce the notation

$$u(q_1 \ldots q_f, F_1 \ldots F_f) \equiv u^*(F_1 \ldots F_f, q_1 \ldots q_f) \tag{67.6}$$

in the interests of symmetry.

It is of interest to note that our originally postulated relations (57.2) between the probability amplitudes $\psi(q,t)$ and $\phi(p,t)$, appearing respectively in coordinate or in momentum representations, can now be regarded as special cases of the above more general relations for transforming to other than coordinate languages. Thus, in the case of a system of one degree of freedom, our previous relations (57.2),

$$\psi(q,t) = h^{-\frac{1}{2}f} \int \phi(p,t)e^{(2\pi i/h)pq}\, dq$$

and

$$\phi(p,t) = h^{-\frac{1}{2}f} \int \psi(q,t)e^{-(2\pi i/h)pq}\, dq,$$

can now be regarded as special cases of (67.4) and (67.5), where the transformation functions

$$u(p,q) = h^{-\frac{1}{2}f}e^{(2\pi i/h)pq}$$

and

$$u(q,p) = u^*(p,q) = h^{-\frac{1}{2}f}e^{-(2\pi i/h)pq}$$

are seen in agreement with (64.28) to be appropriately normalized eigensolutions for states characteristic of the momentum p.

It is also of interest to note the formal possibility of regarding the eigensolutions for states characteristic of q itself as playing the role of transformation functions. Thus in the case of a system of one degree freedom, equation (67.5) would assume the form

$$\psi(q',t) = \int \psi(q,t)\delta(q-q')\,dq = \psi(q,t)_{q=q'}$$

if we substitute our previous eigensolution (64.24) for a state characteristic of the coordinate eigenvalue q'. The result is seen to be valid although trivial.

(*b*) **Operators in general.** When a coordinate representation $\psi(q, t)$ is being employed, the expectation value for an observable quantity $F(q,p)$ will be given by the relation

$$\bar{F} = \int \psi^*(q,t)\mathbf{F}\psi(q,t)\,dq, \tag{67.7}$$

assuming the operator $\mathbf{F}$ to have a known form, in accordance with § 55, suitable for operating on a function of the coordinates q. Substituting the transformation equation (67.1), we can also write this relation in the form

$$\bar{F} = \sum_{n,k} \int a^*(n,t)u^*(n,q)\mathbf{F}a(k,t)u(k,q)\,dq,$$

and this can also be expressed in the simpler form

$$\bar{F} = \sum_{n,k} a^*(n,t)F_{nk}\,a(k,t), \tag{67.8}$$

where the quantities F_{nk} are the elements of the Hermitian matrix defined by

$$F_{nk} = \int u^*(n,q)\mathbf{F}u(k,q)\,dq. \tag{67.9}$$

Defining the *transformed operator* $\mathbf{F}^{(a)}$, suitable for use in the $a(n,t)$ representation, by

$$\mathbf{F}^{(a)}a(n,t) = \sum_k F_{nk}\,a(k,t), \tag{67.10}$$

we see that the new expression for the expectation value $\bar{F}$ as given by (67.8) has a form

$$\bar{F} = \sum_n a^*(n,t)\mathbf{F}^{(a)}a(n,t), \tag{67.11}$$

which is the appropriate analogue of the original form of expression for $\bar{F}$ as given by (67.7) in the $\psi(q,t)$ representation.

This illustrates the general possibility of transforming such operators to any desired language. It also shows, as already remarked, that the operators in the quantum mechanics which correspond to observable quantities may assume a variety of forms. These include, in addition to the simple differential operators which we often have in the case of coordinate representations, operators whose action is expressible as in (67.10) by summation with the components of appropriate matrices, and operators where such summation is replaced by integration. The possibility of reducing the expression for the action of an operator to a simple form, merely involving multiplication and differentiation, must be regarded as the exception rather than the rule. The frequent simplicity of operator formulation in coordinate language, together with the physical insight afforded by such language, makes the use of coordinate representations specially convenient in the quantum mechanics.

(c) The Schroedinger equation in general. As the most fundamental relation in our development of the quantum mechanics, we have the Schroedinger equation, which can be written in the q-language in the form

$$\mathbf{H}\psi(q,t)+\frac{h}{2\pi i}\frac{\partial\psi(q,t)}{\partial t}=0. \tag{67.12}$$

This equation can be readily re-expressed in an analogous form in any other mode of representation.

Introducing the transformation equation (67.1) into (67.12), and noting that the operator **H** in the case of an isolated system will depend on the coordinates alone, we obtain

$$\sum_k a(k,t)\mathbf{H}u(k,q)+\frac{h}{2\pi i}\sum_k\frac{\partial a(k,t)}{\partial t}u(k,q)=0.$$

Multiplying throughout by $u^*(n,q)$ and integrating over all values of the coordinates q, we then obtain, with the help of the normalization and orthogonality of the eigenfunctions $u(k,q)$, the result

$$\sum_k a(k,t)\int u^*(n,q)\mathbf{H}u(k,q)\,dq+\frac{h}{2\pi i}\frac{\partial a(n,t)}{\partial t}=0,$$

and this can be re-expressed in the simpler form

$$\sum_k H_{nk}a(k,t)+\frac{h}{2\pi i}\frac{\partial a(n,t)}{\partial t}=0, \tag{67.13}$$

where the quantities H_{nk} are the components of the Hermitian matrix defined by

$$H_{nk}=\int u^*(n,q)\mathbf{H}u(k,q)\,dq. \tag{67.14}$$

Noting our previous definition (67.10) of the appropriate form for operators in the $a(n,t)$ representation, we see that (67.13) is the immediate analogue, in our present language, of the original Schroedinger equation (67.12). The relation given by (67.13) may hence be called the *transformed Schroedinger equation*.

(*d*) **The Hermitian matrices corresponding to observable quantities.** The above-defined Hermitian matrices, which correspond to observable quantities F with components

$$F_{nk} = \int u^*(n,q)\mathbf{F}u(k,q)\,dq \tag{67.15}$$

in any particular mode of representation specified by the choice of eigenfunctions $u(k,q)$, prove sufficiently important to warrant a special examination of their properties.

Owing to the Hermitian character of the operators $\mathbf{F}$, which correspond to observable quantities in the quantum mechanics, it will be immediately seen from the definition of Hermitian operators (55.25) that the matrices under consideration will themselves be Hermitian with

$$F_{nk} = F^*_{kn}. \tag{67.16}$$

It will also be noted, in accordance with our general expression (57.4) for the expectation value of a quantity, that any diagonal component of such a matrix,

$$F_{kk} = \int u^*(k,q)\mathbf{F}u(k,q)\,dq, \tag{67.17}$$

will give the expectation value

$$F_{kk} = \bar{F}_k \tag{67.18}$$

for the observable F when the system is actually in the state specified by the single eigenfunction $u(k,q)$ then involved.

Let us now consider two different Hermitian operators $\mathbf{F}$ and $\mathbf{G}$ corresponding to two different observable quantities F and G, and let us consider their action on the members of a set of eigenfunctions $u(k,q)$ of the kind that we have been considering. In accordance with our general possibility for developing functions of the coordinates q, we can evidently write as special cases

$$\mathbf{F}u(k,q) = \sum_n u(n,q)F_{nk} \quad \text{with} \quad F_{mk} = \int u^*(m,q)\mathbf{F}u(k,q)\,dq \tag{67.19}$$

$$\text{and} \quad \mathbf{G}u(l,q) = \sum_m u(m,q)G_{ml} \quad \text{with} \quad G_{nl} = \int u^*(n,q)\mathbf{G}u(l,q)\,dq.$$

Combining these expressions, we then obtain

$$\int [\mathbf{F}u(k,q)]^*\mathbf{G}u(l,q)\,dq = \sum_{n,m} F^*_{nk}G_{ml}\int u^*(n,q)u(m,q)\,dq,$$

or, in accordance with the Hermitian character of the operators $\mathbf{G}$ and

F (see (55.25) and (55.26)),

$$\int u(l,q)[\mathbf{GF}u(k,q)]^*\,dq = \int u^*(k,q)\mathbf{FG}u(l,q)\,dq = \sum_n F^*_{nk}G_{nl},$$

or finally, by (67.16),

$$[FG]_{kl} = \sum_n F_{kn}G_{nl}. \tag{67.20}$$

We thus see that the matrix elements corresponding to the product **FG** of two operators **F** and **G** are related to the matrix elements for the two operators themselves in accordance with the rule for matrix multiplication. Hence (67.20) can also be expressed in matrix language in the form

$$\|FG\| = \|F\|\,\|G\|. \tag{67.21}$$

Such matrices were of importance for the formulation of the so-called matrix mechanics by Heisenberg, Born, and Jordan,† which antedated the wave mechanics of Schroedinger. In accordance with preceding discussions, it will be readily seen that the matrices q_{mn}, p_{mn}, and H_{mn} corresponding to the coordinates, momenta, and energy of a system will have the following properties when obtained with the help of a set of eigenfunctions $u_n(q)$ corresponding to the energy eigenvalues E_n for the system. The matrices for the coordinates and momenta will turn out to be Hermitian and to be subject to commutation rules that result from the commutation properties of their operators as given by (55.38); and the energy matrix will turn out to be diagonal, with eigenvalues E_n corresponding to the eigenfunctions $u_n(q)$ employed, when it is evaluated by treating the coordinates and momenta in the classical Hamiltonian as Hermitian matrices obeying the above commutation rules and the ordinary rules of matrix manipulation. This, then, explains the original procedure of Heisenberg in accordance with which the eigenvalues of energy for a system were to be obtained by purely algebraic means by finding matrices for the coordinates, momenta, and energy having the above properties. In practice the procedure proved to be a feasible one only in a few simple cases.

To complete this brief discussion of the matrices corresponding to observable quantities, a generalization of some interest must be mentioned. The matrices so far defined by expressions of the form

$$F_{nk} = \int u^*(n,q)\mathbf{F}u(k,q)\,dq \tag{67.22}$$

are independent of the time t except in so far as the operator **F** might itself be an explicit function of t. A more general *time-dependent matrix* corresponding to the observable quantity F may be defined by

$$F_{\alpha\beta}(t) = \int \psi^*_\alpha(q,t)\mathbf{F}\psi_\beta(q,t)\,dq, \tag{67.23}$$

† *Zeits. f. Phys.* **33**, 879 (1925); **34**, 858 (1925); **35**, 557 (1925–6).

where ψ_α and ψ_β are any pair of possible solutions of the Schroedinger equation for the system under consideration in their full time dependence. These matrices will be Hermitian with

$$F_{\alpha\beta}(t) = F^*_{\beta\alpha}(t), \tag{67.24}$$

and their diagonal elements

$$F_{\alpha\alpha}(t) = \bar{F}_\alpha(t) \tag{67.25}$$

will give the expectation values for the observable F as functions of time when the system is carrying out the behaviour corresponding to the different possible solutions ψ_α of the Schroedinger equation.

Differentiating (67.23) with respect to the time, on the assumption that $\mathbf{F}$ has no explicit dependence on t, and making use of the Schroedinger equation (57.5), we obtain

$$\begin{aligned} \frac{dF_{\alpha\beta}(t)}{dt} &= \int \left[\frac{\partial\psi^*_\alpha}{\partial t}\mathbf{F}\psi_\beta + \psi^*_\alpha\mathbf{F}\frac{\partial\psi_\beta}{\partial t}\right] dq \\ &= \frac{2\pi i}{h}\int [(\mathbf{H}\psi_\alpha)^*\mathbf{F}\psi_\beta - \psi^*_\alpha\mathbf{F}\mathbf{H}\psi_\beta]\, dq \\ &= \frac{2\pi i}{h}\int [\psi_\beta(\mathbf{F}\mathbf{H}\psi_\alpha)^* - \psi^*_\alpha\mathbf{F}\mathbf{H}\psi_\beta]\, dq \\ &= \frac{2\pi i}{h}\int \psi^*_\alpha(\mathbf{H}\mathbf{F} - \mathbf{F}\mathbf{H})\psi_\beta\, dq, \end{aligned} \tag{67.26}$$

where the last two forms of writing are consequences of the Hermitian character of the Hamiltonian operator $\mathbf{H}$ for the energy of the system and of the operator $\mathbf{F}$ for the observable quantity of interest. (See (55.25) and (55.26).)

The result given by (67.26) can also evidently be expressed in the forms

$$\frac{dF_{\alpha\beta}(t)}{dt} = \frac{2\pi}{ih}[FH - HF]_{\alpha\beta} = \frac{2\pi}{ih}[F, H]_{\alpha\beta}$$

and

$$\frac{d\|F(t)\|}{dt} = \frac{2\pi}{ih}\|FH - HF\| = \frac{2\pi}{ih}\|F, H\|. \tag{67.27}$$

Comparing with our previous classical equation (11.4) for the rate of change in the *exact value* of a function F of the coordinates and momenta, we see that the quantities $\frac{2\pi}{ih}[F, H]_{\alpha\beta}$ and $\frac{2\pi}{ih}\|F, H\|$ may be regarded in the present connexion as the quantum mechanical analogues of the classical Poisson bracket $\{F, H\}$. Furthermore, comparing the above with our previous equation (63.3) for the rate of change in the *expectation value* of a function F, we see that the diagonal terms appearing in the present equations are equivalent in import to the earlier equation.

(e) **Unitary transformations between different quantum mechanical representations.** In the preceding parts of this section we have examined the consequences of transforming the treatment of a quantum mechanical problem from a coordinate representation, in which the state of the system is given by a probability amplitude $\psi(q,t)$ for the coordinates q, to some other representation, in which the state of the system would be given by a probability amplitude $a(k,t)$ for the index k labelling the eigenvalues and eigenfunctions $u(k,q)$ for some other kind of observable quantity or quantities. The results obtained are evidently general in character since they merely depend on the assumption that the eigenfunctions $u(k,q)$ form a complete, orthogonal set suitable for the expansion of functions of the coordinates. Hence results of similar form will be obtained if we transform to any other such representation, in which the state of the system would be given by a probability amplitude, say $b(r,t)$, for the index r labelling some other set of eigenfunctions, say $v(r,q)$. It will be useful and instructive to examine the transformation between two such modes of representation, corresponding to the amplitudes $a(k,t)$ and $b(r,t)$, since our previous expressions have been affected by the lack of symmetry arising from a formalism which was chosen, on the one hand, to give explicit recognition to the continuous range of possible eigenvalues for the coordinates q, and, on the other hand, to the possibility of labelling individual eigenfunctions $u(k,q)$, corresponding to some chosen observable or observables, with a specific index k.

To carry out the investigation we may begin by considering the interrelation between any two different complete sets of independent, normalized, orthogonal eigenfunctions $u(k,q)$ and $v(r,q)$, which are suitable for the expansion of functions of the coordinates q, in some problem under consideration. Since the two sets of eigenfunctions $u(k,q)$ and $v(r,q)$ are by hypothesis both of them suitable for the expansion of functions of q, it is evident that the members of either set may be expressed by an expansion in terms of the members of the other set. Using for the time being the shortened notation u_k and v_r to designate the members of the two sets, these expansions may be written in the forms

$$v_r = \sum_k u_k S_{kr} \quad \text{and} \quad u_k = \sum_r v_r S^{-1}_{rk}, \tag{67.28}$$

where S_{kr} and S^{-1}_{rk} are the coefficients needed for the two expansions, and the particular order of writing has been chosen to agree with that customary in matrix multiplication. From the normalization and orthogonality of the two sets of eigenfunctions we see that these coefficients, or components of the transformation matrices, will have the values

$$S_{lr} = \int u_l^* v_r \, dq \quad \text{and} \quad S_{sk}^{-1} = S_{ks}^* = \int v_s^* u_k \, dq. \tag{67.29}$$

Furthermore, in accordance with the foregoing we note the successions of relations

$$\delta_{rs} = \int v_r^* v_s \, dq = \sum_{k,l} \int u_k^* S_{kr}^* u_l S_{ls} \, dq = \sum_k S_{kr}^* S_{ks}$$

and $$\delta_{kl} = \int u_l^* u_k \, dq = \sum_{r,s} \int v_s^* (S_{sl}^{-1})^* v_r S_{rk}^{-1} \, dq = \sum_r S_{kr}^* S_{lr},$$

which gives us the pair of relations

$$\sum_k S_{kr}^* S_{ks} = \delta_{rs} \quad \text{and} \quad \sum_r S_{kr}^* S_{lr} = \delta_{kl}. \tag{67.30}$$

A *transformation matrix*, having components S_{kr}, which imply the existence of those for the inverse transformation S_{rk}^{-1} in the manner given by the second equation (67.29), and which are subject to the relations (67.30), is called *unitary*; and the transformations effected by it are called *unitary transformations*.†

With the help of the foregoing we may now consider the transformation relations between the probability amplitudes $a(k, t)$ and $b(r, t)$, corresponding to the two modes of representation. Since the state of the system as expressed by $\psi(q, t)$ can be expanded in either set of eigenfunctions $u(k, q)$ or $v(r, q)$, we can write the equations

$$\psi(q, t) = \sum_k a(k, t) u(k, q) = \sum_r b(r, t) v(r, q). \tag{67.31}$$

Substituting from (67.28), this then gives us

$$\sum_r b(r, t) v(r, q) = \sum_{k,r} a(k, t) v(r, q) S_{rk}^{-1}$$

† It is sometimes illuminating to regard the elements of the above transformation matrices as correlated with *unitary transformation operators* $\mathbf{S}$ and $\mathbf{S}^{-1}$, which establish a linear correspondence between the two sets of functions u and v in accordance with the expressions

$$\mathbf{S} u_r = v_r \quad \text{and} \quad \mathbf{S}^{-1} v_k = u_k \quad (\text{for all } r \text{ and } k).$$

Equations (67.29) can then be expressed in the form

$$S_{lr} = \int u_l^* \mathbf{S} u_r \, dq \quad \text{and} \quad S_{sk}^{-1} = S_{ks}^* = \int v_s^* \mathbf{S}^{-1} v_k \, dq.$$

It will be noted that these unitary operators are not Hermitian and are thus quite different in character from the quantum mechanical operators $\mathbf{F}$ that correspond to observable quantities. It will be noted, however, that the unitary operators $\mathbf{S}$ and $\mathbf{S}^{-1}$, by which we transform at a given time between two quantum mechanical modes of representation a and b, have a character similar to the unitary operators $\mathbf{U}(t)$ and $\mathbf{U}(-t)$, introduced later in § 96 (*d*), by which we transform with a given quantum mechanical mode of representation between two different times 0 and t.

for any arbitrary state, and hence, from the independence of the eigenfunctions $v(r,q)$,

$$b(r,t) = \sum_k S_{rk}^{-1}a(k,t) = \sum_k a(k,t)S_{kr}^*$$
$$\text{and, similarly,} \quad a(k,t) = \sum_r S_{kr}\,b(r,t) = \sum_r b(r,t)(S_{rk}^{-1})^* \tag{67.32}$$

as the relations between the two kinds of probability amplitudes. This will then permit us to re-express a state of the system in either language.

It will also be of interest to consider the transformation relations between the matrix components

$$F_{nm}^{(a)} = \int u^*(n,q)\mathbf{F}u(m,q)\,dq \quad \text{and} \quad F_{sr}^{(b)} = \int v^*(s,q)\mathbf{F}v(r,q)\,dq, \tag{67.33}$$

which would correspond in the two modes of representation to the same observable quantity F. Substituting from (67.28) into the second of the above equations, we obtain

$$F_{sr}^{(b)} = \sum_{l,k} \int u^*(l,q)S_{ls}^*\,\mathbf{F}u(k,q)S_{kr}\,dq,$$

which gives us, in accordance with the first of equations (67.33),

$$F_{sr}^{(b)} = \sum_{l,k} F_{lk}^{(a)}\,S_{ls}^*S_{kr} = \sum_{l,k} S_{sl}^{-1}F_{lk}^{(a)}S_{kr}$$
$$\text{and, similarly,} \quad F_{nm}^{(a)} = \sum_{s,r} F_{sr}^{(b)}(S_{sn}^{-1})^*S_{rm}^{-1} = \sum_{s,r} S_{ns}\,F_{sr}^{(b)}\,S_{rm}^{-1}, \tag{67.34}$$

$$\text{or in matrix language} \qquad \|F^{(b)}\| = \|S^{-1}F^{(a)}S\|$$
$$\text{and} \qquad \|F^{(a)}\| = \|SF^{(b)}S^{-1}\|, \tag{67.35}$$

as the relations between the Hermitian matrix components and matrices corresponding to an observable quantity F in the two modes of representation. This will then permit us to re-express, in either language, relations which depend on these components, such as the generalized Schroedinger equation or the expression for the computation of expectation values.

The completely symmetrical character of the above transformation relations depends on the circumstance that our formalism now gives no recognition to the possibility that the indices k and r appearing in the probability amplitudes $a(k,t)$ and $b(r,t)$ might really correspond in the one case to a continuous and in the other case to a discrete manifold of values. The general features of the transformation theory are, however, most clearly appreciated with such a symmetrical if not completely descriptive formalism.

To complete this brief treatment of unitary transformations, it is to be pointed out that the successive application of such transformations will itself be a unitary transformation. To see this let us now consider the successive transformations

$$v_r = \sum_k u_k S_{kr} \quad \text{and} \quad w_c = \sum_r v_r T_{rc}, \tag{67.36}$$

together with the inverse transformations

$$u_k = \sum_r v_r S_{rk}^{-1} \quad \text{and} \quad v_r = \sum_c w_c T_{cr}^{-1}. \tag{67.37}$$

We must now show that the direct transformation from the representation corresponding to the eigenfunctions u_k to that corresponding to w_c would have unitary character. Combining the equations (67.36), we have

$$\left.\begin{aligned} w_c &= \sum_{k,r} u_k S_{kr} T_{rc} = \sum_k u_k U_{kc} \\ \text{with} \qquad U_{kc} &= \sum_r S_{kr} T_{rc}, \end{aligned}\right\} \tag{67.38}$$

and, similarly, from (67.37)

$$\left.\begin{aligned} u_k &= \sum_{c,r} w_c T_{cr}^{-1} S_{rk}^{-1} = \sum_c w_c U_{ck}^{-1} \\ \text{with} \qquad U_{ck}^{-1} &= \sum_r T_{cr}^{-1} S_{rk}^{-1}, \end{aligned}\right\} \tag{67.39}$$

where U_{kc} and U_{ck}^{-1} are the components of the new transformation matrices. With the help of the unitary properties of the individual matrix components S_{kr} and T_{rc} as given by (67.29) and (67.30), however, the new components are also seen to have the properties

$$U_{ck}^{-1} = \sum_r T_{rc}^* S_{kr}^* = \sum_r S_{kr}^* T_{rc}^* = U_{kc}^* \tag{67.40}$$

$$\left.\begin{aligned} &\text{and} \quad \sum_k U_{kc}^* U_{kd} = \sum_{k,r,s} S_{kr}^* T_{rc}^* S_{ks} T_{sd} = \sum_{r,s} \delta_{rs} T_{rc}^* T_{sd} = \sum_s T_{sc}^* T_{sd} = \delta_{cd}, \\ &\text{and, similarly,} \qquad \sum_c U_{kc}^* U_{lc} = \delta_{kl}, \end{aligned}\right\} \tag{67.41}$$

which completes the demonstration of the unitary character of the new transformation matrices.

(*f*) **Concluding remarks on the general quantum mechanical language provided by the transformation theory.** As an important consequence of the transformation theory, we are now provided with a very general kind of language for expressing the principles of the quantum mechanics. Thus our original postulates for the quantum mechanics, as summarized in § 57 in the specific languages corresponding to the coordinates q or to the momenta p, may now be re-expressed in the general language corresponding to an index n by which the eigenstates for any kind of observable quantity or suitable set of such quantities can be labelled.

For the relation between the probability $W(n, t)$ of finding the system in the state labelled by n and the probability amplitude $a(n,t)$ for that state, we have

$$W(n,t) = a^*(n,t)a(n,t), \tag{67.42}$$

where the values of the probability amplitude $a(n,t)$ can be referred back if desired to an expression of the state of the system in coordinate language with the help of (67.2).

For the transformation to a different language, involving probability amplitudes $b(r,t)$ for states of a different kind r, we have

$$a(k,t) = \sum_r S_{kr}\, b(r,t) \quad \text{and} \quad b(r,t) = \sum_k a(k,t) S^*_{kr}, \tag{67.43}$$

where the components of the unitary transformation matrix S_{kr} depend on the two kinds of states in a manner which can be referred back to the properties of the system expressed in coordinate language with the help of (67.29).

For the action on the probability amplitude $a(n,t)$, of an operator $\mathbf{F}^{(a)}$ corresponding in the present language to an observable quantity F, we have

$$\mathbf{F}^{(a)} a(n,t) = \sum_k F_{nk}\, a(k,t), \tag{67.44}$$

where the components of the Hermitian matrix F_{nk} can be referred back to the properties of the system expressed in coordinate language with the help of (67.9).

For the mean or expectation value of an observable quantity F we have

$$\bar{F} = \sum_{n,k} a^*(n,t) F_{nk}\, a(k,t). \tag{67.45}$$

And for the change in state with time we have the generalized Schroedinger equation

$$\sum_k H_{nk}\, a(k,t) + \frac{h}{2\pi i} \frac{\partial a(n,t)}{\partial t} = 0, \tag{67.46}$$

where the H_{nk} are components of the Hermitian matrix corresponding to the energy of the system.

By comparison it will be seen that these five equations are indeed the analogues in general language of the five equations (57.1–5), which we took as the original postulates of the quantum mechanics. These equations thus provide a generalized statement of the principles of the quantum mechanics, which can be regarded as valid in any quantum mechanical language that may be of interest for a particular problem. The form of the equations may seem a somewhat inappropriate one when the specific language desired is actually one that corresponds to observables exhibiting continuous rather than discrete spectra of eigenvalues, and reference back to coordinate language may often be desirable in treating a specific problem. Nevertheless, the provision of a generalized language is very important since we can now deduce consequences which will be valid in any quantum mechanical language that may later be chosen as of interest, and since the generalized language is one which often gives a specially good insight into the characteristic features of the quantum mechanics. For this reason we shall often use the above generalized language in developing the fundamental principles of the quantum statistical mechanics.

68. The method of variation of constants

The possibility of obtaining a transformed Schroedinger equation, appropriate for use with any mode of representing the state of a system, as discussed in the last section, finds a specially important application when the state of the system is represented by an expansion in terms of so-called *unperturbed energy eigenfunctions*, which correspond to a Hamiltonian operator which differs from the true Hamiltonian for the system by a small term which is regarded as a perturbation. This method of treatment was first developed by Dirac,† and may be called the method of variation of constants, since the constant coefficients, such as the c_k in equation (66.12), which would appear if the state of the system were expanded in terms of the *true energy eigenfunctions* for the system must now be replaced by coefficients which are allowed to *vary with the time*.

The method of variation of constants is very important for the quantum mechanics in gaining an insight into the changes that take place in a system with time, since our natural interest often lies in the *nearly steady states* (unperturbed energy eigenstates) which a system can exhibit, and since the differential equations for the change in such states with time which the method provides can be readily integrated by an approximate method. Hence we can now afford to devote a section to the method, even though we shall again discuss it in Chapter XI when we undertake our general consideration of the changes that take place in quantum mechanical systems with time. We shall develop the method *ab initio* without reference to the general features of the transformation theory.

(*a*) **Derivation of the differential equations.** Let us consider a quantum mechanical system which can be treated with the help of a Hamiltonian operator

$$\mathbf{H} = \mathbf{H}^0 + \mathbf{V}, \tag{68.1}$$

† Dirac, *Proc. Roy. Soc.* A, **112**, 661 (1926); **114**, 243 (1927).

which can be regarded as the sum of two terms, the *unperturbed Hamiltonian* $\mathbf{H}^0$ corresponding to a nearly precise expression for the energy of the system, and a *perturbation term* $\mathbf{V}$ corresponding to the remainder necessary for an actually exact expression of the energy. For example, in the case of a dilute gas the term $\mathbf{H}^0$ could correspond to an expression for the internal kinetic energy of the system regarded as composed of non-interacting molecules, and the term $\mathbf{V}$ to the added potential energy needed for an expression that also allows for the effects of interaction at times of collision. Or in the case of an atom in equilibrium with radiation the term $\mathbf{H}^0$ could correspond to the sum of the energies of the atom and the electromagnetic field regarded as having no effect on each other, and the term $\mathbf{V}$ to the effects arising from the actual interaction made evident by the fact that the atom could absorb and emit radiation.

Without real loss in generality we can consider our system as isolated with $\mathbf{H}^0$ and $\mathbf{V}$ not explicitly dependent on the time.

Using the unperturbed Hamiltonian $\mathbf{H}^0$, we can then obtain a set of eigenfunctions $u_n(q)$ for the system with the help of the equation determining characteristic states

$$\mathbf{H}^0 u_n(q) = E_n^0 u_n(q), \tag{68.2}$$

where the quantities E_n^0 are the eigenvalues corresponding to the operator $\mathbf{H}^0$, degeneracy being allowed for by the possibility that the eigenvalues for different values of the index n might actually turn out to be equal. These eigenvalues E_n^0 may be conveniently called the *unperturbed energy eigenvalues* for the system. In case the effect of the perturbation term $\mathbf{V}$ is sufficiently small, they would be nearly equal to the actual possible eigenvalues E_n of the true energy of the system.

Since the eigenfunctions $u_n(q)$ will form a complete, normalized, orthogonal set, any state of the system can be conveniently represented, in accordance with our previous discussion, as an expansion in terms of these functions. In the absence of the perturbation term $\mathbf{V}$, this expansion would have at all times the form

$$\psi(q,t) = \sum_k c_k u_k(q) e^{-(2\pi i/h)E_k^0 t}$$

in accordance with our previous equation (66.12), where the c_k are constant coefficients. For the actual system, however, in the presence of the perturbation $\mathbf{V}$, we must use a more general form of expansion

$$\psi(q,t) = \sum_k c_k(t) u_k(q) e^{-(2\pi i/h)E_k^0 t}, \tag{68.3}$$

where we now allow the coefficients $c_k(t)$ to vary with the time in the manner actually demanded by the Schroedinger equation of motion. The expansion given by (68.3) is, of course, entirely possible, in accordance with the completeness of the set $u_k(q)$, and differs in form from our earlier general expansion (66.7) only because it is now convenient to separate out from the total time dependence the periodic part corresponding to the exponential term in (68.3).

Substituting (68.3) into the Schroedinger equation

$$\mathbf{H}\psi(q,t)+\frac{h}{2\pi i}\frac{\partial\psi(q,t)}{\partial t}=0, \tag{68.4}$$

making use of the complete expression for the energy operator (68.1), and also making use of the expression (68.2) by which the eigenfunctions $u_k(q)$ are defined, we obtain

$$\sum_k c_k(t)E_k^0 u_k(q)e^{-(2\pi i/h)E_k^0 t}+\sum_k c_k(t)\mathbf{V}u_k(q)e^{-(2\pi i/h)E_k^0 t}+$$
$$+\frac{h}{2\pi i}\sum_k\frac{dc_k(t)}{dt}u_k(q)e^{-(2\pi i/h)E_k^0 t}-\sum_k c_k(t)u_k(q)E_k^0 e^{-(2\pi i/h)E_k^0 t}=0.$$

Noting that the first and last terms of this expression will cancel, and multiplying through by $u_n^*(q)e^{+(2\pi i/h)E_n^0 t}$ and integrating over the full range of values for the coordinates q, we then obtain, with the help of the normalization and orthogonality of the eigenfunctions $u_k(q)$, the result

$$\sum_k c_k(t)e^{\frac{2\pi i}{h}(E_n^0-E_k^0)t}\int u_n^*(q)\mathbf{V}u_k(q)\,dq+\frac{h}{2\pi i}\frac{dc_n(t)}{dt}=0,$$

and this can be rewritten in the final form

$$\frac{dc_n(t)}{dt}=-\frac{2\pi i}{h}\sum_k V_{nk}\,e^{\frac{2\pi i}{h}(E_n^0-E_k^0)t}c_k(t), \tag{68.5}$$

where the quantities V_{nk} are defined by

$$V_{nk}=\int u_n^*(q)\mathbf{V}u_k(q)\,dq, \tag{68.6}$$

and are seen to be the components of a Hermitian matrix with

$$V_{nk}=V_{kn}^*. \tag{68.7}$$

Equation (68.5) gives the desired expression for the transformed Schroedinger equation, which in principle will allow us to calculate the change with time in any selected coefficient $c_n(t)$ in the expansion (68.3) from information as to the instantaneous values of all the coefficients $c_k(t)$. A knowledge of the values of these coefficients as a function of the time is of course important, since in accordance with our original expansion (68.3) it is evident that the probability at any time of finding the system in the state indicated by the index n would be given by

$$W_n(t) = c_n^*(t)c_n(t), \tag{68.8}$$

provided, of course, that the solution is normalized to give

$$\sum_k W_k(t) = \sum_k c_k^*(t)c_k(t) = 1. \tag{68.9}$$

(*b*) **Approximate integration for a special case.** The set of differential equations given by (68.5) will in general be infinite in number. Nevertheless, their approximate integration will be simple in case we start the system off at time $t_0 = 0$ in a special state, with all the coefficients $c_k(t_0)$ equal to zero except one, say $c_n(t_0)$, which we put equal to unity; i.e.

$$c_n(t_0) = 1, \qquad c_k(t_0) = 0 \quad (k \neq n). \tag{68.10}$$

Under these circumstances the system would be definitely in the state indicated by the index n at the time $t = t_0 = 0$.

So far we have made no use of the smallness of the perturbation term $\mathbf{V}$ and have introduced no approximation. If, however, we now assume a successful separation of the energy expression into two parts such that the effect of the term $\mathbf{V}$ in causing the coefficients to vary with the time will be small, we can carry out an approximate integration by treating the coefficients as nearly retaining the constant values (68.10) over a short time interval in the neighbourhood of $t = 0$.

For the rate of increase in the coefficient $c_m(t)$ for any state m other than that in which the system starts, we then obtain from (68.5)

$$\frac{dc_m(t)}{dt} = -\frac{2\pi i}{h} V_{mn}\, e^{\frac{2\pi i}{h}(E_m^0 - E_n^0)t},$$

and this can evidently be integrated to give

$$c_m(t) = V_{mn} \frac{e^{\frac{2\pi i}{h}(E_m^0 - E_n^0)t} - 1}{E_n^0 - E_m^0} \tag{68.11}$$

as an approximately correct expression for the coefficient $c_m(t)$ in the neighbourhood of the time $t = 0$. Multiplying (68.11) by the corresponding expression for the conjugate complex quantity $c_m^*(t)$, we can write, in accordance with (68.8),

$$W_m(t) = c_m^*(t)c_m(t) = |V_{mn}|^2 \frac{2 - e^{\frac{2\pi i}{h}(E_m^0 - E_n^0)t} - e^{-\frac{2\pi i}{h}(E_m^0 - E_n^0)t}}{(E_n^0 - E_m^0)^2},$$

or, by a simple transformation,

$$W_m(t) = |V_{mn}|^2 \frac{4 \sin^2\{(\pi/h)(E_m^0 - E_n^0)t\}}{(E_m^0 - E_n^0)^2}, \tag{68.12}$$

and have thus obtained an expression, which should be valid in the neighbourhood of $t = 0$, for the probability of finding the system in a state m different from the state n in which it started.

Our treatment has not yet given us an expression for the decrease in the probability $W_n(t)$ of finding the system in the original state n, since it has been based on the approximation $c_n(t) = 1$ over the short time interval involved in the integration. Nevertheless, if we now go to the next stage of approximation and substitute the values of $c_m(t)$ given by (68.11) into (68.5), we can then compute the rate of change $c_n(t)$ with the time and thus obtain an expression for $W_n(t)$. The computation is a little long, however, and we can obtain the same result directly from the fact that the total probability of finding the system in some state must always be equal to unity as given by (68.9). We can then write at once.

$$W_n(t) = 1 - \sum_m |V_{mn}|^2 \frac{4\sin^2\{(\pi/h)(E_m^0 - E_n^0)t\}}{(E_m^0 - E_n^0)^2} \qquad (68.13)$$

for the probability of finding the system still in its original state n, in the neighbourhood of the time $t = 0$, the summation being over all states $m \neq n$.

In accordance with the expression given by (68.12), we see, on account of the appearance of the so-called 'resonance denominator' $(E_m^0 - E_n^0)^2$, that there will be an appreciable tendency for excitation of a quantum state m only when the corresponding unperturbed energy E_m^0 lies close to the unperturbed energy E_n^0 of the initial state n. Exact equality is not demanded since E_n^0 and E_m^0 are not true energy levels for the actual system. Starting the system off at $t = 0$ in the state n would correspond to a distribution over the various possible values of the true energy, with a relatively high probability of finding values close to E_n^0 if a measurement were made; and this relatively high probability of finding such values would be maintained in time in accordance with the quantum mechanical analogue of the principle of the conservation of energy as discussed in § 63 (*c*).

The foregoing shows the possibility of integrating the differential equations (68.5) for the coefficients $c_k(t)$ that determine the state of the system, provided that we start the system off in a specially simple state with one of these coefficients $c_n(t)$ equal to unity, and that we make use of a simple method of approximate calculation. It is evident, however, that at least in principle we could give a similar treatment starting the system off in any desired state, and that by successive approximations we could achieve any desired accuracy in the calculation.

This must now complete our account of the general features of the quantum mechanics.

VIII

SOME SIMPLE APPLICATIONS OF QUANTUM MECHANICS

69. Simple one-dimensional solutions

In the preceding chapter we have developed the fundamental principles and general methods for a non-relativistic quantum mechanics. In the present chapter we shall give a brief and partial account of some simple applications. These will be chosen either in order to illustrate the general nature of the quantum mechanical treatment of the atoms and molecules composing the systems of usual interest in statistical mechanics, or in order to furnish material specifically needed for our later statistical studies. The results to be obtained in the chapter are all well known, and for the most part the treatment will be given in a condensed form with references to more complete treatments. The results of most immediate interest for statistical mechanics are obtained at the end of the present section, where we discuss the correlation between quantum mechanical states and extensions in the classical phase space, in § 71 where we discuss the number of eigensolutions lying in any given energy range for the case of a particle in a container, in § 75 where we discuss the effect of spin on the number of eigensolutions for a particle, and in § 76 where we discuss the enumeration of eigensolutions for systems composed of similar particles. The treatment in these places will be somewhat more complete.

(*a*) **Solutions in regions of constant potential.** In the present section we shall consider solutions of the Schroedinger equation which can be regarded as corresponding to the one-dimensional motion of a particle of mass m along the x-axis. We shall first treat the simple case of solutions in regions of *constant potential* $V(x)$. Such solutions often make it possible to obtain considerable insight into the nature of quantum mechanical results with a minimum of computation.

Since any solution $\psi(x, t)$ of the general Schroedinger equation (57.5) can be expressed, in accordance with the methods of § 66 (*c*), in the form

$$\psi(x, t) = \sum_k c_k u_k(x) e^{-(2\pi i/h)E_k t}, \tag{69.1}$$

as a superposition of steady state solutions each multiplied by a suitable coefficient c_k, it will be sufficient to investigate the solutions of the time-free Schroedinger equation

$$\mathbf{H}u(x) = Eu(x), \tag{69.2}$$

which provide the different eigenfunctions $u_k(x)$ and energy eigenvalues E_k to be used in making such superpositions.

The quantity E in the above equation is some particular possible eigenvalue of the energy, and the Hamiltonian operator $\mathbf{H}$ will be given in agreement with (56.5) by

$$\mathbf{H} = \frac{1}{2m}\mathbf{p}_x^2+V(x)$$
$$= -\frac{h^2}{8\pi^2 m}\frac{\partial^2}{\partial x^2}+V(x), \qquad (69.3)$$

where for our present simplified considerations the potential $V(x)$ will be taken as a constant independent of x, except for the possibility of sudden changes in value in going from one region of constant potential to another. Substituting (69.3) in (69.2), we then obtain

$$\frac{\partial^2 u}{\partial x^2} = -\frac{8\pi^2 m}{h^2}(E-V)u \qquad (69.4)$$

as the equation for determining the eigensolutions corresponding to different allowable values of the energy E.

This equation has two types of solution according as the constant E is greater or less than the constant V.

If the *actual energy E is greater than the potential V*, the solution will be trigonometric and can be written quite generally, either with the help of sines and cosines in the form

$$u = A\sin\frac{2\pi}{h}\sqrt{\{2m(E-V)\}}x+B\cos\frac{2\pi}{h}\sqrt{\{2m(E-V)\}}x, \qquad (69.5)$$

or, with the help of imaginary exponents, in the form

$$u = A'e^{\frac{2\pi i}{h}\sqrt{\{2m(E-V)\}}x}+B'e^{-\frac{2\pi i}{h}\sqrt{\{2m(E-V)\}}x}, \qquad (69.6)$$

where A and B or A' and B' are the pair of arbitrary constants that correspond to the second-order character of the original differential equation (69.4). The *oscillatory nature* of these solutions is characteristic for particles with a total energy E greater than the potential V, and is in agreement with the use of the name *wave mechanics* as an alternative for quantum mechanics.

These two different forms, in which an eigensolution for the case $E > V$ can be written, are of course entirely equivalent and can be transformed into each other. The first form (69.5) is, however, the convenient one to take when we wish to use such eigensolutions in making a superposition of steady state solutions (69.1) to correspond to a situation where there are equal probabilities of finding the particle moving either in the positive or negative x-direction, e.g. a particle between reflecting walls. And the second form (69.6) is the convenient one to take in dealing with situations where there can be a net probability for motion in a given direction, e.g. a particle which has chances of reflection or transmission at a place where there is a sudden change in potential from one constant value to another.

To see the reason for this difference in the two forms of solution, we note, in accordance with our previous expression (58.4), that the probability current corresponding to any stationary state solution, of the form $\psi(x,t) = u(x)e^{-(2\pi i/h)Et}$, would be given by

$$S_x = \frac{h}{4\pi im}\left(\psi^*\frac{\partial\psi}{\partial x} - \psi\frac{\partial\psi^*}{\partial x}\right) = \frac{h}{4\pi im}\left(u^*\frac{\partial u}{\partial x} - u\frac{\partial u^*}{\partial x}\right). \tag{69.7}$$

This at once shows, however, that the first form of solution (69.5) will be convenient for situations where there is no preferential flow, since with the constants A and B both chosen as *real* the probability current would be zero, $S_x = 0$. On the other hand, the second form of solution (69.6) will be convenient for situations where this is a net flow. Thus, for example, with the constant B' set equal to zero, we readily find from (69.7)

$$S_x = |A'|^2\sqrt{\left(\frac{2(E-V)}{m}\right)} = |A'|^2 v, \tag{69.8}$$

where v is the classical expression for the velocity of a particle with total energy E and potential energy V.

Turning now to a consideration of equation (69.4) for the case when the *actual energy E is less than the potential V*, we see that the solution will then be *exponential* in character and can be written in the general form

$$u = Ce^{\frac{2\pi}{h}\sqrt{\{2m(V-E)\}}x} + De^{-\frac{2\pi}{h}\sqrt{\{2m(V-E)\}}x}, \tag{69.9}$$

where C and D are now the pair of arbitrary constants corresponding to the second-order character of the original differential equation. It is specially noteworthy that this solution contains no implication that the probability $u^*u\,dx$ for finding the particle in a region where the total energy E is less than the potential V must be zero. Indeed it is a characteristic feature of the quantum mechanics that particles can penetrate and under suitable circumstances actually traverse classically forbidden regions.

Although the above forms of eigensolution only hold in regions where the potential V is a constant, it is nevertheless possible to allow sudden changes in the potential, from one constant value, say V_1, holding over a given range of x to another value, say V_2, holding in a neighbouring range, and then connect the appropriate expressions for the forms of solution in the contiguous regions. In this way, introducing a succession of stepwise changes in potential if necessary, it is possible to get an approximate representation of many kinds of potential field and thus to treat a wide variety of problems.

In passing from one region of constant potential V_1 to another region of constant potential V_2, it is evident from the form of our differential equation for the eigenfunctions (69.4) that there will be a sudden change at the division point in the value of the second derivative $\partial^2 u/\partial x^2$. For the eigenfunction u itself, however, and for its first derivative $\partial u/\partial x$ it is evident that we can still require continuity at that point. Indeed, if we did not demand this measure of continuity it would be possible to construct solutions such that the probability density and probability current would have to correspond to a source or sink at the division point, which would in general be physically unallowable.

The requirement of continuity for u and $\partial u/\partial x$ at the boundary between any pair of regions is sufficient to give the proper connexion between the forms of solution holding on the two sides of a boundary where a sudden change in potential takes place. For example, if to the left of the point $x = 0$ we have an expression of the form (69.5) holding with the constant potential $V_1 < E$, and to the right an expression of the form (69.9) with the constant potential $V_2 > E$, it is evident that the constants for the two forms of expression would be connected by the relations

$$B = C+D$$

and

$$A\sqrt{(E-V_1)} = (C-D)\sqrt{(V_2-E)}. \tag{69.10}$$

And similarly the connexion between expressions of the forms (69.6) and (69.9) holding on the two sides of $x = 0$ would be given by

$$A'+B' = C+D$$

and

$$i(A'-B')\sqrt{(E-V_1)} = (C-D)\sqrt{(V_2-E)}. \tag{69.11}$$

In both cases it will be seen that the number of independent constants which can be arbitrarily chosen to describe the solution will be only two, in agreement with the second-order character of the original equation (69.4), which governs the solution over the whole range in x.

Making use of the indicated methods of treatment, insight as to the nature of quantum mechanical results can be obtained in a considerable variety of different specific situations.

Thus it is easy to show that an oscillatory expression such as (69.5) with A and B real holding in a region where V_1 is less than E will always be connected with a *diminishing exponential expression* in a neighbouring region where V_2 remains *permanently greater* than E, so that there will be an ever-decreasing probability of finding the particle as we proceed farther and farther into the classically forbidden region. It can also be shown, for the limiting case where the potential V_2 goes to infinity, that the solution for u will fall to zero at the boundary and that there will be no probability of finding the particle outside the classically permitted region; this situation can be regarded as providing a model for a *perfectly reflecting wall.*

For the case of a particle in a region where V_1 is less than E, bounded on *each side* by regions where the potential remains permanently higher than E, it can be shown that the conditions for a diminishing exponential form of expression at both sides can only be met with discrete values for E. By treating this so-called problem of the quantization of a particle in a trough, we thus obtain a simple illustration of the occurrence of *discrete energy spectra* in the quantum mechanics.

Studies can also be made of the behaviour of particles when regions in which the potential V_1 is less than E are connected with regions of *only limited extent* where the potential V_2 is greater than E. In this way considerable understanding can be obtained for the phenomena of *penetration through potential barriers* in general, of *resonance penetration,* and of *weakened quantization* for a particle between two such barriers.†

(*b*) **Approximate solutions in regions of varying potential.** In case a more accurate treatment of one-dimensional problems is needed than is provided by the above possibility of representing any actual distribution of potential by a series of constant potentials, use can be made of a more closely approximate solution of the Schroedinger equation

$$\frac{\partial^2 u}{\partial x^2} = -\frac{8\pi^2 m}{h^2}\{E-V(x)\}u \tag{69.12}$$

which has been specially studied by Wentzel, Kramers, and Brillouin. For regions where $E > V(x)$, this solution can be written in the trigonometric form

$$u(x) = \frac{A}{\sqrt[4]{\{2m(E-V)\}}}\sin\frac{2\pi}{h}\int^{x}\sqrt{\{2m(E-V)\}}\,dx + \\ +\frac{B}{\sqrt[4]{\{2m(E-V)\}}}\cos\frac{2\pi}{h}\int^{x}\sqrt{\{2m(E-V)\}}\,dx, \tag{69.13}$$

† For a specially thorough and valuable study of the method, see Schroedinger, *Berl. Ber., Phys.-Math. Klasse,* 1929, p. 668.

and in regions where $E < V(x)$ in the *exponential* form

$$u(x) = \frac{C}{\sqrt[4]{\{2m(V-E)\}}} e^{\frac{2\pi}{h}\int^{x} \sqrt{\{2m(V-E)\}}\,dx} + \frac{D}{\sqrt[4]{\{2m(V-E)\}}} e^{-\frac{2\pi}{h}\int^{x} \sqrt{\{2m(V-E)\}}\,dx}, \tag{69.14}$$

where A, B, C, and D are constants. By substituting these expressions into (69.12) it will be seen that they do furnish a good approximation provided $|E-V|$ is large enough or $|\partial V/\partial x|$ and $|\partial^2 V/\partial x^2|$ are small enough. This condition makes the present treatment inapplicable to discontinuous changes in potential such as just treated in § 69 (a). The solutions fail in the neighbourhood of points where $E = V$, and we have a transition between the two forms of expression. If at a point $x = a$ we pass from an 'exponential region' where $E < V$ to a 'trigonometric region' where $E > V$, it is found† that exponential solutions holding to the left of the point a can be satisfactorily connected with trigonometric solutions to the right of the point a in the manner indicated by the abbreviated formalism

$$\frac{C}{\sqrt[4]{(-)}} e^{-\frac{2\pi}{h}\int_{x}^{a} \sqrt{(-)}\,dx} \underset{x=a}{\longleftrightarrow} \frac{2C}{\sqrt[4]{(-)}} \cos\left\{\frac{2\pi}{h}\int_{a}^{x} \sqrt{(-)}\,dx - \tfrac{1}{4}\pi\right\}, \tag{69.15}$$

$$\frac{C'}{\sqrt[4]{(-)}} e^{\frac{2\pi}{h}\int_{x}^{a} \sqrt{(-)}\,dx} \underset{x=a}{\longleftrightarrow} \frac{-C'}{\sqrt[4]{(-)}} \sin\left\{\frac{2\pi}{h}\int_{a}^{x} \sqrt{(-)}\,dx - \tfrac{1}{4}\pi\right\}, \tag{69.16}$$

and, if at a point $x = b$ we pass from a region of trigonometric to one of exponential solutions, the connexions may be expressed in the form

$$\frac{2C''}{\sqrt[4]{(-)}} \cos\left\{\frac{2\pi}{h}\int_{x}^{b} \sqrt{(-)}\,dx - \tfrac{1}{4}\pi\right\} \underset{x=b}{\longleftrightarrow} \frac{C''}{\sqrt[4]{(-)}} e^{-\frac{2\pi}{h}\int_{b}^{x} \sqrt{(-)}\,dx}, \tag{69.17}$$

$$\frac{C'''}{\sqrt[4]{(-)}} \sin\left\{\frac{2\pi}{h}\int_{x}^{b} \sqrt{(-)}\,dx - \tfrac{1}{4}\pi\right\} \underset{x=b}{\longleftrightarrow} \frac{-C'''}{\sqrt[4]{(-)}} e^{\frac{2\pi}{h}\int_{b}^{x} \sqrt{(-)}\,dx}, \tag{69.18}$$

where the C's are arbitrary constants. This provides all that is necessary for using the so-called W.K.B. approximate method of solution for the general one-dimensional wave equation.‡

It will be of interest to apply the method to the problem of the quantized behaviour of a particle in a region $a < x < b$ where $E > V$, bounded on both sides by regions $-\infty < x < a$ and $b < x < \infty$ where

† See, for example, Pauli, *Handbuch der Physik*, xxiv/1, second edition, p. 171

‡ For another method of approximation, which is perhaps even more satisfactory, see Langer, *Phys. Rev.* **51**, 669 (1937).

$E < V$ over the entire range. From a classical point of view the particle would be strictly confined between the points $x = a$ and $x = b$, since its total energy E would be less than the potential energy V needed for positions outside that range. In the quantum mechanics, on the other hand, some penetration of the particle into the classically forbidden regions would be possible, since the probability density would not go abruptly to zero on passing the points a and b. Nevertheless, in order to prevent the probability density from going to infinity at $x = \pm\infty$, it is evident that we can only permit descending exponential solutions outside the points a and b. Hence we are restricted in the present problem to solutions of the forms (69.15) and (69.17), and within the range a to b must have

$$C\cos\left\{\frac{2\pi}{h}\int\limits_a^x \sqrt{\{2m(E-V)\}}\,dx - \tfrac{1}{4}\pi\right\} = C''\cos\left\{\frac{2\pi}{h}\int\limits_x^b \sqrt{\{2m(E-V)\}}\,dx - \tfrac{1}{4}\pi\right\} \tag{69.19}$$

holding for all points x. It will be readily seen, however, that this relation can be satisfied for arbitrary values of x only if we have

$$\frac{2\pi}{h}\int\limits_a^b \sqrt{\{2m(E-V)\}}\,dx = n\pi + \tfrac{1}{2}\pi$$

and

$$C'' = (-1)^n C, \tag{69.20}$$

where n is an integer.

The first of these expressions may be rewritten in the forms

$$2\int\limits_a^b \sqrt{\{2m(E-V)\}}\,dx = \oint p\,dx = (n+\tfrac{1}{2})h, \tag{69.21}$$

where p denotes the classical momentum of the particle and the integration is over a complete period for the classical motion of the particle from a to b and back. In the case of a system of more than one degree of freedom, such that the wave equation corresponding to (69.12) can be treated by the method of separation of variables, the result given by (69.21) can be applied separately to the coordinate and momentum, q_i and p_i, for each degree of freedom that corresponds classically to an oscillatory motion. The approximations involved in obtaining (69.21) become negligible at sufficiently large values of n.

The result given by (69.21) is of interest in showing the connexion between the *older quantum theory* and the *present quantum mechanics.* In the older quantum theory the allowed values of the energy for quasi-periodic steady states of motion were to be obtained, in accordance with the Wilson-Sommerfeld rule of quantization, by setting the phase integrals $\int p_i\,dq_i$ equal to Planck's constant h multiplied by an integral quantum number n. In the quantum mechanics we now see that a better approximation is obtained in oscillatory situations by taking half-integral quantum numbers $n+\frac{1}{2}$. (Cf. § 72, eq. (72.3).) However, in simple one-dimensional rotational situations, not involving particle spin, we shall still find integral quantum numbers. (Cf. § 73, remarks in connexion with eq. (73.27).)

In accordance with the result given by (69.21) we can regard each of the successive energy eigenstates, obtained by taking successive values of the integer n, as correlated with the classical states which would lie in a region of the phase space of approximate area h. This is an illustration of our previous remark, end of § 20, that a region in the phase space of magnitude h^f, in the case of a system of f degrees of freedom, could be regarded as an approximate classical analogue for a precisely defined quantum mechanical state. We shall note further illustrations of this correlation in the course of the present chapter.

The volumes of the phase space to be associated with specified quantum mechanical states can be taken as assuming the precise values h^f, as we approach the correspondence principle limit where the quantum mechanics and classical mechanics lead to concordant conclusions, i.e. in the above case as we go to large values of n. This will be of interest in a later chapter, § 84, when we discuss the agreement, at the correspondence principle limit, between the basic hypotheses for the classical and for the quantum statistics.

70. Particle in free space

We may next turn to a consideration of the behaviour of a free particle in three-dimensional space. Since any solution of the complete Schroedinger equation $\psi(q,t)$ can be expressed as a superposition of steady state solutions,

$$\psi(q,t) = \sum_k c_k\,u_k(q)e^{-(2\pi i/h)E_k t}, \tag{70.1}$$

we may first consider the nature of the eigenfunctions $u_k(q)$. These will be solutions of the time-free Schroedinger equation

$$\mathbf{H}u = Eu, \tag{70.2}$$

where E is a particular eigenvalue of the energy and $\mathbf{H}$ is the Hamiltonian operator.

For the case of a particle in free space we may take the potential V

as a constant, which we can set equal to zero without loss of generality. Using a coordinate representation corresponding to the Cartesian coordinates x, y, z, the Hamiltonian operator then assumes the simple form

$$\mathbf{H} = \frac{1}{2m}(\mathrm{p}_x^2+\mathrm{p}_y^2+\mathrm{p}_z^2) = -\frac{h^2}{8\pi^2 m}\left(\frac{\partial^2}{\partial x^2}+\frac{\partial^2}{\partial y^2}+\frac{\partial^2}{\partial z^2}\right),$$

where m is the mass of the particle. And substituting into (70.2), we obtain

$$\frac{\partial^2 u}{\partial x^2}+\frac{\partial^2 u}{\partial y^2}+\frac{\partial^2 u}{\partial z^2} = -\frac{8\pi^2 m}{h^2}Eu \tag{70.3}$$

as the equation for determining the eigensolutions $u(x,y,z)$ corresponding to different values of E.

A general solution of this equation can evidently be written in the form

$$u = C\sin(ax+\alpha)\sin(by+\beta)\sin(cz+\gamma), \tag{70.4}$$

where a, b, c, α, β, γ, and C are constants, provided the first three of these constants are connected by the relation

$$a^2+b^2+c^2 = \frac{8\pi^2 m}{h^2}E, \tag{70.5}$$

thus reducing the number of independent constants to six in agreement with the form of the original equation (70.3).

A solution of the form (70.4) is, however, not very convenient for studying situations of any great physical interest, since taking the constants occurring in the sine terms as *real* we see that it then merely corresponds to a uniform sinusoidal distribution of the probability density u^*u for finding the particle somewhere in space, and to a zero value for the probability current $\mathbf{S}$ in any direction.

A form of solution which is better adapted for giving insight into situations of physical interest is obtained by considering solutions which are simultaneously characteristic of the three components of momentum p_x, p_y, p_z and of the energy E. Since the operators for all of these observables commute with each other it is evident, in accordance with the discussion of § 64 (d), that such solutions should be possible. As the four equations which must be satisfied by such a solution we evidently have

$$\mathrm{p}_x u = p_x u, \qquad \mathrm{p}_y u = p_y u, \qquad \mathrm{p}_z u = p_z u, \qquad \mathbf{H}u = Eu,$$

or, making use of a coordinate representation,

$$\frac{h}{2\pi i}\frac{\partial u}{\partial x} = p_x u, \qquad \frac{h}{2\pi i}\frac{\partial u}{\partial y} = p_y u, \qquad \frac{h}{2\pi i}\frac{\partial u}{\partial z} = p_z u,$$
$$-\frac{h^2}{8\pi^2 m}\left(\frac{\partial^2 u}{\partial x^2}+\frac{\partial^2 u}{\partial y^2}+\frac{\partial^2 u}{\partial z^2}\right) = Eu. \tag{70.6}$$

As a solution satisfying all of these equations, we then have

$$u(x,y,z) = Ce^{\frac{2\pi i}{h}(p_x x+p_y y+p_z z)} \tag{70.7}$$

with

$$E = \frac{1}{2m}(p_x^2+p_y^2+p_z^2)$$

and C a constant.

Multiplying (70.7) by the appropriate exponential time factor, we obtain

$$\psi(x,y,z,t) = Ce^{\frac{2\pi i}{h}(p_x x+p_y y+p_z z-Et)} \tag{70.8}$$

as the corresponding time-dependent solution. And introducing the wave numbers σ_x, σ_y, σ_z and frequency ν defined by

$$p_x = h\sigma_x, \qquad p_y = h\sigma_y, \qquad p_z = h\sigma_z, \qquad E = h\nu, \tag{70.9}$$

we are once more led to the *de Broglie waves* for a free particle

$$\psi(x,y,z,t) = Ce^{2\pi i(\sigma_x x+\sigma_y y+\sigma_z z-\nu t)} \tag{70.10}$$

already obtained in § 61.

Solutions of the form (70.8) may be superposed to give wave packets which will give an approximate representation of the kinematical behaviour of a single particle as already studied in § 62. They may also be used directly to represent an infinite stream of similar non-interacting particles of a given energy, having the uniform constant probability density

$$W = \psi^*\psi = |C|^2 \tag{70.11}$$

and the constant components of probability current

$$\begin{aligned} S_x &= \frac{h}{4\pi i m}\left(\psi^*\frac{\partial\psi}{\partial x}-\psi\frac{\partial\psi^*}{\partial x}\right) = |C|^2\frac{p_x}{m}, \\ S_y &= \frac{h}{4\pi i m}\left(\psi^*\frac{\partial\psi}{\partial y}-\psi\frac{\partial\psi^*}{\partial y}\right) = |C|^2\frac{p_y}{m}, \\ S_z &= \frac{h}{4\pi i m}\left(\psi^*\frac{\partial\psi}{\partial z}-\psi\frac{\partial\psi^*}{\partial z}\right) = |C|^2\frac{p_z}{m}. \end{aligned} \tag{70.12}$$

This latter use is valuable in the study of problems involving collision with a scatterer, where a solution in Cartesian coordinates to represent the oncoming particles can be combined with a solution in polar coordinates around the scatterer to represent the deflected particles.

71. Particle in a container

(*a*) **The energy eigenvalues and eigenfunctions.** We now turn to a consideration of a particle inside a container rather than completely free in space. This will be a first step towards our later consideration of systems, commonly treated in statistical mechanics, such as an assembly of gas molecules in a vessel.

Since any solution of the Schroedinger equation for such a system can be regarded as given by a superposition of eigensolutions $u_k(q)$, corresponding to the different eigenvalues E_k for the energy of the particle, i.e. by

$$\psi(q,t) = \sum_k c_k u_k(q) e^{-(2\pi i/h)E_k t}, \tag{71.1}$$

we may at once turn our attention to these eigensolutions. As the equation determining these eigensolutions we have, in agreement with (60.4),

$$\mathbf{H}u(q) = Eu(q). \tag{71.2}$$

The quantity E occurring in this equation is some particular possible eigenvalue for the energy of the system, and the Hamiltonian operator $\mathbf{H}$ for a single simple particle of mass m will evidently be given in agreement with (56.5) by

$$\begin{aligned}\mathbf{H} &= \frac{1}{2m}(\mathrm{p}_x^2+\mathrm{p}_y^2+\mathrm{p}_z^2)+V(x,y,z) \\ &= -\frac{h^2}{8\pi^2 m}\left(\frac{\partial^2}{\partial x^2}+\frac{\partial^2}{\partial y^2}+\frac{\partial^2}{\partial z^2}\right)+V(x,y,z),\end{aligned} \tag{71.3}$$

where x, y, and z are Cartesian coordinates for the position of the particle and V is the classical expression for the potential energy of the particle as a function of position. Substituting (71.3) in (71.2), we then obtain

$$\frac{\partial^2 u}{\partial x^2}+\frac{\partial^2 u}{\partial y^2}+\frac{\partial^2 u}{\partial z^2} = -\frac{8\pi^2 m}{h^2}(E-V)u \tag{71.4}$$

as the equation for determining the different eigensolutions $u(x,y,z)$ in question.

For simplicity let us now consider that the container for the particle is a rectangular box having one corner at the origin of coordinates $x = y = z = 0$, and having the length, breadth, and height $x = l_1$, $y = l_2$, and $z = l_3$. Inside the box the potential V would be a constant and this may be conveniently taken as the starting-point for energy measurements, so that we may put $V = 0$ inside the container. On the other hand, at the walls of the container, $x = 0$, $x = l_1$, etc., we shall take the potential as suddenly increasing to infinity in agreement with the *perfectly reflecting* character which we shall wish to ascribe to the walls.

Taking $V = 0$ inside the box, we may then evidently write as a general solution of (71.4) in the interior the same expression that we found for the case of the free particle

$$u = C\sin(ax+\alpha)\sin(by+\beta)\sin(cz+\gamma), \tag{71.5}$$

where a, b, c, α, β, γ, and C are constants, the first three being connected by the relation

$$a^2+b^2+c^2 = \frac{8\pi^2 m}{h^2}E, \tag{71.6}$$

thus reducing the number of independent constants to six, as would be expected from the form of (71.4).

We must next consider the boundary conditions which would be imposed on this solution by the presence of the walls of the box. Since by hypothesis V goes to infinity at the walls, it is evident from the form of (71.4) that we must require u itself to be zero at the walls, since otherwise its second derivative would become $\pm\infty$ and we should have an overwhelming probability of finding the particle outside rather than inside the box. To obtain the result $u = 0$ at $x = 0$, $y = 0$, and $z = 0$ we must set the constants

$$\alpha = \beta = \gamma = 0, \tag{71.7}$$

and to obtain that result at $x = l_1$, $y = l_2$, and $z = l_3$ we must take the constants

$$a = \frac{n_1\pi}{l_1}, \qquad b = \frac{n_2\pi}{l_2}, \qquad c = \frac{n_3\pi}{l_3}, \tag{71.8}$$

where n_1, n_2, and n_3 are integers.

Under these circumstances our eigensolution (71.5) reduces to the form

$$u = C\sin\frac{n_1\pi x}{l_1}\sin\frac{n_2\pi y}{l_2}\sin\frac{n_3\pi z}{l_3} \tag{71.9}$$

and the corresponding eigenvalue of the energy has—in accordance with (71.6) and (71.8)—the value

$$E = \frac{h^2}{8m}\left(\frac{n_1^2}{l_1^2}+\frac{n_2^2}{l_2^2}+\frac{n_3^2}{l_3^2}\right). \tag{71.10}$$

The solution can be normalized to unity by taking the constant

$$C = \sqrt{\left(\frac{8}{l_1 l_2 l_3}\right)}. \tag{71.11}$$

We can thus obtain a complete set of normalized orthogonal eigenfunctions $u_{n_1 n_2 n_3}(x,y,z)$, corresponding to different choices of the quantum numbers n_1, n_2, and n_3, and to the different possible discrete values of the energy $E_{n_1 n_2 n_3}$. In case any two of the lengths l_1, l_2, l_3 are commensurable, degeneracy will occur since different eigensolutions would then correspond to the same value of the energy. Having obtained the set of eigenfunctions, any solution for the state of the system can then be expressed as a superposition of such eigenfunctions each multiplied by a suitable coefficient and exponential time factor.

(*b*) **The number of eigensolutions in a given range of energy.** In our later statistical mechanical work we shall be specially interested in the number of different eigensolutions or steady quantum states, having eigenvalues of the energy lying within a given range, say E to $E+\Delta E$.

To determine this number let us consider a system of Cartesian coordinates with axes for plotting the values of the three quantum numbers n_1, n_2, n_3, and let us construct a cubical lattice consisting of points whose coordinates are integral values of these quantities. Each such point with positive values of n_1, n_2, and n_3 will then correspond to a particular eigensolution or quantum state of the system. Furthermore, in the interior of such a lattice there will be one unit cube for each lattice point. Hence, if E is not too small, we can calculate the total number of quantum states corresponding to values of the energy less than E by considering the volume

$$V = \iiint dn_1\,dn_2\,dn_3 \tag{71.12}$$

over the range of quantum numbers given by

$$\frac{n_1^2}{l_1^2}+\frac{n_2^2}{l_2^2}+\frac{n_3^2}{l_3^2} \leqslant \frac{8mE}{h^2}, \tag{71.13}$$

which, in accordance with (71.10), will include all quantum states with energy eigenvalues less than E. From the known formula (see Appendix II) for the volume of an ellipsoid this will give us

$$V = \tfrac{4}{3}\pi\left(\frac{8mE}{h^2}\right)^{\frac{3}{2}} l_1 l_2 l_3 = \tfrac{4}{3}\pi\left(\frac{8mE}{h^2}\right)^{\frac{3}{2}} v, \tag{71.14}$$

where $v = l_1 l_2 l_3$ is the spatial volume of the container for the molecule. And since the positive values of the quantum numbers n_1, n_2, n_3 with which we are concerned are limited to one octant, we shall then have

$$\tfrac{1}{8}V = \tfrac{4}{3}\pi\frac{(2mE)^{\frac{3}{2}}}{h^3}v \tag{71.15}$$

as the number of eigensolutions or quantum states with eigenvalues of the energy less than E. By differentiation we then obtain

$$G = \frac{4\pi v}{h^3} m\sqrt{(2mE)}\,\Delta E \tag{71.16}$$

as the desired expression for the number of eigensolutions or quantum states G corresponding to the energy range E to $E+\Delta E$. In case the particles have a spin, which will be treated in § 75, the number of such states will be twice as great:

$$G = \frac{8\pi v}{h^3} m\sqrt{(2mE)}\,\Delta E. \tag{71.17}$$

We shall find these results very important for our later statistical considerations. By comparing the number of eigensolutions in the range ΔE as given by (71.16) with the volume of the phase space $4\pi v m\sqrt{(2mE)}\,\Delta E$, which would correspond classically to that range, we see for large quantum numbers that we can regard each eigensolution for this system of three degrees of freedom as correlated with an extension in the phase space of magnitude h^3. This is an example of the general possibility for such correlations as discussed at the end of § 69 (*b*).

72. Particle in a Hooke's law field of force

As our next simple mechanical system we shall consider a particle in a potential field $V = kx^2$, which from a classical point of view would give us a linear harmonic oscillator. This will also be a step towards the treatment of systems of physical interest since the oscillations of actual atomic systems can be treated with some degree of success by similar methods.

In accordance with the suggested form, $V = kx^2$, for the potential function, the time-free Schroedinger equation

$$\mathbf{H}u = Eu, \tag{72.1}$$

which determines the energy eigenvalues E and eigenfunctions u for the problem, can be written as

$$\frac{\partial^2 u}{\partial x^2} + \frac{8\pi^2 m}{h^2}(E - kx^2)u = 0. \tag{72.2}$$

The eigenvalues of E which will lead to solutions of this equation, such that $u(x)$ does not go to infinity as x goes to infinity, are found to be given by the simple formula

$$E_n = (n+\tfrac{1}{2})h\sqrt{(k/2\pi^2 m)} = (n+\tfrac{1}{2})h\nu \quad (n = 0, 1, 2, 3, \ldots), \tag{72.3}$$

where ν is an abbreviation for the classical frequency with which such a particle would oscillate. Furthermore, the complete set of normalized orthogonal eigenfunctions, which are solutions of (72.2) for the different values of n, are found to be expressible in the form†

$$u_n = C_n\, e^{-\frac{1}{2}\xi^2} H_n(\xi), \tag{72.4}$$

† For a detailed derivation of this solution of (72.2), together with a table of Hermite polynomials $H_n(\xi)$, see, for example, Pauling and Wilson, *Introduction to Quantum Mechanics*, New York, 1935, pp. 67–81.

where for simplicity we have made the substitution

$$\xi = \sqrt[4]{\left(\frac{8\pi^2mk}{h^2}\right)}x, \tag{72.5}$$

the Hermite polynomials $H_n(\xi)$ are defined by

$$H_n(\xi) = (-1)^n e^{\xi^2}\frac{d^n e^{-\xi^2}}{d\xi^n}, \tag{72.6}$$

and the normalizing factor C_n, leading to the value unity when the square of an individual eigenfunction is integrated over all values of the *original variable* x, is given by

$$C_n = \left\{\frac{1}{2^n n!}\sqrt[4]{\left(\frac{8mk}{h^2}\right)}\right\}^{\frac{1}{2}}. \tag{72.7}$$

These results are of interest in connexion with the oscillatory behaviour of diatomic molecules, and the behaviour of the modes of vibration by which we can represent the thermal energy of an elastic solid or the electromagnetic energy of the radiation in a hollow enclosure. In connexion with (72.3), it is specially interesting to note the appearance of 'half-quantum numbers' in disagreement with the integral quantum numbers prescribed by the older quantum theory. In accordance with this result, the lowest oscillatory quantum state would exhibit a so-called zero-point energy $\frac{1}{2}h\nu$. The superiority of the new result in the case of diatomic molecules was first observationally demonstrated by the work of Mulliken.†

By comparing the expression for energy given by (72.3) with the classical expression $E\sqrt{(2\pi^2m/k)}$ for the area in the phase space that would correspond classically to values of the energy less than E, we see for large values of n that we can regard each eigensolution for this system of one degree of freedom as correlated with an extension in the phase space of magnitude h. This is another example of the general possibility for such correlations as discussed at the end of § 69 (*b*).

73. Particle in a central field of force

As our next mechanical system we shall take a simple particle in a central field of force. In such a field of force the angular momentum around any axis—if it has a definite value at some initial time—will remain fixed, in accordance with the treatment of § 63 (*d*), so long as the system is not disturbed; and we shall be interested in knowing the possible eigenvalues which the angular momentum can assume. We shall commence by considering some general properties of the operators

† Mulliken, *Phys. Rev.* **25**, 259 (1925).

for the components of angular momentum which are of interest without reference to the kind of force field acting on the particle.

(*a*) **Operators for the components of angular momentum.** If we consider a particle with the Cartesian coordinates x, y, z and corresponding components of momentum p_x, p_y, p_z, the classical expressions for its components of angular momentum around the x, y, and z coordinate axes would be

$$\begin{aligned} M_x &= yp_z - zp_y, \\ M_y &= zp_x - xp_z, \\ M_z &= xp_y - yp_x. \end{aligned} \tag{73.1}$$

Making use of our rules for obtaining the corresponding operators suitable for use in coordinate language, we at once obtain as the corresponding operators

$$\begin{aligned} \mathbf{M}_x &= \frac{h}{2\pi i}\left(y\frac{\partial}{\partial z} - z\frac{\partial}{\partial y}\right), \\ \mathbf{M}_y &= \frac{h}{2\pi i}\left(z\frac{\partial}{\partial x} - x\frac{\partial}{\partial z}\right), \\ \mathbf{M}_z &= \frac{h}{2\pi i}\left(x\frac{\partial}{\partial y} - y\frac{\partial}{\partial x}\right), \end{aligned} \tag{73.2}$$

where no ambiguity arises as to the order of factors, since non-conjugate coördinates and momenta commute.† Furthermore, by squaring and adding we obtain for the operator for the square of the total angular momentum with respect to the origin

$$\begin{aligned} \mathbf{M}^2 &= \mathbf{M}_x^2 + \mathbf{M}_y^2 + \mathbf{M}_z^2 \\ &= \left(\frac{h}{2\pi i}\right)^2 \left(x^2\frac{\partial^2}{\partial y^2} - 2xy\frac{\partial^2}{\partial x \partial y} + y^2\frac{\partial^2}{\partial x^2} + y^2\frac{\partial^2}{\partial z^2} - 2yz\frac{\partial^2}{\partial y \partial z} + z^2\frac{\partial^2}{\partial y^2} + \right. \\ &\qquad \left. + z^2\frac{\partial^2}{\partial x^2} - 2zx\frac{\partial^2}{\partial z \partial x} + x^2\frac{\partial^2}{\partial z^2} - 2x\frac{\partial}{\partial x} - 2y\frac{\partial}{\partial y} - 2z\frac{\partial}{\partial z}\right). \end{aligned} \tag{73.3}$$

For some purposes it is more convenient to express these operators in the language of a set of polar coordinates r, θ, ϕ which can be introduced in accordance with the equations

$$x = r\sin\theta\cos\phi, \qquad y = r\sin\theta\sin\phi, \qquad z = r\cos\theta. \tag{73.4}$$

† Treating these operators (73.2) as the components of a vector **M**, they have the important property that the changes in any scalar function of position under an infinitesimal rotation $\delta\boldsymbol{\omega}$ may be simply expressed in terms of them by

$$\delta F(x,y,z) = -\frac{2\pi i}{h}(\delta\boldsymbol{\omega}\,.\,\mathbf{M})F(x,y,z),$$

where $(\delta\boldsymbol{\omega}\,.\,\mathbf{M})$ is the inner product of the two vectors. See for this, and its connexion with the commutation rules (73.7) and (75.1), Pauli, *Handbuch der Physik*, xxiv/1, second edition, pp. 176 ff.

By a straightforward but somewhat tedious process of substitution the above operators are then found to assume the forms

$$\mathbf{M}_x = \frac{h}{2\pi i}\left(-\sin\phi\frac{\partial}{\partial\theta}-\cot\theta\cos\phi\frac{\partial}{\partial\phi}\right),$$

$$\mathbf{M}_y = \frac{h}{2\pi i}\left(\cos\phi\frac{\partial}{\partial\theta}-\cot\theta\sin\phi\frac{\partial}{\partial\phi}\right), \tag{73.5}$$

$$\mathbf{M}_z = \frac{h}{2\pi i}\frac{\partial}{\partial\phi}$$

and

$$\mathbf{M}^2 = \left(\frac{h}{2\pi i}\right)^2\left\{\frac{1}{\sin\theta}\frac{\partial}{\partial\theta}\left(\sin\theta\frac{\partial}{\partial\theta}\right)+\frac{1}{\sin^2\theta}\frac{\partial^2}{\partial\phi^2}\right\}. \tag{73.6}$$

It will be of interest, also in connexion with our later introduction of the spin of the electron and other fundamental particles, to consider the commutation properties of the above operators. It will be readily found that the operators for the different components of angular momentum do not commute with each other, but have the commutators

$$[\mathbf{M}_x, \mathbf{M}_y] = -\frac{h}{2\pi i}\mathbf{M}_z,$$

$$[\mathbf{M}_y, \mathbf{M}_z] = -\frac{h}{2\pi i}\mathbf{M}_x, \tag{73.7}$$

$$[\mathbf{M}_z, \mathbf{M}_x] = -\frac{h}{2\pi i}\mathbf{M}_y.$$

On the other hand, it will be seen that the operator for the square of the total angular momentum with respect to the origin commutes with any of the operators for the component of angular momentum around a particular axis.

In accordance with the above, we may then conclude that it would not be possible for the components of angular momentum around two different axes to exhibit precisely defined values at the same time, except in the special case when all components, and thus M^2, vanish. On the other hand, however, it would be possible for the square of the total angular momentum with respect to the origin and the component around any selected axis to exhibit simultaneously defined values.

(*b*) **The eigenfunctions and eigenvalues corresponding to angular momentum.** We may now inquire into the eigenfunctions and eigenvalues corresponding to a component of angular momentum around a particular axis and to the total angular momentum with respect to the origin.

Let us first consider the angular momentum around some definite direction which we may take as the z-axis. Taking the variables r, θ, and ϕ, the eigenfunctions $u(r,\theta,\phi)$, corresponding to an eigenvalue M_z for the z-component of angular momentum, would then be given, in accordance with (64.2), by

$$\mathbf{M}_z u(r,\theta,\phi) = M_z u(r,\theta,\phi). \tag{73.8}$$

Substituting the expression for the operator $\mathbf{M}_z$ given by (73.5), this gives us the equation

$$\frac{h}{2\pi i}\frac{\partial}{\partial\phi}u(r,\theta,\phi) = M_z u(r,\theta,\phi) \tag{73.9}$$

for determining the desired eigenfunctions.

This equation can easily be treated by the method of *separation of variables* by assuming a solution of the form

$$u(r,\theta,\phi) = v(r,\theta)\Phi(\phi), \tag{73.10}$$

where the function of ϕ alone must evidently satisfy the equation

$$\frac{h}{2\pi i}\frac{\partial}{\partial\phi}\Phi(\phi) = M_z\Phi(\phi). \tag{73.11}$$

And this has the general form of solution

$$\Phi(\phi) = Ce^{(2\pi i/h)M_z\phi}, \tag{73.12}$$

where C is a constant.

For such a solution, however, to be a satisfactory constituent of an eigenfunction $u(r,\theta,\phi)$, which could be used in predicting the probability of finding the particle at any point r,θ,ϕ, it will be appreciated that $\Phi(\phi)$ must have the same values for ϕ and $\phi+2\pi$. Hence the constituent *eigenfunctions* $\Phi(\phi)$ must have the special form

$$\Phi(\phi) = Ce^{im\phi}, \tag{73.13}$$

with

$$m = 0, \pm 1, \pm 2, \pm 3,..., \tag{73.14}$$

and with

$$M_z = m\frac{h}{2\pi}, \tag{73.15}$$

where m—not to be confused with the mass of the particle—may be called the quantum number for the allowed eigenvalues M_z of the z-component of angular momentum. The constant C may be given the value

$$C = \frac{1}{\sqrt{(2\pi)}} \tag{73.16}$$

in case it is desired to normalize to unity for integration over the range 0 to 2π.

Turning now to the total angular momentum with respect to the origin, we shall have the characteristic equation

$$\mathbf{M}^2 u(r,\theta,\phi) = M^2 u(r,\theta,\phi). \tag{73.17}$$

Substituting the expression for the operator $\mathbf{M}^2$ given by (73.6), and the solution characteristic of M_z provided by (73.10), we can then obtain solutions simultaneously characteristic of total angular momentum squared and its z-component from the equation

$$\left(\frac{h}{2\pi i}\right)^2\left\{\frac{1}{\sin\theta}\frac{\partial}{\partial\theta}\left(\sin\theta\frac{\partial}{\partial\theta}\right)+\frac{1}{\sin^2\theta}\frac{\partial^2}{\partial\phi^2}\right\}v(r,\theta)\Phi(\phi) = M^2 v(r,\theta)\Phi(\phi). \quad (73.18)$$

This equation can also evidently be treated by the method of separation of variables if we write the solution in the form

$$u(r,\theta,\phi) = v(r,\theta)\Phi(\phi) = R(r)\Theta(\theta)\Phi(\phi), \quad (73.19)$$

which, by substituting the specific expression for $\Phi(\phi)$ given by (73.13), will require

$$\frac{1}{\sin\theta}\frac{\partial}{\partial\theta}\left\{\sin\theta\frac{\partial\Theta(\theta)}{\partial\theta}\right\}-\frac{m^2}{\sin^2\theta}\Theta(\theta)+\left(\frac{2\pi}{h}\right)^2 M^2\Theta(\theta) = 0. \quad (73.20)$$

The allowable solutions of this latter equation can be shown to be of the form†

$$\Theta(\theta) = CP_l^m(\cos\theta), \quad (73.21)$$

with

$$l = |m|, |m|+1, |m|+2,\ldots, \quad (73.22)$$

and with the allowed eigenvalues for the square of the total angular momentum

$$M^2 = l(l+1)\left(\frac{h}{2\pi}\right)^2, \quad (73.23)$$

where

$$P_l^m(\cos\theta) = \frac{1}{2^l\, l!}\sin^{|m|}\theta\,\frac{d^{|m|+l}(\cos^2\theta-1)^l}{d(\cos\theta)^{|m|+l}} \quad (73.24)$$

is the *associated Legendre function* of order m and degree l, and l may be called the quantum number for total angular momentum. The constant C can be given the value

$$C = \sqrt{\frac{(2l+1)(l-|m|)!}{2(l+|m|)!}} \quad (73.25)$$

in case it is desired to normalize to unity when the square of (73.21) is multiplied by $\sin\theta\, d\theta$ and integrated from $\theta = 0$ to π.

Returning to (73.19), we may now write

$$\psi(r,\theta,\phi,t_0) = u(r,\theta,\phi) = R(r)P_l^m(\cos\theta)e^{im\phi}, \quad (73.26)$$

as an expression for the probability amplitude giving a state of the system at a time t_0 such that the particle would exhibit the component of angular momentum around the z-axis

$$M_z = m\frac{h}{2\pi} \qquad (m = 0, \pm1, \pm2, \pm3,\ldots) \quad (73.27)$$

† For a detailed derivation of this solution of (73.20), together with a table of normalized Legendre functions $P_l^m(\cos\theta)$, see, for example, Pauling and Wilson, *Introduction to Quantum Mechanics*, New York, 1935, chap. v.

and the total angular momentum with respect to the origin

$$M^2 = l(l+1)\left(\frac{h}{2\pi}\right)^2 \qquad (l = |m|, |m|+1, |m|+2,\ldots). \qquad (73.28)$$

Comparing (73.27) with the classical expression $\oint M_z\, d\phi = 2\pi M_z$ for the phase integral corresponding to rotation around the z-axis, we see in this simple situation that the successive eigenvalues of M_z would be determined by the original Wilson-Sommerfeld rule of setting the phase integral equal to Planck's constant h multiplied by an integer. And noting that (73.28) would give eigenvalues for the absolute magnitude of the total angular momentum which would approach $|M| = lh/2\pi$ at large values of l, we see that the successive values of $|M|$ would then be separated by the amount demanded by the Wilson-Sommerfeld rule. In both cases we see, for large quantum numbers, that we can regard the eigensolutions for the single degree of freedom in question as associated with an extension in the classical phase space of magnitude h. This is still another example of the general possibility for such correlations as discussed at the end of § 69 (b), this time for angular momentum eigensolutions instead of for energy eigensolutions.

(c) **Steady states in a central field of force.** So far we have made no assumption which would require our particle to remain permanently in any particular eigenstate of angular momentum. We may now introduce the assumption that the field of force, which would be acting on the particle from a classical point of view, is spherically symmetrical around the origin of coordinates. We shall then be able to find steady states corresponding to permanently fixed eigenvalues for all three quantities—component of angular momentum M_z, total angular momentum squared M^2, and energy E.

Assuming such a radial field of force, the Hamiltonian operator can then be written in the form

$$\begin{aligned}\mathbf{H} &= \frac{1}{2m}(\mathrm{p}_x^2+\mathrm{p}_y^2+\mathrm{p}_z^2)+V(r)\\ &= -\frac{h^2}{8\pi^2 m}\nabla^2+V(r)\\ &= -\frac{h^2}{8\pi^2 m}\left\{\frac{1}{r^2}\frac{\partial}{\partial r}\left(r^2\frac{\partial}{\partial r}\right)+\frac{1}{r^2\sin\theta}\frac{\partial}{\partial\theta}\left(\sin\theta\frac{\partial}{\partial\theta}\right)+\frac{1}{r^2\sin^2\theta}\frac{\partial^2}{\partial\phi^2}\right\}+V(r),\end{aligned} \qquad (73.29)$$

where $V(r)$ is the spherically symmetrical potential field, and we have introduced in the last form of writing the known expression for the Laplacian operator in spherical coordinates.

It will be seen on examination, however, that this Hamiltonian operator commutes both with that for the z-component of angular momentum given by (73.5) and with that for the total angular momentum squared given by (73.6). Under these circumstances we can then conclude in the first place, in accordance with § 64 (d), that we can have states simultaneously characteristic of all three quantities M_z, M^2, and E; and can conclude in the second place, in accordance with (63.3), that these angular momenta will then be conserved with time so long as the system is left undisturbed.

To examine such states, permanently characteristic of all three observables, we have

$$\mathbf{H}u(r,\theta,\phi) = Eu(r,\theta,\phi) \tag{73.30}$$

as the general equation for a steady state of constant energy E. And substituting for the operator $\mathbf{H}$ the expression given by (73.29), this can be written in the specific form

$$\left\{\frac{1}{r^2}\frac{\partial}{\partial r}\left(r^2\frac{\partial}{\partial r}\right)+\frac{1}{r^2\sin\theta}\frac{\partial}{\partial\theta}\left(\sin\theta\frac{\partial}{\partial\theta}\right)+\frac{1}{r^2\sin^2\theta}\frac{\partial^2}{\partial\phi^2}\right\}u(r,\theta,\phi)+ \\ +\frac{8\pi^2m}{h^2}\{E-V(r)\}u(r,\theta,\phi) = 0. \tag{73.31}$$

For $u(r,\theta,\phi)$, however, we may now substitute the specific form

$$u(r,\theta,\phi) = R(r)\Theta(\theta)\Phi(\phi) \tag{73.32}$$

already made characteristic for the angular momenta M_z and M^2. Doing so, and noting the form of $\Phi(\phi)$ given by (73.13), we obtain

$$\left\{\frac{1}{r^2}\frac{\partial}{\partial r}\left(r^2\frac{\partial}{\partial r}\right)+\frac{1}{r^2\sin\theta}\frac{\partial}{\partial\theta}\left(\sin\theta\frac{\partial}{\partial\theta}\right)-\frac{m^2}{r^2\sin\theta}\right\}R(r)\Theta(\theta)+ \\ +\frac{8\pi^2m}{h^2}\{E-V(r)\}R(r)\Theta(\theta) = 0.$$

And noting the property of $\Theta(\theta)$ given by (73.20), this becomes

$$\left\{\frac{1}{r^2}\frac{\partial}{\partial r}\left(r^2\frac{\partial}{\partial r}\right)-\left(\frac{2\pi}{h}\right)^2\frac{M^2}{r^2}\right\}R(r)+\frac{8\pi^2m}{h^2}\{E-V(r)\}R(r) = 0.$$

By substitution of (73.23), this then gives the desired result

$$\frac{1}{r^2}\frac{\partial}{\partial r}r^2\frac{\partial R(r)}{\partial r}-\frac{l(l+1)}{r^2}R(r)+\frac{8\pi^2m}{h^2}\{E-V(r)\}R(r) = 0 \tag{73.33}$$

as an equation for determining the form of the remaining function

$R(r)$, in states of energy E, and angular momenta $M_z = mh/2\pi$ and $M^2 = l(l+1)h/2\pi$. Writing

$$\chi(r) = rR(r), \tag{73.34}$$

this result can also be expressed in the simpler form

$$\frac{d^2\chi(r)}{dr^2}+\frac{8\pi^2 m}{h^2}\left\{E-\frac{h^2}{8\pi^2 m}\frac{l(l+1)}{r^2}-V(r)\right\}\chi(r) = 0, \tag{73.35}$$

which is similar to that for a one-dimensional problem and may be treated by the approximate methods of § 69 (*b*).

The actual solutions of this equation will depend on the form of $V(r)$; and the conditions on E necessary to secure an allowable form for use as a constituent of the eigenfunction $u(r,\theta,\phi)$ will determine the spectrum of energy eigenvalues E. The energy levels will, in general, be degenerate since more than one value of the quantum number m can correspond to any given value of l except for the case of $l = 0$. As a special condition on the allowability of eigensolutions it should be stated that $\chi(r) = rR(r)$ has to go to zero at $r = 0$, since otherwise it can be shown that the Hermitian character of the Hamiltonian would be lost at this singular point.† Having found a satisfactory solution for $R(r)$, we then have all that is necessary for a knowledge of the eigenfunction

$$u(r,\theta,\phi) = CP_l^m(\cos\theta)e^{im\phi}R(r) = CP_l^m(\cos\theta)e^{im\phi}\frac{\chi(r)}{r} \tag{73.36}$$

which corresponds to specified possible values of E, M_z, and M^2. It will be noted, in agreement with earlier discussions, that these eigensolutions for the three-dimensional problem may be regarded, on going to high quantum numbers, as each associated with an extension of magnitude h^3 in the classical phase space.

74. Two interacting particles

In the foregoing applications we have considered the behaviour of a single particle in a steady potential field. Let us now consider the somewhat more complicated case of two *non-identical* interacting particles of masses m_1 and m_2. This will provide methods useful in treating the two-particle problems presented by the hydrogen atom, by the diatomic molecule, and in the theory of collisions and scattering. We may again adopt the method of studying steady state solutions since any solution could then be obtained by superposition.

The energy eigensolutions $u(q)$ corresponding to the steady states of

† See Pauli, *Handbuch der Physik*, xxiv/1, second edition, pp. 123–4.

our present system will now be a function of the Cartesian coordinates x_1, y_1, z_1 and x_2, y_2, z_2 for each of the two particles, and the equation determining them can evidently be written in the form

$$\frac{1}{m_1}\left(\frac{\partial^2}{\partial x_1^2}+\frac{\partial^2}{\partial y_1^2}+\frac{\partial^2}{\partial z_1^2}\right)u+\frac{1}{m_2}\left(\frac{\partial^2}{\partial x_2^2}+\frac{\partial^2}{\partial y_2^2}+\frac{\partial^2}{\partial z_2^2}\right)u+\frac{8\pi^2}{h^2}(E-V)u = 0, \quad (74.1)$$

where m_1 and m_2 are the masses of the two particles, E is the total energy of the combined system, and V is the potential regarded as a function of the coordinates of the two particles.

(*a*) **Separation into external and internal equations.** If we now assume that the potential V can be regarded as the sum of two terms V_{ext} and V_{int}, one dependent on the coordinates for the centre of gravity of the two particles and the other on the coordinate differences between the two particles, it proves possible *to separate* the above equation into two equivalent equations depending separately on the coordinates of the centre of gravity of the system and on the coordinate differences. To accomplish this we may introduce the new variables X, Y, Z,

$$\begin{aligned} X &= \frac{m_1 x_1+m_2 x_2}{m_1+m_2}, \\ Y &= \frac{m_1 y_1+m_2 y_2}{m_1+m_2}, \\ Z &= \frac{m_1 z_1+m_2 z_2}{m_1+m_2}, \end{aligned} \quad (74.2)$$

which correspond classically to the position of the centre of gravity of the system as a whole, and x, y, z,

$$\begin{aligned} x &= x_2-x_1, \\ y &= y_2-y_1, \\ z &= z_2-z_1, \end{aligned} \quad (74.3)$$

which correspond classically to the differences between the coordinates of the two particles.

On making the indicated substitutions, equation (74.1) is found to assume the form

$$\frac{1}{m_1+m_2}\left(\frac{\partial^2}{\partial X^2}+\frac{\partial^2}{\partial Y^2}+\frac{\partial^2}{\partial Z^2}\right)u+\frac{1}{\mu}\left(\frac{\partial^2}{\partial x^2}+\frac{\partial^2}{\partial y^2}+\frac{\partial^2}{\partial z^2}\right)u+ \\ +\frac{8\pi^2}{h^2}\{E-V_{\text{ext}}(X,Y,Z)-V_{\text{int}}(x,y,z)\}u = 0, \quad (74.4)$$

where we have now explicitly indicated that the potential V is assumed expressible as a sum of two parts depending on the external and internal coordinates, and the symbol μ has been introduced for the so-called *reduced mass* of the system

$$\mu = \frac{m_1 m_2}{m_1+m_2}. \tag{74.5}$$

We readily see, however, that equation (74.4) will be satisfied by a solution of the form

$$u(X,Y,Z,x,y,z) = u_{\text{ext}}(X,Y,Z)u_{\text{int}}(x,y,z), \tag{74.6}$$

provided the new functions u_{ext} and u_{int} are taken as solutions of the separate equations

$$\frac{1}{m_1+m_2}\left(\frac{\partial^2}{\partial X^2}+\frac{\partial^2}{\partial Y^2}+\frac{\partial^2}{\partial Z^2}\right)u_{\text{ext}}+\frac{8\pi^2}{h^2}(E_{\text{ext}}-V_{\text{ext}})u_{\text{ext}} = 0 \tag{74.7}$$

and

$$\frac{1}{\mu}\left(\frac{\partial^2}{\partial x^2}+\frac{\partial^2}{\partial y^2}+\frac{\partial^2}{\partial z^2}\right)u_{\text{int}}+\frac{8\pi^2}{h^2}(E-E_{\text{ext}}-V_{\text{int}})u_{\text{int}} = 0, \tag{74.8}$$

where E_{ext} is a constant. This is evidently satisfactory since by multiplying the first of these equations by u_{int} and the second by u_{ext} and adding, we reproduce the original equation (74.4) with the suggested form of solution (74.6).

It will be noted that the first of the two equations is the same as that for a single particle with a total mass m_1+m_2, equal to the sum of the masses of the two particles, with the total energy E_{ext}, and moving in a potential field $V_{\text{ext}}(X,Y,Z)$. This equation can hence be treated by the same methods as for a single particle, and the allowable values of E_{ext} can give a continuous or discrete spectrum according to the nature of the external field V_{ext}.

(*b*) **Separation of internal equation in case of central forces.** The second of the above equations (74.8) can also be given a simple treatment in case the internal potential V_{int} is assumed to be solely a function of the classical expression for the distance between the two particles $r = \sqrt{(x^2+y^2+z^2)}$. To treat the problem under these circumstances we may first transform to polar coordinates r, θ, and ϕ in accordance with the equations

$$\begin{aligned} x &= r\sin\theta\cos\phi, \\ y &= r\sin\theta\sin\phi, \\ z &= r\cos\theta. \end{aligned} \tag{74.9}$$

Classically this corresponds to taking polar coordinates for the second particle with the first particle at the origin. Introducing these new coordinates into (74.8), we obtain

$$\left\{\frac{1}{r^2}\frac{\partial}{\partial r}\left(r^2\frac{\partial}{\partial r}\right)+\frac{1}{r^2\sin\theta}\frac{\partial}{\partial\theta}\left(\sin\theta\frac{\partial}{\partial\theta}\right)+\frac{1}{r^2\sin^2\theta}\frac{\partial^2}{\partial\phi^2}\right\}u_{\text{int}}+$$
$$+\frac{8\pi^2\mu}{h^2}\{E_{\text{int}}-V_{\text{int}}(r)\}u_{\text{int}}=0, \quad (74.10)$$

where we have explicitly indicated the assumed dependence of V_{int} on r alone, and have introduced the symbol E_{int} in accordance with

$$E_{\text{int}} = E-E_{\text{ext}}. \quad (74.11)$$

It will be immediately noticed that equation (74.10) for our two particles has the same form as our earlier equation (73.31) for a single particle in a spherically symmetrical field, with the mass of the single particle m replaced by the reduced mass μ for the two particles. This will make it possible to make use of results obtained in the preceding section. To do this we may treat equation (74.10) by the method of *separation of variables* by assuming a solution of the form

$$u_{\text{int}}(r,\theta,\phi) = R(r)\Theta(\theta)\Phi(\phi), \quad (74.12)$$

where we take Φ, Θ, and R as solutions of the individual equations

$$\frac{d^2\Phi}{d\phi^2}+m^2\Phi = 0, \quad (74.13)$$

$$\frac{1}{\sin\theta}\frac{d}{d\theta}\left(\sin\theta\frac{d\Theta}{d\theta}\right)+\beta\Theta-\frac{m^2}{\sin^2\theta}\Theta = 0, \quad (74.14)$$

$$\frac{1}{r^2}\frac{d}{dr}\left(r^2\frac{dR}{dr}\right)+\frac{8\pi^2\mu}{h^2}\{E_{\text{int}}-V_{\text{int}}(r)\}R-\frac{\beta}{r^2}R = 0, \quad (74.15)$$

with β and m^2 as constants. This separation is evidently satisfactory since the original equation (74.10) will evidently be reproduced with the suggested form of solution (74.12), if we multiply the above three equations by $R\Theta/r^2\sin^2\theta$, $R\Phi/r^2$, and $\Theta\Phi$ respectively and then add.

(*c*) **Solutions of the separate equations.** The three separate equations which we have thus obtained are closely related to equations which we have already encountered in the preceding section, owing to the similarity in form of (74.10) with (73.31) mentioned above.

As a satisfactory solution of the first of the above equations we may evidently take, in agreement with (73.13) and (73.14),

$$\Phi(\phi) = Ce^{im\phi} \quad (74.16)$$

with

$$m = 0, \pm 1, \pm 2, \pm 3,..., \quad (74.17)$$

and shall have in agreement with (73.15)

$$M_z = m\frac{h}{2\pi} \tag{74.18}$$

as the eigenvalues for the z-component of the angular momentum of the pair of particles.

As a satisfactory solution of the second of the above equations (74.14) we may evidently take, in agreement with (73.21) and (73.22),

$$\Theta(\theta) = CP_l^m(\cos\theta) \tag{74.19}$$

with $$l = |m|, |m|+1, |m|+2,..., \tag{74.20}$$

and shall have, in agreement with (73.23),

$$M^2 = \beta\left(\frac{h}{2\pi}\right)^2 = l(l+1)\left(\frac{h}{2\pi}\right)^2 \tag{74.21}$$

as the allowed eigenvalues for the total angular momentum squared. The normalizing factors for (74.16) and (74.19) are, of course, the same as already given by (73.16) and (73.25).

Finally, substituting the value for β given by (74.21) into (74.15), we can rewrite the third of our separated equations in the form

$$\frac{1}{r^2}\frac{d}{dr}\left(r^2\frac{dR}{dr}\right) + \frac{8\pi^2\mu}{h^2}\{E - V(r)\}R - \frac{l(l+1)}{r^2}R = 0, \tag{74.22}$$

where for simplicity we have now dropped the subscripts from E_{int} and V_{int}. Putting

$$R(r) = \frac{\chi(r)}{r}, \tag{74.23}$$

this can be more simply written as

$$\frac{d^2\chi}{dr^2} + \frac{8\pi^2\mu}{h^2}\left\{E - V(r) - \frac{l(l+1)h^2}{8\pi^2\mu r^2}\right\}\chi = 0, \tag{74.24}$$

which is similar in form to that for a one-dimensional problem and may be treated by the approximate methods discussed in § 69 (b).

The actual solutions of (74.22) and (74.24) which will be obtained will depend, of course, on the form of the potential $V(r)$. For the solutions to represent a permanently associated pair of particles they may be normalized to give

$$\int\limits_0^\infty |R|^2 r^2\,dr = \int\limits_0^\infty |\chi|^2\,dr = 1. \tag{74.25}$$

And for the study of collisions they can be normalized to represent the desired probability current. For the solutions to be allowable it is necessary for $\chi(r) = rR(r)$ to go to zero at $r = 0$, as already mentioned in connexion with (73.35).

Having obtained a suitable solution, we may then write the complete eigensolution of (74.10), corresponding to a steady state of a system of two particles, in the form

$$u(r,\theta,\phi) = CP_l^m(\cos\theta)e^{im\phi}R(r) = CP_l^m(\cos\theta)e^{im\phi}\frac{\chi(r)}{r}. \quad (74.26)$$

(*d*) **Indicated nature of applications.** We may now indicate the nature of some of the applications which can be made of this result. The possible applications differ in accordance with the assumptions made as to the form of the potential $V(r)$.

If we assume that the potential corresponds to the Coulomb attraction of an electron of charge $-e$ and the core of an atom of effective charge Ze, the potential would be of the form

$$V(r) = -\frac{Ze^2}{r}. \quad (74.27)$$

Substituting this value into (74.22), it is found possible to obtain exact eigensolutions of that equation. In the case $E < 0$, corresponding to a bound electron, the eigensolutions can be expressed in terms of the so-called associated Laguerre polynomials of degree $(n-l-1)$ and order $(2l+1)$, where n is a new so-called *total quantum number*; and the corresponding allowed eigenvalues of E are given by

$$E_n = -\frac{2\pi^2\mu Z^2e^4}{n^2h^2} \quad (74.28)$$

with $$n = l+1,\, l+2,\, l+3,\ldots.$$

This result is in agreement with the known energy levels of the hydrogen atom when the fine structure, due to spin and relativistic effects, is neglected. In the case $E > 0$, corresponding to a free electron, the eigensolutions can be expressed in terms of hypergeometric functions and the energy levels give a continuous spectrum.

A second possibility of applying the foregoing results is presented by the rotation and vibration of diatomic molecules, provided that the two atoms are of a character to permit their idealization as simple particles without spin. Under these circumstances we can use equation (74.24) in a very simple manner to obtain a zero-order approximation for the lower energy levels of such a system. We first assume that the potential function can be represented with sufficient approximation by taking a Hooke's law of force between the atoms in the neighbourhood of some equilibrium distance r_e, and writing

$$V = k(r-r_e)^2 = kx^2. \tag{74.29}$$

Returning to equation (74.24), we then replace r^2 by the constant value r_e^2 in the denominator of the last term, since χ will be small and $V(r)$ large except in the neighbourhood of $r = r_e$. Substituting x for $r-r_e$, equation (74.24) can then be written as

$$\frac{d^2\chi(x)}{dx^2}+\frac{8\pi^2\mu^2}{h^2}\left\{E-\frac{l(l+1)h^2}{8\pi^2\mu r_e^2}-kx^2\right\}\chi(x) = 0, \tag{74.30}$$

which we recognize by comparison with (72.2) to be that for a linear harmonic oscillator with $E-l(l+1)h^2/8\pi^2\mu r_e^2$ as the parameter whose eigenvalues must now be chosen so as to secure acceptable eigen-solutions for $\chi(x)$. In accordance with (72.3), we can then write for the energy eigenvalues of such a system

$$E_{nl} = (n+\tfrac{1}{2})h\sqrt{\left(\frac{k}{2\pi^2\mu}\right)}+\frac{l(l+1)h^2}{8\pi^2\mu r_e^2}, \tag{74.31}$$

or, using more nearly the usual notation of band spectroscopists,

$$E_{vJ} = (v+\tfrac{1}{2})h\nu_e+\frac{J(J+1)h^2}{8\pi^2 I_e}, \tag{74.32}$$

where v and J are quantum numbers that take the values 0, 1, 2, 3,..., and the constants ν_e and I_e may be called the equilibrium values for vibration frequency and moment of inertia. This gives a good representation of non-electronic energy levels for diatomic molecules in singlet Σ states, i.e. those for which the model is appropriate, and forms a satisfactory basis for higher order approximations that can be made. Noting the relation of l to m given by (74.20), we see for any given value of the quantum number J—in our present notation—that the quantum number m which would specify the angular momentum of the molecule around a given axis could assume the values

$$m = 0, \pm 1, \pm 2,..., \pm J. \tag{74.33}$$

Hence the multiplicity or quantum weight of a level specified by v and J would be given by

$$g_{vJ} = 2J+1. \tag{74.34}$$

Further possibilities of applying (74.26) arise in the study of collisions between particles, which can be undertaken by setting up a steady state solution to represent the colliding and scattered particle (i.e. an incoming plane wave and an outgoing spherical wave). For a treatment of these problems see, for instance, *The Theory of Atomic Collisions*, by Mott and Massey (Oxford, 1933).

75. Particles with spin

(a) The spin variables and operators. A complete and satisfactory account of all the electronic energy levels, found in the analysis of atomic spectra, has only been made possible by ascribing to the electron, as originally proposed by Uhlenbeck and Goudsmit, in addition to the properties of position and momentum which correspond to the classical observables x, y, z and p_x, p_y, p_z, a further property called *electron spin*, which corresponds to the appearance of certain new non-classical observables s_x, s_y, s_z called the components of spin with reference to the axes indicated. This new property of electron spin has the general nature of an intrinsic angular momentum, together with an associated magnetic moment, which must be ascribed to the electron as permanent attributes without reference to its orbital motion. Furthermore, it has now been found that similar spin properties must be assigned to all the fundamental material particles, electrons, positrons, protons, and neutrons whose existence at present seems necessary.

From the analysis of spectral and other data it is found that the component of intrinsic angular momentum parallel to any given axis can only exhibit the eigenvalues $\pm h/4\pi$, in contrast with the integral multiples of the Bohr unit of angular momentum $h/2\pi$ found in cases of ordinary rotation as studied in §§ 73 and 74. For the magnitude of the total intrinsic angular momentum we then have

$$\sqrt{3}(h/4\pi) = \sqrt{\{\tfrac{1}{2}(1+\tfrac{1}{2})\}}h/2\pi.$$

The ratio of intrinsic magnetic moment to intrinsic angular momentum is found to be different for the different fundamental particles. For the electron the ratio is $-e/mc$, where $-e$ is the charge and m is the mass of the electron. It is interesting, although perhaps not very fundamental, to note that this is the ratio which would be calculated classically for a sphere of mass m and uniform density revolving on its axis and carrying a surface charge $-e$. For the positron the absolute magnitude of the ratio is presumably the same as for the electron. For the proton the ratio is about $2{\cdot}7e/Mc$, where M is the mass of the proton. For the neutron the ratio appears to be of the same order as for the proton.

Although from a classical point of view a charged particle revolving around its own axis should exhibit an intrinsic angular momentum with an associated magnetic moment, it is not profitable for two reasons to try in detail to treat the spin of a particle as due to this kind of revolutional motion. In the first place, if the phenomenon could be handled by the methods of § 73 as a simple example of axial revolution, we should expect to find all integral multiples of $h/2\pi$ as the allowed eigenvalues of intrinsic angular momentum around any selected axis, while the actual analysis of spectral energy levels has shown that $+h/4\pi$ and $-h/4\pi$ are the only eigenvalues which really do occur. In the second place, if the spin could be regarded as a revolution, we should have to treat not only the components of intrinsic angular momentum but also the conjugate angular variables for the axial orientation of the electron as observable quantities. It is immediately evident, however, that the orientation of the electron around its own axes would presumably defy observation; and observable quantities corresponding to such orientation do not appear either in the original non-relativistic Pauli† theory of the spin which we shall consider, nor in the later more fundamental relativistic treatment of the electron given by Dirac.‡

We are now ready to undertake the theoretical treatment of this new phenomenon. Corresponding to the fact that the only eigenvalues for the components of intrinsic angular momentum have half the magnitude of the Bohr unit of angular momentum $h/2\pi$, it is customary to introduce three new observables, the so-called spin variables s_x, s_y, and s_z, which have as their only eigenvalues $\pm\frac{1}{2}$. If one of these observables exhibits a definite eigenvalue, say, for example, $s_x = +\frac{1}{2}$, this then means that the particle is in a state such that the component of intrinsic angular momentum in the positive x-direction would exhibit the precise value $\frac{1}{2}(h/2\pi)$. Since this amount of angular momentum would go to zero as h goes to zero, it has no classical analogue, and we regard the new observables s_x, s_y, and s_z as *non-classical*.‖

In addition to these new observables we may also introduce *three new* operators, s_x, s_y, and s_z, which are regarded as correlated with the corresponding observables s_x, s_y, and s_z in the usual quantum mechanical manner; and we must now consider the properties of these operators. For the *commutation rules* applying to the new operators we take††

$$\begin{aligned} \mathrm{s}_y\,\mathrm{s}_z-\mathrm{s}_z\,\mathrm{s}_y &= i\mathrm{s}_x, \\ \mathrm{s}_z\,\mathrm{s}_x-\mathrm{s}_x\,\mathrm{s}_z &= i\mathrm{s}_y, \\ \mathrm{s}_x\,\mathrm{s}_y-\mathrm{s}_y\,\mathrm{s}_x &= i\mathrm{s}_z \end{aligned} \tag{75.1}$$

† Pauli, *Zeits. f. Phys.* **43**, 601 (1927).

‡ Dirac, *The Principles of Quantum Mechanics*, second edition, Oxford, 1935, chapter xii.

‖ As emphasized by Bohr, *Atomtheorie und Naturbeschreibung*, Berlin, 1931, observations of sufficient accuracy to give a direct separation between the intrinsic angular momentum of an electron and that due to its orbital motion cannot be made.

†† See foregoing reference to Pauli.

in analogy with the commutation rules (73.7) applying to the components of ordinary angular momentum. And, in accordance with the fact that the squares of the new observables have $+\frac{1}{4}$ as their only eigenvalue, we also take

$$\mathrm{s}_x^2 = \mathrm{s}_y^2 = \mathrm{s}_z^2 = \tfrac{1}{4}. \tag{75.2}$$

This agrees with the consideration that the characteristic equation $\mathrm{s}_x^2\psi = s_x^2\psi$, applying, for example, to the eigenvalues and eigenfunctions of the observable s_x^2, should necessarily reduce to the form $\mathrm{s}_x^2\psi = \frac{1}{4}\psi$.

The commutation rules given by (75.1) can be expressed in another form with the help of some straightforward algebra.† We first obtain the preliminary result

$$\begin{aligned} i(\mathrm{s}_x\mathrm{s}_y+\mathrm{s}_y\mathrm{s}_x) &= (i\mathrm{s}_x)\mathrm{s}_y+\mathrm{s}_y(i\mathrm{s}_x) \\ &= (\mathrm{s}_y\mathrm{s}_z-\mathrm{s}_z\mathrm{s}_y)\mathrm{s}_y+\mathrm{s}_y(\mathrm{s}_y\mathrm{s}_z-\mathrm{s}_z\mathrm{s}_y) \\ &= -\mathrm{s}_z\mathrm{s}_y^2+\mathrm{s}_y^2\mathrm{s}_z \\ &= -\tfrac{1}{4}\mathrm{s}_z+\tfrac{1}{4}\mathrm{s}_z \\ &= 0, \end{aligned}$$

where the second form of writing comes from the first of equations (75.1), and the next to the last form from (75.2). This result, which can also be expressed in the form

$$\mathrm{s}_x\mathrm{s}_y = -\mathrm{s}_y\mathrm{s}_x, \tag{75.3}$$

can be described by saying that the two operators *anticommute*. Substituting (75.3), together with the two further relations given by symmetry considerations, into (75.1), we then obtain

$$\begin{aligned} \mathrm{s}_x\mathrm{s}_y &= -\mathrm{s}_y\mathrm{s}_x = \tfrac{1}{2}i\mathrm{s}_z, \\ \mathrm{s}_z\mathrm{s}_x &= -\mathrm{s}_x\mathrm{s}_z = \tfrac{1}{2}i\mathrm{s}_y, \\ \mathrm{s}_y\mathrm{s}_z &= -\mathrm{s}_z\mathrm{s}_y = \tfrac{1}{2}i\mathrm{s}_x \end{aligned} \tag{75.4}$$

as the desired alternative expression for the commutation relations (75.1).

Since the three new operators which we have introduced do not commute with each other, we can determine only one of the observables s_x, s_y, and s_z at any given time. On the other hand, since the new operators are assumed to commute with those for position, we can simultaneously determine the values of x, y, z and any one of the new observables that we desire. By convention this latter is usually chosen as s_z for the component of spin parallel to the z-axis. Making this choice, we may then introduce—in the spirit of our previous develop-

† See Dirac, *The Principles of Quantum Mechanics*, second edition, Oxford, 1935, § 19.

ment of the quantum mechanics—probability amplitudes, which can be written in the forms

$$\psi(x, y, z, s_z; t_0) = u(x, y, z, s_z) = u(q, s_z), \tag{75.5}$$

for specifying, at any selected time t_0, the state of our system in the x, y, z, s_z language. These amplitudes will be functions of the continuous variables of position x, y, z and of the two discrete values of $s_z = \pm\frac{1}{2}$ which are possible. For the actual probability of finding the particle in a specified range $dxdydz$ with a specified value of s_z, we shall have

$$W(x, y, z, s_z)\, dxdydz = u^*(q, s_z)u(q, s_z)\, dxdydz. \tag{75.6}$$

And for the total probability of finding the particle somewhere and with one or the other of the two possible values of the spin variable, we shall have the summation of two terms:

$$\sum_{s_z=-\frac{1}{2}}^{s_z=+\frac{1}{2}} \iiint u^*(q, s_z)u(q, s_z)\, dxdydz = 1. \tag{75.7}$$

We must next inquire into the action of our operators s_x, s_y, s_z on probability amplitudes of the above kind. In accordance with the spirit of our previous development of the quantum mechanics (see § 67 (*b*)), we shall assume that each operator s_k can be correlated in any selected language with an appropriate Hermitian matrix, and the action of the operator then represented by an expression of the general form

$$\begin{gathered} \mathrm{s}_k\, u(q, m) = \sum_n [s_k]_{mn}\, u(q, n) \\ m, n = \pm\tfrac{1}{2}, \end{gathered} \tag{75.8}$$

where $[s_k]_{mn}$ is an element of the Hermitian matrix of two rows and two columns which corresponds in the language employed to the operator s_k $(k = x, y, z)$.

We may now investigate the specific form of these matrices in the s_z language. The matrix correlated with s_z will be specially simple in this language, since the action of the operator should then reduce merely to multiplication by s_z itself. This leads at once to the form

$$[s_z]_{mn} = \begin{pmatrix} \frac{1}{2} & 0 \\ 0 & -\frac{1}{2} \end{pmatrix}, \tag{75.9}$$

which does give us, on substitution into (75.8), the desired results

$$\mathrm{s}_z u(q, +\tfrac{1}{2}) = +\tfrac{1}{2}u(q, +\tfrac{1}{2}),$$
$$\mathrm{s}_z u(q, -\tfrac{1}{2}) = -\tfrac{1}{2}u(q, -\tfrac{1}{2}).$$

Turning next to the matrix corresponding to the operator s_x, we can apply the general principle (see (67.20)) that matrices will satisfy the same algebraic relations as the operators to which they correspond provided matrix multiplication is used in place of operator multiplication. The second of equations (75.4) can then be written in the form

$$\sum_n [s_z]_{mn}[s_x]_{nl} = -\sum_n [s_x]_{mn}[s_z]_{nl},$$

which, by considering the cases $m = l = \pm\frac{1}{2}$ and making use of (75.9), gives us

$$[s_x]_{\frac{1}{2}\frac{1}{2}} = [s_x]_{-\frac{1}{2}-\frac{1}{2}} = 0.$$

Furthermore, writing the first of equations (75.2) in the form

$$\sum_n [s_x]_{mn}[s_x]_{nm} = \tfrac{1}{4},$$

we then obtain

$$[s_x]_{\frac{1}{2}-\frac{1}{2}}[s_x]_{-\frac{1}{2}\frac{1}{2}} = \tfrac{1}{4},$$

which, in accordance with the Hermitian character of the matrix, can only be satisfied by

$$[s_x]_{\frac{1}{2}-\frac{1}{2}} = \tfrac{1}{2}e^{i\phi} \quad \text{and} \quad [s_x]_{-\frac{1}{2}\frac{1}{2}} = \tfrac{1}{2}e^{-i\phi},$$

where ϕ is an arbitrary phase. For example, the arbitrary phase ϕ can be taken as zero, which now gives us

$$[s_x]_{mn} = \begin{pmatrix} 0 & \frac{1}{2} \\ \frac{1}{2} & 0 \end{pmatrix} \tag{75.10}$$

as the desired expression for the matrix in question.

Turning finally to the matrix corresponding to the operator s_y, we can re-express the second of equations (75.4) in the form

$$[s_y]_{mn} = 2i \sum_l [s_x]_{ml}[s_z]_{ln},$$

which, with the help of (75.9) and (75.10), leads at once to

$$[s_y]_{mn} = \begin{pmatrix} 0 & -\frac{1}{2}i \\ \frac{1}{2}i & 0 \end{pmatrix} \tag{75.11}$$

as the expression for the remaining matrix. This then completes the apparatus necessary for the non-relativistic treatment of problems involving spin.

(*b*) **Applications involving spin.** The most familiar necessity for application of the foregoing theoretical apparatus for spin arises in the study of electronic energy levels and their correlation with the data of atomic spectra. For this purpose several considerations must be taken into account. Firstly, in treating the total angular momentum, and its conservation in fields of suitable symmetry, both spin and orbital momentum must be included. Secondly, in treating energy levels involving more than a single electron, only those solutions which are antisymmetric both in coordinate and spin variables are to be considered, as will be discussed more completely in § 76. And finally, the Hamiltonian for an electron must be modified by the inclusion of terms which correspond classically to the interaction of its magnetic moment with any imposed external magnetic field, and with the internal electromagnetic field which it encounters in its orbital motion.

Making due use of these several considerations, it then becomes possible to give a good account of electronic energy levels. For example, by adding to the Hamiltonian for the electron a term, corresponding to its magnetic energy in an external magnetic field $\mathcal{H}_x$, $\mathcal{H}_y$, $\mathcal{H}_z$, of the form

$$H_1 = \mu(s_x \mathcal{H}_x + s_y \mathcal{H}_y + s_z \mathcal{H}_z),$$

where μ is the appropriate constant corresponding to the electron's intrinsic magnetic moment, it is possible to give a treatment of the Zeeman effect for the hydrogen atom. The suggested considerations also find their application in studying the energy levels of molecules as well as atoms. Nevertheless, any detailed treatment of the effect of spin phenomena in the fields of atomic and molecular spectra would have to be too elaborate for inclusion here.

For the considerations of statistical mechanics, the most important consequence of the existence of particle spin lies in the simple fact that we must now expect twice as many solutions of Schroedinger's equation for a particle having this property as for a simple particle without it. If, for example, we have a steady state solution for a particle of the form

$$\psi(x, y, z, \tfrac{1}{2}; t) = u(x, y, z, \tfrac{1}{2})e^{-(2\pi i/h)Et},$$

where E is the energy of the particle and the spin variable has the specified value $s_z = +\frac{1}{2}$, we must also expect a related steady state solution of similar form,

$$\psi'(x, y, z, -\tfrac{1}{2}; t) = u'(x, y, z, -\tfrac{1}{2})e^{-(2\pi i/h)E't},$$

where the spin variable has its other possible eigenvalue $s_z = -\frac{1}{2}$. And, indeed, in most cases of interest we may even expect u and u' to have the same dependence on the coordinates x, y, and z and E and E' to be equal, since the interaction of the intrinsic magnetic moment of the particle with the surrounding electromagnetic field will either be absent, negligible, or at least on the average independent of the value of s_z.

Such a simple doubling in the number of eigensolutions is of frequent importance for statistical mechanics and is easily taken into account. We have already called attention in § 71 (*b*) to the doubling that thus arises in the number of eigensolutions for a particle in a container.

76. Systems of two or more like particles

In §74 we have already given a treatment to systems consisting of two *dissimilar* particles of masses m_1 and m_2, and must now turn in the present section to some important new considerations that must be kept in mind in the quantum mechanical treatment of systems containing two or more *entirely similar* particles. These new considerations may be regarded as finding their philosophical justification in a combination of the operational point of view, which would confine the formalism of physics to the treatment of conceivably possible observations, with the quantum mechanical conclusion as to theoretical limitations on observability, which may preclude the possibility of any knowledge as to the separate individual behaviour of two entirely similar particles.

As an illustration, it is evident in the case of collisions between entirely similar particles that the Heisenberg uncertainty principle would make it theoretically impossible to follow the separate motions of the two particles throughout the course of a close encounter. Hence our formalism for treating such collisions should be devised so as to introduce no possibility of distinguishing as to which particle is which after such an encounter has taken place. And it has indeed been found, for example in the case of collisions of alpha particles with alpha particles, that correct conclusions can only be drawn when the formalism does allow for this consideration.†

As another illustration, it is evident, in treating the steady states of a system such as a box containing two identical particles, that the system could be maintained in its assumed steady state only if we refrain from interfering with the constant value of its energy by making observations on the whereabouts of the two particles. Hence the formalism for treating such steady states should be devised so as to be entirely symmetrical in the predictions which it makes as to the behaviour of the two particles. And here also it has been found, for example in studying the steady states of a helium atom with its two entirely similar electrons, that correct conclusions can only be drawn from a formalism which does provide symmetrical predictions for the two particles.‡

We may now turn to a more detailed examination of the nature of the new considerations, beginning with the case of only two similar particles.

† See Oppenheimer, *Phys. Rev.* **32**, 361 (1928); and Mott, *Proc. Roy. Soc.* A, **125**, 222 (1929); ibid. **126**, 259 (1930).

‡ See Heisenberg, *Zeits. f. Phys.* **39**, 499 (1926).

(a) Symmetric and antisymmetric solutions for pairs of like particles. To study the case of two similar particles let us start with Schroedinger's equation for the system

$$\mathbf{H}\psi + \frac{h}{2\pi i}\frac{\partial \psi}{\partial t} = 0 \tag{76.1}$$

and take

$$\psi = \psi(q_1 q_2, t) \tag{76.2}$$

as some actual solution of this equation, where q_1 represents the observables—say coordinates and spin—describing the first of the two particles, and q_2 represents the observables describing the second of the two particles.

Such an expression as (76.2) would not in general have to exhibit any symmetry, in the observables q_1 and q_2 for the two particles, merely in order to be a correct solution of equation (76.1). For illustration, let us consider a system constructed by introducing two simple similar particles into a box, q_1 representing for specificity the coordinates for the first particle put into the box and q_2 for the second particle so introduced. As a conceivable steady state solution of Schroedinger's equation for this system we could then evidently write

$$\psi(q_1 q_2, t) = u_k(q_1)u_l(q_2)e^{-\frac{2\pi i}{h}(E_k+E_l)t}, \tag{76.3}$$

where for simplicity we neglect the energy of interaction between the two particles and take u_k and u_l as eigensolutions for a single particle in the box corresponding to the respective energies E_k and E_l. Such an expression would be a formally correct solution of Schroedinger's equation for the system. It would, nevertheless, be quite unsymmetrical with respect to the coordinates q_1 and q_2 for the two particles, and would permit the unsymmetrical assertion that the first particle has the energy E_k and the second the energy E_l. Such an assertion, moreover, would be quite impossible of observational verification since all we could possibly hope to show would be that one of the two particles that we had put into the box—either the first or the second—had the energy E_k and the other the energy E_l.

As a consequence of considerations such as the above, we now see that a formally *correct* solution of Schroedinger's equation of the form (76.2) would not necessarily be an allowable solution, from a physical point of view, if the situation is such that we must demand symmetrical predictions for the two particles. Nevertheless, starting with any correct solution of Schroedinger's equation for two similar particles, we shall find it easily possible to construct allowable solutions which do have the necessary symmetry properties.

Owing to the fact that the two particles under consideration have by hypothesis entirely similar properties, it is evident that the Hamiltonian operator **H** for the system will depend in an entirely symmetrical manner on the observables q_1 and q_2 for the two particles. Hence, if $\psi(q_1 q_2, t)$ is a correct solution of Schroedinger's equation, it is evident that $\psi(q_2 q_1, t)$—which differs therefrom only by an interchange in the occurrence of the observables for the two particles within the form of expression—would also be a correct solution. Furthermore, since both of the operators **H** and $(h/2\pi i)\partial/\partial t$ occurring in the Schroedinger equation are linear, it is evident that any linear combination of these two solutions would also be a correct solution of Schroedinger's equation.

Two such linear combinations are of special interest. The first of these is the so-called *symmetric solution*

$$\psi_s(q_1 q_2, t) = \frac{1}{\sqrt{2}}\{\psi(q_1 q_2, t)+\psi(q_2 q_1, t)\}, \tag{76.4}$$

where $1/\sqrt{2}$ proves to be the correct normalizing factor if the original solution $\psi(q_1 q_2, t)$ was normalized to unity. This solution is characterized by complete invariance to an interchange in the indices for the two particles, i.e.

$$\psi_s(q_1 q_2, t) = \psi_s(q_2 q_1, t). \tag{76.5}$$

The second linear combination of interest is the so-called *antisymmetric solution*,

$$\psi_a(q_1 q_2, t) = \frac{1}{\sqrt{2}}\{\psi(q_1 q_2, t)-\psi(q_2 q_1, t)\}. \tag{76.6}$$

It is characterized by a change in sign which occurs when the indices for the two particles are interchanged, i.e.

$$\psi_a(q_1 q_2, t) = -\psi_a(q_2 q_1, t). \tag{76.7}$$

As we shall immediately show, *either* of these solutions has the desired property of giving symmetrical predictions as to the two particles.

(*b*) **Properties of the symmetric and antisymmetric solutions.** We must now consider the properties of our symmetric and antisymmetric solutions in some detail. First of all we must show that they do give symmetrical predictions as to the two particles.

We may begin by inquiring into the direct predictions which the two solutions would provide as to the probability of finding different possible values for the observables q_1 and q_2 for the two particles. Assuming appropriate normalization, these predictions could be directly obtained in the two cases with the help of the *probability densities*

$$W_s(q_1 q_2, t) = \psi_s^*(q_1 q_2, t)\psi_s(q_1 q_2, t)$$

and

$$W_a(q_1 q_2, t) = \psi_a^*(q_1 q_2, t)\psi_a(q_1 q_2, t). \tag{76.8}$$

We immediately see, however, since $\psi_s(q_1 q_2, t)$ is completely invariant to an interchange of indices for the two particles, and since $\psi_a(q_1 q_2, t)$ merely changes its sign on such an interchange, that the two probability densities are themselves invariant to any interchange in particle indices. Hence they would both provide predictions which would have the desired character of being the same for one of the two particles as for the other.

For example, if we consider for simplicity a particle without spin and let $dq^{(\alpha)}$ and $dq^{(\beta)}$ denote two infinitesimal ranges in which the coordinates q_1 and q_2 for the particles might fall, we should compute *equal* probabilities of the general form

$$W(q_1 q_2, t)\, dq_1^{(\alpha)} dq_2^{(\beta)} = W(q_1 q_2, t)\, dq_2^{(\alpha)} dq_1^{(\beta)} \tag{76.9}$$

for finding the first in $dq^{(\alpha)}$ and the second in $dq^{(\beta)}$ and for finding the first in $dq^{(\beta)}$ and the second in $dq^{(\alpha)}$, since the probability density $W(q_1 q_2, t)$ is itself invariant to an interchange of indices, both for the case of the symmetric and of the antisymmetric solution. We thus see, since our particles are not distinguishable on the basis of any intrinsic properties, that our formalism is really only adapted for calculating the total probability

$$W(q_1 q_2, t)(dq_1^{(\alpha)} dq_2^{(\beta)} + dq_2^{(\alpha)} dq_1^{(\beta)}) \tag{76.10}$$

of finding one particle in the range $dq^{(\alpha)}$ and the other in the range $dq^{(\beta)}$. Analogous calculations also show that the probability currents are symmetrical in the two particles.

We thus see for the case of two completely similar particles that the symmetric and the antisymmetric solutions both have the property of giving predictions which are entirely symmetrical with respect to the two particles and provide no means of distinguishing one of the particles from the other. We also see for the case of only two particles that these are the only forms of solution which would have this property.

We must next emphasize the further important property of these solutions in that they always maintain their original particular symmetry character as time proceeds. This is an evident consequence of their construction in the form $\psi(q_1 q_2, t) \pm \psi(q_2 q_1, t)$ from individual solutions each of which has the time dependence demanded by the

Schroedinger equation. The necessity for the property can also be seen if we think of the application of Schroedinger's equation

$$\frac{\partial \psi}{\partial t} = -\frac{2\pi i}{h} \mathbf{H}\psi \tag{76.11}$$

to a direct calculation of the change with time in the total expression for ψ. Since the operator $\mathbf{H}$ is itself symmetrical in q_1 and q_2, it is evident that the change in ψ with time will have to have the same symmetry properties as ψ itself. As a consequence of these considerations we now see that a pair of particles originally in a symmetric state would continue permanently in states having the symmetric property, and a pair originally in an antisymmetric state would always maintain the antisymmetric property.

As to the symmetry character of the different kinds of particles that actually occur in nature we must of course rely on observation. It is found in the first place that all the *fundamental material* particles—*electrons*, *protons*, and *neutrons*—occur only in *antisymmetric* states; and this may be a consequence of some underlying necessity of particle structure that we do not now understand. On the other hand, the *non-material* particles—*photons*—which can be introduced for the treatment of radiation are found to occur only in *symmetric* states. Furthermore, the *complex material nuclei* and the *atoms* as a whole, when treated as simple particles, are found to occur only in *symmetric* or *antisymmetric* states according as they are constructed respectively from an even or an odd number of fundamental particles, nuclei being regarded as built only from protons and neutrons. Thus, for example, alpha particles and nitrogen nuclei when treated as individual particles exhibit symmetric properties even though the fundamental material particles are all antisymmetric.

To complete our general discussion of symmetric and antisymmetric solutions we must also investigate the predictions which they allow in a situation where the separate behaviour of the two particles is at least approximately known, so that the two particles can be distinguished one from the other over a certain length of time. In such a situation we must, of course, still use either a symmetric or an antisymmetric solution, as the case may be, since the appropriate form of solution is determined for all time by the nature of the particles. Nevertheless, our predictions for the two particles must not now contradict the possibility that one particle could be known to be carrying out a behaviour quite different from that of the other.

To investigate such a situation let us consider two solutions of Schroedinger's equation $\psi_k(q,t)$ and $\psi_l(q,t)$ for a single particle in the potential field of interest, and let there be practically no overlapping of the solutions during the time of interest, ψ_k being nearly zero for those values of q where ψ_l is large and vice versa. As possible symmetric and antisymmetric solutions for two particles we could then take

$$\psi_s(q_1 q_2, t) = \frac{1}{\sqrt{2}}\{\psi_k(q_1,t)\psi_l(q_2,t)+\psi_k(q_2,t)\psi_l(q_1,t)\} \tag{76.12}$$

and
$$\psi_a(q_1 q_2, t) = \frac{1}{\sqrt{2}}\{\psi_k(q_1,t)\psi_l(q_2,t)-\psi_k(q_2,t)\psi_l(q_1,t)\}, \tag{76.13}$$

where, on account of the failure of the solutions to overlap, we shall have

$$\psi_k^*(q,t)\psi_l(q,t) \approx \psi_l^*(q,t)\psi_k(q,t) \approx 0 \tag{76.14}$$

when the index for either particle 1 or 2 is attached to the observables q.

Noting (76.14), we then see that the probability density would assume the same form both for the symmetric and the antisymmetric case, namely,

$$\begin{aligned} W_s(q_1 q_2, t) &= W_a(q_1 q_2, t) \\ &= \tfrac{1}{2}\{|\psi_k(q_1,t)|^2|\psi_l(q_2,t)|^2+|\psi_k(q_2,t)|^2|\psi_l(q_1,t)|^2\}. \end{aligned} \tag{76.15}$$

This shows us that any differences in the laws of behaviour for particles having symmetric and antisymmetric properties do not become operative so long as the wave functions for the two particles do not overlap.

Furthermore, if we now compute the probability of finding one particle in a range $dq^{(k)}$, where ψ_k is appreciably different from zero, and a second particle in a range $dq^{(l)}$, where ψ_l is appreciably different from zero, we shall obtain both in the symmetric and the antisymmetric case, in accordance with the expression for such probabilities given by (76.10),

$$\begin{aligned} &\tfrac{1}{2}\{|\psi_k(q_1,t)|^2|\psi_l(q_2,t)|^2+|\psi_k(q_2,t)|^2|\psi_l(q_1,t)|^2\}(dq_1^{(k)}\,dq_2^{(l)}+dq_2^{(k)}\,dq_1^{(l)}) \\ &\quad = \tfrac{1}{2}|\psi_k(q_1,t)|^2|\psi_l(q_2,t)|^2\,dq_1^{(k)}\,dq_2^{(l)}+\tfrac{1}{2}|\psi_k(q_2,t)|^2|\psi_l(q_1,t)|^2\,dq_2^{(k)}\,dq_1^{(l)}, \end{aligned} \tag{76.16}$$

where two terms in the product disappear since ψ_k is practically zero in the range $dq^{(l)}$ and ψ_l in the range $dq^{(k)}$. Examining (76.16), however, we now see that, so long as the solutions do not overlap, we can rewrite our expression for the probability of finding one particle in $dq^{(k)}$ and the other in $dq^{(l)}$ in the form

$$W(q^{(k)}q^{(l)},t)\,dq^{(k)}dq^{(l)} = |\psi_k(q^{(k)},t)|^2|\psi_l(q^{(l)},t)|^2\,dq^{(k)}dq^{(l)}, \tag{76.17}$$

where the superscript (k) now denotes that particle whose behaviour follows the solution ψ_k, and (l) the particle following the solution ψ_l.

This then satisfactorily demonstrates, in accordance with the spirit of the correspondence principle, that our formalism is such that two particles whose intrinsic properties are identical can nevertheless be distinguished from each other by their behaviour in the classical limit where quantum mechanical uncertainties connected with the overlapping of probability amplitudes can be neglected. Incidentally it also shows that our choice of $1/\sqrt{2}$ as the normalizing factor in (76.4) and (76.6) was the correct one to make.

(*c*) **Treatment of more than two like particles.** We must now make a brief statement concerning the treatment of more than two particles having identical intrinsic properties. Here again we shall wish to demand that the *allowed* solutions of the Schroedinger equation should be such as to lead to symmetrical predictions for the different particles involved in a situation where quantum mechanical limitations on observability would prevent the maintenance of any knowledge as to which particle was which. It is evident that such a demand would continue to be satisfied with more than two particles either by a *symmetric solution* $\psi_s(q_1 q_2 q_3 \dots q_n, t)$ for n particles, having the property

$$\mathbf{P}\psi_s(q_1 q_2 q_3 \dots q_n, t) = \psi_s(q_1 q_2 q_3 \dots q_n, t), \tag{76.18}$$

where the application of the operator $\mathbf{P}$ represents any permutation of particle indices, or by an *antisymmetric solution* $\psi_a(q_1 q_2 q_3 \dots q_n, t)$ having the property

$$\mathbf{P}\psi_a(q_1 q_2 q_3 \dots q_n, t) = \mp\psi_a(q_1 q_2 q_3 \dots q_n, t), \tag{76.19}$$

where we have the negative or positive sign according as the permutation is odd or even. In either case the probability densities would evidently be invariant to any interchange in particle indices.

As soon as we go to three or more particles, however, it also proves possible to find further types of solution, having more complicated symmetry characters than the above, which on appropriate combination also lead to probability densities which are invariant to a permutation of particle indices. It seems difficult to eliminate these other possibilities on a purely theoretical basis. Nevertheless, if they existed we should sometimes find a pair of particles of a given kind which exhibited none but antisymmetric solutions, and at other times a pair of the same kind of particles which exhibited none but symmetric solutions. For example, a helium atom with the ordinary antisymmetric solution, corresponding to two electrons in the K-shell with their spins *antiparallel*, could appear, after collision with a third electron, with two electrons in the K-shell with *parallel* spins, and this new atom would then exhibit none but symmetric states. Hence solutions other than the symmetric and the antisymmetric could lead to the unexpected result that all pairs of particles would not have to have identical properties even though constructed from single particles having identical properties.

We must, of course, ultimately appeal to observation to determine the symmetry character of the solutions that do correspond to natural phenomena. So far no evidence has been found for solutions other than the completely symmetric or completely antisymmetric. With any number of similar particles in the system, the different kinds of particle actually occurring in nature fall either into the symmetric or antisymmetric class as already described above in the case of pairs of particles.

If $\psi(q_1 q_2 q_3 \dots q_n, t)$ is any formally correct solution of Schroedinger's equation for a system of n particles, which has been normalized to unity, the corresponding allowable symmetric and antisymmetric solutions would be given by

$$\psi_s(q_1 \dots q_n, t) = \frac{1}{\sqrt{n!}} \sum_{\mathbf{P}} \mathbf{P}\psi(q_1 \dots q_n, t)$$

and

$$\psi_a(q_1 \dots q_n, t) = \frac{1}{\sqrt{n!}} \sum_{\mathbf{P}} (\mp 1)\mathbf{P}\psi(q_1 \dots q_n, t), \tag{76.20}$$

where $1/\sqrt{n!}$ is the proper normalizing factor, the operator $\mathbf{P}$ is regarded as permuting the different particle indices $1, \dots, n$, and we take a summation over all possible independent permutations including that of identity. In the symmetric case the $n!$ terms are all taken with the positive sign, and in the antisymmetric case with negative or positive sign according as the permutation is odd or even.

(*d*) **Further properties of symmetric and antisymmetric solutions—case of small interaction between particles. Pauli exclusion principle.** In case we have a system of n similar particles at such a dilution that the energy of interaction between the particles is small, it is often advantageous to treat the n particles as being distributed among the various unperturbed energy eigensolutions $u_k(q)$, $u_l(q)$, $u_m(q)$, etc., which would be found for a single particle, using an unperturbed Hamiltonian $\mathbf{H}^0$ for the system that neglects interaction. At any instant we could then regard the state of the system as describable with the help of a superposition of products of the general form $u_k(q_1)u_l(q_2)u_m(q_3) \dots u_r(q_n)$.

In case the particles belong to the symmetric class, we may make our superpositions by first combining products such as the above to give *symmetric eigenfunctions* of the form

$$U_s(q_1 q_2 q_3 \dots q_n) = \sum_{\mathbf{P}} \mathbf{P} u_k(q_1) u_l(q_2) u_m(q_3) \dots u_r(q_n), \tag{76.21}$$

where a summation is taken over all the $n!$ permutations of the particle indices. Such eigenfunctions would evidently themselves be symmetric, i.e.

$$\mathbf{P} U_s(q_1 q_2 q_3 \dots q_n) = U_s(q_1 q_2 q_3 \dots q_n) \tag{76.22}$$

for any interchange of indices. Their linear superposition, when multiplied by suitable coefficients $c_{klm\dots}$, which can incidentally take care of normalization, would then let us express any instantaneous state of the system, and by allowing these coefficients to change with the time, as in the method of the variation of constants (see § 68), we could express any solution of Schroedinger's equation for our system of symmetric particles in its full time dependence.

In case the particles belong to the antisymmetric class, we may first combine products such as $u_k(q_1)u_l(q_2)u_m(q_3)\dots u_r(q_n)$ to give *antisymmetric eigenfunctions* of the form

$$U_a(q_1 q_2 q_3 \dots q_n) = \sum_{\mathbf{P}} (\mp 1) \mathbf{P} u_k(q_1) u_l(q_2) u_m(q_3) \dots u_r(q_n), \tag{76.23}$$

where the negative or positive sign is to be used according as the permutation is odd or even. Such eigenfunctions would evidently themselves be antisymmetric, i.e.

$$\mathbf{P} U_a(q_1 q_2 q_3 \dots q_n) = \mp U_a(q_1 q_2 q_3 \dots q_n), \tag{76.24}$$

and they could be superposed in the manner described above to express any solution of Schroedinger's equation for a system of antisymmetric particles. It is sometimes convenient to note that an antisymmetric eigenfunction of the form (76.23) could also be expressed as the determinant

$$U_a(q_1 q_2 q_3 \dots q_n) = \begin{vmatrix} u_k(q_1) & u_l(q_1) & u_m(q_1) & \cdot & \cdot & u_r(q_1) \\ u_k(q_2) & u_l(q_2) & u_m(q_2) & \cdot & \cdot & u_r(q_2) \\ u_k(q_3) & u_l(q_3) & u_m(q_3) & \cdot & \cdot & u_r(q_3) \\ \cdot & \cdot & \cdot & \cdot & \cdot & \cdot \\ \cdot & \cdot & \cdot & \cdot & \cdot & \cdot \\ u_k(q_n) & u_l(q_n) & u_m(q_n) & \cdot & \cdot & u_r(q_n) \end{vmatrix}. \tag{76.25}$$

This form of writing makes it very easy to see that these antisymmetric eigenfunctions would change sign for any single interchange of a pair of particle indices, since this would be equivalent to the interchange of two rows of the determinant.

In setting up the products $u_k(q_1)u_l(q_2)u_m(q_3)...u_r(q_n)$ to be used in constructing a *symmetric* eigenfunction with the help of (76.21), it is evident that the same individual solution, for example $u_k(q) = u_l(q)$, might occur more than once so that a number of particles could be in the same 'unperturbed quantum state'. In constructing an *antisymmetric* eigenfunction, however, with the help of (76.23) or (76.25), it is immediately evident that any eigenfunction $U_a(q_1...q_n)$ would automatically be equal to zero if the same individual solution, for example $u_k(q)$, should occur more than once in the original product, since this would make two columns identical in the determinant (76.25). Hence for the antisymmetric case no two particles could ever be found in the same 'unperturbed quantum state'.

This result, which holds for all the fundamental material particles—electrons, protons, and neutrons—may be regarded as a generalization of the original Pauli exclusion principle describing the fact that we never find more than one electron in an atom with a given set of values for all four quantum numbers, including that for the spin, or, in other words, neglecting the direction of spin, never more than a pair of electrons in any single atomic orbit. The great importance of this principle in explaining the facts of spectroscopy and chemistry is well known.

It will be seen not only that the antisymmetric property of solutions implies the exclusion principle, but, vice versa, that the exclusion principle can only be satisfied by antisymmetric solutions since only these have the property of vanishing with two particles in the same 'quantum state'. It is also interesting to note in the case of antisymmetric solutions that two similar particles with the same direction of spin could never occupy the same position in space, since the probability amplitude $\psi_a(q_1 q_2, t)$ describing the situation would then have to be equal both to $\pm\psi_a(q_2 q_1, t)$ which can only be satisfied by the value zero.

It may be remarked that the exclusion principle does not have to be interpreted as meaning that all the electrons in the world 'must keep in a mysterious communication in order not to get into symmetric states', since we have already seen in connexion with equation (76.15) that the differences between symmetric and antisymmetric solutions will automatically only come into play when the wave functions for the different electrons under consideration overlap.

(*e*) **Enumeration of eigensolutions.** We may now conclude this long treatment of systems containing similar particles by emphasizing some considerations which will be of special importance in our later statistical applications. In making these applications to systems of similar particles we shall find it convenient to regard the state of the system at any time as expressed by a superposition of eigenfunctions of the kind discussed above, and by considering the behaviour of ensembles of such systems we shall be led to assign equal *a priori* probabilities to the different eigenstates that correspond to these eigenfunctions. Hence, in calculating the probability of finding the system in any given specified condition, we shall wish to count the number of eigenstates that correspond to the specified condition. In carrying out such enumeration we encounter two important differences between quantum statistics and classical statistics.

In the first place, since a system of similar particles can never change from states of one symmetry class to those of another, but depending on the kind of particle must remain permanently in states which are either symmetric or antisymmetric, it is evident that our enumeration of the number of eigenstates corresponding to a specified condition of the system must include only those states which have the appropriate symmetry. This quantum mechanical limitation to a particular class of *accessible states*† has no immediate analogy in the classical statistics.

In the second place, in counting the number of eigenstates that correspond to a specified steady condition of the assemblage, the question arises as to whether a new state will be obtained by the interchange of similar particles between the individual eigenfunctions describing them; for example, by the change of $u_k(q_1)u_l(q_2)$ into $u_k(q_2)u_l(q_1)$. In accordance with all of the foregoing discussion, it is evident, however, that such an interchange does not lead to a new eigenstate. In the case of symmetric solutions, the interchange would not even have any effect on the eigenfunctions; and in the case of antisymmetric ones it would only change the sign of the eigenfunctions, which, as we have seen above, has no physical meaning or consequences. This also is different from the classical situation where the interchange of similar particles between different individual states was regarded as leading to a new state of the system as a whole.

Some additional remarks concerning this latter difference between classical and quantum statistics will be clarifying. In the classical mechanics, the procedure of taking the interchange of similar particles between different individual states as leading to a new state of the

† The useful term 'accessible' was first employed by R. H. Fowler, *Statistical Mechanics*, Cambridge University Press, 1929.

system as a whole was justified, not because the particles were thought of as carrying labels which would distinguish between them, but because it would in principle be possible to think of an observer who could follow the motions of the individual particles—without disturbing them—and actually determine whether two similar particles had interchanged roles or not. In the quantum mechanics, however, the possibility of following the behaviour of the individual particles is limited by the Heisenberg uncertainty principle, and this can make it impossible to determine whether such an interchange has taken place. Moreover, such a limitation is regarded in the quantum mechanics, not as an accident due to an unsatisfactory choice of the tools of observation, but as a limitation in principle which can make the question of such interchange meaningless. As a result, in the quantum statistics we do not count as different two states which cannot at least by conceivable observations be distinguished one from the other.

The foregoing remarks will also make it possible to differentiate between cases where it is necessary to keep the above considerations in mind in counting eigenstates and cases where the enumeration does not involve any of these special alterations in principle arising from the quantum mechanics. In a general way it may be said that the situations actually met with in counting eigenstates fall into three classes.

In the first class we meet systems of interacting particles having symmetric properties. In determining the probability of an equilibrium condition of such a system we must then count none but symmetric states and realize that a mere interchange in particles does not lead to a new eigenstate. In the second class we have systems of similar particles having antisymmetric properties, and must include only antisymmetric states without allowance for particle interchange in our enumerations. Finally, in the third class of situations, the significance of which is sometimes overlooked, we meet systems composed of elements, having similar properties, which are nevertheless distinguishable in principle the one element from another. For example, in treating the specific heat of solids it is convenient to take the various possible modes of elastic vibration as the similar elements to be considered, and these would be distinguishable one from the other, even with the same intrinsic frequency of vibration, by spatial location or orientation. Such possibilities of distinguishing one element from another also arises in connexion with the modes of electromagnetic vibration in a hollow enclosure, in connexion with isotopic particles, and in connexion with intrinsically similar particles which are kept in separate containers and not allowed to interact. In such cases, even though the individual eigensolutions $u_k(q_1)$, $u_l(q_2)$, etc., for the different elements may be the same in functional form, it is evident that we must still count it as leading to a new eigenstate when we interchange the elements among such individual solutions.

As we shall later see, the above three classes of situation correspond respectively to the occurrence in the quantum mechanics of the so-called Bose-Einstein, Fermi-Dirac, and Maxwell-Boltzmann types of statistical situation.

IX

STATISTICAL ENSEMBLES IN THE QUANTUM MECHANICS

77. Introduction of statistical methods in the classical and in the quantum mechanics

We are now ready to undertake our investigation of statistical quantum mechanics. We may begin by making some remarks on the reasons for introducing statistical methods into classical and into quantum mechanical considerations, and on a notation which we shall find useful in the quantum statistical mechanics.

In the classical mechanics we can regard the state of a system at any time as specified by the values of its coordinates and momenta, q and p, and can regard the future behaviour of the system as uniquely determined by its state at some initial time. Nevertheless, in spite of this theoretical possibility of exact predictions as to behaviour, we often encounter situations in the classical mechanics such that a precise treatment of behaviour with time would not be appropriate. Such situations arise, even in the treatment of simple mechanical systems, when our knowledge of the initial state is too inexact to justify its idealization as some particular precise state, and arise quite commonly in the treatment of complicated mechanical systems, both because our knowledge of initial state is then made necessarily inexact by empirical difficulties, and because the calculation of precise behaviour is made impracticable by computational difficulties.

Under such conditions we may then find it useful not to treat the behaviour of any single system, but to consider instead the properties of a collection or ensemble of systems, each of the same character as the one of actual interest, but distributed in the coordinate momentum phase space with a continuous range of values for their coordinates and momenta, q and p. From a study of the average behaviour of the systems in a properly chosen ensemble we may then feel able to draw conclusions as to the average or expected behaviour of a single system of interest.

Quite similar situations are also encountered in the quantum mechanics. To be sure, the state of the system must then be regarded as specified at any time by a probability amplitude, such as $\psi(q,t)$, $\phi(p,t)$, or $a(k,t)$, which will in general only permit statements as to the probability of finding different values of the coordinates and momenta q and p or other quantities of interest. Such probability amplitudes, nevertheless, will themselves change with the time in the case of an isolated system in the definite manner dictated by the Schroedinger equation, so that in principle it is also possible in the quantum mechanics to make precise predictions as to the changes in the state of a system as time proceeds. Nevertheless, here too, just as in the case of the classical mechanics, we may encounter situations where it would not be practicable nor profitable to try to treat the precise state of a system as time proceeds. And under such circumstances we may again resort, as in the classical mechanics, to the study of an ensemble of systems of the same kind as the one of interest but distributed over a range of possible states. From the average behaviour of the systems in such an ensemble we may then be assisted as before in drawing conclusions as to the average or expected behaviour of a single system of interest.

In accordance with the above we see that the classical and the quantum forms of statistical mechanics are both suitable for making predictions merely as to the average or expected values of the coordinates and momenta or other properties of a given system of interest. In the classical statistical mechanics this arises because we then treat any single system as being in an average state for the systems in the ensemble as a whole. In the quantum statistical mechanics it arises for a double reason, not only because we treat any single system as being in an average state for the systems of the ensemble, but also because the specification of the quantum mechanical state of a single system would in general itself merely permit predictions as to the average or expectation values of its coordinates and momenta or other properties.

As we proceed we shall find it helpful to have a *notation* which gives specific recognition to this twofold reason for the appearance of probabilities and averages in the quantum statistical mechanics. For this purpose we shall continue to use the letter W to denote a probability arising because of a spread in the predictions associated with the specification of a single quantum mechanical state; and shall use the letter P to denote a probability arising in addition because of the introduction of an ensemble of systems distributed over a range of such states. Furthermore, we shall also find it convenient to use a *single* bar to denote a mean value which has been obtained by the single process of averaging over the range of possibilities corresponding to the quantum mechanical state of a single system, and to use a *double* bar to denote a mean value which has been obtained by the process of averaging over the systems of an ensemble. In general the double bar will then correspond to the result of taking a mean first over the range of possibilities presented by each member of the ensemble and then over the members of the ensemble. Thus

$$\bar{F} = \int \psi^*(q,t)\mathbf{F}\psi(q,t)\,dq$$

would denote the mean value of the observable F for a system in the quantum mechanical state ψ. And

$$\bar{\bar{F}} = \frac{1}{N}\sum_{\alpha=1}^{N} \bar{F}_{(\alpha)}$$

would denote the mean value of that observable for the N systems $\alpha = 1, 2,..., N$ of an ensemble. We shall use the double bar, however, in the precise sense of denoting a mean for the members of an ensemble without reference to the necessity or not for first taking a mean for each member. Thus in the case of a system whose state could be given by a probability amplitude of the form $a(k,t)$, the symbol

$$\overline{\overline{a^*(m,t)a(n,t)}} = \frac{1}{N}\sum_{\alpha=1}^{N} a^*_{(\alpha)}(m,t)a_{(\alpha)}(n,t)$$

would denote the mean value, over all the systems of the ensemble, of a quantity directly given for each system in the ensemble. We shall later see that this particular mean is a very important one.

78. The density matrix in quantum statistical mechanics

In the classical statistical mechanics it is convenient to represent the state of a system of f degrees of freedom by the position of a point in a qp-phase space of $2f$ dimensions, and then to represent an ensemble of such systems by a 'cloud' of phase points distributed with the *density* ρ. Assuming that this density has been normalized to unity, so that we have

$$\int ... \int \rho\, dq_1 ... dq_f\, dp_1 ... dp_f = 1, \tag{78.1}$$

we can then calculate the mean value—for the systems in the ensemble—of any function $F(q,p)$ of the coordinates and momenta with the help of the equation

$$\bar{\bar{F}} = \int ... \int F(q,p)\rho\, dq_1 ... dq_f\, dp_1 ... dp_f. \tag{78.2}$$

It has been shown by von Neumann† that a quantity called the

† von Neumann, *Göttinger Nachrichten*, 1927, p. 245. See also Dirac, *Proc. Cambridge Phil. Soc.* **25**, 62 (1929); **26**, 376 (1930); **27**, 240 (1930).

density matrix with components ρ_{nm} can be introduced in the quantum mechanics to play a role somewhat similar to that of the density ρ in the classical mechanics. Since the specific expression of this matrix will depend on the particular language or quantum mechanical representation which is being used, it will be desirable to use the generalized language provided by the transformation theory (see § 67 (f)) in defining the matrix. For this purpose let us regard the state of each system, in an ensemble of similar non-interacting quantum mechanical systems, as represented by a *generalized probability amplitude* $a(n,t)$. This quantity can be regarded as providing the coefficients for expanding the probability amplitude in coordinate language in the form

$$\psi(q,t) = \sum_k a(k,t)u(k,q), \tag{78.3}$$

where the $u(k,q)$ are any desired complete set of orthogonal, normalized eigenfunctions. Using the general $a(n,t)$ language, the density matrix can then be defined by its component elements

$$\rho_{nm} = \frac{1}{N}\sum_{\alpha=1}^{N} a^*_{(\alpha)}(m,t)a_{(\alpha)}(n,t) = \overline{\overline{a^*_m a_n}}, \tag{78.4}$$

where we take a mean of the products $a^*(m,t)a(n,t)$ for all the systems of the ensemble $\alpha = 1, 2,..., N$, and the order of the indices for ρ_{nm} has been chosen to agree with the convention usually made in this connexion.

Since

$$W_n = a^*_n a_n$$

would evidently give the probability of finding an individual system described by (78.3) in the characteristic state corresponding to the eigenfunction $u(n,q)$, we can write

$$P_n = \rho_{nn} = \overline{\overline{W_n}} = \overline{\overline{a^*_n a_n}} \tag{78.5}$$

as an expression for the probability that a system chosen at random from the ensemble would be found in the state n.† In accordance with (78.5), it will be seen that the diagonal elements of the density matrix are necessarily non-negative in any quantum mechanical language.

As a consequence of the normalization and orthogonality of the $u(k,q)$ in (78.3), and of the normalization of $\psi(q,t)$ itself, we can write for each system in the ensemble

$$\int \psi^*\psi\, dq = \sum_k a^*_k a_k = 1,$$

† Later, in Chapter XII, we shall find it convenient to use the symbols P_n and ρ_{nn} respectively in the specific senses of the 'coarse-grained' and 'fine-grained' probabilities for finding the eigenstate n in the ensemble.

and hence for the ensemble as a whole

$$\sum_k P_k = \sum_k \rho_{kk} = \sum_k \overline{\overline{W_k}} = \sum_k \overline{\overline{a_k^* a_k}} = 1. \tag{78.6}$$

This result, which gives unit probability as the chance of finding a system chosen from the ensemble in one or another of the possible states denoted by k, may be regarded as the quantum mechanical analogue of the classical expression for normalization given by (78.1). It will be noted that the classical operation of integrating the density ρ over the whole of phase space is replaced in the quantum mechanics by the process of taking the sum over all the *diagonal elements* of the density matrix ρ_{nm}, thus obtaining the so-called *trace* of that matrix.

It is also possible to use the density matrix to calculate the mean value of any observable F for the systems in the ensemble. For a single system in the ensemble, the mean value of the observable would be given, in agreement with § 67 (*b*), by

$$\begin{aligned}\bar{F} &= \int \psi^* \mathbf{F} \psi \, dq \\ &= \sum_{m,n} \int a_m^* u_m^* \mathbf{F} a_n u_n \, dq \\ &= \sum_{m,n} F_{mn} a_m^* a_n,\end{aligned} \tag{78.7}$$

where F_{mn} is defined by

$$F_{mn} = \int u_m^* \mathbf{F} u_n \, dq. \tag{78.8}$$

Hence, noting the definition of the density matrix as given by (78.4), we can then write

$$\overline{\overline{F}} = \sum_{m,n} F_{mn} \overline{\overline{a_m^* a_n}} = \sum_{n,m} F_{mn} \rho_{nm} \tag{78.9}$$

for the mean value of the observable for all the systems in the ensemble.

This result may be regarded as the quantum mechanical analogue of the classical expression for obtaining the mean value of any function of the coordinates and momenta given by (78.2). Noticing that the matrix elements F_{mn} and ρ_{nm} are combined in (78.9) in the manner corresponding to matrix multiplication, we can also rewrite this expression in the form

$$\overline{\overline{F}} = \sum_m [F\rho]_{mm} = \sum_n [\rho F]_{nn} \tag{78.10}$$

as the sum of the diagonal elements of the matrix given by $[F\rho]_{kl}$ which is itself the product of the two matrices given by F_{kn} and ρ_{nl}. Hence, in this quantum mechanical analogue to the classical equation (78.2), we again see that the integrand over all of phase space of a classical quantity is replaced in the quantum mechanics by the trace of the corresponding quantum mechanical matrix.

79. Transformation of the density matrix from one quantum mechanical language to another

In introducing the density matrix by the equation of definition (78.4) given above, we have used the quantum mechanical mode of representation corresponding to the probability amplitude $a(n,t)$, which furnishes the coefficients in the expansion

$$\psi(q,t) = \sum_k a(k,t)u(k,q). \tag{79.1}$$

It is evident, however, since we have made no use of any specific properties of the eigenfunctions $u(k,q)$, that we should have obtained results having a similar form and leading to the same physical predictions if we had used any other mode of representation, corresponding, say, to the probability amplitude $b(r,t)$ and the eigenfunctions $v(r,q)$ in the expansion

$$\psi(q,t) = \sum_r b(r,t)v(r,q). \tag{79.2}$$

It will prove useful later if we now investigate the *unitary transformation* between two such languages in agreement with our general treatment of unitary transformations in § 67 (*e*).

Using a prime to denote the density matrix in the new language provided by the $v(r,q)$, we may evidently write at once, in accordance with our previous equation (67.32) for transforming between the probability amplitudes $b(r,t)$ and $a(k,t)$,

$$\begin{aligned} \rho'_{sr} &= \overline{\overline{b_r^* b_s}} = \sum_{k,l} \overline{\overline{a_k^* S_{kr} a_l S_{ls}^*}} \\ &= \sum_{k,l} \rho_{lk} S_{ls}^* S_{kr} = \sum_{k,l} S_{sl}^{-1} \rho_{lk} S_{kr} \end{aligned} \tag{79.3}$$

as the equation of transformation for the components of the density matrix from the old to the new language. The expression for transforming the density matrix is hence of the same form as that given by (67.34) for transforming the matrix corresponding to an ordinary observable from one quantum mechanical representation to another.

We can also use the relations for a unitary transformation to show that the new language is equivalent—as it must be—to the old for the purpose of drawing physical conclusions. This equivalence may be regarded as guaranteed by the general principle of the *invariance of the trace of a matrix* under unitary transformation of its representation, since, as we have seen above, physical conclusions are drawn in the present connexion by taking the trace of some appropriate matrix. Nevertheless, it may be informing to examine the matter in detail.

We may first show that the new density matrix, in the language corresponding to the new eigenfunctions v_r, would be properly normalized. For the total probability in that language of finding a system chosen from the ensemble in one or another of the characteristic states corresponding to the v_r, we shall have

$$\sum_r \rho'_{rr} = \sum_r \overline{\overline{W'_r}}. \tag{79.4}$$

In accordance with (79.3) and (67.30), however, we can write

$$\sum_r \rho'_{rr} = \sum_{r,k,l} \rho_{lk}\, S^*_{lr} S_{kr} = \sum_{k,l} \rho_{lk}\, \delta_{lk} = \sum_k \rho_{kk}, \tag{79.5}$$

which demonstrates that we *shall* have the same normalization, e.g. to unity, in the new language as in the old.

We may also show that the new language provides, as it must, the appropriate expressions for the mean values of observables. Using that language, the mean value of an observable F for the systems of the ensemble should be given by

$$\overline{\overline{F}} = \sum_{s,r} F'_{rs}\, \rho'_{sr}, \tag{79.6}$$

where

$$F'_{rs} = \int v^*_r\, \mathbf{F} v_s\, dq \tag{79.7}$$

would be the matrix components corresponding to the operator $\mathbf{F}$ in our present language. Making use, however, of previous relations given by (79.3), (79.7), (67.29), and (67.28), we can write

$$\begin{aligned}\sum_{s,r} F'_{rs}\, \rho'_{sr} &= \sum_{s,r,k,l} F'_{rs}\, \rho_{lk}\, S^*_{ls}\, S_{kr} \\ &= \sum_{s,r,k,l} \rho_{lk} \int v^*_r (S^{-1}_{rk})^* \mathbf{F} v_s\, S^{-1}_{sl}\, dq \\ &= \sum_{k,l} \rho_{lk} \int u^*_k\, \mathbf{F} u_l\, dq \\ &= \sum_{k,l} \rho_{lk}\, F_{kl},\end{aligned} \tag{79.8}$$

which shows, in view of (78.9) and (79.6), that we shall obtain—as we must—the same mean values for an observable in either language.

The two results (79.5) and (79.8), which can be written in the forms

$$\sum_r \rho'_{rr} = \sum_k \rho_{kk}$$

and

$$\sum_r [F'\rho']_{rr} = \sum_k [F\rho]_{kk},$$

are particular cases of the general invariance of the trace of a matrix towards unitary transformation, which was mentioned above.

In the case of a *finite* Hermitian matrix, in accordance with a well-known theorem, it would necessarily be possible to make a unitary transformation to a representation which will put the matrix itself into diagonal form. It will be reasonable to assume the same possibility for the density matrix even though it has an infinite number of components. Having made such a transformation, the diagonal terms of the density matrix may then be called its eigenvalues. Since we have already seen, by (78.5) and (78.6), that the diagonal elements of the density matrix are necessarily non-negative in any language and have a sum equal to unity, we may now state the general principle that the *eigenvalues of the density matrix are necessarily non-negative with a sum equal to unity,*

$$\rho_d \geqslant 0, \qquad \sum_d \rho_d = 1. \tag{79.9}$$

Among the possibilities of transforming the density matrix from one quantum mechanical language to another, we have, of course, the possibility of transforming from the original generalized language, corresponding to the probability amplitude $a(n, t)$ in terms of which we defined the density matrix ρ_{nm}, to coordinate language, corresponding to the familiar probability amplitude $\psi(q, t)$. It will be of interest to show that the transformation to this language does lead to the expected form of expression for the density matrix. It will be sufficient for this purpose to consider a one-dimensional case.

For the components of the transformed density matrix we shall have

$$\rho'_{sr} = \sum_{k,l} \rho_{lk}\, S^*_{ls} S_{kr}, \tag{79.10}$$

in accordance with (79.3), where the components of the transformation matrix will have the values, in agreement with (67.29),

$$S^*_{ls} = \int v^*_s\, u_l\, dq \quad \text{and} \quad S_{kr} = \int u^*_k\, v_r\, dq, \tag{79.11}$$

$u_l(q)$ and $u^*_k(q)$ being eigenfunctions for the states determining the untransformed language, and $v^*_s(q)$ and $v_r(q)$ being eigenfunctions for the states determining the transformed language. In order that the transformation shall actually be to a coordinate representation, these latter eigenfunctions must be taken as cases of our previous solution (64.24), characteristic of eigenvalues of the coordinates,

$$v^*_s = \delta(q-q_s) \quad \text{and} \quad v_r = \delta(q-q_r), \tag{79.12}$$

where we use the delta function of Dirac. Substituting into (79.11), we then obtain

$$S^*_{ls} = \int u_l(q)\delta(q-q_s)\, dq = u_l(q_s)$$

$$\text{and} \qquad S_{kr} = \int u^*_k(q)\delta(q-q_r)\, dq = u^*_k(q_r), \tag{79.13}$$

where $u_l(q_s)$ and $u^*_k(q_r)$ are the values of the eigenfunctions, determining the original language, at the points q_s and q_r respectively. Substituting into (79.10), and remembering the original definition for ρ_{lk}, as given by (78.4), we then obtain

$$\rho'_{sr} = \sum_{k,l} \rho_{lk}\, u_l(q_s) u_k^*(q_r) = \sum_{k,l} \overline{\overline{a_k^*(t) u_k^*(q_r) a_l(t) u_l(q_s)}}$$
$$= \overline{\overline{\psi^*(q_r, t)\psi(q_s, t)}}, \tag{79.14}$$

where the last form of writing is made possible by the form of expansion (78.3) for the probability amplitude $\psi(q, t)$ in terms of the eigenfunction $u_k(q)$.

We thus see that the density matrix does have the expected form in coordinate language in terms of the probability amplitudes $\psi^*(q_r, t)$ and $\psi(q_s, t)$ for the coordinate eigenvalues q_r and q_s. And the treatment could easily be extended to systems of more than one degree of freedom.

In agreement with this finding it is evident, of course, that we could have started with a definition of the density matrix in coordinate language. Since our considerations were going to be very general in character, however, it seemed best to start with a definition in the very general language provided by the transformation theory as discussed in § 67 (f). Furthermore, this general language, by giving no specific recognition to the circumstance that continuous as well as discrete spectra of eigenstates must actually be considered, really provides a simpler formalism for the treatments that we must undertake, than coordinate language with its definite recognition of the circumstance that the eigenvalues actually do exhibit in that case a continuous spectrum.

80. Density matrix corresponding to a pure state

The preceding sections have shown the use of the density matrix ρ_{nm} in describing the distribution of the quantum mechanical systems which compose a statistical ensemble. A case of particular interest arises when all the systems in the ensemble are in the same state. We shall then say that the ensemble is in a pure state; namely, that state which is common to all the members of the ensemble. In making statistical applications, an ensemble in a *pure state* can be regarded as representing an individual system whose exact quantum mechanical state is known; and an ensemble in an appropriate *mixed state* can be regarded as representing a system whose complete possible quantum mechanical specification is not known.

In the case of an ensemble in a pure state, the density matrix will evidently reduce to

$$\rho_{nm} = \overline{\overline{a_m^* a_n}} = a_m^* a_n, \tag{80.1}$$

since at any given time all the systems in the ensemble will have the same values of a_m^* and a_n. The density matrix for a pure state will or can, of course, still be normalized to unity with

$$\sum_m \rho_{mm} = 1; \tag{80.2}$$

and equations of the same form as (78.9),

$$\overline{\overline{F}} = \sum_{n,m} F_{mn} \rho_{nm}, \tag{80.3}$$

can then still be used to calculate the mean value of any observable F, where a double bar over the F is now no longer really necessary since the mean is the same for each member of the ensemble.

When an ensemble is in a pure state the density matrix will satisfy the special relation

$$\sum_k \rho_{nk} \rho_{km} = \sum_k a_k^* a_n a_m^* a_k = a_m^* a_n = \rho_{nm}, \tag{80.4}$$

since $\sum_k a_k^* a_k$ will be equal to unity. Hence we may take the equality of the square of the density matrix with itself,

$$[\rho^2]_{nm} = \rho_{nm}, \tag{80.5}$$

as a *necessary* requirement for a pure state. By considering the possibility of transforming to a representation in which the density matrix is diagonal, it will be seen that this requirement can only be satisfied if the eigenvalues of the density matrix are all equal to 0 or 1. From this it is evident that one and only one of the eigenvalues will be unity, since their sum also has to be equal to unity by (79.9). This means that all the systems in the ensemble must be in the same state, and hence (80.5) is a *sufficient* as well as a necessary requirement for a pure state. In the general case $\rho_{nm}-[\rho^2]_{nm}$ will always be a non-negative matrix, since the eigenvalues of ρ_{nm} can in no case be greater than unity.

The distinction between pure and mixed states is of considerable importance for statistical mechanics and its applications to the interpretation of thermodynamics.

If we have the *maximal knowledge* of a system allowed by the quantum mechanics, the system at any time of interest will be in a perfectly definite quantum mechanical state, and can then be represented by an ensemble in a pure state, each member of the ensemble being in the same state as the system itself. The only uncertainties as to its properties and the only needs for taking averages will then arise because of the inherently statistical character of the quantum mechanics itself, for which there is no classical analogy. On the other hand, if our knowledge of a system is less than maximal so that we represent it by an ensemble in a mixed state, then the system itself might be in any one of the various quantum mechanical states represented in the ensemble. Further uncertainties and needs for taking averages will then be present, of the same kind as encountered in the classical mechanics, due to the distribution over different states. The distinction between pure and mixed states is thus related to the necessity for taking two kinds of averages which we have already emphasized earlier.

When we come to the statistical mechanical interpretation of thermodynamics we shall find it possible to relate the entropy of systems to the statistical properties of the ensembles suitable for their representation. The distinction between pure and mixed states will then be important, since we shall find the rational zero-point for the entropy of a system to be that amount of entropy which it has when the corresponding representative ensemble is in a pure state.

81. The analogue of Liouville's theorem in quantum statistical mechanics

We have previously shown that the density matrix ρ_{nm} for an ensemble of quantum mechanical systems can to some degree at least be regarded as the quantum analogue of the density ρ for a classical ensemble. We may now investigate the rate of change of the elements ρ_{nm} with time. We shall thus obtain the quantum analogue of *Liouville's theorem* giving the dependence on time of the classical density ρ.

Although an equation for the time dependence of the elements ρ_{nm} of the density matrix can be written in a general form applicable in any quantum mechanical language, the specific content of that form will depend on the particular mode of representation employed. In practice we are often interested in representing the state of a system, either by the coefficients for an expansion in terms of the true steady state solutions for the system, or by the coefficients for an expansion in terms of a set of approximately steady state solutions which correspond to the Hamiltonian of the system after neglecting a small perturbation term. For this reason we shall first treat the dependence of ρ_{nm} on time for these special cases, and postpone the general treatment until the end of the present section.

(*a*) Time dependence of density matrix in language provided by true energy states. Let us consider a system with the Hamiltonian operator

H. With the help of this operator we can determine a set of eigenfunctions $u_n(q)$ corresponding to the various possible eigenvalues E_n of the energy of the system by considering the allowable solutions of the equation

$$\mathbf{H}u_n(q) = E_n u_n(q). \tag{81.1}$$

Any state of the system can then be expressed as an expansion in terms of these eigensolutions, having the form

$$\psi(q,t) = \sum_k c_k u_k(q) e^{-(2\pi i/h)E_k t}, \tag{81.2}$$

where the c_k are *constant coefficients*, and the *probability amplitude* $a(n,t)$ corresponding to the energy level language is seen to be given by

$$a(n,t) = c_n e^{-(2\pi i/h)E_n t}. \tag{81.3}$$

In accordance with this expression, the density matrix with our present mode of representation will have the components

$$\rho_{nm} = \overline{\overline{a_m^* a_n}} = \overline{\overline{c_m^* c_n}} e^{-\frac{2\pi i}{h}(E_n-E_m)t}, \tag{81.4}$$

where we take a mean over all the systems of the ensemble. Since the coefficients c_n are constants, the dependence of the density matrix on time will then be given by

$$\frac{\partial \rho_{nm}}{\partial t} = -\frac{2\pi i}{h}(E_n \rho_{nm} - \rho_{nm} E_m). \tag{81.5}$$

The form of the time dependence of the density matrix is hence very simple in our present language. We shall see later that (81.5) is a special case of the general formula for time dependence in any quantum mechanical language, and that this general formula furnishes a natural analogue for Liouville's theorem in the classical mechanics.

(*b*) **Time dependence of density matrix in language provided by unperturbed energy states.** Let us next consider that the Hamiltonian operator for our system can be treated as the sum of two parts,

$$\mathbf{H} = \mathbf{H}^0 + \mathbf{V}, \tag{81.6}$$

the *unperturbed Hamiltonian* operator $\mathbf{H}^0$ which corresponds to a nearly complete expression for the energy of the system and the *perturbation* operator $\mathbf{V}$ which corresponds to the remainder. With the help of the unperturbed Hamiltonian we can now determine a set of *unperturbed energy eigenfunctions* $u_n(q)$ corresponding to the various possible eigenvalues E_n^0 for the energy of the unperturbed system, by considering the allowable solutions of the equation

$$\mathbf{H}^0 u_n(q) = E_n^0 u_n(q). \tag{81.7}$$

Any state of the system can then be expressed as an expansion in terms of these unperturbed eigenfunctions, having the form

$$\psi(q,t) = \sum_k c_k(t)u_k(q)e^{-(2\pi i/h)E_k^0 t}, \tag{81.8}$$

where the c_k must now be allowed to vary with the time in the manner actually demanded by the Schroedinger equation, since the $u_k(q)$ and E_k^0 are not now quite the true steady state solutions and energy levels for the actual system. The *probability amplitude* $a(n,t)$ corresponding to our present unperturbed energy level language is seen to be given by

$$a(n,t) = c_n(t)e^{-(2\pi i/h)E_n^0 t}. \tag{81.9}$$

In accordance with this expression the density matrix will now have the components

$$\rho_{nm} = \overline{\overline{a_m^* a_n}} = \overline{\overline{c_m^*(t)c_n(t)}}e^{-\frac{2\pi i}{h}(E_n^0-E_m^0)t}, \tag{81.10}$$

where we again take a mean over all the systems of the ensemble. Differentiating with respect to t, the time dependence of the density matrix will now be given by

$$\frac{\partial\rho_{nm}}{\partial t} = -\frac{2\pi i}{h}\overline{\overline{c_m^* c_n}}\,e^{-\frac{2\pi i}{h}(E_n^0-E_m^0)t}(E_n^0-E_m^0)+ \\ +\overline{\overline{\left(c_m^*\frac{\partial c_n}{\partial t}+c_n\frac{\partial c_m^*}{\partial t}\right)}}e^{-\frac{2\pi i}{h}(E_n^0-E_m^0)t}. \tag{81.11}$$

In our consideration of the method of variation of constants in § 68, however, we have already obtained for the time dependence of the coefficients c_n the expression

$$\frac{\partial c_n}{\partial t} = -\frac{2\pi i}{h}\sum_k V_{nk}\,e^{-\frac{2\pi i}{h}(E_k^0-E_n^0)t}c_k, \tag{81.12}$$

with
$$V_{nk} = \int u_n^*(q)\mathrm{V}u_k(q)\,dq \tag{81.13}$$

and
$$V_{nk} = V_{kn}^*. \tag{81.14}$$

Using these results in connexion with (81.11), we can then express the time dependence of the density matrix in the desired form

$$\frac{\partial\rho_{nm}}{\partial t} = -\frac{2\pi i}{h}(E_n^0\,\rho_{nm}-\rho_{nm}\,E_m^0)-\frac{2\pi i}{h}\sum_k(V_{nk}\,\rho_{km}-\rho_{nk}\,V_{km}). \tag{81.15}$$

This expression, as compared with (81.5), contains an additional group of terms because of the fact that E_n^0 and E_m^0 are now unperturbed rather than true energy levels for the system.

The present language is one of those most frequently used in statistical mechanics. The language is specially advantageous when the effects of the perturbation operator **V** actually are small, since approximate integrations can then be carried out (see Chapter XI) which will let us study the transitions of a system between nearly steady states. For example, in the case of a dilute gas $\mathbf{H}^0$ can correspond to the energy of the molecules, considered as not affecting each other, and **V** to their energy of interaction. This then provides an appropriate language for treating the transitions between those nearly steady states of the gas which would persist as permanently steady states in the absence of collisions.

(c) **Time dependence of density matrix in general language.** To obtain a more general treatment of the time dependence of the density matrix, let us now consider an expansion for the state of our system in the quite general form

$$\psi(q,t) = \sum_k a_k(t)u_k(q), \tag{81.16}$$

where the $u_k(q)$ may now be any complete set of normalized, orthogonal eigenfunctions for the system, and the $a_k(t)$ are the corresponding probability amplitudes.

For the rate of change of $a_n(t)$ with time we then have the generalized Schroedinger equation (67.13)

$$\frac{\partial a_n}{\partial t} = -\frac{2\pi i}{h}\sum_k H_{nk}\,a_k, \tag{81.17}$$

with

$$H_{nk} = \int u_n^*\,\mathbf{H}u_k\,dq \tag{81.18}$$

and

$$H_{nk} = H_{kn}^*. \tag{81.19}$$

With this apparatus the rate of change of ρ_{nm} with time in our present quite general language assumes the form

$$\frac{\partial\rho_{nm}}{\partial t} = \frac{\partial}{\partial t}\overline{\overline{a_m^*\,a_n}} = -\frac{2\pi i}{h}\sum_k (H_{nk}\overline{\overline{a_m^*\,a_k}} - H_{mk}^*\overline{\overline{a_k^*\,a_n}}),$$

which can be more simply written as

$$\frac{\partial\rho_{nm}}{\partial t} = -\frac{2\pi i}{h}\sum_k (H_{nk}\,\rho_{km} - \rho_{nk}\,H_{km}). \tag{81.20}$$

Two remarks may be made in connexion with this result, which gives a general expression for the quantum analogue of Liouville's theorem.

In the first place, it will be immediately appreciated that our previous expressions (81.5) and (81.15) for the rate of change of the density matrix with time are really special cases of this more general result. They differ in apparent form from (81.20) merely because they contain the explicit expressions for H_{nk} and H_{km} that correspond respectively to the languages provided by the true and by the unperturbed energy states for the system.

In the second place, if we rewrite (81.20) in the form

$$\frac{\partial \rho_{nm}}{\partial t} = -\frac{2\pi}{ih}\sum_k (\rho_{nk}H_{km}-H_{nk}\rho_{km}) = -\frac{2\pi}{ih}[\rho, H]_{nm}, \quad (81.21)$$

or, in matrix language,

$$\frac{\partial \|\rho\|}{\partial t} = -\frac{2\pi}{ih}\|\rho H-H\rho\| = -\frac{2\pi}{ih}\|\rho, H\|,$$

we see, by comparison with our previous equation (19.9) for the dependence of the classical density ρ on time, that the quantities $(2\pi/ih)[\rho, H]_{nm}$ and $(2\pi/ih)\|\rho, H\|$ appearing in the above equations may be regarded in the present connexion as the analogues of the classical Poisson bracket $\{\rho, H\}$. The similarity will be noted with previous analogues for Poisson brackets as given by (63.3) and (67.27). It will also be noted that we take the Poisson bracket or its analogue with a *positive sign* to obtain the time dependence of the exact or expectation value of a function of the coordinates and momenta for a single system, and with a *negative sign* to obtain the time dependence of the density or density matrix for an ensemble of such systems.

82. Conditions for statistical equilibrium

For the purposes of statistical mechanics we shall be specially interested in ensembles which are in statistical equilibrium; that is, having such a distribution of systems among the different possible quantum states that the components of the density matrix ρ_{nm} do not change with time. In analogy with the classical mechanics this result can be achieved either by taking an initial distribution with the density matrix set equal to a constant, or by taking a distribution with the density matrix set equal to some function of the matrix for any constant of motion for the system such as its energy. We may give separate attention to these two possibilities.

(*a*) **Density a constant.** Let us begin by considering the possibility of setting the density matrix equal to a constant. The component elements of the density matrix will then be given by

$$\rho_{nm} = \rho_0\,\delta_{nm}, \quad (82.1)$$

the non-diagonal elements of the matrix all being equal to zero, and the diagonal elements all equal to the same quantity ρ_0.

We may readily show that the distribution given by (82.1) would persist as time proceeds. To see this, we have as our general equation (81.20) for the dependence of the density matrix on time

$$\frac{\partial \rho_{nm}}{\partial t} = -\frac{2\pi i}{h} \sum_k (H_{nk}\rho_{km} - \rho_{nk}H_{km}). \tag{82.2}$$

And substituting (82.1), we obtain

$$\begin{aligned}\frac{\partial \rho_{nm}}{\partial t} &= -\frac{2\pi i}{h}\rho_0 \sum_k (H_{nk}\delta_{km} - \delta_{nk}H_{km}) \\ &= -\frac{2\pi i}{h}\rho_0(H_{nm} - H_{nm}) \\ &= 0.\end{aligned} \tag{82.3}$$

We thus find, by starting with the initial distribution $\rho_{nm} = \rho_0 \delta_{nm}$, that we do obtain an ensemble which retains that same distribution independent of time. We shall see later that the properties of such a permanent uniform distribution will be very important in setting up the fundamental postulate of the quantum statistics as to equal *a priori* probabilities and random *a priori* phases.

(*b*) **Density a function of a constant of the motion.** We may now turn to the more general possibility of securing statistical equilibrium, by setting the density matrix $\|\rho\|$ equal to a function of a matrix $\|A\|$ corresponding to any constant of the motion for the kind of system under consideration.

For this purpose we may begin by taking

$$F_{nm}(t) = \int u_n^*(q) e^{(2\pi i/h)\mathbf{H}t}\mathbf{F}e^{-(2\pi i/h)\mathbf{H}t}u_m(q)\,dq \tag{82.4}$$

as a general expression, in the language given by the eigenfunctions $u_n(q)$, for the elements of the *time-dependent matrix* $\|F(t)\|$, corresponding to any dynamical variable for the system. The symbols $\mathbf{F}$ and $\mathbf{H}$ in this expression denote the operators in coordinate language which correspond to the variable of interest F and to the energy of the system H, and the exponential operators $e^{(2\pi i/h)\mathbf{H}t}$ and $e^{-(2\pi i/h)\mathbf{H}t}$ may be regarded if desired as a shorthand expression for the corresponding series expansions, as will be discussed in more detail in § 96. It will be noted that the equation of definition (82.4) is a particular case of our previous general expression (67.23) for time-dependent matrices, with

$$\psi^*(q,t) = u_n^*(q)e^{(2\pi i/h)\mathbf{H}t}$$

and

$$\psi(q,t) = e^{-(2\pi i/h)\mathbf{H}t}u_m(q)$$

as the two—formally correct—solutions of the Schroedinger equation which are now of interest.

Differentiating (82.4) with respect to the time t, assuming that **H** and **F** have no explicit dependence on t, we obtain

$$\begin{aligned}\frac{dF_{nm}(t)}{dt} &= \frac{2\pi i}{h}\int u_n^*(q)e^{(2\pi i/h)\mathbf{H}t}(\mathbf{HF}-\mathbf{FH})e^{-(2\pi i/h)\mathbf{H}t}u_m(q)\,dq \\ &= \frac{2\pi i}{h}[HF(t)-F(t)H]_{nm} \\ &= \frac{2\pi i}{h}\sum_k \{H_{nk}F_{km}(t)-F_{nk}(t)H_{km}\},\end{aligned} \tag{82.5}$$

where the first form of writing depends on the commutability of the operator **H** with any function of itself, and the second on the general relation between quantum mechanical operators and the corresponding matrices. It will be noted that (82.5) is a particular case of our previous equation (67.27).

Equation (82.5) gives a general expression in our present language for the time dependence of the matrix elements corresponding to any dynamical quantity F. We may now define a *constant of the motion* A as being a quantity which has an operator **A** which commutes with that for the energy **H**,

$$\mathbf{HA} = \mathbf{AH}. \tag{82.6}$$

In accordance with (82.5), we shall then have, for such a quantity, matrix elements which do not change with time and hence can be regarded at all times as given by the simplified expression

$$A_{nm}(t) = A_{nm} = \int u_n^*(q)\mathbf{A}u_m(q)\,dq. \tag{82.7}$$

Returning now to Liouville's theorem (81.20) for the rate of change in the density matrix with time,

$$\frac{\partial\rho_{nm}}{dt} = -\frac{2\pi i}{h}\sum_k (H_{nk}\rho_{km}-\rho_{nk}H_{km}), \tag{82.8}$$

we see that this rate of change will be zero for an ensemble set up with the density matrix put equal to any function of the matrix for a constant of the motion,

$$\rho_{nm} = [f(A)]_{nm}, \tag{82.9}$$

since the right-hand side of (82.8) will then be zero in accordance with the commutativity expressed by (82.6). Hence we can now achieve statistical equilibrium in this quite general way. As in the classical statistical mechanics, we shall be specially interested in cases of statistical equilibrium which are obtained by taking the density as a function of the energy,

$$\rho_{nm} = [f(H)]_{nm}. \tag{82.10}$$

83. The uniform, microcanonical, and canonical ensembles in the quantum mechanics

We may now consider in some detail the properties of different kinds of ensemble used in the quantum mechanics, which may be regarded as the analogues of the uniform, microcanonical, and canonical ensembles familiar in the classical mechanics.

(*a*) **The uniform ensemble.** The first of these is the uniformly distributed ensemble with the elements of the density matrix given by

$$\rho_{nm} = \rho_0 \delta_{nm}, \tag{83.1}$$

where ρ_0 is a constant. As we have already seen in the preceding section, such a distribution, if once set up, would remain unaltered as time proceeds, and hence the uniform ensemble has the important property of being in *statistical equilibrium*.

In the case of such uniform ensembles, since there will be equal probabilities of finding a system chosen from the ensemble in any one of an infinite number of different states, it is not profitable to try to normalize the density matrix to unity. Instead, we may, if desired, take

$$\rho_{nn} = \overline{\overline{W}}_n = \rho_0, \tag{83.2}$$

with ρ_0 finite, as expressing the *relative probability* of finding a system chosen at random from the ensemble in state n, such relative probabilities in this case being equal for all the different states.

An important property of the uniform ensemble lies in the circumstance that the density matrix for this particular ensemble would have just the same form of expression in different quantum mechanical representations. To show this we have as our general formula (79.3) for transforming the density matrix to a new quantum mechanical language

$$\rho'_{sr} = \sum_{k,l} \rho_{lk} S^*_{ls} S_{kr},$$

and, substituting (83.1), this gives us

$$\rho'_{sr} = \sum_{k,l} \rho_0 \delta_{kl} S^*_{ls} S_{kr} = \rho_0 \sum_l S^*_{ls} S_{lr} = \rho_0 \delta_{sr}, \tag{83.3}$$

where the last form of writing depends on the fundamental property given by (67.30) for the matrix elements S^*_{ls} and S_{kr} that lead to unitary transformations.

In accordance with (83.3), we now see that an ensemble set up to have a uniform distribution in one particular quantum mechanical language would not only retain a uniform distribution for all time in the original language, but would also have a uniform distribution, and with the same magnitude of the constant ρ_0, in any language. This is the quantum mechanical analogue of the classical finding, in § 22 (*a*), that a uniform ensemble would not only be in statistical equilibrium but would be distributed with the same density ρ in the phase spaces corresponding to any choice of classical canonical variables.

To complete this discussion of uniform ensembles in the quantum mechanics, a certain point of difference, as compared with uniform ensembles in the classical mechanics, must now be considered. In the classical mechanics it was clear that a uniform distribution of systems in the phase space, with

$$\rho = \text{const.}, \tag{83.4}$$

would necessarily imply the same number of systems in each of the groups of states corresponding to the equal regions $\delta q_1 \ldots \delta p_f$ into which we might think of the phase space as divided. In the quantum mechanics, however, the uniform ensemble, expressed by

$$\rho_{nm} = \rho_0 \delta_{nm}, \tag{83.5}$$

cannot be regarded as carrying necessary implications as to the numbers of systems in different states, owing to the new possibility for the superposition of quantum mechanical states.

To investigate this we may return to our original definition (78.4) for the density matrix ρ_{nm}, and express this in a form to show the dependence of the density matrix on the magnitudes and phases of the probability amplitudes a_m^* and a_n for the different states of the systems in the ensemble. For this purpose we write

$$\begin{aligned} \rho_{nm} &= \overline{\overline{a_m^* a_n}} \\ &= \overline{\overline{r_n r_m e^{i(\phi_n - \phi_m)}}} \\ &= \overline{\overline{r_n r_m \{\cos(\phi_n - \phi_m) + i\sin(\phi_n - \phi_m)\}}}, \end{aligned} \tag{83.6}$$

where r_n and r_m are the absolute magnitudes, and ϕ_n and ϕ_m are the phases, of the probability amplitudes for the states n and m of a system in the ensemble, and the double bar indicates that we are to take a mean of the quantity indicated for all the systems in the ensemble. For the case of a uniform ensemble, the expression given by (83.6) must then reduce to

$$\overline{\overline{r_n r_m \{\cos(\phi_n - \phi_m) + i\sin(\phi_n - \phi_m)\}}} = \rho_0 \delta_{nm} \tag{83.7}$$

for any choice of states n and m.

It is evident, however, that this result would not be sufficient to give unique information as to the distribution of the systems of the ensemble in different states, since (83.7) could evidently be satisfied by taking the magnitudes r_n, r_m, etc., for the different systems in any manner which would lead to the mean results

$$\overline{\overline{r_n^2}} = \overline{\overline{r_m^2}} = ... = \rho_0, \tag{83.8}$$

and by taking the phases ϕ_n, ϕ_m, etc., in any manner which would lead to the mean results

$$\overline{\overline{r_n r_m \cos(\phi_n - \phi_m)}} = 0 \quad \text{and} \quad \overline{\overline{r_n r_m \sin(\phi_n - \phi_m)}} = 0, \tag{83.9}$$

when n and m are not the same. Both of these results could be achieved in a great variety of ways. In particular it is to be noted that (83.9) would be satisfied by any random selection of phases which would make the above means zero on account of the equal weighting for positive and negative values of $\cos(\phi_n - \phi_m)$ and $\sin(\phi_n - \phi_m)$ when n and m are not the same. Hence the fulfilment of (83.7) would not be sufficient to describe the superposition of different eigenstates n, m, etc., which might prevail for any single system in the ensemble.

In accordance with the foregoing, it is customary to describe the uniform ensemble as corresponding to *equal probabilities* and *random phases* for the different eigenstates which define the quantum mechanical representation that is being employed, where the term *random phases* means that there is no special selection of phases for the different states of the different members of the ensemble which would lead to non-diagonal terms in the density matrix. As a consequence of our earlier discussion, it is evident that these properties of equal probabilities and random phases for different eigenstates in the case of any uniform ensemble would persist both as time proceeds and on transformation to other quantum mechanical languages.

It may be specially emphasized that the mere prescription of equal probabilities for different eigenstates in a particular quantum mechanical language, without also prescribing random phases in that language, would not be sufficient to secure these important properties of the maintenance of equal probabilities for different eigenstates as time proceeds, and the similar maintenance on transformation to other quantum mechanical languages. If we had contented ourselves with defining the uniform ensemble in a particular quantum mechanical language by

$$\rho_{nm} = \rho_0 \delta_{nm} + B_{nm}(1 - \delta_{nm}),$$

we should have indeed secured equal probabilities ρ_{nn} for the different states n determining that language. Nevertheless, our proof of statistical equilibrium for the ensemble, as given by (82.3), would no longer be possible since in general $\|H\|$ and $\|B\|$ would not commute, and our proof of invariance against transformation, as given by (83.3), would no longer be possible, since in general we should not have a reduction in form which would permit us to use the simple property of transformation matrices,

$$\sum_l S^*_{ls} S_{lr} = \delta_{sr}.$$

Just as in the classical statistics, the quantum mechanical uniform ensemble cannot be regarded as a representative ensemble for a system concerning the condition of which we have any information, since the ensemble gives equal probabilities for all possible different states. As in the classical statistics, however, we shall find the properties of the uniform ensemble very important when we come to set up the fundamental postulatory basis for statistical mechanics.

(*b*) **The microcanonical ensemble.** We must next consider the analogue of the classical microcanonical ensemble, which will be defined—also in the quantum mechanics—in a manner to give uniform probability of finding members of the ensemble in energy states lying within a narrow energy range and zero probability for finding members of the ensemble outside that range. To obtain such an ensemble it will be necessary to set the density matrix equal to an appropriate function of the energy matrix for the kind of system under consideration. It will be advantageous to begin by making some preliminary general remarks as to the nature of such functional relationships between two matrices.

The possibility of setting one matrix equal to a function f of another depends on the possibility of regarding the function f as having a form which can be represented with sufficient accuracy by a power series in its argument. With the help of the operations of matrix multiplication and addition it will then be possible to obtain explicit expressions for the elements of the first matrix in terms of those of the other. Thus we may relate the density matrix to the energy matrix by the expression

$$\|\rho\| = f\|H\|,$$

provided we regard this as equivalent to an expansion of the form

$$\|\rho\| = a_0+a_1\|H\|+a_2\|H\|^2+a_3\|H\|^3+..., \tag{83.10}$$

where the quantities a_0, a_1, a_2, a_3, etc., are constant coefficients. As explicit expressions for the elements of the density matrix in terms of the elements of the energy matrix, we can then write, in accordance with the rules for matrix multiplication,

$$\rho_{nm} = a_0\delta_{nm}+a_1 H_{nm}+a_2\sum_k H_{nk}H_{km}+a_3\sum_{k,l} H_{nk}H_{kl}H_{lm}+..., \tag{83.11}$$

and this can also be expressed in the abbreviated form

$$\rho_{nm} = [f(H)]_{nm} \tag{83.12}$$

if we so desire.

The above expressions assume a very simple form if we make use of the mode of representation provided by the energy language. The energy matrix will then be diagonal with elements given by

$$\begin{aligned} H_{nm} &= \int u_n^*(q)\mathbf{H}u_m(q)\,dq \\ &= E_m \int u_n^*(q)u_m(q)\,dq \\ &= E_m\,\delta_{nm}, \end{aligned} \tag{83.13}$$

owing to the circumstance that $u_m(q)$ will now be an energy eigenfunction, and owing to the normalization and orthogonality of such eigenfunctions. Substituting expressions of the form (83.13) into (83.11), we can then write

$$\begin{aligned} \rho_{nm} &= a_0\,\delta_{nm} + a_1\,E_m\,\delta_{nm} + a_2 \sum_k E_k\,E_m\,\delta_{nk}\,\delta_{km} + \\ &\qquad\qquad + a_3 \sum_{k,l} E_k\,E_l\,E_m\,\delta_{nk}\,\delta_{kl}\,\delta_{lm} + \ldots \\ &= a_0\,\delta_{nm} + a_1\,E_m\,\delta_{nm} + a_2\,E_m^2\,\delta_{nm} + a_3\,E_m^3\,\delta_{nm} + \ldots, \end{aligned} \tag{83.14}$$

and this can also be expressed in the abbreviated form

$$\rho_{nm} = f(E_m)\delta_{nm}. \tag{83.15}$$

In the light of this preliminary discussion we can now easily define the quantum mechanical *microcanonical ensemble* by requiring a form for the function $f(E_m)$ such that the density matrix expressed in the energy language will be given by

$$\begin{aligned} \rho_{nm} &= \rho_0\,\delta_{nm} \quad (E_m \text{ in range } E \text{ to } E+\delta E), \\ \rho_{nm} &= 0 \qquad\quad (E_m \text{ not in range } E \text{ to } E+\delta E), \end{aligned} \tag{83.16}$$

where ρ_0 is a constant, and E to $E+\delta E$ denotes some particular small range of interest in the energy of the systems under consideration. This simple form of expression holds, of course, only in the *energy language*, and more complicated forms of expression in which the density matrix would exhibit non-diagonal as well as diagonal terms would be found in general on transformation to other modes of representation.

In accordance with the above definition there would be equal probabilities,

$$\rho_{nn} = \overline{\overline{W_n}} = \rho_0, \tag{83.17}$$

for finding a system selected at random from the ensemble in each of the eigenstates n for an eigenvalue of the energy E_n lying in the range

E to $E+\delta E$, and zero probability for finding a system with energy outside that range. In the case of degeneracy, each eigenstate corresponding to the same energy is to be treated as a separate state, having its appropriate quota of systems.

In accordance with § 82 (*b*), the microcanonical ensemble has the important property of being in *statistical equilibrium*, since the density matrix has been set up as a function of the energy matrix, and our actual interest will lie in conservative systems for which the energy will be a constant of the motion. Hence the expression for the density matrix, as given in the energy language by (83.16), will remain unaltered as time proceeds, and the probability of finding the different energy states will remain permanently equal to ρ_0 or 0, according as the energy does or does not lie in the range E to $E+\delta E$. This is in agreement with the quantum mechanical form of the principle of the conservation of energy, which makes the relative probabilities of finding different values of the energy remain constant with time for each individual system in the ensemble.

The microcanonical distribution will prove useful, in the quantum mechanics as in the classical mechanics, as a representative ensemble for an individual system in a steady condition of equilibrium with its energy content sufficiently well determined to be assigned to the specified range E to $E+\delta E$. Since a determination of the energy of a quantum mechanical system of interest would be subject to the Heisenberg uncertainty,

$$\Delta E \approx \frac{h}{\Delta t}, \tag{83.18}$$

where Δt is the time available for observation, it will be appropriate to regard δE as *large* compared with ΔE. In the quantum mechanics there will be even less reason than in the classical mechanics for considering surface ensembles with all the systems having precisely the same energy, since a precise determination of energy would necessitate an infinite time of observation and in the absence of degeneracy would put the system into some single energy eigenstate.

(*c*) **The canonical ensemble.** It is also useful in the quantum statistics to have an analogue of the classical canonical ensemble. This may be defined in general by setting the density matrix $\|\rho\|$ equal to a function of the energy matrix $\|H\|$, for the systems under consideration, having the form

$$\|\rho\| = e^{\frac{\psi-\|H\|}{\theta}}, \tag{83.19}$$

where ψ and θ are parameters which determine the particular distribution of interest, and we regard the expression given as equivalent to the series expansion

$$\|\rho\| = e^{\psi/\theta}\left\{1-\frac{\|H\|}{\theta}+\frac{1}{2!}\frac{\|H\|^2}{\theta^2}-\frac{1}{3!}\frac{\|H\|^3}{\theta^3}+...\right\}. \tag{83.20}$$

In accordance with the rules for matrix multiplication, we can then write as an expression for the elements of the density matrix

$$\rho_{nm} = e^{\psi/\theta}\left\{\delta_{nm}-\frac{H_{nm}}{\theta}+\frac{1}{2!}\frac{1}{\theta^2}\sum_k H_{nk}H_{km}-\frac{1}{3!}\frac{1}{\theta^3}\sum_{k,l} H_{nk}H_{kl}H_{lm}+...\right\}, \tag{83.21}$$

valid for any language in which the H_{nm} are expressed.

In the *energy language* the elements of the energy matrix will assume the simple form

$$H_{nm} = \int u_n^*(q)\mathbf{H}u_m(q)\,dq = E_m\,\delta_{nm}, \tag{83.22}$$

and the above expression will reduce to

$$\rho_{nm} = e^{\psi/\theta}\left\{1-\frac{E_m}{\theta}+\frac{1}{2}\frac{E_m^2}{\theta^2}-\frac{1}{3!}\frac{E_m^3}{\theta^3}+...\right\}\delta_{nm}.$$

This can then be conveniently rewritten as

$$\rho_{nm} = e^{\frac{\psi-E_m}{\theta}}\,\delta_{nm}. \tag{83.23}$$

The equation of definition for the canonical ensemble thus assumes a very simple form in the language corresponding to the energy eigenvalues E_m for the system.

In accordance with the above definition of the canonical ensemble, the probability for finding a system selected at random from the ensemble in any specified energy eigenstate n would be given by

$$\rho_{nn} = \overline{\overline{W}}_n = e^{\frac{\psi-E_n}{\theta}}, \tag{83.24}$$

where each eigenstate is to be treated as a separate state in the case of degeneracy. Since the total probability for finding a system in one or another state will be taken as normalized to unity,

$$\sum_n \rho_{nn} = \sum_n \overline{\overline{W}}_n = \sum_n e^{\frac{\psi-E_n}{\theta}} = 1, \tag{83.25}$$

the *distribution parameters* ψ and θ will be subject to the relation

$$e^{-\psi/\theta} = \sum_n e^{-E_n/\theta}, \tag{83.26}$$

where we take a summation over all energy eigenstates n. Owing to a definition obtained by setting the density matrix equal to a function of the energy matrix for the systems under consideration, the canonical ensemble will have the important property of being in *statistical equili-*

brium, in the case of conservative systems, and the distribution given in the energy language by (83.23) will remain unaltered as time proceeds.

The close analogy between the properties of the quantum canonical ensemble as defined above and of the classical canonical ensemble as defined in § 22 (*c*) will be noted. As was shown for the classical mechanics by Gibbs, the canonical ensemble will prove especially useful in the quantum mechanics when we come to consider the relations between statistical mechanics and thermodynamics, and the canonical distribution will then be found to provide the appropriate representative ensemble for a system of interest having a specified temperature.

The foregoing examples of the uniform, microcanonical, and canonical ensembles have been selected to illustrate the general nature of ensembles in the quantum mechanics because of their definite character and of their usefulness for our later considerations. It will be appreciated, however, in the quantum statistics as in the classical statistics, that we are in no way limited to ensembles which are in statistical equilibrium, and that ensembles where the distribution is changing with time will be of great importance when we come to consider systems of interest which are not themselves in a steady condition of macroscopic equilibrium.

It may be well to call attention at this point to the consideration that ensembles in the quantum mechanics are, of course, to be constructed, in the case of systems containing essentially similar particles, in such a manner as to exclude non-accessible states not having the symmetry properties known to occur in nature. This circumstance, for which there was no classical analogy, provides the explanation for important differences between classical and quantum statistical results, as we shall see later. It is also to be noted in the case of dynamical variables having no classical analogy—e.g. spin—that ensembles must be constructed so as to include the states that correspond to the different possible eigenvalues of such quantities.

84. The fundamental hypothesis of equal *a priori* probabilities and random *a priori* phases for the quantum mechanical states of a system

We must now turn to a consideration of the fundamental hypothesis of the quantum statistical mechanics. In the quantum as in the classical case, the necessity for some postulate, in addition to those sufficient for an exact mechanics, arises when we desire to give a reasonable treatment to the properties and behaviour of some *system of interest* in a condition which is not sufficiently specified to give a precise determination of state. Under such circumstances we then resort to an appropriately chosen *representative ensemble* of systems, of similar structure to the one of actual interest, and regard the average properties and average behaviour of the systems in this ensemble as providing estimates as to what may reasonably be expected for the actual system of interest. In setting up such a representative ensemble we shall, of course, take the members of the ensemble as being in states which agree with our partial knowledge of the precise state of the actual system of interest. We need some hypothesis, however, to guide us in choosing the probabilities and phases for different states that agree equally well with that partial knowledge.

For this purpose we now introduce, as a postulate, the hypothesis of *equal* a priori *probabilities and random* a priori *phases for the quantum mechanical states of a system.* This hypothesis may be taken in a general way as signifying, in the absence of precise knowledge of the probability amplitudes $a_n = r_n e^{i\phi_n}$ for the different eigenstates n of a system, that there is no *a priori* reason, provided, for example, by the quantum mechanics itself, for assuming other than equal importance for the probabilities, $W_n = r_n^2$, or other than random values for the phases ϕ_n of states n which agree equally well with our knowledge of the actual condition of the system.

We can obtain a more adequate idea of the meaning of the hypothesis by applying it in a specific example. For this purpose let us consider a situation in which our partial knowledge of the state of the system is regarded as obtained from an *approximate measurement* of some observable quantity F which is a property of the system. As the equations determining the eigenstates of the system corresponding to this observable, we can write—in coordinate language for specificity—

$$\mathbf{F}u_k(q) = F_k u_k(q), \tag{84.1}$$

where $\mathbf{F}$ is the quantum mechanical operator corresponding to the observable F, and the different possible eigenvalues of this quantity and associated eigensolutions are denoted by F_k and $u_k(q)$. With the help of these eigensolutions, any state of the system could then be expressed by a summation of the form

$$\psi = \sum_k a_k u_k = \sum_k r_k e^{i\phi_k} u_k, \tag{84.2}$$

where the complex quantities $a_k = r_k e^{i\phi_k}$ are the probability amplitudes for the different eigenstates k at any time of interest.

A precise measurement of the quantity F, showing that it had a particular eigenvalue F_k, would now tell us—neglecting degeneracy—that the system was in a particular one of the above states k. The properties and behaviour of the system could then be treated by the precise methods of the quantum mechanics, by setting the probability for this particular state equal to unity with

$$a_k^* a_k = r_k^2 = 1, \tag{84.3}$$

and by taking the phase ϕ_k for this state at random, since we immediately see that its value has no effect on the properties of the system at the time of measurement, and see, from the fact that the generalized Schroedinger equation (67.13) would reduce to

$$\frac{\partial a_n}{\partial t} = -\frac{2\pi i}{h} \sum_k H_{nk} a_k = -\frac{2\pi i}{h} H_{nk} r_k e^{i\phi_k} \tag{84.4}$$

with only one surviving term of the summation, that its value would have no effect on the probabilities $a_n^* a_n$ which we should predict for any state n at a later time. Cf. § 96 (c). In view of the approximate character of our actual measurement of the quantity F, however, we shall actually have to turn to the methods of statistical mechanics and construct an appropriate representative ensemble for the treatment of the system.

For this purpose we might now proceed by regarding our partial information as to the state of the system as equally well satisfied by any one of a group of G_κ neighbouring states k which have eigenvalues F_k in substantial agreement with our approximate measurement of F. In accordance with our hypothesis as to *a priori* probabilities and phases, we could then construct an appropriate representative ensemble by taking equal mean probabilities $\overline{\overline{r_k^2}}$ and random phases ϕ_k for the different states k that lie in this group of G_κ neighbouring states, and zero probabilities for states lying outside that group. This would then give us

$$\rho_{kl} = \overline{\overline{a_l^* a_k}} = \overline{\overline{r_k r_l e^{i(\phi_k - \phi_l)}}} = \begin{cases} \rho_0 \delta_{kl} & \text{(state } k \text{ in group } G_\kappa), \\ 0 & \text{(state } k \text{ not in group } G_\kappa), \end{cases} \tag{84.5}$$

as a specific expression for the density matrix of the representative ensemble at the time of measurement, where the quantity ρ_0 is a constant which gives the equal mean values for the probabilities of the states that do lie in the group G_κ, and any possible non-diagonal terms of the matrix drop out as a result of averaging over the random phases for those states.

It would also be possible to proceed by regarding our approximate measurement as telling us that the value of F was almost certainly in the immediate neighbourhood of a particular eigenvalue F_0, with a decreasing chance for values more and more removed therefrom. In accordance with our hypothesis as to equal *a priori* probabilities and random *a priori* phases, we could then construct an appropriate representative ensemble by taking the density matrix as given, for example, by

$$\rho_{kl} = \rho_0 e^{-\frac{(F_k-F_0)^2}{2\sigma^2}} \delta_{kl}, \tag{84.6}$$

where ρ_0 and σ are constants. With a small value of the dispersion σ, the probability for different states k would then be concentrated among those for which F_k was nearly equal to F_0. The two proposals (84.5) and (84.6), as to an appropriate ensemble for representing our approximate knowledge of the quantity F for the system of interest, would lead to practically identical conclusions in the situations commonly treated by statistical methods.

As previously mentioned, we shall regard the introduction of the hypothesis of equal *a priori* probabilities and random *a priori* phases as making a definite addition to our postulatory basis, which can be ultimately justified only by the agreement between the results to be calculated with its help and the results actually obtained by observation and experiment. Nevertheless, certain remarks may be made beforehand as to the reasonable character of the postulate that we have chosen.

First of all, emphasis may be laid on our previous finding, see § 83 (*a*), that a uniform ensemble, if once set up, would retain for all time a distribution which could be described by saying that it corresponds to equal probabilities and random phases for the different eigenstates that define the quantum mechanical language employed. We may hence conclude that the quantum mechanics does not itself have a character which would make one eigenstate inherently more probable than another, or would make any particularly ordered relation between the phases of different states inherently more probable than a completely random relation. In view of this character of the quantum mechanics itself it would then seem quite arbitrary to adopt any procedure other than the assignment of equal probabilities and random phases to those states that do correspond equally well with the approximate knowledge of state that has been obtained.

This may be seen more clearly if we return once more to our previous example in which our observation of the state of the system was regarded as insufficient to distinguish between the members of a group of G_κ neighbouring eigenstates, corresponding to eigenvalues F_k, all of which are in substantial agreement with our approximate measurement of an observable quantity F. With regard to the probabilities of the different states in the group G_κ, it is then immediately evident that it would be arbitrary to weight one of these states more highly than another, in constructing the appropriate representative ensemble, since the quantum mechanics itself has not provided any reason for thinking that one state is inherently more probable than another, and our actual knowledge is equally well represented by any one of the states in the group. And with regard to the phases of the different states in the group G_κ, it is also evident that it would be arbitrary to proceed otherwise than by the assignment of random phases, since the quantum mechanics has not itself provided any reason for thinking that any particular arrangement of phases is inherently more probable than a random one, and the observation actually made on the system has not been of a kind to give any information as to the actual phases of the different possible eigenstates k.

To see that an approximate measurement of the quantity F would not give information as to the phases ϕ_k of the amplitudes a_k for the eigensolutions u_k that correspond to the eigenvalues F_k of that quantity, let us take the actual state of the system, at the time of our approximate measurement thereof, as expressed by an expansion of the form

$$\psi = \sum_k a_k u_k = \sum_k r_k e^{i\phi_k} u_k, \tag{84.7}$$

where the quantities $a_k = r_k e^{i\phi_k}$ are the actual probability amplitudes at that time for the different eigenstates k. For the probabilities of finding the different possible values F_k of the quantity F we should then have, in accordance with (66.9),

$$W(F_k) = |a_k|^2 = r_k^2, \tag{84.8}$$

where the phases ϕ_k, whatever they may be, disappear from the expression. It is hence evident that a measurement of F, of any degree of accuracy, would give us no information as to the phases ϕ_k and would be equally well satisfied by any values of those quantities. It would hence indeed be arbitrary to take these phases in any but a random way in constructing an appropriate representative ensemble.

In connexion with the foregoing remarks in justification of the assignment of random phases to states that agree equally well with our approximate knowledge of the condition of the system of interest, it will be well to emphasize the important point, that such states, which agree equally well with our knowledge, will themselves, of course, be eigenstates for the particular kind of quantity which was approximately measured in obtaining that knowledge. Hence the assignment of random phases to these states will lead to the disappearance of non-diagonal terms in the density matrix for the representative ensemble, when the matrix is expressed in the language corresponding to the quantity measured. On transformation to other modes of representation, non-diagonal terms may, of course, be expected in the density matrix, since each of the states, agreeing equally well with our approximate knowledge, would then be expressed in general by a summation over some new kind of eigensolutions each multiplied by a coefficient which would be definitely given as to magnitude and phase. It is also to be noted that the assignment of equal probabilities and random phases is, of course, to be regarded as applying at the time of our approximate measurement of the state of the system of interest, and that as time proceeds the distribution of the representative ensemble will then change in whatever way is demanded by equation (81.20), which gives the quantum analogue of Liouville's theorem. In general this actually leads, as time proceeds, to a more uniform distribution over different states and a less random distribution of phases as we shall see in Chapters XI and XII.

As a further consideration, giving a preliminary partial justification for our postulate as to equal *a priori* probabilities and random *a priori* phases for the quantum mechanical states of a system, it is to be pointed out that this hypothesis would stand in agreement, at the correspondence principle limit, with our previous classical hypothesis as to equal *a priori* probabilities for equal extensions in the phase space. To discuss this we note as a first point, in accordance with the transformation properties of the density matrix for a completely uniform ensemble as given by (83.3), that the prescription of equal *a priori* probabilities and random *a priori* phases for the eigenstates which correspond to any particular quantum mechanical language, would necessitate the same equal *a priori* probabilities for all quantum mechanical states corresponding to any language. We also note, as a second point, that there is a good one-to-one correlation, as we approach the classical limit, between different quantum mechanical states for a system of f degrees of freedom and regions of extension h^f in the phase space which would contain the phase points for classical systems that exhibit properties in substantial agreement with those of the quantum mechanical system in the state considered. Such a correlation can be seen from §§ 62 (*a*) and 63 (*e*) of Chapter VII, which show the possibility of obtaining a quantum mechanical state at the classical limit, which could be described as a wave packet giving a fairly permanent concentration of probability density within a region of extension h^f moving along a classical trajectory; and similar correlations are given in §§ 69, 71, 72, and 73 of Chapter VIII, where we have found —for large quantum numbers, i.e. at the correspondence limit—several examples of the association of eigenstates of energy or of angular momentum, with corresponding regions of extension h^f in the phase space, such that the classical systems thereby defined would have substantially the same values of those quantities. As a consequence of the two points noted above, we now see that our postulate, of equal *a priori* probabilities and random *a priori* phases for different quantum mechanical states, is equivalent at the correspondence principle limit to the assumption of equal *a priori* probabilities for different regions of equal extension h^f in the classical phase space, and hence indeed in agreement with the classical postulate as to equal *a priori* probabilities in general for equal regions, owing to the small size of the extensions h^f from the point of view of practical measurement.

In connexion with the above relation of quantum mechanical probabilities and phases to classical probabilities, it is interesting to note that a partial statement of this relation taken in the reverse sense that equal probabilities for equal regions in the classical phase space ought to imply equal probabilities for the allowed states of the older quantum theory, was often made basic in early attempts to found a satisfactory quantum statistics. It is also of interest to note that we can secure an appropriate symmetry between the statements of the quantum mechanical and classical postulates by saying that the dual quantum mechanical hypothesis, of equal *a priori* probabilities *and* of random *a priori* phases for quantum mechanical states, is equivalent at the correspondence principle limit to the dual classical hypothesis, of equal *a priori* probabilities for equal regions in the classical coordinate space *and* of equal *a priori* probabilities for equal regions in the classical momentum space.

This discussion of the fundamental postulate of the quantum statistical mechanics has been so long that it may be well to conclude the present section by stating this postulate once more in the specific form, that the construction of an appropriate *representative ensemble*, to correspond to the knowledge we have gained as to the condition of some *system of interest* by an *approximate measurement* of some quantity F, is to be obtained by assigning equal probabilities and random phases, at the time of measurement, to those eigensolutions, characteristic of the quantity F, which agree equally well with the approximate knowledge of state given by the measurement.

85. Validity of statistical quantum mechanics

It will be evident from the discussion given in this chapter that statistical methods can be employed in the quantum mechanics for predicting the properties and behaviour of a system in a partially specified quantum mechanical state, in a manner which is essentially similar to that by which they are employed in the classical mechanics for predicting the properties and behaviour of a system in a partially specified classical state. In both cases the procedure consists in setting up a representative ensemble of systems of the same structure as the one of actual interest, and then using the average properties and behaviour of the systems in this ensemble as giving reasonable estimates for quantities pertaining to the actual system of interest. And in both cases the representative ensemble is selected in such a manner as to agree with our partial knowledge as to the precise state of the actual system of interest, and is otherwise constructed in accordance with a reasonable and non-arbitrary postulate as to the *a priori* likelihood of different possibilities.

In changing from the classical to the quantum mechanical application of this procedure, there are, of course, certain differences which must be considered. In the case of systems composed of similar particles we have the new feature that non-accessible states, having symmetry properties now appreciated as not occurring in nature, must be excluded. Furthermore, in the case of dynamical variables having no classical analogue—e.g. spin—the ensembles must now be constructed so as to allow for the states that correspond to the different possible eigenvalues of those quantities. More important, it has to be recognized that our considerations will in general now have a twofold statistical character, on the one hand for the reason that the quantum mechanics itself will in general only provide statistical predictions for systems in precisely specified states, and on the other hand for the reason—essentially involved in the development of what we name statistical mechanics—that we shall actually wish to treat systems in states which are not precisely specified. Moreover, as will be considered more fully in Chapter XI, it is to be noted that an ensemble which has been set up to represent a given quantum mechanical state of interest can be regarded as strictly suitable for this purpose only for the time interval between one measurement and the next, owing to the disturbing effects of observation which we now recognize as not necessarily negligible. Finally, there is, of course, a somewhat striking change in mathematical formalism when we change from classical to quantum mechanics, and this is reflected in the apparatus appropriate for the statistical extensions given to those two disciplines, as is made evident by the change from the classical phase density ρ to the quantum mechanical density matrix ρ_{nm}. Nevertheless, allowing for such changes in subject-matter, it is evident that there is a quite similar logical structure for the general framework within which the classical and quantum statistical methods are applied.

On account of this fundamental similarity between the logical presuppositions for the classical and quantum mechanical developments, it will be possible also in the quantum case to take the point of view as to the validity of statistical considerations which was expressed in § 25. This point of view may be summarized, in the present connexion, as follows. The methods of statistical quantum mechanics are to be regarded as truly statistical in character, giving results that can be expected on the average on repeated experiments with the system in question in the condition defined by its partial specification of state, but not results that can be precisely expected in a single trial. The methods lead to calculated fluctuations around the averages which turn out to be small in the case of typical applications, and in other cases can be compared with empirical findings. The methods have to be based, on account of their statistical character, on some hypothesis as to the *a priori* likelihood of different possibilities; the postulate actually introduced for this purpose is the only non-arbitrary one that can be selected, and agrees at the correspondence principle limit with that selected for the classical statistics. The methods developed do have, as far as is known, the *a posteriori* justification of agreement with experimental findings. It will thus be seen that an entirely similar point of view as to the validity of statistical mechanics is adopted, for the purposes of this book, in the quantum as in the classical case.

In this connexion it is of interest to inquire into the possibility of adopting a different point of view as to the validity of the quantum statistics which would be based on some tenable quantum analogue of the original ergodic hypothesis. In the classical mechanics the introduction of the ergodic hypothesis, that a system in the course of time would pass through every point in the phase space corresponding to classical states compatible with its energy, made it possible to conclude, as we have seen in § 25, that the probability of finding a given system of interest at a random time in any specified state would then be equal to the probability of finding a system picked at random from the corresponding microcanonical ensemble in that state, and hence that the time average for the properties of any system of interest would be equal to the averages of those same properties for the members of the ensemble.

This method of attempting to validate the classical statistics proved unsatisfactory when it became recognized that classical systems, except for the trivial case of a single degree of freedom, do not have the required ergodic character. It might be felt at first sight, however, that quantum mechanical systems would perhaps be found to have the necessary character, owing to the circumstance that quantum mechanical states correspond to finite regions in the phase space instead of merely to points, or to the circumstance that quantum considerations have in general a less deterministic character than classical ones. Nevertheless, in spite of feelings of disappointment with the outcome,† it is found that quantum mechanical systems also do not have the needed ergodic character and that the proposed method of validating statistical mechanics is also unsatisfactory in the quantum case.

To discuss this, let us consider a given system of interest for which we merely specify the energy E_0 with an uncertainty ΔE, related to the time Δt available for its measurement, by the Heisenberg expression

$$\Delta E \Delta t \approx h. \tag{85.1}$$

And let us consider the treatment of this system with the help of an appropriate microcanonical ensemble of systems, having a density matrix which can be described in the language corresponding to the energy eigensolutions $u_n(q)$ for the system by

$$\rho_{nm} = \begin{cases} \rho_0 \delta_{nm} & (E_n \text{ in range } E \text{ to } E+\delta E), \\ 0 & (E_n \text{ not in range } E \text{ to } E+\delta E), \end{cases} \tag{85.2}$$

where we choose the range E to $E+\delta E$ in such a way as to include the specified energy E_0, and assume conditions such that δE can be taken as large compared with the uncertainty ΔE in this energy, and yet at the same time small enough so that the ensemble can be regarded as appropriately representing a system of energy E_0. We must now inquire into the possible existence of some quantum mechanical ergodic principle, which would make the probability at a random time, of finding our system of interest in a specified quantum mechanical state, the

† Compare Schroedinger, *Ann. der Phys.* **83**, 956 (1927).

same as the probability of finding a system picked at random from the above ensemble in that state, or let us say in a practically identical state on account of the range in the possible energies E to $E+\delta E$.

To investigate this we must consider the nature of the eigenfunctions $u_n(q)$, which describe the states of equal probability in the above ensemble, and which can also be used by superposition to describe any possible state for the system of interest. In accordance with their character, as normalized eigensolutions of the Schroedinger differential equation

$$\mathbf{H}u(q) = Eu(q), \tag{85.3}$$

these eigenfunctions for the case of a confined system of f degrees of freedom can be expected to have the form

$$u_n(q) = u_n(q_1 \dots q_f, E, C_2 \dots C_f), \tag{85.4}$$

where the index n is to be regarded as designating some particular set of values for the energy E and for the further $f-1$ parameters $C_2 \dots C_f$ which determine the different possible eigenfunctions. In the case of the confined systems, ordinarily treated in statistical mechanics, in which the component particles are enclosed in walls or held together by their own forces, we can expect that suitable eigenfunctions $u_n(q)$ will be strictly possible only for discrete values of the energy E and the other parameters $C_2 \dots C_f$. Nevertheless, for the situations that now interest us, we may take the energy range E to $E+\delta E$ as lying high enough and the number of degrees of freedom as large enough, so that the eigenvalues for the f quantities E, $C_2 \dots C_f$ can be treated as having nearly continuous spectra. The parameters $C_2 \dots C_f$ are to be regarded as designating the values of physical properties of the system, in addition to its energy, which remain constant with time for a system in a given steady state. By adopting an appropriate set of eigenfunctions, constructed with the help of linear combinations of any original set, we can relate some of these 'constants of the motion' in a simple manner to such well-recognized properties of the system as its components of linear and of angular momentum. In general, however, in the case of a system of many degrees of freedom, most of these $f-1$ constants of the motion will have no simple familiar interpretation, as is also true for most of the constants of the motion that appear in the classical treatment of a complicated system. These remarks as to the nature of the eigenfunctions, that interest us in statistical applications, are well illustrated—except for the extreme simplicity of the system—by the eigenfunctions for a particle in a box, as treated in the preceding chapter in § 71.

With the help of this description of the nature of the eigenfunctions $u_n(q)$, we now see also in the quantum mechanics that there could be no ergodic principle which would have the desired character. On the one hand, in accordance with the description of the microcanonical ensemble by

$$\rho_{nm} = \begin{cases} \rho_0 \delta_{nm} & (E_n \text{ in } E \text{ to } E+\delta E), \\ 0 & (E_n \text{ not in } E \text{ to } E+\delta E), \end{cases} \tag{85.5}$$

we should have equal probabilities of picking a system from the ensemble in any state n for which the energy lies in the range E to $E+\delta E$. On the other hand, in accordance with the possibility of describing any possible state of the actual system of interest by a summation, over states in this energy range, of the form

$$\psi(q,t) = \sum_n c_n u_n(q) e^{-(2\pi i/h)E_n t}, \tag{85.6}$$

we should have the probabilities $|c_n|^2$ for finding the system itself in the different states n. There is no reason, however, which would require equal magnitudes for the coefficients c_n in any particular case; and we have seen above that the different states n, although all corresponding to approximately the same energy E, would be characterized by very different possibilities for the remaining $f-1$ constants of motion $C_2 \ldots C_f$. Hence, also in the quantum mechanics, we cannot regard a microcanonical ensemble as giving a necessarily perfect representation of a particular system of interest, but can only regard it as giving the best representation available in the absence of knowledge of the system except as to energy.

It is interesting to note the similar reasons for the failure of a classical or of a quantum system of f degrees of freedom to have ergodic character. In the classical mechanics the ergodic hypothesis has to be abandoned, as seen in § 25, since there would be $2f-2$ constants of the motion—in addition to energy and position along the trajectory—to which arbitrary values could be assigned. In the quantum mechanics the analogous hypothesis has to be abandoned, as seen above, since there would still be $f-1$ or half as many constants of the motion—in addition to energy—to which arbitrary values could be assigned. In both cases such an assignment could then lead to values of properties of the system of interest, which could not change with time, and which might be very different from the values for such properties predicted as probable from the corresponding microcanonical ensemble.

It will, of course, again be appreciated, in the quantum as in the classical statistics, that the failure to find any exact ergodic principle does not mean that we must usually expect great deviations between the properties of an actual system of interest and the corresponding representative ensemble. In the case of simple constants of the motion, such as the components of linear and angular momentum of the system as a whole, we should presumably regard ourselves as knowing the values of such quantities, and hence should make use of a somewhat more limited ensemble than the general microcanonical one. And in the case of constants of the motion, having a very complicated physical significance, the limitation to particular values would often have but little effect on the distribution of values for the quantities of actual interest. It will, of course, also again be appreciated, as in the classical case, that the assumption of ergodic character for the behaviour of quantum systems would actually prevent the proposed methods from exhibiting their full statistical usefulness in allowing for the occasional occurrence of systems exhibiting quite unusual properties.

It is of interest to realize in this latter connexion, however, that there would be a somewhat reduced possibility in the quantum mechanics for the conceptual construction of systems exhibiting exceptional properties, since the number of constants of motion to which arbitrary values could be assigned is decreased by a half—as already noted above—on changing from the classical to the quantum mechanical treatment of a system of f degrees of freedom. The reason for this decrease lies in the quantum mechanical elimination of the possibility for the simultaneous assignment of precise values to conjugate variables. Thus our classical example (§ 25) of a gas with all its molecules permanently moving parallel to the x-axis between plane walls would not be possible in the quantum mechanics. This would be prevented by the impossibility of simultaneously specifying values for the y and z coordinates of molecules, which would place them on separate tracks, together with zero values for the conjugate momenta p_y and p_z, which would secure a permanent motion in those tracks. Hence, speaking somewhat loosely, we can say that there are only half as many possibilities, in the quantum mechanics as compared with the classical mechanics, for systems to exhibit exceptional behaviour.

We may best summarize this chapter, on the general nature of statistical methods in the quantum mechanics, by saying that the proposed methods provide the only possible non-arbitrary procedure that we can employ when it becomes necessary to treat systems in partially specified quantum mechanical states.

X

THE MAXWELL-BOLTZMANN, EINSTEIN-BOSE, AND FERMI-DIRAC DISTRIBUTIONS

86. The microcanonical ensemble as representing a system in equilibrium

We are now ready to commence the application of the methods of statistical quantum mechanics which were developed in the preceding chapter. As in the classical statistics, we shall usually be interested in applying these methods to idealized models for actual physical-chemical systems of common occurrence such as gases, liquids, solids, or enclosures filled with thermal radiation. These systems can be regarded as composed of large numbers of similar subsystems such as atoms, molecules, electrons, photons, modes of vibration, or other *constituent elements*.

In the present chapter we shall study the equilibrium conditions for such systems, when the molecules or other elements composing the system can be taken as interacting only weakly with each other, so that the energy of the system as a whole can be regarded as practically equal to the sum of the energies of the individual elements treated as independent of one another. As a typical example of such systems we may take a dilute gas, the degree of dilution being great enough so that the constituent molecules can be regarded for the most part as free rather than as engaged in the process of collision. Another example arises in studying the thermal properties of a solid under circumstances such that its thermal energy can be regarded as distributed among the different modes of elastic vibration of which the solid is capable; these modes of vibration then play the role of the weakly interacting elements out of which the system is taken as constructed. A similar example occurs in studying the distribution of radiant energy between the modes of electromagnetic vibration in a hollow enclosure.

The treatment to be given to such situations in the present chapter will be based on the use of the microcanonical ensemble as providing an appropriate representative ensemble for a system in a steady condition of phenomenological equilibrium. The present study will thus be similar to the analogous classical treatment of equilibrium given in Chapter IV. Depending, however, on the quantum mechanical symmetry properties of the elements composing the system, we shall now be led to three different possible types of distribution of such elements among their own individual states, rather than merely to the Maxwell-Boltzmann distribution as in the classical equilibrium of systems composed of weakly interacting elements.

We must now give detailed consideration to the nature of the systems of interest and representative ensembles that will concern us in the present chapter. As in the corresponding classical studies of Chapter IV, we may regard the system of interest as being enclosed in a massive container of volume v, and may assume that the system cannot interchange energy with the walls of the container but can adjust its linear and angular momentum by interaction with these (approximately) stationary walls. Referring for simplicity to axes at rest with respect to the system as a whole, we may then assume that our knowledge of the condition of the system is limited to a specification of its energy E_0, with an uncertainty ΔE which would be connected with the time Δt available for observation by the Heisenberg uncertainty relation

$$\Delta E \Delta t \approx h. \tag{86.1}$$

As a consequence of this partial specification of the state of the system of interest we must evidently resort to the methods of statistical mechanics in order to study the expected properties of the system. For this purpose, in accordance with the discussions of the preceding chapter, we may now introduce, as an appropriate representative ensemble for our system of interest, a microcanonical distribution of systems (§ 83 (*b*)) all having approximately the same energy as that specified for the system itself. As a specific expression for the density matrix, defining this ensemble in the energy language, we may take

$$\rho_{\mu\nu} = \begin{cases} \rho_0 \delta_{\mu\nu} & (E_\mu \text{ in range } E \text{ to } E+\delta E), \\ 0 & (E_\mu \text{ not in range } E \text{ to } E+\delta E), \end{cases} \tag{86.2}$$

where it will now be convenient to use Greek letters for the indices that designate the different energy eigenstates for the system as a whole, where we choose the range E to $E+\delta E$ so as to include the specified energy E_0, and where we assume conditions such that δE can be taken as large compared with the uncertainty ΔE in that energy, and yet at the same time as small enough so that it can be regarded from a phenomenological point of view as an infinitesimal quantity.

In accordance with the known property of statistical equilibrium for microcanonical ensembles, we see—in the quantum mechanics as in the classical mechanics—that a system for which the energy alone is specified must be considered as exhibiting no preferential tendency for change in any particular direction. Hence, from the phenomenological point of view, such a system may be regarded as in a steady state of equilibrium. Furthermore, in accordance with the character of the distribution described by (86.2), we see that the microcanonical ensemble appropriate for such a system would give equal probabilities for each of the different energy eigenstates μ for which the corresponding eigenvalue of energy E_μ would lie in the specified range E to $E+\delta E$.

87. Specification of condition for a system composed of weakly interacting elements

We shall next have to undertake a consideration of the different conditions, which will be of importance for our statistical considerations, in the case of a system composed of weakly interacting elements. In the present quantum treatment, as in the corresponding classical one of Chapter IV, we shall first specify these conditions by stating the distribution of the elements composing the system among the different elementary states which they can themselves assume, and shall then determine the number of states for the system as a whole which would correspond to each such specified condition. To understand these matters it will be desirable to begin with a general presentation of the interrelation between eigensolutions for the system as a whole and eigensolutions for its constituent elements.

(*a*) **Relation of eigensolutions for system to eigensolutions for its component elements.** The different possible energy eigensolutions for the system as a whole will themselves have to be solutions of the *time-free Schroedinger equation* for the system, which was first discussed in § 60. For our present purposes it will be convenient to write this equation with the slightly altered notation

$$\mathbf{H}U_\mu(q) = E_\mu\, U_\mu(q), \tag{87.1}$$

where we now use a capital U with a Greek index to designate an eigensolution for the system as a whole, and reserve the small u with Latin indices for later use to designate eigensolutions for a single molecule or other kind of element out of which the system is composed. The symbol $\mathbf{H}$ in the above equation is the Hamiltonian operator for the system expressed in coordinate language, and E_μ is an allowed eigenvalue for the energy of the system, which will assure the appropriate character as an eigensolution for the function of the coordinates $U_\mu(q)$, where the single letter q is to be regarded as symbolizing the total collection of coordinates for all the molecules or other elements composing the system.

Let us now consider the circumstance that our present interest lies in systems composed of *weakly interacting* molecules or other *constituent elements*. If there are n such elements in all, it will then be possible to express the Hamiltonian operator **H**, corresponding to the energy of the whole system, in the form

$$\mathbf{H} = \mathbf{H}_1+\mathbf{H}_2+\mathbf{H}_3+...+\mathbf{H}_n+\mathbf{V}, \tag{87.2}$$

where $\mathbf{H}_1 ... \mathbf{H}_n$ are the Hamiltonian operators that correspond separately to the energies of the n constituent elements, taken as not acting on each other but as acted on by any general potential field—such as that of the walls—which may be present, and **V** is the operator that corresponds to the remaining energy, arising from the actual presence of interaction. If this interaction is sufficiently weak, however, we can—for our present purposes—neglect the term **V** in comparison with the others and write

$$\mathbf{H} = \mathbf{H}_1+\mathbf{H}_2+\mathbf{H}_3+...+\mathbf{H}_n \tag{87.3}$$

as a suitable approximation. Substituting (87.3) into (87.1), and using the letters $q_1 ... q_n$ to symbolize the collection of coordinates for each of the n separate elements, we then obtain

$$(\mathbf{H}_1+\mathbf{H}_2+...+\mathbf{H}_n)U_\mu(q_1, q_2,..., q_n) = E_\mu\, U_\mu(q_1, q_2,..., q_n) \tag{87.4}$$

as a new expression for the equation that must be satisfied by the different eigensolutions $U_\mu(q_1, q_2,..., q_n)$ for the system as a whole. It will be appreciated as we proceed that we are thus assuming sufficiently weak interaction so that these unperturbed energy eigenstates can be taken as substantially the same as the true energy eigenstates that would be used in constructing a microcanonical ensemble for the representation of equilibrium.

The above equation may now be compared with the equations which would determine the eigensolutions—say $u_k(q_1)$, $u_l(q_2)$,..., $u_s(q_n)$—for the n separate elements which compose the system. These equations will evidently be of the forms

$$\begin{aligned} \mathbf{H}_1\, u_k(q_1) &= E_k\, u_k(q_1), \\ \mathbf{H}_2\, u_l(q_2) &= E_l\, u_l(q_2), \\ &\cdots\cdots \\ &\cdots\cdots \\ \mathbf{H}_s\, u_s(q_n) &= E_s\, u_s(q_n), \end{aligned} \tag{87.5}$$

where $\mathbf{H}_1$, $\mathbf{H}_2$,..., $\mathbf{H}_n$ are the Hamiltonian operators for the n separate elements, expressed respectively in terms of the coordinates q_1, q_2,..., q_n for the elements, and E_k, E_l,..., E_s are possible eigenvalues of the energy respectively for the n elements. Comparing equation (87.4) with equations (87.5), it then becomes easy to express any eigenfunction $U_\mu(q_1, q_2,..., q_n)$ for the system as a whole in the terms of eigenfunctions $u_k(q_1)$, $u_l(q_2)$,..., $u_s(q_n)$ for its separate elements, provided we pay due attention to the symmetry restrictions that must be imposed on the eigensolutions for systems containing indistinguishable particles, as already discussed in § 76 (*d*).

In the case of a system composed of n distinguishable elements, *no symmetry restrictions* will have to be imposed, and it is evident that the *energy eigensolutions* for the system will then be of the general form

$$U_\mu(q_1, q_2,..., q_n) = u_k(q_1)u_l(q_2)... u_s(q_n), \tag{87.6}$$

where we must regard any change in the specified elements nos. 1, 2, ..., n assigned to the elementary eigenstates k, l,..., s, or any change in the selection of eigenstates, as leading to a new eigensolution μ for the system as a whole. The corresponding *eigenvalues of energy* will be given by

$$E_\mu = E_k+E_l+...+E_s. \tag{87.7}$$

In the case of a system composed of n indistinguishable particles of a character requiring *symmetric solutions*, the *energy eigensolutions* for the system will, with suitable normalization, evidently be of the general form

$$U_\mu(q_1, q_2,..., q_n) = \sum_{\mathbf{P}} \mathbf{P}u_k(q_1)u_l(q_2)... u_s(q_n), \tag{87.8}$$

where we take a sum over all permutations $\mathbf{P}$ of the particle indices 1, 2,..., n, and must regard each different selection of elementary eigensolutions k, l,..., s as leading to a different eigensolution μ for the system as a whole. The corresponding *eigenvalues of energy* will again be expressed by

$$E_\mu = E_k+E_l+...+E_s. \tag{87.9}$$

In the case of a system composed of n indistinguishable particles of a character requiring *antisymmetric solutions*, the *energy eigensolutions* for the system will, again with suitable normalization, evidently be of the general form

$$U_\mu(q_1, q_2,..., q_n) = \sum_{\mathbf{P}} \mp\mathbf{P}u_k(q_1)u_l(q_2)... u_s(q_n), \tag{87.10}$$

where we now take a sum over all permutations $\mathbf{P}$ of the particle indices 1, 2,..., n using the negative or positive sign according as the permutation is odd or even, and must regard each possible selection of elementary eigensolutions k, l,..., s as leading to a different eigensolution

μ for the system as a whole. The corresponding *eigenvalues of energy* will once more be expressed by

$$E_\mu = E_k + E_l + ... + E_s. \tag{87.11}$$

We thus obtain three typical examples of the different possible ways in which the eigensolutions for a system as a whole can depend on the eigensolutions for its constituent elements. The first of these examples (87.6) would apply in the case of a system of *n distinguishable elements*, such as spatially located and oriented modes of oscillation, and will lead to a type of statistical result which can be appropriately designated as *Maxwell-Boltzmann*. The second example (87.8) applies in the case of a system of *n indistinguishable particles subject to symmetric restrictions*, such as nuclei and atoms composed of an even number of fundamental particles, or photons treated as particles, and will lead to the so-called *Einstein-Bose* type of statistical result. The third example (87.10) applies in the case of a system of *n indistinguishable particles subject to antisymmetric restrictions*, such as the fundamental material particles, electrons, protons, and neutrons, or nuclei and atoms composed of an odd number of such particles, and will lead to the so-called *Fermi-Dirac* type of statistical result.

More complicated cases of systems composed of more than a single type of element can evidently be handled by simple extensions of the above treatment.

(*b*) **Method of specifying different conditions.** We may now give detailed attention to the method of specifying the conditions of such systems which will be of importance for our statistical considerations. These specifications will depend on the nature of the energy spectra for the individual elements composing the system.

We shall assume, for the time being for simplicity, that our system of interest will be composed of n constituent elements of only a single kind, and shall take such elements as having a known *spectrum of energy eigenvalues*, the same for each individual element without reference to the possibility or impossibility of maintaining a distinction between them. Thus our n elements could be n oscillators, all having the same energy spectrum because of the same intrinsic frequency ν, but distinguishable from each other by spatial location or orientation. Or the n elements could be n particles, all having the same energy spectrum because of the same mass m and spin s, but not permanently distinguishable from each other on account of their free motion inside a common container.

To obtain an appropriate approximate description of the energy spectrum for a single constituent element in the potential field applying to the system as a whole, we shall regard the total possible range in energy ϵ as divided up into a succession of small ranges ϵ to $\epsilon+\Delta\epsilon$, each range being identified by an index κ and being of a width to correspond to the approximate accuracy with which we shall later wish to specify different conditions of the system. We shall then describe the spectrum by stating the *number of energy eigensolutions* g_κ which fall in each such *range* $\Delta\epsilon_\kappa$. In most cases the energy spectrum will be nearly enough continuous so that this description can be regarded as given by an expression of the form

$$g_\kappa = f(\epsilon_\kappa)\Delta\epsilon_\kappa, \tag{87.12}$$

where $f(\epsilon_\kappa)$ is a continuous function of the energy ϵ_κ that locates the range. As an example of such expressions we have our previous finding for particles as given by (71.16).

With the help of such a description of the energy spectrum for a single constituent element we shall then *specify the different conditions* of the system as a whole that concern us by stating the *number of constituent elements* n_κ, which lie in each of the *groups* of g_κ eigenstates, that correspond to our succession of *ranges* in energy $\Delta\epsilon_\kappa$.

It will be noted, in the quantum treatment as in the analogous classical treatment, that our specifications of *condition* will not in general be sufficient to specify a *precise state* of the system, this being, of course, a characteristic feature of statistical mechanical considerations. It will also be noted that our method of specifying different conditions does not itself have to make any direct reference to the three different types of relation which may actually be found for the dependence of the eigensolutions for the system on the eigensolutions for its elements.

(*c*) **Number of eigenstates corresponding to a specified condition of the system.** We are now finally ready to consider the number of eigenstates for the system as a whole that would correspond to any particular condition of the system as specified by the method described above. The form of expression, describing the dependence of the number of eigenstates G for the system on the number of elements n_κ in each group of g_κ elementary eigenstates, will be different for different types of relation between eigensolutions for the system and eigensolutions for its elements. We shall hence give separate treatment to the three types of such relation described in the first part of this section.

In the *Maxwell-Boltzmann* case, arising when the system is composed of n distinguishable elements, it is evident, from the relation between the two kinds of eigenstates given by (87.6), that each particular assignment of the n elements to their possible elementary eigenstates would correspond to a different eigenstate for the system as a whole. To determine the consequences of this we note that our total collection of n elements could evidently be divided into quotas containing n_1, $n_2,\ldots, n_\kappa,\ldots$ members in

$$\frac{n!}{n_1!\,n_2!\ldots n_\kappa!\ldots}$$

different ways, and that the n_κ elements in any such quota could be assigned to g_κ different eigenstates in $g_\kappa^{n_\kappa}$ different ways. Hence we can at once take

$$G = \prod_\kappa \frac{n!}{n_\kappa!} g_\kappa^{n_\kappa}, \tag{87.13}$$

in the Maxwell-Boltzmann case, as the desired expression for the number of eigenstates G that would correspond to a condition of the system specified by the n_κ in the manner that we have described.

In the *Einstein-Bose* case, arising when the system is composed of n indistinguishable particles and the eigensolutions must be symmetric, it is evident, from the relation between the two kinds of eigenstates given by (87.8), that each possible way of selecting occupied states for our n particles would correspond to a different eigenstate for the system as a whole, and that there is no restriction on taking a given eigenstate as occupied by more than a single particle. To determine the consequences of this, let us fix our attention on the n_κ particles assigned to a specified group of g_κ states, and let us consider the possibilities for making permutations in a linear array of $n_\kappa+g_\kappa-1$ objects, which can be regarded as the n_κ particles together with the $g_\kappa-1$ partitions which would be sufficient to separate the length available for the array into g_κ cells. Since the total number of permutations for the $n_\kappa+g_\kappa-1$ objects would be equal to $(n_\kappa+g_\kappa-1)!$, and since the permutations of the particles among themselves, or of the partitions among themselves, would not be significant, we should then have a total of

$$\frac{(n_\kappa+g_\kappa-1)!}{n_\kappa!\,(g_\kappa-1)!}$$

different ways of selecting occupied states from the group under consideration. Hence we can now evidently take

$$G = \prod_\kappa \frac{(n_\kappa+g_\kappa-1)!}{n_\kappa!\,(g_\kappa-1)!} \tag{87.14}$$

in the Einstein-Bose case, as the desired expression for the number of eigenstates G that would correspond to a condition of the system specified by the n_κ.

Finally, in the *Fermi-Dirac* case, arising when the system is composed of n indistinguishable particles and the eigensolutions must be antisymmetric, it is evident, from the relation between the two kinds of eigenstates given by (87.10), that each possible way of selecting occupied states for our n particles would correspond to a different eigenstate for the system as a whole, but that it would now be impossible to have more than a single particle in any individual state, since—in agreement with our previous discussion of the Pauli exclusion principle—we see from the form of (87.10) that this would make the corresponding eigensolution U_μ equal to zero. As a consequence it is evident that the number of particles n_κ assigned to any group of elementary eigenstates cannot now be greater than the number of states g_κ in the group. If the n_κ particles were distinguishable from one another they could evidently be assigned in

$$g_\kappa(g_\kappa-1)(g_\kappa-2)\dots(g_\kappa-n_\kappa+1) = \frac{g_\kappa!}{(g_\kappa-n_\kappa)!}$$

different ways without putting more than a single particle into a given elementary state. Hence, making due allowance for their actual indistinguishability, we can now take

$$G = \prod_\kappa \frac{g_\kappa!}{n_\kappa!\,(g_\kappa-n_\kappa)!}, \tag{87.15}$$

in the Fermi-Dirac case, as the desired expression for the number of eigenstates G that would correspond to a condition of the system specified by the n_κ.

This must now complete our necessarily somewhat complicated discussion of the specification of condition for quantum mechanical systems composed of weakly interacting elements.

88. The probabilities for different conditions of the system

We may now combine the results of the two preceding sections. In the first of these sections, § 86, it has been shown that a quantum mechanical system, in a steady state of phenomenological equilibrium with a specified energy, can be appropriately represented by a microcanonical ensemble, and that such an ensemble gives equal probabilities for each of the eigenstates μ for which the eigenvalue of energy E_μ

would fall in a narrow range E to $E+\delta E$ selected as including the energy specified for the system. In the second of these sections, § 87, we have then obtained—for the three possible types of systems composed of n weakly interacting elements—the expressions (87.13), (87.14), and (87.15) for the number G of such eigenstates μ that would correspond to a condition of the system specified by the numbers of elements n_κ assigned to the different groups of g_κ elementary eigenstates that interest us. Combining these results, we may now write the following expressions for the probabilities P of finding our system in a condition specified by the n_κ: in the *Maxwell-Boltzmann* case

$$P = \text{const.} \prod_\kappa \frac{n!}{n_\kappa!} g_\kappa^{n_\kappa}, \tag{88.1}$$

in the *Einstein-Bose* case

$$P = \text{const.} \prod_\kappa \frac{(n_\kappa+g_\kappa-1)!}{n_\kappa!\,(g_\kappa-1)!}, \tag{88.2}$$

and in the *Fermi-Dirac* case

$$P = \text{const.} \prod_\kappa \frac{g_\kappa!}{n_\kappa!\,(g_\kappa-n_\kappa)!}, \tag{88.3}$$

where the factor denoted by 'const.' would in every case have a value independent of the particular assignment n_κ considered. The expressions apply, of course, to conditions such that the energy of the system would lie in the selected range E to $E+\delta E$, the probabilities for other conditions being zero.

For our further purposes it will be more convenient to take the logarithms of these probabilities. Assuming for the time being that the numbers n, n_κ, g_κ, and also $g_\kappa-n_\kappa$ in the Fermi-Dirac case, are all large compared with unity, we then obtain, with the help of the Stirling approximation for factorials (28.3), the approximate results, in the *Maxwell-Boltzmann* case

$$\log P = n\log n + \sum_\kappa \{n_\kappa \log g_\kappa - n_\kappa \log n_\kappa\} + \text{const.}; \tag{88.4}$$

in the *Einstein-Bose* case

$$\log P = \sum_\kappa \{(n_\kappa+g_\kappa)\log(n_\kappa+g_\kappa) - n_\kappa \log n_\kappa - g_\kappa \log g_\kappa\} + \text{const.}; \tag{88.5}$$

and in the *Fermi-Dirac* case

$$\log P = \sum_\kappa \{(n_\kappa-g_\kappa)\log(g_\kappa-n_\kappa) - n_\kappa \log n_\kappa + g_\kappa \log g_\kappa\} + \text{const.} \tag{88.6}$$

It will be noted, except for the term $n\log n$ in (88.4) which can be included in the constant term when it is not necessary to consider variations in the number of elements n, that the three expressions approach identical form with $g_\kappa \gg n_\kappa$, that is, with many states g_κ available in the different energy ranges $\Delta\epsilon_\kappa$ for the n_κ elements assigned to them. As we shall see later, this will be true under conditions of high temperature and great dilution where the three types of quantum mechanical statistical results come into correspondence principle agreement with the classical Maxwell-Boltzmann results.

89. Condition of maximum probability. The three distribution laws

With the help of the foregoing expressions for the probabilities of different conditions we can now determine the most probable condition, for a system of the kind that we are considering, by examining the effect of varying the assignment of elements n_κ to the different groups of g_κ states, keeping the total number of elements n constant (except in the special case of photons), and keeping the total energy E constant since all our conditions must correspond to an energy restricted to a narrow range E to $E+\delta E$. Making use of (88.4–6), this is seen to lead to the following variational equations for determining the desired conditional maximum of $\log P$ and hence of P itself, where the three equations in the first group apply respectively to the Maxwell-Boltzmann, Einstein-Bose, and Fermi-Dirac cases:

$$-\delta \log P = 0 = \begin{cases} \sum\limits_\kappa \{\log n_\kappa - \log g_\kappa + 1\}\,\delta n_\kappa, \\ \sum\limits_\kappa \{\log n_\kappa - \log(n_\kappa + g_\kappa)\}\,\delta n_\kappa, \\ \sum\limits_\kappa \{\log n_\kappa - \log(g_\kappa - n_\kappa)\}\,\delta n_\kappa, \end{cases} \tag{89.1}$$

$$\delta n = \sum_\kappa \delta n_\kappa = 0, \tag{89.2}$$

$$\delta E = \sum_\kappa \epsilon_\kappa\,\delta n_\kappa = 0. \tag{89.3}$$

In setting up the last of these equations we assume the ordinary case of a nearly continuous energy spectrum for the elements composing the system. This makes it possible to have a large number of states g_κ, as assumed above, in an energy range ϵ_κ to $\epsilon_\kappa+\Delta\epsilon_\kappa$ which is narrow enough to be treated as differential. Combining the above equations for the conditional maximum of $\log P$, using the Lagrange method of undetermined multipliers, we then obtain for the three cases respectively

$$\sum_\kappa \left\{\log\frac{n_\kappa}{g_\kappa}+\alpha+\beta\epsilon_\kappa\right\}\delta n_\kappa = 0,$$

$$\sum_\kappa \left\{\log\frac{n_\kappa}{n_\kappa+g_\kappa}+\alpha+\beta\epsilon_\kappa\right\}\delta n_\kappa = 0, \tag{89.4}$$

$$\sum_\kappa \left\{\log\frac{n_\kappa}{g_\kappa-n_\kappa}+\alpha+\beta\epsilon_\kappa\right\}\delta n_\kappa = 0,$$

where α and β are the undetermined constants.

Since the variations δn_κ can now be treated as arbitrary, these equations can only be satisfied when the coefficients of the δn_κ are individually equal to zero, and solving for the n_κ this then leads for the three cases respectively to the desired expressions: in the *Maxwell-Boltzmann* case

$$n_\kappa = \frac{g_\kappa}{e^{\alpha+\beta\epsilon_\kappa}}, \tag{89.5}$$

in the *Einstein-Bose* case

$$n_\kappa = \frac{g_\kappa}{e^{\alpha+\beta\epsilon_\kappa}-1}, \tag{89.6}$$

and in the *Fermi-Dirac* case

$$n_\kappa = \frac{g_\kappa}{e^{\alpha+\beta\epsilon_\kappa}+1}. \tag{89.7}$$

Depending on the type of quantum mechanical system considered, we thus obtain the three different distribution laws which describe the equilibrium distribution of the constituent elements of a system over their own individual states. When convenient, the three distribution laws may be expressed in the combined form

$$n_\kappa = \frac{g_\kappa}{e^{\alpha+\beta\epsilon_\kappa}\pm 1(0)}, \tag{89.8}$$

where we take 0, -1, or $+1$ according as we are interested in the Maxwell-Boltzmann, Einstein-Bose, or Fermi-Dirac case.

The derivations which we have given for these relations have been obtained with the help of the Stirling approximation for factorials, and this may be regarded to some extent as a defect, especially in the Fermi-Dirac case where we have had to assume $g_\kappa-n_\kappa$ large compared with unity and yet shall actually wish to make applications to cases (conduction electrons) where the number of states g_κ and number of particles n_κ would be nearly equal. Nevertheless, we shall be able to show later, in §§ 113 and 114, that these approximate expressions for the most probable numbers of elements in different states in a microcanonical ensemble can be taken as the exact expressions for the mean numbers of elements in different states in a canonical or in a grand canonical ensemble.

In developing the derivations we have assumed, in accordance with (89.2), that the total number of elements n was to be kept constant in carrying out the variations which determine the most probable condition. In our later application of the Einstein-Bose result to photons treated as particles (see § 93 (a)) we shall wish to remove this restriction, since there is no conservation law for the number of photons in a system. It will be seen from the method by which equations (89.4) have been obtained that this can be accomplished by setting the undetermined constant α equal to zero in the final result (89.6).

It will be noted, in agreement with the remarks made at the end of § 88, that all three distribution laws approach the same form when $g_\kappa \gg n_\kappa$, since $e^{\alpha+\beta\epsilon_\kappa}$ would then be large compared with unity.

90. Distribution in systems containing constituent elements of more than one kind

For simplicity the foregoing treatments of the Maxwell-Boltzmann, Einstein-Bose, and Fermi-Dirac distributions were carried out assuming systems composed of only a single kind of constituent element. It is immediately evident, however, that similar methods could be applied to systems composed of large numbers of different kinds of weakly interacting elements, say n of one kind, m of another kind, and so forth.

In such a case any condition of the system of interest could be specified by giving the numbers of elements n_κ, m_λ, etc., that fall respectively in the different groups of elementary eigenstates g_κ, g_λ, etc., that could be set up for each kind of constituent element. The probability P for any such a condition would now evidently be given by a product

$$P = P_1 P_2 \ldots, \tag{90.1}$$

where the factors P_1, P_2, etc., would be respectively associated with the elements of the different kinds n, m, etc., in number, and would have one or another of the forms (88.1–3), according to the particular symmetry character for the kind of element involved. And the condition of maximum probability would now be determined by a system of simultaneous variational equations of the form

$$-\delta \log P = -\sum_\kappa \frac{\partial \log P_1}{\partial n_\kappa}\,\delta n_\kappa - \sum_\lambda \frac{\partial \log P_2}{\partial m_\lambda}\,\delta m_\lambda - \ldots = 0, \tag{90.2}$$

$$\delta n = \sum_{\kappa} \delta n_{\kappa} = 0,$$
$$\delta m = \sum_{\lambda} \delta m_{\lambda} = 0, \tag{90.3}$$
$$\cdots\cdots\cdots$$
$$\cdots\cdots\cdots$$

$$\delta E = \sum_{\kappa} \epsilon_{\kappa} \delta n_{\kappa} + \sum_{\lambda} \epsilon_{\lambda} \delta m_{\lambda} + \ldots = 0. \tag{90.4}$$

Combining these equations by the method of undetermined multipliers, we should then obtain an expression of the form

$$\sum_{\kappa}\left(-\frac{\partial \log P_1}{\partial n_{\kappa}} + \alpha_n + \beta\epsilon_{\kappa}\right)\delta n_{\kappa} + \sum_{\lambda}\left(-\frac{\partial \log P_2}{\partial m_{\lambda}} + \alpha_m + \beta\epsilon_{\lambda}\right)\delta m_{\lambda} + \ldots = 0, \tag{90.5}$$

where α_n, α_m, etc., and β would now be undetermined multipliers. Since the variations δn_{κ}, δm_{λ}, etc., could then be treated as independent, this would necessitate the value zero for the coefficients of δn_{κ}, δm_{λ}, etc., in the above equation.

Hence, comparing (90.5) with the earlier equations (89.4) appearing at the analogous stage of our previous treatment of equilibrium in systems containing only a single kind of constituent element, it is now seen that we shall be led in the general case for each kind of constituent element to a distribution law of the general form

$$n_{\kappa} = \frac{g_{\kappa}}{e^{\alpha+\beta\epsilon_{\kappa}} \pm 1(0)}, \tag{90.6}$$

where we take the Maxwell-Boltzmann, Einstein-Bose, or Fermi-Dirac specific form, depending on the character of the elements considered. It will also be noted, as can be concluded from an examination of the steps leading to (90.5), that the constant β in expressions of the form (90.6) would have the same value for all the different kinds of constituent element out of which our total system is constructed. This character of the constant β, as a parameter applying at equilibrium to all constituents of the system, is of importance in connexion with the relation of β to the temperature T of the system, which will be obtained in the next section.

91. Evaluation of constants in the distribution laws

We must now consider the values of the two constants α and β, which were introduced as arbitrary multipliers in the course of the foregoing derivations, and which still appear in the final expressions for the three forms of distribution law

$$n_\kappa = \frac{g_\kappa}{e^{\alpha+\beta\epsilon_\kappa}\pm 1(0)}. \tag{91.1}$$

These constants are to be regarded as adjustable parameters, whose values depend on the number and kind of elements composing the system, and on the equilibrium condition considered.

(*a*) **Value of constant** α. A formal expression for evaluating the constant α, for any assigned value of β, can be immediately obtained by summing (91.1) over all the different groups of g_κ elementary eigenstates which have been set up. In this way we must obtain the total number of elements n in the system, and hence can write the equality

$$n = \sum_\kappa \frac{g_\kappa}{e^{\alpha+\beta\epsilon_\kappa}\pm 1(0)}. \tag{91.2}$$

In principle this equation could then be solved for α, assuming β known, and assuming the information as to the character of the energy spectrum of the elements needed for the determination of the g_κ and ϵ_κ.

In the Maxwell-Boltzmann case, or in the other two cases under conditions such that the added term ± 1 in the denominator becomes negligible, we can at once obtain from equations (91.2) the explicit solution

$$e^\alpha = \frac{1}{n}\sum_\kappa g_\kappa e^{-\beta\epsilon_\kappa}, \tag{91.3}$$

which is the quantum analogue of our previous classical equation (32.3) for evaluating the equivalent constant. In general, however, in the Einstein-Bose and Fermi-Dirac cases, no simple explicit solutions for the constant α are available, and special methods of treatment have to be devised as will be discussed later.

In any case it will be noted that α, as was true for the equivalent classical constant, depends not only on the structure of the system but also on the particular value assigned to β, and hence as we shall see on the temperature of the system.

(*b*) **Value of constant** β. In order to proceed to an understanding of the other constant β, it will now be profitable, as in the analogous classical development of § 32 (*b*), to take a portion of our system as consisting of a dilute monatomic gas, composed of n particles, of mass m, and let us say for specificity particles not having spin, all enclosed in a container of volume v. Our system is thus provided with an adjunct, which can be made to serve the functions of a classical perfect gas thermometer, and to lead to an interpretation of β.

Making use of the result (71.16) already calculated for the number of eigensolutions for a particle in a container, we may write

$$g_\kappa = \frac{4\pi v}{h^3} m\sqrt{(2m\epsilon_\kappa)}\,\Delta\epsilon_\kappa \tag{91.4}$$

as an expression for the number of elementary eigenstates g_κ that would be available to the particles, composing our gas, in any small energy range $\Delta\epsilon_\kappa$. Substituting this into the distribution law (91.1), and changing for convenience to a differential form of expression, we can then write

$$dn = \frac{4\pi v}{h^3} m\sqrt{(2m\epsilon)}\frac{d\epsilon}{e^{\alpha+\beta\epsilon}\pm 1} \tag{91.5}$$

for the number of particles in our thermometer which would have energies in the range $d\epsilon$, where we are to take the positive or negative sign in the denominator depending on the symmetry character of the particles. Integrating over all possible values of the energy, we also obtain an expression of the form

$$n = \frac{4\pi v}{h^3} m\sqrt{(2m)} \int\limits_0^\infty \frac{\epsilon^{\frac{1}{2}}\,d\epsilon}{e^{\alpha+\beta\epsilon}\pm 1} \tag{91.6}$$

for the total number of particles of the gas.

We are now in a position to understand the possibility of using such a container full of gas as a classical perfect gas thermometer providing an evaluation of β. For this purpose, with any given number of particles n, we have only to choose the volume of the container v and the mass of the particles m large enough. This will have the following consequences. In the first place, by taking the mass of the particles m sufficiently large, it will be possible—in accordance with the correspondence principle—to regard the particles as obeying the classical mechanics. In the second place, by taking the volume v sufficiently large, it will then be possible to regard the gas as perfect and hence as satisfying the consequences derived by the classical statistical mechanics for perfect gases. In the third place, by taking v and m sufficiently large, it will be seen from the form of (91.6), for any given number of particles n, that the parameter e^α can be made so great as to be very large compared with unity; this will then permit us to rewrite the distribution law (91.5) in the form

$$dn = \frac{4\pi v e^{-\alpha}}{h^3} m\sqrt{(2m\epsilon)}e^{-\beta\epsilon}\,d\epsilon = \text{const.}\, e^{-\beta\epsilon}\epsilon^{\frac{1}{2}}\,d\epsilon, \tag{91.7}$$

which now agrees with our previous classical expression (31.3) applied to the distribution of the particles of a perfect gas. In the classical case, however, we have already shown in § 32 (*b*) that the parameter β in such a distribution expression would be connected with the temperature T of the perfect gas by the relation $\beta = 1/kT$, where k is Boltzmann's constant. Hence, since the quantity β introduced in the quantum treatment was shown in § 90 to have the same value for all constituents of the system, and has now been shown equal to the classical β for a constituent of the system so chosen as to obey the classical statistics, we can now write, also in the quantum statistics,

$$\beta = \frac{1}{kT} \tag{91.8}$$

as a connexion between the statistical mechanical parameter β and the phenomenological temperature of the system T as measured on a perfect gas thermometer.

As in the classical statistics, we may regard this interpretation of β in terms of T as providing increased insight into the character of the statistical results that we obtain. It may also again be remarked that we shall later present in §§ 122 and 131 what may seem to be a more fundamental method of introducing the thermodynamic notion of temperature into statistical mechanical considerations.

92. Maxwell-Boltzmann systems

Although it is the primary purpose of this book to consider the principles of statistical mechanics rather than their applications, we may now conclude the present chapter with sections giving a partial account of the methods employed, and results obtained, in the three cases of systems exhibiting the Maxwell-Boltzmann, Einstein-Bose, and Fermi-Dirac types of statistical result. We shall begin by considering systems composed of weakly interacting elements that are permanently distinguishable from each other, and hence systems having eigensolutions not subject to symmetry restrictions and having the Maxwell-Boltzmann type of equilibrium distribution.

(*a*) **Mean energy of oscillators of frequency ν.** The Maxwell-Boltzmann systems, of major interest from the point of view of applications, may be regarded as consisting of sets of harmonic oscillators all having the same frequency ν, but distinguishable from each other by permanent spatial location or orientation. It will be desirable to obtain an expression for the mean energy of such a set of oscillators under equilibrium conditions.

In accordance with the quantum mechanical properties of such oscillators, as illustrated by the treatment given in § 72 to a particle in a Hooke's law field of force, we can take the energy spectrum for our oscillators as expressed by

$$\epsilon_\kappa = (\kappa+\tfrac{1}{2})h\nu, \tag{92.1}$$

with $\kappa = 0, 1, 2, 3, \ldots.$

And hence for the number of eigenstates in the energy range

$$\Delta\epsilon_\kappa = h\nu\,\Delta\kappa,$$

we can take

$$g_\kappa = \Delta\kappa. \tag{92.2}$$

Substituting in the Maxwell-Boltzmann form of the distribution law (89.5),

$$n_\kappa = g_\kappa e^{-\alpha-\beta\epsilon_\kappa},$$

we can then write

$$n_\kappa = e^{-\alpha-\beta\epsilon_\kappa}\Delta\kappa = e^{-\alpha-\beta(\kappa+\frac{1}{2})h\nu}\Delta\kappa, \tag{92.3}$$

as an expression for the number of oscillators at equilibrium which would have values of the quantum number κ falling in the range $\Delta\kappa$.

With the help of this distribution law we can then evidently calculate the mean energy of such oscillators by the expression

$$\bar{\epsilon} = \frac{\sum\limits_{\kappa=0}^{\infty} e^{-\alpha-\beta(\kappa+\frac{1}{2})h\nu}(\kappa+\frac{1}{2})h\nu\,\Delta\kappa}{\sum\limits_{\kappa=0}^{\infty} e^{-\alpha-\beta(\kappa+\frac{1}{2})h\nu}\,\Delta\kappa}, \tag{92.4}$$

where we now use a single bar to denote a mean value for the oscillators composing the system of interest. To evaluate this expression we may choose equal values for the different ranges $\Delta\kappa$, and then cancel the factors $\Delta\kappa$, $e^{-\alpha}$, and $e^{-\beta h\nu/2}$ from all terms in the numerator and denominator. This then leads to

$$\begin{aligned}\bar{\epsilon} &= h\nu\frac{\sum\limits_{\kappa=0}^{\infty} e^{-\kappa\beta h\nu}\kappa}{\sum\limits_{\kappa=0}^{\infty} e^{-\kappa\beta h\nu}} + \frac{h\nu}{2} \\ &= h\nu\,\frac{e^{-\beta h\nu}+2e^{-2\beta h\nu}+3e^{-3\beta h\nu}+\ldots}{1+e^{-\beta h\nu}+e^{-2\beta h\nu}+e^{-3\beta h\nu}+\ldots} + \frac{h\nu}{2} \\ &= \frac{h\nu}{e^{\beta h\nu}-1} + \frac{h\nu}{2},\end{aligned}$$

or finally to

$$\bar{\epsilon} = \frac{h\nu}{e^{h\nu/kT}-1} + \frac{h\nu}{2}, \tag{92.5}$$

where the next to the last form is obtained by performing the indicated division, and the last form is obtained by substituting for β its known expression (91.8) in terms of the temperature T of the system. We thus obtain the desired expression in the quantum statistical mechanics for the mean energy $\bar{\epsilon}$ to be ascribed to oscillators of frequency ν in a system which has come to equilibrium at temperature T.

(*b*) **Application to the specific heat of solids.** The use of the above result in obtaining an understanding of the specific heat of solids has been very important. Empirically, the heat capacity of a crystal composed of n atoms of an elementary substance is known to drop from the classical value

$$C_v = 3nk, \tag{92.6}$$

as given by Dulong and Petit, towards zero as we go to sufficiently low temperatures.

The general nature of the explanation for this phenomenon was first given by Einstein,† who appreciated that the $3n$ oscillators corresponding to the n atoms of a crystalline solid should be assigned the mean energy which they would have on the basis of the quantum theory,‡ rather than the classical value $3nkT$. In carrying this out he assumed somewhat too schematically that the same frequency ν could be used for all $3n$ oscillators, but obtained nevertheless an expression for heat capacity which had the qualitatively correct behaviour of dropping from the classical value $3nk$ to zero.

The more complete theory for the specific heats of crystalline solids given by Debye|| regards the thermal energy of a solid as distributed among the various modes of elastic vibration of different frequencies of which the solid is capable. We shall give further consideration to this theory in Chapter XIV, when we come to a consideration of the thermodynamic properties of crystals. See § 137.

(*c*) **Application to radiation.** The foregoing result, as to the mean energy of oscillators at equilibrium, also has an important application in connexion with the equilibrium distribution of radiation in a hollow enclosure which has come to thermal equilibrium. For this purpose we may regard the radiant energy in such an enclosure as resident in the various modes of electromagnetic vibration which the enclosure would present.

For the number of such modes of vibration inside a hollow enclosure

† Einstein, *Ann. der Phys.* **22**, 180 (1907).

‡ For this purpose Einstein used only the first term of (92.5), since the term $h\nu/2$ was not given by the older quantum theory. This makes no difference, however, after differentiation by T to obtain heat capacities.

|| Debye, *Ann. der Phys.* **39**, 789 (1912). See also Born and von Kármán, *Phys. Zeits.* **13**, 297 (1912); and further articles in volumes **14** and **15**.

of volume v, and lying in the frequency range ν to $\nu+d\nu$, we may take the known expression

$$dZ = \frac{8\pi v\nu^2}{c^3}\,d\nu. \tag{92.7}$$

And for the mean energy of a mode of vibration of frequency ν in a system at temperature T, we may take the expression (92.5)

$$\bar{\epsilon} = \frac{h\nu}{e^{h\nu/kT}-1}+\frac{h\nu}{2} \tag{92.8}$$

as obtained for harmonic oscillators of frequency ν.

In combining these two expressions in the case of radiant energy we shall arbitrarily drop the term $\frac{1}{2}h\nu$—known to be appropriate in mechanical considerations—since its retention would lead in the present case to an infinite density of electromagnetic energy even in empty space at the absolute zero of temperature. Combining the two expressions with this elimination, and dividing by the volume v, we are then led to the well-known *Planck radiation law*

$$du = \frac{8\pi h\nu^3}{c^3}\,\frac{1}{e^{h\nu/kT}-1}\,d\nu, \tag{92.9}$$

which gives the density of radiant energy in the range ν to $\nu+d\nu$ in a hollow enclosure which has come to thermal equilibrium at temperature T. It will be noted from the form of (92.9) that, if we should let h go to zero, we should then have, in accordance with the correspondence principle, the Rayleigh-Jeans law $du = 8\pi\nu^2/c^3\,.\,kT\,d\nu$ which corresponds to a classical treatment of the modes of oscillation in the enclosure.

The above provides a simple and formally satisfactory derivation of the Planck radiation law which is very instructive. A more elaborate treatment of the problem would necessitate a development of the quantum mechanical theory of fields, which at the present time can hardly be regarded as leading to a deeper insight into the physical nature of the problem.

93. Einstein-Bose systems

We may now turn our attention to systems of weakly interacting constituent elements which exhibit the Einstein-Bose type of equilibrium distribution. This will be the case for systems of particles which cannot be permanently distinguished the one from another, and which exhibit eigensolutions that must be *symmetric* in character.

The general treatment of Einstein-Bose distributions is mathematically somewhat complicated, owing to the circumstance—already noted in § 91 (*a*)—that simple explicit solutions for the constant α are not ordinarily available. For this reason we shall first consider the mathematically simple case of photons where this difficulty does not arise.

(*a*) **Application to radiation.** The treatment of the electromagnetic energy in a hollow enclosure, which has come to thermal equilibrium, as an Einstein-Bose system in which photons are regarded as taking the place of more ordinary particles, has proved quite interesting. Such a treatment was the original purpose of the statistics introduced by Bose† before the development of the quantum mechanics. To make such an application is formally at least quite simple.

For the number of photons n_κ in the g_κ elementary eigenstates corresponding to an energy range ϵ_κ to $\epsilon_\kappa+\Delta\epsilon_\kappa$, we may take

$$n_\kappa = \frac{g_\kappa}{e^{\beta\epsilon_\kappa}-1}, \tag{93.1}$$

in agreement with our assumption that photons are to be treated as particles obeying the Einstein-Bose distribution law (89.6), and in agreement with the consideration already mentioned in § 89, that the constant α is to be taken as zero in applying this equation to photons, since their total number in an isolated system does not have to be regarded as conserved. Taking the energy of a photon as related to the corresponding frequency ν by

$$\epsilon = h\nu, \tag{93.2}$$

substituting

$$\beta = 1/kT, \tag{93.3}$$

and changing to a differential mode of expression, the equation given by (93.1) may then be rewritten in the more convenient form

$$dn = \frac{1}{e^{h\nu/kT}-1}\,dg, \tag{93.4}$$

where dn is now to be taken as the number of photons that would be present at equilibrium in the dg elementary eigenstates that would be in the frequency range ν to $\nu+d\nu$.

To make use of this expression we must now make some assumption as to the number of eigenstates for a photon that would lie in a given frequency range. For this purpose we may regard the relation between photons and electromagnetic waves as roughly similar to the relation between ordinary particles and probability waves. On this basis we

† Bose, *Zeits. f. Phys.* **27**, 384 (1924).

may then associate each eigenstate for a photon in an enclosure with a corresponding steady state solution of the electromagnetic wave equation, just as we have previously associated each eigenstate for a particle in a container with a steady state solution of the Schroedinger wave equation. Doing so, we can then put the number of elementary eigenstates dg, in range ν to $\nu+d\nu$, for a photon in an enclosure of volume v, equal to the known number of modes of electromagnetic vibration in such an enclosure:

$$dg = \frac{8\pi v\nu^2}{c^3}\,d\nu. \tag{93.5}$$

Combining (93.4) with (93.5), multiplying by the energy per photon $h\nu$, and dividing by v so as to obtain the energy density, we are then led once more to the *Planck radiation law*

$$du = \frac{8\pi h\nu^3}{c^3}\,\frac{1}{e^{h\nu/kT}-1}\,d\nu, \tag{93.6}$$

for the density of radiant energy in the range ν to $\nu+d\nu$, in a hollow enclosure, which has come to thermal equilibrium at temperature T. Here it may be noted that we should have been led to the Wien law $8\pi h\nu^3/c^3 . e^{-h\nu/kT}\,d\nu$ if we had treated the photons as statistically independent and applied the Maxwell-Boltzmann rather than the Einstein-Bose distribution law, since the (-1) in the denominator would then have been missing.

We are thus able to give a derivation of the Planck law, either from the point of view of the preceding section treating radiation as electromagnetic waves, or from the point of view of the present section treating radiation as photons, provided in each case we make appropriate quantum theoretic generalizations of the classical treatment. Thus, also in the case of electromagnetic as well as in the case of mechanical phenomena, we encounter the wave-particle duality which plays so fundamental a role in quantum theory. The existence of a formal equivalence between results obtained from oscillators obeying Maxwell-Boltzmann relations and particles obeying Einstein-Bose relations has been specially emphasized by Dirac. In conclusion, it is to be remarked that both of the above derivations for the Planck law may appear at the present stage of theory to be somewhat arbitrarily guided by the known character of the result to be obtained and the heuristic application of the correspondence principle.

(*b*) **Useful integrals in the Einstein-Bose case.** It will now be of interest to give some consideration to the mathematical methods that can be employed in the general treatment of Einstein-Bose systems of particles, when simple explicit solutions for the parameter α, in the distribution law

$$n_\kappa = \frac{g_\kappa}{e^{\alpha+\beta\epsilon_\kappa}-1}, \tag{93.7}$$

are not available.

For this purpose it proves convenient to consider integrals of the general form

$$U(\alpha,\rho) = \frac{1}{\Gamma(\rho+1)}\int\limits_0^\infty \frac{z^\rho\,dz}{e^{\alpha+z}-1}, \tag{93.8}$$

where α is the parameter mentioned, and ρ is a number which in our applications will be either $\frac{1}{2}$ or $\frac{3}{2}$. We shall only have to consider positive values of α, since with α negative it is evident that the distribution law (93.7) could lead to negative values of the numbers of particles n_κ in eigenstates of sufficiently low energy ϵ_κ. With α positive, integrals of the form (93.8) can be treated by carrying out the indicated division of z^ρ by $(e^{\alpha+z}-1)$ in the integrand. This then gives a series of terms which can be handled by known formulae of integration (see Appendix II). In this way we readily obtain

$$U(\alpha,\rho) = e^{-\alpha}+\frac{e^{-2\alpha}}{2^{\rho+1}}+\frac{e^{-3\alpha}}{3^{\rho+1}}+\frac{e^{-4\alpha}}{4^{\rho+1}}+\dots \tag{93.9}$$

which will converge rapidly when $\alpha \gg 1$.

(*c*) **Values of parameter α, energy E, and pressure p.** We may now make use of such integrals in investigating the values of the parameter α, energy E, and pressure p for gases composed of particles exhibiting Einstein-Bose distributions at equilibrium.

For this purpose we may begin by writing, in accordance with our previous treatment of § 71 (*b*),

$$g_\kappa = \frac{4\pi vg}{h^3} m\surd(2m\epsilon_\kappa)\,\Delta\epsilon_\kappa \tag{93.10}$$

as an explicit expression for the number of elementary eigensolutions for a particle of mass m, in a container of volume v, in the energy range ϵ_κ to $\epsilon_\kappa+\Delta\epsilon_\kappa$. A factor g has now been explicitly introduced in this expression to allow for different possibilities for the intrinsic angular momentum of the particles considered. It is equal to the number of eigenstates of angular momentum that would be possible with each eigenstate of kinetic energy ϵ. Substituting (93.10) in the Einstein-Bose distribution law (93.7), and changing to a differential mode of expression, we then have

$$dn = \frac{4\pi vg}{h^3} m\sqrt{(2m)} \frac{\epsilon^{\frac{1}{2}}\, d\epsilon}{e^{\alpha+\beta\epsilon}-1} \tag{93.11}$$

for the number of particles in the energy range ϵ to $\epsilon+d\epsilon$.

Integrating (93.11) over all possible values of ϵ, so as to obtain the total number of particles n, substituting $\beta = 1/kT$, and putting $\frac{1}{2}\sqrt{\pi} = \Gamma(\frac{1}{2}+1)$, we can then readily obtain the result

$$\frac{nh^3}{vg(2\pi mkT)^{\frac{3}{2}}} = \frac{1}{\Gamma(\frac{1}{2}+1)} \int\limits_0^\infty \frac{z^{\frac{1}{2}}\, dz}{e^{\alpha+z}-1} = U(\alpha, \tfrac{1}{2}), \tag{93.12}$$

in terms of an integral of the type introduced above. This equation shows that the value of the *parameter* α will depend on the nature and condition of the gas, as expressed by a combination of the quantities n, v, m, T appearing at the left. For this important combination of quantities it will be convenient to introduce the symbol

$$y = U(\alpha, \tfrac{1}{2}) = \frac{nh^3}{vg(2\pi mkT)^{\frac{3}{2}}}. \tag{93.13}$$

It will be seen from the above equations that large values of α will correspond to small values of y, or in more physical terms to small values of the concentration n/v combined with large values for the mass of the particles m and for the temperature of the system T.

Multiplying (93.11) by ϵ and integrating over all values, we obtain for the total (kinetic) *energy* of the gas

$$E = \frac{4\pi vg}{h^3} m\sqrt{(2m)} \int\limits_0^\infty \frac{\epsilon^{\frac{3}{2}}\, d\epsilon}{e^{\alpha+\beta\epsilon}-1}, \tag{93.14}$$

and again substituting $\beta = 1/kT$, this can be re-expressed in the form

$$E = \tfrac{3}{2}kT \frac{vg(2\pi mkT)^{\frac{3}{2}}}{h^3} \frac{1}{\Gamma(\frac{3}{2}+1)} \int\limits_0^\infty \frac{z^{\frac{3}{2}}\, dz}{e^{\alpha+z}-1}. \tag{93.15}$$

Hence, noting (93.8) and (93.12), we can also express the energy of the gas in a form

$$E = \tfrac{3}{2}nkT \frac{U(\alpha, \frac{3}{2})}{U(\alpha, \frac{1}{2})}, \tag{93.16}$$

which depends on integrals of the type introduced above.

Finally, in order to investigate the pressure of such a gas, we note on the one hand, in accordance with our original derivation of the eigenfunctions for a particle in a container, see (71.10), that the energy for such eigenfunctions would be of the form

$$\epsilon = \frac{h^2}{8m}\left(\frac{n_1^2}{l_1^2}+\frac{n_2^2}{l_2^2}+\frac{n_3^2}{l_3^2}\right), \tag{93.17}$$

where n_1, n_2, and n_3 are integral quantum numbers, and l_1, l_2, and l_3 are the linear dimensions of the assumed rectangular container. On the other hand, it can be shown, see § 97, that a particle, started off in any eigenstate, as defined by the above quantum numbers, would remain in that same state during any sufficiently slow adiabatic change in the dimensions of the container. Hence, noting the above inverse dependence on the square of the linear dimensions, and considering the average effect for different states, we can take

$$E = \frac{\text{const.}}{v^{\frac{2}{3}}}, \tag{93.18}$$

in the case of a gas containing many such particles, as an expression for the dependence of energy on volume when reversible adiabatic changes in the volume are to be considered. In accordance with the definition of pressure in terms of the work done in a reversible adiabatic expansion, we can then take

$$p = -\frac{\partial E}{\partial v} = \frac{2}{3}\frac{\text{const.}}{v^{\frac{5}{3}}} = \frac{2}{3}\frac{E}{v} \tag{93.19}$$

as an expression—in the quantum as well as in the classical mechanics—for the dependence of the pressure p of a dilute gas on its energy E and volume v, under equilibrium conditions. Combining (93.19) with (93.16), we then obtain

$$p = \frac{nkT}{v}\frac{U(\alpha, \frac{3}{2})}{U(\alpha, \frac{1}{2})} \tag{93.20}$$

as the desired expression for the *pressure* of an Einstein-Bose gas in terms of integrals of the type that we have introduced.

(*d*) **Case of slight degeneration.** We may now investigate the values of α, E, and p for an Einstein-Bose gas under conditions where α is large enough so that the series expansion (93.9), for the integrals that we have introduced, will converge rapidly. These may be called conditions of slight degeneration since the properties of the gas will then approach those for a classical perfect gas.

Under such conditions, with the help of equations (93.9) and (93.13), we can then write as a series expansion for the quantity y, which conveniently characterizes the condition of the gas,

$$y = U(\alpha, \tfrac{1}{2}) = \frac{nh^3}{vg(2\pi mkT)^{\frac{3}{2}}} = e^{-\alpha}+\frac{e^{-2\alpha}}{2^{\frac{3}{2}}}+\frac{e^{-3\alpha}}{3^{\frac{3}{2}}}+\dots \tag{93.21}$$

which can be readily solved to give

$$e^{-\alpha} = y\left\{1-\frac{1}{2^{3/2}}y+\left(\frac{1}{4}-\frac{1}{3^{3/2}}\right)y^2-...\right\}. \tag{93.22}$$

Furthermore, we can also make use of the series expansion (93.9) to rewrite our expression for total energy (93.16) in the form

$$E = \tfrac{3}{2}nkT\frac{1}{y}\left(e^{-\alpha}+\frac{e^{-2\alpha}}{2^{5/2}}+\frac{e^{-3\alpha}}{3^{5/2}}+...\right), \tag{93.23}$$

which, by substituting (93.22), leads to the result

$$E = \tfrac{3}{2}nkT(1-0{\cdot}1768y-0{\cdot}0033y^2-...). \tag{93.24}$$

For the pressure of the gas we then obtain

$$p = \frac{nkT}{v}(1-0{\cdot}1768y-0{\cdot}0033y^2-...). \tag{93.25}$$

In accordance with these results, noting the dependence of y on n/v, m, and T, we see that the properties of our Einstein-Bose gas would approach those for an ordinary perfect gas as we increase the dilution, particle mass, and temperature of the gas. As we approach this limit, however, the energy and pressure would always be less than those for a perfect gas.

Nevertheless, these theoretical deviations are not important under ordinary circumstances. For example, in the case of molecular hydrogen, which should act as an Einstein-Bose gas, the value of y under standard conditions of pressure and temperature would be only of the order 10^{-5}. Hence the slight degeneration present under such circumstances would be entirely too small to detect. At very low temperatures and high densities the deviations of Einstein-Bose gases from the classical behaviour of perfect gases would of course become important but would then be difficult to distinguish from deviations due to strong instead of weak particle interaction.† At the present time the application of the Einstein-Bose results to photons is the most important one that we have.

† For the treatment of highly degenerate Einstein-Bose gases, it is not always legitimate to replace sums by integrals as we have done (see Uhlenbeck, *Over statistische methoden in de Theorie der Quanta*, 's-Gravenhage, 1927, p. 70). For deviations between classical and quantum behaviour in the case of transport phenomena, see Uehling and Uhlenbeck, *Phys. Rev.* **43**, 552 (1933); Uehling, *Phys. Rev.* **46**, 917 (1934); Massey and Mohr, *Proc. Roy. Soc.* A, **141**, 434 (1933), A, **144**, 188 (1933).

94. Fermi-Dirac systems

We may now turn to the consideration of systems of weakly interacting constituent elements which exhibit the Fermi-Dirac type of equilibrium distribution. This will be the case for systems of particles which cannot be permanently distinguished one from another, and which exhibit eigensolutions that must be *antisymmetric* in character.

(*a*) **Useful integrals in the Fermi-Dirac case.** Explicit solutions for the parameter α appearing in the Fermi-Dirac distribution law

$$n_\kappa = \frac{g_\kappa}{e^{\alpha+\beta\epsilon_\kappa}+1} \tag{94.1}$$

are in general not available. And it now proves convenient to make the general treatment depend on integrals of the general form

$$V(\alpha,\rho) = \frac{1}{\Gamma(\rho+1)}\int\limits_0^\infty \frac{z^\rho\,dz}{e^{\alpha+z}+1}, \tag{94.2}$$

where α is the parameter mentioned, and ρ is a number which in our applications will be either $\frac{1}{2}$ or $\frac{3}{2}$.

These integrals are similar in form to the previous integrals $U(\alpha,\rho)$ introduced for the treatment of Einstein-Bose systems, but have $+1$ instead of -1 in the denominator of the integrand. In evaluating these integrals it will not now be possible to confine the consideration to positive values of α, since the distribution given by (94.1) does not now become physically impossible for negative values of α.

When α is positive the above integral can easily be treated by performing the indicated division by the term $(e^{\alpha+z}+1)$. This then gives a series of terms which can be handled by known formulae of integration (see Appendix II). In this way we readily obtain

$$V(\alpha,\rho) = e^{-\alpha}-\frac{e^{-2\alpha}}{2^{\rho+1}}+\frac{e^{-3\alpha}}{3^{\rho+1}}-\frac{e^{-4\alpha}}{4^{\rho+1}}+\dots \tag{94.3}$$

which will converge rapidly if $\alpha \gg 1$.

When α is negative the foregoing series is no longer useful. An approximate treatment by Sommerfeld† then leads to the result

$$V(\alpha,\rho) = \frac{(-\alpha)^{\rho+1}}{\Gamma(\rho+2)}\left[1+2\left\{\frac{(\rho+1)\rho c_2}{\alpha^2}+\frac{(\rho+1)\rho(\rho-1)(\rho-2)c_4}{\alpha^4}+\dots\right\}\right], \tag{94.4}$$

where the constants c_2, c_4, etc., have the values

$$c_\nu = 1-\frac{1}{2^\nu}+\frac{1}{3^\nu}-\frac{1}{4^\nu}+\dots. \tag{94.5}$$

† Sommerfeld, *Zeits. f. Phys.* **47**, 1 (1928).

This series converges rapidly when $-\alpha \gg 1$. The approximation employed in obtaining it involves an error of the order of e^{α}.

(*b*) **Values of parameter α, energy E, and pressure p.** We may now make use of such integrals in investigating the values of the parameter α, energy E, and pressure p for gases composed of particles exhibiting Fermi-Dirac distributions at equilibrium. For this purpose we may begin by re-expressing our distribution law (94.1) in the more convenient differential form

$$dn = \frac{4\pi vg}{h^3} m\sqrt{(2m)} \frac{\epsilon^{\frac{1}{2}}\, d\epsilon}{e^{\alpha+\beta\epsilon}+1}, \tag{94.6}$$

where dn denotes the number of particles in the energy range ϵ to $\epsilon+d\epsilon$. The method of obtaining this form of expression is similar to that given in detail for the analogous expression (93.11) in the Einstein-Bose development and the quantities involved have a similar significance.

Integrating (94.6) over all possible values, so as to obtain the total number of particles n, substituting $\beta = 1/kT$, and putting $\frac{1}{2}\sqrt{\pi} = \Gamma(\frac{1}{2}+1)$, we can then readily obtain the result

$$\frac{nh^3}{vg(2\pi mkT)^{\frac{3}{2}}} = \frac{1}{\Gamma(\frac{1}{2}+1)} \int\limits_0^\infty \frac{z^{\frac{1}{2}}\, dz}{e^{\alpha+z}+1} = V(\alpha, \tfrac{1}{2}), \tag{94.7}$$

in terms of an integral of the type that we have introduced. This equation shows also in the Fermi-Dirac case that the value of the *parameter* α will depend on the nature and condition of the gas, as expressed by a combination of quantities, which we can again denote by the symbol

$$y = V(\alpha, \tfrac{1}{2}) = \frac{nh^3}{vg(2\pi mkT)^{\frac{3}{2}}}. \tag{94.8}$$

It will again be seen from the present two equations that large values of α will correspond also in the Fermi-Dirac case to small values of y, or in more physical terms to small values of the concentration n/v combined with large values for the mass of the particles m, and for the temperature of the system T.

Multiplying (94.6) by ϵ, and integrating over all values, we now obtain for the total (kinetic) *energy* of the gas

$$E = \frac{4\pi vg}{h^3} m\sqrt{(2m)} \int\limits_0^\infty \frac{\epsilon^{\frac{3}{2}}\, d\epsilon}{e^{\alpha+\beta\epsilon}+1}, \tag{94.9}$$

and again substituting $\beta = 1/kT$, this can be rewritten in the form

$$E = \tfrac{3}{2}kT\frac{vg(2\pi mkT)^{\frac{3}{2}}}{h^3}\frac{1}{\Gamma(\frac{3}{2}+1)}\int\limits_0^\infty \frac{z^{\frac{3}{2}}\,dz}{e^{\alpha+z}+1}. \tag{94.10}$$

Hence, noting (94.2) and (94.7), we can also express the energy of the gas in a form

$$E = \tfrac{3}{2}nkT\frac{V(\alpha, \frac{3}{2})}{V(\alpha, \frac{1}{2})}, \tag{94.11}$$

which depends on integrals of the type that we have introduced.

Finally, in order to obtain an expression for the *pressure* of the gas, it is evident that we can again use the relation (93.19),

$$p = \frac{2}{3}\frac{E}{v}, \tag{94.12}$$

and hence write

$$p = \frac{nkT}{v}\frac{V(\alpha, \frac{3}{2})}{V(\alpha, \frac{1}{2})}. \tag{94.13}$$

(c) **Case of slight degeneration.** We may now investigate the values of α, E, and p for a Fermi-Dirac gas under conditions where α is large enough so that the first of the two series expansions (94.3), for the integrals $V(\alpha, \rho)$, will converge rapidly. These may be called conditions of slight degeneration since the properties of the gas will then approach those for a classical perfect gas.

Under such conditions, with the help of equations (94.3) and (94.8), we can then write as a series expansion for the quantity y, which conveniently characterizes the condition of the gas,

$$y = V(\alpha, \tfrac{1}{2}) = \frac{nh^3}{vg(2\pi mkT)^{\frac{3}{2}}} = e^{-\alpha} - \frac{e^{-2\alpha}}{2^{\frac{3}{2}}} + \frac{e^{-3\alpha}}{3^{\frac{3}{2}}} - \dots \tag{94.14}$$

which can readily be solved to give

$$e^{-\alpha} = y\left\{1 + \frac{1}{2^{\frac{3}{2}}}y + \left(\frac{1}{4} - \frac{1}{3^{\frac{3}{2}}}\right)y^2 + \dots\right\}. \tag{94.15}$$

Furthermore, we can now also make use of the series expansions (94.3) to rewrite our expression for total kinetic energy (94.11) in the form

$$E = \tfrac{3}{2}nkT\frac{1}{y}\left(e^{-\alpha} - \frac{e^{-2\alpha}}{2^{\frac{5}{2}}} + \frac{e^{-3\alpha}}{3^{\frac{5}{2}}} - \dots\right), \tag{94.16}$$

which, by substituting (94.15) and computing the numerical coefficients, leads to the result

$$E = \tfrac{3}{2}nkT(1 + 0{\cdot}1768y - 0{\cdot}0033y^2 + \dots). \tag{94.17}$$

For the pressure of the gas, in agreement with (94.12), we shall then obtain

$$p = \frac{nkT}{v}(1+0{\cdot}1768y-0{\cdot}0033y^2+...). \tag{94.18}$$

These results, except for signs, are similar to those for the Einstein-Bose case. In accordance therewith, noting the dependence of y on n/v, m, and T, we see that the properties of our Fermi-Dirac gas would approach those for an ordinary perfect gas as we increase the dilution, particle mass, and temperature of the gas. As we approach this limit, however, the energy and pressure of the Fermi-Dirac gas would always be higher than those for a perfect gas. Nevertheless, for particles of the mass of ordinary atoms at familiar concentrations and temperatures, the deviations from perfect gas behaviour would lie well below detection. (See remarks in § 93 (*d*) concerning Einstein-Bose gases.)

Only in the case of stellar interiors,† and in the case of the conduction electrons in a metal, which can be treated in first approximation as an electron gas, do we find a large enough concentration or small enough particle mass for the new theory to become important. We then find, however, even at the high temperatures of stellar interiors and up to ordinary temperatures for metals, that the gas can be in a condition of extreme degeneration.

(*d*) **Case of extreme degeneration.** To treat the case of extreme degeneration in Fermi-Dirac gases (i.e. α large and negative) we must make use of expansions of the form (94.4). Doing so, equation (94.8) then becomes

$$\begin{aligned} y = V(\alpha,\tfrac{1}{2}) &= \frac{nh^3}{vg(2\pi mkT)^{\frac{3}{2}}} \\ &= \frac{4(-\alpha)^{\frac{3}{2}}}{3\sqrt{\pi}}\left[1+2\left\{\frac{(\frac{3}{2})(\frac{1}{2})c_2}{\alpha^2}+\frac{(\frac{3}{2})(\frac{1}{2})(-\frac{1}{2})(-\frac{3}{2})c_4}{\alpha^4}+...\right\}\right]. \end{aligned} \tag{94.19}$$

Introducing for c_2 the exact value

$$c_2 = 1-\frac{1}{2^2}+\frac{1}{3^2}-\frac{1}{4^2}+... = \frac{\pi^2}{12}, \tag{94.20}$$

this can then be solved to give for α, to the indicated order of approximation,

$$-\alpha = \left(\frac{3\sqrt{\pi}}{4}y\right)^{\frac{2}{3}}\left\{1-\frac{\pi^2}{12}\left(\frac{4}{3\sqrt{\pi}y}\right)^{\frac{4}{3}}...\right\}. \tag{94.21}$$

Furthermore, again using expansions of the form (94.4), we find that the expression for the total energy (94.11) now becomes, to the same order of approximation,

$$E = \tfrac{3}{5}nkT(-\alpha)\left(1+\frac{\pi^2}{2\alpha^2}...\right), \tag{94.22}$$

† See Fowler, *Monthly Notices*, **87**, 114 (1926–7).

which, with the help of our values for α and y, finally reduces to

$$E = n\frac{3h^2}{10m}\left(\frac{3n}{4\pi vg}\right)^{\frac{2}{3}} + n\frac{\pi^2 mk^2}{2h^2}\left(\frac{4\pi vg}{3n}\right)^{\frac{2}{3}} T^2 \dots \tag{94.23}$$

For the pressure of the gas, in agreement with (94.13), we shall then obtain

$$p = \frac{n}{v}\frac{h^2}{5m}\left(\frac{3n}{4\pi vg}\right)^{\frac{2}{3}} + \frac{n}{v}\frac{\pi^2 mk^2}{3h^2}\left(\frac{4\pi vg}{3n}\right)^{\frac{2}{3}} T^2 \dots \tag{94.24}$$

In accordance with these results we see in the case of extreme degeneration that a Fermi-Dirac gas would have a residual zero-point energy and pressure even at the absolute zero of temperature. It would also have a heat capacity at low temperatures

$$C_v = \frac{n\pi^2 mk^2}{h^2}\left(\frac{4\pi vg}{3n}\right)^{\frac{2}{3}} T, \tag{94.25}$$

which would be proportional to the temperature, and would be small per particle in the case of large densities n/v and small particle mass m.

(*e*) **Remarks on applications to conduction electrons.** It will not be out of place to make a few remarks concerning the really great advances in our understanding of the properties of metallic conductors which have resulted from the foregoing development of quantum statistics.

To explain the properties of such substances, we may regard the outer valence electrons, provided by the atoms of a metal, as behaving in first approximation as a free electron gas, with the repulsive forces between them neutralized on the average by the positive ions which then compose the lattice. This picture of metallic structure thus agrees in a general way with that employed by Drude in the classical theory of metallic conduction, and leads to similar explanations for the non-electrolytic character of metallic conduction, for the high thermal and electrical conductivities exhibited by metals, and for the Wiedemann-Franz law as to the ratio between these conductivities. The picture cannot agree in all details with the classical one, however, since we now know that electrons, being subject to the Pauli exclusion principle, would have to exhibit the properties of a Fermi-Dirac gas rather than those of an ordinary perfect gas as assumed by Drude.

To determine the importance of this change in point of view, we must obtain an idea as to the values of the important quantity

$$y = \frac{nh^3}{vg(2\pi mkT)^{\frac{3}{2}}} \tag{94.26}$$

which are to be expected in the case of typical metals. We may calculate this quantity for the case of *silver at room temperature*, assuming one free electron per atom which gives us an electron density of approximately $5{\cdot}9\times10^{22}$ per cubic centimetre. Taking $g = 2$ corresponding to the two directions of electron spin, together with the approximate values $n/v = 5{\cdot}9\times10^{22}$ cm.$^{-3}$, $h = 6{\cdot}5\times10^{-27}$ gm. cm.^{2} sec.$^{-1}$, $m = 9{\cdot}0\times10^{-28}$ gm., $k = 1{\cdot}4\times10^{-16}$ gm. cm.^{2} sec.$^{-2}$ deg.$^{-1}$, and $T = 300$ deg., we then obtain for the above quantity the large value

$$y = 2{,}200. \tag{94.27}$$

And, substituting in (94.21), this then gives us the large negative value

$$\alpha = -205 \tag{94.28}$$

for the parameter which determines the degree of degeneration of our gas. On account of high electron concentration and low electron mass, large negative values for α are thus found in general for typical metals even up to considerable temperatures. We must hence conclude that the Fermi-Dirac gas, by which we represent the conduction electrons of a metal, will be in a condition of extreme degeneration, with properties quite different from those of an ordinary perfect gas.

To get a good qualitative idea of the properties of such a gas, it will be helpful to note that there would be a great tendency under conditions of extreme degeneration for most of the electrons to accumulate in the eigenstates of lowest energy, filling each such state with the one electron allowed by the Pauli exclusion principle. The reason for this tendency can be seen with the help of our previous equation (94.1)

$$n_\kappa = \frac{g_\kappa}{e^{\alpha+\beta\epsilon_\kappa}+1}, \tag{94.29}$$

connecting the number of particles n_κ in a given energy range ϵ_κ to $\epsilon_\kappa+\Delta\epsilon_\kappa$ with the number of states g_κ in that range. With α large and negative, this is found to make the number of electrons n_κ sensibly equal to the number of states g_κ available, for all energies ϵ_κ from zero up to a value high enough to include nearly the total number of electrons.

This finding, that most of the electrons would reside in the practically completely filled eigenstates of lowest energy, provides an immediate solution for the major difficulty of the classical Drude theory as to the negligible contribution actually made by conduction electrons to the total specific heat of a metal. This negligible contribution to total specific heat is now explained by the consideration that most of the electrons cannot be lifted to adjoining states of higher energy by rise in temperature since these states are already filled. Only the electrons in the layer between filled and empty states can be so affected. Quantitatively this latter is seen to be unimportant, however, since our previous formula (94.25) for the heat capacity C_v of a degenerate Fermi-Dirac gas, applied to the case of the conduction electrons in silver at room temperature, gives the small value

$$C_v = 0{\cdot}023nk \tag{94.30}$$

instead of the classically expected value

$$C_v = \tfrac{3}{2}nk. \tag{94.31}$$

Our new idea, as to the existence of only a small fraction of the conduction electrons in the layer between filled and empty states, also provides the basis for Pauli's brilliant treatment of the magnetic properties of metals.† With the discovery that the electron has an intrinsic magnetic moment for itself, it might be supposed at first sight that the presence of free conduction electrons would make metals highly paramagnetic, since the electrons acting as elementary magnets could all orient themselves in the direction of any applied magnetic field. From the point of view of our present theory it is evident, however, that only the electrons in the boundary layer would be provided with unfilled states of nearly the same energy which could be occupied after the reorientation. This consideration, together with an allowance for the diamagnetic effect resulting according to the quantum theory from moving charges, has been found to give a good account of the magnetic susceptibilities of the alkali metals.

Our present point of view also provides the basis for Sommerfeld's fairly complete theory of electrical and thermal conductivities and of thermo-electric effects, together with such further elaborations as those of Houston, Bloch, Peierls, Nordheim, Brillouin, and Wilson.‡ It also provides the basis for the treatments of electron emission from metals as carried out by Fowler, Nordheim, Wentzel, and others. It has, however, provided no immediate understanding for the still unexplained phenomena of super-conductivity at low temperatures.

† Pauli, *Zeits. f. Phys.* **41**, 81 (1927).

‡ See Sommerfeld and Bethe, *Handbuch der Physik*, xxiv/2, second edition, Berlin, 1933.

XI
THE CHANGE IN QUANTUM MECHANICAL SYSTEMS WITH TIME

95. Dynamical reversibility in the quantum mechanics

In the preceding chapter we have considered possibilities of applying statistical mechanics to quantum systems which are in a steady condition of phenomenological equilibrium. In this chapter we may now commence the treatment of systems which are not in a steady condition. In the present section we shall first consider the status of dynamical reversibility in the quantum mechanics. In the next two sections, §§ 96 and 97, we shall then undertake the mathematical problem of obtaining integrals for Schroedinger's equation for the change in state of a system with time, both in the case of isolation and in the case when changes are made in some external parameter for the system such as its volume. In § 98 we shall then turn to more physical questions by studying the nature of the approximate observations by which in practice it might actually be profitable to follow the changes in a quantum mechanical system with time. In §§ 99 and 100, with the help of representative ensembles to correspond to approximate observations on the nearly steady states of a system, we shall then study time-proportional transitions in general, and the special case of transitions by molecular collision taking place with a probability proportional to the time. Finally, in § 101 we shall consider the general case of changes with time in an ensemble of isolated members. Just as our consideration of the effect of molecular collisions in changing the state of classical systems made it possible to proceed to a study of Boltzmann's H-theorem, so the considerations of the present chapter will then make it possible to study the natural analogue of that theorem in the quantum mechanics.

We may begin by investigating the possibility of dynamical reversibility in the quantum mechanics, that is, in the case of any quantum mechanical system which is changing with the time, the possibility of specifying the conditions for a second system of entirely similar structure which would be changing with the time in the reverse manner. The question is of considerable importance for statistical mechanics and thermodynamics, since it is evident that the actual phenomenological irreversibilities of macroscopic behaviour would assume quite a different aspect from that in classical theory, if the principle of dynamical reversibility had to be abandoned in the quantum mechanics.

We must first define what we can reasonably mean in the quantum mechanics by two systems which are changing with time in the reverse manner. As in the corresponding classical treatment of Chapter V, we may here regard ourselves as concerned for simplicity with isolated conservative systems, since we take it as possible to look at any system from a point of view which would make this true. Let us denote the first system by the letter S, and—in accordance with the limited possibilities of prediction provided by the quantum mechanics—take $W(q,t)\,dq$ and $W(p,t)\,dp$ as giving the probabilities at time t of finding the coordinates and momenta for this system in the ranges dq and dp, and $\overline{F}(q,p,t)$ as the expectation value at time t for any function of these coordinates and momenta. The second system may then be denoted by the letter S' and analogous symbols $W'(q,t)\,dq$, $W'(p,t)\,dp$, and $\overline{F}'(q,p,t)$ be taken for the probabilities and expectation values for this second system of similar structure. Using this notation, we shall then say that the behaviour of the second system is the reverse of that of the first if it satisfies the conditions

$$\begin{aligned} W'(q,t) &= W(q,-t), \\ W'(p,t) &= W(-p,-t), \\ \overline{F}'(q,p,t) &= \overline{F}(q,-p,-t). \end{aligned} \tag{95.1}$$

In accordance with these equations, the second system would then exhibit at time t the same probability for specified values of the coordinates, the same probability for specified values of the momenta taken with reversed sign, and the same expectation value for any function of the coordinates and reversed momenta, as would be exhibited by the first system at time $-t$. These conditions are thus the natural analogue of the classical conditions for the reversal of motion, in which the two systems would exhibit *precisely* the same values of their coordinates, reversed momenta, and functions thereof at times t and $-t$ respectively.

To show the actual possibility of such reversed behaviour in the quantum mechanics we must consider Schroedinger's equation which governs the change of quantum mechanical systems with time. In applying this equation we may restrict our considerations for simplicity to isolated, conservative systems and take the Hamiltonian operator $\mathbf{H}$, for the system under investigation, as obtainable from the classical expression $H(q,p)$ for the energy in the simple manner

$$\mathbf{H} = \mathbf{H}\left(q, \frac{h}{2\pi i}\frac{\partial}{\partial q}\right),$$

where the classical expression involves only *even* powers of the momenta. This will make the operator **H** 'real', so that it will be unaffected in passing from an equation in which this operator occurs to the corresponding complex conjugate equation.

For the first of our two systems S we may then write Schroedinger's equation (57.5) in the form

$$\mathbf{H}\psi(q,t)+\frac{h}{2\pi i}\frac{\partial\psi(q,t)}{\partial t}=0, \tag{95.2}$$

and the corresponding complex conjugate equation, with the help of the last of the above assumptions, in the simple form

$$\mathbf{H}\psi^*(q,t)-\frac{h}{2\pi i}\frac{\partial\psi^*(q,t)}{\partial t}=0. \tag{95.3}$$

The expressions $\psi(q,t)$ and $\psi^*(q,t)$ occurring in the above equations are to be regarded as giving, for the first of the two systems S, the actual dependence of the probability amplitude and its complex conjugate on the coordinates q and time t.

For the second of our two systems S' we shall then consider the possibility of taking the probability amplitude $\psi'(q,t)$ and its complex conjugate $\psi'^*(q,t)$ as determined by the equations

$$\psi'(q,t)=\psi^*(q,-t) \quad \text{and} \quad \psi'^*(q,t)=\psi(q,-t). \tag{95.4}$$

For this to be a possible specification it is evident that these new quantities must also satisfy Schroedinger's equation and its complex conjugate equation

$$\mathbf{H}\psi'(q,t)+\frac{h}{2\pi i}\frac{\partial}{\partial t}\psi'(q,t)=0$$

and

$$\mathbf{H}\psi'^*(q,t)-\frac{h}{2\pi i}\frac{\partial}{\partial t}\psi'^*(q,t)=0.$$

Substituting, however, from (95.4), the first of these equations can be written in the form

$$\mathbf{H}\psi^*(q,-t)-\frac{h}{2\pi i}\frac{\partial}{\partial(-t)}\psi^*(q,-t)=0$$

and the second in the form

$$\mathbf{H}\psi(q,-t)+\frac{h}{2\pi i}\frac{\partial}{\partial(-t)}\psi(q,-t)=0,$$

which are at once seen to be valid by comparison with (95.2) and (95.3).

We have thus found that (95.4) does give a possible specification for the probability amplitude and its complex conjugate for the second system S'. We shall now show that this specification satisfies the three requirements given by (95.1) for a behaviour of system S' which is the reverse of that of the first system S.

Making use of the relation (57.1) between probability density and probability amplitude

$$W(q,t) = \psi^*(q,t)\psi(q,t),$$

we see that the specification given by (95.4) immediately satisfies the first of the three requirements for reverse behaviour

$$W'(q,t) = W(q,-t). \tag{95.5}$$

This shows that the second system would exhibit at time t the same probability for specified values of the coordinates q as the first system exhibits at time $-t$.

Furthermore, making use of the transformation relation (57.2) between probability amplitudes for coordinates and momenta, we evidently obtain, with the help of (95.4),

$$\begin{aligned}\phi'(p,t) &= h^{-\frac{1}{2}f}\int \psi'(q,t)e^{-\frac{2\pi i}{h}[p_1q_1+\ldots+p_fq_f]}\,dq\\ &= h^{-\frac{1}{2}f}\int \psi^*(q,-t)e^{+\frac{2\pi i}{h}[(-p_1)q_1+\ldots+(-p_f)q_f]}\,dq\\ &= \phi^*(-p,-t),\end{aligned}$$

and similarly $\qquad \phi'^*(p,t) = \phi(-p,-t).$

This then gives at once the next of our three requirements

$$W'(p,t) = W(-p,-t), \tag{95.6}$$

which shows that the second system would exhibit at time t the same probability for specified momenta taken with reversed sign as the first system exhibits at time $-t$.

Finally, making use of the expression (57.4), which makes it possible to calculate expectation values $\bar{F}$ from a knowledge of the corresponding operator $\mathbf{F}\left(q, \frac{h}{2\pi i}\frac{\partial}{\partial q}\right)$, we obtain, with the help of (95.4),

$$\begin{aligned}\bar{F}'(q,p,t) &= \int \psi'^*(q,t)\mathbf{F}\left(q, \frac{h}{2\pi i}\frac{\partial}{\partial q}\right)\psi'(q,t)\,dq\\ &= \int \psi(q,-t)\mathbf{F}\left(q, \frac{h}{2\pi i}\frac{\partial}{\partial q}\right)\psi^*(q,-t)\,dq\\ &= \int \psi(q,-t)\left[\mathbf{F}\left(q, \frac{h}{2\pi i}\frac{\partial}{\partial(-q)}\right)\psi(q,-t)\right]^*dq\\ &= \int \psi^*(q,-t)\mathbf{F}\left(q, \frac{h}{2\pi i}\frac{\partial}{\partial(-q)}\right)\psi(q,-t)\,dq\\ &= \bar{F}(q,-p,-t),\end{aligned} \tag{95.7}$$

where the fourth form of writing is justified in accordance with (55.25) by the Hermitian character of the operator **F**. This result then gives the last of our three requirements for reversed behaviour by showing that the second system would exhibit at time t the same expectation value for any function of the coordinates and momenta as exhibited for the same function of coordinates and reversed momenta by the first system at time $-t$.

We thus find that the principle of dynamical reversibility would hold in the quantum mechanics in much the same way as in the classical mechanics. Hence the introduction of the quantum mechanics—at least in its present form—cannot be regarded as throwing any new kind of light on the problem of the actual phenomenological irreversibility of thermodynamic processes. Just as in the classical mechanics, this irreversibility will have to be explained by considering the probable behaviour of a collection or ensemble of systems rather than from consideration of the purely mechanical behaviour of a single system.

96. Integration of Schroedinger equation for changes with time in an isolated system

(*a*) **Introduction.** Before proceeding to more physical considerations, it will first be profitable to make a preliminary disposition of purely mathematical problems connected with the integration of Schroedinger's differential equation for the change in the state of a quantum mechanical system with time. In the present section we shall consider isolated systems, and in the next section systems where some external parameter such as volume is itself changed with time.

In the case of an isolated quantum mechanical system it is of course evident that we could in any case write Schroedinger's equation for the dependence of any state $\psi(q,t)$ on the time in the quite general integrated form

$$\psi(q,t) = \sum_k c_k u_k(q) e^{-(2\pi i/h)E_k t},$$

where the c_k are constant coefficients whose squares $|c_k|^2$ give the probabilities for the different true energy eigenstates k for the system corresponding to the possible eigensolutions $u_k(q)$ and eigenvalues of energy E_k. Such an expansion in terms of true energy solutions is not usually convenient, however, for treating the temporal behaviour of a system, since we are not ordinarily interested in the fact that there would be no change with time in the relative probabilities for the truly steady states of an isolated system—whose close determination would indeed take an exceedingly long time—but are, on the other hand, interested in the actual change with time in the probabilities for states of the system which could be closely determined by observations taking a length of time short compared with that needed for appreciable change.

For this reason we shall now begin by considering the integration of Schroedinger's equation, by the method of variation of constants, when the state of the system is expanded in terms of nearly steady states on which observations might be made. We shall then follow by considering a general method of integration when the state of the system is expanded in terms of any kind of eigenstates that we might wish to observe.

(*b*) **Expansion of state in terms of unperturbed energy eigenstates and integration by the method of variation of constants.** The integration of the Schroedinger equation by the method of variation of constants has already been discussed in § 68, and it will now merely be necessary to recite certain results in a form available for use in the present chapter. To apply the method we consider the *Hamiltonian operator* **H** for the system to be expressed as the sum of two terms

$$\mathbf{H} = \mathbf{H}^0 + \mathbf{V}, \tag{96.1}$$

where the *unperturbed Hamiltonian operator* $\mathbf{H}^0$ corresponds to a nearly precise expression for the energy of the system, and the *perturbation operator* **V** to the remainder necessary for an exact expression of the energy. The unperturbed Hamiltonian may then be used to determine a set of normalized orthogonal eigenfunctions $u_k(q)$ for the system with the help of the characteristic equations

$$H^0 u_k(q) = E_k^0 u_k(q), \tag{96.2}$$

where the E_k^0 and $u_k(q)$ may be called the unperturbed energy eigenvalues and unperturbed energy eigenfunctions for the system. Since these eigenfunctions are taken as providing a complete set, the probability amplitude for the state of a system could then be expressed as a function of time t by a summation of the form

$$\psi(q,t) = \sum_k c_k(t) u_k(q) e^{-(2\pi i/h)E_k^0 t}, \tag{96.3}$$

where the probability coefficients $c_k(t)$, which would be constants if we had made an expansion in terms of true energy eigenfunctions, are now allowed to vary in the manner actually demanded by the quantum mechanics. As an actually appropriate expression for the time dependence of these probability coefficients, we have obtained (68.5) the specialized form of the Schroedinger equation

$$\frac{dc_n(t)}{dt} = -\frac{2\pi i}{h}\sum_k V_{nk}\, e^{\frac{2\pi i}{h}(E_n^0-E_k^0)t} c_k(t), \tag{96.4}$$

where the V_{nk} are the elements of an Hermitian matrix defined by

$$V_{nk} = \int u_n^*(q)\mathbf{V}u_k(q)\, dq, \tag{96.5}$$

with

$$V_{nk} = V_{kn}^*. \tag{96.6}$$

The differential equations (96.4) for change with time will in general form an infinite set and contain so far no approximation. By solving them, we could in principle determine for any case of interest the values of the coefficients $c_n(t)$ as a function of time. This would then give us the probability

$$W_n(t) = c_n^*(t)c_n(t) \tag{96.7}$$

of finding the system at time t in any state of interest n, provided we normalize to unity with

$$\sum_k W_k(t) = \sum_k c_k^*(t)c_k(t) = 1. \tag{96.8}$$

Although the precise solution of equations (96.4) would be complicated, we can easily obtain an approximate integration by assuming that the separation of the true Hamiltonian into two parts as given by (96.1) has actually been carried out in such a way as to make the effect of the perturbation operator **V** really small. Under these circumstances the matrix components V_{nk} in (96.4) will be small and the coefficients $c_n(t)$ will be changing but slowly with the time. Hence we can make an approximate integration over a short time interval, in the neighbourhood of $t = 0$, by assigning to the coefficients $c_k(t)$ on the right-hand side of (96.4) the values $c_k(0)$ which they have at time $t = 0$. Doing so we then obtain in the neighbourhood of $t = 0$ the simple result

$$c_n(t) = c_n(0) + \sum_k V_{nk}\frac{e^{\frac{2\pi i}{h}(E_n^0-E_k^0)t}-1}{E_k^0-E_n^0} c_k(0), \tag{96.9}$$

as can readily be verified by redifferentiation. By resubstituting this result into (96.4), we could then obtain a higher stage of approximation, and so on to any desired accuracy. We shall leave the consideration of the exact integration of Schroedinger's equation, however, for the following very general treatment, where we consider the expansion of state in terms of any kind of eigenfunctions that may be desired.

A specially simple case of interest arises when the system is known at time $t = 0$ to be in a particular state k with

$$W_k(0) = c_k^*(0)c_k(0) = 1, \tag{96.10}$$

and with all other coefficients $c_n(0)$ equal to zero. For any state n not the same as k, equation (96.9) now reduces to

$$c_n(t) = \frac{V_{nk}\, e^{\frac{2\pi i}{h}(E_n^0-E_k^0)t}-1}{E_k^0-E_n^0}\, c_k(0). \tag{96.11}$$

Noting (96.10), the probability of finding the system in state n at any time in the neighbourhood of $t = 0$ then becomes

$$W_n(t) = c_n^*(t)c_n(t) = |V_{nk}|^2 \frac{2-e^{\frac{2\pi i}{h}(E_n^0-E_k^0)t}-e^{-\frac{2\pi i}{h}(E_n^0-E_k^0)t}}{(E_n^0-E_k^0)^2},$$

or by a simple transformation to trigonometric form

$$W_n(t) = 4|V_{nk}|^2 \frac{\sin^2\frac{\pi}{h}(E_n^0-E_k^0)t}{(E_n^0-E_k^0)^2}. \tag{96.12}$$

In accordance with this result we see, on account of the appearance of the so-called resonance denominator $(E_n^0-E_k^0)^2$, that the probability of finding the system in state n will continue to increase with the time t the longer, the closer the unperturbed energy E_n^0 of the system is to the original unperturbed energy E_k^0. Exact equality of E_n^0 with E_k^0 is not necessary for a transition to take place, since the above are only unperturbed rather than true energy levels for the actual system. Starting the system off at time $t = 0$ in the state k would not be equivalent to putting it in a steady state of definite energy from which no transitions could occur, but would correspond to a distribution over various possible values of the true energy with a relatively high probability of finding values close to E_k^0 if a measurement were made. This relatively high probability of finding values of the energy in the neighbourhood of E_k^0 will then be preserved in time in accordance with the quantum analogue of the principle of the conservation of energy as discussed in § 63 (*c*).

With the help of (96.12) we can also calculate, at times near $t = 0$, the probability of still finding the system in the original state k, since the fact that the total probability of finding the system in one or another state must remain equal to unity will at once permit us to write

$$W_k(t) = 1-4\sum_n |V_{nk}|^2 \frac{\sin^2\frac{\pi}{h}(E_n^0-E_k^0)t}{(E_n^0-E_k^0)^2}. \tag{96.13}$$

This result can also be verified by direct calculation starting from the differential equations (96.4), provided we carry the computation to the next stage of approximation beyond that employed above.

This provides an account of the approximate integration of the Schroedinger equation, by the method of variation of constants, which will be sufficient for the purposes of the present chapter. The results obtained are expressed in terms of the probability coefficients $c_n(t)$ for a set of unperturbed energy eigenstates of the system and are only valid provided the unperturbed Hamiltonian which determines these eigenstates is sufficiently close to the true Hamiltonian $\mathbf{H}$ for the system, and provided the time interval over which the integration is taken is not too long. The results are of special interest because of the possibility for a simple understanding of the character of the unperturbed energy eigenstates, or *nearly steady states*, which furnish the quantum mechanical language employed, and because of their later application in studying—with the help of statistical considerations—transitions which occur with a probability that is proportional to the time between groups of neighbouring nearly steady states.

(*c*) **Expansion of state in terms of general eigenfunctions and integration as a Taylor's series in the time.** We may now turn to a consideration of the more general problem of obtaining a formally exact integration for the Schroedinger equation expressed in the generalized language corresponding to any set of eigenfunctions for the system. This will give us a very powerful and actually simple mathematical formalism for treating the changes in a quantum mechanical system with time, but nevertheless a formalism which may not seem very transparent from the point of view of an intuitive appreciation of the character of the results.

In accordance with our discussion of the quantum mechanical transformation theory in § 67, we may give a very general expression for the quantum mechanical state of a system in the form

$$\psi(q,t) = \sum_k a_k(t)u_k(q), \tag{96.14}$$

where the $u_k(q)$ are any desired complete set of normalized orthogonal eigenfunctions for the system, and the $a_k(t)$ may be called the generalized probability amplitudes for the different eigenstates k. Making use of this formalism, the content of Schroedinger's equation can then be expressed, in accordance with (67.13), in the general form

$$\frac{da_n(t)}{dt} = -\frac{2\pi i}{h}\sum_k H_{nk}\,a_k(t), \tag{96.15}$$

where the elements of the Hermitian matrix H_{nk} are related to the Hamiltonian operator $\mathbf{H}$ for the system by the expression

$$H_{nk} = \int u_n^*(q)\mathbf{H}u_k(q)\,dq, \tag{96.16}$$

with

$$H_{nk} = H_{kn}^*. \tag{96.17}$$

In order to obtain an integrated expression for the generalized equation (96.15), it will only be necessary to consider the values of the successive derivatives of the probability amplitude $a_n(t)$ with respect to t at some initial time $t = 0$, and then use these to give a Taylor's expansion for the amplitude $a_n(t)$ at any desired later time $t = t$. For the derivative at $t = 0$ we can write at once from (96.15)

$$\left(\frac{da_n}{dt}\right)_{t=0} = -\frac{2\pi i}{h}\sum_k H_{nk}\,a_k(0). \tag{96.18}$$

For the second derivative we can then write

$$\begin{aligned}\left(\frac{d^2a_n}{dt^2}\right)_{t=0} &= -\frac{2\pi i}{h}\sum_l H_{nl}\left(\frac{da_l}{dt}\right)_{t=0}\\ &= \left(-\frac{2\pi i}{h}\right)^2\sum_{l,k} H_{nl}\,H_{lk}\,a_k(0)\\ &= \left(-\frac{2\pi i}{h}\right)^2\sum_k H_{nk}^2\,a_k(0),\end{aligned} \tag{96.19}$$

where the first two forms of writing are made possible by the successive application of equation (96.15), and the last form by the rules for matrix multiplication. Proceeding in this manner, we should then evidently obtain as a general expression for the Nth derivative

$$\left(\frac{d^Na_n}{dt^N}\right)_{t=0} = \left(-\frac{2\pi i}{h}\right)^N\sum_k H_{nk}^N\,a_k(0). \tag{96.20}$$

Making use of expansion in the form of a Taylor's series, we can then write for the probability amplitude $a_n(t)$ at any desired time t the expression

$$\begin{aligned}a_n(t) = a_n(0)+\left(-\frac{2\pi i}{h}t\right)\sum_k H_{nk}\,a_k(0)+\frac{1}{2!}\left(-\frac{2\pi i}{h}t\right)^2\sum_k H_{nk}^2\,a_k(0)+\\ +\dots+\frac{1}{N!}\left(-\frac{2\pi i}{h}t\right)^N\sum_k H_{nk}^N\,a_k(0)+\dots,\end{aligned} \tag{96.21}$$

thus obtaining the desired integration of the generalized Schroedinger equation.

This result, moreover, can be written in quite a compact form if we introduce an exponential operator $e^{-(2\pi i/h)\mathbf{H}t}$ which we regard as equivalent to the series expansion

$$e^{-(2\pi i/h)\mathbf{H}t} = \sum_{N=0}^{\infty} \frac{1}{N!}\left(-\frac{2\pi i}{h}t\right)^N \mathbf{H}^N. \tag{96.22}$$

Considering the matrix elements corresponding to the successive terms of this series, we then see that the integrated Schroedinger equation (96.21) can be re-expressed in the form

$$a_n(t) = \sum_k [e^{-(2\pi i/h)\mathbf{H}t}]_{nk}\, a_k(0). \tag{96.23}$$

Furthermore, we may give an immediate demonstration of the satisfactory character of this solution of Schroedinger's equation, since we obtain by redifferentiation

$$\begin{aligned} \frac{da_n(t)}{dt} &= -\frac{2\pi i}{h}\sum_k [e^{-(2\pi i/h)\mathbf{H}t}\mathbf{H}]_{nk}\, a_k(0) \\ &= -\frac{2\pi i}{h}\sum_k [\mathbf{H}e^{-(2\pi i/h)\mathbf{H}t}]_{nk}\, a_k(0) \\ &= -\frac{2\pi i}{h}\sum_{k,l} H_{nl}[e^{-(2\pi i/h)\mathbf{H}t}]_{lk}\, a_k(0) \\ &= -\frac{2\pi i}{h}\sum_l H_{nl}\, a_l(t), \end{aligned} \tag{96.24}$$

where the second form of writing is made possible by the commutativity of the operator **H** with any function of itself, the third form of writing comes from the rules for matrix multiplication, and the last form of writing comes from a further application of the original formula (96.23). It will be noted that this last form does agree with the generalized Schroedinger equation (96.15).

(*d*) **Change with time regarded as a unitary transformation.** It will be convenient to introduce the symbol

$$\mathbf{U}(t) = e^{-(2\pi i/h)\mathbf{H}t} \tag{96.25}$$

to denote the important operator defined by (96.22). Our solution of the generalized Schroedinger equation (96.23) can then be written in the very simple form

$$a_n(t) = \sum_k U_{nk}(t)a_k(0). \tag{96.26}$$

The consequences of going from time $t = 0$ to time $t = t$ can now be conveniently spoken of as the result of a transformation with the help of the matrix elements $U_{nk}(t)$ corresponding to the transformation operator $\mathbf{U}(t)$. We now proceed to consider the properties of such a transformation.

We may first show the possibility† of obtaining the *inverse operator* to $\mathbf{U}(t)$ which will permit us to calculate the earlier probability amplitudes $a_l(0)$ in terms of the later ones $a_n(t)$. In accordance with the equation of definition (96.25) we can evidently write for any two times t_1 and t_2

$$\mathbf{U}(t_1)\mathbf{U}(t_2) = \mathbf{U}(t_1+t_2), \tag{96.27}$$

and hence in particular may write

$$\mathbf{U}(-t)\mathbf{U}(t) = \mathbf{U}(0) = \mathbf{I}, \tag{96.28}$$

where $\mathbf{I}$ is the identity operator, with the corresponding matrix elements

$$I_{lk} = \sum_n U_{ln}(-t)U_{nk}(t) = \delta_{lk}. \tag{96.29}$$

Multiplying (96.26) by $U_{ln}(-t)$ and summing over n, we then obtain

$$\begin{aligned}\sum_n U_{ln}(-t)a_n(t) &= \sum_k \sum_n U_{ln}(-t)U_{nk}(t)a_k(0)\\ &= \sum_k \delta_{lk}\, a_k(0),\end{aligned}$$

which gives

$$a_l(0) = \sum_n U_{ln}(-t)a_n(t). \tag{96.30}$$

We thus see that $\mathbf{U}(-t)$, with the corresponding matrix elements $U_{ln}(-t)$, is the desired inverse operator.

We may next obtain a simple relation connecting the matrix elements corresponding to the two operators $\mathbf{U}(t)$ and $\mathbf{U}(-t)$. Making use of the series expansion (96.22), we can write as a matrix element corresponding to $\mathbf{U}(t)$

$$U_{nk}(t) = \sum_{N=0}^{\infty} \frac{1}{N!}\left(-\frac{2\pi it}{h}\right)^N \int u_n^* \mathbf{H}^N u_k \, dq, \tag{96.31}$$

and hence by the ordinary rules can write for its complex conjugate

$$U_{nk}^*(t) = \sum_{N=0}^{\infty} \frac{1}{N!}\left(+\frac{2\pi it}{h}\right)^N \int u_n (\mathbf{H}^N u_k)^* \, dq. \tag{96.32}$$

Similarly, we can evidently write as a matrix element corresponding to $\mathbf{U}(-t)$

$$\begin{aligned}U_{kn}(-t) &= \sum_{N=0}^{\infty} \frac{1}{N!}\left(+\frac{2\pi it}{h}\right)^N \int u_k^* \mathbf{H}^N u_n \, dq\\ &= \sum_{N=0}^{\infty} \frac{1}{N!}\left(+\frac{2\pi it}{h}\right)^N \int u_n (\mathbf{H}^N u_k)^* \, dq,\end{aligned} \tag{96.33}$$

† For certain general restrictions implied by the existence of the inverse operator see the discussion of Pauli, *Handbuch der Physik*, xxiv/1, second edition, Berlin, 1933, p. 142.

where the second form of expression is made possible by the Hermitian character of the operator $\mathbf{H}^N$. Comparing (96.32) and (96.33), we then obtain the desired relation connecting the two kinds of matrix elements

$$U_{kn}(-t) = U^*_{nk}(t), \tag{96.34}$$

showing incidentally that the operator $\mathbf{U}(t)$ is not Hermitian in character.

Finally, with the help of (96.34), we may now express an important summation property for the matrix elements corresponding to $\mathbf{U}(t)$ by rewriting (96.29) in the form

$$\sum_n U^*_{nl}(t)U_{nk}(t) = \delta_{lk}. \tag{96.35}$$

Similarly, we can obtain

$$\sum_k U^*_{nk}(t)U_{mk}(t) = \delta_{nm}. \tag{96.36}$$

In accordance with these summation properties, and with the existence of the inverse operator as demonstrated above, we may now describe the transformation brought about by the operator $\mathbf{U}(t)$ as having *unitary character*. It is to be emphasized, however, that this unitary transformation, to the same quantum mechanical language at a later time, is not to be confused with our previous possibility, see § 67 (*e*), for a unitary transformation to a different quantum mechanical language at a given time.

(*e*) **Application to the calculation of probabilities as a function of time.** We may now apply the foregoing apparatus to the important problem of calculating the probability $W_n(t)$ of finding a system at time $t = t$ in a given state n from a specification of its initial state at an earlier time $t = 0$.

In accordance with our fundamental equation (96.26) we can write

$$a_n(t) = \sum_k U_{nk} a_k(0), \tag{96.37}$$

where $a_n(t)$ is the probability amplitude at time $t = t$ for the final state n, and the $a_k(0)$ are the various probability amplitudes at time $t = 0$ for the initial states k, and where we shall now often find it convenient to write U_{nk} in place of the more explicit $U_{nk}(t)$. Similarly, we can write

$$a^*_n(t) = \sum_l U^*_{nl} a^*_l(0) \tag{96.38}$$

for the complex conjugate amplitude. Hence, multiplying (96.37) by (96.38), we now obtain for the probability of interest

$$W_n(t) = a^*_n(t)a_n(t) = \sum_{l,k} U^*_{nl} U_{nk} a^*_l(0)a_k(0). \tag{96.39}$$

Or summing separately for the cases $l = k$ and $l \neq k$, we can also write this in the form

$$W_n(t) = \sum_k |U_{nk}|^2 W_k(0) + \sum_{l \neq k} U^*_{nl} U_{nk} a^*_l(0) a_k(0). \tag{96.40}$$

The general character, precise validity, and formal simplicity of the results obtained by the present methods may well be emphasized. In this connexion we note the simple form of the fundamental expression (96.26) obtained in the present treatment,

$$a_n(t) = \sum_k U_{nk} a_k(0), \tag{96.41}$$

for the precise connexion between probability amplitudes, at times $t = 0$ and $t = t$, in the general language corresponding to any selection of quantum mechanical eigenstates. This may be contrasted with the considerably more complicated form of the analogous expression (96.9) obtained in the preceding treatment,

$$c_n(t) = c_n(0) + \sum_k V_{nk} \frac{e^{\frac{2\pi i}{h}(E_n^0 - E_k^0)t} - 1}{E_k^0 - E_n^0} c_k(0), \tag{96.42}$$

for the approximate connexion between probability coefficients, at times $t = 0$ and $t = t$ sufficiently close together, and in the special language corresponding to some set of unperturbed energy eigenstates. It is also of interest to compare the general connexion between probabilities at $t = 0$ and $t = t$ given by (96.39) with the special connexion given in the preceding treatment by (96.12), where it is to be remembered that (96.12) applies only when the system starts out in a single eigenstate k at $t = 0$. Nevertheless, it must also be remarked that the simple final form of the general and precise results of the present treatment has been obtained in part by the device of introducing simple symbols for quantities which in practical applications would really be of a complicated character, and that the development of a satisfactory physical intuition for the significance of the present results may seem difficult.

In studying the temporal behaviour of ensembles of isolated systems, we shall first make use of the results of the preceding treatment, and shall be led to the easily appreciated—but special and only approximately valid—notion of transitions between different conditions which take place with a probability proportional to the time. In our final studies of temporal behaviour, however, we shall make use of the results of the present treatment, and shall be led to less transparent, but nevertheless general and precise methods for treating the changes in ensembles with time.

97. Integration of Schroedinger equation when an external parameter is varied

(*a*) **Probability amplitudes for energy states that depend on an external parameter.** The preceding section dealt with the integration of the Schroedinger equation in the case of an isolated system. We must now consider a method of procedure which can be used in treating the temporal behaviour of a system having an external parameter which is itself changed with time in some prescribed manner.† We shall not have to make use of the results obtained in this section in the present chapter, but shall need them at a later place, in § 124 of Chapter XIII, when we discuss adiabatic changes in the external coordinates for a thermodynamic system.

Let us consider a system having a Hamiltonian operator,

$$\mathbf{H}\left(q, \frac{h}{2\pi i}\frac{\partial}{\partial q}, a\right),$$

with a form dependent on the value of some external parameter a—for example, the volume of the system—which we can later regard as varying with the time. For any particular value of the parameter a we can then determine a set of energy eigenfunctions $u_k(q,a)$ with the help of the equations

$$\mathbf{H}\left(q, \frac{h}{2\pi i}\frac{\partial}{\partial q}, a\right)u_k(q,a) = E_k(a)u_k(q,a), \tag{97.1}$$

where the quantities $E_k(a)$ are the eigenvalues of the energy of the system with that value of the parameter a. These eigenfunctions may be taken as forming a complete, normalized, orthogonal set. For the given value of the parameter a, the solution of Schroedinger's equation for the system,

$$\mathbf{H}\left(q, \frac{h}{2\pi i}\frac{\partial}{\partial q}, a\right)\psi(q,t)+\frac{h}{2\pi i}\frac{\partial}{\partial t}\psi(q,t) = 0, \tag{97.2}$$

could then be expressed, as a superposition of truly steady state solutions, in the form

$$\psi(q,t) = \sum_k c_k\, u_k(q,a)e^{-(2\pi i/h)E_k(a)t}, \tag{97.3}$$

where the quantities c_k are a set of constants which will in general be complex numbers.

Let us now consider that the external parameter a instead of being held constant is made to change with time in accordance with some

† The method of treatment will be similar to that of Güttinger, *Zeits. f. Phys.* **73**, 169 (1931–2).

prescribed law. The Hamiltonian operator itself would then be a function of the time through its dependence on a, and we should have to write Schroedinger's equation in the form, appropriate for non-conservative systems,

$$\mathbf{H}\left[q, \frac{h}{2\pi i}\frac{\partial}{\partial q}, a(t)\right]\psi(q,t)+\frac{h}{2\pi i}\frac{\partial}{\partial t}\psi(q,t) = 0. \tag{97.4}$$

As a form of solution for this equation we shall then find it profitable to take the superposition expressed by

$$\psi(q,t) = \sum_k c_k(t)u_k(q,a)e^{-(2\pi i/h)E_k(a)t}, \tag{97.5}$$

where at each time t we use the steady state solutions that correspond to the instantaneous value of a that prevails, and now allow the quantities $c_k(t)$ to change with time in whatever manner is actually demanded by Schroedinger's equation. This proposed solution may also be expressed in a convenient form by combining the two factors giving an explicit dependence on t, by substituting

$$C_k(t) = c_k(t)e^{-(2\pi i/h)E_k(a)t}. \tag{97.6}$$

This then gives us
$$\psi(q,t) = \sum_k C_k(t)u_k(q,a). \tag{97.7}$$

The quantities $C_k(t)$ are then the probability amplitudes for the different states k that correspond to the instantaneous forms of the eigenfunctions $u_k(q,a)$.

To determine how these probability amplitudes depend on the time we may now substitute the proposed solution (97.7) in Schroedinger's equation (97.4). Noting (97.1), this gives us

$$\sum_k C_k(t)E_k(a)u_k(q,a)+\frac{h}{2\pi i}\frac{\partial}{\partial t}\sum_k C_k(t)u_k(q,a) = 0, \tag{97.8}$$

or carrying out the indicated differentiation this can be written in the form

$$\frac{2\pi i}{h}\sum_k C_k(t)E_k(a)u_k(q,a)+\sum_k \dot{C}_k(t)u_k(q,a)+\dot{a}\sum_k C_k(t)\frac{\partial u_k(q,a)}{\partial a} = 0, \tag{97.9}$$

where the last term takes care of the change with time in the eigenfunctions $u_k(q,a)$ which we are using. Multiplying through by $u_n^*(q,a)$ and integrating over the configuration space, making use of the normalization and orthogonality of the eigenfunctions $u_k(q,a)$, we can then obtain

$$\dot{C}_n(q,a)+\frac{2\pi i}{h}C_n(t)E_n(a)+\dot{a}\sum_k C_k(t)\int u_n^*(q,a)\frac{\partial u_k(q,a)}{\partial a}\,dq = 0. \tag{97.10}$$

Before making use of this equation it will be desirable to consider the nature of the integrals appearing in the last term.

Let us first consider the case $k = n$. We should then evidently be able to write

$$\int u_n^*(q,a)\frac{\partial u_n(q,a)}{\partial a}\,dq+\int u_n(q,a)\frac{\partial}{\partial a}u_n^*(q,a)\,dq = 0, \tag{97.11}$$

since the eigenfunctions are normalized to unity for all values of the parameter a. Hence the integral of interest could in any case only be a pure imaginary quantity

$$\int u_n^*(q,a)\frac{\partial u_n(q,a)}{\partial a}\,dq = ip_n(a), \tag{97.12}$$

where $p_n(a)$ is real. This result will then make it possible to choose the phase factor for our eigenfunctions so that the integral in question would vanish. To see this, consider the new eigenfunctions

$$U_n(q,a) = u_n(q,a)e^{-i\int^a p_n(a)\,da}, \tag{97.13}$$

which would also be solutions of (97.1). Our integral would then take the form

$$\begin{aligned}\int U_n^*(q,a)\frac{\partial}{\partial a}U_n(q,a)\,dq &= \int u_n^*(q,a)e^{+i\int^a p_n(a)\,da}\frac{\partial}{\partial a}\Big[u_n(q,a)e^{-i\int^a p_n(a)\,da}\Big]\,dq\\ &= \int u_n^*(q,a)\Big[\frac{\partial u_n(q,a)}{\partial q}-ip_n(a)u_n(q,a)\Big]\,dq\\ &= ip_n(a)-ip_n(a) = 0,\end{aligned} \tag{97.14}$$

with the help of (97.12). For simplicity we shall assume that our eigenfunctions have been chosen in such a way as to secure this result. Hence the summation in (97.10) will only have to be taken over states where $k \neq n$.

Let us now consider the values of such integrals when k is not equal to n. Differentiating equation (97.1) with respect to a, we can write

$$\begin{aligned}\frac{\partial}{\partial a}\mathbf{H}\Big(q,\frac{h}{2\pi i}\frac{\partial}{\partial q},a\Big)u_k(q,a)+\mathbf{H}\Big(q,\frac{h}{2\pi i}\frac{\partial}{\partial q},a\Big)\frac{\partial}{\partial a}u_k(q,a)&\\ = \frac{\partial E_k(a)}{\partial a}u_k(q,a)+E_k(a)\frac{\partial}{\partial a}u_k(q,a).&\end{aligned} \tag{97.15}$$

And multiplying this through by $u_n^*(q,a)$ and integrating over configuration space, we then obtain from the normalization and orthogonality of the eigenfunctions, and from the Hermitian character of the operator **H**,

$$\left[\frac{\partial H}{\partial a}\right]_{nk} + E_n \int u_n^*(q,a)\frac{\partial}{\partial a}u_k(q,a)\,dq = E_k \int u_n^*(q,a)\frac{\partial}{\partial a}u_k(q,a)\,dq, \tag{97.16}$$

where the indicated matrix components are defined by

$$\left[\frac{\partial H}{\partial a}\right]_{nk} = \int u_n^*(q,a)\frac{\partial}{\partial a}\mathbf{H}\left(q, \frac{h}{2\pi i}\frac{\partial}{\partial q}, a\right)u_k(q,a)\,dq. \tag{97.17}$$

Solving (97.16) then allows us to calculate the desired integrals from

$$\int u_n^*(q,a)\frac{\partial}{\partial a}u_k(q,a)\,dq = \frac{[\partial H/\partial a]_{nk}}{E_k - E_n}. \tag{97.18}$$

Substituting (97.18) into (97.10), and remembering that we have to sum only over states $k \neq n$, we may now write our equation for the time dependence of the probability amplitudes in the form

$$\dot{C}_n(t) + \frac{2\pi i}{h}C_n(t)E_n + \dot{a}\sum_{k\neq n}\frac{C_k(t)}{E_k - E_n}\left[\frac{\partial H}{\partial a}\right]_{nk} = 0. \tag{97.19}$$

(*b*) **Gradual change in parameter.** We are now ready to consider the effect of a very slow change in the parameter a on the probability for finding the system in different states n. For this purpose it is convenient to transform (97.19) by substituting

$$C_n(t) = F_n(t)e^{-\frac{2\pi i}{h}\int_0^t E_n dt}. \tag{97.20}$$

Since the quantities $F_n(t)$ have the same moduli as the $C_n(t)$, they will be just as useful to us in determining the probabilities for the different states n. Introducing (97.20) into (97.19), this can now be written in the simplified form

$$\dot{F}_n + \dot{a}\sum_{k\neq n}\frac{F_k e^{\frac{2\pi i}{h}\int_0^t (E_n - E_k)\,dt}}{E_k - E_n}\left[\frac{\partial H}{\partial a}\right]_{nk} = 0, \tag{97.21}$$

or, introducing the abbreviation

$$E_n - E_k = h\nu_{nk},$$

we have

$$\dot{F}_n = \dot{a}\sum_{k\neq n}\left[\frac{\partial H}{\partial a}\right]_{nk}\frac{F_k}{h\nu_{nk}}e^{2\pi i\int_0^t \nu_{nk}\,dt}. \tag{97.22}$$

We now wish to use this equation to calculate the change in F_n when the parameter a changes by a prescribed amount, say from a_0 to $a_0 + \Delta a$, at a vanishingly small rate $\dot{a}$. Without loss in significance we can take this rate as constant and write

$$\Delta a = \dot{a}T, \tag{97.23}$$

where T is the total time involved in the change. With a fixed change Δa in the parameter, we shall then be interested in the limiting case when

$$\dot{a} \to 0 \quad \text{and} \quad T \to \infty. \tag{97.24}$$

Integrating (97.22), we have

$$F_n(t) - F_n(0) = \dot{a} \sum_{k \neq n} \int_0^T \left[\frac{\partial H}{\partial a}\right]_{nk} \frac{F_k}{h\nu_{nk}} e^{2\pi i \int_0^t \nu_{nk}\,dt'}\,dt, \tag{97.25}$$

and must now estimate the value of the right-hand side of this equation at the limit $T \to \infty$.

To do this we note that the integrals involved on the right-hand side of (97.25) can be re-expressed with the help of a partial integration in the form

$$J_{nk} = \int_0^T \left[\frac{\partial H}{\partial a}\right]_{nk} \frac{F_k}{h\nu_{nk}} e^{2\pi i \int_0^t \nu_{nk}\,dt'}\,dt$$

$$= \left[\frac{\partial H}{\partial a}\right]_{nk} \frac{F_k}{h\nu_{nk}} \frac{e^{2\pi i \int_0^t \nu_{nk}\,dt'}}{2\pi i\nu_{nk}}\Bigg|_0^T -$$

$$- \int_0^T \frac{e^{2\pi i \int_0^t \nu_{nk}\,dt'}}{2\pi i\nu_{nk}} \frac{d}{dt}\left\{\left[\frac{\partial H}{\partial a}\right]_{nk} \frac{F_k}{h\nu_{nk}}\right\}\,dt. \tag{97.26}$$

We shall treat this expression on the assumption that

$$\nu_{nk} \neq 0, \text{ and is continuous,} \tag{97.27}$$

throughout the process. The first term on the right-hand side of (97.26) is then at once seen to be finite since it is the difference of two quantities which are themselves recognized to be finite. To study the second term we may rewrite it, with the help of (97.22) and with the help of the circumstance that certain quantities therein depend on time only through their dependence on a, in the form

$$-\dot{a} \int_0^T \frac{e^{2\pi i \int_0^t \nu_{nk}\,dt'}}{2\pi i\nu_{nk}} \left[\frac{\partial H}{\partial a}\right]_{nk} \frac{1}{h\nu_{nk}} \sum_{l \neq k} \left[\frac{\partial H}{\partial a}\right]_{kl} \frac{F_l}{h\nu_{kl}} e^{2\pi i \int_0^t \nu_{kl}\,dt'}\,dt -$$

$$-\dot{a} \int_0^T \frac{e^{2\pi i \int_0^t \nu_{nk}\,dt'}}{2\pi i\nu_{nk}} \frac{\partial}{\partial a}\left\{\left[\frac{\partial H}{\partial a}\right]_{nk} \frac{1}{h\nu_{nk}}\right\} F_k\,dt.$$

In accordance with the assumption expressed by (97.27), none of the factors in the above integrands will be infinite. The integrals will then be of the order of the integration interval T, and hence the second term in (97.26) is itself of the order $\dot{a}T = \Delta a$ which is finite. Thus the integrals J_{nk} are bounded even when T goes to infinity.

Hence, returning to (97.25), we can now write

$$F_n(t) = F_n(0)+O(\dot{a}), \tag{97.28}$$

where $O(\dot{a})$ indicates a quantity of the order of magnitude $\dot{a}$. Squaring both sides of this equation, and noting from (97.20) that F_n has the same modulus as the probability amplitude C_n for the state n, we then obtain in general

$$|C_n(t)|^2 = |C_n(0)|^2+O(\dot{a}), \tag{97.29}$$

and

$$|C_n(t)|^2 = O(\dot{a}^2) \tag{97.30}$$

in the special case where the initial probability for the state n is equal to zero.

Considering the limit when $\dot{a} \to 0$, we thus obtain the *general quantum mechanical principle that the probabilities for finding a system in its energy eigenstates n do not change with time when changes are made in external parameters at a vanishingly slow rate.* This is the quantum mechanical analogue of Ehrenfest's *so-called adiabatic principle* familiar in the older quantum theory.

We have derived the principle with the help of the assumption expressed by (97.27). It is evident, however, that the principle would still be true if ν_{nk} should pass through zero during the process without too high an order of contact. Hence we may ascribe validity to the principle in all but very special cases.

(*c*) **Abrupt change in parameter.** We must also treat the effect of an abrupt change, in the external parameter a, on the probabilities for finding a system in its different energy eigenstates. For this purpose we may return to our original equation (97.8) for the time dependence of the probability amplitudes C_k and write this in the form

$$\frac{d}{dt}\sum_k C_k(t)u_k(q,a) = -\frac{2\pi i}{h}\sum_k C_k(t)E_k(a)u_k(q,a). \tag{97.31}$$

Integrating over a time interval 0 to t, during which a change is made in the parameter from the value a_0 to $a_0+\Delta a$, this gives the general relation

$$\sum_k C_k(a_0+\Delta a)u_k(q,a_0+\Delta a)$$
$$= \sum_k C_k(a_0)u_k(q,a_0)-\frac{2\pi i}{h}\int_0^t \sum_k C_k(t)E_k(a)u_k(q,a)\,dt. \tag{97.32}$$

In the case of an abrupt change, however, it is evident that the last term will go to zero as we make the time interval 0 to t shorter and shorter, since the integrand involved is certainly finite. Hence we shall have

$$\sum_k C_k(a_0+\Delta a)u_k(q,a_0+\Delta a) = \sum_k C_k(a_0)u_k(q,a_0). \tag{97.33}$$

Multiplying by $u_n^*(a_0+\Delta a)$, and integrating over the coordinate space, this then gives us

$$C_n(a_0+\Delta a) = \sum_k S_{nk}\, C_k(a_0), \tag{97.34}$$

where the matrix elements indicated are defined by

$$S_{nk} = \int u_n^*(q,a_0+\Delta a)u_k(q,a_0)\, dq. \tag{97.35}$$

It is to be noted that these quantities are the components of the *transformation matrix* connecting the two kinds of eigenfunctions which we have introduced as specially appropriate before and after the change has been made in the parameter a. To see this let us put

$$u_n^*(a_0+\Delta a) = \sum_m S_{nm}\, u_m^*(a_0) \tag{97.36}$$

as the development of the new eigenfunctions in terms of the old. Multiplying by $u_k(a_0)$, and integrating over the coordinate space, we then indeed do obtain

$$\int u_n^*(a_0+\Delta a)u_k(a_0)\, dq = \sum_m S_{nm} \int u_m^*(a_0)u_k(a_0)\, dq$$
$$= S_{nk} \tag{97.37}$$

in agreement with the definition (97.35).

It will also be noted that the components of this matrix satisfy the two necessary conditions for a *unitary transformation*

$$\sum_m S_{km}^* S_{nm} = \delta_{kn}$$

and

$$\sum_m S_{mk}^* S_{mn} = \delta_{kn}. \tag{97.38}$$

To see this we may use (97.36) to write

$$\int u_n^*(a_0+\Delta a)u_k(a_0+\Delta a)\, dq = \sum_{l,m} S_{kl}^* S_{nm} \int u_m^*(a_0)u_l(a_0)\, dq,$$

which, from the normalization and orthogonality of the eigenfunctions, gives us

$$\delta_{kn} = \sum_{l,m} S_{kl}^*\, S_{nm}\, \delta_{lm} = \sum_m S_{km}^* S_{nm},$$

verifying the first of equations (97.38), and the second can be verified by similar methods.

As already remarked, we shall need the results of this method of treating Schroedinger's equation for a non-isolated system at a later place, in § 124 of Chapter XIII. For the remainder of the present chapter, however, it will only be necessary to consider the methods of integrating Schroedinger's equation in the case of an isolated system, which were discussed in the preceding section.

98. Observation and specification of state in studying the change of quantum mechanical systems with time

(*a*) **Complementarity restrictions on observations.** Having made a suitable disposition of the purely mathematical problem of integrating the Schroedinger equation for the change in the quantum mechanical state of a system with time, we may now turn to the more physical question of considering the nature of the observations and specifications of state that would be theoretically possible or actually appropriate in investigating the temporal behaviour of such a system. We encounter quantum mechanical aspects of this problem which are quite different from anything that had to be considered in the classical mechanics.

In the classical mechanics, in following the temporal behaviour of a system, it was regarded as theoretically possible to make precise and instantaneous observations of the classical state of a system, at any time of interest, without in any way interfering with its further behaviour, and then to use the result obtained for the prediction of future states. In the quantum mechanics, in order to treat the temporal behaviour of a system, it is also regarded as theoretically possible to make a precise observation of quantum mechanical state and to use the result for the prediction of future states and of corresponding expectation values for quantities of interest. Nevertheless, it is no longer regarded as always possible to make precise observations without appreciably affecting the further behaviour of the system, nor to carry them out instantaneously, since the quantum mechanics recognizes that the act of observation may involve an uncontrollable interaction between system and measuring instrument, and that an interval of time may be needed to secure precision in the measurement of energy or quantities related thereto. Both of these new aspects of the situation, due to the complementarity features characteristic of the quantum mechanics, have consequences for the kind of observations which may now be regarded as appropriate in studying the temporal behaviour of systems.

We may first consider the possible effects of observation on the further behaviour of a system of interest. Such effects arise from the circumstance that the measurement of one kind of dynamical variable is connected in the quantum mechanics with uncontrollable changes in other variables which stand thereto in a complementary relation. Such changes, moreover, become more and more serious the greater we try to make the precision of measurement. As a consequence, it may then become inappropriate in the quantum mechanics to make a precise determination of one kind of quantity when the temporal behaviour of the system which is being studied is also dependent on a complementary kind of quantity.

These remarks may be illustrated by an example of the kind typically treated by the methods of statistical mechanics. In studying the tendency for a sample of gas of approximately given energy content to distribute itself uniformly between two connecting containers, it is clear that the quantum mechanics would allow us in principle to approach a very precise observation of the initial positions of the component molecules, for instance by using a battery of microscopes illuminated by exceedingly hard γ-rays. Nevertheless, the result of a method of measurement so 'violent' as this would be to make large and uncontrolled changes in the kinetic energies of the molecules, and thus to affect the actual rate of flow from one container to another. Hence, if we desire to study the rate of flow when the gas has a given fairly well determined energy content, it is evident that we shall wish to employ 'gentler' though less precise methods of observing the initial spatial distribution, for instance by measuring the colours, or densities of the gas in the two containers.

We may now turn to a discussion of the circumstance that instantaneous observations are not generally feasible in the quantum mechanics. Limits on the duration of observation in the quantum mechanics arise as a consequence of complementarity relations involving the time, as given by the Heisenberg expression

$$\Delta E \Delta t \geqslant h, \tag{98.1}$$

which connects the order of magnitude of the uncertainty ΔE in the energy E of a system with the uncertainty Δt in the time t at which an observation is made on the system. In accordance with this expression, if we make an observation on the condition of a system compatible with a specification of energy of precision ΔE, it will then be necessary in making the observation to employ an interval of time τ which will satisfy

$$\tau \geqslant \frac{h}{\Delta E}. \tag{98.2}$$

As a consequence we now see that the time interval τ devoted to observation cannot be taken too short in the quantum mechanics, if we desire a given precision in our definition of the energy of a system. On the other hand, it will also be noted that the time interval τ cannot be taken too long, if we desire to study a given kind of change which would itself take place in a period denoted by T, since it is evident that it will also be necessary for the time of observation τ to satisfy the relation

$$\tau \ll T \tag{98.3}$$

in order to carry out the study at all. This imposition of upper and lower limits on the length of time τ devoted to observation has an important influence on the nature of the observations appropriate in studying the change of quantum mechanical systems with time.

These remarks may also be illustrated by an example of the kind typically treated by the methods of statistical mechanics. In studying the tendency for the molecules of an enclosed sample of gas to acquire their equilibrium distribution of kinetic energy as a result of collisions, the *classical* mechanics would allow us in principle to make a precise assay of the instantaneous distribution. Nevertheless, it is evident in the *quantum* mechanics that we must not try to make such an observational assay too precise, since the time τ needed for this purpose would become longer the smaller we make the uncertainty ΔE in the measured energies, and we must keep this time short compared with the time T which we take as an approximate expression for that involved in the redistribution of energies by collision.

It will be seen from the foregoing discussion that the complementarity features characteristic of the quantum mechanics have the general effect of making the use of too accurate methods of observation inappropriate in studying the changes in quantum mechanical systems with time.

(*b*) **Approximate specifications of state in the quantum mechanics.** Paying attention to the general character of the observational process and to the special quantum mechanical characteristics of observations discussed above, we may now consider the specifications of condition that might be appropriately made in applying the principles of mechanics to obtain insight into the temporal behaviour of quantum mechanical systems. As in the case of the classical mechanics, it is of course possible in the quantum case to start with a precise specification of the initial state of a system of interest and then apply the principles of the exact quantum mechanics—i.e. make use of Schroedinger's equation—to predict the precise state at any later time together with the corresponding expectation values for quantities of interest. Nevertheless, there are a number of reasons which often make it possible to obtain a better insight into the temporal behaviour of quantum systems by starting out with approximate specifications of initial state and then using the principles of statistical quantum mechanics to investigate the temporal behaviour of a suitable representative ensemble for the system of interest.

In the first place it will be recognized, just as in the classical mechanics, that the actual observations which we make on the initial states of systems are for practical reasons usually only approximate in character, since actual measuring instruments provide in general less precision than is theoretically possible. Hence, for this reason, approximate specifications of state would have a better correspondence with the real situations to be studied.

In the second place, again as in the classical mechanics, it will be appreciated in the case of very complicated systems composed of many molecules that the application of the methods of exact mechanics might itself become very complicated and difficult. Hence also for this reason it may be more appropriate to use the methods of statistical quantum mechanics with the accompanying approximate specifications of state.

In addition to such 'classical' reasons, special 'quantum' reasons, making approximate specifications of state more suitable than exact ones in obtaining a theoretical insight into the temporal behaviour of systems, can arise from the previously discussed circumstance that approximate methods of observation may be more appropriate in the study of changes with time than exact ones on account of complementarity relations. This can be most simply illustrated if we consider observations that might be made on unperturbed energy eigenstates of a system in studying its transitions from one such state to another. In making such observations, limitations would be imposed on the precise determination of energy, as was seen above, since the time τ devoted to observation would have to be reduced to an interval short compared with the period T characterizing the transitions.† In the case of an almost continuous energy spectrum, this can lead to an uncertainty in energy ΔE wide enough to include a number of unperturbed energy

† The actual study of the periods T which would be involved in such transitions will be carried out in the next section.

eigenstates. This would then make it appropriate in studying such rates of transition to regard the initial state of the system as only approximately specified, and to represent the initial condition by an ensemble of similar systems suitably distributed over the different unperturbed eigenstates that could agree with the knowledge available. This method of procedure will be discussed in more detail in the next section.† Similar reasons for approximate specifications may arise in cases where it would be natural to use other quantum mechanical languages than those provided by unperturbed energy eigenfunctions in order to have states corresponding to the kind of initial observation of interest. The more general method of procedure then appropriate will also be discussed in what follows in § 101.

Attention has already been called, at the beginning of the book, to the circumstance that the idea of the precise state of a system is in any case an abstract limiting concept, and hence that the methods of exact mechanics apply to rather highly idealized situations, while the methods of statistical mechanics apply to conceptual situations involving less drastic abstraction from reality. As indicated in the foregoing, the appropriateness of the less abstract methods of statistical mechanics becomes even more evident in the quantum than in the classical mechanics, since the very process of obtaining the knowledge of precise state necessary for the application of exact mechanics might now actually interfere with those features of the behaviour which it was desired to study. Indeed, even the treatment of such a well-known problem as the transitions of a simple atomic system back and forth between the levels of a discrete and of a continuous spectrum of states actually involves considerations of an essentially statistical mechanical character, as we shall see in the next section. The special circumstance, that the applications of the exact quantum mechanics itself are so frequently concerned with statistical quantities, such as expectation values, has had a tendency to obscure the circumstance that an added incorporation of ideas, really germane to statistical mechanics, is made in the course of the treatment of such problems.

(*c*) **Approximate specification of unperturbed energy eigenstates.** It will now be profitable to give more detailed treatment to the principles and formalism to be applied in the specification of states for quantum mechanical systems that are changing with time. We may begin by considering specifications that would be appropriate when observations

† The above arguments justifying such a procedure were first clearly expressed by Pauli, *Sommerfeld Festschrift*, Leipzig, 1928, p. 30.

are made on the unperturbed energy eigenstates of a system, as defined by equations of the form

$$\mathbf{H}^0 u_k(q) = E_k^0 u_k(q), \tag{98.4}$$

where $\mathbf{H}^0$ is some suitable unperturbed Hamiltonian operator for the system.

In accordance with our previous discussion, it would in any event be desirable to limit the accuracy of such observations to the extent demanded by the restrictions on the time used in observation which were expressed by (98.2) and (98.3). These restrictions may be rewritten in the combined form

$$\Delta E \geqslant \frac{h}{\tau} \gg \frac{h}{T}, \tag{98.5}$$

where ΔE expresses the extent to which the energy of the system must be regarded as undetermined, τ is the time interval that can be suitably devoted to the observation, and T is the period characterizing the change from one condition of interest to another. As a consequence of this uncertainty in energy, combined with the approximate agreement between true and unperturbed energy levels for the system, it then becomes necessary to conclude that an observation on the system of the kind under consideration could not lead to a perfectly precise determination of its unperturbed energy eigenstate.

In the special case of an initial observation, such that only a single discrete energy level corresponding to the observation happens to be within the range of uncertainty ΔE, it would then indeed be appropriate to predict the further behaviour of the system by regarding its initial condition as precisely specified by the indicated eigenstate. Under such circumstances, no immediate necessity for the use of statistical mechanics would arise.

On the other hand, in the case of an initial observation such that a number of eigenstates corresponding to the observation are found to have energies within the range ΔE that has to be allowed, and, in general, when the observations are not sufficiently accurate to distinguish between states, it would no longer be justifiable to regard the initial condition of the system as precisely specified by any particular eigenstate. Under such circumstances it would then be necessary to turn to the methods of statistical mechanics to obtain a reasonable prediction as to the future behaviour of the system.

To employ the methods of statistical mechanics in this connexion a suitable representative ensemble would have to be set up to correspond to our actual knowledge of the initial condition of the system. For this purpose let us take the result of the initial observation as showing that the actual state of the system might be equally well represented by any one of a group of G_κ unperturbed energy eigenstates k, lying within the energy range ΔE that describes the accuracy of the observation. In accordance with the fundamental hypothesis of equal *a priori* probabilities and random *a priori* phases for quantum mechanical states, as discussed in Chapter IX, the initial condition of the representative ensemble might then be obtained by assigning equal mean values to the squares of the probability amplitudes $|a_k|^2$ for the various states k that lie in the group G_κ and random values to their phases. Noting that the probability amplitude a_k for the state k would be related to the corresponding probability coefficients c_k by the simple expression

$$a_k = c_k\, e^{-(2\pi i/h)E_k^0 t}, \tag{98.6}$$

and that exponents would cancel in diagonal terms, the initial state of the ensemble would then be described by the density matrix

$$\rho_{kl} = \overline{\overline{c_l^* c_k}} = \begin{cases} \dfrac{1}{G_\kappa}\delta_{kl} & \text{(state } k \text{ in group } G_\kappa), \\ 0 & \text{(state } k \text{ not in group } G_\kappa). \end{cases} \tag{98.7}$$

By studying the temporal behaviour of such an ensemble of systems it would then be possible to make reasonable predictions as to the later condition of the system of interest itself, as we shall see in detail in the next section.

(*d*) **Approximate specification of eigenstates in general.** We must now turn to a consideration of specifications of initial condition that could be used in the general case of observations on the characteristic eigenstates for any kind of quantity F, instead of observations restricted merely to unperturbed energy eigenstates. Attention has already been given in § 84 to the methods of specifying condition that would be appropriate in such a general case, in order to illustrate the nature of the hypothesis of equal *a priori* probabilities and random *a priori* phases.

If $\mathbf{F}$ is the operator corresponding to the quantity F on which initial observation is made, the eigenvalues F_k and eigenfunctions $u_k(q)$, for the states k now of importance, would be determined by equations of the form

$$\mathbf{F}u_k(q) = F_k\, u_k(q). \tag{98.8}$$

We shall be primarily interested in situations such that the initial observation is not accurate enough to place the system definitely in any single one of these eigenstates k. In accordance with our fundamental hypothesis as to equal *a priori* probabilities and random *a priori* phases, and in agreement with our previous equation (84.5), it would then be appropriate to represent the initial condition of the system by an ensemble with a density matrix of the form

$$\rho_{kl} = \overline{\overline{a_l^* a_k}} = \begin{cases} \dfrac{1}{G_\kappa}\delta_{kl} & \text{(state } k \text{ in group } G_\kappa\text{)}, \\ 0 & \text{(state } k \text{ not in group } G_\kappa\text{)}, \end{cases} \tag{98.9}$$

if we regard the observation as equally well represented by any one of a group of G_κ eigenstates k. Or, in agreement with (84.6), it would be appropriate to represent the initial condition by an ensemble with the density matrix

$$\rho_{kl} = \overline{\overline{a_l^* a_k}} = \rho_0 e^{-\frac{(F_k - F_0)^2}{2\sigma^2}} \delta_{kl} \tag{98.10}$$

if we wish to emphasize a decreasing probability for values of the quantity F more and more removed from a most probable value F_0.

As a new aspect of such situations, which we now appreciate as applying in the case of an initial observation from which we wish to make predictions as to further behaviour, it may be emphasized that the discussion of this section has provided us with additional quantum mechanical reasons making approximate specifications of state often specially desirable on account of the necessity for avoiding too 'violent' and too 'time-consuming' methods of measurement.

(*e*) **Remarks on the assignment of equal probabilities and random phases.** We may now conclude this long section, on the observation and specification of state in studying the temporal behaviour of quantum mechanical systems, with some general remarks on the assignment of initial probabilities and phases in constructing a suitable representative ensemble for a system of interest.

In the first place it is to be emphasized that the assignment of equal probabilities and of random phases to certain states, as illustrated for example by (98.9), is to be regarded as definitely correlated with the nature of the initial observation. In accordance with our fundamental hypothesis as to *a priori* probabilities and phases, the assignment in question is to be made to quantum mechanical states that agree equally well with our partial knowledge of actual state. Hence this equality of probability and randomness of phase will apply only in the quantum mechanical language that is determined by the kind of quantity measured, and on transformation to other quantum mechanical languages this characteristic of the representative ensemble will in general no longer be apparent.

In the second place it is to be noted that the assignment is also to be regarded as definitely correlated with the time at which the initial observation is made. As time proceeds the initial assignment of probability and phase for eigenstates in the language corresponding to the original observation will in general be lost. Indeed, as we shall see later, we can usually expect a tendency for the distribution to spread over many states together with a tendency for decrease in the randomness of phase. In this connexion it may be mentioned that the original complete randomness of phase will play a primary role in deriving this conclusion.

One further point will be of importance. Although our representative ensemble will be set up to exhibit equal probabilities and random phases for certain states in the quantum mechanical language corresponding to the quantity which we regard as approximately observed, nevertheless it is evident that transformation to a language corresponding to a quantity which is only slightly different therefrom may produce only a slight disturbance in the equality of probability and randomness of phase. Indeed, in the frequently typical case of an approximate observation, which could be equally well represented by many states, the complete invariance towards transformation of the density matrix for a uniform ensemble would make us expect wide possibilities for transformation to other languages without greatly disturbing the equality and randomness mentioned. These remarks are of interest in connexion with the circumstance that the performance of a measurement which is only approximate in character is not really sufficient for a unique assertion as to the precise quantity which we should regard ourselves as observing. We can now see, however, that slight changes in this respect cannot be expected to have a critical effect in determining the nature of our conclusions.

99. Time-proportional transitions

Quantum mechanical transitions between conditions described by unperturbed energy eigenfunctions, which have a probability of occurrence proportional to the time, are of great importance for an understanding of atomic and molecular behaviour and for statistical mechanical applications. Such time-proportional transitions are found to occur when at least one of the two quantum mechanical conditions, between which the transition takes place, is to be described by a collection of states belonging to a practically continuous spectrum of unperturbed energy eigenstates that correspond to some suitable unperturbed Hamiltonian $\mathbf{H}^0$. We are now ready to treat the theory of such transitions with the help of the approximate integration of Schroedinger's equation by the method of variation of constants as discussed in § 96 (*b*), together with the help of the principles to be used in setting up an initial representative ensemble as particularly discussed in § 98 (*c*).

(*a*) **Transition from a discrete state to a continuous spectrum of states.** Let us first consider a quantum mechanical system which has a non-degenerate, discrete, unperturbed energy eigenstate k with a value for its unperturbed energy E_k^0 that happens to lie in the range overlapped by a nearly continuous spectrum of non-degenerate, unperturbed energy levels E_n^0, which correspond to an observably different condition of the system. For example, the discrete state might be one for an undissociated molecule and the overlapping continuous states correspond to dissociation, or the discrete state might correspond to an excited atom and the continuous states to unexcited atom plus radiation.

Let us then consider that we are interested in making approximate observations on the condition of the system which would determine its actual unperturbed energy eigenstate, within an *energy range* ΔE subject in any case to the lower limit prescribed by our previous expression (98.5). Let us assume that our initial observation can be well represented by taking the system at time $t = 0$ as actually being in the discrete state k, and let us then calculate the probability of finding it at a later time t in a condition corresponding to a group of G_ν eigenstates n, belonging to the continuous spectrum and having a range ΔE of unperturbed energies E_n^0 which overlap the original energy E_k^0.

In accordance with the initial observation we can take the original condition of the system at $t = 0$ as specified by setting the quantum mechanical probability for the state k equal to unity,

$$W_k(0) = c_k^*(0)c_k(0) = 1. \tag{99.1}$$

And, in accordance with our previous result (96.12), we can then take the probability for finding the system at time t, in any particular state n, belonging to the continuous spectrum as given by

$$W_n(t) = 4|V_{nk}|^2 \frac{\sin^2 \frac{\pi}{h}(E_n^0 - E_k^0)t}{(E_n^0 - E_k^0)^2}, \tag{99.2}$$

where V_{nk} is the matrix component, corresponding to the perturbation which leads the system to change from one unperturbed state to another, and where the time 0 to t must be short enough to justify the method of variation of constants. Summing up over all such states n that lie in the group of G_ν states included in the energy range ΔE of interest, we can now write

$$P_\nu(t) = \sum_n W_n(t) = 4\sum_n |V_{nk}|^2 \frac{\sin^2\frac{\pi}{h}(E_n^0-E_k^0)t}{(E_n^0-E_k^0)^2} \tag{99.3}$$

for the total probability of finding the system at time t in one or another of these G_ν states, provided the time t is short enough to justify such a direct addition of probabilities.

To evaluate this expression for the probability of finding that transition has occurred to the final condition of interest we may now make use of our assumption that the levels n form a practically continuous spectrum. We may then write

$$dn = \sigma_n\, dE_n^0 \tag{99.4}$$

as a suitable expression for the number of levels in the range dE_n^0, where the factor σ_n can be treated as a continuous function of E_n^0. Substituting this expression in (99.3) and replacing summation by integration over the energy range ΔE which we have introduced as corresponding to the accuracy of our observation, and which we may take as including E_k^0 at its centre, we then obtain

$$P_\nu(t) = 4\int\limits_{E_k^0-\frac{1}{2}\Delta E}^{E_k^0+\frac{1}{2}\Delta E} |V_{nk}|^2 \frac{\sin^2\frac{\pi}{h}(E_n^0-E_k^0)t}{(E_n^0-E_k^0)^2}\,\sigma_n\, dE_n^0. \tag{99.5}$$

In order to treat this integral it will be convenient to make a change of variables by substituting

$$y = \frac{\pi}{h}(E_n^0-E_k^0)t, \tag{99.6}$$

where E_k^0 is of course a constant, and we treat t as a constant parameter which denotes some particular time of interest for observation. The integral then assumes the form

$$P_\nu(t) = \frac{4\pi}{h}t\int\limits_{-\frac{\pi}{2}\frac{t\Delta E}{h}}^{+\frac{\pi}{2}\frac{t\Delta E}{h}} |V_{nk}|^2\sigma_n \frac{\sin^2 y}{y^2}\, dy. \tag{99.7}$$

To evaluate the integral we now note that the integrand will be large only in a narrow range where y is approximately equal to zero, that is, where E_n^0 is approximately equal to E_k^0. This will make it approximately correct to treat $|V_{nk}|^2$ and σ_n as constants, assigning to them the values that they have at the middle of the range where $E_n^0 = E_k^0$. It will also make it approximately valid to take the limits of integration as from minus to plus infinity, provided the time t is long enough so that

$$\tfrac{1}{2}\pi \frac{t\Delta E}{h} \gg \tfrac{1}{2}\pi$$

or

$$t\Delta E \gg h. \tag{99.8}$$

The integral then assumes the form

$$P_\nu(t) = \frac{4\pi}{h}|V_{nk}|^2\sigma_n t \int_{-\infty}^{+\infty} \frac{\sin^2 y}{y^2}\, dy. \tag{99.9}$$

From a known formula of integration we now obtain

$$P_\nu(t) = \frac{4\pi^2}{h}|V_{nk}|^2\sigma_n t, \tag{99.10}$$

as the desired expression for the probability of finding that a transition has occurred at the time t from the original discrete state k to the collection of states n. We see that such a transition would take place with a probability proportional to the time.

It will be noted that this result has been obtained directly from the quantum mechanics proper without any necessity of resorting to statistical quantum mechanics. This arises from the circumstance that we have regarded the system as starting out in a definitely specified quantum mechanical state, and hence have not had to introduce any ensemble to represent its initial condition. It will be necessary to employ the methods of statistical mechanics, however, in treating the inverse problem of transition from the continuous spectrum back to the discrete state, as we shall see presently.

The expression given by (99.10) is of course only valid when the approximations made in its derivation are justified. For this to be the case the time t must be short enough to make the probability that transition has occurred small, since this assumption was involved in developing the method of variation of constants in order to integrate (96.4), and was again involved above in adding the probabilities for different states n in order to obtain (99.3). On the other hand, the time t must be long enough to satisfy the relation (99.8),

$$t\Delta E \gg h, \tag{99.11}$$

which was used in justifying the limits of integration $\pm\infty$ in (99.9).

Furthermore, if we desire an observational test of our predictions, it is evident that t must be of the order of the time T necessary for an appreciable change to occur, which by (99.10) can be taken as

$$T \approx \frac{h}{4\pi^2|V_{nk}|^2\sigma_n}. \tag{99.12}$$

Since (99.11) must at least be satisfied by such a time T, we have

$$\Delta E \gg 4\pi^2|V_{nk}|^2\sigma_n, \tag{99.13}$$

as an evaluation of the extent to which the energy of the system must be left undetermined. On the other hand, ΔE must, of course, also be small enough so as to permit a distinction of the discrete level k from its neighbours. It will be seen that the success of the method depends on the separation of the Hamiltonian $\mathbf{H}$ for the system into an unperturbed Hamiltonian $\mathbf{H}^0$ and a sufficiently small perturbation term $\mathbf{V}$.

There are a number of important atomic processes which can be described as a transition from a nearly stationary discrete state to a continuous spectrum of states of overlapping energy, and which are actually found to take place with a probability proportional to the time. These include the transition of atoms from excited states to lower levels with the emission of radiation into a narrow range of the possible continuous radiation spectrum, the radioactive decay of excited nuclei with the emission of alpha particles into a narrow range of the possible continuous spectrum of kinetic energies, the dissociation of excited atoms into ion plus electron (Rosseland-Auger effect), the dissociation of excited molecules into smaller constituents (predissociation), and certain scattering processes occurring in the case of atomic collisions which can also be described in the above language.

The actual detailed treatment of these processes is often more complicated than indicated above. In particular a simple, appropriate separation of the Hamiltonian into an unperturbed Hamiltonian plus a perturbation term may not be feasible. Nevertheless, by proper handling, differential equations of the type (96.4) can often be obtained which depend on the elements of a suitable Hermitian matrix V_{nk}, even though this may not be immediately related to a simple perturbation operator $\mathbf{V}$. The fundamental aspects of the calculation are then not greatly altered.

(*b*) **Transition from the continuous spectrum back to the discrete state.** In the preceding we have investigated the time-proportional transitions from a discrete state k to a condition which could be described by a group of states n belonging to a continuous spectrum and lying in a range ΔE of unperturbed energies E_n^0 which overlaps the original energy E_k^0. We now wish to consider the inverse process of transition back to the discrete state.

For this purpose we shall now assume that the initial condition of the system at time $t = 0$ would be one in which the system would certainly be found in one or another of the G_ν individual states n which we take as lying in the specified range ΔE. We then have unit probability at the start,

$$\sum_n W_n(0) = \sum_n c_n^*(0)c_n(0) = 1, \tag{99.14}$$

for finding the system in one or another of the states n, where we sum over all the G_ν states, together with zero probability, i.e.

$$c_k(0) = 0, \tag{99.15}$$

for being found in the discrete state k. Making use of our general formula (96.9) for the values of probability coefficients as a function of time, and noting (99.15), we can then write

$$c_k(t) = \sum_n V_{kn} \frac{e^{\frac{2\pi i}{h}(E_k^0-E_n^0)t}-1}{E_n^0-E_k^0} c_n(0) \tag{99.16}$$

together with

$$c_k^*(t) = \sum_{n'} V_{kn'}^* \frac{e^{-\frac{2\pi i}{h}(E_k^0-E_{n'}^0)t}-1}{E_{n'}^0-E_k^0} c_{n'}^*(0) \tag{99.17}$$

as expressions for the probability coefficients at a later time t corresponding to the discrete state k, where we take summations for n and n' over all the G_ν states originally occupied. Multiplying (99.16) and (99.17), we then have

$$\begin{aligned} W_k(t) &= c_k^*(t)c_k(t) \\ &= \sum_{n',n} V_{kn'}^* V_{kn} \frac{e^{-\frac{2\pi i}{h}(E_k^0-E_{n'}^0)t}-1}{E_{n'}^0-E_k^0} \frac{e^{\frac{2\pi i}{h}(E_k^0-E_n^0)t}-1}{E_n^0-E_k^0} c_{n'}^*(0)c_n(0) \end{aligned} \tag{99.18}$$

for the probability of finding the system back in state k at time $t = t$.

The above summation could evidently be evaluated only on the basis of some definite assignment of values for the initial coefficients $c_{n'}^*(0)$ and $c_n(0)$. This assignment, moreover, could actually be carried out in a great variety of ways, all compatible with the initial condition described by (99.14); and a great variety of actually different results could be obtained. Hence we shall now be interested in turning to the general point of view adopted in statistical mechanics, and making some estimate as to what might be expected as to the average behaviour of the system in a situation such as we are considering. Such an appeal to statistical methods is indeed made definitely necessary by the circumstance that our original condition is not sufficiently well defined for a precise specification of initial state.

To carry out the application of statistical methods we may now calculate the mean behaviour of the systems in an appropriate representative ensemble for the system of actual interest. For this purpose, in accordance with the postulate of equal *a priori* probabilities and random *a priori* phases, and in agreement with the discussion leading to (98.7), we may take the initial state of the representative ensemble for our system at time $t = 0$ as described by the density matrix

$$\rho_{nn'} = \overline{\overline{c_{n'}^*(0)c_n(0)}} = \begin{cases} \dfrac{1}{G_\nu}\delta_{n'n} & \text{(state } n \text{ in group } G_\nu), \\ 0 & \text{(state } n \text{ not in group } G_\nu), \end{cases} \tag{99.19}$$

where the double bar, as usual, signifies a mean for all the members of the ensemble. Combining with (99.18), we shall then obtain the simplified expression

$$\begin{aligned} P_\kappa(t) &= \overline{\overline{W_k(t)}} \\ &= \frac{1}{G_\nu}\sum_n |V_{kn}|^2 \frac{e^{-\frac{2\pi i}{h}(E_k^0-E_n^0)t}-1}{E_n^0-E_k^0}\,\frac{e^{\frac{2\pi i}{h}(E_k^0-E_n^0)t}-1}{E_n^0-E_k^0} \\ &= \frac{4}{G_\nu}\sum_n |V_{kn}|^2 \frac{\sin^2\frac{\pi}{h}(E_k^0-E_n^0)t}{(E_k^0-E_n^0)^2}, \end{aligned} \tag{99.20}$$

for the probability of finding a system in the ensemble in state k at time t, where the summation is over all states n in the group G_ν.

This summation may now be handled in the same manner as the similar expression (99.3), occurring in the preceding example, by taking $dn = \sigma_n\,dE_n$ as the number of states in the range dE_n, replacing summation by integration, and introducing suitable approximations. We are thus led to

$$P_\kappa(t) = \frac{1}{G_\nu}\frac{4\pi^2}{h}|V_{kn}|^2\sigma_n t, \tag{99.21}$$

as a suitable approximate expression for the probability of finding that a transition has occurred at time t from the condition described by the original collection of G_ν states in the range ΔE back to the discrete state k, where V_{kn} and σ_n are to be given the values which they assume when $E_n^0 = E_k^0$.

This result may now be compared with our previous expression (99.10) giving the probability for the inverse transition from state k to the group G_ν. Since we can take the number of states G_κ corresponding to the initial condition as unity, this may be written in a form similar to (99.21),

$$P_\nu(t) = \frac{1}{G_\kappa}\frac{4\pi^2}{h}|V_{nk}|^2\sigma_n t. \tag{99.22}$$

We see that both kinds of transition would take place with a probability proportional to the time. Furthermore, since the Hermitian character of the matrix elements V_{nk} will give us

$$|V_{nk}|^2 = |V_{kn}|^2, \tag{99.23}$$

we note that the probabilities per unit time for the two transitions differ only by the factors G_ν and G_κ which give the number of states in the two conditions of interest. The approximations necessary for the validity of (99.21) are similar to those which were previously discussed in connexion with the relation now given by (99.22).

(*c*) **Transition from one group of continuous states to another.** The foregoing treatments which are due to Pauli† are sufficient to illustrate the general character of time-proportional transitions. With systems composed of many molecules or other elements, however, as is often the case of typical interest in statistical mechanical studies, we can expect an unperturbed Hamiltonian operator for the system to provide energy levels all of which will in general be practically continuous rather than discrete in character. Hence we may now supplement the discussion of transitions between a discrete state and a group of continuous states by a discussion of transitions between two such groups of continuous unperturbed energy eigenstates.

To carry out the treatment we may take G_κ, G_λ, G_μ, etc., as being the 'weights' or numbers of individual states k, l, m, etc., occurring in the different groups of interest. In any particular problem we shall take the different groups as corresponding to observably different situations, but as each corresponding to the *same given range* in energy E to $E+\Delta E$, which in any case cannot be taken too narrow owing to the finite time τ allowable for observation, as already discussed in § 98 (*c*).

We must now investigate the probability for transition between the conditions specified by one such group of states k and another of states n. For the sake of generality we shall now allow for the possibility that the energy levels E_k^0 and E_n^0 of the kind present in the two spectra may

† Pauli, *Sommerfeld Festschrift*, Leipzig, 1928, p. 30.

be degenerate, and shall denote the individual unresolved states for a given level by symbols of the form

$$kr \ (r = 1, 2, ..., g_k) \quad \text{and} \quad ns \ (s = 1, 2, ..., g_n), \tag{99.24}$$

where g_k and g_n are the respective multiplicities of the two kinds of level. The symbols $c_{kr}(t)$ and $c_{ns}(t)$ will be used as expressions for the probability coefficients for the different individual eigenstates in the two groups. We may then take the initial condition of our system at time $t = 0$ as corresponding to

$$\sum_{k,r} W_{kr}(0) = \sum_{k,r} c_{kr}^*(0)c_{kr}(0) = 1, \tag{99.25}$$

together with

$$c_{ns} = 0, \tag{99.26}$$

where the summation is taken in (99.25) over all the G_κ eigenstates in the first group, and (99.26) must hold for all G_ν eigenstates in the second group. This will then put our system at time $t = 0$ definitely in the condition corresponding to the group of states k, and we may now calculate the probability for finding it in the group of states n at a later time $t = t$.

Making use of our general formula (96.9) for probability coefficients as a function of time, and noting (99.26), we can write

$$c_{ns}(t) = \sum_{k,r} V_{ns,kr} \frac{e^{\frac{2\pi i}{h}(E_n^0 - E_k^0)t} - 1}{E_k^0 - E_n^0} c_{kr}(0) \tag{99.27}$$

as an expression for the value at time t of the probability coefficient for any particular state ns in the second group of G_ν states, where $V_{ns,kr}$ is the indicated element of the perturbation matrix and the summation is to be taken over all G_κ states of the first group. Similarly, for the complex conjugate of this quantity we can write

$$c_{ns}^*(t) = \sum_{k',r'} V_{ns,k'r'}^* \frac{e^{-\frac{2\pi i}{h}(E_n^0 - E_{k'}^0)t} - 1}{E_{k'}^0 - E_n^0} c_{k'r'}^*(0). \tag{99.28}$$

Multiplying (99.27) and (99.28), and summing over all G_ν states ns in the second group, we then obtain as the total probability for finding the system at time t in the condition corresponding to these G_ν states

$$\begin{aligned}\sum_{n,s} W_{ns}(t) &= \sum_{n,s} c_{ns}^*(t)c_{ns}(t) \\ &= \sum_{n,s}\sum_{k',r'}\sum_{k,r} V_{ns,k'r'}^* V_{ns,kr} \frac{e^{-\frac{2\pi i}{h}(E_n^0 - E_{k'}^0)t} - 1}{E_{k'}^0 - E_n^0} \times \\ &\qquad \times \frac{e^{\frac{2\pi i}{h}(E_n^0 - E_k^0)t} - 1}{E_k^0 - E_n^0} c_{k'r'}^*(0)c_{kr}(0).\end{aligned} \tag{99.29}$$

As in the earlier example provided by (99.18), this expression could be given a specific evaluation only on the basis of some specific assignment of values to the initial coefficients $c_{kr}(0)$. This we do not have, however, since our knowledge of the initial situation is only sufficient to give us (99.25), which could be satisfied in many different ways leading to many actually different results. Hence, as previously in § 99 (*b*), we must resort to methods of an essentially statistical mechanical character, and try to find what we might expect on the average. To do this we may again calculate the mean behaviour in an appropriate representative ensemble, set up in agreement with the postulate of equal *a priori* probabilities and random *a priori* phases, see (98.7), so as to have the initial density matrix, at time $t = 0$,

$$\rho_{kr,k'r'} = \overline{\overline{c^*_{k'r'}(0)c_{kr}(0)}} = \begin{cases} \dfrac{1}{G_\kappa}\delta_{k'r',kr} & \text{(state } kr \text{ in group } G_\kappa), \\ 0 & \text{(state } kr \text{ not in group } G_\kappa), \end{cases} \tag{99.30}$$

where the double bar once more signifies a mean for all the members of the ensemble. Combining with (99.29) and changing to trigonometric form, equation (99.29) is then seen to give

$$P_\nu(t) = \sum_{n,s} \overline{\overline{W}}_{ns}(t) = \frac{4}{G_\kappa}\sum_{n,s}\sum_{k,r} |V_{ns,kr}|^2 \frac{\sin^2\frac{\pi}{h}(E^0_n - E^0_k)t}{(E^0_n - E^0_k)^2} \tag{99.31}$$

for the probability of finding a system in the ensemble in the group of G_ν states at time t, where the summations are now over all states ns and kr of the two groups.

To evaluate this summation we may now take

$$dn_{kr} = \sigma_{kr}\, dE^0_k$$

and

$$dn_{ns} = \sigma_{ns}\, dE^0_n \tag{99.32}$$

as expressions for the numbers of individual states of the kind indicated in the energy ranges dE^0_k and dE^0_n. Substituting in (99.31), and replacing an appropriate part of the summation by integration, this then gives

$$P_\nu(t) = \frac{4}{G_\kappa}\sum_{s=1}^{g_n}\sum_{r=1}^{g_k} |V_{ns,kr}|^2 \sigma_{ns}\sigma_{kr} \int\limits_E^{E+\Delta E}\int\limits_E^{E+\Delta E} \frac{\sin^2\frac{\pi}{h}(E^0_n - E^0_k)t}{(E^0_n - E^0_k)^2}\, dE^0_n\, dE^0_k, \tag{99.33}$$

where certain factors have been taken outside the integral signs since they may be regarded as constant over the range E to $E+\Delta E$ selected as corresponding to observation.

To perform the integrations indicated in (99.33), we may first integrate with respect to E_n^0 holding E_k^0 constant. In doing so, we may then again assume, as in the treatment of § 99 (*a*), that for a time t which is not too short the quantity $(E_n^0 - E_k^0)$ can be assumed to vary over the whole range from minus to plus infinity. Evaluating the first integration with the help of this assumption, and then carrying out the second integration between the actual limits given, we then obtain the desired expression

$$P_\nu(t) = \frac{1}{G_\kappa}\frac{4\pi^2}{h}\sum_{s=1}^{g_n}\sum_{r=1}^{g_k}|V_{ns,kr}|^2\sigma_{ns}\,\sigma_{kr}\,\Delta E\,t \qquad (99.34)$$

for the probability of finding a system of the ensemble at time t in one or another of the G_ν individual states belonging to the final condition. We again obtain a probability of transition which increases linearly with time. The approximations made in the derivation are similar to those already discussed in § 99 (*a*); and although transitions to unperturbed energy states lying outside the energy range E to $E+\Delta E$ are not absolutely prevented by the quantum mechanical form of the energy principle, they will nevertheless take place with a very small probability owing to the effect of the resonance denominator in (99.33).

In a similar manner, starting out at time $t = 0$ with a representative ensemble corresponding to a system in the condition given by the group of G_ν states of the kind n, we should be led to an expression of the form

$$P_\kappa(t) = \frac{1}{G_\nu}\frac{4\pi^2}{h}\sum_{s=1}^{g_n}\sum_{k=1}^{g_k}|V_{kr,ns}|^2\sigma_{ns}\,\sigma_{kr}\,\Delta E\,t \qquad (99.35)$$

for the average probability of finding the condition corresponding to the G_κ states of the kind k, at $t = t$. We note that there will be a simple relation between the probabilities for the two inverse transitions, owing to the equality

$$|V_{kr,ns}|^2 = |V_{ns,kr}|^2 \qquad (93.36)$$

arising from the Hermitian character of the perturbation operator **V**.

(*d*) **General formulation of transition probabilities.** It will now be desirable for later use to give a somewhat more general formulation to the foregoing results on time-proportional transitions. In accordance with all of the expressions—(99.21), (99.22), (99.34), and (99.35)—which we have obtained for such transitions, it will be seen that we can represent the average probability of finding a system in a condition ν at time $t = t$, which starts in a condition κ at time $t = 0$, by an expression of the form

$$P_\nu(t) = \frac{1}{G_\kappa} T_{\kappa\nu} t, \tag{99.37}$$

where G_κ is the number of individual unperturbed eigenstates k corresponding to the condition κ, and $T_{\kappa\nu}$ is an abbreviation for certain factors that we do not now need to express in detail. Similarly, for the inverse transition we shall have an expression of the form

$$P_\kappa(t) = \frac{1}{G_\nu} T_{\nu\kappa} t, \tag{99.38}$$

where we can write
$$T_{\nu\kappa} = T_{\kappa\nu} \tag{99.39}$$

on account of the Hermitian character of the perturbation matrices V_{nk} and V_{kn} which determine the probabilities of transition. The two expressions (99.37) and (99.38) give the probabilities at time $t = t$ of finding a system in the group of states G_ν or G_κ in a representative ensemble which has been set up initially at time $t = 0$ to correspond to a system of interest in the group of states G_κ or G_ν respectively.

We may now consider somewhat more general possibilities for the initial condition of the system of interest by assuming that the original observation at time $t = 0$ is not such as to place the system definitely in a single group of states G_κ, but such that we assign the probabilities P_κ, P_λ, P_μ, etc., to different groups of states G_κ, G_λ, G_μ, etc., in number, each of which corresponds to the range in energy ΔE limiting our observations. The initial condition of our representative ensemble can then be described by taking P_κ, P_λ, P_μ, etc., as the total probabilities for each of these groups, and by assigning equal probabilities and random phases to the individual states within each such group.

To treat this more general case it will now be convenient to define a new quantity characterizing transitions by the expression

$$A_{\kappa\nu} = \frac{1}{G_\kappa G_\nu} T_{\kappa\nu}, \tag{99.40}$$

with
$$A_{\kappa\nu} = A_{\nu\kappa} \tag{99.41}$$

from the previous equality (99.39) which results from the Hermitian character of perturbation matrices. Making use of (99.37), we can then evidently write for the rate of transition in the representative ensemble from condition κ to ν
$$Z_{\kappa\nu} = A_{\kappa\nu} G_\nu P_\kappa, \tag{99.42}$$

and making use of (99.38), can write for the rate of inverse transition from condition ν to κ the symmetrical expression

$$Z_{\nu\kappa} = A_{\nu\kappa} G_{\kappa} P_{\nu}, \tag{99.43}$$

where these new quantities are to be understood in the sense that $Z_{\kappa\nu}$ and $Z_{\nu\kappa}$ express those contributions to the total rates of change in P_{κ} and P_{ν} which are to be ascribed respectively to transitions in the ensemble from condition κ to ν and from condition ν to κ.

It will be noted that the validity of the above expressions depends on the assignment of random phases to the states in the original representative ensemble. It is this circumstance which permits us to set the probability for transition from condition κ to ν in the ensemble under consideration equal to a product of the probability for such a transition, in an ensemble where κ is the only condition represented, multiplied by the probability P_{κ} for that condition in the actual ensemble considered. Hence these expressions only apply at the time of an initial observation which can be represented in the manner described.

The results given by (99.42) and (99.43), together with the equality of $A_{\kappa\nu}$ and $A_{\nu\kappa}$ given by (99.41), will prove very important in deriving the quantum mechanical H-theorem in the next chapter.

100. The probabilities for transition by collision in Fermi-Dirac and Einstein-Bose gases

The changes in condition of a dilute gas as a result of molecular interaction or collision provide important special cases of quantum mechanical transitions which take place—at least in first approximation—with a probability which is proportional to the time. We shall devote the present section to the treatment of such transitions, applying methods similar to those developed in the preceding section both to the case of a system of Fermi-Dirac particles, characterized by antisymmetric eigenfunctions, and to the case of a system of Einstein-Bose particles, characterized by symmetric eigenfunctions. We shall thus be able to obtain the quantum analogues of the classical expressions for the probability of collisions which were used in deriving the classical form of Boltzmann's H-theorem. In the next chapter we shall use these quantum analogues in deriving a quantum form of H-theorem.

In order to treat these problems, using the language of the method of variation of constants, we shall regard the true Hamiltonian operator $\mathbf{H}$ for the gas as separated into two parts

$$\mathbf{H} = \mathbf{H}^0 + \mathbf{V}, \tag{100.1}$$

where the unperturbed Hamiltonian $\mathbf{H}^0$ corresponds to the energy of the particles neglecting their interaction, and the perturbation operator $\mathbf{V}$ corresponds to the remaining energy associated with the forces of interaction between them. Equations of the form

$$\mathbf{H}^0 U_k(q_1 q_2 \cdots q_n) = E_k^0 U_k(q_1 q_2 \cdots q_n), \tag{100.2}$$

where $q_1 \cdots q_n$ represent the coordinates for the n particles composing the system, will then determine the unperturbed eigenfunctions U_k and unperturbed eigenvalues of the energy E_k^0 for the different eigenstates k to be used in describing the condition of the system. And equations of the form

$$V_{nk} = \int U_n^* \mathbf{V} U_k \, dq \tag{100.3}$$

will determine the elements of the perturbation matrix involved in the transitions from one eigenstate k to another eigenstate n which result from the interaction or collision of particles.

It will first be necessary for us to obtain more specific expressions for these elements of the perturbation matrix. This will involve somewhat lengthy and complicated calculations, which will be carried out separately for the Fermi-Dirac and Einstein-Bose cases. We start with the Fermi-Dirac case, which is the simpler of the two.

(*a*) **Perturbation matrix for the interaction of Fermi-Dirac particles.** In accordance with our previous treatment as given in § 76 (*d*), the antisymmetric unperturbed eigenfunctions for a Fermi-Dirac gas, composed of n similar particles in a container, would be of the form

$$\begin{aligned} U_a(q_1 \cdots q_n) &= A \sum_{\mathbf{P}} (\mp 1)^{\mathbf{P}} u_k(q_1) u_l(q_2) u_m(q_3) \cdots u_r(q_n) \\ &= A \sum_{\mathbf{P}_\alpha} (\mp 1)^{\mathbf{P}_\alpha} \prod_k u_k(q_\alpha), \end{aligned} \tag{100.4}$$

where the second form of writing will serve as a convenient abbreviation. The quantity A occurring in the above expression is an appropriate normalizing factor, the symbols $u_k \ldots u_r$ represent individual eigenfunctions for a single particle in the container in question, the symbols $q_1 \cdots q_n$ stand for the coordinates of the n particles, and the formalism indicates that we are to take a summation over all permutations $\mathbf{P}$ of the particle indices $\alpha = 1, \ldots, n$, using the minus or plus sign according as the permutation is odd or even. As already discussed in § 76 (*d*), it is evident that the individual eigenfunctions $(u_k \ldots u_r)$ appearing above would all have to be different from each other in order that the total expression given by the summation should not reduce to zero. Hence the summation expressed by (100.4) will contain $n!$ terms all of which are different.

We must begin by determining the magnitude of the factor A, which will be necessary in order that such an eigenfunction U_a for the system as a whole shall be properly normalized to unity, that is, in order to secure the result

$$1 = \int \cdots \int U_a^*(q_1 \cdots q_n) U_a(q_1 \cdots q_n)\, dq_1 \cdots dq_n$$

$$= A^*A \int \cdots \int \sum_{P_\alpha} (\mp 1)^{P_\alpha} \prod_k u_k^*(q_\alpha) \sum_{P_\alpha} (\mp 1)^{P_\alpha} \prod_k u_k(q_\alpha)\, dq_1 \cdots dq_n. \tag{100.5}$$

To calculate this value let us first consider a particular term, say

$$u_k(q_1)u_l(q_2)u_m(q_3) \ldots u_r(q_n),$$

of the $n!$ different terms occurring in the *second* summation appearing in the above integral. Then it is evident that there will be one and only one term occurring in the *first* summation, namely

$$u_k^*(q_1)u_l^*(q_2)u_m^*(q_3) \ldots u_r^*(q_n),$$

which has the same assignment of coordinate and eigenfunction indices in the different factors that make up the term. Hence, making use of the orthogonality of the different individual eigenfunctions $u_k \ldots u_r$, it is evident that each of the $n!$ terms in the second summation will be multiplied by only one term in the first summation which will lead to a result other than zero on integration. And, furthermore, making use of the assumed previous normalization of the individual eigenfunctions, it is evident that such a pair of terms will give the result unity on integration. As a consequence we at once see that the normalization (100.5) can only be secured by assigning to the normalizing factor A a magnitude such that

$$A^*A = \frac{1}{n!} \quad \text{or} \quad |A| = \frac{1}{(n!)^{\frac{1}{2}}}. \tag{100.6}$$

We are now ready to proceed to the determination of the elements of the perturbation matrix, corresponding to a transition from one such eigenfunction for the system as a whole to another. To make the calculation we shall assume that the perturbation operator, corresponding to the mutual energy of interaction between the particles, would be given by an expression of the form

$$\mathbf{V}(q_1 \cdots q_n) = \tfrac{1}{2} \sum_{\beta \neq \gamma} V(q_\beta, q_\gamma), \tag{100.7}$$

where $V(q_\beta, q_\gamma)$ expresses the mutual energy of two particles as a function of their coordinates, and we take the appropriate expression for the result of summing over all pairs of particles. The expression neglects, for our present case of a *dilute* gas, the necessity of including terms depending on the coordinates for more than one pair of particles at a time.

Using the above expression for the perturbation operator, we should then have an expression of the form

$$\begin{aligned} V_{a'a} &= \tfrac{1}{2}\int \dots \int U^*_{a'}(q_1 \dots q_n) \sum_{\beta \neq \gamma} V(q_\beta, q_\gamma) U_a(q_1 \dots q_n)\, dq_1 \dots dq_n \\ &= \tfrac{1}{2} A'^* A \int \dots \int \sum_{P_\alpha} (\mp 1)^{P_\alpha} \prod_{k'} u^*_{k'}(q_\alpha) \sum_{\beta \neq \gamma} V(q_\beta, q_\gamma) \times \\ &\qquad \times \sum_{P_\alpha} (\mp 1)^{P_\alpha} \prod_{k} u_k(q_\alpha)\, dq_1 \dots dq_n \end{aligned} \quad (100.8)$$

for the element of the perturbation matrix corresponding to the transition from the state corresponding to one eigenfunction U_a to the state corresponding to a different one $U_{a'}$.

To evaluate this expression we note in the first place, as a result of the orthogonality of the individual eigenfunctions, that the expression would in any case reduce to zero if more than two of the eigenfunctions $u_{k'}(q)$ appearing in $U_{a'}$ should differ from any eigenfunction $u_k(q)$ in U_a. Two eigenfunctions in U_a could be replaced by different ones in $U_{a'}$, however, without getting a null result, since they could be assigned the coordinates q_β and q_γ which appear in $V(q_\beta, q_\gamma)$. We shall consider that $u_k(q)$ and $u_l(q)$ in U_a are replaced by $u_m(q)$ and $u_n(q)$ in $U_{a'}$. We can then write (100.8) in the more explicit form

$$\begin{aligned} V_{mn,kl} &= \tfrac{1}{2} A'^* A \int \dots \int \sum_{P} (\mp 1)^{P} u^*_m(q_1) u^*_n(q_2) u^*_o(q_3) \dots u^*_r(q_n) \times \\ &\quad \times \sum_{\beta \neq \gamma} V(q_\beta, q_\gamma) \sum_{P} (\mp 1)^{P} u_k(q_1) u_l(q_2) u_o(q_3) \dots u_r(q_n)\, dq_1 \dots dq_n, \end{aligned} \quad (100.9)$$

where the individual eigenfunctions $u_o \dots u_r$ belonging to the total eigenfunctions U_a are matched pair by pair with the eigenfunctions $u^*_o \dots u^*_r$ in $U^*_{a'}$.

To continue with the evaluation, let us now consider a particular term, say

$$u_k(q_1) u_l(q_2) u_o(q_3) \dots u_r(q_n), \quad (100.10)$$

of the different terms appearing in the *second* summation over the different permutations of the particle indices in (100.9). There are in all $n!$ such terms.

Having chosen this term, let us next consider a particular assignment of values to β and γ, say

$$\beta = 1, \qquad \gamma = 2, \quad (100.11)$$

which can lead to a non-vanishing result on integration. There are in all two such assignments ($\beta = 1$, $\gamma = 2$) and ($\beta = 2$, $\gamma = 1$) which would lead to the same final result.

Having made the above assignment for β and γ, let us finally pick out the terms in the *first* summation over the permutations of particle indices, which would lead to a non-vanishing result. These are seen to be

$$+u_m^*(q_1)u_n^*(q_2)u_o^*(q_3)\ldots u_r^*(q_n)$$

and

$$-u_m^*(q_2)u_n^*(q_1)u_o^*(q_3)\ldots u_r^*(q_n). \tag{100.12}$$

Taking account of the numbers of possibilities given above for different choices which would lead to an end result of the same magnitude, and making use of the normalization of the individual eigenfunctions $u_0 \ldots u_r$, we now see that (100.9) will reduce to

$$V_{mn,kl} = \tfrac{1}{2}A'^*An!\,2\iint [u_m^*(q_1)u_n^*(q_2)-u_m^*(q_2)u_n^*(q_1)]V(q_1,q_2)u_k(q_1)u_l(q_2)\,dq_1\,dq_2. \tag{100.13}$$

Noting the magnitudes of the normalizing factors A' and A given by (100.6), and using I_1 to denote the so-called 'direct integral'

$$I_1 = \iint u_m^*(q_1)u_n^*(q_2)V(q_1,q_2)u_k(q_1)u_l(q_2)\,dq_1\,dq_2, \tag{100.14}$$

and I_2 for the so-called 'exchange integral'

$$I_2 = \iint u_m^*(q_2)u_n^*(q_1)V(q_1,q_2)u_k(q_1)u_l(q_2)\,dq_1\,dq_2, \tag{100.15}$$

we then obtain as the desired expression for the square of the absolute magnitude of the above matrix element

$$|V_{mn,kl}|^2 = |I_1 - I_2|^2. \tag{100.16}$$

In accordance with our development of the method of variation of constants—note, for example, equation (96.12)—it will be seen that this is the quantity which would determine the probability of a collisional transition in which the numbers of particles in the individual eigenstates k, l, m, and n change in the manner indicated by

$$\begin{pmatrix} n_k = n_l = 1 \\ n_m = n_n = 0 \end{pmatrix} \rightarrow \begin{pmatrix} n_k = n_l = 0 \\ n_m = n_n = 1 \end{pmatrix}, \tag{100.17}$$

where, in agreement with the Pauli exclusion principle, the only possible occupation numbers are zero and one. It will be noted, moreover, that our derivation of (100.16) only applies to a case where the initial situation is that described by (100.17), and that the value of $|V_{mn,kl}|^2$ would reduce to zero with any other choice for the initial occupation numbers.

Hence we may now write as an entirely general expression for this important quantity, in the case of Fermi-Dirac particles,

$$|V_{nm,kl}|^2 = |I_1 - I_2|^2 n_k n_l (1-n_m)(1-n_n), \tag{100.18}$$

where the initial occupation numbers appear in the added factors in such a way as to give 1 or 0, according as the initial situation does or does not agree with that given in (100.17).

(*b*) **Perturbation matrix for the interaction of Einstein-Bose particles.** We must next determine the analogous expression for the square of the perturbation matrix in the case of Einstein-Bose particles. The computation is somewhat more complicated than for the Fermi-Dirac case, but will be carried out so as to run parallel to the treatment just given which will help to make it understandable.

In accordance with the discussion given in §76 (*d*), the symmetric eigenfunction for an Einstein-Bose gas composed of n similar particles may be expressed in the form

$$U_s(q) = A \sum_{\mathbf{P}} \mathbf{P} u_k(q_1) \ldots u_k(q_{n_k}) u_l(q_{n_k+1}) \ldots u_l(q_{n_k+n_l}) \ldots u_r(q_n)$$
$$= A \sum_{\mathbf{P}} \mathbf{P} \prod_k [u_k(q)]^{n_k}, \tag{100.19}$$

where it is hoped that the second manner of writing will serve as a convenient and informative abbreviation. The quantity A, occurring above, again represents an appropriate normalizing factor, the symbols $u_k \ldots u_r$ represent individual eigenfunctions for a single particle, the symbols $q_1 \ldots q_n$ stand for the coordinates of the n particles, and the formalism indicates that we are to take a summation over all permutations $\mathbf{P}$ of the particle indices $1,\ldots, n$. The form of writing in (100.19) gives specific expression to the possibility of now having more than one particle assigned to the same individual eigenfunction (e.g. n_k to u_k), without a reduction of the whole eigenfunction to zero.

We must again begin by determining, for the present case, the magnitude of the factor A which will secure the proper normalization,

$$1 = \int \ldots \int U_s^*(q_1 \ldots q_n) U_s(q_1 \ldots q_n)\, dq_1 \ldots dq_n$$
$$= A^*A \int \ldots \int \sum_{\mathbf{P}} \mathbf{P} \prod_k [u_k^*(q)]^{n_k} \sum_{\mathbf{P}} \mathbf{P} \prod_k [u_k(q)]^{n_k}\, dq_1 \ldots dq_n. \tag{100.20}$$

To calculate this value let us first consider a particular term, say

$$u_k(q_1) \ldots u_k(q_{n_k}) u_l(q_{n_k+1}) \ldots u_l(q_{n_k+n_l}) \ldots u_r(q_n),$$

of the $n!$ successive terms which make up the *second* summation appearing in the above integral. In accordance with the normalization and orthogonality of the individual eigenfunctions, it is then evident that if we multiply this by the similar term

$$u_k^*(q_1)\dots u_k^*(q_{n_k})u_l^*(q_{n_k+1})\dots u_l^*(q_{n_k+n_l})\dots u_r^*(q_n)$$

occurring in the *first* summation and then integrate we shall obtain the result unity. Furthermore, it is evident that we can carry out $n_k!\,n_l!\dots n_r!$ permutations of the particle indices in this latter term before we reach a situation where the orthogonality of the different individual eigenfunctions would lead to the result zero. Hence, since there are in all $n!$ successive terms in the second summation over the permutations $\mathbf{P}$, we see that the normalization (100.20) is to be obtained by assigning to the factor A a magnitude such that

$$A^*A = \frac{1}{n!\,n_k!\,n_l!\dots n_r!} \quad \text{or} \quad |A| = \frac{1}{(n!\,n_k!\,n_l!\dots n_r!)^{\frac{1}{2}}}. \tag{100.21}$$

We are now ready to proceed to the determination of the elements of the perturbation matrix corresponding to the transition from one such eigenfunction U_s for the system as a whole to another $U_{s'}$. In the present case these elements will be given by an expression of the form

$$V_{s's} = \tfrac{1}{2}\int\dots\int U_{s'}^*(q_1\dots q_n)\sum_{\beta\neq\gamma}V(q_\beta,q_\gamma)U_s(q_1\dots q_n)\,dq_1\dots dq_n$$

$$= \tfrac{1}{2}A'^*A\int\dots\int\sum_{\mathbf{P}}\mathbf{P}\prod_{k'}[u_{k'}^*(q)]^{n_{k'}}\sum_{\beta\neq\gamma}V(q_\beta,q_\gamma)\sum_{\mathbf{P}}\mathbf{P}\prod_{k}[u_k(q)]^{n_k}\,dq_1\dots dq_n, \tag{100.22}$$

where we again take $\frac{1}{2}\sum_{\beta\neq\gamma}V(q_\beta,q_\gamma)$ as an appropriate expression for the perturbation operator.

To evaluate the above, we again begin by noting that the result would reduce to zero, as a consequence of the orthogonality of the individual eigenfunctions, if more than two eigenfunctions $u_k(q)$ are changed to different eigenfunctions $u_{k'}(q)$ when we pass from U_s to $U_{s'}$. Two such eigenfunctions, however, let us say u_k and u_l, could be replaced by different ones, say u_m and u_n, in passing from U_s to $U_{s'}$, without leading to a null result since these could be the eigenfunctions to which we assign the coordinates q_β and q_γ appearing in $V(q_\beta,q_\gamma)$. Making such a replacement, the transition from U_s to $U_{s'}$ could then be described by the formalism

$$\begin{aligned} n_k &\to n_k-1 \qquad & n_m &\to n_m+1 \\ n_l &\to n_l-1 \qquad & n_n &\to n_n+1. \end{aligned} \tag{100.23}$$

And we can then write the element of the perturbation matrix in the more explicit form

$$V_{mn,kl} = \tfrac{1}{2}A'^*A \int \ldots \int \sum_{\mathrm{P}} \mathrm{P}[u_k^*(q)]^{n_k-1}[u_l^*(q)]^{n_l-1} \times \\ \times [u_m^*(q)]^{n_m+1}[u_n^*(q)]^{n_n+1}[u_o^*(q)]^{n_o} \ldots [u_r^*(q)]^{n_r} \times \\ \sum_{\beta \neq \gamma} V(q_\beta, q_\gamma) \sum_{\mathrm{P}} \mathrm{P}[u_k(q)]^{n_k}[u_l(q)]^{n_l}[u_m(q)]^{n_m} \times \\ \times [u_n(q)]^{n_n}[u_o(q)]^{n_o} \ldots [u_r(q)]^{n_r}\, dq_1 \ldots dq_n, \quad (100.24)$$

where the individual eigenfunctions from o to r are matched pair by pair in the expressions for U_s and $U_{s'}^*$.

To continue with the evaluation, let us now consider a particular term, say

$$u_k(q_1) \ldots u_k(q_{n_k})u_l(q_{n_k+1}) \ldots u_l(q_{n_k+n_l})u_o(q_{n_k+n_l+1}) \ldots u_r(q_n), \quad (100.25)$$

of the successive terms appearing in the *second* summation over the different permutations of particle indices in (100.24). There are in all $n!$ such successive terms, although some of them can be equal to one another.

Having chosen this term, let us next consider a particular assignment of values to β and γ, say

$$\beta = 1, \qquad \gamma = n_k+1, \quad (100.26)$$

which can lead to a non-vanishing result on integration. There are evidently $2n_k n_l$ such assignments in all, since β could be taken anywhere in the range $1, \ldots, n_k$ with γ in the range $n_k+1, \ldots, n_k+n_l$, or vice versa.

Having made the above assignment for β and γ, let us finally count the number of terms appearing in the *first* summation over the different permutations of particle indices in (100.24) which would lead to a non-vanishing result on integration. Evidently there will be

$$(n_k-1)!\,(n_l-1)!\,(n_m+1)!\,(n_n+1)!\,n_o! \ldots n_r! \quad (100.27)$$

such terms, in which q_1 appears in $u_m^*(q)$ with q_{n_k+1} in $u_n^*(q)$, and the same number with q_1 in $u_n^*(q)$ and q_{n_k+1} in $u_m^*(q)$.

Hence, taking account of the numbers of possibilities given above for the different choices which would lead to an end result of the same magnitude, and making use of the normalization of the individual eigenfunctions, we now see that (100.24) will reduce to

$$V_{mn,kl} = \tfrac{1}{2}A'^*An!\,2n_k n_l(n_k-1)!\,(n_l-1)!\,(n_m+1)!\,(n_n+1)!\,n_o! \ldots n_r! \times \\ \times \iint [u_m^*(q_1)u_n(q_2)+u_m^*(q_2)u_n^*(q_1)]V(q_1,q_2)u_k(q_1)u_l(q_2)\,dq_1\,dq_2. \quad (100.28)$$

Introducing the magnitudes of A' and A as given by (100.21), and using our previous symbols I_1 and I_2 for the 'direct' and 'exchange integrals'

(100.14) and (100.15), this then gives us for the desired square of the magnitude of this quantity

$$|V_{mn,kl}|^2 = \frac{[n!\,n_k\,n_l(n_k-1)!\,(n_l-1)!\,(n_m+1)!\,(n_n+1)!\,n_o!\dots n_r!]^2|I_1+I_2|^2}{(n!)^2(n_k-1)!\,(n_l-1)!\,(n_m+1)!\,(n_n+1)!\,n_k!\,n_l!\,n_m!\,n_n!\,(n_o!)^2\dots(n_r!)^2}, \tag{100.29}$$

which is easily seen to reduce to the final result

$$|V_{mn,kl}|^2 = |I_1+I_2|^2 n_k\,n_l(1+n_m)(1+n_n). \tag{100.30}$$

This is the quantity which will determine in the Einstein-Bose case the probability of collisional transitions in which a pair of particles leave the elementary states k and l to appear in the states m and n.

By comparison with (100.18), it will be seen that the two formulae could be expressed in the combined form

$$|V_{mn,kl}|^2 = |I_1 \pm I_2|^2 n_k\,n_l(1\pm n_m)(1\pm n_n), \tag{100.31}$$

where the upper signs apply to the Einstein-Bose and the lower to the Fermi-Dirac case, and the symbols n_k, n_l and n_m, n_n denote the numbers of particles originally present in the states from and to which transition occurs. In the Einstein-Bose case there would be no limit on the number of particles in such states, but in the Fermi-Dirac case the possible values would be limited to 0 and 1. In both cases the probability of transition would be proportional to the numbers of particles in the states from which transition occurs. In the Einstein-Bose case the transition would be favoured by particles already present in the states to which transition is to occur, but in the Fermi-Dirac case transition would be prevented if either of these states were already occupied.†

(c) **Time-proportional collision probabilities.** Having made the foregoing necessary investigations of the pertinent matrix elements, we are now ready to study the probabilities for different specified kinds of collision to take place in a Fermi-Dirac or in an Einstein-Bose gas. We shall first have to decide on a reasonable procedure for specifying different kinds of collision in a way that would correspond to conceptual observational possibilities. With such a specification we shall then find—at least in first approximation—that the probability for any given kind of collision to occur would be proportional to the time.

In order to see what is involved in making a reasonable specification of a given kind of collision we must first pay attention to the fact that the individual eigenfunctions $u_k(q)$ for a single particle in a container

† For the method of generalization of these results to ternary and higher order interactions see Jordan, *Zeits. f. Phys.* **45**, 766 (1927).

of any considerable size would correspond—except for the very lowest energies—to a practically continuous spectrum of unperturbed energies ϵ_k. Hence, in a finite time of observation τ, such as we should be willing to allow for studying the initial condition of the system before the probability of transition itself became too large, it is evident, in agreement with the discussion of § 98, that we could not determine the precise eigenstates occupied by the molecules since this would give a more precise knowledge of their energy than would be allowable in accordance with the Heisenberg uncertainty principle in the form

$$\tau\Delta\epsilon \geqslant h. \tag{100.32}$$

For this reason it is not feasible to specify the different kinds of collisions taking place in a gas by stating the exact states k, l and m, n from and to which transition takes place.

To meet this consideration we may proceed by treating our continuous spectrum of states for a single particle as divided into groups of states each corresponding to the *same energy range* $\Delta\epsilon$, which would be allowed by the Heisenberg relation (100.32), for the time of observation. We may use Greek letters κ, λ, μ, ν,... to designate these different groups of states in place of our previous Latin letters for the precise states, and may take the 'weight' or number of individual precise states k, l, m, n,... in the different groups of energy width $\Delta\epsilon$ as designated by the symbols g_κ, g_λ, g_μ, g_ν,... . The condition of the system can now be specified in a manner corresponding with observational possibilities by giving the numbers of particles n_κ, n_λ, n_μ, n_ν,... in these different groups of states, and the different possible kinds of collision can be specified by stating the groups of states, say κ, λ and μ, ν, from and to which transition takes place.

We now turn to the computation of the probabilities for such transitions to take place. For this purpose it will be desirable to carry through once more a complete calculation for the specific case under consideration, rather than to try to substitute into our previous formulae for probabilities of transition.

We may take the initial condition of the gas at $t = 0$ as one in which there are n_κ, n_λ, n_μ, and n_ν particles in the indicated groups of states, and take the transition as specified by the formalism

$$\begin{aligned} n_\kappa &\rightarrow n_\kappa - 1 \qquad & n_\mu &\rightarrow n_\mu + 1 \\ n_\lambda &\rightarrow n_\lambda - 1 \qquad & n_\nu &\rightarrow n_\nu + 1. \end{aligned} \tag{100.33}$$

The numbers of particles present at $t = 0$ in states other than those in the groups κ, λ, μ, and ν need not be specified since they will not be involved in the computation. The method of variation of constants must now be applied to calculate the probability of finding at a later time $t = t$ that the transition (100.33) has occurred.

In order to apply the method of variation of constants, which deals with precisely defined states, we shall use the letters i and f to designate precise states of the gas which would agree respectively with the initial and final conditions of the gas, which were more broadly specified above by the occupation numbers

$$(n_\kappa, n_\lambda, n_\mu, n_\nu) \quad \text{and} \quad (n_\kappa - 1, n_\lambda - 1, n_\mu + 1, n_\nu + 1),$$

for the groups of particle eigenstates of interest. Employing the symbols $c_i(t)$ and $c_f(t)$ for the corresponding probability coefficients, we shall then take unit probability at time $t = 0$,

$$\sum_i W_i(0) = \sum_i c_i^*(0)c_i(0) = 1, \tag{100.34}$$

for finding the system in one or another of the states i which correspond to the specified initial condition of the gas, and zero probability, i.e.

$$c_f(0) = 0, \tag{100.35}$$

for finding the system in any state f corresponding to the final condition of interest.

We must now compute the probability for finding the system in one or another of the states f at a later time $t = t$. Making use of the approximate integration of Schroedinger's equation (96.9), we can write for the value at time t of the probability coefficient for any particular state f

$$c_f(t) = \sum_i V_{fi} \frac{e^{\frac{2\pi i}{h}(E_f^0 - E_i^0)t} - 1}{E_i^0 - E_f^0} c_i(0), \tag{100.36}$$

where V_{fi} is the element of the perturbation matrix corresponding to the transition from i to f, and the summation is taken over all the initial states i. Multiplying by the appropriate similar expression for the complex conjugate quantity,

$$c_f^*(t) = \sum_{i'} V_{fi'}^* \frac{e^{-\frac{2\pi i}{h}(E_f^0 - E_{i'}^0)t} - 1}{E_{i'}^0 - E_f^0} c_{i'}^*(0), \tag{100.37}$$

and summing over all states f which agree with the final condition of interest, we then obtain

$$\sum_f W_f(t) = \sum_f c_f^*(t)c_f(t)$$

$$= \sum_f \sum_{i'} \sum_i V_{fi'}^* V_{fi} \frac{e^{-\frac{2\pi i}{h}(E_f^0 - E_{i'}^0)t} - 1}{E_{i'}^0 - E_f^0} \frac{e^{\frac{2\pi i}{h}(E_f^0 - E_i^0)t} - 1}{E_i^0 - E_f^0} c_{i'}^*(0)c_i(0) \tag{100.38}$$

as an expression for the desired probability for finding the system in the final condition of interest at time $t = t$.

To obtain an exact evaluation of this expression it would be necessary to have some precise assignment of values for the coefficients $c_i(0)$ and $c_{i'}^*(0)$. This, however, we do not have, since our only knowledge of these initial values is merely given by the fact that the total probability for finding the system in one or another of the states i must be unity, as shown by (100.34), and this can be satisfied in many different ways. Hence, as in our previous treatment of similar cases in § 99, we must now resort to methods of an essentially statistical mechanical character and try to find what we might expect on the average. To do this we may calculate the mean behaviour in an appropriate representative ensemble, set up in agreement with the postulate of equal *a priori* probabilities and random *a priori* phases—see (98.7)—so as to have the initial density matrix, at time $t = 0$,

$$\rho_{ii'} = \overline{\overline{c_{i'}^*(0)c_i(0)}} = \begin{cases} \dfrac{1}{G_i}\delta_{i'i} & \text{(state } i \text{ in group } G_i\text{)}, \\ 0 & \text{(state } i \text{ not in group } G_i\text{)}, \end{cases} \tag{100.39}$$

where G_i is the total number of precise states that correspond to our specified initial condition. Combining with (100.38), we then obtain, after reduction to trigonometric form,

$$P_f(t) = \sum_f \overline{\overline{W}}_f(t) = \frac{4}{G_i}\sum_f \sum_i |V_{fi}|^2 \frac{\sin^2\frac{\pi}{h}(E_f^0-E_i^0)t}{(E_f^0-E_i^0)^2} \tag{100.40}$$

as an expression for the probability of finding a system in the ensemble in the group of G_f states that correspond to the final condition of interest, where the summations are over all states i and f in the two groups.

To treat this expression let us begin by considering a particular initial state i, which may be defined by giving the numbers of particles n_k, n_l, n_m, n_n in each of the $g_\kappa, g_\lambda, g_\mu, g_\nu$ individual eigenstates k, l, m, n which go to make up the groups of interest. Any final state f associated with this initial state may then be specified by giving the particular states k, l and m, n involved in the transition. Hence the summation over f corresponds to a quadruple summation over k, l, m, n. Introducing the expression for $|V_{fi}|^2$ in terms of n_k, n_l, n_m, and n_n given by (100.31), replacing $(E_f^0-E_i^0)$ by $(\epsilon_m+\epsilon_n-\epsilon_k-\epsilon_l)$, and re-expressing the summation over f, we then obtain

$$P_f(t) = \frac{4}{G_i}\sum_i \sum_{k,l,m,n} |I_1 \pm I_2|^2 n_k n_l (1\pm n_m)(1\pm n_n)\frac{\sin^2\frac{\pi}{h}(\epsilon_m+\epsilon_n-\epsilon_k-\epsilon_l)t}{(\epsilon_m+\epsilon_n-\epsilon_k-\epsilon_l)^2}, \tag{100.41}$$

where the upper signs are for the Einstein-Bose case and the lower ones for the Fermi-Dirac case.

To simplify this expression we must introduce further approximations. We take the 'direct and exchange integrals' I_1 and I_2 as being practically the same for any set of individual states picked from the groups κ, λ, μ, ν; and we assign to n_k, n_l, n_m, n_n for each individual eigenstate the mean values which they have for their respective groups:

$$n_k = \frac{n_\kappa}{g_\kappa}, \qquad n_l = \frac{n_\lambda}{g_\lambda}, \qquad n_m = \frac{n_\mu}{g_\mu}, \qquad n_n = \frac{n_\nu}{g_\nu}. \tag{100.42}$$

This then permits us to take the factors containing these quantities outside the summation over k, l, m, and n, and to regard the summation over all initial states i as cancelled by the factor $1/G_i$, which is the reciprocal of the total number of initial states. We then obtain

$$P_f(t) = 4|I_1 \pm I_2|^2 \frac{n_\kappa}{g_\kappa}\frac{n_\lambda}{g_\lambda}\Big(1 \pm \frac{n_\mu}{g_\mu}\Big)\Big(1 \pm \frac{n_\nu}{g_\nu}\Big) \sum_{k,l,m,n} \frac{\sin^2 \frac{\pi}{h}(\epsilon_m + \epsilon_n - \epsilon_k - \epsilon_l)t}{(\epsilon_m + \epsilon_n - \epsilon_k - \epsilon_l)^2}. \tag{100.43}$$

The summation still remaining in this expression may now be evaluated by introducing an approximate integration in the familiar manner already employed in the preceding section. Considering the summation over n as to be taken first, we can introduce

$$dn = \frac{g_\nu}{\Delta\epsilon} d\epsilon_n \tag{100.44}$$

as an expression for the number of eigenstates of the kind n in the range $d\epsilon_n$, since g_ν is by definition the number in the interval $\Delta\epsilon$ determined by our time of observation τ. Replacing summation by integration, we can then write for a given time t

$$\sum_{k,l,m,n} \frac{\sin^2 \frac{\pi}{h}(\epsilon_n + \epsilon_m - \epsilon_k - \epsilon_l)t}{(\epsilon_n + \epsilon_m - \epsilon_k - \epsilon_l)^2}$$

$$= \frac{g_\nu}{\Delta\epsilon} t \sum_{k,l,m} \int\limits_{\epsilon}^{\epsilon+\Delta\epsilon} \frac{\sin^2 \frac{\pi}{h}(\epsilon_n + \epsilon_m - \epsilon_k - \epsilon_l)t}{(\epsilon_n + \epsilon_m - \epsilon_k - \epsilon_l)^2 t^2} \, d[(\epsilon_n + \epsilon_m - \epsilon_k - \epsilon_l)t]. \tag{100.45}$$

In carrying out the integration indicated in this expression, ϵ_n is to be treated as a variable and ϵ_m, ϵ_k, ϵ_l, and t are to be treated as parameters. On account of the presence of the resonance denominator $(\epsilon_n + \epsilon_m - \epsilon_k - \epsilon_l)^2 t^2$ it will then be seen for sufficiently large values of t

that important contributions to the integral will be made only when we have the approximate relation

$$\epsilon_n \approx \epsilon_k + \epsilon_l - \epsilon_m. \tag{100.46}$$

This will have two consequences. In the first place it will restrict the possibilities for appreciable transition to cases where the energies for the groups of states κ, λ, μ, ν satisfy within small limits the equation of conservation

$$\epsilon_\mu + \epsilon_\nu = \epsilon_\kappa + \epsilon_\lambda. \tag{100.47}$$

And in the second place it will permit us to substitute for the integral occurring in (100.45) the value π^2/h, which it would have if the limits of integration for the variable $(\epsilon_n + \epsilon_m - \epsilon_k - \epsilon_l)t$ were taken as extending all the way from minus to plus infinity. In place of (100.45) we can then write

$$\sum_{k,l,m,n} \frac{\sin^2 \frac{\pi}{h}(\epsilon_n + \epsilon_m - \epsilon_k - \epsilon_l)t}{(\epsilon_n + \epsilon_m - \epsilon_k - \epsilon_l)^2} = \frac{g_\kappa g_\lambda g_\mu g_\nu}{\Delta\epsilon} \frac{\pi^2}{h} t, \tag{100.48}$$

where the factors g_κ, g_λ, g_μ are introduced by the later summation over those states.

Substituting (100.48) into (100.43), we then obtain for the desired expression for the probability of finding a system in the ensemble at time t in the final condition of interest

$$P_f(t) = \frac{4\pi^2}{h} \frac{|I_1 \pm I_2|^2}{\Delta\epsilon} n_\kappa n_\lambda (g_\mu \pm n_\mu)(g_\nu \pm n_\nu) t. \tag{100.49}$$

Thus in first approximation, for times t which are neither too long nor too short, we do have that the probability for finding a system in the final condition specified will grow proportionally to the time t.

Taking $Z_{\kappa\lambda,\mu\nu}$ as the average frequency per unit time which we can expect for collisions in which a pair of particles are thrown from regions κ, λ to μ, ν, we can also write our result in the form

$$Z_{\kappa\lambda,\mu\nu} = A_{\kappa\lambda,\mu\nu} n_\kappa n_\lambda (g_\mu \pm n_\mu)(g_\nu \pm n_\nu), \tag{100.50}$$

where $A_{\kappa\lambda,\mu\nu}$ is introduced as an abbreviation for the factors containing the integrals I_1 and I_2 whose values depend on the particular regions κ, λ and μ, ν in the manner given by the equations of definition (100.14) and (100.15). The upper and lower signs in this expression refer respectively to the Einstein-Bose and Fermi-Dirac cases. In using this expression, it should be noted, in accordance with (100.47), that it applies to cases satisfying the relation

$$\epsilon_\mu + \epsilon_\nu = \epsilon_\kappa + \epsilon_\lambda, \tag{100.51}$$

and that probabilities of transition for cases where this is not satisfied can be taken as practically zero.

Owing to the Hermitian character of the operators involved in obtaining the integrals I_1 and I_2, it will also be seen that we can write the very important relation

$$A_{\mu\nu,\kappa\lambda} = A_{\kappa\lambda,\mu\nu} \tag{100.52}$$

for the factor which would determine the frequency of collisions that are the inverse of those which we have been considering; i.e. collisions in which particles are thrown from μ, ν to κ, λ instead of from κ, λ to μ, ν.

It may be felt that the lengthy investigation of collision probabilities which has been given in the three parts of the present section is hardly justified, in view of the many approximations which have been introduced. These include not only the usual kind of approximation employed in connexion with the method of variation of constants but, in addition, further approximations necessitated by the special nature of the application. Nevertheless, we shall be able to obtain a derivation of the quantum mechanical H-theorem with the help of the above results which will be the analogue of Boltzmann's original derivation of that theorem based on the classical behaviour of colliding molecules. This will give us insight into the physical character of statistical mechanics. Our final derivation of the H-theorem will be based on a general treatment of changes in quantum mechanical ensembles with time made with the help of the exact integration of the Schroedinger equation. To this treatment we now turn.

101. General treatment of changes in ensembles with time

We may now bring the chapter to a close by giving a general and formally precise treatment to temporal changes in ensembles, which would hold in any quantum mechanical language as well as in that provided by unperturbed energy eigenstates, and which would be valid over a long as well as over a short interval of time. We shall be specially interested in the probability at time $t = t$ for finding a member of the ensemble in any particular state n, when the state of the ensemble itself has been specified at some initial time $t = 0$. We can base the discussion on the exact integration of the generalized Schroedinger equation as given in § 96.

If we consider any single system, we have seen, (96.37), that the probability amplitude $a_n(t)$ for any state n of the system at time $t = t$

can be calculated from the probability amplitudes $a_k(0)$ for the various states k at time $t = 0$ by the expression

$$a_n(t) = \sum_k U_{nk}\, a_k(0), \tag{101.1}$$

where the U_{nk} are matrix elements corresponding to the operator $\mathbf{U}(t)$ and to the language provided by the states k. With the help of this integration of the generalized Schroedinger equation we have then seen, (96.40), that the probability for finding this system in state n at time t would be given by

$$a_n^*(t)a_n(t) = \sum_k |U_{nk}|^2 a_k^*(0)a_k(0) + \sum_{l \neq k} U_{nl}^*\, U_{nk}\, a_l^*(0)a_k(0), \tag{101.2}$$

where the first term is summed over all states k and the second term is summed over all states l and all states k which are not the same.

Hence, if we now turn our interest from a single system to an ensemble of systems, it is evident, from the significance of the density matrix and from its original definition by

$$\rho_{nm} = \overline{\overline{a_m^*\, a_n}} \tag{101.3}$$

as a mean over all members of the ensemble, that we can write

$$\rho_{nn}(t) = \sum_k |U_{nk}|^2 \rho_{kk}(0) + \sum_{l \neq k} U_{nl}^*\, U_{nk}\, \rho_{kl}(0) \tag{101.4}$$

as an expression for the probability $\rho_{nn}(t)$ of finding a member of the ensemble in state n at time t, in terms of the initial probabilities $\rho_{kk}(0)$ for such states in the ensemble and in terms of the initial non-diagonal elements for the density matrix $\rho_{kl}(0)$.

This expression assumes a specially simple form if we now regard the ensemble as having been originally set up to represent the results of an initial observation made on a system of interest at time $t = 0$. Taking this observation as being a measurement of some quantity F of which the states k are characteristic, we have then seen, especially from the discussion in § 98 (d), that the fundamental postulate of statistical mechanics will lead us to the initial assignment of equal probabilities and random phases to those states k which agree equally well with the observation. Hence, in accordance with the assignment of *random phases* (see §§ 83 (a) and 84), the non-diagonal terms of the initial matrix will be zero,

$$\rho_{kl} = 0 \quad (l \neq k), \tag{101.5}$$

and the expression for the probability of finding a member of the ensemble in state n at time t will reduce to the simple form

$$\rho_{nn}(t) = \sum_k |U_{nk}|^2 \rho_{kk}(0). \tag{101.6}$$

In making statistical mechanical applications of this result, our ultimate interest will ordinarily lie in its relation to the probability $P_\nu(t)$ of finding a member of the ensemble in a group of G_ν states n between which our approximate observations do not distinguish. We shall find the result very important for our final general derivation of the quantum mechanical H-theorem in the next chapter.

The validity of the special form of expression given by (101.6) is of course dependent on the assumption of random phases for the probability amplitudes $a_k(0)$ for the various states k at the initial time $t = 0$. In this connexion it is of interest to examine what happens to the original randomness of phase as time proceeds. For this purpose we may use (101.1) to give

$$\rho_{nm}(t) = \overline{\overline{a_m^*(t)a_n(t)}} = \sum_{l,k} U_{ml}^* U_{nk} \overline{\overline{a_l^*(0)a_k(0)}}, \tag{101.7}$$

as a general expression for the density matrix at the later time $t = t$. If the original phases of the states k, l were random, this would then become

$$\rho_{nm}(t) = \sum_k U_{mk}^* U_{nk} \rho_{kk}(0), \tag{101.8}$$

which would not in general reduce to zero for n not equal to m. Hence we must conclude that the initial randomness of phase would in general be lost as time proceeds. As we shall see in the next chapter, however, this can be regarded as compensated by an increased randomness of distribution over possible states.

We are now ready to apply the results of this chapter to the derivation of the H-theorem.

XII

THE QUANTUM MECHANICAL *H*-THEOREM

A. DERIVATION OF THEOREM

102. Definition of H for a gas

We are now ready to proceed to a discussion of the quantum mechanical H-theorem. For this purpose we shall first adopt a point of view similar to the original one of Boltzmann, defining a quantity H which directly characterizes the condition of a gas, and showing the tendency for this quantity to decrease with time towards an equilibrium value as a result of molecular collisions. We shall then turn to a more general point of view similar to that of Gibbs, defining a quantity $\bar{\bar{H}}$ which would characterize the condition of the representative ensemble for any system of interest, and showing the general tendency for this quantity to decrease with time towards a steady value where the ensemble would give a suitable representation for the condition of equilibrium. The order of treatment will thus be the same as that of the classical discussion in Chapter VI. We shall now regard it as desirable, however, to give only a brief consideration to the effect of collisions in changing the quantity H for a given sample of gas, and devote most of our attention to the more general behaviour of the quantity $\bar{\bar{H}}$ for a representative ensemble of systems and to the conclusions that can be drawn therefrom.

We must begin by defining a suitable quantum mechanical quantity H that would characterize the condition of a sample of gas. In agreement with the classical expression (47.9), which relates the quantity H to the number of states G—defined by equal volumes in the phase space—that correspond to the specified condition of the gas, it also proves profitable in the quantum statistics to introduce a similar definition

$$H = -\log G, \tag{102.1}$$

where G now denotes the number of quantum mechanical states of the gas that correspond to its specified condition.

In order to give specific content to this expression, let us now consider a sample of gas consisting of n similar particles enclosed in a container of volume v. In accordance with our previous discussion in § 100 (c), the condition of such a gas could be appropriately specified by regarding the eigenstates of energy for a single particle in the container as divided up into groups of g_κ neighbouring states, with a range in energy $\Delta\epsilon$ related to the time available for observation, and by then stating the number of particles n_κ assigned to each such group κ. For the number of unperturbed energy eigenstates G for the gas as a whole, that would correspond to such a specified condition, we could then write, in agreement with (87.14) and (87.15), in the case of an Einstein-Bose gas

$$G = \prod_\kappa \frac{(n_\kappa + g_\kappa - 1)!}{n_\kappa!\,(g_\kappa - 1)!}, \tag{102.2}$$

and in the case of a Fermi-Dirac gas

$$G = \prod_\kappa \frac{g_\kappa!}{n_\kappa!\,(g_\kappa - n_\kappa)!}, \tag{102.3}$$

where we take products $\prod$ over all groups of elementary eigenstates κ. Taking the logarithms of these quantities, using the Stirling approximation for factorial numbers as in § 88, and substituting in (102.1), we can then write the desired explicit expressions for H in the combined form

$$H = \sum_\kappa \{n_\kappa \log n_\kappa - (n_\kappa \pm g_\kappa)\log(g_\kappa \pm n_\kappa) \pm g_\kappa \log g_\kappa\}, \tag{102.4}$$

where the upper signs refer to the Einstein-Bose and the lower signs to the Fermi-Dirac case.

These expressions agree with those originally introduced in this connexion for the two cases respectively by Einstein† and by Fermi.‡ The expressions are not precise for those groups of states κ where the use made of the Stirling approximation is unjustified. This is not important, however, since we shall use the expressions to obtain an insight into the special mechanism by which the decrease in H with time takes place in gases as a result of molecular collisions, and shall give more rigorous treatment to the generalized H-theorem later.

At the correspondence principle limit, where the conclusions drawn from the quantum mechanics should approach the classical ones, the above expressions come into agreement with Boltzmann's expression for H. The attainment of this limit will be favoured by high dilution, large particle mass, and large energy for the gas as a whole; and these factors will have the general effect of making the numbers of particles n_κ in most of the groups of states κ small compared with the numbers of elementary states g_κ in the group. The expressions given by (102.4) then approach‖

$$H = \sum_\kappa \{n_\kappa \log n_\kappa - n_\kappa \log g_\kappa\}. \tag{102.5}$$

† Einstein, *Berl. Ber.* 1925, p. 3, equation (29 *a*).

‡ Fermi, *Zeits. f. Phys.* **36**, 903 (1926), equation (10).

‖ This approximation is not close enough for a correct calculation of the Sackur-Tetrode expression for the entropy of monatomic gases. See § 135 (*b*).

This agrees, in the manner to be expected, with the classical expression (47.7) for Boltzmann's H,

$$H = \sum_i \{n_i \log n_i - n_i \log \delta v_\mu\}, \tag{102.6}$$

provided we express volumes δv_μ in the μ-space in terms of units of magnitude h^r, where r is the number of degrees of freedom for the kind of molecule involved.

103. Change of H with time as a result of collisions

We must now consider the effect of collisions in leading to a decrease with time in the value of the above defined expression for the quantity H.† In accordance with our previous investigation of collisions in § 100, with the help of statistical methods we have seen in the case of a sample of gas in a condition specified by taking n_κ, n_λ, n_μ, n_ν, etc., as the numbers of particles in different possible groups of g_κ, g_λ, g_μ, g_ν, etc., elementary states, that we can take (100.50),

$$Z_{\kappa\lambda,\mu\nu} = A_{\kappa\lambda,\mu\nu}\, n_\kappa n_\lambda (g_\mu \pm n_\mu)(g_\nu \pm n_\nu), \tag{103.1}$$

as a reasonable expression for the expected number of collisions per unit time in which a pair of particles would be thrown from groups κ, λ to μ, ν. The upper signs in this expression refer to the Einstein-Bose and the lower to the Fermi-Dirac case. The quantity $A_{\kappa\lambda,\mu\nu}$ is a coefficient satisfying the equality

$$A_{\kappa\lambda,\mu\nu} = A_{\mu\nu,\kappa\lambda} \tag{103.2}$$

on account of its relation to the elements of an Hermitian perturbation matrix. And the value of this coefficient is practically zero except for collisions which satisfy in close approximation the energy relation

$$\epsilon_\mu + \epsilon_\nu = \epsilon_\kappa + \epsilon_\lambda. \tag{103.3}$$

In agreement with (103.1), we may now write, as an appropriate expression for the rate of change in the number of particles in group κ,

$$\frac{dn_\kappa}{dt} = -\sum_{\lambda,(\mu\nu)} A_{\kappa\lambda,\mu\nu}\, n_\kappa n_\lambda (g_\mu \pm n_\mu)(g_\nu \pm n_\nu) + \\ + \sum_{\lambda,(\mu\nu)} A_{\mu\nu,\kappa\lambda}\, n_\mu n_\nu (g_\kappa \pm n_\kappa)(g_\lambda \pm n_\lambda), \tag{103.4}$$

where we sum over all groups λ and over all pairs of groups $(\mu\nu)$, and make a double inclusion of those terms in the summation for which $\lambda = \kappa$.

With the help of this expression we can now investigate the rate of change with time in the quantity H, which itself depends on the

† The treatment follows that of Pauli, *Sommerfeld Festschrift*, Leipzig, 1928, p. 30.

numbers of particles n_κ in the different groups. Starting with the explicit expression (102.4) which we have given,

$$H = \sum_\kappa \{n_\kappa \log n_\kappa - (n_\kappa \pm g_\kappa)\log(g_\kappa \pm n_\kappa) \pm g_\kappa \log g_\kappa\}, \tag{103.5}$$

and differentiating with respect to the time, we obtain

$$\frac{dH}{dt} = \sum_\kappa \{\log n_\kappa - \log(g_\kappa \pm n_\kappa)\} \frac{dn_\kappa}{dt}. \tag{103.6}$$

Substituting (103.4), this then gives us

$$\frac{dH}{dt} = -\sum_\kappa \sum_{\lambda,(\mu\nu)} A_{\kappa\lambda,\mu\nu}\, n_\kappa n_\lambda (g_\mu \pm n_\mu)(g_\nu \pm n_\nu) \log \frac{n_\kappa}{g_\kappa \pm n_\kappa} +$$

$$+ \sum_\kappa \sum_{\lambda,(\mu\nu)} A_{\mu\nu,\kappa\lambda}\, n_\mu n_\nu (g_\kappa \pm n_\kappa)(g_\lambda \pm n_\lambda) \log \frac{n_\kappa}{g_\kappa \pm n_\kappa}, \tag{103.7}$$

where the summations are over all groups κ and λ, and over all pairs of groups $(\mu\nu)$. Changing to a summation over all pairs of groups $(\kappa\lambda)$, this then becomes

$$\frac{dH}{dt} = -\sum_{(\kappa\lambda),(\mu\nu)} A_{\kappa\lambda,\mu\nu}\, n_\kappa n_\lambda (g_\mu \pm n_\mu)(g_\nu \pm n_\nu) \log \frac{n_\kappa n_\lambda}{(g_\kappa \pm n_\kappa)(g_\lambda \pm n_\lambda)} +$$

$$+ \sum_{(\kappa\lambda),(\mu\nu)} A_{\mu\nu,\kappa\lambda}\, n_\mu n_\nu (g_\kappa \pm n_\kappa)(g_\lambda \pm n_\lambda) \log \frac{n_\kappa n_\lambda}{(g_\kappa \pm n_\kappa)(g_\lambda \pm n_\lambda)}. \tag{103.8}$$

And taking the arithmetical mean of this expression with the equivalent result obtained by interchanging the pair of indices $(\kappa\lambda)$ with the pair $(\mu\nu)$, we finally obtain

$$\frac{dH}{dt} = -\frac{1}{2} \sum_{(\kappa\lambda),(\mu\nu)} A_{\kappa\lambda,\mu\nu}\, n_\kappa n_\lambda (g_\mu \pm n_\mu)(g_\nu \pm n_\nu) \times$$

$$\times \left\{\log \frac{n_\kappa n_\lambda}{(g_\kappa \pm n_\kappa)(g_\lambda \pm n_\lambda)} - \log \frac{n_\mu n_\nu}{(g_\mu \pm n_\mu)(g_\nu \pm n_\nu)}\right\} -$$

$$-\frac{1}{2} \sum_{(\kappa\lambda),(\mu\nu)} A_{\mu\nu,\kappa\lambda}\, n_\mu n_\nu (g_\kappa \pm n_\kappa)(g_\lambda \pm n_\lambda) \times$$

$$\times \left\{\log \frac{n_\mu n_\nu}{(g_\mu \pm n_\mu)(g_\nu \pm n_\nu)} - \log \frac{n_\kappa n_\lambda}{(g_\kappa \pm n_\kappa)(g_\lambda \pm n_\lambda)}\right\}. \tag{103.9}$$

To proceed farther, *as an essential step* we must now introduce the equality between the coefficients for inverse collisions

$$A_{\mu\nu,\kappa\lambda} = A_{\kappa\lambda,\mu\nu} \tag{103.10}$$

which has already been mentioned above. Substituting in (103.9), we then obtain, after some reorganization, the desired form of expression

$$\frac{dH}{dt} = -\frac{1}{2}\sum_{(\kappa\lambda),(\mu\nu)} A_{\kappa\lambda,\mu\nu}[n_\kappa n_\lambda(g_\mu \pm n_\mu)(g_\nu \pm n_\nu) - n_\mu n_\nu(g_\kappa \pm n_\kappa)(g_\lambda \pm n_\lambda)] \times$$
$$\times \log\frac{n_\kappa n_\lambda(g_\mu \pm n_\mu)(g_\nu \pm n_\nu)}{n_\mu n_\nu(g_\kappa \pm n_\kappa)(g_\lambda \pm n_\lambda)}. \qquad (103.11)$$

This result, however, has a value which can only be zero or negative. To see this, we note for any term in the indicated summation that the factor $A_{\kappa\lambda,\mu\nu}$ could not be negative on account of its physical significance, and that expressions of the form $n_\kappa n_\lambda(g_\mu \pm n_\mu)(g_\nu \pm n_\nu)$ could also not be negative, since even in the Fermi-Dirac case—where the negative sign appears—we should have to have $g_\mu \geqslant n_\mu$ on account of the Pauli exclusion principle. Thus the various terms in the summation will all be of the form

$$A(x-y)\log\frac{x}{y}, \qquad (103.12)$$

with none of the quantities A, x, y negative. As a consequence, these terms will themselves be quantities which can only assume zero or positive values. Hence, noting the negative sign for the whole summation in (103.11), we see that this expression for the expected rate of change in H must satisfy the relation

$$\frac{dH}{dt} \leqslant 0. \qquad (103.13)$$

Furthermore, as the necessary and sufficient condition for a null value of this rate of change, it is readily seen that it would be necessary to satisfy

$$\log\frac{n_\kappa}{g_\kappa \pm n_\kappa} + \log\frac{n_\lambda}{g_\lambda \pm n_\lambda} = \log\frac{n_\mu}{g_\mu \pm n_\mu} + \log\frac{n_\nu}{g_\nu \pm n_\nu} \qquad (103.14)$$

for all collisions for which $A_{\kappa\lambda,\mu\nu}$ has an appreciable value, and hence, in general, in accordance with the discussion of § 100 (c), for all collisions approximately satisfying the energy relation

$$\epsilon_\kappa + \epsilon_\lambda = \epsilon_\mu + \epsilon_\nu, \qquad (103.15)$$

already mentioned above. As the simultaneous solution of these two equations, we then obtain an expression of the form

$$\log\frac{n_\kappa}{g_\kappa \pm n_\kappa} + \alpha + \beta\epsilon_\kappa = 0, \qquad (103.16)$$

where α and β are constants independent of κ. Or, solving for n_κ, this can also be written in the more familiar form

$$n_\kappa = \frac{g_\kappa}{e^{\alpha+\beta\epsilon_\kappa} \mp 1}, \qquad (103.17)$$

in agreement with our previous expressions (89.6) and (89.7) for the distributions prevailing at equilibrium in the case of Einstein-Bose and Fermi-Dirac gases.

We thus indeed find, also in the quantum mechanics, that we can define a suitable quantity H to characterize the condition of a gas which will have the desired property of exhibiting a tendency to decrease with time as a result of collisions, unless the observed distribution of the molecules is already such as to satisfy the accepted expression for equilibrium. No elaborate discussion of this finding will be necessary, since the detailed discussion of the classical H-theorem, given in the various parts of § 49, can be readily adapted to the present quantum mechanical analogue. The main points of that discussion may again be mentioned, and it will be seen on reflection that they would apply also in the quantum mechanical case.

The H-theorem is a principle, having a statistical rather than an exact character, which makes it a probable rather than a certain prediction that H will decrease with time as a result of collisions when a gas is observed to be in a condition differing from equilibrium. The theorem is such as to make it plausible to assume that continued observations would show a continued decrease in H with time towards a minimum equilibrium value. The theorem relates to the changes that may be expected in the condition rather than in the precise state of a gas; and the conclusion of statistical mechanics, as to a preferred direction of change in *conditions* of the system, stands in no conflict with the conclusion of exact mechanics, as to dynamical reversibility, which also in the quantum mechanics (§ 95) makes it possible for a system to pass in either direction through a series of precise *states*. As a consequence of the theorem, we can expect the final behaviour of an isolated gas to consist in a succession of fluctuations in the value of H around its minimum, with large fluctuations therefrom occurring very infrequently.

In concluding this discussion it is of interest to note that the foregoing quantum mechanical derivation of the H-theorem is in one respect made very simple by the possibility of introducing a direct equality (103.10) between the probability coefficients for inverse collisions. This equality arises, as we have seen in § 100 (c), on the one hand from the statistical assumptions involved in our treatment of the groups of states between which transitions are taken as occurring, and on the other hand from the Hermitian character of the perturbation operator corresponding to the energy of interaction between the particles.

Analogous simple methods of treatment would be possible also in the case of molecules more complicated than particles.

In the corresponding classical treatment of the H-theorem as given in §48 (b), it was not possible to proceed in general in such a simple manner, since the more complete specification which is given to molecular states in the classical mechanics rules out the existence of inverse collisions except in the special case of spherical molecules. This made it necessary to base the general classical derivation on a somewhat complicated consideration of Boltzmann's closed cycles of corresponding collisions. We shall return to a discussion of questions related to this difference in §116.

104. Definition of $\overline{\overline{H}}$ for a representative ensemble of systems

(a) **Fine-grained and coarse-grained probabilities in the quantum mechanics.** We are now ready to turn to more powerful methods of treating the approach of quantum mechanical systems towards equilibrium which are analogous to the methods introduced into the classical statistics for this purpose by Gibbs. We may begin by defining a suitable quantity $\overline{\overline{H}}$ which would characterize the condition of the representative ensemble for any system of interest.

In the quantum mechanics, as in the classical mechanics, the definition of the quantity $\overline{\overline{H}}$ for a representative ensemble depends on a distinction between what may be called the fine-grained and coarse-grained probabilities for the different possible states of members of the ensemble. The necessity for such a distinction depends, as before, on the circumstance that our actual interest in *statistical mechanics* lies ordinarily in conditions of a system which are not sufficiently well known for a precise specification of state. We must first discuss the character of the two kinds of probabilities.

Let us consider an ensemble in a state which is described by the elements of the density matrix

$$\rho_{nm} = \overline{\overline{a_m^* a_n}}, \tag{104.1}$$

where the quantities a_m^* and a_n denote probability amplitudes for the various states of interest m, n, etc., for a single system, and we take a mean of the indicated product over all members of the ensemble as indicated by the double bar. As the probability of finding a member of the ensemble in a particular state n, we can then write

$$\rho_{nn} = \overline{\overline{W_n}} = \overline{\overline{a_n^* a_n}}. \tag{104.2}$$

This may be called the *fine-grained probability* for the state n since it gives the probability for finding that state if precise observations are made on the different states n of the members of the ensemble.

In the situations of interest for statistical mechanics, however, we regard ourselves as making observations of only limited accuracy, which are in general insufficient to distinguish between neighbouring states n having similar properties. Hence, if we consider a group of G_ν states n between which our measurements do not distinguish, we shall also be interested in the total probability P_ν of finding a member of the ensemble in this group. For this we can evidently write

$$P_\nu = \sum_{n=1}^{G_\nu} \rho_{nn}, \tag{104.3}$$

where we take a sum over all states n in the group. In terms of this expression, for the probability of the whole group of G_ν states, we may then write

$$P_n = \frac{P_\nu}{G_\nu} = \frac{1}{G_\nu} \sum_{n=1}^{G_\nu} \rho_{nn}, \tag{104.4}$$

as an equation defining the *coarse-grained probability* for the states n in the group of G_ν states.

As in the classical mechanics, we thus define the coarse-grained probability for a state as the mean of fine-grained probabilities taken over neighbouring states of nearly identical properties. And it will be seen that our present fine-grained and coarse-grained probabilities, ρ_{nn} and P_n for a quantum mechanical state n, are indeed the natural analogues for our previous probabilities, $\rho\, dq_1 \ldots dp_f$ and $P\, dq_1 \ldots dp_f$ for a classical state defined by the infinitesimal region $dq_1 \ldots dp_f$ in the phase space.

The quantum mechanical language, to be used in setting up our coarse-grained probabilities P_n, and the range and character of states n, to be used in taking the mean, will, of course, be determined by the nature of the observational processes that we have in view. Both the fine-grained and the coarse-grained probabilities will be regarded as normalized to unity with

$$\sum_k \rho_{kk} = 1 \tag{104.5}$$

and

$$\sum_k P_k = 1, \tag{104.6}$$

where we take summations over all possible states k.

(*b*) **General expression for $\overline{\overline{H}}$.** In terms of such coarse-grained probabilities we may now give a general definition of the quantity $\overline{\overline{H}}$ for an ensemble, by the summation

$$\overline{\overline{H}} = \sum_k P_k \log P_k \tag{104.7}$$

taken over all possible states k. As the work of the present chapter proceeds, we shall find that this definition does give us a quantity which can be regarded as measuring deviation from equilibrium and which tends to decrease with time to an equilibrium value.

The defining expression given by (104.7) can also be written in other forms that prove useful.

Since $\log P_k$ will evidently have the same value for all states k that lie in each particular group of G_κ states for which P_k is the mean of ρ_{kk}, and since the summation of P_k or ρ_{kk} over such a group would give the same result, we can also write (104.7) in the form

$$\bar{\bar{H}} = \sum_k \rho_{kk} \log P_k. \tag{104.8}$$

We thus see that $\bar{\bar{H}}$ can be regarded as the mean value in the ensemble of $\log P_k$ for all states k. The symbolism that we have adopted thus agrees with our general convention of using a double bar to indicate a mean value for an ensemble as a whole.

Noting the relation (104.4), between the total probability P_ν for a group of G_ν states and the mean probability P_n for the members of the group, it will be seen that we can also rewrite our defining expression (104.7) in the form

$$\bar{\bar{H}} = \sum_\kappa \sum_{k=1}^{G_\kappa} \frac{P_\kappa}{G_\kappa} \log \frac{P_\kappa}{G_\kappa} = \sum_\kappa P_\kappa \log \frac{P_\kappa}{G_\kappa}, \tag{104.9}$$

where the final result arises from the circumstance that the summation over $k = 1$ to $k = G_\kappa$ would consist in the addition of G_κ equal terms.

In closing these general remarks on the definition of $\bar{\bar{H}}$, we may emphasize the distinction between the two different quantities

$$\sum_k \rho_{kk} \log \rho_{kk} \quad \text{and} \quad \sum_k P_k \log P_k, \tag{104.10}$$

the latter being the one by which $\bar{\bar{H}}$ is actually defined.† Only in special cases will a representative ensemble be in such a state that the two quantities are equal (e.g. an ensemble representing a measurement which has just been made, see § 106 (*a*), equation (106.4)).

† The definition of $\bar{\bar{H}}$ given above is in agreement with that of Pauli, *Sommerfeld Festschrift*, Leipzig, 1928, p. 30. We have, however, perhaps laid more emphasis on the circumstances that the quantity is defined in terms of coarse-grained probabilities, and is primarily characteristic of the condition of a representative ensemble for a system of interest rather than immediately characteristic of the system itself. The definition does *not* agree, on the other hand, with that of Klein, *Zeits. f. Phys.* **72**, 767 (1931), who uses the first of the two quantities (104.10) and thus fails to give clearly appropriate recognition to the main function of statistical mechanics in providing a treatment for the behaviour of systems in conditions that are not sufficiently defined for a precise specification of state. See § 106 (*c*).

It is also of interest to note that our present quantum mechanical definition of the quantity $\bar{\bar{H}}$ by $\sum_k P_k \log P_k$ is indeed the natural analogue of our previous classical definition by $\int \ldots \int P \log P \, dq_1 \ldots dp_f$.

(c) **Relation between $\bar{\bar{H}}$ and H.** It will now be informing to examine the relation between our present quantity $\bar{\bar{H}}$, which characterizes the condition of the representative ensemble for a system of interest, and our previous quantity H, which more directly characterizes the condition of a single system of the kind under discussion. In accordance with (104.9), we can write the definition of $\bar{\bar{H}}$ in the form

$$\bar{\bar{H}} = -\sum_\kappa P_\kappa \log G_\kappa + \sum_\kappa P_\kappa \log P_\kappa, \tag{104.11}$$

where P_κ is the probability of finding a member of the ensemble in an observed condition κ, and G_κ is the number of quantum mechanical states for a single system which would correspond to this condition. On the other hand, in accordance with (102.1), we can write the definition of H in the form

$$H_\kappa = -\log G_\kappa, \tag{104.12}$$

where G_κ is the number of quantum mechanical states that correspond to a condition κ actually specified for a single system. Combining the two expressions, we may then write

$$\begin{aligned} \bar{\bar{H}} &= \sum_\kappa P_\kappa H_\kappa + \sum_\kappa P_\kappa \log P_\kappa \\ &= \bar{\bar{H}}_{\text{syst}} + \sum_\kappa P_\kappa \log P_\kappa, \end{aligned} \tag{104.13}$$

where we use the symbol $\bar{\bar{H}}_{\text{syst}}$ to denote the mean value, for all members of the ensemble, of the quantity H applying in the original form of H-theorem to a single system. For the special case of a system which has been observed to be in a given condition, the above relation would reduce at the instant of observation to

$$\bar{\bar{H}} = H, \tag{104.14}$$

since P_κ would then be unity for the particular condition κ that had been found and otherwise zero.

Equations (104.13) and (104.14) are the quantum mechanical analogues of our previous classical equations (51.33) and (51.34). They will permit similar views in the quantum as in the classical mechanics, concerning the relation between the original form of the H-theorem as already discussed in § 103, and the generalized form of that theorem to the discussion of which we now turn.

105. Change of $\overline{\overline{H}}$ with time by the method of transition probabilities

We are now ready to consider a derivation of the generalized H-theorem, in which we shall use our previous results as to transition probabilities to demonstrate the tendency for $\overline{\overline{H}}$ to decrease with time in the case of an ensemble which has been set up to represent some system of interest. For this purpose we shall assume that we are going to be concerned with approximate observations on the system of interest which can be correlated with groups of G_κ unperturbed energy eigenstates k which correspond to some unperturbed Hamiltonian $\mathbf{H}^0$ for the system. In accordance with the character of our interest, and in agreement with (104.9), we can then take

$$\overline{\overline{H}} = \sum_\kappa P_\kappa \log \frac{P_\kappa}{G_\kappa} \tag{105.1}$$

as an appropriate expression for the value of $\overline{\overline{H}}$ in our representative ensemble, where P_κ is the probability at some time of interest of finding a member of the ensemble in a specified condition κ corresponding to such a group of G_κ unperturbed states. Differentiating (105.1) with respect to the time t, we then obtain as a general expression for the rate of change in $\overline{\overline{H}}$ with time

$$\begin{aligned}\frac{d\overline{\overline{H}}}{dt} &= \sum_\kappa (\log P_\kappa - \log G_\kappa + 1)\frac{dP_\kappa}{dt} \\ &= \sum_\kappa (\log P_\kappa - \log G_\kappa)\frac{dP_\kappa}{dt},\end{aligned} \tag{105.2}$$

where the second form of writing is justified by the consideration that the total probability of finding a member of the ensemble in one or another condition κ cannot change with time.

To make a specific application of this general expression, let us now consider that we are interested in determining the value of $d\overline{\overline{H}}/dt$ at an initial time when the ensemble is set up to represent an initial observation on the condition of the system of interest. In accordance with our fundamental hypothesis as to *a priori* probabilities and phases, the initial state of the ensemble will then be specified in such a manner as to give probabilities for the different states of its members that agree with the information made available by the observation, and also in particular to give *random phases* for all the states that are represented. Thus the ensemble will be in a state to satisfy the conditions necessary for applying the general formulation given to transition probabilities in § 99 (d), and we can take our previous expressions (99.42) and (99.43),

$$Z_{\kappa\nu} = A_{\kappa\nu}\, G_\nu\, P_\kappa \tag{105.3}$$

and

$$Z_{\nu\kappa} = A_{\nu\kappa}\, G_\kappa\, P_\nu, \tag{105.4}$$

as giving those contributions to the total rates of change in P_κ and P_ν which are to be ascribed respectively to transitions in the ensemble from condition κ to ν and from condition ν to κ.

This will then let us write

$$\frac{dP_\kappa}{dt} = -\sum_\nu (A_{\kappa\nu}\, G_\nu\, P_\kappa - A_{\nu\kappa}\, G_\kappa\, P_\nu), \tag{105.5}$$

where we take a summation over all conditions ν, as an expression for the rate of change in P_κ with time. Substituting in (105.2), this now gives us

$$\frac{d\bar{\bar{H}}}{dt} = -\sum_{\kappa,\nu} (A_{\kappa\nu}\, G_\nu\, P_\kappa - A_{\nu\kappa}\, G_\kappa\, P_\nu)\log\frac{P_\kappa}{G_\kappa} \tag{105.6}$$

for the rate of change in $\bar{\bar{H}}$. Furthermore, by taking the arithmetical mean of this expression with the equivalent result obtained by interchanging the indices κ and ν, we can write

$$\frac{d\bar{\bar{H}}}{dt} = -\frac{1}{2}\sum_{\kappa,\nu} (A_{\kappa\nu}\, G_\nu\, P_\kappa - A_{\nu\kappa}\, G_\kappa\, P_\nu)\left(\log\frac{P_\kappa}{G_\kappa} - \log\frac{P_\nu}{G_\nu}\right). \tag{105.7}$$

To proceed farther, *as an essential step* we must now introduce the equality between the coefficients for inverse transitions

$$A_{\kappa\nu} = A_{\nu\kappa}, \tag{105.8}$$

which we have found in § 99 as a consequence of the Hermitian character of perturbation operators and of our statistical assumptions. Substituting above, we can then rewrite (105.7) in the form

$$\frac{d\bar{\bar{H}}}{dt} = -\frac{1}{2}\sum_{\kappa,\nu} A_{\kappa\nu}\, G_\kappa\, G_\nu\left(\frac{P_\kappa}{G_\kappa} - \frac{P_\nu}{G_\nu}\right)\left(\log\frac{P_\kappa}{G_\kappa} - \log\frac{P_\nu}{G_\nu}\right). \tag{105.9}$$

Since the factor $A_{\kappa\nu}\, G_\kappa\, G_\nu$ cannot be negative merely on grounds of physical significance, and since the remaining combination of factors cannot be negative on the combined grounds of significance and form (see remarks made in connexion with (48.13)), we are then led to the expected result

$$\frac{d\bar{\bar{H}}}{dt} \leqslant 0 \tag{105.10}$$

for the rate of change in $\bar{\bar{H}}$ with time.

Furthermore, as the necessary and sufficient requirement for a null value of this rate of change, it will be seen that it would be necessary to have

$$\frac{P_\kappa}{G_\kappa} = \frac{P_\nu}{G_\nu} = \text{const.} \tag{105.11}$$

for all conditions κ, ν, etc., that are represented in the ensemble or that could be reached from those present. Or, making use of the definition of coarse-grained probability P_n in terms of the total probability P_ν for a group of G_ν states, as given by (104.4), we can also take the uniformity of coarse-grained probability

$$P_n = \text{const.} \tag{105.12}$$

for the unperturbed eigenstates n corresponding to such conditions as the requirement for equilibrium. In interpreting this result it is to be remembered, in accordance with the method of transition probabilities, that the different possible states n are all regarded as having unperturbed energies E^0 that lie within a given range E to $E+\Delta E$ having a width determined by the character of our approximate observations. It will be noted that the above requirement for a stationary value of $\bar{\bar{H}}$ is also the requirement for a minimum value of that quantity.

We thus find that we can predict a decreasing value of $\bar{\bar{H}}$ in the case of an ensemble set up to represent an initial approximate observation on the unperturbed energy eigenstates n of a system of interest. We also note that the requirement for a minimum equilibrium value of $\bar{\bar{H}}$, as given above by $P_n = \text{const.}$, would be satisfied with our previous choice of the microcanonical ensemble to represent a system in a steady condition of equilibrium.

The foregoing simple and satisfactory derivation of the generalized quantum mechanical H-theorem is originally due to Pauli.† The derivation is very informing in giving a real insight into the mechanism by which the approach of quantum mechanical systems towards equilibrium takes place.

The derivation involves some measure of approximation, since the method of transition probabilities is based on an approximate rather than on an exact integration of Schroedinger's equation. Furthermore, although the derivation is applicable to systems in general, it assumes that our knowledge of the condition of a system is to be obtained from the particular kind of measurements that correspond to the eigenstates for some unperturbed Hamiltonian $\mathbf{H}^0$ for the system. Moreover,

† Pauli, *Sommerfeld Festschrift*, Leipzig, 1928, p. 30.

although the derivation tells us that $\bar{\bar{H}}$ can be expected to move in the direction of equilibrium at the time an ensemble is set up to represent the results of an initial measurement, it does not provide all the information that would be desirable and possible as to the continued behaviour of the ensemble. The first two of these inadequacies will be removed and the last at least partially remedied by the derivation to be given in the next section.

106. Change of $\bar{\bar{H}}$ with time from the exact integration of the Schroedinger equation

(*a*) **The representative ensemble.** We now turn to a more general, more rigorous, and more complete consideration of the generalized H-theorem. The treatment will be based on results obtained from the exact integration of the Schroedinger equation, rather than on the results as to transition probabilities obtained from the approximate integration of that equation by the method of variation of constants.

In the interests of generality we shall now assume that we are going to be concerned with observations on the system of interest which could be regarded as approximate measurements of the value of any desired kind of quantity F pertaining to the system. Such observations can be taken as giving approximate information as to the probability that the system would actually behave as though in one or another of the precise eigenstates k, that would be determined by equations of the form

$$\mathbf{F}u_k(q) = F_k\, u_k(q), \tag{106.1}$$

where $\mathbf{F}$ is the quantum mechanical operator, and the F_k and $u_k(q)$ are the eigenvalues and eigenfunctions corresponding to the observable F. In accordance with this character of the contemplated observations, it will then be appropriate to treat the behaviour of the system of interest with the help of a representative ensemble described in the language provided by these eigenstates k.

In order to set up such a representative ensemble, let us consider that we make an initial observation on the system of interest, at time $t = 0$, which furnishes values at that time for the probabilities $P_\kappa(0)$ for different groups of G_κ neighbouring eigenstates k between which our approximate measurements cannot distinguish. In accordance with the definition given in § 104 (*a*), we can then take the coarse-grained probabilities $P_k(0)$ for different states k at that time as given by

$$P_k(0) = \frac{P_\kappa(0)}{G_\kappa}. \tag{106.2}$$

Hence, in accordance with our fundamental hypothesis as to equal *a priori* probabilities and random *a priori* phases, we can now describe the initial state of the representative ensemble by the density matrix

$$\rho_{kl}(0) = P_k(0)\,\delta_{kl}. \qquad (106.3)$$

Thus the initial state of the ensemble, at time $t = 0$, is such as to give equal fine-grained and coarse-grained probabilities

$$\rho_{kk}(0) = P_k(0) \qquad (106.4)$$

for any state k of the kind that concerns us, and such as to make the non-diagonal elements of the density matrix equal to zero, with

$$\rho_{kl}(0) = 0 \quad (k \neq l) \qquad (106.5)$$

in the language that is being employed.

To use this ensemble for predicting the result of later observations on the system of interest it will be necessary to have some principle which will describe the temporal behaviour of the coarse-grained probabilities $P_k(t)$ for the different states k involved in our measurements. It is for this reason that we are interested in determining the laws of behaviour for the quantity $\overline{\overline{H}}$, characterizing the distribution of coarse-grained probability in accordance with the definition of § 104 (*b*),

$$\overline{\overline{H}}(t) = \sum_k P_k(t) \log P_k(t). \qquad (106.6)$$

We shall show that the general behaviour of this quantity is to exhibit a tendency to decrease with time, and hence, as we shall examine in some detail in § 109, to proceed to as uniform a distribution of coarse-grained probabilities as is consistent with the energy principle.

(*b*) **Needed results of the exact integration of Schroedinger's equation.** In the interests of rigour we shall base our present demonstration of the tendency for $\overline{\overline{H}}$ to decrease with time on results obtained from the exact integration of the generalized Schroedinger equation for isolated systems. This will also satisfy our interests for a more complete treatment of the temporal behaviour of $\overline{\overline{H}}$ than could be obtained from the use of transition probabilities, since the exact integration provides expressions holding at any time $t = t$ later than the initial time $t = 0$.

In accordance with the exact integration of Schroedinger's equation, as carried out in § 96, and as applied to the behaviour of ensembles in § 101, we can take the fine-grained probability $\rho_{nn}(t)$ for finding a member of the ensemble in any state n at time $t = t$ as given by an equation (101.4) of the form

$$\rho_{nn}(t) = \sum_k |U_{nk}|^2 \rho_{kk}(0) + \sum_{l \neq k} U^*_{nl}\, U_{nk}\, \rho_{kl}(0), \qquad (106.7)$$

where the quantities $\rho_{kl}(0)$ are the elements of the density matrix at an initial time $t = 0$, and the quantities U_{nk} are the elements of the unitary transformation matrix that correspond to the operator $\mathbf{U}(t)$ in the manner discussed in § 96 (*d*). Furthermore, if the density matrix is *diagonal* at the initial time $t = 0$, as will be true in the cases of present interest as a consequence of the hypothesis of random *a priori* phases (see (106.3)), this expression will reduce to the simpler form (101.6)

$$\rho_{nn}(t) = \sum_k |U_{nk}|^2 \rho_{kk}(0). \tag{106.8}$$

This result, together with the important summation properties for the elements of the transformation matrix, which correspond to its unitary character,

$$\sum_n U^*_{nl} U_{nk} = \delta_{kl} \tag{106.9}$$

and

$$\sum_k U^*_{mk} U_{nk} = \delta_{nm}, \tag{106.10}$$

provide those consequences of the exact integration of Schroedinger's equation which will be necessary for the demonstration.

In addition to making use of these quantum mechanical results, it will also be necessary in carrying out the demonstration to make use of the essentially positive character of a certain combination of quantities, as given by the expression

$$x \log x - x \log y - x + y \geqslant 0, \tag{106.11}$$

where x and y are themselves quantities which cannot assume negative values. The validity of this expression has already been demonstrated in § 51 (*b*). The equality sign applies only when x and y are equal, and the combination assumes more and more rapidly increasing positive values as y is made increasingly different from x.

(*c*) **The Klein relation as a necessary lemma.** Before proceeding to our ultimate investigation of the temporal behaviour of the quantity $\sum_k P_k \log P_k$, by which $\bar{\bar{H}}$ is defined, it will first be necessary to deduce a lemma as to the behaviour of the similar quantity $\sum_k \rho_{kk} \log \rho_{kk}$, containing fine-grained instead of coarse-grained probabilities. The relation to be derived was first obtained by Klein.†

To deduce this lemma we begin by defining an auxiliary quantity

$$Q_{kn} = \rho_{kk}(0) \log \rho_{kk}(0) - \rho_{kk}(0) \log \rho_{nn}(t) - \rho_{kk}(0) + \rho_{nn}(t), \tag{106.12}$$

where $\rho_{kk}(0)$ and $\rho_{nn}(t)$ are the fine-grained probabilities for the states k and n respectively at the initial time $t = 0$ and at any later time of

† Klein, *Zeits. f. Phys.* **72**, 767 (1931).

interest $t = t$. In accordance with (106.11) and with the circumstance that the probabilities $\rho_{kk}(0)$ and $\rho_{nn}(t)$ could not themselves be negative on account of their physical significance, Q_{kn} is itself a quantity which cannot be less than zero.

Multiplying the essentially positive quantity Q_{kn} by the essentially positive quantity $|U_{nk}|^2$, and summing over all values of n and of k, we then obtain a result which can be written out in detail in the form

$$\begin{aligned}\sum_{n,k} |U_{nk}|^2 Q_{kn} = \sum_k \sum_n U^*_{nk} U_{nk} \rho_{kk}(0) \log \rho_{kk}(0) - \sum_n \sum_k U^*_{nk} U_{nk} \rho_{kk}(0) \log \rho_{nn}(t) - \\ - \sum_k \sum_n U^*_{nk} U_{nk} \rho_{kk}(0) + \sum_n \sum_k U^*_{nk} U_{nk} \rho_{nn}(t) \\ \geqslant 0. \qquad (106.13)\end{aligned}$$

By applying the consequence of Schroedinger's equation given by (106.8) to the second of the explicit terms given above, and by applying the summation properties for the elements of the transformation matrix given either by (106.9) or (106.10) to the other three terms, we then obtain the result

$$\sum_k \rho_{kk}(0) \log \rho_{kk}(0) - \sum_n \rho_{nn}(t) \log \rho_{nn}(t) - \sum_k \rho_{kk}(0) + \sum_n \rho_{nn}(t) \geqslant 0.$$

Or, since the summations of the probabilities $\rho_{kk}(0)$ and $\rho_{nn}(t)$ over all states will both have the value unity and hence cancel out, this can also be written in the finally desired form

$$\sum_k \rho_{kk}(0) \log \rho_{kk}(0) \geqslant \sum_n \rho_{nn}(t) \log \rho_{nn}(t). \qquad (106.14)$$

The result given by (106.14) may be called the Klein relation. Its validity depends on the diagonal character of the density matrix $\rho_{kl}(0)$ at the time of initial observation $t = 0$, since otherwise it would be necessary to use the complete expression (106.7) rather than the simplified expression (106.8) in evaluating the probabilities $\rho_{nn}(t)$ at time $t = t$. This initial character of the density matrix will be justified in our applications, nevertheless, by the fundamental hypothesis as to *a priori* phases.

It is of interest to compare the Klein inequality as given by (106.14) with the quantum mechanical equality

$$\sum_k [\rho(0) \log \rho(0)]_{kk} = \sum_n [\rho(t) \log \rho(t)]_{nn}, \qquad (106.15)$$

which can be readily obtained as an example of the general invariance of the trace of a matrix towards unitary transformation. The quantities $\sum_k \rho_{kk} \log \rho_{kk}$ and $\sum_k [\rho \log \rho]_{kk}$ become identical when the density matrix ρ_{kl} is itself diagonal.

It is also of interest to make a comparison with the classical equality (51.12)

$$\int \dots \int \rho(0) \log \rho(0)\, dq_1 \dots dp_f = \int \dots \int \rho(t) \log \rho(t)\, dq_1 \dots dp_f \quad (106.16)$$

which played an essential role in the derivation of the classical form of the generalized H-theorem. In the derivation of the quantum form of the theorem the similar role will be played by the Klein inequality (106.14) rather than by the equality (106.15).

The quantum mechanical possibility for the quantity $\sum_k \rho_{kk} \log \rho_{kk}$ to exhibit decreased values at times $t = t$ later than the initial time $t = 0$ may well be contrasted with the classical constancy of the quantity $\int \dots \int \rho \log \rho\, dq_1 \dots dp_f$. Thus the quantum mechanics implies some tendency, as time proceeds, towards a spreading out of fine-grained probability over different states which has no analogue in the classical mechanics. This quantum mechanical tendency towards increased uniformity of distribution over precisely defined states arises, of course, from the statistical features inherent in the exact quantum mechanics itself even when applied directly to a single system. This is illustrated by the finding that a quantum mechanical system started out initially at $t = 0$ with unit quantum mechanical probability, $W_k(0) = 1$, for being in a single particular precise state k must be expected at a later time $t = t$ to exhibit a distribution of such probabilities $W_n(t)$ over various states n, as can be calculated with the help of (96.40).

As often emphasized in the foregoing, such statistical features, which arise in the quantum mechanics proper as applied to precisely specified states, must be kept distinct from those additional statistical features which arise, in what we specifically name statistical mechanics, from the circumstance that we then deal with systems in conditions which are not sufficiently defined for a precise specification of state. More particularly in the present case, the quantum mechanical tendency towards a more uniform distribution of fine-grained probabilities ρ_{kk} must not be confused, as has sometimes been done in the past, with the whole of the statistical mechanical tendency towards a more uniform distribution of coarse-grained probabilities P_k as time proceeds.

(*d*) **Derivation of the generalized H-theorem.** We are now ready to investigate the temporal behaviour of coarse-grained probabilities $P_k(t)$, by deriving the generalized H-theorem showing the tendency for

$$\bar{\bar{H}} = \sum_k P_k \log P_k \quad (106.17)$$

to decrease with time. To carry out the demonstration we may begin by writing

$$\bar{\bar{H}}(0)-\bar{\bar{H}}(t) = \sum_k P_k(0)\log P_k(0) - \sum_n P_n(t)\log P_n(t) \qquad (106.18)$$

as an evident expression for the difference between the value of $\bar{\bar{H}}$ at an initial time $t = 0$ and its value at any later time $t = t$. Moreover, in accordance with the circumstance that our ensemble will be set up initially to correspond to the results of an observation made on the system of interest at that time, it is evident, in agreement with (106.4), that we can substitute

$$\sum_k P_k(0)\log P_k(0) = \sum_k \rho_{kk}(0)\log \rho_{kk}(0) \qquad (106.19)$$

in terms of the initial fine-grained probabilities; and in accordance with the possibility of expressing $\bar{\bar{H}}$ in either of the forms (104.7) or (104.8), it is evident that we can also substitute

$$\sum_n P_n(t)\log P_n(t) = \sum_n \rho_{nn}(t)\log P_n(t), \qquad (106.20)$$

where a partial change to fine-grained probabilities has been made. Doing so, our expression for the difference between the two values of $\bar{\bar{H}}$ becomes

$$\bar{\bar{H}}(0)-\bar{\bar{H}}(t) = \sum_k \rho_{kk}(0)\log \rho_{kk}(0) - \sum_n \rho_{nn}(t)\log P_n(t). \qquad (106.21)$$

To treat this expression, we may first add the Klein relation, as derived above,

$$\sum_k \rho_{kk}(0)\log \rho_{kk}(0) \geqslant \sum_n \rho_{nn}(t)\log \rho_{nn}(t). \qquad (106.22)$$

This gives us

$$\bar{\bar{H}}(0)-\bar{\bar{H}}(t) \geqslant \sum_n \rho_{nn}(t)\log \rho_{nn}(t) - \sum_n \rho_{nn}(t)\log P_n(t). \qquad (106.23)$$

Adding and subtracting quantities which sum up to unity, this can then be rewritten in the form

$$\bar{\bar{H}}(0)-\bar{\bar{H}}(t) \geqslant \sum_n [\rho_{nn}(t)\log \rho_{nn}(t) - \rho_{nn}(t)\log P_n(t) - \rho_{nn}(t) + P_n(t)]. \qquad (106.24)$$

In accordance with (106.11), however, we note that the quantity in square brackets, which is summed over all values of n, has a form which is essentially positive, so that we can write

$$\sum_n [\rho_{nn}(t)\log \rho_{nn}(t) - \rho_{nn}(t)\log P_n(t) - \rho_{nn}(t) + P_n(t)] \geqslant 0, \qquad (106.25)$$

with the equality sign holding only with $\rho_{nn}(t)$ equal to $P_n(t)$ for all states n. This then provides all that is necessary for the demonstration of the H-theorem, since, by combining with (106.24), we are led to the expected expression

$$\bar{\bar{H}}(0)-\bar{\bar{H}}(t) \geqslant 0 \tag{106.26}$$

as a description of the temporal behaviour of the quantity $\bar{\bar{H}}$ in a representative ensemble which is set up at the time of initial observation $t = 0$ for the purpose of making predictions at any later time $t = t$ as to the condition of some system of interest.

In accordance with this result, we see that the quantity $\bar{\bar{H}}$ for such a representative ensemble could never exhibit at a later time a value larger than it had when the ensemble was initially set up. We also note, in accordance with (106.22) and (106.25), that $\bar{\bar{H}}$ could exhibit a later value equal to its initial one only if the Klein relation were actually satisfied by the sign of equality rather than inequality, and if in addition the fine-grained and coarse-grained probabilities, $\rho_{nn}(t)$ and $P_n(t)$, were actually the same for all states n. These two requirements would indeed be met in the case of an ensemble set up to represent a condition of equilibrium with the ρ_{nn} chosen so as to be independent of time and equal to P_n for all states n. Nevertheless, except in the case of equilibrium, the possibility for a value $\bar{\bar{H}}(t)$ as large as the initial one $\bar{\bar{H}}(0)$ must evidently be regarded as very exceptional.

In addition to concluding that $\bar{\bar{H}}(t)$ will in general be smaller than $\bar{\bar{H}}(0)$ in non-equilibrium cases, it also appears reasonable to conclude that the ensemble—at least for a considerable period after the initial time $t = 0$—would in general exhibit successively lower and lower values of $\bar{\bar{H}}(t)$ as we go to later and later times $t = t$. Here, as in the analogous part of the classical treatment, it is evident that our arguments can be only of a qualitative and plausible kind,† unless we should be willing and able to apply the methods of exact mechanics to a specific example.

To apply general, qualitative arguments, let us consider an ensemble which is set up at $t = 0$ to correspond to an initial observation which we take for simplicity as showing that the system of interest might be

† The qualitative arguments given here may be regarded to some extent as supplemented by the recent results of Pauli and Fierz, *Zeits. f. Phys.* **106**, 572 (1937), who have shown, by introducing a somewhat conventional definition of the probabilities of making different kinds of observations, that the long-time-average value of $\bar{\bar{H}}$ for a perfectly isolated system would differ appreciably from the minimum possible value of $\bar{\bar{H}}$ less and less frequently as the size of the groups characteristic of the observations made on the system increases indefinitely. It must be emphasized, however, that the times contemplated by Pauli and Fierz are far longer than any which can be of physical interest or than those which we shall actually consider in §§ 109 and 111, and, further, that these authors purposely neglect those contributions to the decrease of $\bar{\bar{H}}$ with time which are a direct consequence of the quantum mechanical disturbance of the system by observations and which we shall discuss under the term quantum mechanical spreading.

equally well in any one of a single group κ of G_κ neighbouring states k between which our approximate measurements do not distinguish. At the initial time $t = 0$, as a consequence of our rules of procedure, we must then take $\rho_{kk}(0)$ and $P_k(0)$ equal, with the same positive value for all states k in the group κ, and with the value zero for all states not in the group κ. As time proceeds, however, we may expect at least for a considerable interval after $t = 0$ a gradual increase in the probabilities $\rho_{nn}(t)$ and $P_n(t)$ for states outside the original group κ and a decrease in probabilities for states inside that group.

With such a gradual change in the state of the ensemble there would then be two factors operating which would tend to give continually decreasing values of $\bar{\bar{H}}$. In accordance with the first of the two inequalities (106.22), on which the above derivation of the H-theorem has been based, one factor would consist in the circumstance that the occupation of more and more states n would lead to lower and lower values of $\sum_n \rho_{nn} \log \rho_{nn}$. And in accordance with the second of those inequalities (106.25), a second factor would consist in the circumstance that the unequal rates at which ρ_{nn} would grow for different states outside the original group κ and diminish for states inside that group would lead to the development of inequalities between the values of ρ_{nn} and P_n for different states n. These two factors on which the decrease of $\bar{\bar{H}}$ with time depends may well be given descriptive names.

The first of the two factors, the tendency for the occupation of more and more states n to lead to lower and lower values of $\sum_n \rho_{nn} \log \rho_{nn}$, is a specific quantum effect which may be called the *quantum mechanical spreading* of fine-grained probability. There is no similar classical effect since the classical conservation of fine-grained probability ρ, as we follow any moving point in the phase space, leads to a constant value in time for the classical quantity $\int \dots \int \rho \log \rho \, dq_1 \dots dp_f$. The quantum mechanical analogue of this classical result is the constancy of the matrix trace $\sum_k [\rho \log \rho]_{kk}$, expressed by (106.15), and holding in any quantum mechanical language during the undisturbed behaviour of the ensemble. Hence the changes in $\bar{\bar{H}}$ connected with quantum mechanical spreading are not to be thought of as associated with the undisturbed development in time of the systems composing the ensemble, but rather with the irreversible disturbance associated with the act of subsequent observation on some particular kind of state n. It will be appreciated that the effects of quantum mechanical spreading are of predominant importance when the observations on the system approach in definition the limits permitted by the quantum mechanics, and may be expected to be unimportant in the opposite limit of observations so crude that the accompanying disturbance of the system may be neglected.

The second of the two factors on which the decrease in $\bar{\bar{H}}$ with time depends, namely the tendency for inequalities to develop between the fine- and coarse-grained probabilities ρ_{nn} and P_n, is an effect, also present in the classical statistics, which may be called the *inhomogeneous redistribution* of fine-grained probability. In the classical statistics it provided the only mechanism for the decrease in $\bar{\bar{H}}$ with time. In the quantum statistics it will be appreciated that the effects of inhomogeneous redistribution will grow in importance with increasing size of the groups of states κ involved in the observations made on the system, since as the size of these groups increases, on the one hand the distinction between fine- and coarse-grained probabilities becomes more marked, and, on the other hand, the disturbances produced by the observations themselves, i.e. the specific quantum mechanical features of the observations, become less and less important. In fact, it has recently been shown by Pauli and Fierz,† if we consider the limiting case where the groups κ comprise a very large number of states, that the inhomogeneous redistribution is then alone sufficient to assure an ultimate decrease of $\bar{\bar{H}}$ to a value which in general differs from the minimum possible value less and less as the groups κ increase in size.

In view of the foregoing discussion, it is evident, in the case of an ensemble set up to represent the observation of a non-equilibrium condition, that both the factors of quantum mechanical spreading and of inhomogeneous redistribution could be expected to operate at least over a considerable period of time in the direction of lower and lower values of $\bar{\bar{H}}$. Hence we shall now regard ourselves as justified in expecting not only that the quantity

$$\bar{\bar{H}} = \sum_k P_k \log P_k \tag{106.27}$$

can never assume a value larger than its initial one in the case of an ensemble set up to represent an approximate observation of the state k for some system of interest, but also in general that this quantity would tend to assume smaller and smaller values as time proceeds until we approach the minimum value of $\bar{\bar{H}}$ permitted by the nature of the systems composing the ensemble and by such restrictions as are imposed by the conservation of energy. As the final condition of the

† Pauli and Fierz, *Zeits. f. Phys.* **106**, 572 (1937).

ensemble, corresponding to equilibrium for the system of interest, we can then take as uniform a distribution of the coarse-grained probabilities P_k for different states k as is physically possible.

(*e*) **Further discussion of the generalized *H*-theorem.** In concluding this treatment of the generalized H-theorem, several further points of interest may be mentioned.

In the first place it is well to emphasize that the tendency towards a preferred direction of change, which has been found in the foregoing, is a consequence of the circumstance that our ensembles are initially set up in a particular kind of state, with equal fine-grained and coarse-grained probabilities and with random phases for all states k of the kind involved in observation. The directional character of the behaviour then resulting stands in no conflict with the possibility of specifying exact behaviours for a single system which would agree with the principle of dynamical reversibility, also valid in the quantum mechanics, as we have seen in § 95. It will be noted, however, that this directional character is now not only due to the development of inequalities between fine-grained and coarse-grained probabilities as in the classical statistics, but also to the purely quantum effect of spreading which would operate, for example, in the case of a single system started out in a single precisely defined state.

It is of interest in this connexion to note again that the Klein relation (106.14), which controls the general character of the quantum effect of spreading when more than a single initial state is involved, has a validity which definitely depends on the assumption of random phases at the initial time $t = 0$. The assumption of random phases at the later time $t = t$ would make it necessary to reverse the sign of inequality in that expression. It is also of interest to recall once more our earlier finding, as demonstrated by (101.8), that an initial randomness of phase for the states k represented in an ensemble would in general be lost as time proceeds. We now see that the loss in the randomness of the phases of states can be regarded in a sense as compensated by a gain in the randomness of distribution over states.

In connexion with the tendency for $\overline{\overline{H}}$ to decrease with time towards a minimum equilibrium value, inquiry might be made into the rate at which the approach to equilibrium would occur. It is clear that no general quantitative answers to this important question can be given since the rate would depend on the specific character of the system of interest. Nevertheless, it is evident that we can often expect a rapid decrease in $\overline{\overline{H}}$ to start at once as a consequence of the combined effects of quantum mechanical spreading and of inhomogeneous redistribution, and that an important measure of redistribution in the values of the P_k might take place in a short time. In the quantum as in the classical statistics there is no reason for assuming that very long periods of time would be necessary for the approach to equilibrium.

In the foregoing treatment of the H-theorem it was assumed that the representative ensemble, set up to correspond to an initial approximate measurement of a particular kind of quantity pertaining to the system of interest, was then to be used for making reasonable predictions as to the result of a later measurement on the system of just the *same kind*. To give a more general treatment, it will be noted that our derivation of the relation $\bar{\bar{H}}(0)-\bar{\bar{H}}(t) \geqslant 0$ depends in principle, in the first place, on the nature of the initial state of the ensemble as expressed by $\rho_{kl}(0) = P_k(0)\,\delta_{kl}$, and, in the second place, on the summation properties exhibited by the elements of the transformation matrix U_{nk} in accordance with the unitary character of the transformation from $t = 0$ to $t = t$. If, however, we denote by $S_{n'n}$ the elements of the unitary matrix for transforming at a given time from an original quantum mechanical language to another language corresponding to states n', it would also be possible to consider a combined unitary transformation to a later time and to another language, with the help of the matrix elements

$$T_{n'k} = \sum_n S_{n'n}\,U_{nk}, \tag{106.28}$$

which would themselves have unitary character in accordance with the discussion of § 67 (e). Using similar methods to those employed above in deriving the H-theorem, this then makes it possible to obtain an expression of the form

$$\bar{\bar{H}}(0)-\bar{\bar{H}}'(t) = \sum_k P_k(0)\log P_k(0) - \sum_{n'} P_{n'}(t)\log P_{n'}(t) \geqslant 0, \tag{106.29}$$

provided we still assume the same initial character of the ensemble expressed by $\rho_{kl}(0) = P_k(0)\,\delta_{kl}$.

This result is of considerable interest in giving some information as to the probabilities $P_{n'}$ for states of a different kind from those originally approximately observed. The result applies also if we merely transform to a new language at the original time $t = 0$. It demonstrates that we cannot increase $\bar{\bar{H}}$ above its initial value either by going to a later time or by contemplating a different kind of observation. It is to be remarked, however, that investigation shows that there are no longer the same reasons for expecting continually decreasing values of $\bar{\bar{H}}'$ with time if the states n' are of a very different kind from the original states n.

We may close this long section on the generalized H-theorem by laying emphasis on the appropriate and powerful character of the statistical methods of Gibbs, also in the quantum mechanics, in investigating the theory of temporal behaviour.

107. Application of H-theorem to interacting systems

In the foregoing derivation of the H-theorem it has been assumed that the system of interest could be regarded as completely isolated from other systems and from its surroundings. It was this circumstance which made it appropriate to represent the system of interest by an ensemble of members, each of which was itself taken as isolated, and thus made it then possible to base the derivation on results obtained from the integration of the Schroedinger equation in the form applying to isolated systems. It is evident, however, that we can also be interested in situations where a transfer of energy could take place between two or more interacting systems or between an enclosed system and its surroundings.

Such a situation would obviously arise when we wish to treat the behaviour of some system of interest which has been purposely placed in good 'thermal contact' with a heat reservoir. Such a situation would also arise, moreover, even when we wish to treat the behaviour of a practically isolated system, provided the possible fluctuation of energy between the system proper and its walls is not small enough so that its effects can be appropriately neglected. Indeed, as we shall see in the next chapter, the two cases, of a system in good 'thermal contact' with a heat reservoir or in what we shall call 'essential' rather than 'perfect' isolation from its surroundings, will be of primary importance in providing a statistical mechanical explanation of the principles of thermodynamics as applied to the behaviour of ordinary macroscopic systems.

In order to study the consequences of the H-theorem when applied to interacting systems, let us consider two *individual systems* S_1 and S_2 which can be placed in contact with each other in such a way as to form a *combined system* S_{12}, which can itself be regarded as isolated and hence treated by the methods already developed for isolated systems. Let us introduce symbols of the forms k_1, l_1, m_1, etc., and k_2, l_2, m_2, etc., to denote states, of the kind proposed for observation, for the two individual systems S_1 and S_2 respectively. And assuming weak interaction between the individual systems when in contact, let us regard the possible states for the combined system S_{12} as consisting of associated pairs of the above states, which can be denoted by symbols of the form $k_1 k_2$, $k_1 l_2$, $l_1 k_2$, $l_1 l_2$, etc.

At some initial time $t = 0$ we may now regard ourselves as making approximate observations on the states of the two individual systems S_1 and S_2, and if interaction has not already been established between S_1 and S_2 we may regard them as placed in contact with each other at $t = 0$ in order to form the combined system S_{12}. In accordance with our general statistical procedure, the initial conditions of the two systems S_1 and S_2 can then be represented by setting up a pair of ensembles, with the diagonal density matrices

$$\rho_{k_1 l_1}(0) = P_{k_1}(0)\,\delta_{k_1 l_1} \quad \text{and} \quad \rho_{k_2 l_2}(0) = P_{k_2}(0)\,\delta_{k_2 l_2}, \tag{107.1}$$

that correspond to our observational determination of the coarse-grained probabilities P_{k_1} and P_{k_2}, and to the choice of random phases for different states. Furthermore, an appropriate ensemble to give a suitable representation of the initial condition of the combined system S_{12} can evidently be set up by regarding each member of the ensemble for S_1 as combined separately with each member of the ensemble for S_2, thus giving a new ensemble, with a number of members equal to the product of the numbers in the two uncombined ensembles. The initial density matrix for the combined ensemble would then be expressed by

$$\rho_{k_1 k_2, l_1 l_2}(0) = P_{k_1 k_2}(0)\,\delta_{k_1 k_2, l_1 l_2}. \tag{107.2}$$

This matrix would be diagonal in agreement with the random phases in the two original ensembles; and the coarse-grained probabilities in the new ensemble would be given in terms of those for the original ensembles by

$$P_{k_1 k_2}(0) = P_{k_1}(0) P_{k_2}(0) \tag{107.3}$$

as a consequence of the method of construction.

At the initial time $t = 0$ we may then write as expressions for $\bar{\bar{H}}$, corresponding to the distributions that represent the two individual systems S_1 and S_2,

$$\bar{\bar{H}}_1(0) = \sum_{k_1} P_{k_1}(0) \log P_{k_1}(0) \quad \text{and} \quad \bar{\bar{H}}_2(0) = \sum_{k_2} P_{k_2}(0) \log P_{k_2}(0). \tag{107.4}$$

And we may write as an expression for $\bar{\bar{H}}$, corresponding to the distribution that represents the combined system S_{12},

$$\begin{aligned}
\bar{\bar{H}}_{12}(0) &= \sum_{k_1, k_2} P_{k_1 k_2}(0) \log P_{k_1 k_2}(0) \\
&= \sum_{k_1, k_2} P_{k_1}(0) P_{k_2}(0) \{\log P_{k_1}(0) + \log P_{k_2}(0)\} \\
&= \sum_{k_1} P_{k_1}(0) \log P_{k_1}(0) + \sum_{k_2} P_{k_2}(0) \log P_{k_2}(0) \\
&= \bar{\bar{H}}_1(0) + \bar{\bar{H}}_2(0),
\end{aligned} \tag{107.5}$$

where the second form of writing is made possible by (107.3) and the next to the last form by the circumstance that the summation of a probability over all possible states will give unity.

Let us now regard the combined ensemble as left undisturbed by external influences from the time $t = 0$ to some later time $t = t$. In accordance with our derivation of the H-theorem in the simple form for isolated systems, we may then expect a value for $\bar{\bar{H}}_{12}$ at this later time satisfying in any case the relation

$$\bar{\bar{H}}_{12}(t) \leqslant \bar{\bar{H}}_{12}(0), \tag{107.6}$$

and indeed, if we wait long enough, we may expect a tendency for the quantity $\bar{\bar{H}}_{12}$ to decrease to a final equilibrium value.

At this later time $t = t$ we may now consider the possibility of making renewed observations on the two individual systems S_1 and S_2, and also of separating the two systems from each other if that should be desired. At this later time we can write

$$\bar{\bar{H}}_1(t) = \sum_{k_1} P_{k_1}(t) \log P_{k_1}(t) \quad \text{and} \quad \bar{\bar{H}}_2(t) = \sum_{k_2} P_{k_2}(t) \log P_{k_2}(t) \tag{107.7}$$

as explicit expressions for the values of $\bar{\bar{H}}$ corresponding to the distributions of probabilities for states of the two individual systems S_1 and S_2, and can write

$$\bar{\bar{H}}_{12}(t) = \sum_{k_1,k_2} P_{k_1k_2}(t) \log P_{k_1k_2}(t) \tag{107.8}$$

as an explicit expression for the value of $\bar{\bar{H}}$ corresponding to the distribution for the combined system S_{12}. It will not now be possible in general, however, to set $P_{k_1k_2}$ equal to $P_{k_1}P_{k_2}$, since there may be a tendency for particular states k_1 for systems of the kind S_1 to be preferentially associated with particular states k_2 for systems of the kind S_2. Nevertheless, we can in any case write the evident relations

$$\sum_{k_1,k_2} P_{k_1k_2}(t) = 1, \qquad \sum_{k_1} P_{k_1}(t) = 1, \qquad \sum_{k_2} P_{k_2}(t) = 1 \tag{107.9}$$

and

$$P_{k_1}(t) = \sum_{k_2} P_{k_1k_2}(t), \qquad P_{k_2}(t) = \sum_{k_1} P_{k_1k_2}(t), \tag{107.10}$$

as a consequence of the significance of the quantities involved as expressing probabilities. Combining (107.7) and (107.8), we may then write, with the help of (107.9) and (107.10),

$$\begin{aligned}
&\bar{\bar{H}}_{12}(t) - [\bar{\bar{H}}_1(t) + \bar{\bar{H}}_2(t)] \\
&\quad = \sum_{k_1,k_2} P_{k_1k_2}(t) \log P_{k_1k_2}(t) - \sum_{k_1} P_{k_1}(t) \log P_{k_1}(t) - \sum_{k_2} P_{k_2}(t) \log P_{k_2}(t) \\
&\quad = \sum_{k_1,k_2} P_{k_1k_2}(t) \log P_{k_1k_2}(t) - \sum_{k_1,k_2} P_{k_1k_2}(t) \log P_{k_1}(t) - \sum_{k_2,k_1} P_{k_1k_2}(t) \log P_{k_2}(t) \\
&\quad = \sum_{k_1,k_2} [P_{k_1k_2}(t) \log P_{k_1k_2}(t) - P_{k_1k_2}(t) \log P_{k_1}(t) P_{k_2}(t) - P_{k_1k_2}(t) + P_{k_1}(t) P_{k_2}(t)],
\end{aligned} \tag{107.11}$$

where in the last mode of writing we have added and subtracted terms which cancel out on summation. In accordance with the form of (107.11), we thus obtain a summation over terms of an essentially positive character (see (106.11)), and are hence led to the useful result

$$\bar{\bar{H}}_1(t)+\bar{\bar{H}}_2(t) \leqslant \bar{\bar{H}}_{12}(t). \tag{107.12}$$

Collecting together the expressions of special interest as given by (107.5), (107.6), and (107.12), and combining their consequences, we can then write

$$\bar{\bar{H}}_1(0)+\bar{\bar{H}}_2(0) = \bar{\bar{H}}_{12}(0), \tag{107.13}$$

$$\bar{\bar{H}}_{12}(t) \leqslant \bar{\bar{H}}_{12}(0), \tag{107.14}$$

$$\bar{\bar{H}}_1(t)+\bar{\bar{H}}_2(t) \leqslant \bar{\bar{H}}_{12}(t), \tag{107.15}$$

$$\bar{\bar{H}}_1(t)+\bar{\bar{H}}_2(t) \leqslant \bar{\bar{H}}_1(0)+\bar{\bar{H}}_2(0), \tag{107.16}$$

as a summary of results which will concern us in treating the statistical behaviour of a pair of individual systems S_1 and S_2 which can interact with each other, over a time interval $t = 0$ to $t = t$, in the form of a combined system S_{12}.

In accordance with the derivation of the H-theorem for an isolated system, it is seen that the value of $\bar{\bar{H}}_{12}$ for the distribution of probabilities representing the combined system S_{12} can be expected to decrease with time towards the minimum value that would be physically possible. And it is also seen, as a consequence of the above, that the sum of the values of $\bar{\bar{H}}_1$ and $\bar{\bar{H}}_2$ for the distributions representing the two individual systems S_1 and S_2 can never be expected to exhibit a value greater than at the initial time, since $\bar{\bar{H}}_1+\bar{\bar{H}}_2$ is originally equal to $\bar{\bar{H}}_{12}$ and is never again greater than $\bar{\bar{H}}_{12}$. It may be noted, however, that $\bar{\bar{H}}_1$ and $\bar{\bar{H}}_2$ do not in general each individually have to proceed to lower values, since an increase in the one may be compensated by a sufficient decrease in the other.

We shall find the methods and results of this section specially useful in the present chapter when we wish to consider the equilibria of non-isolated systems, and in the next chapter when we wish to explain the thermodynamic consequences of placing systems in contact with each other.

B. RELATION OF H-THEOREM TO BEHAVIOUR AT EQUILIBRIUM

108. Relation to previous studies of equilibrium

As shown in the preceding part of this chapter, the H-theorem proves very informing in studying the mechanism by which the approach of a system towards equilibrium would take place. In the remainder of the chapter we now wish to make use of the theorem in developing more fundamental and more powerful methods of studying the condition of equilibrium itself than have been previously available to us. Before proceeding to that development we may first consider the relation of the H-theorem to our earlier studies of equilibrium.

These earlier studies, as carried out in Chapter X, were based on the use of a microcanonical ensemble, uniformly distributed over a small energy range E to $E+\delta E$, as giving a suitable representation of a system in equilibrium with an energy lying in that range. Making use of the equal probabilities for different precise states in such an ensemble, we then calculated the numbers of such states G which would correspond to different observable conditions of the system of interest, and regarded the probabilities for these conditions as proportional to the corresponding values of G. We then took a maximum of G, under necessary subsidiary conditions as to total number of particles and total energy, as giving a determination of the average (most probable) condition for the system at equilibrium.

The relation of this procedure to the H-theorem may now be readily seen. In accordance with (102.1), we have defined the quantity H for a system in a given specified condition by

$$H = -\log G, \tag{108.1}$$

where G is the number of precise states that correspond to the condition. And in accordance with the H-theorem, we have found a tendency, in the absence of equilibrium, for this quantity to decrease with time. Hence our earlier treatment may now be regarded as correlating the condition of equilibrium with that which might be expected as a highly probable final condition in a system left in isolation to carry out its own behaviour.

Our earlier studies of equilibrium thus continue to appear reasonably satisfactory when examined in the light of our later understanding of the H-theorem. Nevertheless, from a theoretical point of view, it is evident that the determination of equilibrium, as corresponding to the maximum possible value of G and thus to the minimum possible value of H, does not quite give a satisfactory recognition to those fluctuations around the most probable condition which might be expected at equilibrium. And, from a computational point of view, it will be remembered that our previous calculations of the molecular distributions to be expected at equilibrium were actually carried out for simplicity, with a somewhat unsatisfactory introduction of the Stirling approximation for factorial numbers, which, as emphasized by Fowler,† might

† Fowler, *Statistical Mechanics*, second edition, Cambridge, 1936, p. 35.

even involve the use of that approximation for integers as small as zero or one. We shall be able to avoid these unsatisfactory features when we apply the more fundamental methods of treating equilibrium which we are now going to study.

109. The long time behaviour of ensembles representing perfectly isolated systems

In order to develop more fundamental methods of treating equilibria it will be rational to begin by considering the long time behaviour of the representative ensembles that could be set up to correspond to systems of interest started off initially in conditions not corresponding to equilibrium. This will then give us the kind of information appropriate for constructing ensembles to represent the ultimate conditions of equilibrium that would be reached by such systems. It will be convenient to give separate attention to the two cases of systems which are isolated from their surroundings and systems where energy interchange with the surroundings can take place.

In the present section we consider the behaviour of ensembles corresponding to *perfectly isolated* systems. To obtain such an ensemble we may regard ourselves at some initial time $t = 0$ as making an approximate observation on the system of interest which will give us a set of initial values $P_k(0)$ for the coarse-grained probabilities of states k of the kind observed. In accordance with our statistical methods, we may then set up an appropriate ensemble to represent the actual system of interest by taking a collection of similar systems distributed over different states $k, l,\ldots$ with the initial density matrix

$$\rho_{kl}(0) = P_k(0)\,\delta_{kl}. \tag{109.1}$$

On account of the character of the systems of interest now under consideration, it is to be noted that each member of this ensemble must itself be taken as perfectly isolated from its surroundings.

As time proceeds, the distribution of the ensemble will undergo changes away from that described by (109.1), assuming that the initial situation does not itself happen to correspond to equilibrium. This behaviour of the ensemble can be characterized, in accordance with the generalized H-theorem, by the tendency for the quantity $\bar{\bar{H}}$, defined by

$$\bar{\bar{H}} = \sum_k P_k \log P_k \tag{109.2}$$

summed over all states k, to decrease with time to lower values. This tendency arises from the combined effect of two factors, which we have described as the quantum mechanical spreading and the inhomogeneous redistribution of fine-grained probabilities ρ_{kk}. In the case of an ensemble, initially having a diagonal density matrix as in (109.1) and having an arbitrary distribution of coarse-grained probabilities P_k in states k which are not themselves steady energy states, we can regard this tendency as making it practically certain that $\bar{\bar{H}}$ will immediately start to decrease with time, and as making it highly probable that $\bar{\bar{H}}$ will continue to decrease over a long period towards the minimum value which would be physically possible.

As the ensemble carries out such a change with time, however, it is evident that the possible values of the quantities P_k would be subject to restrictions. A quite general restriction immediately arises from the significance of these quantities as probabilities, which makes it necessary that their sum should at all times satisfy the relation

$$\sum_k P_k = 1. \tag{109.3}$$

In addition more specific restrictions arise from the necessity for each member of the ensemble to carry out its behaviour in accordance with the laws of the quantum mechanics.

In considering the restrictions imposed by the laws of mechanics on the possible values of the P_k, we shall regard ourselves as primarily concerned only with those which arise from the principle of the conservation of energy. This character of our concern is due, as in the corresponding classical considerations of § 51 (*c*), in the first place to the circumstance that we shall in any case take the ensemble as only containing members with such specified values for the components of total linear and angular momentum as are of interest, e.g. zero values, and in the second place to the circumstance that we regard the restrictions imposed by the initial state of the ensemble on the values of other 'constants of the motion'—more complicated and physically less significant than energy and momentum—as leading during the course of time to exclusion from otherwise possible states k, distributed at random within the groups of states involved in the determination of coarse-grained probabilities, in a manner that does not have physical consequences of interest.

To study the character of the restrictions imposed on the behaviour of the ensemble by the energy principle, we note once more that each member of the ensemble must itself be regarded as perfectly isolated. Hence, in accordance with the quantum mechanical form of the law of the conservation of energy, each member of the ensemble would at all times have to exhibit unchanged probabilities for the different possible values of its energy. Thus the application of the energy principle to the ensemble as a whole may be taken as requiring constant probability for finding a member of the ensemble in any specified narrow energy range, of width ΔE corresponding to the accuracy of our observations.

In the special case, when our initial observation is made on unperturbed energy states k of the system corresponding to eigenvalues E_k^0 which are close to the true eigenvalues of energy for the system, it is easy to give a simple, closely approximate expression to the above requirements of the energy principle. Under such circumstances the probability becomes exceedingly high that a system in state k would actually exhibit a value of energy very close to the eigenvalue E_k^0. This then makes it possible to write

$$\sum_{E_k^0=E}^{E_k^0=E+\Delta E} {}_k \, P_k = P(E \text{ to } E+\Delta E), \tag{109.4}$$

where we sum over all values of k for which E_k^0 lies in the range E to $E+\Delta E$, as an approximate expression for the probability of finding a member of the system in the energy range E to $E+\Delta E$; and this expression approaches exactness as the observed states k approach the true energy states of the system. The requirements of the energy principle may then be imposed by taking the sum of the P_k, over each such narrow energy range ΔE, as being constant in time, with a value which always remains equal to that which it had when the ensemble was initially set up.

In the more general case, when our initial observation is made on states k of an arbitrary character, an explicit statement of the consequences of the energy principle may be obtained by making use of transformation theory (see (79.3)) to furnish expressions for the constant probabilities ρ'_{nn} of any energy state n, in terms of the elements ρ_{kl} of the density matrix used in describing the ensemble, and in terms of the elements S_{kn} of the unitary matrix used in transforming from the language of observation to that of energy. Such expressions, however, are found to depend on the non-diagonal elements of the density matrix ρ_{kl} and on the specific values of the S_{kn} in a manner that does not permit a simple general expression of the resulting restrictions on the possible values of the coarse-grained probabilities P_k, independent of the nature of the system under consideration. As a rough approximation, in the case of states k of an arbitrary kind, we could assume constancy for expressions of the form (109.4) now taken as summed over states k having expectation values of energy $\bar{E}_k$ which fall in the various ranges E to $E+\Delta E$. This might not be a very exact procedure, however, since the probability for a value of the energy quite different from the expectation value could now be large.

In accordance with the foregoing, it proves simplest in studying the long time behaviour of representative ensembles to consider that the initial observation on the system of interest consists in an approximate determination of unperturbed energy eigenstates, corresponding to some suitable selection of unperturbed Hamiltonian $\mathbf{H}^0$ for the system. This should not be regarded as an inappropriate kind of state to observe, owing to our natural interest in the *nearly steady* conditions of the system to which such states correspond. Furthermore, this procedure should not be regarded as making unnecessary the general treatment given to the H-theorem in § 106, since it was the discussion of this general derivation which made it clear that the tendency for $\bar{\bar{H}}$ to decrease with time could be expected to persist for a considerable period after the initial observation.

Assuming such initial observation on nearly steady states k of the system, we can now describe the long time behaviour of the representative ensemble, for any perfectly isolated system of interest started off in an arbitrary condition, as a tendency for the quantity

$$\bar{\bar{H}} = \sum_k P_k \log P_k \tag{109.5}$$

to decrease with time at least well along in the direction of the minimum value, which would be allowed by the restrictions imposed on the P_k,

$$\sum_k P_k = 1, \tag{109.6}$$

as a consequence of the significance of those quantities as probabilities, and

$$\sum_{E_k^0=E}^{E+\Delta E} {}^k\, P_k = \text{const.}, \tag{109.7}$$

for each narrow range in energy, as a consequence of the energy principle.

Furthermore, applying the usual methods of determining a conditional minimum, we see that the minimum physically possible value of the above quantity $\bar{\bar{H}}$ would require substantially equal values

$$P_k = \text{const.} \tag{109.8}$$

for the probabilities of the different states k which have eigenvalues E_k^0 lying inside of each of the individual ranges E to $E+\Delta E$, into which we take the total possible energy range as divided. Moreover, with the help of our previous equation (105.5) applying to the rate of change of coarse-grained probabilities with time, we see, making use of (105.8), that such a uniform distribution for the probabilities P_k within each energy range ΔE, if once attained, could be expected to exhibit a tendency to persist. Complete permanence would not be expected since this could be concluded from the equations mentioned only with random phases for the various states k.

We are thus provided with a good picture of the long time behaviour of such a representative ensemble, for a *perfectly isolated* system, as consisting in a *tendency* for the coarse-grained probabilities P_k for different nearly steady states k to assume ultimate values which would be nearly constant in time and nearly the same for all states k lying in any one of the narrow ranges E to $E+\Delta E$ into which we divide the total energy. This picture does not in any way contradict the conclusions to which we were led by the simple methods of § 105, but has been obtained from a more complete consideration of long time behaviour in general, and gives attention to the possibility that more than a single energy range ΔE may have been originally populated with members of the ensemble.

In the case of ensembles set up to represent the initial measurement of an *arbitrary* kind of state k, such a simple picture of behaviour is not possible. We see qualitatively, however, that there will be a tendency towards equal probabilities P_k for states k which make similar contributions to the probabilities for different energies.

110. The microcanonical ensemble as representing equilibrium for a perfectly isolated system

In accordance with the foregoing, if we make an initial approximate observation on the possible nearly steady states k of an isolated system of interest and then set up a representative ensemble to predict the future behaviour of that system, we can expect this ensemble to proceed towards an ultimate, approximately steady distribution with the coarse-grained probabilities P_k for different states k appreciably dependent only on the corresponding energy. The ultimate condition approached by a system of interest, however, may be regarded as a condition of equilibrium, substantially unaffected by the character of any approximate measurements that may have been made on the system itself, and substantially unchanging in time. To take cognizance of these considerations in constructing ensembles for the general representation of equilibrium in the case of perfectly isolated systems, it will be appropriate to choose distributions having a density matrix of the general form

$$\rho_{nm} = f(E_n)\,\delta_{nm} \tag{110.1}$$

when expressed in the true energy language, with $f(E_n)$ dependent only on the energies E_n of the various states n.

Such a distribution would, as we know, remain strictly constant as time proceeds—see, for example, (81.5)—and hence give, as expected, a description of equilibrium that would not depend on time. Moreover, with a suitable choice for the form of $f(E_n)$, it could be made to give any distribution of probabilities for different energies that might be needed to correspond to observations originally made on the system which it represents. Furthermore, from the close connexion between nearly steady states k and true energy states n, it would give coarse-grained probabilities P_k for different states k that would appreciably depend only on the corresponding energy. Hence, in the case of an isolated system, such a distribution would meet all requirements appropriate for representing the condition of equilibrium, including those which arise from treating this condition as that which would be ultimately expected if the system is left to its undisturbed behaviour.

In what may be regarded as the typical applications of statistical mechanics to isolated systems, our initial observations of the system of interest may be taken as compatible with and as including a rough determination of energy. Such a determination will not be accurate enough to specify any precise stationary state for the system but will be accurate enough to fix the energy of the system within a range, say, E to $E+\delta E$, which is quite narrow from a macroscopic point of view. Under such typical circumstances it will then be appropriate to represent the condition of equilibrium by giving our general expression for the density matrix (110.1) the special form

$$\rho_{nm} = \begin{cases} \rho_0\,\delta_{nm} & (E_n \text{ in range } E \text{ to } E+\delta E), \\ 0 & (E_n \text{ not in range } E \text{ to } E+\delta E), \end{cases} \tag{110.2}$$

with ρ_0 a constant. Or, paying particular attention to the diagonal terms, we can write

$$P_n = \rho_{nn} = \begin{cases} \rho_0 & (E_n \text{ in range } E \text{ to } E+\delta E), \\ 0 & (E_n \text{ not in range } E \text{ to } E+\delta E), \end{cases} \tag{110.3}$$

with ρ_0 a constant, as an expression for the equilibrium probability of finding the system of interest in any energy state n. In writing this expression we have taken it as appropriate to equate coarse- and fine-grained probabilities P_n and ρ_{nn} since observations are possible in the case of equilibrium and of true energy states which can be carried out with any desired degree of accuracy by taking sufficient time.

We are thus provided, also from the present point of view, with a justification for the *microcanonical ensemble* as representing equilibrium for a system of specified energy, and are hence provided with an apparatus for studying the equilibrium condition for any system for which we can calculate the character of the different energy states n. It will be appreciated, however, that such ensembles would be strictly applicable only in the case of perfectly isolated systems, where a definite specification of energy would be significant.† Nevertheless, such ensembles may also be used in the case of systems in contact with suitably chosen surroundings provided the resulting energy fluctuations have an effect which can be neglected. This is a usual case for systems of many degrees of freedom. Having selected a suitable microcanonical ensemble, the expected properties of the corresponding system of interest can then be studied by taking any kind of average over the members of the ensemble that we may select. Thus we may compute the most probable values of quantities by the methods of Chapter X, or could compute mean values by the methods devised by Darwin and Fowler.

111. The long time behaviour of ensembles representing systems in contact with their surroundings

(*a*) **Probabilities for states of the combined system.** We now turn to a consideration of the long time behaviour of ensembles representing systems in contact with their surroundings. The treatment will be based on our previous application of the *H*-theorem to two interacting systems as carried out in § 107.

Employing the terminology of that section, we shall take S_1 as the *system proper*, and S_2 as the *surroundings* with which it can interact; and shall regard their combination as giving a *combined system* S_{12}, which can itself be treated as isolated. Proceeding as before, we may then denote observed states of the system proper and its surroundings by symbols of the forms k_1, l_1, m_1, etc., and k_2, l_2, m_2, etc., respectively, and may regard the association of pairs of such states for its two parts as giving the possible states $k_1 k_2$, $k_1 l_2$, $l_1 k_2$, etc., for the combined system.

The extent of the surroundings to the system proper, which should

† The application of the methods of statistical mechanics to the internal rearrangements that take place in an atomic nucleus after disturbance—see Bohr, *Science*, **86**, 161 (1937)—provides a good example of a system which could be appropriately treated as perfectly isolated.

be included in order to secure a combined system that can be appropriately treated as isolated, would depend on specific circumstances. In the case of a situation where a study of the consequences of thermal interchange with the surroundings is definitely desired, it might be an appropriate idealization to regard the combined system as consisting of the system proper together with a suitable heat bath in which we could think of the system as being immersed. On the other hand, in the case of a situation where thermal interchange with the surroundings is not desired, but where the system proper can nevertheless interact with the walls of its container or other parts of its immediate environment, it might be necessary to regard the combined system as including more and more extensive surroundings as we treat processes of longer and longer duration.

In accordance with the methods of § 107, we may now consider ourselves at some initial time $t = 0$ as making approximate observations of the condition of the system proper S_1 and its surroundings S_2. Such observations will give us initial values for the coarse-grained probabilities $P_{k_1}(0)$ and $P_{k_2}(0)$ for states of the system proper and its surroundings respectively, and thus will also provide initial values

$$P_{k_1k_2}(0) = P_{k_1}(0)P_{k_2}(0) \tag{111.1}$$

for the coarse-grained probabilities for states of the combined system. With the help of these findings we can then set up a representative ensemble, with the initial diagonal density matrix

$$\rho_{k_1k_2,l_1l_2}(0) = P_{k_1k_2,l_1l_2}(0)\,\delta_{k_1k_2,l_1l_2}, \tag{111.2}$$

in order to make predictions as to the further behaviour of the combined system.

As time proceeds, the distribution of the ensemble will in general undergo changes away from that described by (111.2). In agreement with our previous treatment of isolated systems in § 109, this behaviour of the ensemble can be regarded as characterized by a tendency for the quantity

$$\overline{\overline{H}}_{12} = \sum_{k_1,k_2} P_{k_1k_2} \log P_{k_1k_2} \tag{111.3}$$

to decrease with time, at least well along towards the minimum value that could be attained under the appropriate restrictions on the possible values of the $P_{k_1k_2}$. The first of these restrictions is a simple consequence of the significance of those quantities as probabilities which can be expressed in the form

$$\sum_{k_1,k_2} P_{k_1k_2} = 1. \tag{111.4}$$

The remaining restrictions, which it seems necessary to consider, are a consequence of the energy principle. We may assume for simplicity, as in § 109, that the states under consideration are nearly steady states, which approximate to the true energy states of the combined system, and may assume weak enough interaction between system proper and surroundings so that the sum of the unperturbed energy eigenvalues $E^0_{k_1}+E^0_{k_2}$ will be a close expression for the energy of the combined system when in state $k_1 k_2$. We may then write

$$\sum_{E^0_{k_1}+E^0_{k_2}=E}^{E+\Delta E} {}_{k_1,k_2}\, P_{k_1k_2} = P(E \text{ to } E+\Delta E), \tag{111.5}$$

summed over the indicated states, as a close approximation for the probability of finding a member of the ensemble in any specified energy range of width ΔE corresponding to the accuracy of our observations. The restrictions imposed by the energy principle may now be expressed by requiring constant values in time for such sums taken over each energy range E to $E+\Delta E$ which is appreciably populated.

Applying the usual methods of determining a conditional minimum, we then see that the minimum physically possible value of the quantity $\bar{\bar{H}}_{12}$, as given by (111.3) under the restrictions expressed by (111.4) and (111.5), would require substantially equal values

$$P_{k_1k_2} = \text{const.} \tag{111.6}$$

for the probabilities of the different states $k_1 k_2$ which have energies $E^0_{k_1}+E^0_{k_2}$ lying inside of each one of the individual ranges E to $E+\Delta E$ into which we regard the total possible energy range as divided.

In agreement with our previous discussion of (109.8), this now permits us to describe the long time behaviour of the ensemble, representing a system proper in contact with its surroundings, as a tendency towards the ultimate assumption of probabilities $P_{k_1k_2}$ for states of the combined system, which would be approximately equal for all states $k_1 k_2$ lying inside of any specified narrow energy range E to $E+\Delta E$ that we wish to consider, and which would remain approximately constant in time when once reached.

(*b*) **Number of states of surroundings as a function of energy.** Although the foregoing gives us information as to the probabilities that may be ultimately expected for states $k_1 k_2$ of the combined system, it does not yet provide the information that we desire as to the probabilities for states k_1 of the system proper. In order to obtain such information, it will first be necessary to introduce, by way of digression, a discussion of the distribution of states of the surroundings as a function of energy. This will then permit us to find the number of states $k_1 k_2$, which would lie in any selected energy range E to $E+\Delta E$ with a specified value of k_1 and with any possible value of k_2. From the equal probabilities for all states $k_1 k_2$ of the combined system within any such range, we can then draw the desired conclusions as to the relative probabilities for different states k_1 of the system proper.

The surroundings which we shall usually wish to consider will be some ordinary macroscopic structure of many degrees of freedom—such as a heat bath, the walls of a suitable container, or the environment provided by a laboratory—which will be of sufficient extent and high enough energy content, so that its energy will lie in a range of the spectrum which can be treated as practically continuous. Under such circumstances the number of unperturbed energy eigenstates for a system increases very rapidly with energy in a manner which can be appropriately described by an expression of the form

$$\frac{dG(E)}{dE} = CE^n, \tag{111.7}$$

where $G(E)$ is the number of eigenstates of energy equal to or less than E, C is a constant, and the exponent n proves to be a large number of the order of the number of degrees of freedom of the system which have been excited. In many simple cases the exponent n is found to be a constant. And in general n may be taken as varying only gradually with the energy E, since the main part of the rapid increase in number of states with energy is found to be given by the proportionality of dG/dE to a power of E which is of the order of the number of degrees of freedom of the system.

We may illustrate these remarks as to the character of the above expression for the distribution of unperturbed energy eigenstates by giving detailed evaluations of the distribution in the case of two simple quantum mechanical systems. The evaluations will be found to depend on a known type of definite integral, of the kind studied by Dirichlet, for which we can write the formula†

$$\int \cdots \iint \epsilon_1^{i_1-1}\epsilon_2^{i_2-1}\cdots\epsilon_N^{i_N-1}\, d\epsilon_1 d\epsilon_2 \ldots d\epsilon_N = \frac{\Gamma(i_1)\Gamma(i_2)\ldots\Gamma(i_N)}{\Gamma(i_1+i_2+\ldots+i_N+1)} E^{i_1+i_2+\ldots+i_N},$$

$$\epsilon_1+\epsilon_2+\ldots+\epsilon_N \leqslant E, \tag{111.8}$$

where the integration is taken over all positive values of $\epsilon_1, \epsilon_2, \ldots, \epsilon_N$ for

† This is a special case of equation (11) in Appendix II.

which the sum is less than some specified value E, and $i_1, i_2,..., i_N$ may be any set of numbers greater than zero.

As the first example for which we shall give a detailed evaluation of the distribution of energy states, we choose a system of N weakly interacting harmonic oscillators of frequency ν. In accordance with the energy spectrum for such oscillators as given by (72.3), we can take

$$g(\Delta\epsilon) = \frac{1}{h\nu}\Delta\epsilon \tag{111.9}$$

as an expression for the number of eigenstates for a single oscillator that would lie in any energy range ϵ to $\epsilon+\Delta\epsilon$, provided that $\Delta\epsilon$ is large compared with $h\nu$, a condition which restricts our final result (111.11) to the essentially classical limit where the mean energy of the oscillators is large compared with $h\nu$. For the number of unperturbed eigenstates for the whole system of N oscillators that would have energies less than a specified value E, we may then write

$$G(E) = \left(\frac{1}{h\nu}\right)^N \int ... \int\int d\epsilon_1\, d\epsilon_2 ... d\epsilon_N, \qquad \epsilon_1+\epsilon_2+...+\epsilon_N \leqslant E, \tag{111.10}$$

where the integration is to be taken over all values for which the sum of the energies of the individual oscillators is not greater than the total energy of interest E. Evaluating the integral on the right-hand side of this expression, with the help of (111.8), choosing unity as the value of the quantities $i_1, i_2,..., i_N$, we obtain

$$G(E) = \left(\frac{1}{h\nu}\right)^N \frac{1}{\Gamma(N+1)} E^N. \tag{111.11}$$

And differentiating with respect to E, we then have

$$\frac{dG(E)}{dE} = \left(\frac{1}{h\nu}\right)^N \frac{N}{\Gamma(N+1)} E^{N-1} \tag{111.12}$$

as a specific illustration of our general expression (111.7), with the value of the constant C explicitly expressed, and with the exponent n in this case equal to one less than the number of degrees of freedom of the system.

As a second example of a specific system we may choose a dilute gas composed of N weakly interacting simple similar particles of mass m in a container of volume v. Making use of our previous expression (71.16)

$$g(\Delta\epsilon) = \frac{4\pi vm}{h^3}\sqrt{(2m\epsilon)}\,\Delta\epsilon, \tag{111.13}$$

for the number of states for a single particle in the energy range ϵ to $\epsilon+\Delta\epsilon$, we can then immediately write

$$G(E) = \frac{1}{N!}\left(\frac{2^{\frac{5}{2}}\pi vm^{\frac{3}{2}}}{h^3}\right)^N \int \dots \iint \epsilon_1^{\frac{1}{2}}\epsilon_2^{\frac{1}{2}}\dots\epsilon_N^{\frac{1}{2}}\,d\epsilon_1\,d\epsilon_2\dots d\epsilon_N,$$

$$\epsilon_1+\epsilon_2+\dots+\epsilon_N \leqslant E \tag{111.14}$$

as an expression for the number of states of the N particles of the gas having a total energy equal to or less than E. The factor $1/N!$ in this expression arises from the circumstance that we do not regard the interchange of similar particles in a quantum mechanical system as leading to a new state; the integration is over all values for which the sum of the energies of the particles is not greater than the total energy of interest E; and we assume this latter high enough so that we can neglect the necessity for giving special consideration to portions of the above integral which would correspond to more than a single particle in the same elementary state. Evaluating the right-hand side of (111.14) by another application of the formula of integration (111.8), this time taking $\frac{3}{2}$ as the value of the quantities $i_1, i_2,\dots, i_n$, we obtain

$$G(E) = \frac{1}{N!}\left(\frac{2^{\frac{5}{2}}\pi vm^{\frac{3}{2}}}{h^3}\right)^N \frac{\{\Gamma(\frac{3}{2})\}^N}{\Gamma(\frac{3}{2}N+1)} E^{\frac{3}{2}N}. \tag{111.15}$$

And, differentiating with respect to E, we now have

$$\frac{dG(E)}{dE} = \frac{3}{2}\frac{1}{(N-1)!}\left(\frac{2^{\frac{5}{2}}\pi vm^{\frac{3}{2}}}{h^3}\right)^N \frac{\{\Gamma(\frac{3}{2})\}^N}{\Gamma(\frac{3}{2}N+1)} E^{\frac{3}{2}N-1} \tag{111.16}$$

as a second illustration of our general expression (111.7), again with the constant C given explicit expression, and with the exponent n in this case equal to one less than half the number of degrees of freedom of the system.

These two systems, which we have chosen for the detailed illustration of our general expression (111.7) for the distribution of unperturbed energy states, may be regarded as reasonably typical, although rather specially simple, examples of the kind of surroundings that it will be natural to consider. Moreover, from the character of the general formula of integration (111.8) on which our evaluations of the above distributions have depended, it will be appreciated that similar results may be expected in general in the case of other systems composed of weakly interacting elements. When the mean energy of the elements is large compared with the spacing of their energy levels, and the density of these levels per unit energy range may be taken to vary with a constant power of the energy, it will be seen that our general expression (111.7) for the distribution of the states of the whole system could be expected to hold with the exponent n strictly constant. And in less special cases it will be seen that the exponent n could be reasonably supposed to vary only gradually with the total energy E. This, then, confirms the original remarks which we have made in connexion with our general expression (111.7) for the distribution of states.

In view of the foregoing discussion, we may now conclude this digression by writing, in agreement with (111.7),

$$\frac{dG(E_2)}{dE_2}\Delta E_2 = C_2\, E_2^{n_2}\,\Delta E_2 \tag{111.17}$$

as a reasonable expression for the number of states of the surroundings S_2 which would lie in any small energy range E_2 to $E_2+\Delta E_2$, with C_2 a constant and n_2 a number which does not vary rapidly with the energy E_2 and which is of the order of the number of degrees of freedom of the surroundings. These quite moderate requirements as to the character of the surroundings S_2 will be all that is needed for our investigation of the ultimate probabilities P_{k_1} for different states k_1 of the system proper S_1.

(c) **Probabilities for states of the system proper.** In accordance with (111.6), we have found a tendency for the combined ensemble to assume an ultimate condition such that there would be equal probabilities $P_{k_1k_2}$ for all states $k_1\,k_2$ in each range E to $E+\Delta E$ for the total energy of the system proper plus its surroundings. Hence, if we fix our attention, for the time being, on that portion of the representative ensemble for the combined system having members which lie in one particular such energy range E to $E+\Delta E$, we can set the probability P_{k_1} for any particular state k_1 of the system proper proportional to the number of states k_2 of the surroundings that would have energies $E^0_{k_2}$ lying in the range

$$E-E^0_{k_1} < E^0_{k_2} < E-E^0_{k_1}+\Delta E, \tag{111.18}$$

where $E^0_{k_1}$ is the energy of the system proper in the particular state k_1 under consideration. With the help of the explicit expression given by (111.17) for the number of states of the surroundings lying in such an energy range, we can then write

$$\begin{aligned} P_{k_1} &= \text{const.}\, C_2(E_2)^{n_2}\,\Delta E \\ &= \text{const.}\, C_2(E-E^0_{k_1})^{n_2}\,\Delta E \\ &= \text{const.}(E-E^0_{k_1})^{n_2} \end{aligned} \tag{111.19}$$

as an expression giving the probabilities of finding different states k_1 for the system proper within that portion of the ensemble which we are now considering. Furthermore, since E will be substantially equal to the sum $\bar{\bar{E}}_1+\bar{\bar{E}}_2$ of the mean energies of the system and its surroundings in the narrow range of total energies ΔE, we can rewrite this expression in the form

$$P_{k_1} = \text{const.}(\bar{\bar{E}}_1+\bar{\bar{E}}_2-E^0_{k_1})^{n_2}$$
$$= \text{const.}(\bar{\bar{E}}_2)^{n_2}\left(1+\frac{\bar{\bar{E}}_1-E^0_{k_1}}{\bar{\bar{E}}_2}\right)^{n_2}. \tag{111.20}$$

We shall now be interested in showing that this result can be re-expressed in a very simple form for states of the system proper having energies $E^0_{k_1}$ near enough to the mean energy $\bar{\bar{E}}_1$ for the system proper to satisfy

$$\left|\frac{\bar{\bar{E}}_1-E^0_{k_1}}{\bar{\bar{E}}_2}\right| \ll 1. \tag{111.21}$$

This restriction will have two effects. In the first place, since we can neglect higher powers of the above ratio, it permits us to change into exponential form by making the substitution

$$1+\frac{\bar{\bar{E}}_1-E^0_{k_1}}{\bar{\bar{E}}_2} = e^{(\bar{\bar{E}}_1-E^0_{k_1})/\bar{\bar{E}}_2}. \tag{111.22}$$

In the second place, since we shall evidently have the approximate equality $\bar{\bar{E}}_1-E^0_{k_1} = E^0_{k_2}-\bar{\bar{E}}_2$, where $E^0_{k_2}$ is the energy for a state of the surroundings compatible with the particular state of the system k_1 under consideration and $\bar{\bar{E}}_1$ and $\bar{\bar{E}}_2$ are mean energies in the narrow range of total energy ΔE, it will make it feasible to expand n_2 around the value which it has at $\bar{\bar{E}}_2$ and write

$$n_2 = \bar{n}_2+\frac{d\bar{n}_2}{dE_2}(\bar{\bar{E}}_1-E^0_{k_1}), \tag{111.23}$$

where we use $\bar{n}_2$ and $d\bar{n}_2/dE_2$ as convenient symbols for the values of n_2 and dn_2/dE_2 at $E_2 = \bar{\bar{E}}_2$, and where we drop higher terms in accordance with the assumed smallness of $|\bar{\bar{E}}_1-E^0_{k_1}|$ compared with $\bar{\bar{E}}_2$, and in accordance with our previous finding that n_2 would at most be a slowly varying function of E_2.

Substituting (111.22) and (111.23) in (111.20), retaining terms only to the first order in $(\bar{\bar{E}}_1-E^0_{k_1})$, it will be seen that we can then write

$$P_{k_1} = \text{const.}\, e^{\bar{n}_2\log\bar{\bar{E}}_2+\frac{d\bar{n}_2}{dE_2}(\bar{\bar{E}}_1-E^0_{k_1})\log\bar{\bar{E}}_2+\bar{n}_2\frac{\bar{\bar{E}}_1-E^0_{k_1}}{\bar{\bar{E}}_2}}. \tag{111.24}$$

Or defining two constants C and β by

$$C = \text{const.}\, e^{\bar{\bar{n}}_2 \log \bar{\bar{E}}_2 + \frac{d\bar{\bar{n}}_2}{d\bar{\bar{E}}_2}\bar{\bar{E}}_1 + \frac{\bar{\bar{n}}_2 \bar{\bar{E}}_1}{\bar{\bar{E}}_2}} \tag{111.25}$$

and

$$\beta = \frac{d\bar{\bar{n}}_2}{d\bar{\bar{E}}_2} \log \bar{\bar{E}}_2 + \frac{\bar{\bar{n}}_2}{\bar{\bar{E}}_2}, \tag{111.26}$$

we now obtain

$$P_{k_1} = Ce^{-\beta E^0_{k_1}} \tag{111.27}$$

as the desired simple expression for the ultimate probabilities of states k_1 of the system proper, where it will be remembered that we are confining our attention to states of the system proper having energies close enough to the mean to satisfy (111.21), and are confining our attention for the time being only to a particular selected energy range E to $E+\Delta E$ for the combined system proper and surroundings.

With regard to the circumstance that the simple distribution law (111.27) only applies to states k_1 of the system proper having energies satisfying (111.21), it will be appreciated that the law would, nevertheless, be substantially valid for all states of importance whenever the energy content and number of degrees of freedom for the surroundings are large compared with those quantities for the system proper. To examine this in detail we may return to (111.20), and, by introducing the appropriate factor for the number of states of the system proper that would have energies in a range dE_1, may write

$$\begin{aligned} P(E_1)\,dE_1 &= \text{const.}(E_1)^{n_1}(\bar{\bar{E}}_2)^{n_2}\left(1+\frac{\bar{\bar{E}}_1-E_1}{\bar{\bar{E}}_2}\right)^{n_2} dE_1 \\ &= \text{const.}(\bar{\bar{E}}_1+\bar{\bar{E}}_2-E_2)^{n_1}(\bar{\bar{E}}_2)^{n_2}\left(1+\frac{\bar{\bar{E}}_1-E_1}{\bar{\bar{E}}_2}\right)^{n_2} dE_1 \\ &= \text{const.}(\bar{\bar{E}}_1)^{n_1}(\bar{\bar{E}}_2)^{n_2}\left(1+\frac{\bar{\bar{E}}_2-E_2}{\bar{\bar{E}}_1}\right)^{n_1}\left(1+\frac{\bar{\bar{E}}_1-E_1}{\bar{\bar{E}}_2}\right)^{n_2} dE_1 \\ &= \text{const.}(\bar{\bar{E}}_1)^{n_1}(\bar{\bar{E}}_2)^{n_2}\left(1-\frac{\bar{\bar{E}}_2}{\bar{\bar{E}}_1}\cdot\frac{\bar{\bar{E}}_1-E_1}{\bar{\bar{E}}_2}\right)^{n_1}\left(1+\frac{\bar{\bar{E}}_1-E_1}{\bar{\bar{E}}_2}\right)^{n_2} dE_1, \end{aligned} \tag{111.28}$$

as an expression for the probability of finding the system proper in different energy ranges dE_1. We then indeed see, with $\bar{\bar{E}}_2 \gg \bar{\bar{E}}_1$ and $n_2 \gg n_1$, that the whole range of large probabilities could be covered without contradicting the restriction (111.21),

$$\left|\frac{\bar{\bar{E}}_1-E_1}{\bar{\bar{E}}_2}\right| \ll 1,$$

on which our previous considerations were based, since with $(\bar{\bar{E}}_1 - E_1)$ positive the large factor $\bar{\bar{E}}_2/\bar{\bar{E}}_1$ in the next to the last parenthesis in (111.28) would be operative in leading to low probabilities, and with $(\bar{\bar{E}}_1 - E_1)$ negative the large exponent n_2 for the last parenthesis would so operate. Hence we can regard our simple formula (111.27) as substantially valid for all states of importance provided the surroundings are sufficiently large compared with the system proper.

With regard to the circumstance that the foregoing considerations have all been concerned with members of the combined ensemble lying in one particular energy range, a separate expression of the form (111.27) would have to be applied to each energy range E to $E+\Delta E$ for the combination of system proper and surroundings, the constant C being determined by the probability of finding a member of the combined ensemble in that range, and the constant β being determined in accordance with (111.26) by the mean energy $\bar{\bar{E}}_2$ of the surroundings for members of the combined ensemble in the range mentioned. It will be noted, nevertheless, that the constant β would have substantially the same value for neighbouring ranges E to $E+\Delta E$, since the mean energies $\bar{\bar{E}}_2$ for the surroundings would be practically the same in such neighbouring ranges for the total energy. It will be appropriate, moreover, to concern ourselves with situations where our initial knowledge of condition is such as to give a high concentration of the probability for different values of the total energy in the neighbourhood of some particular range E to $E+\Delta E$. Thus, with an appropriate value for the constant C, it will also be possible to take an expression of the form (111.27) as applying to the whole ensemble.

Hence, by way of summary, we may now describe the long time behaviour of a system of interest S_1 in contact with its surroundings S_2 as characterized by a tendency for the corresponding statistical representation to proceed to an ultimate nearly steady condition, such that the distribution of coarse-grained probabilities for the different unperturbed energy states k_1 of the system proper would be given by the simple formula

$$P_{k_1} = Ce^{-\beta E^0_{k_1}}, \tag{111.29}$$

where C and β would be constants, and the formula would be valid over a range of energies $E^0_{k_1}$, in the neighbourhood of the mean energy $\bar{\bar{E}}_1$ for the system proper which would be related to the mean energy $\bar{\bar{E}}_2$ of the surroundings by

$$|\bar{\bar{E}}_1 - E^0_{k_1}| \ll \bar{\bar{E}}_2, \tag{111.30}$$

this range being wide enough, in the case of surroundings very much larger than the system, to include substantially all energies of importance.

(*d*) **The concepts of thermal equalization, essential isolation, and essentially adiabatic process.** In closing this discussion of the behaviour that can be expected for a system which interacts with its surroundings, it will prove convenient to distinguish three idealized types of such interaction which provide useful abstractions. These may be denoted by the terms thermal equalization, essential isolation, and essentially adiabatic process, and, assuming the allowance of sufficient time, will all three be taken as leading to or involving a distribution of probabilities P_{k_1} for states k_1 of the system proper which can be substantially described by the simple formula (111.29).

In a situation which we idealize as a case of *thermal equalization* we shall take the system proper as purposely placed in good thermal contact with a very large heat reservoir, having a large capacity for furnishing or taking up energy, and a mean energy and number of degrees of freedom which could be taken as going to infinity. Under such limiting circumstances it will be seen from the foregoing discussion that we could regard our formula (111.29), for the ultimate probabilities of different states k_1 of the system proper, as approaching validity over the entire range of energies $E^0_{k_1}$.

In a situation which we idealize as a case of *essential isolation* we shall take the system proper as not purposely placed in thermal contact with its surroundings, and on the average as having no net interchange of energy with the outside in a series of similar experiments, but nevertheless as *necessarily in contact* with immediate surroundings, such as the walls of a container or portions of the laboratory environment. Such a situation could be represented by a combined ensemble for the system and its surroundings, with the possibility for any representative of the system proper to adjust its energy by interchange with the associated representative for the surroundings, but with the special proviso that the surroundings are so chosen as to secure *constant mean energy* for the representatives of the system proper. Since this special feature of the surroundings would not disturb our previous considerations as to the long time behaviour of the combined ensemble, the ultimate probabilities for different states k of the system proper would then be described by equation (111.29) over a range of energies $E^0_{k_1}$ around the mean $\bar{\bar{E}}_1$ having a width which would depend on the extent of the surroundings that interact with the system. In the situations to which we apply the idealization of essential isolation we regard the extent of the surroundings as sufficient to justify a use of the simple exponential law (111.29) as providing a sufficiently valid complete description of the ultimate probability distribution.

Finally, in a situation which we idealize as an *essentially adiabatic process*, we shall take the system proper as having the limited kind of contact with its immediate surroundings which was described above but as subjected to changes in the value of some macroscopic external parameter, such as volume, which describes its condition. An essentially adiabatic process would thus be one in which the mean energy ascribed to the system proper would not be changed by 'thermal flow' from the surroundings, but might be changed by 'mechanical action' from the outside. Furthermore, in situations to which we apply the idealization of essentially adiabatic process, we regard the extent of the surroundings that interact with the system proper as sufficient, so that by carrying out the process extremely slowly it would be justifiable at each state of the change to use the simple exponential law (111.29) as though it gave a completely valid description of the probability distribution.

The three abstract idealizations of thermal equalization, essential isolation, and essentially adiabatic process which we have thus introduced are seen to be quite different in character from the previous concept of perfect isolation as treated in §§ 109 and 110. Several remarks may be made with regard to the actual situations in which it would be appropriate to apply the new concepts.

It is at once evident that the idealization which we have called thermal equalization would provide suitable statistical representation for the processes which ensue when a system is immersed in a large heat bath, and further discussion of this case will not be needed.

With regard to the applicability of the concept of essential isolation somewhat more discussion is necessary. It will be appreciated on reflection, however, that this idealization would often provide suitable statistical representation for practical situations in which there is no intention of permitting any flow of energy to or from a system of interest, and in which the system would ordinarily be spoken of as isolated, but in which it would not be appropriate to neglect the possibility for fluctuations to take place in the energy of the system proper, through interaction with actual surroundings, to an extent sufficient to make the simple exponential law (111.29) a natural one to take as giving an approximately valid description of the ultimate probabilities for different states.

Some of the factors which tend to make this procedure quite frequently appropriate may be mentioned here. When the system proper is actually a very small one, the extent of the surroundings that must be considered as effectively concerned will tend to stand in a large ratio to the extent of the system proper, thus giving the conditions necessary for a wide range of validity for (111.29). When the system proper is actually a very large one of many degrees of freedom, it turns out—as we shall appreciate more fully in what follows—that the statistical results obtained tend to be relatively independent of the exact form of law chosen for the distribution of ultimate probabilities provided it gives—as does (111.29)—equal probabilities for states of equal energy and a high concentration in the neighbourhood of a particular energy. Furthermore, whatever the actual size of the system proper, it will be appreciated with any given degree of separation between system proper and surroundings, that the extent of the environment which would have to be regarded as effectively interacting with the system proper would in general increase as we go to longer and longer times in considering our ultimate probability distributions.

It will also be appreciated, in the case of a system having some contact with its surroundings, that the concept of perfect isolation would often be a clearly inexact idealization to employ, that the best possible idealization would depend on the actual degree of isolation achieved together with the length of time allowed for interaction with the environment in too complicated a manner for ready determination or use, and that the concept of essential isolation would in any case provide a distribution law for ultimate probabilities which would have some range of substantial validity in the neighbourhood of the mean energy. This often makes the concept of essential isolation really the most reasonable one to use in developing general ideas as to the long time behaviour of the so-called isolated systems commonly encountered in nature or employed in laboratory or engineering technique.

With regard to the applicability of the concept of essentially adiabatic process as an idealization for actual processes, it will be appreciated that the necessary points have been covered in the foregoing paragraphs on the applicability of the concept of essential isolation, since we merely have to add the idea of making changes in the external parameters for systems which are otherwise left in essential isolation. We shall find the idea of essentially adiabatic processes specially important when we come to the statistical explanation of the principles of thermodynamics in the next chapter.

In conclusion it may be remarked that the concepts of perfect isolation and thermal equalization have often been employed in carrying out statistical mechanical considerations, but that the concepts of essential isolation and essentially adiabatic process are newly introduced here.† This has been done not only in the interest of obtaining closer and less abstract idealizations for certain actual situations, but also, as we shall see in § 124, in order to overcome difficulties in the older treatment of reversible adiabatic processes.

112. The canonical ensemble as representing equilibrium for a system in a heat bath or in essential isolation

(*a*) **The canonical distribution.** In accordance with the foregoing discussion of systems in contact with sufficiently extensive surroundings, we have found a tendency for the coarse-grained probabilities P_k for different unperturbed energy eigenstates k of the system proper to assume an ultimate, quasi-permanent distribution which would be described by

$$P_k = Ce^{-\beta E_k^0}, \tag{112.1}$$

where C and β are constants, and E_k^0 is the unperturbed energy for the state k. Such a distribution may be expected both in cases of thermal contact, where surroundings in the nature of a very large heat reservoir are purposely introduced, and also in cases of essential isolation.

Taking cognizance of this finding, it will now be appropriate to choose as a suitable general representation for the condition of equilibrium—in such cases both of thermal contact and of essential isolation—the distribution precisely defined in the *true energy language* by the density matrix

$$\rho_{nm} = Ce^{-\beta E_n}\delta_{nm}, \tag{112.2}$$

where C and β are parameters having values independent of the state n, and the quantities E_n are the true energies for the various energy eigenstates n. Such a distribution—together with an appropriate distribution to represent the surroundings giving a combined density matrix diagonal in the energy language—would, as we know, remain strictly *constant in time* and hence give as expected a description of equilibrium that would be strictly independent of time. Furthermore, from the

† The importance of the effect of the surroundings of a system on its equilibrium behaviour, even in those cases where we should ordinarily speak of the system as isolated, has also been especially emphasized from a somewhat different point of view by Epstein in his 'Critical Appreciation of Gibbs' Statistical Mechanics' published in *Commentary on the Scientific Writings of J. Willard Gibbs*, vol. ii, Yale University Press, 1936.

close connexion between nearly steady states k and true energy states n, the distribution defined by (112.2) would give coarse-grained probabilities P_k for different states k in substantial agreement with (112.1). Hence, in such cases of thermal contact and of essential isolation, the distribution defined by (112.2) would meet all the appropriate requirements for representing the condition of equilibrium, including those which arise from treating this condition as that which would be ultimately expected in the course of time after an initial approximate measurement of the nearly steady states of the system of interest and of its surroundings.

The distribution defined by (112.2) gives an ensemble which is the natural quantum mechanical analogue of the *canonical ensemble* of Gibbs, and is commonly denoted by this same name. By introducing new quantities defined by

$$\beta = \frac{1}{\theta} \quad \text{and} \quad C = e^{\psi/\theta}, \tag{112.3}$$

the distribution can also be expressed in the familiar form

$$\rho_{nm} = e^{\frac{\psi - E_n}{\theta}} \delta_{nm}, \tag{112.4}$$

which has already been mentioned in § 83 (*c*), and which will prove specially convenient in the next chapter in treating the relations between statistical mechanics and thermodynamics. The relation of the constants β and θ to the equilibrium temperature T of the system will be discussed later, from an immediate phenomenological point of view in § 114 (*c*) of the present chapter and from a thermodynamic point of view in the next chapter.

In accordance with (112.2), the equilibrium probability for finding any given state n of energy E_n will be given by

$$P_n = \rho_{nn} = Ce^{-\beta E_n}, \tag{112.5}$$

where, as mentioned in connexion with (110.3), it will now be appropriate to drop the distinction between coarse-grained probabilities P_n and fine-grained probabilities ρ_{nn}. Summing over all states n, we can then write

$$\sum_n Ce^{-\beta E_n} = 1 \tag{112.6}$$

as a valid expression of the normalization of the ensemble, and

$$\sum_n Ce^{-\beta E_n} E_n = \bar{\bar{E}} \tag{112.7}$$

as a necessary expression for the mean energy $\bar{\bar{E}}$ of the members of the ensemble. This thus gives two equations for determining the parameters C and β. With an appropriate selection for these quantities we see that the ensemble would represent equilibrium for an essentially isolated system with any desired specification for its mean energy.

(*b*) **Justification for the canonical ensemble.** The foregoing method of introducing the canonical ensemble—as giving a good statistical description of the ultimate equilibrium condition approached by systems having sufficient interaction with their surroundings—is somewhat different from the methods that have usually been employed in introducing that ensemble as appropriately descriptive of equilibrium. It will hence be of interest to make a few remarks as to various arguments that have been and may be made in connexion with the introduction of the canonical ensemble.

A somewhat common point of view has sometimes been that the condition of equilibrium for ordinary systems would actually be best represented by a microcanonical ensemble, with all its members having energies practically identical with some precise value ascribed as that of the system of interest. A justification for substituting a canonical ensemble of appropriate mean energy, in place of the microcanonical ensemble, has then been based firstly on the circumstance that a high concentration around the mean energy would make the average properties of the canonical ensemble practically identical with those of a microcanonical ensemble in a case of many degrees of freedom, and secondly on the practical ground that the canonical distribution is mathematically the easier of the two to handle.† From our present point of view, such a justification for the canonical ensemble would seem quite roundabout, since the original selection of the microcanonical ensemble as fundamental would appear based on ideas as to perfect isolation and as to the precise determination of energy which would not usually be appropriate. It is perhaps to be emphasized, however, in the ordinary cases of interest with systems of many degrees of freedom, that the two ensembles do lead to substantially the same conclusions as to the condition of equilibrium.

A formal method of introducing a canonical distribution for the energy states n of a system of interest can be obtained‡ by requiring a minimum of the quantity $\bar{\bar{H}}$ defined in terms of energy states for the system itself by

$$\bar{\bar{H}} = \sum_n P_n \log P_n, \tag{112.8}$$

† See, for example, Lorentz, *Abhandlungen über theoretische Physik*, Teubner, 1907, in particular § 79, p. 289. Compare also the remark of Gibbs, *Elementary Principles of Statistical Mechanics*, Yale University Press, 1902, p. 116.

‡ See, for example, Pauli, *Handbuch der Physik*, xxiv/1, second edition, Berlin 1933, p. 151.

under the subsidiary conditions of a constant total probability for one or another state

$$\sum_n P_n = 1, \tag{112.9}$$

and of a constant value for the mean energy $\bar{\bar{E}}$

$$\sum_n P_n E_n = \bar{\bar{E}}. \tag{112.10}$$

Making use of the usual methods of determining a conditional minimum, we should then indeed be led to the distribution

$$P_n = Ce^{-\beta E_n} \tag{112.11}$$

with C and β constants, in agreement with that for a canonical ensemble. Furthermore, with the help of the concept of essentially isolated system, as discussed in § 111 (d), it will be seen that we can regard this method of attaining a canonical distribution as actually descriptive of the ultimate results which would be expected on the average in the case of a system left to itself in contact with sufficiently extensive surroundings which could adjust the energy of the system to different values without affecting the mean value that would be found in successive trials.

A different kind of justification for the choice of the canonical ensemble as representing equilibrium was given by the discovery of Gibbs† that such a selection makes it possible to obtain very simple statistical mechanical analogues for the fundamental quantities and equations of thermodynamics. These analogues appeared to Gibbs himself more satisfactory than those which he obtained from the choice of the microcanonical distribution‡ as a representation of equilibrium. As we shall see in the next chapter, it can now be made clearly evident that the canonical distribution must be regarded as the one which provides the appropriate mechanical analogues for the quantities and equations of thermodynamics.

Somewhat deeper justification for the choice of the canonical ensemble as representing a system in thermal equilibrium was given by three further discoveries by Gibbs. The first|| of these is the immediately evident result, see the form of (112.4), that a canonical distribution for a system consisting of two parts implies a canonical distribution for each of those parts with the same value of the parameter θ, which, as we shall see later, can be related to temperature. This can be inter-

† Gibbs, *Elementary Principles of Statistical Mechanics*, Yale University Press, 1902, p. 44.

‡ Gibbs, loc. cit., chap. xiv.

|| Gibbs, loc. cit., p. 36.

preted as corresponding to the circumstance that a system in thermal equilibrium can be regarded as composed of parts which are themselves in thermal equilibrium with the same temperature as the whole. The second† of these discoveries of Gibbs was that a microcanonical distribution for a combined system consisting of a small system proper immersed in a large heat bath would imply a canonical distribution for the states of the system proper. This result agrees with our somewhat more general finding as to the ultimate condition of the representative ensemble for a system in contact with extensive surroundings, and gives a similar justification for the canonical distribution as representing equilibrium. The third‡ of the above-mentioned discoveries of Gibbs was the finding that an arbitrary representative ensemble would assume canonical form with the expected value of θ, as a consequence of repeated interaction of its members with those of a canonical ensemble representing a large heat bath. We shall give a special demonstration of this principle in the next chapter, see § 128 (*c*); the result may be interpreted as giving a satisfactory method of representing the process of temperature adjustment. In addition to these discoveries, it should also be mentioned, as emphasized by Gibbs,|| that the microcanonical ensemble could hardly be regarded as giving a satisfactory representation for a system in equilibrium with a heat bath, since it would give no recognition to the fluctuations in energy which would actually be possible.

It will be appreciated that our own method of introducing the canonical ensemble—as representing the ultimate condition of equilibrium to be expected for a system in contact with a heat reservoir or in essential but not perfect isolation from its surroundings—stands in no kind of conflict with the justifications for the use of that ensemble that were provided by the work of Gibbs. It may be emphasized, however, that our considerations give a more complete account of the processes by which equilibrium is achieved, and give due recognition to the possible effect of the walls and other surroundings in adjusting the energy of a system proper to different values. In the work of Gibbs the abstract idealization was definitely introduced that the walls of the container should have no effect.†† We shall see, nevertheless, in the next chapter in § 124 that the effect of the walls and other immediate surroundings must be definitely introduced in order to supply a deficiency in the Gibbs treatment of adiabatic processes.

† Gibbs, loc. cit., p. 183.

‡ Gibbs, loc. cit., p. 161.

|| Gibbs, loc. cit., p. 180.

†† Gibbs, loc. cit., p. 164.

In conclusion it may be well to emphasize that physical-chemical systems of interest, under usual conditions of laboratory or engineering practice, must be regarded as making sufficient contact with their surroundings so that canonical distributions will be appropriate in a fundamental treatment of their conditions of equilibrium. In this connexion it is significant again to point out, in the case of a small system of interest having a small number of degrees of freedom, that the surroundings with which it must interact to secure substantial validity of the canonical distribution will themselves only have to be of moderate extent, and, in the case of a large system of interest of many degrees of freedom, that the differences between the microcanonical and canonical distribution as a description of equilibrium tend to become unimportant. It should also be noted again that it is very fortunate, from a computational point of view, that we can feel justified in selecting the canonical instead of the microcanonical distribution as giving the usually appropriate representation of equilibrium, since the calculation of mean values proves to be much easier in the canonical than in the microcanonical case.

C. SPECIFIC EXAMPLES OF EQUILIBRIUM

113. Equilibrium in Maxwell-Boltzmann systems

We are now ready to make use of the canonical distribution, as giving an appropriate description of equilibrium under ordinary circumstances, in order to treat some specific examples. We shall begin by considering a simple system of the Maxwell-Boltzmann type, which can be regarded as composed of n weakly interacting elements, having similar properties, but permanently distinguishable from each other. As an illustration we might take a system of harmonic oscillators all having the same fundamental frequency but distinguishable from one another by permanent spatial location or orientation.

We may regard the different energy states for a single element as designated by the value of an index or quantum number i and may take the corresponding energy for the element as denoted by the symbol ϵ_i. Assuming weak interaction between the elements, we can then take the possible energy states for the system as a whole as specified by a group of such quantum numbers $i_1, i_2,..., i_n$ for the n separate elements, and may take the total energy for such a state of the whole system as given by a sum of the form

$$E_{i_1,i_2,...,i_n} = \epsilon_{i_1}+\epsilon_{i_2}+...+\epsilon_{i_n}. \tag{113.1}$$

In accordance with our expression for the probabilities of different states in a canonical ensemble, as given by (112.5), we can then write

$$P_{i_1,i_2,\ldots,i_n} = Ce^{-\beta(\epsilon_{i_1}+\epsilon_{i_2}+\ldots+\epsilon_{i_n})}, \tag{113.2}$$

where C and β are constant parameters, as an expression for the equilibrium probability for any specified state of the whole system that we may wish to consider.

Our actual interest is now going to lie in the probabilities for different states of the separate elements. Fixing our attention for the moment on the first of these elements corresponding to the index 1, and noting that the summation of (113.2) over all possible states must lead to the value unity, we may then write

$$P_{i_1} = \frac{Ce^{-\beta\epsilon_{i_1}} \sum\limits_{i_2\ldots i_n} e^{-\beta(\epsilon_{i_2}+\ldots+\epsilon_{i_n})}}{C \sum\limits_{i_1\ldots i_n} e^{-\beta(\epsilon_{i_1}+\ldots+\epsilon_{i_n})}} \tag{113.3}$$

as an expression for the probability of finding element number 1 in the state i_1. With the help of simple cancellations we then obtain

$$P_i = \frac{e^{-\beta\epsilon_i}}{\sum\limits_i e^{-\beta\epsilon_i}} \tag{113.4}$$

as a general expression for the equilibrium probability of finding any element—on which we wish to fix our attention—in the state i, the summation in the denominator being over all possible states i. It will be noted that such a probability for any particular element is independent of the states of other elements of the system.

With the help of this result we may now investigate the mean number of elements $\bar{\bar{n}}_i$ which would be found in a specified state i. In accordance with the independence of the probabilities for the states of the different elements, we can evidently write

$$(P_i)^{n_i}(1-P_i)^{n-n_i} \tag{113.5}$$

for the probability of finding a specified set of n_i elements in the state i, with the remaining $n-n_i$ elements not in that state. Furthermore, such a set of n_i elements could evidently be chosen in

$$\frac{n!}{n_i!\,(n-n_i)!} \tag{113.6}$$

equivalent different ways. Combining (113.5) and (113.6), we can then write

$$P(n_i) = \frac{n!}{n_i!\,(n-n_i)!}(P_i)^{n_i}(1-P_i)^{n-n_i} \tag{113.7}$$

for the total probability $P(n_i)$ of finding precisely n_i elements in state i.

And this will of course permit us to compute the mean number $\bar{\bar{n}}_i$ in that state.

The simplest way to perform the computation is to consider the summation of (113.7) over all possible values of n_i from 0 to n. This gives us

$$\sum_{n_i=0}^{n} P(n_i) = \sum_{n_i=0}^{n} \frac{n!}{n_i!\,(n-n_i)!}(P_i)^{n_i}(1-P_i)^{n-n_i} = 1, \tag{113.8}$$

where the value unity arises in the first instance from the circumstance that the total probability for finding one or another number of elements in state i must evidently add up to 1. It will also be noted, however, that the value of this summation would be equal to unity without reference to the value of P_i, since the second form of expression is seen to be the binomial expansion of $[P_i+(1-P_i)]^n$ which is evidently unity for any value of P_i. Hence, differentiating the second term of (113.8) with respect to P_i, it will be readily seen that we can obtain the equation

$$\sum_{n_i=0}^{n} \left[\frac{n_i}{P_i}P(n_i) - \frac{(n-n_i)}{(1-P_i)}P(n_i)\right] = 0, \tag{113.9}$$

where we re-express the result of differentiating the two factors containing P_i in terms of $P(n_i)$ itself. Clearing of fractions and rearranging, this then gives us

$$\sum_{n_i=0}^{n} P(n_i)n_i = nP_i \sum_{n_i=0}^{n} P(n_i). \tag{113.10}$$

Since $P(n_i)$ is the probability of finding n_i elements in state i, the left-hand side of the above expression is seen to be the mean number of elements in state i, and the summation on the right-hand side is seen to reduce to the factor unity. With the help of the expression (113.4) already found for P_i we can then write

$$\bar{\bar{n}}_i = nP_i$$

or

$$\bar{\bar{n}}_i = n\frac{e^{-\beta\epsilon_i}}{\sum_i e^{-\beta\epsilon_i}} \tag{113.11}$$

as the desired expression at equilibrium for the mean number of elements in any particular state i that would be found in the members of a canonical ensemble representing the condition of equilibrium in a Maxwell-Boltzmann system composed of n similar but distinguishable elements.

It is also easy to extend the use of the above apparatus to a calculation of the deviations that can be expected from the mean number in any state. Differentiating (113.8) a second time with respect to P_i, we readily obtain after some simplification the result

$$\frac{\overline{\overline{(n_i^2)}}-(\overline{\overline{n}}_i)^2}{(\overline{\overline{n}}_i)^2}=\frac{1}{\overline{\overline{n}}_i}-\frac{1}{n}$$

or

$$\frac{\overline{\overline{(n_i-\overline{\overline{n}}_i)^2}}}{(\overline{\overline{n}}_i)^2}=\frac{1}{\overline{\overline{n}}_i}-\frac{1}{n} \tag{113.12}$$

as an expression for the fractional mean square deviation from the mean number in any state i. It will be noted that this becomes small for those states that are highly populated.

If we consider a group of g_κ states all of substantially the same energy ϵ_κ, an expression for the mean number of elements having this energy can be written, in agreement with (113.11), in the familiar form

$$\overline{\overline{n}}_\kappa = g_\kappa e^{-\alpha-\beta\epsilon_\kappa}, \tag{113.13}$$

where α is an appropriate constant. Comparing with our earlier equation (89.5), we now see that our exact expression for the mean number of elements in any condition κ in a canonical ensemble is the same as our previous approximate expression for the most probable number of elements in that condition in the corresponding microcanonical ensemble. Hence our previous results as to the equilibrium properties of Maxwell-Boltzmann systems, as obtained in Chapter X, can now be taken over without substantial alteration.

114. Equilibrium in Einstein-Bose and Fermi-Dirac gases

(*a*) **Derivation of the distribution laws.** We may now turn to the use of the canonical ensemble as representing equilibrium for a system composed of n weakly interacting particles, having similar properties and not permanently distinguishable from one another on account of their free motion inside a common container. Such a system will be an Einstein-Bose or a Fermi-Dirac gas, according as the eigenfunctions which describe the states of the system are symmetric or antisymmetric to the interchange of particle indices as discussed in § 76.

We may regard the different possible energy states for a single particle of the system in the container under consideration as denoted by the letters k, l, m, etc., and may take the corresponding eigenvalues of energy for the particle as denoted by ϵ_k, ϵ_l, ϵ_m, etc. Assuming weak interaction between the particles, we can then take the possible states for the system as a whole as specified by the numbers of particles n_k, n_l, n_m, etc., in the different elementary eigenstates, without any specification as to which particles are selected for the different states. Furthermore, we can take the total energy for any such state of the system as given by a sum of the form

$$E_{n_k,n_l,n_m,\dots} = n_k\epsilon_k+n_l\epsilon_l+n_m\epsilon_m+\dots \tag{114.1}$$

for all n particles. In accordance with our expression for the probabilities of different states in a canonical ensemble, as given by (112.5), we can then write

$$P_{n_k,n_l,n_m,\dots} = Ce^{-\beta(n_k\epsilon_k+n_l\epsilon_l+n_m\epsilon_m+\dots)}, \tag{114.2}$$

where C and β are constant parameters, as an expression for the equilibrium probability for any specified state of the whole system that we may wish to consider.

Our actual interest is now going to lie in the probabilities for finding different numbers of particles in the different possible elementary states. Fixing our attention for the moment on a particular one of these states, say k, we can evidently immediately write

$$P(n_k) = Ce^{-\beta n_k\epsilon_k}\sum_{n_l,n_m,\dots} e^{-\beta(n_l\epsilon_l+n_m\epsilon_m+\dots)} \tag{114.3}$$

as a correct expression for the probability $P(n_k)$ of finding n_k particles in state k provided we take the indicated summation over all possible further assignments n_l, n_m, etc., subject to the necessary restriction

$$n_l+n_m+\dots = n-n_k. \tag{114.4}$$

The restriction expressed by this relation makes any precise evaluation of (114.3) very difficult to carry out and dependent in a specific manner on the energy levels for the particles under consideration. This makes it advantageous to introduce a simple approximate method of treatment which, as we shall see later, proves to be closely correct. For this purpose we now replace our original expression (114.2), for the probabilities of different states in an ensemble of members each composed of a fixed number of particles n, by an expression for the probabilities of different states in an ensemble of members containing all different possible numbers of particles n. To this expression we assign the form

$$P_{n_k,n_l,n_m,\dots} = Ce^{-\alpha n-\beta(n_k\epsilon_k+n_l\epsilon_l+n_m\epsilon_m+\dots)}, \tag{114.5}$$

where C, α, and β are constants, and where we now regard the total number of particles

$$n = n_k+n_l+n_m+\dots \tag{114.6}$$

as a variable which can assume any value from zero to infinity.

We thus replace our original *canonical ensemble* of members each composed of the same number of particles n by a so-called *grand canonical ensemble*, consisting of a weighted collection of canonical ensembles, one for each possible value for the total number of particles n, and all with the same value for the significant parameter β. Under the circumstances of ordinary interest, however, the members of such an ensemble will be highly concentrated in the neighbourhood of a particular value for the total number of particles, as we shall show in the next part of this section. With an appropriate choice of the constant α this concentration can be made to occur at any value for the total number of particles n that we desire. In the present instance we regard this introduction of the grand canonical ensemble merely as a convenient computational device. In a later place, § 140, we shall show that the grand canonical ensemble can actually have a theoretical significance under appropriate circumstances.

Making a fresh start, with the help of the expression for the probabilities of different states in our grand canonical ensemble as given by (114.5), substituting for n from (114.6), and noting that the summation of probabilities over all states must be equal to unity, we can now evidently write

$$P(n_k) = \frac{Ce^{-(\alpha+\beta\epsilon_k)n_k} \sum\limits_{n_l,n_m,\ldots} e^{-(\alpha+\beta\epsilon_l)n_l-(\alpha+\beta\epsilon_m)n_m-\ldots}}{C \sum\limits_{n_k,n_l,n_m,\ldots} e^{-(\alpha+\beta\epsilon_k)n_k-(\alpha+\beta\epsilon_l)n_l-(\alpha+\beta\epsilon_m)n_m-\ldots}} \tag{114.7}$$

as an expression for the probability of finding n_k particles in state k. And, since there is now no fixed total number of particles to put a limitation on the indicated summations, we obtain by simple cancellation the desired simple expression

$$P(n_k) = \frac{e^{-(\alpha+\beta\epsilon_k)n_k}}{\sum\limits_{n_k} e^{-(\alpha+\beta\epsilon_k)n_k}} \tag{114.8}$$

for the probability of finding n_k particles in a designated state k.

We may evidently use this expression to determine the mean number of particles $\bar{\bar{n}}_k$ or the mean square number $(\overline{\overline{n_k^2}})$ that would be found in state k. It now becomes necessary, however, to give separate treatment to the Einstein-Bose case where there would be no limits on the possible number of particles in a given state in agreement with the character of symmetric eigenfunctions, and to the Fermi-Dirac case where, as we know, the possible number of particles in a given state could only be zero or one, in agreement with the Pauli exclusion principle applying with antisymmetric eigenfunctions.

To perform the desired computations in the Einstein-Bose case, it proves convenient to consider the expression

$$\sum_{n_k=0}^{\infty} e^{-(\alpha+\beta\epsilon_k)n_k} = \frac{1}{1-e^{-(\alpha+\beta\epsilon_k)}}, \tag{114.9}$$

where the summation is taken over all values of n_k from zero to infinity, in agreement with the considerations that the grand canonical ensemble places no limitation on the total number of particles in a system, and that the symmetrical character of the eigenfunctions places no limitation on the number in any particular state; and where the left-hand side is seen to be the correct binomial expansion of the right. Differentiating (114.9) with respect to $(\alpha+\beta\epsilon_k)$, we obtain the result

$$\sum_{n_k=0}^{\infty} e^{-(\alpha+\beta\epsilon_k)n_k} n_k = \frac{e^{-(\alpha+\beta\epsilon_k)}}{[1-e^{-(\alpha+\beta\epsilon_k)}]^2}, \tag{114.10}$$

and differentiating a second time, we obtain

$$\sum_{n_k=0}^{\infty} e^{-(\alpha+\beta\epsilon_k)n_k} n_k^2 = \frac{e^{-(\alpha+\beta\epsilon_k)}}{[1-e^{-(\alpha+\beta\epsilon_k)}]^2} + 2\frac{[e^{-(\alpha+\beta\epsilon_k)}]^2}{[1-e^{-(\alpha+\beta\epsilon_k)}]^3}. \tag{114.11}$$

Furthermore, making use of the original expression (114.9), it will be seen that we can rewrite these results in the forms

$$\frac{\sum_{n_k=0}^{\infty} e^{-(\alpha+\beta\epsilon_k)n_k} n_k}{\sum_{n_k=0}^{\infty} e^{-(\alpha+\beta\epsilon_k)n_k}} = \frac{1}{e^{\alpha+\beta\epsilon_k}-1}, \tag{114.12}$$

and

$$\frac{\sum_{n_k=0}^{\infty} e^{-(\alpha+\beta\epsilon_k)n_k} n_k^2}{\sum_{n_k=0}^{\infty} e^{-(\alpha+\beta\epsilon_k)n_k}} = \frac{1}{e^{\alpha+\beta\epsilon_k}-1} + 2\left(\frac{1}{e^{\alpha+\beta\epsilon_k}-1}\right)^2. \tag{114.13}$$

Comparing with the expression for the probability $P(n_k)$ for finding n_k particles in the state k, as given by (114.8), we now see that (114.12) gives

$$\bar{\bar{n}}_k = \frac{1}{e^{\alpha+\beta\epsilon_k}-1} \tag{114.14}$$

as an expression for the mean number of particles in state k. And we see that (114.13) gives

$$\overline{\overline{(n_k^2)}} = \bar{\bar{n}}_k + 2(\bar{\bar{n}}_k)^2 \tag{114.15}$$

as an expression for the mean square number in that state. This latter can be rewritten in the form

$$\frac{\overline{\overline{(n_k-\bar{\bar{n}}_k)^2}}}{(\bar{\bar{n}}_k)^2} = \frac{1}{\bar{\bar{n}}_k} + 1, \tag{114.16}$$

which gives an expression, for the fractional mean square deviation from the mean number in any state, that may be compared with the previous (classical) Maxwell-Boltzmann expression as given by (113.12). We thus obtain the desired results in the case of Einstein-Bose particles.

In the Fermi-Dirac case the possible numbers of particles in any given state k could only be zero or one, as we have already noted. This makes it easy in this case to compute results analogous to those obtained above. Starting with the expression, given by (114.8),

$$P(n_k) = \frac{e^{-(\alpha+\beta\epsilon_k)n_k}}{\sum\limits_{n_k} e^{-(\alpha+\beta\epsilon_k)n_k}}, \tag{114.17}$$

for the probability of finding n_k particles in state k, the summation in the denominator will now include only the two terms for $n_k = 0$ and $n_k = 1$. Hence it will be immediately seen that we shall now obtain the same result for the mean and for the mean square number of particles in the state k

$$\overline{\overline{n}}_k = \overline{\overline{(n_k^2)}} = \frac{e^{-(\alpha+\beta\epsilon_k)}}{1+e^{-(\alpha+\beta\epsilon_k)}}. \tag{114.18}$$

These findings may be rewritten in the more familiar forms

$$\overline{\overline{n}}_k = \frac{1}{e^{\alpha+\beta\epsilon_k}+1}, \tag{114.19}$$

and

$$\frac{\overline{\overline{(n_k-\overline{\overline{n}}_k)^2}}}{(\overline{\overline{n}}_k)^2} = \frac{1}{\overline{\overline{n}}_k} - 1. \tag{114.20}$$

We thus also obtain the desired results in the case of Fermi-Dirac particles.

(*b*) **Investigation of approximation.** The foregoing expressions, both for the Einstein-Bose and for the Fermi-Dirac case, were obtained with the help of a grand canonical ensemble containing members with all possible values for the total number of particles n. We must now demonstrate the validity of this method of procedure† by showing for the ordinary cases of interest that there would be a high concentration of the members of the ensemble around the most probable number of particles $\tilde{n}$.

For this purpose we may start with our expression (114.5),

$$P_{n_k,n_l,n_m,\ldots} = Ce^{-\alpha n-\beta(n_k\epsilon_k+n_l\epsilon_l+n_m\epsilon_m+\ldots)}, \tag{114.21}$$

† Compare Pauli, *Zeits. f. Phys.* **41**, 81 (1927).

for the probability of finding a member of the grand canonical ensemble in the indicated state. With the help of this formula we can then evidently write

$$P(n) = Ce^{-\alpha n} \sum_{n_k+n_l+\ldots=n} e^{-\beta(n_k\epsilon_k+n_l\epsilon_l+\ldots)}, \tag{114.22}$$

as an expression for the probability of finding a member of the ensemble with any selected value for the total number of particles n, provided the summation is taken as indicated over all possible states for a system composed of that number of particles. It will be convenient to rewrite this expression in the abbreviated form

$$P(n) = Ce^{-\alpha n-\gamma(n)}, \tag{114.23}$$

where γ is defined as a function of n by

$$e^{-\gamma(n)} = \sum_{n_k+n_l+\ldots=n} e^{-\beta(n_k\epsilon_k+n_l\epsilon_l+\ldots)}. \tag{114.24}$$

Differentiating (114.23) with respect to n, we obtain

$$\frac{\partial P(n)}{\partial n} = -Ce^{-[\alpha n+\gamma(n)]}\frac{\partial}{\partial n}[\alpha n+\gamma(n)] = 0 \tag{114.25}$$

as the condition for a maximum value of the probability $P(n)$. Hence, if we develop $\alpha n+\gamma(n)$ around the most probable value $\tilde{n}$ for the total number of particles, we see that the second term of the expansion will be missing, and that we shall obtain

$$\alpha n+\gamma(n) = \alpha\tilde{n}+\gamma(\tilde{n})+\frac{1}{2}\left(\frac{\partial^2\gamma}{\partial n^2}\right)_{n=\tilde{n}}(n-\tilde{n})^2+\ldots \tag{114.26}$$

as an appropriate approximation. Substituting in (114.23), we then have

$$P(n) = Ce^{-\alpha\tilde{n}-\gamma(\tilde{n})-\frac{1}{2}\left(\frac{\partial^2\gamma}{\partial n^2}\right)_{n=\tilde{n}}(n-\tilde{n})^2} \tag{114.27}$$

as an approximate expression for the probability of finding a member of the ensemble with a total of n particles. As a suitable approximation for the mean square deviation we shall then have

$$\frac{\int\limits_0^\infty P(n)(n-\tilde{n})^2\,dn}{\int\limits_0^\infty P(n)\,dn} = \frac{\int\limits_0^\infty e^{-\frac{1}{2}\left(\frac{\partial^2\gamma}{\partial n^2}\right)_{n=\tilde{n}}(n-\tilde{n})^2}(n-\tilde{n})^2\,dn}{\int\limits_0^\infty e^{-\frac{1}{2}\left(\frac{\partial^2\gamma}{\partial n^2}\right)_{n=\tilde{n}}(n-\tilde{n})^2}\,dn}. \tag{114.28}$$

Changing the variable of integration to $(n-\tilde{n})$, and noting with $\tilde{n}$ large that it will then be a good approximation to take the limits of integration as running from minus to plus infinity, we then have

$$\overline{\overline{(n-\tilde{n})^2}} = \frac{\int\limits_{-\infty}^{+\infty} e^{-\frac{1}{2}\left(\frac{\partial^2\gamma}{\partial n^2}\right)_{n=\tilde{n}}(n-\tilde{n})^2} (n-\tilde{n})^2\, d(n-\tilde{n})}{\int\limits_{-\infty}^{+\infty} e^{-\frac{1}{2}\left(\frac{\partial^2\gamma}{\partial n^2}\right)_{n=\tilde{n}}(n-\tilde{n})^2}\, d(n-\tilde{n})}, \tag{114.29}$$

which from known formulae of integration (Appendix II) gives us

$$\overline{\overline{(n-\tilde{n})^2}} = \frac{1}{(\partial^2\gamma/\partial n^2)_{n=\tilde{n}}} \tag{114.30}$$

as the desired result.

To investigate the magnitude which can be expected for this mean square deviation, we may turn to the definition of $\gamma(n)$ as given by (114.24),

$$e^{-\gamma} = \sum_{n_k+n_l+...=n} e^{-\beta(n_k\epsilon_k+n_l\epsilon_l+...)}, \tag{114.31}$$

and rewrite this for any value of n in the form

$$e^{-\gamma} = \sum_{\epsilon_1,\epsilon_2,...\epsilon_n} \frac{n_k!\, n_l!\, n_m! \ldots}{n!} e^{-\beta(\epsilon_1+\epsilon_2+...+\epsilon_n)}, \tag{114.32}$$

where we take a sum over all possible values ϵ_1, ϵ_2,..., ϵ_n for the energies of n particles. The symbols n_k, n_l, n_m,... in this expression denote the numbers of particles in each group to which the same elementary state is assigned in any particular term of the summation. Hence the factor introduced in front of the exponential term correctly allows for the circumstance that a mere interchange in particles does not lead to a different state of the system as a whole.

At a later time (see, for example, § 135 (a)) we shall have occasion to obtain explicit evaluations for expressions of the form (114.32), to which the name 'sum-over-states' may be applied. For our present purposes it is merely necessary to call attention to the appearance of factorial n in the denominator of each term of the summation. Using the Stirling approximation for factorial numbers, and taking logarithms, this then gives

$$\gamma = n \log n + ... \tag{114.33}$$

as the 'leading term' of γ. And by differentiation we then see that the denominator in (114.30) will at least have the order of magnitude

$$\left(\frac{\partial^2\gamma}{\partial n^2}\right)_{n=\tilde{n}} \approx \frac{1}{\tilde{n}}, \tag{114.34}$$

in the absence of fortuitous cancellations. Combining with (114.30), we can then take

$$\frac{\overline{\overline{(n-\tilde{n})^2}}}{(\tilde{n})^2} \approx \frac{1}{\tilde{n}}, \tag{114.35}$$

as giving an idea of the order of magnitude of the fractional mean square deviation from the most probable number of particles for the members of our grand canonical ensemble. Since the fractional deviation goes to zero as the most probable number of particles $\tilde{n}$ becomes large, this then justifies our approximate method of calculation in the case of ordinary physical-chemical systems of interest where the actual number of particles is exceedingly high.

A somewhat different method of showing the tendency for the members of a grand canonical ensemble to be concentrated in the neighbourhood of a particular composition will be discussed in § 141 (*e*).

(*c*) **Further discussion of the Einstein-Bose and Fermi-Dirac distribution laws.** For purposes of comparing the results of this section with our previous equilibrium results as obtained in Chapter X, it will now be convenient to fix our attention on any selected group of g_κ elementary states all of which may be regarded as corresponding to substantially the same value of energy ϵ_κ. In accordance with (114.14) and (114.19), we can then write

$$\bar{\bar{n}}_\kappa = \frac{g_\kappa}{e^{\alpha+\beta\epsilon_\kappa} \mp 1} \tag{114.36}$$

as a convenient combined expression for the mean number of particles that would be present at equilibrium with the energy ϵ_κ, where the upper sign refers to the Einstein-Bose and the lower to the Fermi-Dirac case. Comparing with our previous equation (89.8), we now see that our closely exact expression for the mean number of particles in any condition κ in a canonical ensemble is the same as our previous approximate expression for the most probable number in that condition in the corresponding microcanonical ensemble. Hence our previous results as to the equilibrium properties of Einstein-Bose and Fermi-Dirac systems, as obtained in Chapter X, can now be taken over without substantial alteration.

Among these earlier results we obtained in § 91 (*b*) the important possibility of relating the parameter β—appearing in the expressions for the Maxwell-Boltzmann, Einstein-Bose, and Fermi-Dirac distributions—to the temperature T of the system by the relation

$$\beta = \frac{1}{kT}, \tag{114.37}$$

where k is Boltzmann's constant. It will be immediately evident that this possibility still persists in our present mode of treatment. From the general form of expression,

$$P_n = Ce^{-\beta E_n}, \tag{114.38}$$

for the distribution of states in a canonical ensemble, it will be seen, in the case of a system consisting of parts to which separate energies can be ascribed, that each part can be regarded as represented at equilibrium by a canonical distribution with the same value of β. This then makes it possible, as before, to take a part of any system of interest as consisting of a gas sufficiently dilute, so that it can be treated as a classical perfect-gas thermometer, and so that the distribution of particles described by (114.36) will assume its classical form. Using this distribution, it then again becomes possible to obtain the previous relation of the constant β to the pressure p and hence to the phenomenologically defined temperature T of the dilute gas.

To conclude this discussion of equilibrium in Einstein-Bose and Fermi-Dirac gases, this may be taken as a convenient place to consider a question, arising in the case of gases composed of real molecules, which was not discussed in Chapter X. In the above derivation of the Einstein-Bose and Fermi-Dirac distribution laws we have spoken of the gases for simplicity as consisting of particles. It is evident, however, that the derivation given would also apply to gases composed of real molecules, provided the dilution is high enough so that the interaction between molecules can be treated as weak, provided we use the Einstein-Bose or Fermi-Dirac distribution according as the molecules are composed of an even or odd total number of fundamental particles (nuclei being treated as composed of protons and neutrons), and finally provided we recognize that the different states of the molecules to which the distribution laws apply must be specified by giving their internal as well as their external state. This latter circumstance now makes it of interest to inquire into the relative numbers of molecules that would be in different internal states without reference to their external state of motion.

To examine this question it now becomes convenient to regard the different possible states of a molecule as specified by a pair of indices ki referring to external and internal conditions respectively, and to take the total energy of the molecule ϵ_{ki} in any such state as given by the sum of its external kinetic energy ϵ_k and its internal energy ϵ_i,

$$\epsilon_{ki} = \epsilon_k + \epsilon_i. \tag{114.39}$$

In the classical mechanics this additivity of energies would lead to statistical independence of the distributions over states k and states i, and hence to the familiar Boltzmann ratio for the numbers of molecules in different internal states i. We shall see, however, that for degenerate

Einstein-Bose and Fermi-Dirac gases the internal and external distributions are not independent.

In accordance with (114.14) and (114.19), we may write

$$\bar{\bar{n}}_{ki} = \frac{1}{e^{\alpha+\beta\epsilon_k+\beta\epsilon_i}\mp 1} \tag{114.40}$$

as an expression for the mean number of molecules which would be in any such state ki at equilibrium, where the upper sign refers to the Einstein-Bose and the lower to the Fermi-Dirac case. Furthermore, in accordance with (71.16), we may write

$$dg = \frac{4\pi v}{h^3} m\sqrt{(2m\epsilon_k)}\, d\epsilon_k, \tag{114.41}$$

as an expression for the number of eigenvalues of kinetic energy which would fall in any range ϵ_k to $\epsilon_k+d\epsilon_k$, where v is the volume of the container and m the mass of a molecule. Combining with (114.40), we can now write

$$d\bar{\bar{n}}_i = \frac{4\pi v}{h^3} m\sqrt{(2m)} \frac{\epsilon_k^{\frac{1}{2}}\, d\epsilon_k}{e^{[\alpha+\beta\epsilon_i]+\beta\epsilon_k}\mp 1} \tag{114.42}$$

as an expression for the mean number of molecules in the internal state i and in the external energy range $d\epsilon_k$.

Considering the integration of this expression over all possible values of the kinetic energy ϵ_k from zero to infinity, making use of our previous expressions for the result of such integrations as given by (93.12) and (94.7), and replacing β by $1/kT$, we now obtain

$$\bar{\bar{n}}_i = \frac{v(2\pi mkT)^{\frac{3}{2}}}{h^3} U([\alpha+\epsilon_i/kT], \tfrac{1}{2})$$

and

$$\bar{\bar{n}}_i = \frac{v(2\pi mkT)^{\frac{3}{2}}}{h^3} V([\alpha+\epsilon_i/kT], \tfrac{1}{2}) \tag{114.43}$$

as respective expressions in the Einstein-Bose and Fermi-Dirac cases for the mean number of particles in any internal state i. The functions U and V in these expressions are the definite integrals originally introduced in §§ 93 (*b*) and 94 (*a*) for the treatment of the two cases. It will be noted, however, that they now depend on the argument $[\alpha+\epsilon_i/kT]$ instead of simply on α as in our earlier considerations. We may make use of these expressions to obtain the desired information as to the relative numbers of molecules that would be present in different internal states i.

Considering cases where the gas degeneration is sufficiently small to permit the use of the series expansion for U and V, which are given by (93.9) and (94.3), and fixing our attention on any pair of internal states i and j, we readily obtain

$$\frac{\bar{\bar{n}}_i}{\bar{\bar{n}}_j} = \frac{e^{-\epsilon_i/kT}}{e^{-\epsilon_j/kT}} \frac{1 \pm \frac{1}{2^{\frac{3}{2}}} e^{-\alpha-\epsilon_i/kT} + \frac{1}{3^{\frac{3}{2}}} e^{-2\alpha-2\epsilon_i/kT} \pm \ldots}{1 \pm \frac{1}{2^{\frac{3}{2}}} e^{-\alpha-\epsilon_j/kT} + \frac{1}{3^{\frac{3}{2}}} e^{-2\alpha-2\epsilon_j/kT} \pm \ldots}, \tag{114.44}$$

where the upper and lower signs refer respectively to the Einstein-Bose and Fermi-Dirac cases, as an expression for the ratio of the equilibrium numbers of molecules that can be expected in two different internal states i and j at the temperature T. It is of interest to note that this approaches the familiar *Boltzmann ratio*,

$$\frac{\bar{\bar{n}}_i}{\bar{\bar{n}}_j} = \frac{e^{-\epsilon_i/kT}}{e^{-\epsilon_j/kT}}, \tag{114.45}$$

as α gets larger and the degree of gas degeneration gets smaller. For all ordinary gases under laboratory conditions, α is large enough so that deviations from the Boltzmann ratio are quite negligible. This conclusion is of great importance in the study of physical-chemical equilibria.†

A case, however, where the general formula (114.43) leads to results radically different from the Boltzmann ratio is provided by Pauli's treatment of the paramagnetic susceptibility of metals which has already been mentioned in § 94 (*e*). Treating the conduction electrons of a metal as a Fermi-Dirac gas, the second form of (114.43) may be used to compute the ratio of the numbers of electrons with magnetic moment oriented parallel and antiparallel to an external field. Because of the high degeneracy of the electron gas at ordinary temperatures, this ratio is much nearer unity than the Boltzmann ratio and gives correspondingly smaller paramagnetic susceptibility.

115. Equilibrium in general in physical-chemical systems

In accordance with the relation (114.37) between the parameter β and the temperature T,

$$\beta = \frac{1}{kT}, \tag{115.1}$$

which was seen to be justified in the preceding section, we may now

† Deviations from the Boltzmann ratio have been discussed for the Einstein-Bose case by Lewis and Mayer, *Proc. Nat. Acad.* **15**, 208 (1929), and for the Fermi-Dirac case by Brillouin, *Les Statistiques Quantiques*, Paris, 1930, p. 150.

express the *canonical distribution* (112.5) for the equilibrium probabilities P_n for the various energy states of any system in the form

$$P_n = Ce^{-E_n/kT}, \tag{115.2}$$

where C is an appropriate normalizing factor. It will be appreciated that this now provides a fundamental apparatus for studying the equilibrium properties, at a *specified temperature* T, for any kind of system for which we can determine the energy states n corresponding to different possible eigenvalues E_n.

It is of interest to compare the above distribution with the *microcanonical distribution* (110.3), for the probabilities of different energy states, which can be written in the form

$$P_n = \begin{cases} \rho_0 & (E_n \text{ in range } E \text{ to } E+\delta E) \\ 0 & (E_n \text{ not in range } E \text{ to } E+\delta E), \end{cases} \tag{115.3}$$

where ρ_0 is a constant. This provides an apparatus, as was discussed in § 110, for studying the equilibrium properties of a system of *specified energy*.

In investigating the ordinary problems of physical-chemical equilibria which arise, it is usually preferable to employ the apparatus provided by the canonical distribution (115.2) rather than that provided by the microcanonical distribution (115.3). This is due in the first place to the consideration that we are usually interested in systems which have sufficient contact with their surroundings so that they can be better regarded as having a specified temperature than as having a specified energy; and is due in the second place to the circumstance that it is much easier to give accurate mathematical treatment to the canonical than to the microcanonical distribution. Nevertheless, as previously remarked and as already illustrated by examples, the two methods lead to substantially identical results in the case of systems of many degrees of freedom, since the canonical distribution then becomes highly concentrated around a particular value of the energy, in agreement with our physical experience that the fluctuations in energy are unimportant for an ordinary macroscopic system immersed in a temperature bath.

With the help of the canonical distribution described by (115.2), it would now be possible to proceed at once to further studies of equilibrium phenomena, including equilibrium between regions of different potential energy, equilibrium between condensed and gaseous phases, and equilibrium in chemical reactions. We shall regard it as preferable, however, to postpone such studies until after we have presented a statistical mechanical explanation of the principles of thermodynamics. This will make it possible to introduce the idea of temperature into our considerations from a more fundamental thermodynamic point of view, and to express the results obtained in the kind of language commonly employed by physical chemists.

116. The principle of detailed balance in the quantum mechanics

We now conclude the present chapter by using the studies, which we have made in connexion with the H-theorem, to throw light on the mechanism responsible for the maintenance of steady properties in the case of a quantum mechanical system in a state of macroscopic equilibrium. It will be found that the maintenance of such properties can be regarded as guaranteed by a principle which is the quantum analogue of the classical principle of detailed balance, which was discussed in § 50 (c) of Chapter VI. To carry out the investigation we shall consider the frequency of transition between different conditions of the system of interest, making use of the method of transition probabilities as giving a good insight into the mechanism involved, and as being sufficiently exact for that purpose.

Let us fix our attention on two possible conditions of the system of interest—or, if necessary, of the system proper and its surroundings—which we denote by the letters κ and ν. Let us regard these conditions as defined by two groups of neighbouring nearly steady states, G_κ and G_ν in number, which lie in the same energy range E to $E+\Delta E$, but which correspond to observably different situations. Let us then regard ourselves as making observations on such conditions, when the system is in equilibrium, in order to determine the probabilities for transition between them.

In accordance with the circumstances, that equilibrium for the system corresponds in any case to an ensemble giving the same probabilities for different states of the same energy, and that our observations are made on nearly true energy states, we can take the probabilities P_κ and P_ν for finding the two conditions as proportional to the corresponding numbers of states G_κ and G_ν. This then permits us to write

$$\frac{P_\kappa}{P_\nu} = \frac{G_\kappa}{G_\nu} \tag{116.1}$$

for the ratio of the two probabilities. Furthermore, in accordance with the circumstances, that representative ensembles for systems in equilibrium are diagonal in the energy language, and that our observations would give no information as to the phases of the states in G_κ or G_ν, we have the necessary conditions for applying the method of transition probabilities, and can take our previous expressions (99.42) and (99.43),

$$Z_{\kappa\nu} = A_{\kappa\nu} G_\nu P_\kappa \tag{116.2}$$

and

$$Z_{\nu\kappa} = A_{\nu\kappa} G_\kappa P_\nu \tag{116.3}$$

with

$$A_{\kappa\nu} = A_{\nu\kappa} \tag{116.4}$$

as giving the numbers of transitions per unit time, $Z_{\kappa\nu}$ and $Z_{\nu\kappa}$, which we can expect from condition κ to ν and from ν to κ respectively. Combining the foregoing equations, we can then obtain

$$Z_{\kappa\nu} = Z_{\nu\kappa}. \tag{116.5}$$

This result shows that we may expect, *on the average, for a system at equilibrium* the same frequency of transition from the condition κ to ν as from the condition ν to κ. This then means that we can regard the steady state of affairs at equilibrium as maintained by a direct balance between the rates of opposing processes; that is, the transitions from κ to ν do not have to be thought of as balanced with the help of some indirect route such as ν to λ to κ. This is the *quantum mechanical* form of the *principle of detailed balance.*

The derivation of this principle in the quantum mechanics, as given above, is seen to be rather more general and more direct than the corresponding classical derivation as given in Chapter VI. This arises partly from the general character of the formulation in terms of transition probabilities (cf. § 99), which makes it unnecessary to consider the special nature of the processes taking place in the system of interest. In addition, the early introduction of statistical assumptions, especially the assumption of random phases as a necessary prerequisite for the formulation in terms of transition probabilities, may be regarded as including that grouping of states with their reverse states, which was taken as a necessary special step in the classical treatment of this question.

The principle of detailed balance is often very useful in obtaining an insight into the behaviour of complicated systems. A specific application, which is of use in the theory of chemical kinetics, will make its importance clearer. Consider a gas containing molecules which can exist in various states of kinetic energy ϵ_k and internal energy ϵ_i, and let the condition of the gas as a whole be specified by the numbers of molecules of the different kinds present in the different individual states which are possible. Consider, then, two conditions of the gas which differ in that the first condition has a pair of molecules with high kinetic energy and low internal excitation, and the second has a corresponding pair of molecules of average kinetic energy, one of the molecules, however, now being in a highly excited internal state so that we can regard it as *chemically activated* for some reaction of interest. Assuming collisional mechanism for the transitions between these two conditions, the principle of detailed balance will then allow us to equate the numbers of collisions in unit time which would lead at equilibrium to the inverse transitions between the above two conditions. This is of importance in chemical kinetics since it permits a determination of the rate of chemical activation from the more easily calculated frequency of deactivating collisions which would prevail at equilibrium.

XIII

STATISTICAL EXPLANATION OF THE PRINCIPLES OF THERMODYNAMICS

117. Introduction

(*a*) **Thermodynamic system and representative ensemble.** In the present chapter we are now ready to undertake the important task of using the methods of statistical mechanics to provide an explanation for the principles of thermodynamics. The methods of statistical mechanics are specially devised for treating the behaviour, which can be expected on the average, in the case of mechanical systems in conditions corresponding to an incomplete specification of precise state. And the principles of thermodynamics are devised for giving a phenomenological account of the gross behaviour of macroscopic physical systems in conditions corresponding to the specification of a limited number of thermodynamic variables such as volume, pressure, energy, temperature, or entropy. The desired explanation of thermodynamics depends on showing that the science of statistical mechanics provides an appropriate interpretation for such non-mechanical variables as temperature and entropy, and provides predictions as to the average behaviour of systems of many degrees of freedom in substantial agreement with the predictions of thermodynamics.

In accordance with the methods developed in preceding chapters, the properties of a *thermodynamic system*, whose condition is described by the values of a limited number of thermodynamic variables, may be studied with the help of the average properties of an appropriately chosen *representative ensemble* of systems, of similar constitution to the one of actual interest. In a general way it may be said that the appropriate choice of representative ensemble depends on taking a distribution of the members of the ensemble over their different possible individual states, which agrees, on the one hand, with our knowledge of the thermodynamic variables that have been measured, and which conforms, on the other hand, with the hypotheses of equal *a priori* probabilities and of random *a priori* phases on which we have based our deductions of statistical results.

(*b*) **The nature of thermodynamic variables.** We must now consider the nature of the *thermodynamic variables*, the values of which determine the selection of an appropriate representative ensemble. These variables may be conveniently grouped into three classes.

As the first class of such quantities we shall take *external parameters*, having values which we regard as definitely established by external agencies without reference to the internal condition of the system of interest. They include typically such quantities as the volume of the system determined, for example, by the position of a piston, the intensity of fields of force acting on the system due to the presence of neighbouring bodies, and the coordinates locating the position of heat reservoirs which we may wish to bring into thermal contact with the system of interest. These external parameters depend either directly or indirectly on the quantities designated as *external coordinates* by Gibbs† in the classical development of statistical mechanics. As in the classical statistics, we can take the values of these parameters as precisely determined. This is made possible, in the first place, as in the classical development, by the consideration that these quantities are not dependent on the imperfectly known precise internal state of the thermodynamic system of interest; and is made possible, in the second place, for the purposes of a quantum treatment, by the consideration that the external parameters will be chosen so as to depend on the gross properties of macroscopic external bodies, which are unaffected by the limitations imposed by the Heisenberg uncertainty principle. Owing to this precise determination it will be possible to assign the same values, as are exhibited by the system of interest, to the external parameters of all the systems in the representative ensemble.

As the second class of thermodynamic variables we shall take *mechanical quantities*, having values dependent on the condition of the system, which we regard as determined by gross measurements that are not sufficient to give a knowledge of the precise state of our system. As typical examples of such variables we have the pressure exerted by the system and its total energy content. In general, the values of these mechanical quantities cannot be regarded as precisely determined for the particular thermodynamic system of interest. In this connexion, however, we find a certain difference between the classical point of view and the quantum mechanical point of view which we must now adopt. In the classical statistics a quantity such as pressure was taken as necessarily somewhat indeterminate since the instantaneous force exerted on the walls of a container by the molecules composing a system

† Gibbs, *Elementary Principles in Statistical Mechanics*, Yale University Press, 1902. See, for example, chapter xiii, p. 152. It will be noticed that the treatment given in the present chapter is in many places merely the quantum mechanical transcription of the classical treatment of the relation between thermodynamics and statistical mechanics given by Gibbs.

would be subject to fluctuations, depending on the unknown values of internal coordinates and momenta. But a quantity such as the total energy of a system was regarded as subject in principle to precise determination without important effect on the subsequent behaviour of that system. In the quantum statistics, however, we shall generally have to regard both the pressure and the energy of a system as not precisely determined. In the case of pressure, this will be true not only for the classical reason that the necessary knowledge as to internal coordinates and momenta is lacking, but also for the added quantum mechanical reason that the necessary knowledge as to the simultaneous values of such complementary variables could not even be theoretically obtained. And in the case of energy we shall have to take this quantity also as only approximately determined in general, since its precise measurement would involve an infinite length of time and would put the system in a steady quantum mechanical state, thus drastically affecting its subsequent behaviour. In agreement with their approximate determination, we shall represent the values of the mechanical quantities characterizing a thermodynamic system by the average values of these quantities in the representative ensemble.

As the third class of thermodynamic variables we shall take the essentially *non-mechanical thermodynamic quantities*, which the science of thermodynamics has specially introduced on its own level of scientific abstraction for the treatments which it gives. These quantities are temperature and entropy, and the various subsidiary variables which can be defined with their help. We shall find later (§ 122) that the statistical mechanical analogues of temperature and entropy are provided by quantities which characterize the distribution of the representative ensemble appropriate for the thermodynamic system of interest. Further discussion of these quantities will have to be postponed until that time.

(*c*) **Energy, work, and heat.** For the purposes of thermodynamics it has been found essential to distinguish two different kinds of process by which the internal energy of a system may be changed. These processes are the performance of work by the system on its surroundings and the flow of heat from the surroundings into the system, positive and negative values being possible for either of those quantities.

The concepts of the *internal energy* of a thermodynamic system, and of the *transfer of energy* between system and surroundings, presuppose a sufficiently distinct, actual or ideal, boundary between system and surroundings so that the energy belonging to the system can be distinguished—at least with suitable approximation—from that of the surroundings. Under these circumstances the internal energy of the system of interest can then be correlated, as noted above, with the average energy of the systems in the appropriate representative ensemble.

The concept of the *work* performed by a thermodynamic system on its surroundings depends on the possibility of changing the values of external parameters for the system, which have the nature of generalized coordinates. In the thermodynamic treatment, when a small variation is made in the value of such an external coordinate, the work done can be set equal to the product of that variation by the corresponding generalized force exerted by the system on its surroundings, and can be regarded as equal to the (potential) energy thereby transferred to external bodies. As a simple typical example, when the volume of a system is varied by the amount δv, we set the work done and energy thereby transferred equal to $p\,\delta v$, where the generalized force p is the pressure exerted by the system. In the corresponding statistical treatment we can assign to all the systems in the representative ensemble the same values for the external coordinates as those of the thermodynamic system of interest, and can represent the corresponding generalized forces by their average values in that ensemble. The work done and the energy thereby transferred to external bodies can then be represented by its average value for the ensemble, as given by the product of the definite value of the displacement with the average value of the corresponding generalized force.

The concept of the *heat* transferred between a thermodynamic system and its surroundings depends on the notion of thermal contact of the system with a so-called heat reservoir, the transfer of energy then taking place in a manner which cannot be followed in detail by the macroscopic measurements of thermodynamics since it is dependent on the unknown internal states of the two bodies involved. In the statistical mechanical treatment of such processes it is evident that we shall in general have to introduce two representative ensembles, one to correspond to the thermodynamic system of interest and the other to the heat reservoir. By considering each system in the ensemble for the thermodynamic system in thermal contact *separately* with each system in the ensemble for the heat reservoir, we can now create a third ensemble—of higher order in the number of systems—to correspond to the process of heat transfer. We may then represent the heat flow into our thermodynamic system by the average energy decrease in the heat reservoirs in this third ensemble.

118. The energy principle for ensembles

We may now prepare the way for a presentation of the statistical interpretation of the familiar first law of thermodynamics by considering the quantum mechanical analogue of the energy principle as applied to an ensemble of systems. In its quantum form this principle requires, in the case of an ensemble of *isolated systems*, that the probability for finding a member of the ensemble with any specified value of energy should remain constant independent of time. As a particular consequence of this principle it is then evident that the *mean energy* for the members of such an ensemble would be *constant in time*. This restricted form of the principle will be of special interest to us in the explanation of thermodynamic phenomena.

It is evident that the truth of this principle is immediately guaranteed by the laws of the quantum mechanics itself, since these have shown that the mean or expectation value of the energy for each individual system in the ensemble would itself be independent of the time. Nevertheless it may be instructive to give a derivation of the principle with the help of the formalism that has been specially devised for the treatment of statistical mechanical problems.

In accordance with (78.9), we can write

$$\bar{\bar{E}} = \sum_{n,m} H_{mn}\,\rho_{nm} \tag{118.1}$$

for the mean energy of the systems in an ensemble, where the H_{mn} are elements of the matrix corresponding to the Hamiltonian operator $\mathbf{H}$, and the ρ_{nm} are the elements of the density matrix which describe the distribution of the ensemble. Furthermore, for the rate of change in the density matrix with time we can write, in accordance with (81.20),

$$\frac{\partial\rho_{nm}}{\partial t} = -\frac{2\pi i}{h}\sum_k (H_{nk}\,\rho_{km} - \rho_{nk}\,H_{km}). \tag{118.2}$$

Combining (118.1) and (118.2), we then obtain

$$\begin{aligned}\frac{d\bar{\bar{E}}}{dt} &= -\frac{2\pi i}{h}\sum_{m,n,k}(H_{mn}\,H_{nk}\,\rho_{km} - H_{km}\,H_{mn}\,\rho_{nk}) \\ &= -\frac{2\pi i}{h}\sum_{m,n,k}(H_{mn}\,H_{nk}\,\rho_{km} - H_{mn}\,H_{nk}\,\rho_{km}) \\ &= 0,\end{aligned} \tag{118.3}$$

where the second form of writing results from the possibility of changing our choice of letters for dummy indices over which we take a summation.

We thus see indeed that the mean energy in an ensemble of isolated systems would be independent of the time.

119. The analogue of the first law of thermodynamics

We are now ready to consider the first law of thermodynamics and its statistical mechanical analogue. In the classical thermodynamics this law was customarily written in the form

$$\Delta E = Q - W, \tag{119.1}$$

where ΔE is the increase in the energy of a system that accompanies the absorption of heat Q from the surroundings and the performance of work W on the surroundings. It may be regarded as an expression of the classical principle of the conservation of energy written for the purposes of thermodynamics in a form which distinguishes the two different modes of energy transfer, between surroundings and system, denoted by the terms heat and work.

In order to obtain the analogue of the above equation in a quantum statistical mechanics, it is evident that the quantities appearing therein will have to be represented by averages provided by ensembles of systems appropriate for representing the processes of transfer of heat and performance of work. The considerations of the two preceding sections have provided the necessary concepts for obtaining the desired modified form of equation.

The thermodynamic system itself, together with the external bodies on which it can do work, may be represented by an ensemble of systems having the mean internal energy $\bar{\bar{E}}_1$ in the parts corresponding to the thermodynamic system proper and the mean energy $\bar{\bar{V}}_1$ in the bodies external thereto. Furthermore, the heat reservoir involved in the process may be represented by an ensemble of systems having the mean energy $\bar{\bar{E}}_2$. By considering a representative for each member in the first ensemble as placed in appropriate thermal contact separately with each member of the second, we can then construct a new ensemble suitable for studying both the performance of work and the transfer of heat. The mean energy of this final ensemble can evidently be written as

$$\bar{\bar{E}}_{12} = \bar{\bar{E}}_1 + \bar{\bar{V}}_1 + \bar{\bar{E}}_2. \tag{119.2}$$

Each member of this final ensemble, however, is itself an isolated system composed of one representative for the thermodynamic system proper, one for the bodies external thereto upon which work can be done, and one for the heat reservoir. Hence, in accordance with the preceding section on the energy principle for ensembles, the mean energy $\overline{\overline{E}}_{12}$ cannot change with the time, and we shall have

$$\Delta\overline{\overline{E}}_{12} = \Delta\overline{\overline{E}}_1+\Delta\overline{\overline{V}}_1+\Delta\overline{\overline{E}}_2 = 0 \tag{119.3}$$

for any processes that take place.

This result can now be rewritten in the desired form—analogous to the usual first law equation—

$$\Delta\overline{\overline{E}} = \overline{\overline{Q}}-\overline{\overline{W}}, \tag{119.4}$$

where $\Delta\overline{\overline{E}} = \Delta\overline{\overline{E}}_1$ is the mean increase in the internal energy of the representatives of the thermodynamic system proper, $\overline{\overline{Q}} = -\Delta\overline{\overline{E}}_2$ is the mean energy transferred from the heat reservoirs, and $\overline{\overline{W}} = \Delta\overline{\overline{V}}_1$ is the mean work done on external bodies.

In making this interpretation of the first law of thermodynamics, it will be noted that we have actually selected the *mean values* of quantities in an appropriate representative ensemble as the kind of averages to be correlated with the *precise values* which would be assigned to such quantities in thermodynamic considerations. It would, of course, also be possible to select some other kind of average instead of the mean for this purpose. Nevertheless, in the typical situations to which thermodynamics may be applied, the different common kinds of average which might be selected are actually found to have substantially identical values on account of the negligible character of fluctuations around the mean. It is hence possible and proves convenient, as we shall see also in the case of the second law of thermodynamics, to confine attention to simple arithmetical means as the averages to be used in the interpretation of thermodynamic relations.

It is of interest in this connexion to remark that the success of the thermodynamic procedure, of assigning exact values to quantities which cannot really be regarded as precisely defined, depends on the actually negligible character of the fluctuations which can be expected in the values of such quantities in a series of repetitions of the same thermodynamic experiment. When this condition is not satisfied, the methods of thermodynamics must be suitably modified, as will be discussed in the next chapter in § 141.

120. The canonical ensemble as representing thermodynamic equilibrium

We must now undertake a series of statistical mechanical investigations which will ultimately put us in a position (§ 130) to state the statistical mechanical analogue of the second law of thermodynamics.

The first problem to be treated is that of investigating the appropriate ensemble for representing a system in thermodynamic equilibrium.

In picking out a suitable ensemble for this purpose, it is to be noted that the systems, ordinarily treated by thermodynamic methods, are in actuality never in perfect isolation, but are either purposely placed in thermal contact with some other system such as a heat reservoir, or are necessarily in contact with their immediate surroundings such as the walls and piston of a cylinder. Hence, in accordance with §§ 111 and 112, see in particular the discussions in §§ 111 (*d*) and 112 (*b*), it is evident that the condition of *thermodynamic equilibrium* for the systems of usual thermodynamic interest can be best represented by a *canonical ensemble,* since this has been found to give the most appropriate description of equilibrium in the case of systems in thermal contact with their surroundings or in essential rather than perfect isolation therefrom.

In agreement with (112.4), we may write the density matrix for a canonical ensemble, when expressed in the energy language, in the form

$$\rho_{nm} = e^{\frac{\psi - E_n}{\theta}} \delta_{nm}, \tag{120.1}$$

where ψ and θ are constants and E_n is the energy eigenvalue for the state n. In accordance with this density matrix we may then take

$$P_n = \rho_{nn} = e^{\frac{\psi - E_n}{\theta}} \tag{120.2}$$

as an expression for the *equilibrium probability* of finding the system of interest in any particular state n, on successive repetitions of the same condition of equilibrium. Since such probabilities could be determined, in the case of equilibrium and of true energy states, with any desired degree of accuracy by separating the system from its surroundings and taking sufficient time for the observation of its state, we regard ourselves as proceeding to the limit where the observed or coarse-grained probability P_n can be set equal to the fine-grained probability ρ_{nn}.

Important properties of the canonical distribution are expressed by the following equations. Since the total probability of finding one or another state will be *normalized to unity,* we have

$$\sum_n P_n = \sum_n e^{\frac{\psi - E_n}{\theta}} = 1 \tag{120.3}$$

when summed over all states n. For the *mean energy* of the members of the ensemble we have

$$\overline{\overline{E}} = \sum_n P_n E_n = \sum_n e^{\frac{\psi - E_n}{\theta}} E_n. \tag{120.4}$$

And for the *value of the quantity* $\bar{\bar{H}}$, corresponding to the probabilities for the different energy states n, we can write

$$\bar{\bar{H}} = \sum_n P_n \log P_n = \sum_n e^{\frac{\psi - E_n}{\theta}} \frac{\psi - E_n}{\theta} = \frac{\psi - \bar{\bar{E}}}{\theta}. \tag{120.5}$$

Owing to the constancy of ψ and θ, the first of the above equations, (120.3), can also be expressed in the form

$$Z = e^{-\psi/\theta} = \sum_n e^{-E_n/\theta}, \tag{120.6}$$

where the useful quantity Z, which is thus defined, may be given for convenience the name of *sum-over-states*.† We shall find this quantity important for our later work.

The immediately foregoing expressions apply to the different individual states n which the system of interest can exhibit. In the *case of degeneracy*, if we wish for convenience to group together G_k individual states, all of which have the same energy E_k, we can rewrite these expressions in the forms

$$P_k = G_k e^{\frac{\psi - E_k}{\theta}}, \tag{120.7}$$

$$\sum_k G_k e^{\frac{\psi - E_k}{\theta}} = 1, \tag{120.8}$$

$$\bar{\bar{E}} = \sum_k G_k e^{\frac{\psi - E_k}{\theta}} E_k, \tag{120.9}$$

$$\bar{\bar{H}} = \sum_k G_k e^{\frac{\psi - E_k}{\theta}} \frac{\psi - E_k}{\theta} = \frac{\psi - \bar{\bar{E}}}{\theta}, \tag{120.10}$$

$$Z = e^{-\psi/\theta} = \sum_k G_k e^{-E_k/\theta}. \tag{120.11}$$

In using such a canonical ensemble to represent a given system of interest, in a condition of thermodynamic equilibrium, it will be noted that the expression for the normalization of the ensemble as given by (120.3) or (120.8), together with the expression for the mean energy of the members of the ensemble as given by (120.4) or (120.9), provide two equations for the determination of the two *distribution parameters* ψ and θ. Hence the ensemble may be regarded as corresponding to a given system of interest in a *condition of equilibrium* with a specified value for the *mean energy* $\bar{\bar{E}}$ which is to be expected. In using such an ensemble for the interpretation of thermodynamic phenomena, this mean value of energy would then be correlated—as in our previous

† This agrees with the form assumed by Planck's so-called 'Zustandsumme' when consideration is given to the states for a system as a whole. See Planck, *Theory of Heat*, translated by Brose, London, 1932, p. 253, equation (419).

discussion of the first law—with the precise value of energy which would be assigned in thermodynamic considerations.

In this connexion it is of importance to note that the canonical distribution actually does have a character, under usual circumstances, such that the fluctuations in energy around the mean would be small. The tendency towards concentration into states of neighbouring energy is brought about by the combined effects of the form of the distribution law (120.2) with the negative of E_n appearing in the exponent, and of the rapid increase with energy in the number of possible energy eigenstates which occurs in the case of systems of many degrees of freedom. Taking specified values of ψ and θ, the first of the above-mentioned factors then makes systems of relatively high energy improbable, and the second factor makes systems of relatively low energy improbable. In agreement with the remarks made at the end of the preceding section, it will be appreciated that this tendency for the systems in a representative canonical ensemble to be concentrated in neighbouring states of a similar character is very important for the statistical mechanical explanation of the ordinary principles of thermodynamics, since otherwise it would be difficult to account for the actual success of those principles when applied from a point of view which assigns precise values to such macroscopic quantities as energy.

A more complete consideration of the fluctuations to be expected at thermodynamic equilibrium, in the values of energy and other quantities pertaining to a system of interest, will be given in § 141 at the end of the next chapter. When such fluctuations become sufficiently large, the methods of thermodynamics must either be modified or be supplemented by more directly statistical considerations.

121. Relation connecting the values of $\overline{\overline{H}}$ in neighbouring canonical ensembles

We have seen in the preceding section that the canonical distribution, representing a system in thermodynamic equilibrium, can be described in the energy language by the expression

$$P_n = e^{\frac{\psi - E_n}{\theta}}, \tag{121.1}$$

giving the probabilities for finding the system in the different steady states n, corresponding to the energy eigenvalues E_n. It will be noted that any particular canonical distribution is directly dependent on the values of the two distribution parameters ψ and θ. It will also be appreciated, however, that the distribution can be taken as indirectly dependent on the values assigned to any *external coordinates* $a_1, a_2, a_3,\ldots,$ etc., which describe the system, since a change in these—for example, a change in volume—would affect the eigenvalues E_n for the energies of the different states n.

In the present section we wish to consider the effect on the important quantity $\overline{\overline{H}}$ of making small variations in the parameters and external coordinates $\psi, \theta, a_1, a_2, a_3,\ldots$ in such a way as to secure a *neighbouring canonical distribution* suitable for representing the same system in a neighbouring condition of thermodynamic equilibrium. As the expression for $\overline{\overline{H}}$, corresponding to the probabilities for the true energy states n, we can write in the case of a canonical ensemble

$$\overline{\overline{H}} = \sum_n e^{\frac{\psi-E_n}{\theta}} \frac{\psi-E_n}{\theta} = \frac{\psi-\overline{\overline{E}}}{\theta}, \tag{121.2}$$

as given by our earlier equation (120.5). Hence for the variation in $\overline{\overline{H}}$, when we change to a neighbouring canonical ensemble, we can at once take

$$\delta\overline{\overline{H}} = \frac{\delta\psi}{\theta} - \frac{\delta\overline{\overline{E}}}{\theta} - \frac{\psi-\overline{\overline{E}}}{\theta^2}\delta\theta. \tag{121.3}$$

We shall be interested, however, in putting this expression into a more significant form.

Since we are making a change to a new canonical ensemble, it is evident that our variations must be so made as to preserve the truth of the relation

$$\sum_n e^{\frac{\psi-E_n}{\theta}} = 1, \tag{121.4}$$

which makes the total probability for finding the system in one or another state n equal to unity. Hence the variations which we make in the quantities $\psi, \theta, a_1, a_2, a_3,\ldots$ must be taken in any case in such a way as to satisfy

$$\sum_n e^{\frac{\psi-E_n}{\theta}} \left\{\frac{\delta\psi}{\theta} - \frac{1}{\theta}\left(\frac{\partial E_n}{\partial a_1}\delta a_1 + \frac{\partial E_n}{\partial a_2}\delta a_2 + \frac{\partial E_n}{\partial a_3}\delta a_3 + \ldots\right) - \frac{\psi-E_n}{\theta^2}\delta\theta\right\} = 0. \tag{121.5}$$

It will now be convenient for any state n to define the *generalized external forces* $A_1, A_2, A_3,\ldots$ which correspond to the external coordinates $a_1, a_2, a_3,\ldots$ by the equations

$$A_1 = -\frac{\partial E_n}{\partial a_1}, \qquad A_2 = -\frac{\partial E_n}{\partial a_2}, \qquad A_3 = -\frac{\partial E_n}{\partial a_3}, \ldots. \tag{121.6}$$

Introducing these expressions, and making use of the general method of computing mean values, (121.5) can then be rewritten in the form

$$\frac{\delta\psi}{\theta}+\frac{1}{\theta}(\bar{\bar{A}}_1\,\delta a_1+\bar{\bar{A}}_2\,\delta a_2+\bar{\bar{A}}_3\,\delta a_3+...)-\frac{\psi-\bar{\bar{E}}}{\theta^2}\delta\theta = 0, \qquad (121.7)$$

where $\bar{\bar{A}}_1$, $\bar{\bar{A}}_2$, $\bar{\bar{A}}_3$,... denote the mean values of the external forces calculated over the members of the ensemble, and δa_1, δa_2, δa_3,... denote the variations made in the external coordinates, the same, of course, for each member of the ensemble in accordance with our original discussion of external parameters.

Combining (121.7) with our previous expression (121.3) for the variation in $\bar{\bar{H}}$, we now obtain the desired result

$$-\delta\bar{\bar{H}} = \frac{\delta\bar{\bar{E}}}{\theta}+\frac{1}{\theta}(\bar{\bar{A}}_1\,\delta a_1+\bar{\bar{A}}_2\,\delta a_2+\bar{\bar{A}}_3\,\delta a_3+...). \qquad (121.8)$$

We shall consider the application of this important relation to the interpretation of thermodynamics in the next section.

122. Statistical mechanical analogues of entropy, temperature, and free energy

We have already proposed statistical mechanical analogues for the thermodynamic quantities appearing in the first law of thermodynamics. Corresponding to the energy of the system, we take the mean energy $\bar{\bar{E}}$ of the members of a representative ensemble appropriate for the thermodynamic system of interest; corresponding to the heat absorbed, we take the mean energy $\bar{\bar{Q}}$ transferred between members in a combined ensemble representing the system of interest and the heat bath; and corresponding to the work done, we take the mean work $\bar{\bar{W}}$ performed by the members of the ensemble by the forces which they exert on external bodies. We have seen that these quantities satisfy an equation of the same form as that for the ordinary first law of thermodynamics,

$$\Delta\bar{\bar{E}} = \bar{\bar{Q}}-\bar{\bar{W}}. \qquad (122.1)$$

In the present section we wish to propose statistical mechanical analogues for the important thermodynamic quantities entropy, temperature, and free energy which appear in considerations of the second law of thermodynamics. In following sections we shall then show that the proposed analogues do have properties appropriate for representing these quantities, and finally, in § 130, we shall feel able to make a complete statement of the statistical mechanical form of the second law of thermodynamics.

The statistical mechanical analogues of entropy and temperature are provided by comparing the form of the statistical mechanical relation, derived in the preceding section,

$$-\delta\bar{\bar{H}} = \frac{\delta\bar{\bar{E}}}{\theta}+\frac{1}{\theta}(\bar{\bar{A}}_1\,\delta a_1+\bar{\bar{A}}_2\,\delta a_2+\bar{\bar{A}}_3\,\delta a_3+...), \tag{122.2}$$

with the form of the familiar thermodynamic relation

$$\delta S = \frac{\delta E}{T}+\frac{1}{T}(A_1\,\delta a_1+A_2\,\delta a_2+A_3\,\delta a_3+...). \tag{122.3}$$

The first of these equations expresses the variation in the statistical mechanical quantity $\bar{\bar{H}}$, corresponding to the probabilities for energy states, when we pass from the canonical ensemble which represents a system in one condition of thermodynamic equilibrium to that which represents it in a neighbouring condition of thermodynamic equilibrium; while the second of the equations expresses the variation in the thermodynamic quantity, the entropy of the system S, when we pass from the one condition of equilibrium to the other.

It has already seemed appropriate to correlate the thermodynamic quantities energy E, external forces A_1, A_2, A_3,..., and external coordinates a_1, a_2, a_3,... for the system of interest, with the mean energy $\bar{\bar{E}}$, mean external forces $\bar{\bar{A}}_1$, $\bar{\bar{A}}_2$, $\bar{\bar{A}}_3$,..., and external coordinates a_1, a_2, a_3,... for the corresponding representative ensemble, in accordance with the expressions

$$\begin{array}{lll} E \rightleftharpoons \bar{\bar{E}}, & A_1 \rightleftharpoons \bar{\bar{A}}_1, & a_1 \rightleftharpoons a_1 \\ & A_2 \rightleftharpoons \bar{\bar{A}}_2, & a_2 \rightleftharpoons a_2 \\ & \cdots\cdots & \cdots\cdots \end{array} \tag{122.4}$$

Hence the similarity in form of (122.2) and (122.3) now makes it reasonable to propose the further correlations of the thermodynamic quantities *entropy* S and *temperature* T with the statistical mechanical quantities $-\bar{\bar{H}}$ and θ, in accordance with the expressions

$$S \rightleftharpoons -k\bar{\bar{H}}, \qquad T \rightleftharpoons \frac{\theta}{k}, \tag{122.5}$$

where a constant k, with the dimensions of energy over temperature, has to be introduced to allow for difference in units. As indicated by the notation, this constant will actually turn out to be equal to Boltzmann's k, the perfect gas constant per molecule (see § 135 (c)). By making use of the correlations (122.4) and (122.5), the thermodynamic relation (122.3) is indeed seen to agree with the statistical mechanical relation (122.2).

Having introduced the analogues for entropy and temperature, we can now at once find the analogue for *free energy*. To do this we have only to compare our previous statistical mechanical relation (120.5), which can be solved for the parameter ψ in the form

$$\psi = \overline{\overline{E}}+\theta\overline{\overline{H}}, \tag{122.6}$$

with the thermodynamic equation

$$A = E-TS \tag{122.7}$$

by which the free energy A of a thermodynamic system was defined by Helmholtz.† Noting (122.5), it will then be seen that we can make the immediate correlation

$$A \rightleftharpoons \psi. \tag{122.8}$$

It may also be noted, in accordance with (120.6), that the free energy can be correlated with the sum-over-states Z by the equation

$$A \rightleftharpoons -kT\log Z = -kT\log\sum_n e^{-E_n/kT}. \tag{122.9}$$

This result is often very useful in making practical calculations of this important quantity A which characterizes a system in thermodynamic equilibrium.

The above correlations, of the two essential thermodynamic quantities temperature T and entropy S with the statistical mechanical quantities θ and $\overline{\overline{H}}$, have been obtained for the special case of equilibrium by comparing the change in S when a thermodynamic system is changed from one condition of thermodynamic equilibrium to another, with the change in $\overline{\overline{H}}$, when the corresponding representative ensemble is changed from one canonical distribution to another, i.e. from one state of statistical equilibrium to another. Furthermore, the expression for $\overline{\overline{H}}$, used in making the above correlations, is the special one that would correspond to contemplated observations on true energy states carried out with a limiting accuracy such that coarse-grained probabilities P_n and fine-grained probabilities ρ_{nn} would come into agreement. Hence it will now be desirable to make some further remarks as to possibilities that might be available also in the absence of equilibrium for the correlation of temperature and entropy with statistical mechanical quantities, and as to the role played in any such correlations by the type of observation that we have in mind.

Let us first consider the concept of temperature. In a case of equilibrium the science of thermodynamics ascribes a unique and precise value to the temperature T of a system as a whole, which could be empirically determined by any suitable thermometer inserted into any

† We use the letter A to designate the free energy of system *as originally defined by Helmholtz*. This quantity, $A = E-TS$, should be carefully distinguished from the thermodynamic potential, $F = E-TS+pv$, which is often called free energy by chemists.

part of the system considered. This corresponds to the statistical mechanical consideration that a system of interest, in a condition of equilibrium with a specified mean energy, is to be represented by a canonical ensemble with a definite value for the distribution parameter θ which can be immediately correlated with T.

In the absence of equilibrium the idea of the temperature of a system is regarded in thermodynamic treatments as losing its unique and precise character, since a thermometer in different parts of the system or thermometers of different constructions would in general give different indications. This corresponds to the statistical mechanical consideration, that a system which is not in equilibrium would have to be represented by an ensemble which is not in equilibrium—rather than by a canonical distribution—so that no distribution parameter θ would be available for purposes of correlation. Nevertheless, in so far as it remains valid in the absence of equilibrium to apply the concept of temperature in an approximate manner to parts of the whole system or to partial aspects of its behaviour, it also remains possible to find the appropriate statistical mechanical representation of the situation. For example, in a case of heat conduction from one part of a system to another to which temperatures T can be approximately assigned, the conditions of these parts may be described by nearly canonical distributions with corresponding values of θ, and the condition of the system as a whole may then be represented by a combined ensemble constructed in the general manner which was first discussed in § 107.

Let us now turn to the concept of entropy. In a case of equilibrium the science of thermodynamics ascribes a precise value to the entropy S of a system, which could be empirically determined by evaluating the integral $\int dQ/T$ for any reversible process (see § 130) by which the system could be taken from its selected standard condition of thermodynamic equilibrium to the condition of interest. This corresponds to the statistical mechanical consideration that we have decided to correlate the entropy S for a system in thermodynamic equilibrium with the precise value of $\overline{\overline{H}} = \sum P_n \log P_n$ that would be calculated from the (exact) probabilities P_n for the true energy states n in the canonical ensemble which we take as representing the equilibrium.

In the absence of equilibrium the idea of the entropy of a system is still regarded in thermodynamic treatments as retaining at least an approximately valid character. This is made possible by the consideration that the entropy of a system is equal to the sum of the entropies of its parts, and that the system as a whole can be treated as consisting of parts, or of components, approximately in conditions such as could be reached by reversible processes from a standard condition. In accordance with this circumstance, we shall now regard the entropy S for any system as correlated in general with quantities pertaining to the corresponding representative statistical ensemble by

$$S \rightleftharpoons -k\bar{\bar{H}} = -k \sum_n P_n \log P_n, \qquad (122.10)$$

where an additive constant may be included if desired to allow for choice of standard condition, and where the quantities P_n are the coarse-grained probabilities for those states n which are involved in the kind of observations on the system that concern us.

In case we are concerned with an initial approximate observation on the condition of a system which is changing with time, the entropy S would be correlated with the coarse-grained probabilities P_n, which we determine by our initial approximate measurement of states of the kind n, and which we may then use in setting up a representative ensemble for the system with the initial density matrix $\rho_{nm} = P_n \delta_{nm}$. It will be noted, in accordance with the principles of the quantum mechanics, that an immediate repetition of the same kind of observation would lead to the same value of entropy. It will also be appreciated that the value of entropy thus obtained could be expected to be in consonance with the approximate value that would be assigned in a thermodynamic treatment of the system on the basis of an observation of its condition. In case we are then concerned with a subsequent approximate observation on the condition of such a system, at a time later than that of the initial observation or of a different kind, the entropy S would be correlated with the coarse-grained probabilities P_n which would be predicted for those states n that now interest us, with the help of the representative ensemble that has been set up. Finally, in case we are concerned with the precise observation which might be appropriately made on the true energy state of a system in a condition of equilibrium, the entropy S would be correlated with the coarse-grained probabilities P_n which would be predicted for such energy states n, from the ensemble representing the equilibrium, at the limit of exact measurement where coarse- and fine-grained probabilities come into agreement.

At first sight the circumstance, that the entropy of a system can depend on the kind of observation that is contemplated for performance, might seem to stand in contradiction to the appearance of that quantity in definite thermodynamic expressions. This proves not to be the case, however, since the more general statements of thermodynamics are concerned primarily with inequalities rather than equations which are to be satisfied by the entropy S. It is also to be remarked in this connexion that we shall find it generally appropriate, in the development of thermodynamic analogies which follows, to regard ourselves in the case of a system which is changing with time as concerned with approximate observations on suitable unperturbed energy eigenstates which are nearly steady states for the system, and in the case of a system which is in equilibrium as concerned with precise observations on true energy states. Hence, in the absence of explicit mention, it is to be assumed in what follows that the values of $\bar{\bar{H}}$ and S refer to observations on nearly-steady or quite steady states of the system of interest.

We may now sum up the results of the present section by writing down once more the three correlations

$$S \rightleftharpoons -k\bar{\bar{H}}, \qquad T \rightleftharpoons \frac{\theta}{k}, \qquad A \rightleftharpoons \psi, \tag{122.11}$$

which we take as representing the statistical mechanical interpretation of the thermodynamic quantities entropy, temperature, and free energy. These proposals may be regarded, if desired, as tentative for the time being. The next few sections will be devoted to showing that the statistical mechanical quantities $-\bar{\bar{H}}$ and θ, and thus also by implication ψ, do behave in such a way as to justify the above correlation.

123. Effect on $\bar{\bar{H}}$ of leaving a system in essential isolation

The first consideration, justifying the above proposals for the statistical mechanical interpretation of thermodynamic quantities, will be based on the behaviour that would be expected in the course of time for an ensemble representing a system left to itself in essential isolation, the walls of its container and other surroundings with which it necessarily interacts being in such a condition that in the mean no net transfer of energy to or from the system proper takes place. Such a set-up would appear to provide the appropriate statistical mechanical representation for a system which would be regarded in thermodynamics as left to itself in complete isolation.

In accordance with the discussion of essential isolation given in § 111 (*d*), the ensemble representing the above situation could be expected to change, in the course of a time long enough for extensive interaction with the surroundings, to a distribution of probabilities for the different nearly steady states that concern us which would be practically a canonical distribution. Hence, in accordance with a property of the canonical ensemble shown in § 112 (*b*), the ultimate distribution could be taken as closely corresponding to a minimum value for the quantity $\bar{\bar{H}}$ under the condition of constant mean energy $\bar{\bar{E}}$ for the system proper.

Thus the behaviour in time of an essentially isolated system can be described in the language of statistical mechanics as a tendency for the corresponding quantity $\bar{\bar{H}}$ to change to the minimum value compatible with a constant value for the mean energy $\bar{\bar{E}}$. On the other hand, the behaviour of an isolated system can be described in the language of thermodynamics by the well-known principle that the entropy S for the system would tend to increase with time to the maximum value compatible with its constant energy E. This thus provides further justification for our proposed correlation between S and $-\bar{\bar{H}}$.

In this connexion it will be remembered that the derivation which we actually gave for the H-theorem was dependent on the introduction of the hypotheses of equal *a priori* probabilities and of random *a priori* phases in setting up the initial representative ensemble for the system of interest, and that we had to allow for the possibility of occasional deviations from our conclusions when these assumptions were not near enough valid. Accepting the proposed correlation of S with $-\bar{\bar{H}}$, we must now regard the certainty with which S does increase in a system appreciably displaced from equilibrium as due to the infrequent occurrence of important deviations from the consequences of those hypotheses in the case of actual thermodynamic systems of many degrees of freedom.

124. Effect on $\bar{\bar{H}}$ of adiabatic changes in external coordinates

We shall next wish to obtain a second illustration of the propriety of correlating S with $-\bar{\bar{H}}$ by considering the effect on a canonical ensemble of making an *actual change* in the values of the external coordinates for the systems in the ensemble. Here we shall distinguish the two limiting cases of an extremely slow variation in the values of the coordinates and of an abrupt change in their values. We shall thus obtain analogies for the two thermodynamic cases of reversible and irreversible changes in external coordinates. The treatments to be given will be based on our previous investigation of the effect of such changes on the probabilities for the different states of a single quantum mechanical system as carried out in § 97.

Let us consider an isolated thermodynamic system in a condition of thermodynamic equilibrium together with the corresponding canonical ensemble of members which would represent it in that condition, in accordance with the concept of essential isolation. In the first place, let us then make a small change δa in the value of some external coordinate a, pertaining to the system of interest and to the members of the corresponding ensemble, the change being carried out adiabatically from the thermodynamic point of view and hence appropriately taken as essentially adiabatic from the mechanical point of view, in accordance with the discussion of § 111 (d). In the second place, let us then allow sufficient time so that the system of interest comes to its new condition of thermodynamic equilibrium and the ensemble comes to the corresponding new canonical distribution, which, as we have seen, would be established in the course of sufficient time under conditions of essential isolation.

We must now give separate attention to the two limiting cases when the change δa is made very slowly and when it is made quite abruptly. We may begin with the case of a very slow change.

(a) **Reversible adiabatic change in an external coordinate.** Since the total change in the ensemble is from one given canonical distribution to a neighbouring canonical distribution, we can in any case apply our general expression (121.8) connecting the values of $\overline{\overline{H}}$ in two such ensembles. This gives us, correct to the first order in small quantities,

$$-\delta\overline{\overline{H}} = \frac{\delta\overline{\overline{E}}}{\theta} + \frac{\overline{\overline{A}}}{\theta}\delta a, \tag{124.1}$$

where $\delta\overline{\overline{E}}$ is the change in the mean energy of the systems in the ensemble, $\overline{\overline{A}}$ is the mean external force corresponding to the coordinate a, and θ is the parameter defining the original canonical distribution.

To evaluate this expression, in our present case, let us consider the change in mean energy $\delta\overline{\overline{E}}$ which is to be expected when the change in the coordinate a is made very slowly. Since we have shown, by equation (97.29) of § 97 (b), that the probabilities $|C_n|^2$ for finding a single system in its different energy states n would not be altered by a vanishingly slow rate of change in the coordinate a, we can also conclude that the probabilities,

$$P_n = \overline{\overline{W}}_n = \overline{\overline{|C_n|^2}}, \tag{124.2}$$

for finding different states n in the ensemble would not be altered merely as a direct consequence of the change δa. Furthermore, in accordance with the essentially adiabatic character of the change in which we are interested, we can conclude that the mean energy of the ensemble would not be affected by the subsequent readjustment in probabilities to a new canonical distribution which we assume to take place in the course of time by interaction with the surroundings. Hence, for the total change in the mean energy $\overline{\overline{E}}$, we need only consider that which results from the change in eigenvalues E_n when we change the external coordinate a. This then permits us to write the simple expression

$$\delta\overline{\overline{E}} = \sum_n P_n \frac{\partial E_n}{\partial a}\delta a = -\overline{\overline{A}}\,\delta a, \tag{124.3}$$

where we equate the mean value of $\partial E_n/\partial a$ to the negative of the mean generalized force A in accordance with the definition of that quantity by (121.6).

Combining (124.1) and (124.3), we are then led in the present case to the result

$$\delta\overline{\overline{H}} = 0. \tag{124.4}$$

In accordance with this important result there would be no resulting change in the value of $\overline{\overline{H}}$ if we make small adiabatic changes at a vanishingly slow rate in the external coordinates a for the systems in a canonical ensemble and allow the re-establishment of the corresponding new canonical distribution. Furthermore, by making a succession of such changes, allowing establishment of equilibrium at each stage, it is evident that we could then also make a finite change in external coordinates without alteration in $\overline{\overline{H}}$. And by reversing the change in external coordinates it is evident that we could return the ensemble of systems, including the external bodies acted on by forces, to their original condition. Hence, making use of our proposed correlations of a thermodynamic system in equilibrium with a canonical ensemble of systems, and of entropy S with the negative of $\overline{\overline{H}}$, we are thus provided with satisfactory statistical mechanical analogies for the *constancy of entropy* and accompanying *reversibility of the process*, when the coordinates for a thermodynamic system are adiabatically changed at a rate slow enough to preserve at all times a condition of thermodynamic equilibrium.

In connexion with the demonstration of (124.4), it should be noted that it is the unchanging values, assignable to the probabilities for different states n when the change δa is made slowly enough, rather than the subsequent readjustment to a new canonical distribution, which makes it possible to conclude that the value of $\delta\overline{\overline{H}}$ would be zero. Indeed, since the quantity $\overline{\overline{H}}$ may be regarded as defined in the present connexion by

$$\overline{\overline{H}} = \sum_n P_n \log P_n, \tag{124.5}$$

we see that unchanged values for the probabilities P_n when the change δa is made would leave $\overline{\overline{H}}$ strictly unaltered, and that the subsequent readjustment to a new canonical distribution, corresponding to the new value for the mean energy $\overline{\overline{E}}$, would actually lead to a slight decrease in $\overline{\overline{H}}$. Nevertheless, this subsequent change in $\overline{\overline{H}}$ would only be of the second order, since the expression (124.1), for the dependence of $\delta\overline{\overline{H}}$ on $\delta\overline{\overline{E}}$ and δa, which forms the basis for the above treatment, is itself correct to the first order. It will thus be seen that the analogy which we have given for reversible, adiabatic, thermodynamic processes depends essentially on treating the rate of change in external coordinates, in the first place, as sufficiently slow so that the probabilities P_n for different states n will not be altered as a purely mechanical consequence of the change, and in the second place, as still further sufficiently slow so that the readjustment to an appropriate canonical distribution can be regarded as already completed at each new step of the process. This may be regarded as a statistical mechanical formulation of the thermodynamic idea that reversible processes must be carried out sufficiently slowly so as to preserve mechanical and thermodynamic equilibrium at each stage of the process.

In order to complete this discussion of the choice of a very slow, essentially adiabatic, statistical mechanical process as providing the appropriate analogy for a reversible, adiabatic, thermodynamic process, some remarks must be made as to the role which is thereby assigned—in accordance with the discussions of §§ 111 and 112—to the walls or other immediate surroundings of the system proper in readjusting the representative ensemble to a new canonical distribution after each change δa in an external coordinate. In this respect the treatment is different from that of Gibbs, who regarded it as desirable to introduce the definite, abstract idealization that the walls of the container for a system proper should have no effect on its behaviour.† The justification for the procedure, which we have adopted at this point, has a twofold character.

In the first place, the concept of essential isolation provides an idealization, which appears to correspond somewhat more closely to the actual situations of ordinary interest in thermodynamic studies than would be provided by the alternative concept of perfect isolation. The assumption, in the case of adiabatic processes, that a thermodynamic fluid would merely have no net heat flow from the surroundings on the average in a series of similar experiments, would be a less drastic abstraction away from reality than the alternative assumption that the

† Gibbs, *Elementary Principles of Statistical Mechanics*, Yale University Press, 1902, p. 164.

fluid should have absolutely no interchange of heat, for example, with the walls of a cylinder enclosing it or with a piston exerting mechanical action on it. Furthermore, the assumption that the extent of interaction between representatives of the system proper and of the surroundings would be sufficient for a substantial maintenance of canonical distribution seems reasonable in view of the circumstance that we regard ourselves as starting out with an exact canonical distribution, as then making only an infinitesimal change δa in some external coordinate a, and as then waiting a very long time before a further change is made in that coordinate. The initial stage of a finite change in the coordinate would then in any case be characterized by the necessity for only an infinitesimal readjustment of energies to secure canonical redistribution and by plenty of time for this to occur through interchange with the surroundings, and any later stage would also be so characterized provided we neglect the circumstance that previous readjustments could be subject to small and unknown deviations from strict canonical form.

In the second place, the concept of essential isolation provides an idealization which is more appropriately related to our general programme for the statistical mechanical explanation of thermodynamics than would be provided by the alternative concept of perfect isolation. It is characteristic of this programme that we correlate the definite values assigned to mechanical quantities on the basis of thermodynamics with the mean values which they have in a suitable representative ensemble. Hence, having correlated the energy E of a thermodynamic system with the mean energy $\bar{\bar{E}}$ in the corresponding ensemble, it would then appear appropriate to take a thermodynamic, adiabatic process as one in which this mean energy was not affected by thermal flow rather than as one in which absolutely no readjustment in individual energies could take place. Moreover, having initially introduced canonical distribution as the suitable representation for the condition of thermodynamic equilibrium, it would then seem appropriate and desirable to regard that same kind of distribution as representing the successive conditions of thermodynamic equilibrium through which a system would pass in a finite, reversible, adiabatic change. This could not be achieved using perfect isolation as an idealization, since it is only for a very limited class of mechanical systems (e.g. an harmonic oscillator with slowly varied frequency) that canonical distribution would be retained on the slow variation of an external parameter, unless we do introduce the needed redistribution of energies through interchange with the surroundings. It is to be noted in this connexion that we have already decided to correlate the temperature T of a thermodynamic system at equilibrium with the distribution parameter θ for the corresponding canonical ensemble, and hence, if we did not provide for the re-establishment of canonical distribution at each stage of our analogue, we should have nothing to correspond to the continuous change in temperature which in thermodynamic treatments is always taken as accompanying reversible adiabatic changes.

The re-establishment of canonical distribution at each stage of a sufficiently slow, essentially adiabatic, process assumes especial importance in connexion with our later investigations of the Carnot cycle and of the formulation of the second law in general, which are to be carried out in §§ 129 and 130. In those sections we shall derive two relations,

$$\overline{\overline{W}} \leqslant \overline{\overline{Q}}_1 \frac{\theta_1 - \theta_2}{\theta_1}$$

and

$$-\Delta\overline{\overline{H}} \geqslant \int \frac{d\overline{\overline{Q}}}{\theta},$$

which are to be regarded as the analogues of the well-known thermodynamic relations

$$W \leqslant Q_1 \frac{T_1 - T_2}{T_1}$$

and

$$\Delta S \geqslant \int \frac{dQ}{T},$$

which respectively characterize the Carnot cycle and give a general formulation of the second law. The proofs which we shall give for these relations follow methods which were devised by Gibbs, and the proofs would still be valid of we used the idealization of perfect rather than essential isolation in representing adiabatic processes. Nevertheless, the validity of the above relations would then be restricted in general to irreversible processes covered by the signs of inequality, since the limiting case of reversible processes covered by the signs of equality could only be approached if the condition of thermodynamic equilibrium could be regarded at each stage of the change as correlated with a canonical distribution, as can be secured by employing the idealization of essential rather than perfect isolation. Without the continuous readjustment to canonical distribution made possible by the newer idealization, we should in general encounter unavoidable irreversible adjustments in distribution when ensembles are employed to represent the combined reversible processes of making a very slow adiabatic change in a system of interest, followed by an immersion of the system in a heat bath at a temperature so chosen that no thermal flow takes place.

In view of the foregoing discussion it then appears reasonable to adopt our proposed modification of the methods of Gibbs and make use of the concept of essential rather than of perfect isolation in obtaining a suitable idealization for adiabatic thermodynamic processes.

(*b*) **Irreversible adiabatic change in an external coordinate.** Let us now turn to the case of an abrupt change δa in an external coordinate a, followed by the establishment of a new canonical distribution. Under these circumstances our previous general equation

$$-\delta\overline{\overline{H}} = \frac{\delta\overline{\overline{E}}}{\theta}+\frac{\overline{\overline{A}}}{\theta}\delta a \tag{124.6}$$

would still be valid for the change in $\overline{\overline{H}}$, when the new distribution has established itself, but the change in mean energy $\delta\overline{\overline{E}}$ could no longer be calculated from (124.3), since an abrupt change in a would in general change the probabilities for different states n. To treat the present case it is simplest to begin at once by giving separate consideration to the two successive parts of the process: abrupt change in a—establishment of the new distribution.

In accordance with the treatment of abrupt changes given in § 97 (*c*), the probability amplitudes for the different energy states of a single system, before and after a change δa, would be connected in agreement with (97.34) by the relation

$$C_n(a_0+\delta a) = \sum_k S_{nk}\, C_k(a_0), \tag{124.7}$$

where the quantities

$$S_{nk} = \int u_n^*(a_0+\delta a)u_k(a_0)\, dq \tag{124.8}$$

are the elements of a unitary transformation matrix. Hence, if we now consider an ensemble of such systems, we can take the probabilities before and after the change δa as connected by the relation

$$\overline{\overline{|C_n(a_0+\delta a)|^2}} = \sum_k S_{nk}^*\, S_{nk}\overline{\overline{|C_k(a_0)|^2}}+\sum_{k\neq l} S_{nk}^*\, S_{nl}\,\overline{\overline{C_k^*(a_0)C_l(a_0)}}, \tag{124.9}$$

where the double bar indicates, as usual, a mean taken over all the members of the ensemble. Furthermore, making use of the diagonal character of the original canonical distribution, this can be rewritten in the form

$$\rho_{nn}(a_0+\delta a) = \sum_k |S_{nk}|^2\rho_{kk}(a_0). \tag{124.10}$$

This last formula, however, has just the same form as the relation

$$\rho_{nn}(t) = \sum_k |U_{nk}|^2 \rho_{kk}(0) \tag{124.11}$$

on which we based the derivation of the H-theorem itself in § 106, and the S_{nk} have been shown in § 97 (*c*) to have the same unitary properties as the U_{nk} which were needed in that derivation. Hence we can take the abrupt change in the coordinate as not leading to an increase in $\bar{\bar{H}}$, and in general as leading to a decrease in that quantity.

Turning, then, to the further effects that are to be expected from the subsequent establishment of a new canonical distribution under the conditions of essential isolation, it will be evident that they can also only lead to a decrease in $\bar{\bar{H}}$ for the system proper, since that distribution is characterized by a minimum of the quantity $\bar{\bar{H}}$ with a fixed value for the mean energy $\bar{\bar{E}}$.

Hence the succession of the two processes of making an abrupt change δa in some external coordinate a for the members of a canonical ensemble, and then allowing the establishment of a new canonical distribution under essentially adiabatic conditions, would lead to a small change in $\bar{\bar{H}}$ which can be characterized by the relation

$$\delta\bar{\bar{H}} \leqslant 0, \tag{124.12}$$

where the equality sign could only be expected as an exception, or if the change were actually made slowly instead of abruptly.

The above is of immediate interest in providing the statistical mechanical analogy for the thermodynamic principle, that the entropy of a thermodynamic system originally in equilibrium can only increase as the result of abrupt, adiabatic changes in external coordinates and of the subsequent re-establishment of equilibrium. Furthermore, if we regard any change in external coordinates at a *finite rate* as analysable into a succession of abrupt changes, we are also provided with a statistical mechanical analogy for the thermodynamic phenomenon of the *irreversibility* of changes made in external coordinates at a finite rate.

It is also now of interest to return to the general formula (124.6) which holds in any case for the change in $\bar{\bar{H}}$ when we change from one canonical distribution to a neighbouring one. If we apply this formula to the case of an extremely slow and hence reversible change δa, we can write

$$-\delta\bar{\bar{H}}_{\text{rev}} = \frac{\delta\bar{\bar{E}}_{\text{rev}}}{\theta} + \frac{\bar{\bar{A}}}{\theta}\delta a = 0. \tag{124.13}$$

If, however, we apply it to the same change δa made abruptly, we can write

$$-\delta\overline{\overline{H}}_{\text{irr}} = \frac{\delta\overline{\overline{E}}_{\text{irr}}}{\theta}+\frac{\overline{\overline{A}}}{\theta}da \geqslant 0. \tag{124.14}$$

Combining the two, we obtain

$$\delta\overline{\overline{E}}_{\text{irr}} \geqslant \delta\overline{\overline{E}}_{\text{rev}}. \tag{124.15}$$

This provides the statistical mechanical analogy for the thermodynamic principle that a sudden change in the external coordinates for an isolated system in equilibrium will in general require more work (or deliver less work) than a gradual change by the same amount.

125. Effect on $\overline{\overline{H}}$ of interaction in general

We may now continue our discussion of statistical mechanical analogies for thermodynamic processes by considering the effect on $\overline{\overline{H}}$ if an ensemble representing a given thermodynamic system of interest is combined with a second ensemble representing another such system in a manner to correspond to thermal or mechanical interaction between the two systems of interest. The result needed for the discussion has already been provided by our application of the H-theorem to two interacting systems S_1 and S_2 as carried out in § 107.

At an initial time $t = 0$, let us regard ourselves as making observations on the conditions of the two systems and as setting up the corresponding representative ensembles, with $\overline{\overline{H}}_1(0)$ and $\overline{\overline{H}}_2(0)$ as the initial values of $\overline{\overline{H}}$ for the two distributions. Furthermore, at this same time let us regard ourselves as placing the two systems in interaction with each other, if this has not already been done, and as representing the combined system S_{12} by a combined ensemble, which can be constructed by regarding each member of the ensemble for S_1 as combined separately with each member of the ensemble for S_2, thus giving a new ensemble of higher order in the number of its members.

Let us now regard ourselves as leaving the two systems in interaction with each other until some later time $t = t$, at which time renewed observations might be made on their conditions and the two systems again separated if desired. In accordance with the application of the H-theorem to the ensemble representing the combined system, as carried out in § 107, we may then write (107.16),

$$\overline{\overline{H}}_1(t)+\overline{\overline{H}}_2(t) \leqslant \overline{\overline{H}}_1(0)+\overline{\overline{H}}_2(0), \tag{125.1}$$

as an expression connecting the sum of the quantities $\overline{\overline{H}}_1(t)$ and $\overline{\overline{H}}_2(t)$ for the two systems at the later time $t = t$ with the sum of those quantities at the initial time $t = 0$.

In the interests of a *simplified notation* for use in following sections, it will now be convenient to rewrite the above expression in the form

$$\overline{\overline{H}}_1'+\overline{\overline{H}}_2' \leqslant \overline{\overline{H}}_1+\overline{\overline{H}}_2, \tag{125.2}$$

where we use unprimed letters to denote the initial values of quantities and primed letters to denote later values.

Returning again to our correlation of entropy S with the negative of $\overline{\overline{H}}$, we are thus furnished with a satisfactory statistical mechanical analogue for the thermodynamic principle, that the sum of the entropies of two systems cannot be decreased as a result of allowing a period of interaction between them.

126. Lemma on $\overline{\overline{H}}+\overline{\overline{E}}/\theta$

Before proceeding farther with our statistical mechanical analogies for thermodynamic processes, we must first derive a theorem concerning the value of $\overline{\overline{H}}+\overline{\overline{E}}/\theta$ in a canonical ensemble. Let us consider a given canonical ensemble of systems distributed over energy states n according to the expression

$$P_n = e^{\frac{\psi-E_n}{\theta}}. \tag{126.1}$$

And let us also consider any other distribution for the same systems over energy states n which could, of course, necessarily be described by an expression of the form

$$P_n' = e^{\frac{\psi-E_n}{\theta}+x_n}, \tag{126.2}$$

provided we allow the quantities x_n to assume the needed values. We now wish to compare the value of the quantity $\overline{\overline{H}}+\overline{\overline{E}}/\theta$ as calculated for the original canonical distribution with the value of $\overline{\overline{H}}'+\overline{\overline{E}}'/\theta$ as calculated for the second distribution using therein the value of θ for the original canonical ensemble.

To make the calculation we can write, as a consequence of the significance of the quantities $\overline{\overline{H}}$ and $\overline{\overline{E}}$,

$$\begin{aligned}\left(\overline{\overline{H}}'+\frac{\overline{\overline{E}}'}{\theta}\right)-\left(\overline{\overline{H}}+\frac{\overline{\overline{E}}}{\theta}\right) &= \sum_n P_n'\left(\log P_n'+\frac{E_n}{\theta}\right)-\sum_n P_n\left(\log P_n+\frac{E_n}{\theta}\right)\\ &= \sum_n e^{\frac{\psi-E_n}{\theta}+x_n}\left(\frac{\psi-E_n}{\theta}+x_n+\frac{E_n}{\theta}\right)-\sum_n e^{\frac{\psi-E_n}{\theta}}\left(\frac{\psi-E_n}{\theta}+\frac{E_n}{\theta}\right)\\ &= \sum_n e^{\frac{\psi-E_n}{\theta}+x_n}x_n\\ &= \sum_n e^{\frac{\psi-E_n}{\theta}+x_n}x_n+\sum_n e^{\frac{\psi-E_n}{\theta}}-\sum_n e^{\frac{\psi-E_n}{\theta}+x_n}\\ &= \sum_n e^{\frac{\psi-E_n}{\theta}}[e^{x_n}x_n+1-e^{x_n}],\end{aligned} \tag{126.3}$$

where the third and fourth forms of writing make use of the consideration that the total summation over all values of n for the probabilities expressed by (126.1) and (126.2) must lead to the value unity.

The quantity in the final square brackets, however, is of the form

$$Q = e^x x+1-e^x, \tag{126.4}$$

with the derivative with respect to x

$$\frac{dQ}{dx} = e^x x. \tag{126.5}$$

It is then seen to be a decreasing function of x when x is negative, to have the value zero when $x = 0$, and to be an increasing function of x for x positive. It thus satisfies the condition

$$Q \geqslant 0 \tag{126.6}$$

with the value zero only when

$$x = 0. \tag{126.7}$$

Hence, since all the quantities involved in the summation (126.3) must be equal to or greater than zero, we now obtain the relation between the two quantities

$$\bar{\bar{H}}'+\frac{\bar{\bar{E}}'}{\theta} \geqslant \bar{\bar{H}}+\frac{\bar{\bar{E}}}{\theta}. \tag{126.8}$$

In accordance with this expression the quantity $\bar{\bar{H}}+\bar{\bar{E}}/\theta$, where θ is some constant, has a minimum value when the ensemble is canonically distributed with θ as the parameter of the distribution, and this is the lemma which we desired to prove.

127. The direction of thermal flow as dependent on θ

So far our discussion of analogies for thermodynamic processes has been confined to illustrations of the propriety of regarding the negative of the quantity $\bar{\bar{H}}$ as the analogue of entropy S. We now turn to our first illustration of the propriety of regarding the parameter θ in a canonical ensemble as the analogue of the temperature T in the corresponding thermodynamic system. We shall be able to show that the direction of energy flow between two canonical ensembles which are put in thermal contact is determined by the relative values of their distribution parameters θ, just as energy flow between bodies put in contact is determined by the relative values of their temperatures T.

Let us consider two separate canonical ensembles, representing two different systems each in a condition of thermodynamic equilibrium; and let us consider the result of making contact for a time between members of the two ensembles in such a way as to secure an analogy for thermal interaction. *Before the contact* is made let us use

$$\bar{\bar{H}}_1,\ \bar{\bar{E}}_1,\ \theta_1 \quad \text{and} \quad \bar{\bar{H}}_2,\ \bar{\bar{E}}_2,\ \theta_2 \tag{127.1}$$

to denote the quantities indicated for the two ensembles respectively. And *after the contact* is broken let us use

$$\bar{\bar{H}}_1',\ \bar{\bar{E}}_1' \quad \text{and} \quad \bar{\bar{H}}_2',\ \bar{\bar{E}}_2' \tag{127.2}$$

for the quantities indicated, no parameter θ now being included since in general neither of the systems would now be in equilibrium and neither of the ensembles would be canonically distributed. In accordance with the general result as to the effect of interaction between ensembles as shown by expression (125.2) in § 125, we can write

$$\bar{\bar{H}}_1'+\bar{\bar{H}}_2' \leqslant \bar{\bar{H}}_1+\bar{\bar{H}}_2. \tag{127.3}$$

And, in accordance with the theorem of the preceding section as expressed by (126.8), we can write

$$\bar{\bar{H}}_1+\frac{\bar{\bar{E}}_1}{\theta_1} \leqslant \bar{\bar{H}}_1'+\frac{\bar{\bar{E}}_1'}{\theta_1} \tag{127.4}$$

and

$$\bar{\bar{H}}_2+\frac{\bar{\bar{E}}_2}{\theta_2} \leqslant \bar{\bar{H}}_2'+\frac{\bar{\bar{E}}_2'}{\theta_2}, \tag{127.5}$$

where θ_1 and θ_2 are the parameters for the two original canonical ensembles. Adding all three expressions (127.3–5) together, we then have

$$\frac{\bar{\bar{E}}_1}{\theta_1}+\frac{\bar{\bar{E}}_2}{\theta_2} \leqslant \frac{\bar{\bar{E}}_1'}{\theta_1}+\frac{\bar{\bar{E}}_2'}{\theta_2}$$

or

$$\frac{\bar{\bar{E}}_1'-\bar{\bar{E}}_1}{\theta_1}+\frac{\bar{\bar{E}}_2'-\bar{\bar{E}}_2}{\theta_2} \geqslant 0. \tag{127.6}$$

Let us now consider the case of pure thermal interaction where the work done in making and breaking the thermal contact between the two ensembles is negligible, and the members of the two ensembles that have been combined do no work on each other during the interaction. In accordance with the statistical mechanical analogue of the first law of thermodynamics, § 119, we shall then have

$$\bar{\bar{E}}_1'+\bar{\bar{E}}_2'-\bar{\bar{E}}_1-\bar{\bar{E}}_2 = 0, \tag{127.7}$$

and can take

$$\Delta\bar{\bar{E}}_1 = \bar{\bar{E}}_1'-\bar{\bar{E}}_1 = -(\bar{\bar{E}}_2'-\bar{\bar{E}}_2) \tag{127.8}$$

as representing the heat flow into the first system from the second during the period of contact. Substituting in (127.6), we thus have

$$\frac{\Delta\bar{\bar{E}}_1}{\theta_1}-\frac{\Delta\bar{\bar{E}}_1}{\theta_2} \geqslant 0. \tag{127.9}$$

This result shows that the direction of energy transfer can only be from the ensemble originally having the higher value of the distribution parameter θ into the ensemble having the lower value of θ. Returning to our proposed correlation of the distribution parameter θ with temperature T, we are thus provided with a satisfactory analogy for the thermodynamic principle that heat can only flow from a body of higher temperature to one of lower temperature when thermal contact is made.

128. Effects of various kinds of thermal process

(*a*) **Effect on $\bar{\bar{H}}$ of thermal interaction in general.** We may now consider the effects to be expected in a number of different cases where ensembles are combined in such a way as to represent thermal flow. We shall thus obtain further illustrations of the propriety of correlating entropy S with the negative of the quantity $\bar{\bar{H}}$, and temperature T with the parameter θ.

We may first consider the effect on $\bar{\bar{H}}$ of any kind of thermal interaction between ensembles which may represent thermodynamic systems in any condition of interest, whether one of equilibrium or not. To treat the problem we have only to note that a purely thermal interaction between ensembles is merely a special case of interaction in general, when the work done in making and breaking the thermal contact is negligible, and the members of the two ensembles that have been combined do no work on each other. Hence we can at once apply expression (125.2) derived in the general treatment of interaction. This gives us

$$\bar{\bar{H}}_1'+\bar{\bar{H}}_2' \leqslant \bar{\bar{H}}_1+\bar{\bar{H}}_2, \tag{128.1}$$

which says that the sum of the quantities $\bar{\bar{H}}$ for the two systems after the thermal transfer has been completed cannot be greater than the original sum of those quantities. This is in immediate analogy with the thermodynamic principle that the sum of the entropies of two bodies cannot be decreased by thermal interaction.

(*b*) **Effect on $\bar{\bar{H}}$ of thermal transfer from a canonical ensemble.** We now consider the case in which an ensemble originally arbitrarily distributed, representing some system of interest S_1 in an arbitrary condition, is placed in thermal contact with an ensemble which is itself canonically distributed, representing a body S_2 of specified temperature. Using unprimed and primed letters to refer respectively to conditions before and after thermal contact, we may now write, in accordance with our general expression (128.1),

$$\bar{\bar{H}}_1'+\bar{\bar{H}}_2' \leqslant \bar{\bar{H}}_1+\bar{\bar{H}}_2, \tag{128.2}$$

for the sums of the values of $\bar{\bar{H}}$ for the two ensembles before and after thermal contact, and, in accordance with the lemma of § 126, we may write

$$\bar{\bar{H}}_2+\frac{\bar{\bar{E}}_2}{\theta_2} \leqslant \bar{\bar{H}}_2'+\frac{\bar{\bar{E}}_2'}{\theta_2} \tag{128.3}$$

as an expression applying to the ensemble which is originally canonically distributed with the distribution parameter θ_2. Adding these two expressions, we can then obtain

$$\bar{\bar{H}}_1'-\bar{\bar{H}}_1 \leqslant \frac{\bar{\bar{E}}_2'-\bar{\bar{E}}_2}{\theta_2}$$

or, by the energy principle, since the work done is taken as negligible,

$$-(\bar{\bar{H}}_1'-\bar{\bar{H}}_1) \geqslant \frac{\bar{\bar{E}}_1'-\bar{\bar{E}}_1}{\theta_2}, \tag{128.4}$$

where $(\bar{\bar{E}}_1'-\bar{\bar{E}}_1)$ is the mean energy thermally transferred to the members of the first ensemble.

This provides us with a satisfactory analogy for the thermodynamic principle that the increase in entropy of a system, as a result of making thermal contact with a body in an equilibrium condition, cannot be less than the heat absorbed divided by the original temperature of the body supplying the heat, whatever the initial condition of the system of interest.

(*c*) **Thermal equilibrium as a result of successive contacts.** We may also use the expression which we have just derived to investigate the effect of successive thermal contacts between an ensemble having an arbitrary initial distribution and ensembles which are canonically distributed with the parameter θ_2. For this purpose let us write (128.4) in the form

$$\bar{\bar{H}}_1'+\frac{\bar{\bar{E}}_1'}{\theta_2} \leqslant \bar{\bar{H}}_1+\frac{\bar{\bar{E}}_1}{\theta_2}. \tag{128.5}$$

This shows that the quantity $\bar{\bar{H}}+\bar{\bar{E}}/\theta_2$ for the first ensemble can in general be expected to decrease as a result of making contacts with ensembles canonically distributed with the parameter θ_2. We have shown, however, in accordance with the lemma derived in § 126, expressed by (126.8), that the quantity $\bar{\bar{H}}+\bar{\bar{E}}/\theta_2$, where θ_2 is some constant, would have a minimum value for an ensemble which is itself canonically distributed with the quantity θ_2 as the parameter of distribution. Hence we may conclude that any original ensemble would be given a canonical distribution with the parameter θ_2, as a result of successive thermal contacts with ensembles which are themselves canonically distributed with θ_2 as the parameter.

We are thus provided with a satisfactory analogy for the thermodynamic principle that any system can be brought into a condition of thermodynamic equilibrium at the temperature T as a result of successive contacts with bodies which are themselves in equilibrium at that temperature.

The foregoing is very interesting in showing that we can regard the canonical distribution as not only appropriate for representing an isolated thermodynamic system which has come to equilibrium by itself (see § 120), but as also appropriate for representing a thermodynamic system which has been brought into equilibrium at a prescribed temperature T by contact with a large heat reservoir at that temperature. This latter was one of the most important of the arguments presented by Gibbs justifying the use of the canonical ensemble as a representation of thermodynamic equilibrium.

(*d*) **The limiting case of reversible thermal transfer.** The results of thermal transfer between ensembles of systems are in general irreversible in that the sum of the quantities $\bar{\bar{H}}$ for the two ensembles is found to be decreased in accordance with (128.1) as a result of their interaction. The limiting case of reversible transfer is of special interest.

To study this, let us consider two representative ensembles which are originally each of them canonically distributed, with the parameters θ_1 and θ_2. Let them then be combined in a manner to represent thermal contact, and separated again after there has been on the average only an infinitesimal transfer of energy. Finally, let each of them be allowed to assume its new canonical distribution as a consequence of the merely essential isolation which we have found it appropriate to assume. In accordance with our general relation (121.8), connecting neighbouring canonical distributions, we can then write for the changes in the quantity $\bar{\bar{H}}$ for the two ensembles

$$-\delta\bar{\bar{H}}_1 = \frac{\delta\bar{\bar{E}}_1}{\theta_1}$$

$$\text{and} \qquad -\delta\bar{\bar{H}}_2 = \frac{\delta\bar{\bar{E}}_2}{\theta_2}, \tag{128.6}$$

where the assumed absence of external work appropriate in a case of pure thermal interaction permits us to omit the terms in (121.8) depending on external forces. The absence of external work will also permit us to take $\delta\bar{\bar{E}}_1 = -\delta\bar{\bar{E}}_2$ as the mean energy transferred by thermal action from the second ensemble to the first. Hence, adding the two above equations, we can now write for the sum of the changes in $\bar{\bar{H}}$ for the two ensembles

$$-\delta(\bar{\bar{H}}_1+\bar{\bar{H}}_2) = \delta\bar{\bar{E}}_1\left(\frac{1}{\theta_1}-\frac{1}{\theta_2}\right) = \delta\bar{\bar{E}}_1\left(\frac{\theta_2-\theta_1}{\theta_1\theta_2}\right). \tag{128.7}$$

This result shows us that the change in the sum of the values of $\bar{\bar{H}}$ for the two ensembles for a given mean energy transfer would approach zero as the distribution parameters for the two ensembles approach each other. Furthermore, by taking θ_2 infinitesimally larger or smaller than θ_1, we could approach the limiting possibility of thermal transfer in either direction without change in the combined values of $\bar{\bar{H}}$ for the two ensembles. Equation (128.7) hence provides us with an analogy for the thermodynamic principle of constant total entropy at the limit when heat is transferred between two bodies having the same temperature. We are thus led to an analogy for the *reversibility* of heat transfer under those limiting conditions, as compared with the *irreversibility* of heat transfer to be expected under more general circumstances.

129. Carnot cycle of processes

As a final example of processes carried out on ensembles which are analogous to typical thermodynamic processes, we may treat the case of the Carnot cycle. To do this let us consider an ensemble of systems No. 0 which we can take as representing 'the engine', and further ensembles Nos. 1, 2, 3,... which we can take as representing 'heat reservoirs'. Let us start with the ensembles for the engine and heat reservoirs each canonically distributed with the distribution parameters θ_0, θ_1, θ_2,.... Let us then carry out a cycle of processes, involving external work consequent on the changes made in the external coordinates for the ensemble corresponding to the engine and involving thermal transfer between the ensembles for the engine and the heat reservoirs. Finally, at the end of the cycle, let the ensemble for the engine have the same values of external coordinates and the same mean energy $\bar{\bar{E}}_0$ as at the start, and let it once more assume its original canonical distribution.

In accordance with our general expressions (124.12) and (125.2) for the effects of changing the external coordinates in an ensemble, and of introducing interactions with other ensembles, we can write

$$\bar{\bar{H}}_0'+\bar{\bar{H}}_1'+\bar{\bar{H}}_2'+... \leqslant \bar{\bar{H}}_0+\bar{\bar{H}}_1+\bar{\bar{H}}_2+..., \tag{129.1}$$

and, in accordance with the lemma expressed by (126.8) as to the minimum of $\bar{\bar{H}}+\bar{\bar{E}}/\theta$ for a canonical distribution, we can write

$$\begin{aligned}\bar{\bar{H}}_0+\frac{\bar{\bar{E}}_0}{\theta_0} &\leqslant \bar{\bar{H}}'_0+\frac{\bar{\bar{E}}'_0}{\theta_0},\\ \bar{\bar{H}}_1+\frac{\bar{\bar{E}}_1}{\theta_1} &\leqslant \bar{\bar{H}}'_1+\frac{\bar{\bar{E}}'_1}{\theta_1},\\ \bar{\bar{H}}_2+\frac{\bar{\bar{E}}_2}{\theta_2} &\leqslant \bar{\bar{H}}'_2+\frac{\bar{\bar{E}}'_2}{\theta_2},\\ &\cdots\cdots\end{aligned} \tag{129.2}$$

where the unprimed and primed letters refer respectively to conditions at the beginning and end of the cycle.

Adding the above expressions, we obtain

$$\frac{\bar{\bar{E}}_0}{\theta_0}+\frac{\bar{\bar{E}}_1}{\theta_1}+\frac{\bar{\bar{E}}_2}{\theta_2}+\ldots \leqslant \frac{\bar{\bar{E}}'_0}{\theta_0}+\frac{\bar{\bar{E}}'_1}{\theta_1}+\frac{\bar{\bar{E}}'_2}{\theta_2}+\ldots. \tag{129.3}$$

By hypothesis, however, we are interested in the case where the final mean energy $\bar{\bar{E}}'_0$ of the ensemble corresponding to the engine is equal to the original value $\bar{\bar{E}}_0$. Hence we can rewrite the above in the form

$$\frac{\bar{\bar{E}}_1-\bar{\bar{E}}'_1}{\theta_1}+\frac{\bar{\bar{E}}_2-\bar{\bar{E}}'_2}{\theta_2}+\ldots \leqslant 0, \tag{129.4}$$

which is arranged to show the dependence on the mean energy transferred from the ensembles representing the heat reservoirs. In addition we can write, in accordance with the energy principle as applied to ensembles,

$$\bar{\bar{W}} = \bar{\bar{E}}_1-\bar{\bar{E}}'_1+\bar{\bar{E}}_2-\bar{\bar{E}}'_2+\ldots, \tag{129.5}$$

where $\bar{\bar{W}}$ is the mean external work done by the ensemble representing the engine.

For the case of a simple Carnot cycle, involving only two heat reservoirs, we can combine (129.4) and (129.5) in the form

$$\frac{\bar{\bar{E}}_1-\bar{\bar{E}}'_1}{\theta_1}+\frac{\bar{\bar{W}}-(\bar{\bar{E}}_1-\bar{\bar{E}}'_1)}{\theta_2} \leqslant 0,$$

or, denoting by $\bar{\bar{Q}}_1 = \bar{\bar{E}}_1-\bar{\bar{E}}'_1$ the mean thermal transfer of energy from the ensemble originally distributed with the parameter θ_1, this can be expressed in the form

$$\bar{\bar{W}} \leqslant \bar{\bar{Q}}_1\frac{\theta_1-\theta_2}{\theta_1}. \tag{129.6}$$

We thus obtain a satisfactory analogy for the fundamental thermodynamic relation of Carnot, which connects the work done by a heat engine with the heat absorbed at the higher temperature, and the values of the two temperatures at which heat is taken in and given out.

The equality sign in (129.6) would be valid for the limiting case of a cycle of reversible steps, with heat reservoirs large compared with the energy that they must absorb or receive. To carry out such a *reversible cycle* the external coordinates for the ensemble representing the engine would have to be changed at a vanishingly slow rate, and the ensemble would in any step have to be canonically distributed with sensibly the same value for θ as that for any heat reservoir with which it is in contact. In accordance with the results of §§ 124 (*a*) and 128 (*d*), we should then have the equality sign holding in the first of the expressions (129.1) on which we have based our deduction. Furthermore, we can in any case use the equality sign in the first of the expressions given by (129.2), since the ensemble for the engine returns by hypothesis to its original canonical distribution. And, with sufficiently large heat reservoirs so that their mean energy is changed only infinitesimally, we could also use the equality sign in the remaining expressions (129.2), in accordance with the substantial maintenance of canonical distribution. Thus, under these limiting circumstances, we should have the equality sign holding in all of the expressions on which the deduction has been based, and hence would also have the equality sign in the final expression (129.6) for the work done.

130. The analogue of the second law of thermodynamics

In preceding sections we have discussed the statistical mechanical analogues for a considerable number of thermodynamic processes that involve the second law of thermodynamics. These processes included examples of the reversible and irreversible performance of external work and of the reversible and irreversible transfer of heat, and the situations treated seemed typical of simple thermodynamic processes in general. The treatments all showed the satisfactory character of our proposals to correlate the entropy S of a thermodynamic system with the negative of the quantity $\bar{\bar{H}}$ for the corresponding representative ensemble of systems, and to correlate the temperature T of a thermodynamic system in an equilibrium condition with the distribution parameter θ for the corresponding canonical ensemble. We are now in a position to make a somewhat complete statement of the statistical mechanical analogue of the second law of thermodynamics.

For our purposes we may regard the content of the ordinary second law of thermodynamics as given by the expression

$$\Delta S \geqslant \int \frac{\delta Q}{T}, \tag{130.1}$$

which states that the increase in the entropy of a system, when it changes from one condition to another, cannot be less than the integral of the heat absorbed divided for each increment of heat by the temperature of a heat reservoir appropriate for supplying the increment in question. The equality sign in this expression is to be taken as applying to the limiting case of reversible changes.

The justification for the statistical mechanical analogue for this expression is immediately provided by relations which we have already derived in the present chapter. The change in the ensemble representing the system of interest will be brought about by making changes in the values of the external coordinates for the ensemble, and by permitting the thermal transfer of energy from other ensembles. We have found, however, in §§ 124 (*a*) and (*b*), that the effect on $\overline{\overline{H}}$ of a small change δa in an external coordinate a would be given by

$$-\delta\overline{\overline{H}} \geqslant 0, \tag{130.2}$$

where the equality sign would apply to the limiting case of a reversible change. And we have found in §§ 128 (*b*) and (*d*) that the effect on $\overline{\overline{H}}$ of the transfer of a small amount of thermal energy of magnitude $\delta\overline{\overline{Q}}$ in the mean from a canonical ensemble with the distribution parameter θ to an ensemble representing any system of interest would be given by

$$-\delta\overline{\overline{H}} \geqslant \frac{\delta\overline{\overline{Q}}}{\theta}, \tag{130.3}$$

where the equality sign again applies to the limiting case of reversible transfer between canonical ensembles having sensibly the same value of θ. Moreover, the nature of our derivations was such that it is proper to regard the effects of thermal transfer and of changes in external coordinates as superposable.

Hence, by combining (130.2) and (130.3), we are now led, after integration, to the desired expression

$$-\Delta\overline{\overline{H}} \geqslant \int \frac{\delta\overline{\overline{Q}}}{\theta}, \tag{130.4}$$

where the equality sign is seen to apply to the limiting case of reversible processes of a nature such that all the ensembles and external bodies involved could be restored to their original condition. We at once see that this is the appropriate statistical mechanical analogue for the general statement of the second law of thermodynamics as expressed by (130.1), provided we correlate $\overline{\overline{S}}$ with $-k\overline{\overline{H}}$ and T with θ/k, as has been proposed.

This, then, completes our statistical mechanical justification for both of the two fundamental laws of thermodynamics. We are now provided with statistical expressions (119.4) and (130.4) of just the same form as those used for the thermodynamic statement of the two laws, with the thermodynamic quantities E, Q, W, S, and T replaced by their statistical mechanical analogues $\bar{\bar{E}}$, $\bar{\bar{Q}}$, $\bar{\bar{W}}$, $-k\bar{\bar{H}}$, and θ/k; and further statistical mechanical deductions can hence now be carried out in a form to run parallel to deductions familiar in thermodynamics.

131. Remarks on the statistical explanation of thermodynamics

We may now conclude the present chapter by making a few remarks concerning the nature of the explanation which the methods of statistical mechanics have provided for the principles of thermodynamics.

The fundamental idea in this explanation lies in regarding the thermodynamic behaviour of a single system of interest as equivalent to the mechanical behaviour which would be exhibited on the average by a suitably chosen ensemble of systems of similar structure. The introduction of such an idea, if we are to attempt a mechanical explanation of the behaviour of thermodynamic systems, seems quite reasonable in view of the consideration that the thermodynamic specification of the condition of a system is incomplete enough to comport with a wide variety of specifications having the precision allowed by mechanics. The importance and significance of the idea was appreciated to some extent by Boltzmann, but was first fully and completely understood and applied by Gibbs. The validity of the idea has been made evident by the results which we have given in the present chapter.

In carrying out the implications of this fundamental idea, we have found it necessary to correlate the thermodynamic variables ordinarily used to specify the condition of a thermodynamic system with mechanical quantities applying to the corresponding representative ensemble. In the case of the external coordinates describing a thermodynamic system, we have found it possible—paying due regard to the new requirements imposed by the quantum mechanics—to give all the systems in the ensemble the same values for their external coordinates as those of the system to be represented. In the case of purely mechanical quantities, such as the energy or external forces exhibited by a thermodynamic system, we have found it possible to make a correlation with the average values of these quantities in the representative ensemble. Here it may be remarked that the mean proves the most desirable average to take, and that inevitable quantum mechanical as well as deliberate experimental limitations on the accuracy of our observational information may play a role in the necessity for representing such quantities by their mean values. In the case of the essentially thermodynamic variables, entropy and temperature, considerable investigation was necessary to validate the choice of statistical correlates.

In the case of entropy, except for a constant factor needed to allow for choice of units, it was found satisfactory to correlate this thermodynamic quantity with the negative of the quantity $\overline{\overline{H}}$ for the ensemble used to represent the system of interest. Several remarks may be made in this connexion.

In the first place, it is to be noted, as has been emphasized in connexion with our derivation of the H-theorem, that the ensemble which we choose to represent a mechanical system of interest is determined by the observations we have made on the condition of that system. Thus also the value of $\overline{\overline{H}}$ will be determined by the nature of our knowledge of the condition of the system. Since the value of $\overline{\overline{H}}$ is lower the less exactly the quantum mechanical state of the system is specified, this provides the reason for the statement sometimes made that the entropy of a system is a measure of the degree of our ignorance as to its condition. From a precise point of view, however, it seems more clarifying to emphasize tnat entropy can be regarded as a quantity which is thermodynamically defined with the help of its relation to heat and temperature given by (130.1), and statistically interpreted with the help of the analogous relation (130.4) between the mechanical quantities $\overline{\overline{H}}$, $\overline{\overline{Q}}$, and θ.

An interesting point arises in connexion with the entropy of systems which are not in a condition of thermodynamic equilibrium. In the older thermodynamics it was possible to say that the entropy of a system originally in a condition of equilibrium would increase if it changed spontaneously into a temporarily unstable condition; for example, as a result of opening a valve which would allow flow into an exhausted container. Nevertheless, it was not so easy to make a precise assignment of the value for the entropy in such a non-equilibrium condition, since precise changes had ultimately to be correlated with the help of (130.1) with reversible changes from one condition of equilibrium to another, and this introduced complications. With the help of our correlation $S \rightleftharpoons -k\overline{\overline{H}}$, it is now easier to grasp what is to be meant by entropy in genéral, since the evaluation of $\overline{\overline{H}}$ is in no wise limited to conditions of equilibrium.

There is another important point connected with the correlation

$$S \rightleftharpoons -k\overline{\overline{H}} = -k \sum_n P_n \log P_n. \tag{131.1}$$

In the older thermodynamics entropy was regarded as defined only to the extent of an arbitrary additive constant, and this circumstance was not affected by the interpretation of thermodynamics with the help of the classical statistical mechanics. From the point of view of quantum mechanics, however, it will be noted that the situation has been altered, since the quantity $\overline{\overline{H}}$ is seen to have the value zero when the system of interest is known to be in a single definite quantum mechanical state, and then to have lower but definitely calculable values for any other distribution. Hence the introduction of the correlation (131.1) definitely implies that we choose the additive constant for entropy in such a way that the entropy of a system will be zero for a *pure state*, and greater than zero but with a perfectly definite value for any so-called *mixed state*. This provides an explanation for the term *absolute entropy* which is sometimes used as descriptive of this situation. We shall not make use of this term in the present book, however, since this would imply a selection of states to be designated as pure that was based on an ultimate analysis into nuclear as well as extra-nuclear conditions. In the actual practice of physical chemists the specification of states which are treated as pure is best regarded as determined by conventions which must be applied in a consistent manner throughout the computation. Assuming the consistent application of conventions as to the specification of pure states, the important point to emphasize is that the correlation expressed (131.1) then provides more powerful methods for the calculation of entropies and entropy differences than were classically available. In the next chapter, which gives some specific applications to thermodynamic problems, we shall see that these more powerful methods provide a satisfactory explanation for those actual relations between entropies which have led to various formulations of the so-called Nernst heat theorem or third law of thermodynamics.

As a final point concerning the correlation between entropy S and the statistical mechanical quantity $\overline{\overline{H}}$, as given by (131.1), it is of interest to consider the special case of a system which we regard as being with *equal* probability in one or another of a group of W eigenstates between which we do not distinguish on the basis of our macroscopic measurements. It will be seen that the relation (131.1) would then assume the form

$$S \rightleftharpoons k \log W. \tag{131.2}$$

This has the form of the relation between entropy and probability considered by Boltzmann and Planck, and the quantity W is sometimes spoken of as the *thermodynamic probability*. It is evident, however, that (131.2) is best regarded merely as a special case of the generally satisfactory and understandable expression for entropy given by (131.1).

We may now turn to the correlation of temperature T with the distribution parameter θ which is given by the expression

$$T \rightleftharpoons \frac{\theta}{k}. \tag{131.3}$$

Several remarks may also be made in this connexion.

It is of interest first of all to point out once more that the temperature of a system is in any case a quantity to which we can assign precise meaning only for systems which are in a condition of equilibrium. This corresponds to the circumstance that the parameter θ characterizes only the particular kind of ensembles, with canonical distribution, which we use to represent systems in thermodynamic equilibrium.

It is also of interest to note once more that our introduction of the concept of essential isolation has made the canonical distribution, rather than any form of microcanonical distribution, seem appropriate for the representation of equilibrium in the case of a system which would be regarded as isolated in a thermodynamic treatment. This now removes the motive which led Gibbs to attempt his second and third analogies for thermodynamic quantities using microcanonical ensembles. The analogies thus obtained for temperature were specially unsatisfactory, as Gibbs himself appreciated.

Finally, it may be emphasized that our present method of correlating temperature T with the parameter θ is very satisfactory from the thermodynamic point of view of the role played by $1/T$ as an integrating factor for the heat dQ absorbed by a system in an infinitesimal reversible step. In our treatment of the statistical mechanical analogue of the second law of thermodynamics, as given in § 130, we have seen that $1/\theta$ plays a similar role as an integrating factor for the mean energy $d\overline{\overline{Q}}$ transferred by reversible thermal action to the representative ensemble for the thermodynamic system of interest. Our present method of connecting temperature with statistical mechanics may hence seem more fundamental than our earlier introduction of temperature into the formulae for the Maxwell-Boltzmann, Einstein-Bose, and Fermi-Dirac distributions with the help of the phenomenological properties of a perfect gas thermometer. Of course, it will in any case be necessary to determine the numerical value of the constant k, involved in the relation of the quantities T and θ, by considering the properties of some specific system; and for this purpose at least it is convenient to consider the properties of a perfect gas as will be done in the next chapter.

In concluding this chapter, it is hoped that due appreciation will be felt for the importance and significance of the great achievement of statistical mechanics in providing a fundamental, mechanical interpretation and explanation of the principles of thermodynamics.

XIV
FURTHER APPLICATIONS TO THERMODYNAMICS

132. Thermodynamic quantities in terms of the free energy

It is evident, as shown in the last chapter, that the principles of statistical mechanics are sufficiently fundamental to provide a mechanical explanation for the phenomena of thermodynamics. It may be emphasized, however, that the principles of statistical mechanics are not only more fundamental but are also more powerful than those of thermodynamics. This greater power arises from the circumstance that statistical mechanics gives due consideration to the microscopic, internal structures of the systems to be treated, while thermodynamics concerns itself solely with their macroscopic, phenomenological behaviour. This makes it possible, by applying the methods of statistical mechanics, to obtain definite theoretical values for quantities such as pressures, specific heats, energy, and entropy differences, which would have to enter a purely thermodynamic treatment as parameters with values needing empirical determination.

In this final chapter we shall now consider several specific applications of statistical mechanics to thermodynamic systems which will illustrate this possibility of deepened theoretical treatment. The systems which we choose for this purpose will be the simplest ones of ordinary physical-chemical interest, and, in accordance with the general purposes of this book, we shall make no attempt to give an account of the many further possible and important applications of statistical mechanics to thermodynamics that can be made.

In addition to a consideration of such illustrative applications, it will also be necessary to pay attention to two questions of more fundamental character. In § 140 we shall give an account of the grand canonical ensemble as providing the appropriate statistical apparatus for treating equilibria involving changes in the amount of material which constitutes a system of interest. This is necessary in order to have a justification for the thermodynamic treatment ordinarily given to so-called 'open' rather than 'closed' systems. Finally, in § 141 we shall give treatment to the fluctuations in the properties of a thermodynamic system which can be expected at equilibrium. This is not only important in determining the extent of validity which can be ascribed to the thermodynamic procedure of assigning precise values to thermodynamic quantities, but can also be of direct observational interest under suitable circumstances.

We must begin the chapter by recalling some simple thermodynamic formulae which will be needed. We may confine our attention for the present to situations simple enough so that the different equilibrium conditions of the thermodynamic system under consideration can be specified by its temperature T and its volume v. In accordance with the first law of thermodynamics, we can then write for the change in energy E, when a small change is made in those variables,

$$\begin{aligned} dE &= dQ-dW \\ &= dQ-p\,dv, \end{aligned} \tag{132.1}$$

where p is the pressure which the system exerts. And, in accordance with the second law of thermodynamics, we can then write for the change in the entropy S of the system

$$\begin{aligned} dS &= \frac{dQ}{T} \\ &= \frac{dE+p\,dv}{T}. \end{aligned} \tag{132.2}$$

We shall be specially interested, however, in the relation of different thermodynamic quantities to the free energy of the system, owing to the simple connexion which we have found between this quantity and the statistical mechanical quantity which we have called the sum-over-states. For the free energy of the system A we can write, in accordance with the original definition of Helmholtz,†

$$A = E-TS. \tag{132.3}$$

By differentiating this expression, we then obtain in general

$$dA = dE-T\,dS-S\,dT. \tag{132.4}$$

And, by substituting (132.2), we obtain for the simple kind of situations now under consideration

$$dA = -p\,dv-S\,dT. \tag{132.5}$$

This now provides what is necessary in order to express the various thermodynamic quantities of interest in terms of the free energy A. In accordance with (132.5), we immediately obtain for the pressure and entropy of the system

$$p = -\frac{\partial A}{\partial v} \tag{132.6}$$

and

$$S = -\frac{\partial A}{\partial T}. \tag{132.7}$$

† This quantity is not to be confused with the thermodynamic potential

$$F = E+pv-TS,$$

sometimes called free energy by chemists.

And, making use of the original equation (132.3) itself, we then obtain for the energy of the system

$$E = A - T\frac{\partial A}{\partial T}, \tag{132.8}$$

and hence for its heat capacity at constant volume

$$C_v = -T\frac{\partial^2 A}{\partial T^2}. \tag{132.9}$$

133. Thermodynamic quantities in terms of the sum-over-states

We are now ready to introduce statistical mechanical considerations with the help of the relation (122.9), already derived as giving the appropriate connexion between the thermodynamic quantity free energy A, and the statistical mechanical quantity sum-over-states Z. This relation has the form

$$A = -kT\log Z, \tag{133.1}$$

where the sum-over-states is defined by the expression

$$Z = \sum_n e^{-E_n/kT}, \tag{133.2}$$

the quantities E_n being the energy eigenvalues for all the different steady state solutions n of which the system is capable.†

Substituting (133.1) in our previous purely thermodynamic relations (132.6) to (132.9), we can now write for the pressure, energy, heat capacity, and entropy of the system, in the order named,

$$p = kT\frac{\partial \log Z}{\partial v}, \tag{133.3}$$

$$E = kT^2\frac{\partial \log Z}{\partial T}, \tag{133.4}$$

$$C_v = 2kT\frac{\partial \log Z}{\partial T} + kT^2\frac{\partial^2 \log Z}{\partial T^2}, \tag{133.5}$$

$$S = k\log Z + kT\frac{\partial \log Z}{\partial T} = \frac{E}{T} + k\log Z. \tag{133.6}$$

With the help of these expressions we can now obtain explicit expressions for the quantities listed in the case of any system for which we have the knowledge of structure necessary for calculating the sum-over-states Z as given by (133.2).

† It may again be pointed out that the *sum-over-states* Z provides in the present treatment the computational apparatus which is provided in the treatments of Darwin and Fowler by their *partition functions*.

We have picked pressure, energy, heat capacity, and entropy as a list of thermodynamic quantities which we shall wish to evaluate for various typical systems, since these particular quantities provide the usual starting-point for the thermodynamic computations actually carried out in practice by physical chemists. The expressions for such a list of quantities will, of course, contain no information not already implied by an expression for Z in terms of v and T. Nevertheless, having once obtained a set of expressions for p, E, C_v, and S in terms of v and T, it will then be natural to follow ordinary thermodynamic procedure and calculate further quantities without returning to the statistical mechanical basis from which we have started. Thus, as examples, we may calculate the *free energy* A itself from

$$A = E-TS, \tag{133.7}$$

the *thermodynamic potential* F from

$$F = E+pv-TS, \tag{133.8}$$

or the *heat capacity at constant pressure* C_p from

$$C_p = C_v+T\left(\frac{\partial p}{\partial T}\right)_v\left(\frac{\partial v}{\partial T}\right)_p, \tag{133.9}$$

once we have evaluated the previous expressions in terms of v and T.

The expression for the energy of a system E provided by (133.4) will give values based on an energy zero-point which will itself be determined by the energy zero-point selected for the eigenvalues of energy E_n used in evaluating the sum-over-states in accordance with (133.2). On the other hand, the expression for the entropy of a system S provided by (133.6) will give values, as noted at the end of the preceding chapter, based on the definite zero-point implied, in correlating entropy with statistical mechanical quantities by taking

$$S = -k\bar{\bar{H}} = -k\sum_n P_n \log P_n \tag{133.10}$$

without including any additive constant. This makes the entropy zero for any pure state—that is, when the probability P_n for some particular quantum mechanical state n is known to be unity. We shall have occasion to direct specific attention to these zero-points in what follows.

134. Sum-over-states as dependent on molecular states

It will be seen from the foregoing that definite numerical values can be obtained for quantities of thermodynamic interest as soon as we can obtain an appropriate evaluation of the sum-over-states,

$$Z = \sum_n e^{-E_n/kT}, \tag{134.1}$$

for the system of interest. The quantities E_n occurring in the expression for the sum-over-states Z are the energy eigenvalues for the different steady states that the system as a whole can exhibit. In the case of systems composed of weakly interacting elements, such as a dilute gas where the energy levels of the individual molecules are not greatly affected by collisions, the calculation of Z can be conveniently made to depend on the energy eigenvalues ϵ_k for the individual elements of which the system is composed.

In making such a calculation to determine the dependence of the sum-over-states for the system as a whole on the states of the individual elements of which the system is composed, it is necessary to distinguish between cases where the specification of the state of the system as a whole depends on which particular elements are assigned to their different elementary eigenstates, and cases where the specification is only dependent on the number of elements assigned to different elementary eigenstates. We may begin by treating the first of these two cases.

(*a*) **Case of permanently distinguishable elements.** Let us consider a system composed of n similar, weakly interacting elements; and let each element exhibit the same set of states of energies ϵ_i, but let the elements be distinguishable one from another so that an eigenstate of the system as a whole must be specified by giving the elementary eigenstate for each of the n different elements. In accordance with our previous nomenclature such a system may be called a Maxwell-Boltzmann system of elements. As an example we could take a system of n oscillators all exhibiting the same frequency ν, but distinguishable from each other by permanent spatial location or orientation.

If we now use the indices i_1, i_2, i_3,..., i_n to denote the eigenstates of these individual elements, it is evident for the case of weak interaction that the energy for a state of the system as a whole could be expressed in the form

$$E_{i_1,i_2,...,i_n} = \epsilon_{i_1}+\epsilon_{i_2}+...+\epsilon_{i_n} \tag{134.2}$$

and that each different assignment of values to the indices i_1, i_2,..., i_n would correspond to a different eigenstate of the system as a whole. Hence, from the definition of the sum-over-states given by (134.1), we can now express that quantity in the form

$$Z = \sum_{i_1,i_2,...,i_n} e^{-\frac{\epsilon_{i_1}+\epsilon_{i_2}+...+\epsilon_{i_n}}{kT}}, \tag{134.3}$$

where the summation is to be taken over all possible values of the indices i_1, i_2,..., i_n. It will be immediately seen, however, from the possibility of decomposing the quantity to be summed into a product of factors of the form $e^{-\epsilon_i/kT}$, and from the assumed identity of the energy spectra for the different elements, that this can be rewritten in the form

$$Z = \Big(\sum_i e^{-\epsilon_i/kT}\Big)^n, \tag{134.4}$$

where the quantities ϵ_i are the energy eigenvalues for the different eigenstates i of a single element of the kind under consideration. In carrying out the indicated summation over i, each separate eigenstate is to be considered. In the case of degeneracy, where several eigenstates may have the same eigenvalue of energy, the above formula can be rewritten if desired in the form

$$Z = \Big(\sum_i g_i e^{-\epsilon_i/kT}\Big)^n, \tag{134.5}$$

where g_i denotes the number of eigenstates which happen to have the same energy ϵ_i, and the indices i are now used to distinguish merely between different energy levels. We thus obtain the desired expressions for the sum-over-states for a Maxwell-Boltzmann system composed of n weakly interacting elements.

(*b*) **Case of Einstein-Bose and Fermi-Dirac gases.** We may now turn to the treatment of systems composed of similar elements, where the eigenstates for the system as a whole are to be specified by stating merely the number of elements in their different possible individual eigenstates, without any distinction as to which elements are so assigned. As examples of such systems we have Einstein-Bose and Fermi-Dirac gases which exhibit eigensolutions which are respectively either symmetric or antisymmetric in the indices denoting the particles composing the gas. For the purposes that we now have in view, we shall be specially interested in the sum-over-states when the phenomenon of gas degeneration can be neglected. The effects of this phenomenon are negligible for ordinary gases under usual circumstances, and have already been treated by appropriate methods for the special cases where they become important. In the absence of appreciable degeneration we shall find that the calculation then leads to a very simple result.

Let us consider a dilute gas composed of n similar, weakly interacting molecules, each exhibiting the same set of states of energies $\epsilon_k, \epsilon_l, \epsilon_m, \ldots$.

Using the symbols $\epsilon_1, \epsilon_2, \ldots, \epsilon_n$ to denote possible values for the energies of the n molecules composing the system, we can then write

$$E = \epsilon_1+\epsilon_2+\ldots+\epsilon_n \tag{134.6}$$

as an expression, in the case of weak interaction, for a possible value of the energy of the system as a whole. Noting that the total number of terms on the right-hand side of this expression is equal to n, and using n_k, n_l, n_m,... to denote the numbers of repetitions in which different terms ϵ refer to the same molecular state, it is evident that the above value for E could be obtained in

$$\frac{n!}{n_k!\,n_l!\,n_m!\,...} \tag{134.7}$$

different ways, by merely permuting the assignment of particle indices $1, ..., n$ to the energies ϵ. In the immediately preceding consideration of systems composed of distinguishable elements we regarded such different permutations as leading to different states of the system as a whole. In our present consideration, however, we know from the symmetry properties of the eigensolutions for the system as a whole that such permutations do not lead to new states. Hence we can now evidently write as an appropriate expression for the sum-over-states of a dilute gas composed of n similar molecules

$$Z = \sum_{\epsilon_1, \epsilon_2, ..., \epsilon_n} \frac{n_k!\,n_l!\,n_m!\,...}{n!} e^{-\frac{\epsilon_1+\epsilon_2+...+\epsilon_n}{kT}}, \tag{134.8}$$

where the factor before the exponential term takes care of the circumstance that the indicated summation would introduce the same states more than once. In applying this expression to Einstein-Bose gases, we must sum over all possible values of $\epsilon_1 ... \epsilon_n$. But in applying it to Fermi-Dirac gases, we must eliminate, in accordance with the Pauli exclusion principle, those terms in which two or more energies, $\epsilon_1 ... \epsilon_n$, refer to the same state, since there is then no corresponding eigenstate for the system as a whole.

We must now consider the simplification of this general expression, under conditions of negligible degeneration, when the temperature of the gas is high and its volume large. Under such circumstances most of the terms, which make important contributions in the summation indicated by (134.8), would be such that the same state is not assigned to more than a single molecule, since many different states are available for the molecules without going to such high total energies as to make the exponential term negligibly small. This has two effects. In the first place it permits us to neglect the fact that the summation as written contains a few terms which ought to be excluded to get a precisely correct result in the Fermi-Dirac case, and in the second place it permits us as an approximation to take the factorials $n_k!\, n_l!\, n_m! ...$ as all having the value unity. Hence we can now write as a satisfactory approximation, for either kind of gas, in the absence of appreciable degeneration,

$$Z = \frac{1}{n!} \sum_{\epsilon_1, \epsilon_2, \ldots, \epsilon_n} e^{-\frac{\epsilon_1 + \epsilon_2 + \ldots + \epsilon_n}{kT}}, \tag{134.9}$$

where we now take all possible values ϵ for the energies of the n particles.

Noting once more the possibility of decomposing the quantity to be summed into a product of factors of the form $e^{-\epsilon_k/kT}$, this can now be rewritten in the form

$$Z = \frac{1}{n!}\left(\sum_k e^{-\epsilon_k/kT}\right)^n, \tag{134.10}$$

where the summation is to be taken over all eigenstates k for a single particle or molecule of the kind in question. We thus obtain the desired simple expression for the sum-over-states in the case of a dilute gas of n similar molecules, in the absence of appreciable degeneration. If desired, in the case of multiple energy levels, the formula can also be written in the form

$$Z = \frac{1}{n!}\left(\sum_k g_k e^{-\epsilon_k/kT}\right)^n, \tag{134.11}$$

where the summation is now over energy levels rather than explicitly over individual states.

135. Perfect monatomic gas

(*a*) **Sum-over-states.** We are now ready to apply the methods which we have developed above to investigate the properties of specific thermodynamic systems. We shall first treat the case of a perfect monatomic gas.

Let us consider a gas composed of n similar particles, enclosed in a container of volume v, and having the temperature T. Under the conditions of negligible degeneracy, necessarily holding in the case of perfect gases, we can then write for the sum-over-states

$$Z = \frac{1}{n!}\left(\sum_k e^{-\epsilon_k/kT}\right)^n, \tag{135.1}$$

where the indicated summation is to be taken over the individual states k for a single particle of the kind under consideration. To evaluate the summation, we may make use of our previous expression (71.16) for the number of eigenstates dG for a single particle in the range ϵ to $\epsilon + d\epsilon$. This can be written in the form

$$dG = \frac{2\pi v g}{h^3}(2m)^{\frac{3}{2}}\epsilon^{\frac{1}{2}}\, d\epsilon, \tag{135.2}$$

where m is the mass of the particle, and, for the sake of generality in case the particle has an intrinsic angular momentum, g stands for the number of eigenvalues of the component of angular momentum parallel to a selected axis. With the help of this expression we can now replace summation by integration and write

$$\sum_k e^{-\epsilon_k/kT} = \frac{2\pi vg}{h^3}(2m)^{\frac{3}{2}} \int\limits_0^\infty e^{-\epsilon/kT}\epsilon^{\frac{1}{2}}\,d\epsilon = \frac{vg}{h^3}(2\pi mkT)^{\frac{3}{2}}, \tag{135.3}$$

with the help of the known value for the definite integral. Substituting (135.3) in (135.1), we then obtain for the sum-over-states of a perfect monatomic gas

$$Z = \frac{1}{n!}\left[\frac{vg}{h^3}(2\pi mkT)^{\frac{3}{2}}\right]^n. \tag{135.4}$$

In our actual applications we shall be interested in the logarithm of this quantity. Introducing the Stirling approximation

$$n! = \sqrt{(2\pi n)}\left(\frac{n}{e}\right)^n \tag{135.5}$$

for the factorial of the very large number n, we then obtain

$$\log Z = n\log\frac{vg}{h^3}(2\pi mkT)^{\frac{3}{2}} - n\log n + n - \log\sqrt{(2\pi n)}$$

$$= n\left[\log\frac{evg}{nh^3}(2\pi mkT)^{\frac{3}{2}} - \frac{1}{n}\log\sqrt{(2\pi n)}\right]. \tag{135.6}$$

Hence, except for a term which goes to zero as the number of particles n increases, we can write

$$\log Z = n\log\frac{evg}{nh^3}(2\pi mkT)^{\frac{3}{2}}. \tag{135.7}$$

(*b*) **Thermodynamic quantities.** We may now substitute this expression for $\log Z$ in our previous expressions (133.3) to (133.6) for the pressure p, energy E, heat capacity C_v, and entropy S of any system. We thus obtain for a perfect monatomic gas, with a minimum of computation, the important collection of results

$$p = \frac{nkT}{v}, \tag{135.8}$$

$$E = \tfrac{3}{2}nkT, \tag{135.9}$$

$$C_v = \tfrac{3}{2}nk, \tag{135.10}$$

$$S = nk\log\frac{e^{\frac{5}{2}}vg}{nh^3}(2\pi mkT)^{\frac{3}{2}}. \tag{135.11}$$

The simplicity with which these results have been obtained illustrates the powerful character of our present methods. The first three expressions for pressure, energy, and heat capacity are of course very familiar ones.

The last of the equations gives the so-called Sackur-Tetrode expression for the entropy of a perfect monatomic gas, modified merely by the inclusion of the factor g, to allow for the possibility that the translational states of the particles composing the gas may, from a modern point of view, have a g-fold degeneracy, owing, for instance, to the presence of intrinsic angular momentum. It will be noted, in accordance with the discussion at the end of § 133, that our present treatment of thermodynamic properties from the standpoint of a *quantum* statistical mechanics leads to a perfectly *definite* value for the entropy of the gas in the condition considered, based on a starting-point which would make the entropy zero if the system were known to be in a single definite quantum mechanical state.

(*c*) **Evaluation of constant k.** So far we have assigned no definite numerical value to the constant k which was introduced into the correlations of entropy and temperature with mechanical quantities,

$$S \rightleftharpoons -k\bar{\bar{H}} \quad \text{and} \quad T \rightleftharpoons \frac{\theta}{k}, \tag{135.12}$$

in order to allow for difference in units. With the help of the expression for the pressure of a perfect gas given by (135.8) we now see, however, as indicated by the notation chosen, that this quantity actually is the so-called Boltzmann constant

$$k = \frac{R}{N}, \tag{135.13}$$

where R is the gas constant per mol and N is the number of molecules in a mol.

Using the system of units which depends on the centimetre, gram, second, and degree centigrade, the quantity in question has the value†

$$k = 1{\cdot}380 \times 10^{-16}\ \text{erg/deg}. \tag{135.14}$$

136. Perfect gases composed of more complicated molecules

(*a*) **Sum-over-states.** We must now consider gases composed of similar elements which are more complicated than simple particles. The energy eigenvalues for a single element can then depend not only on its translational state but on its internal state as well, and we have

† Birge, *Phys. Rev. Supplement*, 1, 1 (1929), modified to correspond to the new value $4{\cdot}803 \times 10^{-10}$ e.s.u. for e.

to allow for this in computing the actual value of the sum-over-states. The case of greatest practical interest is provided by gases composed of molecules, containing more than a single atom, which exhibit different internal states of rotation and nuclear vibration. The methods of treatment are general enough, however, to apply also when it becomes necessary to consider the different internal electronic states which can be excited both in the case of molecules and single atoms as well.

Let us consider a gas composed of n similar, chemically non-interacting molecules, enclosed in a container of volume v, and having the temperature T. Under conditions of negligible gas degeneration we can then again write, in accordance with (134.10),

$$Z = \frac{1}{n!}\Big(\sum_m e^{-\epsilon_m/kT}\Big)^n \tag{136.1}$$

as a general expression for the sum-over-states, where the different states of a single molecule in the container are designated by the letter m. In order to evaluate the indicated summation over the states m, however, we must now consider both the translational and the internal condition of the molecule. To carry this out it is convenient to regard the state of the molecule as specified by two letters k and i, referring respectively to the translational and internal conditions. Except for a negligible relativity effect, we can then take the total energy for a state ki as equal to the sum of a term ϵ_k depending solely on the index k in the same way as for simple particles, and a term ϵ_i which depends solely on the internal condition, i.e. we can take

$$\epsilon_{ki} = \epsilon_k + \epsilon_i. \tag{136.2}$$

Our expression for the sum-over-states is then seen to assume the form

$$Z = \frac{1}{n!}\Big(\sum_k e^{-\epsilon_k/kT}\Big)^n \Big(\sum_i e^{-\epsilon_i/kT}\Big)^n. \tag{136.3}$$

For the purposes of practical computation, it is usually more convenient to rewrite this expression in a form

$$Z = \frac{1}{n!}\Big(\sum_k e^{-\epsilon_k/kT}\Big)^n \Big(\sum_i g_i e^{-\epsilon_i/kT}\Big)^n, \tag{136.4}$$

which gives explicit recognition to the possibility that any given internal energy level including the lowest may be degenerate. The second summation is now over different energy levels i, g_i being the number of separate states having the same internal energy ϵ_i. Taking the logarithm of this quantity, and noting our previous value (135.7) for monatomic gases where Z corresponds to the first two factors in (136.4), we then obtain

$$\log Z = n \log \frac{ev}{nh^3}(2\pi mkT)^{\frac{3}{2}} + n \log \sum_i g_i e^{-\epsilon_i/kT} \tag{136.5}$$

as the desired expression for the sum-over-states.

(*b*) **Thermodynamic quantities.** Substituting this expression for Z into equations (133.3) to (133.6), we now obtain for the pressure, energy, heat capacity at constant volume, and entropy of a dilute gas composed of n molecules of mass m, in volume v and at temperature T,

$$p = \frac{nkT}{v}, \tag{136.6}$$

$$E = \tfrac{3}{2}nkT + nkT^2 \frac{\partial}{\partial T} \log \sum_i g_i e^{-\epsilon_i/kT}, \tag{136.7}$$

$$C_v = \tfrac{3}{2}nk + 2nkT \frac{\partial}{\partial T} \log \sum_i g_i e^{-\epsilon_i/kT} + nkT^2 \frac{\partial^2}{\partial T^2} \log \sum_i g_i e^{-\epsilon_i/kT}, \tag{136.8}$$

$$S = nk \log \frac{e^{\frac{5}{2}}v}{nh^3}(2\pi mkT)^{\frac{3}{2}} + nk \log \sum_i g_i e^{-\epsilon_i/kT} + nkT \frac{\partial}{\partial T} \log \sum_i g_i e^{-\epsilon_i/kT}. \tag{136.9}$$

The first of these formulae, (136.6), is that for the *pressure* of a *perfect gas* in agreement with the circumstance that we have treated gases at high enough dilutions and temperatures so that the phenomenon of gas degeneration and the effects of collision on energy levels could be neglected.

The second formula, (136.7), may be regarded as expressing the total *energy* of the gas as a sum of translational and internal parts. This will be better appreciated by carrying out the indicated differentiation in the last term of (136.7) and re-expressing the energy in the form

$$E = \tfrac{3}{2}nkT + n \frac{\sum_i g_i e^{-\epsilon_i/kT} \epsilon_i}{\sum_i g_i e^{-\epsilon_i/kT}}, \tag{136.10}$$

where the first term is a well-known expression for the mean kinetic energy of the n molecules, and the second term is seen to be an appropriate expression for the mean internal energy under conditions where the Maxwell-Boltzmann distribution over internal states i can be taken as holding. The actual value that we obtain for the energy E will depend of course on the energy zero-point that we select. To examine this let us denote the energy assigned to a molecule at rest in its lowest internal state by ϵ_0. Introducing terms which would mutually cancel, we may then rewrite our expression (136.10) for energy as a sum of three terms

$$E = n\epsilon_0 + \tfrac{3}{2}nkT + n\frac{\sum_i g_i e^{-\frac{(\epsilon_i-\epsilon_0)}{kT}}(\epsilon_i-\epsilon_0)}{\sum_i g_i e^{-\frac{(\epsilon_i-\epsilon_0)}{kT}}}, \tag{136.11}$$

which are respectively seen to give the energy of the molecules at rest in the ground-level, their kinetic energy of translation, and the internal energy that has been excited. In the practical computations of physical chemists the energy ϵ_0 assigned to the ground-level may be taken as is convenient for the problem at hand. For example, it may be taken as zero, or as the potential energy necessary to remove a molecule from some condensed phase of interest, or as the energy involved in forming the molecule from its atoms, in accordance with the nature of the problem under consideration.

The third of the formulae given above, (136.8), is that for the *heat capacity* of the gas at constant volume as obtained by differentiating the expression for energy with respect to temperature. The result is of course independent of any assignment made to the energy of the ground-level.

The last of the above formulae (136.9) gives the entropy of the gas based on a starting-point which assigns zero entropy to any pure quantum mechanical state. By comparing with (136.7), it can also be written in the form

$$S = \frac{E}{T} + nk\log\frac{ev}{nh^3}(2\pi mkT)^{\frac{3}{2}} + nk\log\sum_i g_i e^{-\epsilon_i/kT}, \tag{136.12}$$

which sometimes proves convenient. And by introducing terms which mutually cancel, this can be re-expressed in a form

$$S = \frac{E-n\epsilon_0}{T} + nk\log\frac{ev}{nh^3}(2\pi mkT)^{\frac{3}{2}} + nk\log\sum_i g_i e^{-\frac{(\epsilon_i-\epsilon_0)}{kT}}, \tag{136.13}$$

which only involves energies reckoned from the ground-level up. This, then, shows clearly that our expressions for entropy are independent of the selection of energy zero-point, in agreement with the circumstance that they have their own starting-point of zero entropy for any pure state as mentioned above. For the purposes of the physical chemist, it is often more convenient to have the dependence of entropy on pressure rather than on volume, as can be obtained by substituting

$$v = \frac{nkT}{p} \tag{136.14}$$

in accordance with the gas laws. It is also convenient to rewrite the last term of (136.13) in the form

$$nk\log\sum_i g_i e^{-\frac{(\epsilon_i-\epsilon_0)}{kT}} = nk\log\sum_i \frac{g_i}{g_0}e^{-\frac{(\epsilon_i-\epsilon_0)}{kT}}+nk\log g_0, \quad (136.15)$$

where g_0 is the multiplicity of the ground-level, in order to separate out the part $nk\log g_0$ which does not go to zero at low temperatures. Our expression for entropy (136.13) then assumes the form

$$S = \frac{E-n\epsilon_0}{T}+\tfrac{5}{2}nk\log T-nk\log p+nk\log\sum_i \frac{g_i}{g_0}e^{-\frac{(\epsilon_i-\epsilon_0)}{kT}}+$$
$$+nk\frac{\log e(2\pi m)^{\frac{3}{2}}k^{\frac{5}{2}}g_0}{h^3}, \quad (136.16)$$

where we have rearranged in a manner to segregate constant and variable terms.

(c) **Energy and entropy of actual monatomic and diatomic gases.** The application of the foregoing formulae for energy and entropy to actual gases is of course dependent on the requisite knowledge of the energy levels ϵ_i and quantum weights g_i for those states of the gas molecule which will introduce appreciable terms in the summation $\sum_i e^{-\epsilon_i/kT}$ at the temperature under consideration. Such information as to energy levels and quantum weights must be obtained in general from spectroscopic data and its quantum mechanical interpretation.

In making applications it is often necessary to undertake a detailed specific treatment for the particular kind of gas molecule involved, in order to obtain high precision and to take account of all possible complicating factors. We may illustrate the methods, nevertheless, by giving general—although to some extent approximate and over-simplified—treatments for the cases of gases composed of monatomic and of diatomic molecules. In the case of polyatomic gases it is best at the present time not to try to develop general formulae, but to make specific calculations for each particular case, on account of the complex behaviour of molecules containing more than two atoms. We shall be specially interested in treating the expressions for the energy and entropy of monatomic and diatomic gases in a manner similar to that commonly employed in physical-chemical computations.

Let us first consider an *actual monatomic gas*, composed of n atoms, having a ground-level which may be degenerate, and having higher electronic levels which might become appreciably excited. We may then use the general symbols ϵ_i and g_i to denote the energies and degrees of degeneracy of all the different electronic levels i including the ground-level, and may introduce the special symbols ϵ_0 and g_0 to denote when necessary the particular values of those quantities for the ground-level. The energy ϵ_0 assigned to the ground-level may be, for example, the potential energy which a free atom would have with reference to removal from a crystal of the substance at very low temperatures. The degeneracy g_0 of the ground-level may be due to the presence of nuclear spin or of unresolved electronic states.

For the practical computation of the *energy* of such a monatomic gas it then proves convenient to take our previous expression (136.11)

$$E = n\epsilon_0+\tfrac{3}{2}nkT+n\frac{\sum_i g_i e^{-\frac{(\epsilon_i-\epsilon_0)}{kT}}(\epsilon_i-\epsilon_0)}{\sum_i g_i e^{-\frac{(\epsilon_i-\epsilon_0)}{kT}}}, \tag{136.17}$$

where the evaluation of the internal energy as given by the last term will have to be made, with the help of spectroscopic information, by actually summing over all levels i in the numerator and denominator which make an appreciable contribution at the temperature T under consideration. In making these summations, it will be noticed that the term corresponding to the ground-level, $\epsilon_i = \epsilon_0$, disappears from the numerator but not from the denominator of the above expression for internal energy. As a consequence it will be seen that the internal energy would approach zero at temperatures low enough so that the energy $(\epsilon_1-\epsilon_0)$ needed for exciting the first level above the ground-level is large compared with kT. For many actual monatomic gases this means that internal energies can be neglected up to very high temperatures. This is not always possible, however. For example, in the case of the monatomic halogens the two lowest electronic levels form such a close doublet that the excitation of the upper level must be considered at those temperatures where the dissociation of the diatomic into the monatomic gas would be studied.

For the practical computation of the *entropy* of monatomic gases it proves convenient to take our previous equation (136.16)

$$S = \frac{E-n\epsilon_0}{T}+\tfrac{5}{2}nk\log T-nk\log p+nk\log\sum_i \frac{g_i}{g_0}e^{-\frac{(\epsilon_i-\epsilon_0)}{kT}}+$$
$$+nk\log\frac{e(2\pi m)^{\frac{3}{2}}k^{\frac{5}{2}}g_0}{h^3}, \tag{136.18}$$

where the next to the last term would now have to be evaluated on the basis of specific information as to the electronic levels ϵ_i. It will be noted that this term would go to zero at temperatures low enough so that the energy of the first excited level $(\epsilon_1-\epsilon_0)$ would be large compared with kT. Hence the term can often be neglected. For values of this term in the case of the monatomic halogens where it proves important, and for values of the multiplicity g_0 of the ground-level in the case of a number of monatomic gases, reference may be made to Fowler's *Statistical Mechanics*.†

Let us now turn to a consideration of the energy and entropy of *diatomic gases*, composed of molecules capable of existing in various states of internal rotation, vibration, and electronic excitation. Precise values of the energy and entropy of such a gas at any desired temperature could of course be computed in any specific case with the help of the requisite precise spectroscopic data. It will be of advantage, however, also to consider a somewhat general approximate method of treating the energy and entropy of such gases in the temperature range where their rotational levels can be regarded as fully excited.

To obtain such a treatment we may first recall the results which were derived in § 74 (*d*) for a simple model of a diatomic molecule consisting of two dissimilar particles, without spin, which are regarded as held together by an attractive force obeying Hooke's law. For the energy levels of such a model we obtained the approximate expression

$$\epsilon_{Jv} = \frac{J(J+1)h^2}{8\pi^2 I}+(v+\tfrac{1}{2})h\nu, \tag{136.19}$$

where the rotational and vibrational quantum numbers J and v can assume the values

$$J = 0, 1, 2, 3,..., \qquad v = 0, 1, 2, 3,..., \tag{136.20}$$

and I and ν are the moment of inertia and the classical vibration frequency of the molecule when in its equilibrium configuration. Furthermore, for the multiplicity or quantum weight of the various levels we obtained the result

$$g_{Jv} = 2J+1. \tag{136.21}$$

Actual diatomic molecules are of course more complex than the above simple model. They consist not only of a pair of nuclei, but of surrounding electrons as well, and the nuclear particles are not necessarily dissimilar and without spin. Hence there are many complicating factors which must be considered in the treatment of actual molecules. Nevertheless, in agreement with the foregoing, it is often possible to represent

† Fowler, *Statistical Mechanics*, second edition, Cambridge, 1936, pp. 218–20. For the connexion between Fowler's notation and ours, see footnote † on p. 608.

the internal energy of a diatomic molecule with sufficient approximation by the formula

$$\epsilon_{Jve} = \epsilon_0 + \frac{J(J+1)h^2}{8\pi^2 I} + \epsilon_{ve}, \tag{136.22}$$

where the first term ϵ_0 gives the internal energy which we assign to the molecule in its ground-level—including therein the corresponding half-quantum of vibrational energy—the second term gives the rotational energy corresponding to the quantum number J, and the last term ϵ_{ve} represents the energy associated with the state of vibrational *and* electronic excitation denoted by the subscript *ve*. Both of these latter terms will go to zero for the ground-level denoted by $J = v = e = 0$. Furthermore, it will often be possible to represent the quantum weight of any level Jve by the formula

$$g_{Jve} = (2J+1)g_{ve}, \tag{136.23}$$

where $(2J+1)$ is the multiplicity of the rotational level J, and g_{ve} is the multiplicity of the vibrational and electronic state *ve*. It is to be noted that g_{ve} will not necessarily go to unity in the ground-level, since this may be degenerate on account of nuclear spins or unresolved electronic states.

Assuming the satisfactory character of the above expressions for the energy levels and quantum weights of our diatomic molecule, we can then write for the logarithm of a needed summation over the internal states of the molecule i

$$\begin{aligned} \log \sum_i g_i e^{-\frac{(\epsilon_i-\epsilon_0)}{kT}} &= \log \sum_{J,v,e} (2J+1) g_{ve} e^{-\frac{J(J+1)h^2}{8\pi^2 IkT} - \frac{\epsilon_{ve}}{kT}} \\ &= \log \sum_J (2J+1) e^{-\frac{J(J+1)h^2}{8\pi^2 IkT}} + \log \sum_{v,e} g_{ve} e^{-\frac{\epsilon_{ve}}{kT}} \\ &= \log \sum_J (2J+1) e^{-\frac{J(J+1)h^2}{8\pi^2 IkT}} + \log \sum_{v,e} (g_{ve}/g_0) e^{-\frac{\epsilon_{ve}}{kT}} + \log g_0, \end{aligned} \tag{136.24}$$

where in the last form of writing we have factored out the quantum weight g_0 of the ground-level in order to separate the constant term $\log g_0$ from the remainder of the expression which goes to zero at low temperatures. Furthermore, at sufficiently high temperatures so that we can regard the rotational levels as fully excited, that is, at temperatures where

$$\frac{h^2}{8\pi^2 IkT} \ll 1, \tag{136.25}$$

we can replace summation by integration by putting

$$\sum_{J=0}^{\infty} (2J+1)e^{-\frac{J(J+1)h^2}{8\pi^2 IkT}} = \int_0^{\infty} (2J+1)e^{-\frac{J(J+1)h^2}{8\pi^2 IkT}}\, dJ. \qquad (136.26)$$

In the case of a diatomic molecule, where the two nuclei are *dissimilar*, all the levels $J = 0, 1, 2,...$ could be occupied and the above would give us

$$\sum_J (2J+1)e^{-\frac{J(J+1)h^2}{8\pi^2 IkT}} = \frac{8\pi^2 IkT}{h^2} \qquad (136.27)$$

as a reasonable approximation.† In the case of a diatomic molecule where the two nuclei are *similar*, only alternate levels could be occupied, and we can then take

$$\sum_J (2J+1)e^{-\frac{J(J+1)h^2}{8\pi^2 IkT}} = \frac{1}{2}\frac{8\pi^2 IkT}{h^2} \qquad (136.28)$$

as a reasonable approximation. Hence, returning to (136.24), we can now write in general, for the internal states of our molecule at sufficiently high temperatures,

$$\log \sum_i g_i e^{-\frac{(\epsilon_i-\epsilon_0)}{kT}} = \log\frac{8\pi^2 IkT}{h^2} - \log\sigma + \log\sum_{v,e}(g_{ve}/g_0)e^{-\frac{\epsilon_{ve}}{kT}} + \log g_0, \qquad (136.29)$$

where the *symmetry number* σ assumes the value 1 or 2 according as the two nuclei are dissimilar or alike.

Returning to our general expression (136.11), and making use of the result of differentiating (136.29) with respect to T, we can now write for the *energy of a diatomic gas*, at temperatures T high enough so that the rotational levels can be regarded as fully excited,

$$E = n\epsilon_0 + \tfrac{5}{2}nkT + n\frac{\sum_{v,e} g_{ve}\, e^{-\epsilon_{ve}/kT}\epsilon_{ve}}{\sum_{v,e} g_{ve}\, e^{-\epsilon_{ve}/kT}}. \qquad (136.30)$$

The first term $n\epsilon_0$ gives the energy that we ascribe to the n molecules in the ground-level. The second term gives the sum of their translational energy $\frac{3}{2}nkT$, and the additional rotational energy nkT. The last term gives the further energy to be ascribed to the excitation of the vibrational and electronic levels ve. This term would tend towards zero at sufficiently low temperatures, and for higher temperatures could be computed from a knowledge of spectral data.

Turning now to our general expression for entropy (130.16), sub-

† The above approximation could be slightly improved by integrating from $-\frac{1}{2}$ to ∞. For a treatment of rotational levels based on a more precise formula than (136.27), see Giauque and Overstreet, *Journ. Amer. Chem. Soc.* **54**, 1731 (1932).

stituting (136.29), and segregating variable and constant terms, we obtain for the *entropy of a diatomic gas*, at sufficiently high temperatures,

$$S = (E-n\epsilon_0)/T+\tfrac{7}{2}nk\log T-nk\log p+nk\log\sum_{v,e}(g_{ve}/g_0)e^{-\epsilon_{ve}/kT}+$$
$$+nk\log e(2\pi m)^{\frac{3}{2}}k^{\frac{7}{2}}\frac{8\pi^2 I}{h^5}\frac{g_0}{\sigma}, \quad (136.31)$$

where the next to the last term, which is seen to go to zero at low temperatures, would still have to be evaluated from spectral knowledge as to vibrational and electronic excitation levels. The factors in the final constant term, which refer to the specific diatomic molecule under consideration, are the mass m, moment of inertia I, degeneracy of the ground-level g_0, and symmetry number σ. For the values of g_0 and σ in the case of a number of diatomic gases, and for the effect of the next to the last term in (136.31) in the case of NO where it proves important, reference may be made to Fowler's *Statistical Mechanics*.†

137. Crystals composed of a single substance

(*a*) **The modes of vibration of a crystal.** We may now turn our considerations from the gaseous state characterized by the tendency towards a random distribution of component particles, to the crystalline state characterized by the tendency towards a highly ordered arrangement of component particles. We must begin by concentrating our attention on the positions and motions of the particles composing the crystal, leaving the possibilities for actual particles to assume different internal states for later inclusion in the theory.

Let us consider, at first from a classical point of view, a crystal composed of n similar particles of mass m capable of performing small oscillations about a set of equilibrium positions which are determined by the nature of the forces acting between the particles. These equilibrium positions inside the crystal structure may be specified as to their permanent relative spatial arrangement by the set of numbers 1, 2, 3,..., n.

To treat the condition of the crystal we may now introduce, corresponding to each particle, a set of three Cartesian coordinates

$$(x_1 y_1 z_1),\ (x_2 y_2 z_2),\ldots,\ (x_n y_n z_n) \quad (137.1)$$

to denote the position of the oscillating particle associated with each equilibrium position. It may be noted for later reference that the

† Fowler, loc. cit., pp. 220–5. For the connexion between Fowler's notation and ours, see footnote † on p. 608.

indices 1, 2, 3,..., n thus assigned are determined by the permanent relative positions of the equilibrium positions. Hence, in accordance with the discussions of § 76, when we come to the application of quantum mechanics no solutions will have to be eliminated on the basis of symmetry restrictions.

Using the above coordinates, the energy of the crystal could then be expressed in the form

$$E = \tfrac{1}{2}m(\dot{x}_1^2+\dot{y}_1^2+...+\dot{z}_n^2)+V(x_1, y_1,..., z_n)+\text{const.}, \qquad (137.2)$$

where the first term gives the sum of the kinetic energies of the particles, the second term gives their mutual potential energy, and the last term allows for any desired later choice of the energy zero-point. The explicit expression for the potential energy V appearing in this equation would, however, depend in a very complicated way on the coordinates x_1, y_1,..., z_n, and a change to new coordinates will be convenient. Assuming the oscillations of the particles small enough so that the potential energy of the particles could be regarded as a quadratic function of their displacements from the equilibrium positions, a transformation to so-called *normal coordinates* q_1, q_2,..., q_{3n} can now be introduced, following well-known methods,† which will make it possible to rewrite the energy in the form

$$E = \tfrac{1}{2}(m_1\dot{q}_1^2+m_2\dot{q}_2^2+...+m_{3n}\dot{q}_{3n}^2)+\tfrac{1}{2}(b_1q_1^2+b_2q_2^2+...+b_{3n}q_{3n}^2)+E_0, \qquad (137.3)$$

where the m's and b's are constants, and E_0 is the zero-point energy that we assign to the crystal in the absence of oscillational excitation. Or, by introducing the momenta $p_1 ... p_{3n}$ which correspond to our present coordinates, this can also be written in the desired Hamiltonian form

$$H = \tfrac{1}{2}(a_1p_1^2+a_2p_2^2+...+a_{3n}p_{3n}^2)+\tfrac{1}{2}(b_1q_1^2+b_2q_2^2+...+b_{3n}q_{3n}^2)+E_0. \qquad (137.4)$$

The a's and b's appearing in this final expression for the classical Hamiltonian are constants. The coordinates $q_1 ... q_n$ are linear functions of the original coordinates $x_1 ... z_n$. Six of these coordinates can be taken as determining the position and orientation of the crystal as a whole, so that the corresponding coefficients a will be the reciprocals of the mass and moments of inertia of the crystal as a whole and the associated coefficients b will be zero. The remaining coordinates and momenta can evidently be regarded as corresponding to $3n-6$ modes

† See Whittaker, *Analytical Dynamics*, second edition, Cambridge, 1917, p. 177.

of harmonic vibration, and the internal motions of the crystal can then be regarded as a superposition of the vibrations in these modes.

The expression for the Hamiltonian given by (137.4) is of course not quite precise, since it contains the usual approximations involved in the method which we are developing. In particular we may mention that it neglects small terms in the potential energy which would actually be present if we allowed for the interaction between oscillators corresponding to the possibility of energy transfer from one mode to another. Furthermore, when we wish to regard the crystal as in contact with its vapour it is not strictly accurate to treat the particles on the surface as controlled by pure elastic forces. Nevertheless, the treatment made possible by (137.4) will be accurate enough for our purposes, and since $3n$ will actually be a very large number compared with 6, it will also be sufficiently precise to regard $3n$ as the number of modes of harmonic vibration by which we represent the internal motions of the crystal.

The frequencies ν of the different modes of vibration will be determined from a classical point of view by the values of the corresponding constants a and b in accordance with the familiar relation

$$\nu = \frac{\sqrt{(ab)}}{2\pi}. \tag{137.5}$$

The modes of vibration of the lowest frequencies may be regarded as those that would ordinarily be spoken of as mechanical or acoustical, and those of higher frequencies as corresponding to motions of the particles that would ordinarily be spoken of as thermal. There will be an upper limit ν_m for the maximum possible frequency since there are only $3n$ modes of vibration in all.

Since the frequencies of the different modes of vibration in the case of a large crystal will lie close together over the major portion of the spectrum, we can introduce an expression of the general form

$$dz = f(\nu)\,d\nu \tag{137.6}$$

for the number of modes of vibration in the frequency range ν to $\nu+d\nu$. To determine the form of the function $f(\nu)$, it is sufficiently accurate for many purposes, following Debye,† to neglect the particle structure and treat the crystal as a continuous, elastic solid. For the number of modes of vibration in the range ν to $\nu+d\nu$, we should then have an expression of the known form

$$dz = \frac{4\pi v}{C^3}\nu^2\,d\nu, \tag{137.7}$$

† Debye, *Ann. Physik*, **39**, 789 (1912). This important article has already been mentioned as significant for the development of quantum theory.

where v is the volume of the solid and C is a quantity which is determined by the elastic properties of the solid which can be taken as approximately constant. For example, for an *isotropic solid* $1/C^3$ would be given by

$$\frac{1}{C^3} = \frac{1}{c_l^3} + \frac{2}{c_t^3}, \tag{137.8}$$

where c_l and c_t are the velocities of the longitudinal and transverse waves of which the solid is capable. An expression of the form (137.7) could be expected to give a good representation of the spectral distribution for frequencies low enough so that the corresponding wave-lengths are large compared with the distances between neighbouring particles. The expression must in any case be modified at sufficiently high frequencies since there are only a finite number of frequencies $3n$ in all. Taking the expression as holding sufficiently well up to the maximum frequency ν_m, we can write

$$3n = \frac{4\pi v}{C^3} \int_0^{\nu_m} \nu^2 \, d\nu = \frac{4}{3} \frac{\pi v}{C^3} \nu_m^3, \tag{137.9}$$

which gives for the maximum frequency ν_m the value

$$\nu_m = C\left(\frac{9n}{4\pi v}\right)^{\frac{1}{3}}, \tag{137.10}$$

(*b*) **Application of quantum mechanics.** We are now ready to turn to the application of quantum mechanics to our system by considering the allowed eigensolutions of the Schroedinger equation

$$\mathbf{H}u(q_1 \ldots q_{3n}) = Eu(q_1 \ldots q_{3n}), \tag{137.11}$$

where $\mathbf{H}$ is the Hamiltonian operator for the crystal and E an allowed eigenvalue of its energy. Noting the form of the Hamiltonian operator which would be obtained from (137.4) by the substitution $p_i = \frac{h}{2\pi i}\frac{\partial}{\partial q_i}$, we see that this equation can be handled by the method of separation of variables by assuming a solution of the form

$$u(q_1 \ldots q_{3n}) = u_1(q_1)u_2(q_2) \ldots u_{3n}(q_{3n}) = \prod_i u_i(q_i). \tag{137.12}$$

The complete equation (137.11) can then be solved with the help of $3n$ one-dimensional equations each of the form

$$a_i \frac{\partial^2 u_i}{\partial q_i^2} + \frac{8\pi^2}{h^2}\left(E_i - \frac{b_i}{2} q_i^2\right) u_i = 0, \tag{137.13}$$

where the quantities E_i are the energy eigenvalues for the individual modes of vibration. Since each of the individual equations (137.13) is of the form already studied for a simple harmonic oscillator in § 72, their eigensolutions $u_i(q_i)$ will be of the form previously obtained, and the allowed individual eigenvalues of the vibrational energy will be given by expressions of the form

$$E_i = (k_i+\tfrac{1}{2})h\nu_i, \qquad k_i = 0, 1, 2, 3,..., \tag{137.14}$$

where ν_i is the frequency of the ith mode of vibration. The allowed eigenvalues for the energy of the system as a whole may then be expressed in the form

$$E = E_0 + \sum_i E_i, \tag{137.15}$$

where E_0 is the zero-point energy assigned to the crystal and $\sum_i E_i$ is a sum of possible eigenvalues of energy for the $3n$ modes of vibration.

In connexion with the foregoing it is to be noted that none of the eigensolutions for the different modes of vibration are to be eliminated on the basis of symmetry considerations, since the modes of vibration are permanently distinguishable the one from another by their spatial location and orientation inside the crystal. This is in agreement with the fact that the coordinates $q_1 \dots q_{3n}$ for these modes of vibration are functions of the original coordinates $x_1 y_1 z_1 \dots x_n y_n z_n$ which were assigned to the different particles on the basis of their association with the permanently distinguishable equilibrium positions in the crystal.

(*c*) **Sum-over-states for the crystal.** We have now obtained all that is needed for a calculation of the sum-over-states for the crystal,

$$Z = \sum_m e^{-E_m/kT}, \tag{137.16}$$

where the summation is over all possible energy eigenstates m, each individual state to be separately included in the case of degenerate energy levels. Noting the expression for the total energy of the crystal given by (137.15), and allowing for the possibility that the lowest energy level may be degenerate and actually consist of G_0 individual states, we can then rewrite (137.16) in the form

$$Z = G_0 e^{-\frac{E_0}{kT}} \sum e^{-\frac{E_1+E_2+E_3+\dots+E_{3n}}{kT}} \tag{137.17}$$

where the summation $\sum$ is to be taken over all combinations of the possible eigenvalues E_1, E_2, E_3,..., E_{3n} of the $3n$ modes of oscillation. Introducing the expression for the possible eigenvalues for these energies given by (137.14), we obtain

$$\begin{aligned} Z &= G_0 e^{-E_0/kT} \sum_{k_1=0}^{\infty} e^{-(k_1+\frac{1}{2})h\nu_1/kT} \sum_{k_2=0}^{\infty} e^{-(k_2+\frac{1}{2})h\nu_2/kT} \dots \sum_{k_{3n}=0}^{\infty} e^{-(k_{3n}+\frac{1}{2})h\nu_{3n}/kT} \\ &= G_0 e^{-E_0/kT} \left(\frac{e^{-h\nu_1/2kT}}{1-e^{-h\nu_1/kT}}\right)\left(\frac{e^{-h\nu_2/2kT}}{1-e^{-h\nu_2/kT}}\right)\dots\left(\frac{e^{-h\nu_{3n}/2kT}}{1-e^{-h\nu_{3n}/kT}}\right), \end{aligned} \tag{137.18}$$

where $\nu_1, \nu_2,..., \nu_{3n}$ denote the frequencies of the $3n$ modes of vibration, and the second form of writing can be verified by performing the indicated divisions.

For the purposes of calculating thermodynamic quantities we shall be interested in the logarithm of the above quantity, which can be written in the form

$$\log Z = -\frac{E_0}{kT} - \sum_{i=1}^{3n} \log(1-e^{-h\nu_i/kT}) - \sum_{i=1}^{3n} \frac{h\nu_i}{2kT} + \log G_0, \quad (137.19)$$

where the summations are over all $3n$ modes of vibration i. Substituting the expression (137.7) for the number of modes of vibration in the range ν to $\nu+d\nu$, we can then replace these summations by integrations from $\nu = 0$ to the maximum frequency $\nu = \nu_m$, and write

$$\log Z = -\frac{E_0}{kT} - \frac{4\pi v}{C^3} \int_0^{\nu_m} \nu^2 \log(1-e^{-h\nu/kT})\, d\nu - \frac{\pi v}{C^3} \frac{h\nu_m^4}{2kT} + \log G_0, \quad (137.20)$$

where the next to the last term is obtained by evaluating the integral which depends on the residual 'half-quanta' $\frac{1}{2}h\nu$ which our modes of vibration possess in their lowest quantum states.

Finally, it will be profitable to re-express this result by introducing a *dimensionless variable* x, which is defined by

$$x = \frac{h\nu}{kT}, \quad (137.21)$$

together with a quantity Θ, determined by the elastic properties of the crystal, which is defined by

$$\Theta = \frac{h\nu_m}{k} = \frac{hC}{k}\left(\frac{9n}{4\pi v}\right)^{\frac{1}{3}}, \quad (137.22)$$

where the second form of writing results from (137.10). This quantity may be called the *characteristic temperature* for the crystal. Introducing x and Θ into (137.20), we now obtain the desired result

$$\log Z = -\frac{E_0}{kT} - 9n\left(\frac{T}{\Theta}\right)^3 \int_0^{\Theta/T} x^2 \log(1-e^{-x})\, dx - \tfrac{9}{8}n\left(\frac{\Theta}{T}\right) + \log G_0 \quad (137.23)$$

as a general expression for the sum-over-states Z of such a crystal.

We shall also be specially interested in the form of this expression at very low and at very high temperatures where the evaluation of the integral which it contains is simple. *At very low temperatures*, $T \ll \Theta$, we can evidently take

$$\int_0^{\Theta/T} x^2 \log(1-e^{-x})\,dx = \int_0^{\infty} x^2 \log(1-e^{-x})\,dx$$

$$= -\frac{1}{3}\int_0^{\infty} \frac{x^3}{e^x-1}\,dx$$

$$= -\frac{\pi^4}{45}, \tag{137.24}$$

where the second form of writing is obtained with the help of a partial integration. This then gives us

$$\log Z = -\frac{E_0}{kT}+\frac{\pi^4}{5}n\left(\frac{T}{\Theta}\right)^3-\tfrac{9}{8}n\left(\frac{\Theta}{T}\right)+\log G_0. \tag{137.25}$$

At very high temperatures, $T \gg \Theta$, only very small values of x, between 0 and Θ/T, will be involved in the integrand appearing in (137.23), and we can substitute

$$\log(1-e^{-x}) = \log x. \tag{137.26}$$

Doing so, we have an integral of the simple form $\int x^2 \log x\,dx$, and obtain the result

$$\log Z = -\frac{E_0}{kT}+3n\log\left(\frac{T}{\Theta}\right)+n-\tfrac{9}{8}n\left(\frac{\Theta}{T}\right)+\log G_0. \tag{137.27}$$

(*d*) **Thermodynamic properties of the crystal.** With the help of the foregoing expressions for $\log Z$, we can now investigate the thermodynamic properties of the crystal by making use of our general equations (133.3) to (133.6) for pressure, energy, heat capacity, and entropy.

For the *pressure* we may write at once from (137.23), for any temperature, the general expression

$$p = kT\frac{\partial \log Z}{\partial v}$$

$$= -\frac{\partial E_0}{\partial v}-kT\frac{\partial}{\partial \Theta}\left[9n\left(\frac{T}{\Theta}\right)^3\int_0^{\Theta/T} x^2\log(1-e^{-x})\,dx+\tfrac{9}{8}n\left(\frac{\Theta}{T}\right)\right]\frac{\partial \Theta}{\partial v}, \tag{137.28}$$

where, without attempting any more complete analysis, stress may be laid on the specially important term $-\partial E_0/\partial v$.

In the case of the remaining quantities we may begin by considering the low-temperature and high-temperature results separately.

At very low temperatures we must make use of (137.25). Treating Θ, which depends on the elastic properties of the crystal, as independent of the temperature, we then obtain for the *energy* of the crystal

$$E = kT^2\frac{\partial \log Z}{\partial T} = E_0 + \frac{3\pi^4}{5}\frac{nkT^4}{\Theta^3} + \tfrac{9}{8}nk\Theta, \tag{137.29}$$

where the first term is the zero-point energy, the second the energy of excitation of the modes of oscillation, and the last the residual energy in these modes due to their 'half-quanta' of energy when unexcited. For the *heat capacity* at constant volume we obtain, by differentiation,

$$C_v = \frac{12\pi^4}{5}\frac{nkT^3}{\Theta^3}, \tag{137.30}$$

thus giving the third power dependence on temperature first found by Debye. Finally, for the *entropy* we obtain

$$S = \frac{E}{T} + k\log Z = \frac{4\pi^4}{5}nk\left(\frac{T}{\Theta}\right)^3 + k\log G_0, \tag{137.31}$$

where we shall give later consideration to the important term $k\log G_0$ which depends on the degeneracy of the ground-level. We may regard these results with great confidence, since only the modes of vibration of low frequency and hence of wave-length long compared with the distances between neighbouring particles will be appreciably excited at low temperatures, and these modes may well be taken as distributed in accordance with the assumed expression (137.7).

At very high temperatures we must make use of the expression for $\log Z$ given by (137.27). Treating Θ again as independent of T, we obtain for the *energy* of the crystal

$$E = kT^2\frac{\partial \log Z}{\partial T} = E_0 + 3nkT + \tfrac{9}{8}nk\Theta. \tag{137.32}$$

By differentiation we then obtain for the *heat capacity* at constant volume

$$C_v = 3nk, \tag{137.33}$$

in agreement with the original high-temperature expression of Dulong and Petit. Finally, for the *entropy* we obtain

$$S = \frac{E}{T} + k\log Z = 4nk + 3nk\log\frac{T}{\Theta} + k\log G_0, \tag{137.34}$$

where the term $k\log G_0$ again appears. In so far as these high-temperature results depend merely on the total number of modes of oscillation, and not on the assumed distribution of frequencies which has led to terms containing Θ, we may regard them as quite precise.

At intermediate temperatures, between the very low and high ones treated above, we must make use of the full expression for $\log Z$ given by (137.23). Without presenting the details of the calculation, we then find as the expression for *energy*

$$E = E_0 + \frac{9nkT^4}{\Theta^3} \int_0^{\Theta/T} \frac{x^3}{e^x - 1}\, dx + \tfrac{9}{8}nk\Theta, \tag{137.35}$$

and as the expression for *entropy*

$$S = \frac{12nkT^3}{\Theta^3} \int_0^{\Theta/T} \frac{x^3}{e^x - 1}\, dx - 3nk\log(1 - e^{-\Theta/T}) + k\log G_0, \tag{137.36}$$

where use can now be made of existing tables† for the unevaluated integral—the so-called Debye function—which still remains. The dependence of energy on temperature given by (137.35) is actually found to follow the observations within the limit of error in the case of many crystals, provided Θ is chosen so as to secure the best fit, and some measure of agreement has also been found between such values of Θ and those calculated from the appropriate elastic constants.

It should of course be remarked, in connexion with the above results for crystals at high and intermediate temperatures, that a complete treatment would also have to give recognition to the possibility of exciting internal states of the component molecules above the ground-level when the temperature is sufficiently raised.

(*e*) **Remarks on the entropy of crystals.** The foregoing investigation is of special importance because of its bearing on the values which should be ascribed to the entropies of actual crystals as we approach the absolute zero of temperature. At low temperatures we have obtained the expression

$$S = \frac{4\pi^4}{5} \frac{nkT^3}{\Theta^3} + k\log G_0 \tag{137.37}$$

for the entropy of our simple model of a crystal. As the temperature T goes to lower and lower values, this approaches the value

$$S_{T=0} = k\log G_0, \tag{137.38}$$

where the quantity G_0 is the quantum weight or number of independent quantum states which actually correspond to the state of the crystal when its modes of oscillation have not been excited.

In the case of a crystal uniquely constructed from a single kind of

† See, for example, six-place tables by Beattie, *Journ. Math. and Phys.* (Mass. Inst. Tech.) **6**, 1 (1926).

molecule, which itself exhibits no degeneracy in the ground state, this quantum weight G_0 could be taken as unity since the crystal would approach a single definite structure as we make the temperature lower and lower. It would then be consistent to take the entropy of the crystal as approaching at low temperatures the value

$$S_{T=0} = 0, \tag{137.39}$$

in agreement with the formulation originally given to the so-called *third law of thermodynamics* by Planck.

In the case of many crystals, however, composed of a chemically pure substance, it is now quite certain that we cannot regard the crystal as approaching a condition corresponding to a single quantum mechanical state when we extrapolate from the lowest temperatures available in the laboratory. In such cases we shall have to take G_0 as greater than unity, and hence also take

$$S_{T=0} > 0, \tag{137.40}$$

if we desire a starting-point for entropy, consistent with the assignment of zero entropy to any pure state, as is used when the same substance is considered in the gaseous or other non-crystalline form, or as entering into chemical reaction. The presence of multiplicity in the ground-level approached by pure crystalline substances at low temperatures will be due to one or more of the following factors—existence of isotopes, presence of allotropic forms, randomness in the crystal structure, or unresolved degeneracy in the component molecules.

First of all we may consider the possibility that an elementary substance which we regard as pure from the ordinary chemical point of view is in reality a *mixture of isotopes*. To give a strict treatment to crystals of such substances, suitable for use when possible variations in isotopic composition are to be considered, it will then be necessary to take the quantum weight G_0 for the low-temperature state so as to correspond to the different ways in which the isotopic atoms could be arranged in the crystal lattice, since these different arrangements would correspond to macroscopically indistinguishable crystals but to microscopically distinct quantum states. The entropy of such a crystal at low temperatures would then also have to be assigned a value greater than zero, as can be evaluated by the methods to be developed in the next section where we consider mixtures in general. Nevertheless, in giving a treatment to mixtures of isotopes suitable merely for application to ordinary physical-chemical processes where changes in isotopic composition do not occur, it proves possible and convenient to introduce small approximations which then make it feasible to specify the entropies of crystals in a manner which ignores the existence of separate isotopes.†

As our second case, we may consider the possibility that an elementary substance can contain two *allotropic forms* of molecule, for example ortho- and para-hydrogen, which in the absence of a catalyst cannot change readily from the one form into the other. Under such circumstances the crystal on which we actually make low-temperature measurements will ordinarily be a mixture of the two kinds of molecule in their high-temperature ratio, instead of being composed of the single form that would be thermodynamically stable at the absolute zero. It will then again be necessary to assign a weight G_0 greater than unity and an entropy $S_{T=0}$ greater than zero to the actual low-temperature crystal, as can be calculated by the methods for treating mixtures which will be developed in the next section.

As our third case, we now come to the possibility of *randomness in the crystal structure*. This will arise when the atoms or molecules composing the crystal can actually be arranged in a variety of different ways without important disturbance in the main features of the crystal lattice. A simple example is provided by carbon monoxide in which the two atoms of the CO molecule are so similar that crystals apparently form without much regularity of orientation with respect to the two different ends of the molecule.‡ Assuming complete randomness in such a respect for a crystal of n molecules, we should evidently have

$$G_0 = 2^n, \qquad S_{T=0} = nk\log 2 \text{ or } 1{\cdot}38 \text{ calories/(mol degree)} \qquad (137.41)$$

as the contribution to low-temperature entropy. The actual measurements of the entropy differences between crystalline and gaseous carbon monoxide made by Clayton and Giauque gave a value 1·1 cal. per mol per degree for the entropy which had to be assigned to the crystal on account of such randomness, thus indicating some tendency towards regularity of arrangement. Another good example is provided by ice, where, in accordance with the considerations of Pauling,|| there appear to be several nearly equivalent arrangements for the hydrogen atoms in the oxygen lattice. Here the calculated and observed entropy due to the possibilities of random arrangement are respectively 0·805 and 0·82 calories per mol per degree, which is a very close agreement.

As our final category, we now come to the possibility of *unresolved*

† See Giauque and Overstreet, *Journ. Amer. Chem. Soc.* **54**, 1731 (1932).

‡ See Clayton and Giauque, ibid. **54**, 2610 (1932).

|| Pauling, ibid. **57**, 2680 (1935); Giauque and Stout, ibid. **58**, 1144 (1936).

degeneracy in the component molecules, which provide the similar particles out of which the crystal lattice is really built. Here, if we have a crystal composed of n molecules, which—in our actual low-temperature measurements—are themselves in a level corresponding to g_c separate states, it is evident that we must take

$$G_0 = g_c^n, \qquad S_{T=0} = nk \log g_c, \tag{137.42}$$

as giving the contribution to the low-temperature entropy of the crystal arising from that cause. Such unresolved degeneracy will occur whenever the component molecules have one or more eigenstates with energy so close to that of the very lowest that they can all be regarded as equally excited at the low temperatures actually achieved in the laboratory measurements. The examples of such a situation are of several kinds. In the case of crystals where the molecules can still rotate at the lowest temperatures, we can have—as in the example of orthohydrogen—a degeneracy of the lowest rotational state.† In the case of crystals where the molecules presumably cannot rotate, we can have—as in the example of $Gd_2(SO_4)_3 \cdot 8H_2O$—degeneracy involving electronic spin and orbital momentum.‡ Finally, in the case of crystals in general, we can have possibilities for different orientations of nuclear spin when this is present. In this latter case, however, it is often possible to ignore the presence of nuclear spin since its effects may cancel when entropies involving the same element under different circumstances are subtracted.||

It is thus evident from the foregoing that we cannot neglect the term $k \log G_0$ in the low-temperature entropy of crystals, if we wish to measure their entropies from the same starting-point, $S = 0$ for a pure state, that we find simplest in general. Indeed, we must often expect to find

$$S_{T=0} > 0 \tag{137.43}$$

if we extrapolate down from actual low-temperature measurements on the crystal. Furthermore, we must now expect to find cases where chemical reactions between crystals would be accompanied by a change in entropy as we approach $T = 0$,

$$\Delta S_{T=0} \neq 0, \tag{137.44}$$

which cannot be taken equal to zero as demanded by older formulations of the so-called third law of thermodynamics. A list containing a number of known cases, where the change in entropy for such low-

† See Giauque and Johnston, *Journ. Amer. Chem. Soc.* **50**, 3221 (1928); Pauling, *Phys. Rev.* **36**, 430 (1930).

‡ See Giauque, *Journ. Amer. Chem. Soc.* **49**, 1870 (1927).

|| See Gibson and Heitler, *Zeits. f. Phys.* **49**, 465 (1928).

temperature crystalline reactions would not be equal to zero, will be found in Fowler's *Statistical Mechanics*.†

It is often convenient to write the expression for the residual low-temperature entropy of a crystal in general in the form

$$S_{T=0} = k \log G_0 = nk \log g_c, \tag{137.45}$$

where in simple cases g_c will merely be the multiplicity of the ground-level of the component molecules. Discussion of the values of the quantity g_c for a number of specific types of crystal will also be found in Fowler's treatise.‡

138. Mixtures of substances

So far we have been interested in the thermodynamic properties of gases and crystals composed of a single pure substance, and have merely referred to mixtures in connexion with the possibilities for isotopic or allotropic forms of the same chemical substance. We must now consider the thermodynamic properties of mixtures of different substances. We shall give separate treatments to the limiting cases of mixtures in a highly rarefied gas phase where the molecules have an almost negligible action on each other, and mixtures in a highly condensed crystalline or liquid phase where the interactions between molecules become of prime importance.

(*a*) **Gaseous mixtures.** Let us consider a gaseous mixture composed of n molecules of one kind and m of another kind enclosed in a container of volume v. And let us take the dilution and temperature high enough, so that the effect of collisional interaction will not be sufficient to introduce appreciable change in the energy levels of the molecules, and so that the phenomenon of gas degeneration will not be appreciable. To treat the properties of such a mixture we must first obtain an expression for the corresponding sum-over-states,

$$Z = \sum_i e^{-E_i/kT}, \tag{138.1}$$

where the summation is over all energy eigenstates i for the system as a whole.

Following the same line of argument, developed in more detail in § 134 for systems containing only a single kind of molecule, we may then write

$$E = \epsilon_1 + \ldots + \epsilon_n + \epsilon_1' + \ldots + \epsilon_m' \tag{138.2}$$

as the expression for a possible energy of the system as a whole, where

† Fowler, loc. cit., p. 233.

‡ Fowler, loc. cit., pp. 218 ff. For the connexion between Fowler's notation and ours, see footnote † on p. 608.

$\epsilon_1 \ldots \epsilon_n$ are n possible eigenvalues of energy for a molecule of the first kind and $\epsilon'_1 \ldots \epsilon'_m$ are m possible eigenvalues for a molecule of the second kind. Considering the circumstance that a new state of the system as a whole would not be obtained by a mere permutation in which the same states of energies ϵ or ϵ' were assigned different particle indices $1,\ldots, n$ or $1,\ldots, m$, we may then write for the sum-over-states

$$Z = \sum_{\epsilon_1 \ldots \epsilon_n} \sum_{\epsilon'_1 \ldots \epsilon'_m} \frac{n_k!\, n_l! \ldots}{n!} \frac{m_r!\, m_s! \ldots}{m!} e^{-\frac{\epsilon_1+\ldots+\epsilon_n+\epsilon'_1+\ldots+\epsilon'_m}{kT}}, \tag{138.3}$$

where the symbols n_k, n_l,... and m_r, m_s,... denote the numbers of repetitions of the same states of energies ϵ or ϵ' respectively. In the absence of gas degeneration, where these numbers may be regarded as going to unity for most states of the system that are important, we then obtain

$$Z = \frac{1}{n!}\frac{1}{m!}\Big(\sum_k e^{-\epsilon_k/kT}\Big)^n \Big(\sum_r e^{\epsilon'_r/kT}\Big)^m \tag{138.4}$$

as the desired expression of the sum-over-states for a dilute, gaseous mixture of n molecules of one kind with m of another, where the summations are over all individual eigenstates k and r that would be exhibited by a single molecule of the two kinds respectively in a container of the total volume v.

By comparing this expression for the sum-over-states of the mixture with the expression given by (136.1) on which we based the treatment of a single gas, we now obtain the important conclusion that the sum-over-states for our mixture of two gases would be equal to the product of the values of that quantity that would hold separately for each of the two component gases, provided we had each gas alone in the *same volume* v as that of the mixture in order that its molecules should exhibit the same spectrum of eigenstates ϵ_k or ϵ'_r in the pure and mixed conditions, and at the *same temperature* T as that of the mixture in order to secure unchanged exponential factors. This result can be expressed in the form

$$Z(n+m, v, T) = Z(n, v, T)Z(m, v, T), \tag{138.5}$$

where the arguments of the quantities Z indicate the system for which the sum-over-states is to be considered.

It will now be convenient to take logarithms, since $\log Z$ is the quantity immediately appearing in our formulae for the evaluation of thermodynamic quantities. We can then write for a sufficiently dilute mixture of two gases

$$\log Z(n+m, v, T) = \log Z(n, v, T)+\log Z(m, v, T), \tag{138.6}$$

and this result can evidently be immediately extended to any number of components. Turning now to our general formulae (133.3) to (133.6) for the calculation of the quantities of thermodynamic interest—pressure p, energy E, heat capacity at constant volume C_v, and entropy S—we immediately see that we shall have in general for a sufficiently dilute mixture of gases the additive relations

$$\begin{aligned} p(n+m,v,T) &= p(n,v,T)+p(m,v,T),\\ E(n+m,v,T) &= E(n,v,T)+E(m,v,T),\\ C_v(n+m,v,T) &= C_v(n,v,T)+C_v(m,v,T),\\ S(n+m,v,T) &= S(n,v,T)+S(m,v,T), \end{aligned} \tag{138.7}$$

in agreement with well-known facts as to the properties of perfect gases and their mixtures. The quantities on the left-hand side of the above equations designate properties of the mixture of $n+m$ molecules of the two gases, and those on the right-hand side designate properties of the separate gases when present alone at the *same volume and temperature* as that of the whole mixture.

For some purposes it is preferable to express the thermodynamic properties of the mixture in terms of the properties of the separate gases when present alone at the *same pressure and temperature* as the mixture, rather than at the same volume and temperature. To investigate this we may now introduce the indices 1, 2, and 12 to designate the pure gases and their mixture respectively under any conditions of interest. Making use of the simple dependence of Z on v for a single gas as given by (136.5), we may then rewrite our expression (138.6) for the sum-over-states of the mixture in the somewhat generalized form

$$\begin{aligned} &\log Z_{12}(n_1+n_2, v_{12}, T)\\ &\quad = \log Z_1(n_1, v_1, T)+n_1\log\frac{v_{12}}{v_1}+\log Z_2(n_2, v_2, T)+n_2\log\frac{v_{12}}{v_2}, \end{aligned} \tag{138.8}$$

where v_{12} is the volume of the mixture, v_1 and v_2 are any volumes that we wish to choose for the separate gases, and T is the temperature for each of the three systems under consideration. By choosing the volumes v_1 and v_2, in accordance with the expressions

$$\frac{v_{12}}{v_1} = \frac{n_1+n_2}{n_1} \quad\text{and}\quad \frac{v_{12}}{v_2} = \frac{n_1+n_2}{n_2}, \tag{138.9}$$

we shall then have our separate gases present at the same pressures as that of the mixture, and can rewrite (138.8) in the desired form

$$\log Z_{12} = \log Z_1+\log Z_2+n_1\log\frac{n_1+n_2}{n_1}+n_2\log\frac{n_1+n_2}{n_2}, \tag{138.10}$$

where Z_1, Z_2, and Z_{12} now refer to the separate gases and to the mixture all at the same pressure and temperature.

With the help of our general formulae (133.3) to (133.6) for the evaluation of thermodynamic quantities, we can now write

$$p_{12} = p_1 = p_2, \qquad E_{12} = E_1+E_2, \qquad C_{v_{12}} = C_{v_1}+C_{v_2},$$
$$S_{12} = S_1+S_2+n_1 k\log\frac{n_1+n_2}{n_1}+n_2 k\log\frac{n_1+n_2}{n_2}, \tag{138.11}$$

where the subscripts 12 and 1 and 2 now refer respectively to the mixture of gases at a given pressure and temperature, and to the pure gases alone at that same pressure and temperature. It may be remarked that the last two terms in the expression for entropy, which give the so-called *entropy of mixing*, arise from the circumstance that the formula for the evaluation of entropy contains a term depending directly on $\log Z$ as well as on its derivatives.

(*b*) **Mixed crystals.** We may now turn from gaseous to crystalline mixtures and treat the *ideal limiting case*, which would be approached when the molecules of the two kinds of substance become so nearly alike as to the volume they occupy and the forces they exert on their neighbours, that they can be used indiscriminately for constructing crystals of any desired composition without changing the structure or dimensions of the crystal lattice or affecting the elastic properties of the crystal. Assuming these limiting ideal conditions, let us then consider—at a given pressure p and temperature T—the *two pure crystals* consisting of n_1 molecules of the first kind and n_2 of the second kind, and the *mixed crystal* consisting of n_1+n_2 molecules of the two kinds taken together. In accordance with the above assumptions, the volumes of the three crystals would then be additive, with

$$v_1+v_2 = v_{12}. \tag{138.12}$$

And, in accordance with the general treatment which we have given to crystals in § 137—see (137.22) and (137.23)—it is evident that the logarithm of the sum-over-states for each of the three crystals would be given by an expression of the same form

$$\log Z = -\frac{E_0}{kT}-9n\left(\frac{T}{\Theta}\right)^3\int\limits_0^{\Theta/T} x^2\log(1-e^{-x})\,dx-\tfrac{9}{8}n\left(\frac{\Theta}{T}\right)+\log G_0, \tag{138.13}$$

where E_0 is the energy assigned to the crystal in the absence of vibrational excitation, G_0 is the number of actually independent quantum states that correspond to this condition, and Θ is a parameter which has the same value for all three of our idealized crystals since it depends solely on elastic properties and lattice dimensions in accordance with

$$\Theta = \frac{hC}{k}\left(\frac{9n}{4\pi v}\right)^{\frac{1}{3}}. \tag{138.14}$$

As a consequence of the above generally applicable form (138.13) and of the additivity of the values of E_0 in the case of our idealized crystals, it is then evident that we could relate the sum-over-states Z_{12} for the mixed crystal to the corresponding quantities for the two pure crystals by

$$\log Z_{12} = \log Z_1 + \log Z_2 + \log \frac{G_{12}}{G_1 G_2}, \tag{138.15}$$

where G_{12}, G_1, and G_2 denote the numbers of independent quantum states for the three crystals when they are vibrationally unexcited. Furthermore, in typical cases we can evidently expect G_{12} to differ from the product $G_1 G_2$, only because we must take each different mode of arranging the two kinds of molecule in the lattice for the mixed crystal as corresponding, from a microscopic point of view, to a different quantum-mechanical state. We can then put

$$G_{12} = \frac{(n_1+n_2)!}{n_1!\,n_2!} G_1 G_2, \tag{138.16}$$

where the factor expresses the number of such modes of arrangement. Using the Stirling approximation for factorials, omitting negligible terms when n_1 and n_2 are large, we then have as our desired expression for the sum-over-states of the mixed crystal

$$\log Z_{12} = \log Z_1 + \log Z_2 + n_1 \log \frac{n_1+n_2}{n_1} + n_2 \log \frac{n_1+n_2}{n_2}. \tag{138.17}$$

Making use of our general formulae (133.3) to (133.6) for evaluating quantities of thermodynamic interest, we then also have for the pressure p, energy E, heat capacity at constant volume C_v, and entropy S of our mixed crystal in terms of those quantities for the two pure components

$$\begin{gathered} p_{12} = p_1 = p_2, \qquad E_{12} = E_1 + E_2, \qquad C_{v_{12}} = C_{v_1} + C_{v_2}, \\ S_{12} = S_1 + S_2 + n_1 k \log \frac{n_1+n_2}{n_1} + n_2 k \log \frac{n_1+n_2}{n_2}. \end{gathered} \tag{138.18}$$

These expressions are of just the same form as the analogous ones for mixtures of perfect gases, when the properties of the pure gases are related to volumes v_1 and v_2 such that the pressures will be the same as for the mixture as a whole.

The above treatment of mixed crystals is of course a highly idealized one since it applies under the limiting conditions which we have assumed as to negligible effects from the replacement of one kind of molecule by another in the crystal structure. Nevertheless, it provides a good starting-point from which deviations may be considered. The ideal conditions would be very nearly approached in the case of mixtures of isotopes. The corresponding treatment, which was given to gaseous mixtures, has a wider validity since it should apply in general when the gases are sufficiently dilute.

(*c*) **Liquid mixtures.** Some progress along the lines of the above methods can also be made in understanding the properties of liquid solutions. For this purpose we may consider the *ideal limiting case* of two liquids, which are composed of molecules exerting sufficiently similar forces so that there would be no change in total volume on mixing at a given pressure p and temperature T, and so that a molecule in the mixture would have the same forces exerted on it by its neighbours as in the pure liquid form.

For the volume v_{12} of a mixture of n_1 and n_2 molecules of the two kinds we could then write

$$v_{12} = v_1+v_2, \tag{138.19}$$

where v_1 and v_2 are the volumes of the pure component liquids at the given pressure and temperature.

Furthermore, in accordance with our assumptions, we could regard the energy spectra of the individual molecules of the two kinds as unaltered on making the mixture, since in the case of a condensed liquid phase the energy levels of a molecule could depend on environment but only very slightly on total volume. Following the methods used in § 134 for calculating the sum-over-states as dependent on molecular states, assuming temperatures sufficient so that symmetry restrictions can be neglected, we should then be led to take

$$\begin{aligned} Z_1 &= \frac{1}{n_1!}\Big(\sum_{k_1} e^{-\epsilon_{k_1}/kT}\Big)^{n_1}, \\ Z_2 &= \frac{1}{n_2!}\Big(\sum_{k_2} e^{-\epsilon_{k_2}/kT}\Big)^{n_2}, \qquad (138.20) \\ Z_{12} &= \frac{(n_1+n_2)!}{n_1!\,n_2!}\,Z_1\,Z_2, \end{aligned}$$

where k_1 and k_2 denote the energy eigenstates for a single molecule of the two kinds, as giving fair first approximations for the sum-over-states of the two pure liquids and of the mixture.

The factor, by which $Z_1 Z_2$ is multiplied in the third of these expressions can be regarded as roughly justified if we assume the approximate validity of treating our mixture as corresponding to permanent spatial arrangements of the two kinds of molecules, which could be transformed into each other by the interchange of unlike molecules. The factor would then give the number of different quantum mechanical states for the system as a whole that would result from such interchanges.

This approximate picture, by neglecting the effect of total volume on molecular energy levels, and neglecting the interchanges of unlike molecules which could occur by diffusion, provides a rough treatment for mixed liquids similar to the treatment already given to mixed crystals rather than to that given to mixed gases. For a more correct picture, modifications in the direction of gas-like behaviour might be introduced. Nevertheless, since the two limiting cases of permanent crystals and dilute gases have both led to the same dependence of Z_{12} on Z_1 and Z_2 which we have assumed above, we may regard this expression as fairly satisfactory.

Our approximate picture has also given no consideration to the effect of the relative volumes of the two kinds of molecules in the mixed liquid on the number of interchanges that would be possible. If these volumes were equal, the factor $(n_1+n_2)!/n_1!\,n_2!$ under discussion would seem correct. On the other hand, it might still be a good first approximation in a variety of cases. For example, as suggested by a conversation with Professor J. T. Hildebrand, a mixture of two kinds of straight chain organic molecules, differing greatly in length but having similar chemical character along the chain and at the ends, might provide the conditions necessary for the approximate validity of the above equations. If such molecules were arranged end to end in filaments which were then packed parallel in the liquid, each molecule in the mixture would find itself in a similar environment as in the pure liquid, and the total number of arrangements would be given by the above factor. The actual structure of the liquid mixture might tend to approximate these conditions.

Assuming the validity of (138.20) as a good first approximation, we then obtain

$$\log Z_{12} = \log Z_1+\log Z_2+n_1\log\frac{n_1+n_2}{n_1}+n_2\log\frac{n_1+n_2}{n_2}, \qquad (138.21)$$

and are led to thermodynamic expressions similar to those in the case of perfect gases and idealized crystals,

$$p_{12} = p_1 = p_2, \qquad E_{12} = E_1+E_2, \qquad C_{v_{12}} = C_{v_1}+C_{v_2},$$
$$S_{12} = S_1+S_2+n_1 k \log\frac{n_1+n_2}{n_1}+n_2 k \log\frac{n_1+n_2}{n_2}, \tag{138.22}$$

for the values of the pressure, energy, heat capacity, and entropy of the mixture in terms of those same quantities for the pure liquids.

(*d*) **On the definition of the ideal solution.** In view of the similar character of the results which we have obtained in all three above cases of idealized gaseous, crystalline, and liquid mixtures, we are now in a position to propose a general statistical-mechanical definition of the *ideal or perfect solution*, which can be regarded as a standard of reference in treating the properties of actual solutions. Using the indices 1, 2, and 12 to denote in general the two pure phases consisting of n_1 and n_2 molecules of the two kinds and the similar mixed phase made by their combination, and taking conditions such that the pressures and temperatures satisfy

$$p_{12} = p_1 = p_2, \qquad T_{12} = T_1 = T_2, \tag{138.23}$$

we may define the ideal solution as having a volume which satisfies the additive relation

$$v_{12} = v_1+v_2, \tag{138.24}$$

and as having a sum-over-states which satisfies the relation

$$\log Z_{12} = \log Z_1+\log Z_2+n_1 \log\frac{n_1+n_2}{n_1}+n_2 \log\frac{n_1+n_2}{n_2}. \tag{138.25}$$

In accordance with our general formulae (133.3) to (133.6) for the evaluation of thermodynamic quantities, we can then also show that the energy, heat capacity at constant volume, and entropy of our ideal solution would be given by

$$E_{12} = E_1+E_2, \qquad C_{v_{12}} = C_{v_1}+C_{v_2},$$
$$S_{12} = S_1+S_2+n_1 k \log\frac{n_1+n_2}{n_1}+n_2 k \log\frac{n_1+n_2}{n_2}. \tag{138.26}$$

The foregoing relations may be readily extended to any number of components.

It will be of interest to compare the above statistical-mechanical definition of the ideal solution with the thermodynamic definition proposed by Lewis,† which states that the fugacity of each component of the solution would obey the generalized Raoult law

$$f = xf^0, \tag{138.27}$$

† Lewis, *Journ. Amer. Chem. Soc.* **30**, 668 (1908).

where f and f^0 are the fugacities of the component in the mixture and in the pure state respectively, and x is its molal fraction in the mixture.

To show that this relation would also be satisfied if we take the proposed statistical-mechanical definition of a perfect solution, it will first be necessary to recall the following definitions of thermodynamic quantities. For the *thermodynamic potential* F of a system at a given pressure p and temperature T, we have

$$F = E+pv-TS. \tag{138.28}$$

For the *partial molal potential* $\bar{F}$ of any component of the system—say the component 1—we then have

$$\bar{F}_1 = \left(\frac{\partial F}{\partial N_1}\right)_{p,T,N_2,\dots}, \tag{138.29}$$

where the number of mols of the different substances composing the system are denoted by N_1, N_2, etc. And for the *fugacity* f of that component we then have by definition the relation

$$RT\log f_1 = \bar{F}_1+\text{const.} \tag{138.30}$$

Turning now, however, to the proposed statistical-mechanical definition of a perfect solution and its thermodynamic consequences given by (138.26), we can evidently write for the thermodynamic potential F_{12} of a binary solution

$$F_{12} = F_1+F_2-N_1\,RT\log\frac{N_1+N_2}{N_1}-N_2\,RT\log\frac{N_1+N_2}{N_2}, \tag{138.31}$$

where F_1 and F_2 are the potentials of the two components in pure form. Differentiating with respect to N_1, we then obtain

$$\begin{aligned}\bar{F}_1 &= \bar{F}_1^0-RT\log\frac{N_1+N_2}{N_1}-\frac{N_1}{N_1+N_2}RT+\frac{N_1}{N_1}RT-\frac{N_2}{N_1+N_2}RT\\ &= \bar{F}_1^0+RT\log\frac{N_1}{N_1+N_2}\end{aligned} \tag{138.32}$$

as a relation which connects the value of partial molal potential $\bar{F}_1$ for the indicated component in solution with its value $\bar{F}_1^0$ in the pure state. And substituting (138.30), we then obtain

$$f_1 = \frac{N_1}{N_1+N_2}f_1^0 = x_1 f_1^0 \tag{138.33}$$

for the fugacities f_1 and f_1^0 of that component in the solution and pure state respectively, and relations of the same form could evidently be derived for each constituent in the case of a perfect solution composed of any number of components.

We thus see that our statistical-mechanical definition of a perfect or ideal solution will lead to the same results as the earlier thermodynamic definition. The statistical-mechanical viewpoint may have some advantages, however, in making it easier to see under what conditions we may expect the properties of actual solutions to agree with or deviate from those of the ideal solution.

139. Vapour pressures and chemical equilibria

(*a*) **The thermodynamic potentials of crystals and of gases.** We are now ready to use some of the results which have been obtained in this chapter to illustrate the methods of treating physical-chemical equilibria. We shall give brief consideration first to equilibria between crystals and their vapours, and then to chemical equilibria between reacting gases.

In the thermodynamic treatment of physical-chemical equilibria it is often most convenient to consider the *thermodynamic potential* F of the system of interest.† This quantity may be defined in terms of the energy E, pressure p, volume v, temperature T, and entropy S of the system under consideration by the equation

$$F = E+pv-TS. \tag{139.1}$$

And the condition for thermodynamic equilibrium at a given pressure and temperature will then be given by the expression

$$\delta F = 0 \tag{139.2}$$

for any variation in the condition of the system, holding the pressure and temperature constant. The above relations (139.1) and (139.2) are themselves purely thermodynamic. Nevertheless, if we substitute for the quantities composing F values obtained from statistical-mechanical theory, rather than the empirical values which would have to be used in a purely thermodynamic treatment, we shall reap the full reward of our more powerful statistical-mechanical methods.

For the purposes of the present section we shall be interested in the thermodynamic potentials of crystals at temperatures low enough so that we can apply the *low-temperature* formulae for energy and entropy obtained in § 137 (*d*). We can use those formulae with great confidence since their derivation is relatively independent of the assumed form of distribution for the frequencies of modes of vibration, except in the region of wave-lengths—long compared with intermolecular distances—

† This quantity is sometimes called 'free energy' by chemists, although it differs by the term pv from the quantity originally given that name by Helmholtz.

where the form should be valid. Neglecting in the expression for F given by (139.1) the small term pv, which would be inappreciable for a condensed phase at the pressures that will interest us, we can then write—in accordance with our previous expressions (137.29) and (137.31) for the low-temperature energy and entropy of a crystal—

$$F = E_0+\tfrac{9}{8}nk\Theta-\frac{\pi^4}{5}nk\frac{T^4}{\Theta^3}-kT\log G_0 \tag{139.3}$$

as an expression for the thermodynamic potential of a crystal of n molecules at low temperature. Or putting

$$n\epsilon_c = E_0+\tfrac{9}{8}nk\Theta \tag{139.4}$$

as a convenient expression for the total energy of the unexcited crystal including its residual half-quanta of vibrational energy, and introducing in accordance with (137.45) the more convenient expression,

$$k\log G_0 = nk\log g_c, \tag{139.5}$$

for the effect of the multiplicity of the ground state of the crystal, we can write as the desired expression for the *thermodynamic potential of a crystal at low temperatures*

$$F = n\epsilon_c-\frac{\pi^4}{5}nk\frac{T^4}{\Theta^3}-nkT\log g_c, \tag{139.6}$$

where the characteristic temperature Θ, which depends on the elastic properties of the crystal, can be safely regarded as a constant at the temperatures and pressures that will interest us.

Turning next to monatomic gases, we shall be interested in temperatures high enough so that gas degeneration can be neglected. Setting $pv = nkT$ in (139.1), we can then write, in accordance with the expression connecting entropy and energy given by (136.18), for the *thermodynamic potential of a monatomic gas*, consisting of n molecules at pressure p and temperature T,

$$F = n\epsilon_0-\tfrac{5}{2}nkT\log T+nkT\log p-nkT\log\sum_i\frac{g_i}{g_0}e^{-\frac{(\epsilon_i-\epsilon_0)}{kT}}-$$
$$-nkT\log\frac{(2\pi m)^{\frac{3}{2}}k^{\frac{5}{2}}g_0}{h^3}, \tag{139.7}$$

where m is the mass of a single atom, ϵ_0 and g_0 are the energy and the quantum weight of the ground state of the atom, and the term involving summation is over all appreciably excited electronic states i.

Turning finally to diatomic gases, we shall be interested in temperatures high enough so that gas degeneration can be neglected, and so that the rotational energy of the molecules can be regarded as fully excited. Again setting $pv = nkT$ in (139.1), we can then write, in accordance with the expression connecting entropy and energy given by (136.31), for the *thermodynamic potential of a diatomic gas*

$$F = n\epsilon_0 - \tfrac{7}{2}nkT\log T + nkT\log p - nkT\log\sum_{v,e}(g_{ve}/g_0)e^{-\epsilon_{ve}/kT} - nkT\log(2\pi m)^{\frac{3}{2}}k^{\frac{7}{2}}\frac{8\pi^2 I}{h^5}\frac{g_0}{\sigma}, \quad (139.8)$$

where the newly appearing quantities I and σ are the moment of inertia and symmetry number of the molecule, and the term involving summation is now over all appreciably excited vibrational and electronic states ve.

(*b*) **Vapour pressures of crystals.** We can now use the foregoing expressions for thermodynamic potentials to obtain formulae for the vapour pressures of crystals, over a temperature range where the crystals can be treated by low-temperature methods and their vapours can be regarded as perfect gases with translational and rotational energy fully excited. Such a range exists for all substances except hydrogen, where the full excitation of rotational levels only occurs at somewhat high temperatures owing to the low moment of inertia of the molecule.

Noting the condition for equilibrium given by (139.2), it is evident that a crystal and its vapour would be in equilibrium when we have pressures and temperatures such that the thermodynamic potentials would be equal for the same amount of substance in the crystalline and in the gaseous form. Hence, by equating (139.6) first with (139.7) and then with (139.8), and solving for p, we can obtain expressions for the equilibrium or vapour pressure in the respective cases where the substance is monatomic or diatomic in its gaseous phase. Doing so, we then obtain, after some rearrangement, for the *vapour pressure* of a crystal which gives a *monatomic gas*

$$\log p = -\frac{(\epsilon_0-\epsilon_c)}{kT} + \tfrac{5}{2}\log T + \log\sum_i \frac{g_i}{g_0}e^{-\frac{(\epsilon_i-\epsilon_0)}{kT}} - \frac{\pi^4}{5}\left(\frac{T}{\Theta}\right)^3 + \log\frac{(2\pi m)^{\frac{3}{2}}k^{\frac{5}{2}}}{h^3}\frac{g_0}{g_c}, \quad (139.9)$$

and obtain for the *vapour pressure* of a crystal which gives a *diatomic gas*

$$\log p = -\frac{(\epsilon_0-\epsilon_c)}{kT} + \tfrac{7}{2}\log T + \log\sum_{v,e}\frac{g_{ve}}{g_0}e^{-\epsilon_{ve}/kT} - \frac{\pi^4}{5}\left(\frac{T}{\Theta}\right)^3 + \log(2\pi m)^{\frac{3}{2}}k^{\frac{7}{2}}\frac{8\pi^2 I}{h^5}\frac{g_0}{\sigma g_c}, \quad (139.10)$$

where the significance of the various quantities involved has already been noted.

These formulae apply of course in the temperature range which was specified above. With the help of the well-known purely thermodynamic equation of Clapeyron for the change in vapour pressure with temperature, however, we can of course calculate the vapour pressure at any desired temperature if we know it at some one temperature. This equation for the dependence of vapour pressure on temperature may be written in the closely approximate form

$$\frac{d\log p}{dT} = \frac{\Lambda}{RT^2}, \tag{139.11}$$

where Λ is the heat of evaporation per mol at the temperature in question. As an expression for this heat of evaporation as a function of temperature we can write

$$\Lambda = \Lambda_0 + \int\limits_0^T \Delta C_p\, dT, \tag{139.12}$$

where Λ_0 is the heat of evaporation approached at very low temperatures, and ΔC_p is the difference in heat capacities at constant pressure between a mol of the substance in gaseous and condensed form. Furthermore, we can also put

$$\Delta C_p = \left.\begin{matrix} 5/2 \\ 7/2 \end{matrix}\right\} R + \Delta C_p', \tag{139.13}$$

where the first term gives the constant part of this difference for the two cases of monatomic and diatomic gases respectively, and the second term gives the part which varies with temperature. This then gives us

$$\Lambda = \Lambda_0 + \left.\begin{matrix} 5/2 \\ 7/2 \end{matrix}\right\} RT + \int\limits_0^T \Delta C_p'\, dT. \tag{139.14}$$

Substituting (139.14) in (139.11), and integrating, we then obtain

$$\log\frac{p_2}{p_1} = \frac{\Lambda_0}{RT_1} - \frac{\Lambda_0}{RT_2} + \left.\begin{matrix} 5/2 \\ 7/2 \end{matrix}\right\} \log\frac{T_2}{T_1} + \int\limits_{T_1}^{T_2} \frac{1}{RT'^2}\left[\int\limits_0^{T'} \Delta C_p'\, dT''\right] dT' \tag{139.15}$$

as the desired relation between the vapour pressures p_1 and p_2 at temperatures T_1 and T_2. This expression is valid over any temperature range where the vapour can be treated as a perfect gas with a volume very large compared with that of the condensed phase; it can also be used over ranges where there is a transition in state of the condensed phase—say from crystalline to liquid form—provided we give appropriate treatment to the heat of transition.

In order to combine the consequences of (139.15) with our previous expressions (139.9) and (139.10) for vapour pressures in a low-temperature range we must note in the first place that we can obviously take

$$(\epsilon_0-\epsilon_c)/k = \Lambda_0/R. \tag{139.16}$$

We must also note, in the second place, that the two terms which appear next to the last in the expressions mentioned would go to zero at sufficiently low temperatures. It will then be readily seen that we can write as quite general expressions for the *vapour pressure of a monatomic gas*

$$\log p = -\frac{\Lambda_0}{RT}+\tfrac{5}{2}\log T+\int\limits_0^T \frac{1}{RT'^2}\left[\int\limits_0^{T'} \Delta C_p'\,dT''\right]dT' +i_1, \tag{139.17}$$

and for the *vapour pressure of a diatomic gas*

$$\log p = -\frac{\Lambda_0}{RT}+\tfrac{7}{2}\log T+\int\limits_0^T \frac{1}{RT'^2}\left[\int\limits_0^{T'} \Delta C_p'\,dT''\right]dT' +i_2, \tag{139.18}$$

where the *vapour-pressure constants* i_1 and i_2 are given by

$$i_1 = \log\frac{(2\pi m)^{\frac{3}{2}}k^{\frac{5}{2}}}{h^3}\frac{g_0}{g_c} \tag{139.19}$$

and

$$i_2 = \log(2\pi m)^{\frac{3}{2}}k^{\frac{7}{2}}\frac{8\pi^2 I}{h^5}\frac{g_0}{\sigma g_c}. \tag{139.20}$$

A satisfactory comparison between these theoretical values of the vapour-pressure constants i_1 and i_2 and the empirical values for a considerable number of monatomic and diatomic vapours will be found in Fowler's *Statistical Mechanics*.† We do not attempt any corresponding general treatment for polyatomic vapours, since the complexities of polyatomic molecules make specific treatment in each particular case advisable.‡

† Fowler, *Statistical Mechanics*, second edition, Cambridge, 1936, pp. 218–25. The notation used by Fowler may be connected with that used above by the following relations:

For a crystal $\omega_0 = g_c$.
For a monatomic gas

$$f_v(T) = \sum_i g_i e^{-\frac{(\epsilon_i-\epsilon_0)}{kT}}, \qquad f_v(0) = \rho_0 = g_0.$$

For a diatomic gas

$$n_v(T) = \sum_{v,e} g_{ve}\,e^{-\epsilon_{ve}/kT}, \qquad n_v(0) = \nu_0 = g_0, \qquad A = I, \qquad \sigma = \sigma.$$

‡ See Sterne, *Phys. Rev.* **39**, 993 (1932), and **42**, 556 (1932), for treatments of NH_3 and CH_4.

(*c*) **Chemical equilibria in gases.** We may also investigate the conditions for chemical equilibrium in reacting gases with the help of the expressions for the thermodynamic potentials of gases which are provided by statistical-mechanical considerations. Let us consider a mixture of gases A, B, C, D, etc., at some given pressure p and temperature T, and let a chemical reaction of the general type

$$aA+bB+... = cC+dD+... \tag{139.21}$$

be possible, where a mols of A react with b mols of B, etc., to give c mols of C plus d mols of D, etc. In accordance with our general criterion for physical-chemical equilibria (139.2), we shall then have equilibrium with respect to this chemical reaction if the gases involved are present at such partial pressures that there would be no change in the thermodynamic potential of the system if the reaction in question should take place to an infinitesimal extent in either direction. Noting, for the case of perfect gases, that the thermodynamic potential for the system as a whole can be taken as the sum of expressions for the thermodynamic potentials of the individual gases at pressures equal to their partial pressures in the mixture, we can then write the condition for equilibrium in the form

$$aF_A(n,p_A,T)+bF_B(n,p_B,T)+... = cF_C(n,p_C,T)+dF_D(n,p_D,T)+..., \tag{139.22}$$

where the symbols F_A, F_B, F_C, F_D, etc., denote the thermodynamic potentials of the same number of molecules n of the different gases at the temperature of the mixture T, and at pressures p_A, p_B, p_C, p_D, etc., which are equal to the partial pressures of the different gases in the mixture.

With the help of this expression of the thermodynamic conditions for equilibrium, we can now apply statistical mechanics by substituting for the individual potentials F_A, F_B, F_C, F_D, etc., the values computed for them with the help of statistical mechanics. And since these potentials are themselves dependent on the partial pressures of the gases involved, we can then study what combinations of such pressures would correspond to equilibrium. We are thus provided with a general method for applying statistical mechanics to the treatment of chemical equilibria in gases.

We may illustrate this general method by considering the two different types of gaseous dissociation which would be expressed by the chemical reactions

$$AB = A+B \tag{139.23}$$

and

$$A_2 = 2A, \tag{139.24}$$

where diatomic molecules consisting respectively of two different or two similar atoms decompose.

For the first of these reactions we can readily obtain, after some rearrangement and cancellation, with the help of the condition for equilibrium given by (139.22) and the expressions for the thermodynamic potentials of monatomic and diatomic gases given by (139.7) and (139.8),

$$\log K_p = \log\frac{p_A p_B}{p_{AB}}$$
$$= -\frac{(\epsilon_A+\epsilon_B-\epsilon_{AB})}{kT}+\tfrac{3}{2}\log T+\log\sum\frac{g_A}{g_A^0}e^{-\epsilon_A/kT}+\log\sum\frac{g_B}{g_B^0}e^{-\epsilon_B/kT}-$$
$$-\log\sum\frac{g_{AB}}{g_{AB}^0}e^{-\epsilon_{AB}/kT}+\log\left(\frac{2\pi m_A m_B}{m_{AB}}\right)^{\frac{3}{2}}k^{\frac{5}{2}}\frac{1}{8\pi^2 I_{AB} h}\frac{g_A^0 g_B^0}{g_{AB}^0} \quad (139.25)$$

as an expression for the equilibrium constant K_p which gives the relation between equilibrium values of the partial pressures of the three gases. Similarly, for the second of the above reactions we obtain

$$\log K_p = \log\frac{p_A^2}{p_{A_2}} = -\frac{(2\epsilon_A-\epsilon_{A_2})}{kT}+\tfrac{3}{2}\log T+2\log\sum\frac{g_A}{g_A^0}e^{-\epsilon_A/kT}-$$
$$-\log\sum\frac{g_{A_2}}{g_{A_2}^0}e^{-\epsilon_{A_2}/kT}+\log\left(\frac{2\pi m_A^2}{m_{A_2}}\right)^{\frac{3}{2}}k^{\frac{5}{2}}\frac{1}{8\pi^2 I_{A_2} h}\frac{(g_A^0)^2}{g_{A_2}^0}+\log 2, \quad (139.26)$$

where the additional term $\log 2$ arises from the symmetry number $\sigma = 2$ for the molecule A_2. The symbolism in the above expressions will be self-explanatory when compared with the symbolism in the original formulae (139.7) and (139.8) for the thermodynamic potentials of monatomic and diatomic gases.

The expressions make the equilibrium constant for the two kinds of dissociation reaction dependent, at a given temperature, on the energy of dissociation of the molecule in its ground state into atoms in their ground states, the spectra of energy levels for the molecule and atoms, the molecular and atomic masses, the moment of inertia and symmetry number of the molecule, and the quantum weights of the ground states of vibrational and electronic excitation. The symmetry number 2 appears in the equilibrium expression for the case of like atoms in the manner that would be expected from a consideration of the possible frequencies of dissociation and rècombination by collision.

Expressions for equilibrium constants such as the two given above can also be rewritten, if desired, in a form containing empirical heat capacities of the reacting gases instead of summations over the internal energy levels of the atoms and molecules involved. To do this we may make use of the purely thermodynamic equation of van't Hoff for the dependence of equilibrium constants on temperature. In the case of a reaction between perfect gases as given by

$$aA+bB+... = cC+dD+... \tag{139.27}$$

the van't Hoff equation can be written in the form

$$\frac{d\log K_p}{dT} = \frac{d}{dT}\log\frac{p_C^c p_D^d \cdots}{p_A^a p_B^b \cdots} = \frac{\Delta H}{RT^2}, \tag{139.28}$$

where ΔH is the change in 'heat content' $(E+pv)$ when the reaction as written takes place. As an expression for this change in heat content as a function of temperature we can write

$$\Delta H = \Delta H_0 + \int_0^T \Delta C_p\, dT = \Delta H_0 + \Delta C_{p_0} T + \int_0^T \Delta C_p'\, dT, \tag{139.29}$$

where ΔH_0 is the change in heat content approached at very low temperatures, ΔC_p is the change in heat capacity when the reaction as written takes place, and ΔC_{p_0} and $\Delta C_p'$ are introduced as symbols to denote respectively the constant part of this change in heat capacity at temperatures where translation and rotation are fully excited, and the variable part which changes with temperature as vibrational and electronic energies become excited. Substituting (139.29) into (139.28), and integrating between T_1 and T_2, we obtain for the ratio of equilibrium constants at these two temperatures

$$\log\frac{K_2}{K_1} = \frac{\Delta H_0}{RT_1} - \frac{\Delta H_0}{RT_2} + \frac{\Delta C_{p_0}}{R}\log\frac{T_2}{T_1} + \int_{T_1}^{T_2} \frac{1}{RT'^2}\left[\int_0^{T'} \Delta C_p'\, dT''\right] dT'. \tag{139.30}$$

We may now compare this purely thermodynamic expression with the statistical-mechanical expressions given by (139.25) and (139.26). Noting that ΔH_0 would be directly related to the reaction energy for molecules in their ground-levels, noting that $\Delta C_{p_0}/R$ would be a more general expression for the coefficient of $\log T$ previously appearing, and noting that the terms in the statistical-mechanical expressions involving summations over internal levels would tend to zero at low temperatures, we may now conclude, even for more general cases than have been illustrated, that the equilibrium conditions for a gas reaction could be expressed in the form

$$\log K_p = -\frac{\Delta H_0}{RT} + \frac{\Delta C_{p_0}}{R}\log T + \int_0^T \frac{1}{RT'^2}\left[\int_0^{T'} \Delta C_p'\, dT''\right] dT' + C, \tag{139.31}$$

where C is a constant which can be evaluated from a knowledge of the masses, moments of inertia, symmetry numbers, and quantum weights of the ground-levels for the molecules involved in the reaction.

In the case of a number of gas reactions involving monatomic and diatomic molecules, comparisons showing the agreement of the theoretical and empirical values of the constant C will be found in Fowler's *Statistical Mechanics*.† We thus obtain a finally satisfactory outcome for that long development in the theory of physical-chemical equilibria which has gradually shown us how to supplement the mere requirements of the first and second laws of thermodynamics by additional considerations. With this development we may always associate the names—among others—of Nernst, Planck, Sackur and Tetrode, Ehrenfest and Trkal, Lewis, Giauque, and Fowler.

(*d*) **On the status of the so-called third law of thermodynamics.** We may now conclude this section with a few remarks as to the present status of that form of supplementary principle to which the name of third law of thermodynamics has become attached. During the past three decades various attempts have been made to formulate a principle applying to the entropies of chemical substances which would provide consistent entropy zero-points for a given substance when considered in different states of aggregation or of chemical combination. A common formulation has been the statement that all pure crystalline substances, including different crystalline forms of the same substance and including compounds as well as elements, could be assigned zero entropy at the absolute zero of temperature. As a consequence of this assumption it then became possible, by combining with data as to heat capacities and heats of reaction, to determine the conditions for chemical equilibrium at any desired temperature.

In accordance with our previous discussions, see in particular § 137 (*e*), it is now evident, nevertheless, that it would not be correct to assign zero entropy in general to all actual crystals in the states which they approach as we go to the lowest temperatures available in the laboratory. Indeed we have found rather, with the help of the statistical-mechanical interpretation of entropy, that a rational and actually correct and consistent assignment of entropy zero-points can be most easily obtained by taking the entropy of any system zero when it is known to be in a single pure quantum mechanical state. This, as we have seen, makes the low-temperature entropy of crystals equal to

$$S_{T=0} = k \log G_0 = nk \log g_c, \tag{139.32}$$

† Fowler, loc. cit., p. 228.

where $G_0 = g_c^n$ is the quantum mechanical weight for the condition of the crystal actually approached at low temperatures. And in agreement with this we do have definitely known cases, as already mentioned in § 137 (e), where the change in entropy accompanying chemical reaction between crystals would not approach zero at low temperatures. The frequent validity or approximate validity of the so-called third law must now be regarded as due to the fact that the residual entropies of crystals as given by (139.32) do tend to be small and, in addition, to cancel or partially cancel when the entropies of the same substances in different crystalline forms are subtracted.

In view of the theoretical soundness and simplicity of assigning zero-points in such a way that the entropy for any system in a single pure state will be zero, and in view of the fact that this has been shown to give correct evaluations of the constants in the equations for gaseous equilibria, we may now regard the so-called third law as quite properly replaced by this assignment of zero entropy to any system in a single pure state. We thus obtain what appears to be an entirely satisfactory principle.

If actual crystals, when taken to temperatures far below those approached in the laboratory and held for an infinite length of time, should of themselves go over into single pure states, our principle would of course also agree with the assignment of zero entropy to those crystalline forms. Such an alternative statement of the principle—if valid—could in any case, however, be of only hypothetical interest, when we think of the necessities for isotopes to diffuse out into separate crystals, allotropic modifications to change into the most stable form, randomly oriented molecules to line themselves parallel, and all internal degeneracies of the molecules to become resolved, in order that such hypothetical crystalline forms should be realized. Hence we shall do well to refrain from any formulation of the principle other than that of assigning zero entropy to any system in a pure state, as has been justified by the considerations of statistical mechanics.

140. Equilibrium between connected systems

(*a*) **Thermodynamic relations.** Up to this point the thermodynamic formalism considered has been specially adapted to the treatment of equilibrium in so-called 'closed' systems, having a composition which is assumed not to be altered by transfer of matter to or from the outside. Such a formalism is perhaps sufficient for the treatment of any kind of problem that could arise, since it would always be possible to regard the boundary separating a system of interest from its surroundings as taken large enough to include any regions to or from which transfer might occur. Nevertheless, when we are specially interested in the principles governing equilibrium with respect to the transfer of matter, it proves convenient to make use of a thermodynamic formalism specially adapted to the treatment of so-called 'open' systems, where explicit recognition is given to the possibility of changing the composition of a system of interest by the introduction or withdrawal of matter.

To complete our account of the applications of statistical mechanics to thermodynamics, we must now give brief consideration to the statistical-mechanical explanation of the principles governing equilibrium, as to the transfer of matter as well as energy, when a system of interest is placed in contact with some other system to or from which substances of one or more kinds might be transferred. This will then be found to give us a statistical-mechanical justification for the thermodynamic formalism ordinarily used in treating 'open' systems, and will also be found to give us a justification for the thermodynamic principle that entropy is a quantity having extensive magnitude, so that the entropy of a homogeneous substance is to be taken proportional to the amount of the substance considered.

It will prove advantageous if we begin by considering the thermodynamic formalism appropriate for the treatment of 'open' systems. This formalism was first obtained by Gibbs† in his memoir 'On the Equilibrium of Heterogeneous Substances', where special attention was given to the circumstance that the different systems or regions between which equilibrium is considered might themselves be different homogeneous phases of the same substance or substances.

Let us consider a system composed of h *independent kinds of substances or components*, out of which any further substances present can be regarded as formed—by chemical reaction if necessary—and let us take the condition of the system at equilibrium as specified by its energy E, by the values $a_1, a_2, a_3,...$ of any external coordinates such as volume which have to be considered, and by the number of mols $N_1 ... N_h$ for the h different components which the system contains. Treating these quantities as independent variables, we can then take the change δS in the entropy of the system, when we make a small change from a given equilibrium condition to a neighbouring one, as given in accordance with the principles of the calculus by

$$\delta S = \frac{\partial S}{\partial E}\,\delta E + \frac{\partial S}{\partial a_1}\,\delta a_1 + \frac{\partial S}{\partial a_2}\,\delta a_2 + ... + \frac{\partial S}{\partial N_1}\,\delta N_1 + ... + \frac{\partial S}{\partial N_h}\,\delta N_h. \quad (140.1)$$

† Gibbs, *Trans. Conn. Acad.* III (1875–78).

And, in accordance with the relation of change in entropy to the heat reversibly absorbed by a system of constant composition, we can re-write this in the form

$$\delta S = \frac{1}{T}\delta E + \frac{1}{T}(A_1\,\delta a_1 + A_2\,\delta a_2 + \ldots) + \frac{\partial S}{\partial N_1}\delta N_1 + \ldots + \frac{\partial S}{\partial N_h}\delta N_h, \quad (140.2)$$

where T is the temperature of the system and $A_1, A_2, A_3, \ldots$ are the generalized external forces corresponding to the coordinates $a_1, a_2, a_3, \ldots$. In the common case, where the only external coordinate that has to be considered is the volume of the system v, and the corresponding external force is the pressure p, this expression can be rewritten in the less general form

$$\delta S = \frac{1}{T}\delta E + \frac{p}{T}\delta v + \frac{\partial S}{\partial N_1}\delta N_1 + \ldots + \frac{\partial S}{\partial N_h}\delta N_h. \quad (140.3)$$

With the help of the foregoing it is now easy to investigate the conditions for equilibrium when systems of the above kind are connected with each other so as to form a combined system. Designating quantities applying to the individual systems forming the combination by different numbers of accents, and for simplicity taking a case where volume is the only external coordinate involved, we can evidently write

$$\begin{aligned} \delta S = {} & \frac{1}{T'}\delta E' + \frac{p'}{T'}\delta v' + \frac{\partial S'}{\partial N_1'}\delta N_1' + \ldots + \frac{\partial S'}{\partial N_h'}\delta N_h' + \\ & + \frac{1}{T''}\delta E'' + \frac{p''}{T''}\delta v'' + \frac{\partial S''}{\partial N_1''}\delta N_1'' + \ldots + \frac{\partial S''}{\partial N_h''}\delta N_h'' + \\ & + \text{etc.} \end{aligned} \quad (140.4)$$

as an expression for the dependence of the entropy S of the whole combination on the variables characterizing its parts. In accordance with the second law of thermodynamics, however, if the energy and volume of the whole combination are held constant, the condition of equilibrium would be characterized by an adjustment of the entropy of the combination to its maximum possible value. This adjustment would of course have to be carried out with constant values for the total amounts of the various constituents present. Hence the condition of equilibrium for the combined system can now be obtained by setting the above expression for δS equal to zero, under the subsidiary conditions

$$\begin{aligned} \delta E' + \delta E'' + \delta E''' + \ldots &= 0 \\ \delta v' + \delta v'' + \delta v''' + \ldots &= 0 \\ \delta N_1' + \delta N_1'' + \delta N_1''' + \ldots &= 0 \\ \cdots\cdots\cdots\cdots & \\ \cdots\cdots\cdots\cdots & \\ \delta N_h' + \delta N_h'' + \delta N_h''' + \ldots &= 0. \end{aligned} \quad (140.5)$$

This is then immediately seen to lead to

$$T' = T'' = T''' = ..., \tag{140.6}$$

$$p' = p'' = p''' = ..., \tag{140.7}$$

$$\begin{array}{c} \frac{\partial S_1'}{\partial N_1'} = \frac{\partial S_1''}{\partial N_1''} = \frac{\partial S_1'''}{\partial N_1'''} = ... \\ \cdot \quad \cdot \quad \cdot \quad \cdot \quad \cdot \quad \cdot \quad \cdot \quad \cdot \\ \cdot \quad \cdot \quad \cdot \quad \cdot \quad \cdot \quad \cdot \quad \cdot \quad \cdot \\ \frac{\partial S_h'}{\partial N_h'} = \frac{\partial S_h''}{\partial N_h''} = \frac{\partial S_h'''}{\partial N_h'''} = ... \end{array} \tag{140.8}$$

as an explicit description of the condition for equilibrium between the separate systems which are in connexion. The first of these equations (140.6) expresses the condition for thermal equilibrium, the second equation (140.7) expresses the condition for mechanical equilibrium, and the remaining equations (140.8) express the conditions for no transfer of the h different component substances between the individual systems which form the combination.

Similar treatment can of course also be given, when coordinates other than volume are involved, by giving appropriate consideration in each particular case to the nature of these coordinates and of the corresponding forces. This will then give us the conditions for mechanical equilibrium in a form appropriate to the particular case, and will leave the conditions for thermal equilibrium and for equilibrium with respect to the transfer of matter unaltered.

In what follows, we shall be specially interested in the conditions for equilibrium with respect to the transfer of component substances between the different individual systems. For any given component i the equilibrium condition for no transfer can be expressed by the statement that the partial derivative of the entropy with respect to the amount of that substance shall have the same value

$$\left(\frac{\partial S}{\partial N_i}\right)_{E,a_1,a_2,...,N} = \text{const.} \tag{140.9}$$

for each of the individual systems in the combination. In accordance with our original introduction of such quantities in (140.1), it will be seen that these derivatives are to be taken holding the energy, external coordinates, and amounts of other components constant. This has been indicated in the above expression by the addition of subscripts, where the symbol N is to be taken as applying to all components other than the component i. From the tendency for entropy to increase towards a maximum, it will be seen in the absence of equilibrium that there will be a tendency for the substance i to be transferred from those individual systems in which the value of $\partial S/\partial N_i$ is low to those in which it is high.

For comparison with the actual symbolism employed by Gibbs, and for convenience with respect to our later statistical-mechanical considerations, it will be advantageous also to express the foregoing in a somewhat different form. By solving our fundamental equation (140.2), for the variation in energy δE instead of for the variation in entropy δS, we can write

$$\delta E = T\,\delta S-(A_1\,\delta a_1+A_2\,\delta a_2+...)-T\left(\frac{\partial S}{\partial N_1}\right)\delta N_1-...-T\left(\frac{\partial S}{\partial N_h}\right)\delta N_h, \tag{140.10}$$

and this can be expressed in the simpler form

$$\delta E = T\,\delta S-(A_1\,\delta a_1+A_2\,\delta a_2+...)+\mu_1\,\delta N_1+...+\mu_h\,\delta N_h, \tag{140.11}$$

provided we define the quantity μ_i for any component i by

$$\mu_i = \left(\frac{\partial E}{\partial N_i}\right)_{S,a_1,a_2,...,N} = -T\left(\frac{\partial S}{\partial N_i}\right)_{E,a_1,a_2,...,N}. \tag{140.12}$$

The above quantities μ_i were called by Gibbs the *potentials* of the different component substances constituting a system. It will be seen from the foregoing that equilibrium with respect to the transfer of any component i from one system to another can also be expressed by requiring the same value

$$\mu_i = \text{const.} \tag{140.13}$$

for the potential of that component in each of the individual systems. It will also be noted in the absence of equilibrium that there would be a tendency for substances to be transferred from systems where their potential is high to those where it is low.

The relation of these potentials to partial derivatives of the energy or of the entropy of a system are given by the definition (140.12). It will also be convenient to have their relation to partial derivatives of the Helmholtz *free energy* of the system

$$A = E-TS. \tag{140.14}$$

Varying the quantities in this equation, we have

$$\delta A = \delta E-T\,\delta S-S\,\delta T, \tag{140.15}$$

which by combining with (140.11) gives us

$$\delta A = -S\,\delta T-(A_1\,\delta a_1+A_2\,\delta a_2+...)+\mu_1\,\delta N_1+...+\mu_h\,\delta N_h. \tag{140.16}$$

We hence see that the potential for any component i can also be expressed by

$$\mu_i = \left(\frac{\partial A}{\partial N_i}\right)_{T,a_1,a_2,\ldots,N}, \tag{140.17}$$

as the partial derivative of free energy with respect to the number of mols of the component i, holding temperature, external coordinates, and the other components constant.

This form of expression for the potentials, together with the familiar principle that the change in free energy accompanying a change in state at a given temperature is equal to the isothermal reversible work necessary to bring about the change, is convenient in drawing conclusions as to the factors controlling the values of the μ_i in specific cases. In accordance therewith—if we start out with a system consisting of the actual system of interest in which the value of μ_i is desired, together with some of the pure substance i at the same temperature T as that of the system of interest but in its 'standard' state of zero free energy—the potential μ_i of this substance in the system of interest will be equal to the isothermal reversible work per mol necessary to transfer a small portion of the substance from its 'standard' state into the system of interest, with the temperature and external coordinates for the latter held fixed.

The above method of defining the potentials μ_i for the components of a system makes it possible to draw an important qualitative conclusion, which will be of use in our later statistical-mechanical considerations, in § 141 (*e*), when we come to investigate the fluctuations in composition that would be expected in the grand canonical ensembles that we shall introduce for the representation of open thermodynamic systems. If we consider a system maintained at constant temperature and volume, and provided with a semipermeable membrane through which any chosen component i could be reversibly introduced, say, in gaseous form, it is evident from our general knowledge of the factors controlling the 'escaping tendency' of substances from a system that we can usually expect the work per mol necessary to transfer the component i from its standard state into the system—and hence also the corresponding potential μ_i—to increase markedly as we increase the amount of that component already present in the system, and to be much less affected at constant volume by similar changes that might be made in the amounts of the other components in the system. As a consequence we shall regard it as plausible to assume in typical cases that $\delta\mu_i/\delta N_i$ would be a positive quantity greater than zero if we change the amount of any component i by some given small increment δN_i and then adjust the amounts of the other components j of the system so as to leave their potentials μ_j unaltered, temperature and volume being held constant throughout.† Only under special circumstances—e.g. with the component i present in coexisting phases—could we expect the above quantity $\delta\mu_i/\delta N_i$ to approach zero.

It will be of interest to consider a simple quantitative example of the above qualitative conclusion. For this purpose let us take a mixture of perfect gases at temperature T and volume v, and fix our attention on one of the gases i in this mixture. Let p_i^0 be the pressure of this component in its standard state at that temperature, and p_i its partial pressure in the mixture. In accordance with the work necessary for reversible transfer from the standard state into the mixture, we can then evidently write for the potential μ_i in question

$$\mu_i = RT\log\frac{p_i}{p_i^0}+RT, \tag{140.18}$$

where the first term is the reversible work per mol necessary to raise the pressure of the component from p_i^0 to p_i, and the second term is the work per mol necessary for its reversible introduction through a semi-permeable membrane into the mixture. Re-expressing the partial pressure p_i, with the help of the gas laws, in terms of the volume v of the mixture and the number of mols N_i of the component under consideration, we can rewrite this in the form

$$\mu_i = RT\log\frac{N_i RT}{p_i^0 v}+RT. \tag{140.19}$$

Hence, in this special case, we see that the potential μ_i of each component i would depend only on the number of mols of that component present, and can write the simple result

$$\frac{\delta\mu_i}{\delta N_i} = \frac{RT}{N_i} \tag{140.20}$$

for the change in any particular potential μ_i with the corresponding number of mols N_i, the potentials μ_j of the remaining components being kept constant.

(*b*) **The grand canonical ensemble.** We are now ready to consider the statistical-mechanical apparatus that will be needed in explaining the

† This conclusion as to the dependence of potentials on composition at constant temperature and *volume* is not to be confused with the so-called Gibbs-Duhem relation holding at constant temperature and *pressure*, which can be expressed by the equation

$$N_1\,d\mu_1+...+N_h\,d\mu_h = 0.$$

foregoing thermodynamic method of treating equilibrium with respect to the transfer of matter to or from a system of interest. Since we now contemplate the possibility of changes in the composition of a system of interest by amounts which will not be exactly known, it now proves desirable to introduce the idea of representative ensembles composed of members which can differ not only in state but also in the amounts of material of various kinds which they contain. Using the language of Gibbs,† such ensembles may be given the name of *grand ensembles* and the name of *petit ensembles* may be used to designate our previous kind of ensemble where all the member systems are composed of the same numbers of molecules of the different kinds needed for the construction of the system.

Since we shall be interested in equilibria, we may proceed at once to the definition of the grand canonical ensemble which provides the appropriate apparatus for treating equilibrium not only towards the transfer of energy but also towards the transfer of matter when systems are placed in contact. Let us consider a system composed of *h independent kinds of substances or components*, out of which any further substances in the system can be regarded as composed. Let us use the symbols n_1, n_2,..., n_h to designate the numbers of molecules of these components in any member of the corresponding representative ensemble. And let us use the symbol E_m to designate the energy of such a member in any energy characteristic state m, where it will be noted that the spectrum of such values will depend on the composition of the particular member of the ensemble under consideration. We may then define the *grand canonical ensemble* corresponding to such a system by the formula

$$P_{n_1\ldots n_h,m} = e^{\frac{\Omega+\mu_1 n_1+\ldots+\mu_h n_h-E_m}{\theta}}, \tag{140.21}$$

where $P_{n_1\ldots n_h,m}$ is the probability for finding a member of the ensemble with the composition described by the numbers of molecules $n_1 \ldots n_h$ and in an energy state m which is possible for this composition, and where Ω, θ, and $\mu_1 \ldots \mu_n$ are adjustable parameters.

We regard the above distribution as applying to all possible values—from zero to infinity—for the numbers of molecules $n_1 \ldots n_h$, and to all possible values for the energy E_m. We shall, however, still regard the external coordinates, a_1, a_2, a_3, etc., such, for example, as volume, for the different members of the ensemble as all having the same values as

† Gibbs, *Elementary Principles in Statistical Mechanics*, Yale University Press, 1902, see chapter xv. The treatment of grand ensembles given in the present section is a condensed quantum mechanical transcription of the classical treatment given by Gibbs rather than a complete and detailed development.

pertain in the system of interest itself. In typical cases there will then be a high concentration of probability in the neighbourhood of the mean composition and mean energy for the members of the ensemble, as we shall see later, § 141 (*e*).

Important properties of the grand canonical distribution are expressed by the following equations. Since the total probability for finding one or another number of molecules and one or another value of energy will be *normalized to unity*, we have

$$\sum_{n_1 \dots n_h, m} e^{\frac{\Omega+\mu_1 n_1+\dots+\mu_h n_h-E_m}{\theta}} = 1, \tag{140.22}$$

when summed over all compositions $n_1 \dots n_h$ and over all corresponding states m. For the *mean energy* of the members of the ensemble, we have

$$\overline{\overline{E}} = \sum_{n_1 \dots n_h, m} e^{\frac{\Omega+\mu_1 n_1+\dots+\mu_h n_h-E_m}{\theta}} E_m. \tag{140.23}$$

For the *mean numbers of molecules* for the different kinds of components, we have

$$\begin{aligned} \overline{\overline{n}}_1 &= \sum_{n_1 \dots n_h, m} e^{\frac{\Omega+\mu_1 n_1+\dots+\mu_h n_h-E_m}{\theta}} n_1 \\ &\cdot \quad \cdot \quad \cdot \quad \cdot \quad \cdot \quad \cdot \quad \cdot \quad \cdot \\ &\cdot \quad \cdot \quad \cdot \quad \cdot \quad \cdot \quad \cdot \quad \cdot \quad \cdot \\ \overline{\overline{n}}_h &= \sum_{n_1 \dots n_h, m} e^{\frac{\Omega+\mu_1 n_1+\dots+\mu_h n_h-E_m}{\theta}} n_h. \end{aligned} \tag{140.24}$$

And for the *value of the quantity* $\overline{\overline{H}}$, which we define in analogy to that for a petit ensemble, we have

$$\begin{aligned} \overline{\overline{H}} &= \sum_{n_1 \dots n_h, m} e^{\frac{\Omega+\mu_1 n_1+\dots+\mu_h n_h-E_m}{\theta}} \frac{\Omega+\mu_1 n_1+\dots+\mu_h n_h-E_m}{\theta} \\ &= \frac{\Omega+\mu_1 \overline{\overline{n}}_1+\dots+\mu_h \overline{\overline{n}}_h-\overline{\overline{E}}}{\theta}. \end{aligned} \tag{140.25}$$

It will be noted that the grand canonical ensemble can be regarded as consisting of a weighted collection of petit canonical ensembles, one for each possible composition. The grand ensemble will therefore itself be in a condition of equilibrium corresponding to that which could be regarded as reached as the ultimate result of long-continued essentially isolated behaviour. It will also be noted that the above expressions (140.22–4) provide in all $h+2$ equations which could be solved for the parameters Ω, θ, $\mu_1 \dots \mu_h$, describing the distribution, in terms of the mean energy $\overline{\overline{E}}$ and mean numbers of molecules $\overline{\overline{n}}_1 \dots \overline{\overline{n}}_h$ of the members of the ensemble. By correlating these mean values with the definite values of energy and composition which would be assigned to a system of interest from the point of view of thermodynamics, it then becomes possible, as we shall see, to regard the ensemble as representing thermodynamic equilibrium for a system of specified energy and composition.

(c) **Correlation of thermodynamic and statistical-mechanical quantities.** Assuming this representation of thermodynamic equilibrium, the complete correlation of thermodynamic with statistical-mechanical quantities can now be obtained—as in our previous treatment of petit canonical ensembles—by considering the variation in $\bar{\bar{H}}$, when we pass from any given grand canonical distribution to a similar neighbouring distribution, corresponding to a change in the system of interest from one condition of thermodynamic equilibrium to a similar neighbouring condition, with slightly altered values for the energy, composition, and external coordinates pertaining thereto. Making use of the expression for $\bar{\bar{H}}$ given by (140.25), we immediately obtain for the variation in question

$$\delta\bar{\bar{H}} = \frac{\delta\Omega}{\theta}-\frac{\Omega}{\theta^2}\delta\theta+\sum_{i=1}^{h}\left(\frac{\bar{\bar{n}}_i}{\theta}\delta\mu_i+\frac{\mu_i}{\theta}\delta\bar{\bar{n}}_i-\frac{\mu_i\bar{\bar{n}}_i}{\theta^2}\delta\theta\right)-\frac{\delta\bar{\bar{E}}}{\theta}+\frac{\bar{\bar{E}}}{\theta^2}\delta\theta. \quad (140.26)$$

We shall be interested, however, in putting this expression into a more significant form.

Since we are making a change to a new grand canonical distribution, our variation must be made in such a way as to preserve the validity of equation (140.22) which makes the total probability of finding a member of the ensemble in some possible condition equal to unity. Hence the above variation must be such as to satisfy

$$\sum_{n_1\ldots n_h,m} e^{\frac{\Omega+\mu_1 n_1+\ldots+\mu_h n_h-E_m}{\theta}}\left\{\frac{\delta\Omega}{\theta}-\frac{\Omega}{\theta^2}\delta\theta+\sum_{i=1}^{h}\left(\frac{n_i}{\theta}\delta\mu_i-\frac{\mu_i n_i}{\theta^2}\delta\theta\right)-\right.$$
$$\left.-\frac{1}{\theta}\left(\frac{\partial E_m}{\partial a_1}\delta a_1+\frac{\partial E_m}{\partial a_2}\delta a_2+\ldots\right)+\frac{E_m}{\theta^2}\delta\theta\right\} = 0, \quad (140.27)$$

where the change in the eigenvalues E_m has been appropriately expressed in terms of their dependence on the external coordinates a_1, a_2, etc., such, for example, as volume. Defining the external forces A_1, A_2, etc., corresponding to these coordinates as in (121.6), and making use of the general method of computing mean values, the above equation can be rewritten in the form

$$\frac{\delta\Omega}{\theta}-\frac{\Omega}{\theta^2}\delta\theta+\sum_{i=1}^{h}\left(\frac{\bar{\bar{n}}_i}{\theta}\delta\mu_i-\frac{\mu_i\bar{\bar{n}}_i}{\theta^2}\right)\delta\theta+\frac{1}{\theta}(\bar{\bar{A}}_1\delta a_1+\bar{\bar{A}}_2\delta a_2+\ldots)+\frac{\bar{\bar{E}}}{\theta^2}\delta\theta = 0. \quad (140.28)$$

And combining with (140.26), this then gives us

$$\delta\overline{\overline{H}} = -\frac{\delta\overline{\overline{E}}}{\theta}+\sum_{i=1}^{h}\frac{\mu_i}{\theta}\delta\overline{\overline{n}}_i-\frac{1}{\theta}(\overline{\overline{A}}_1\,\delta a_1+\overline{\overline{A}}_2\,\delta a_2+...). \tag{140.29}$$

For the purpose of making the desired correlations, this result may now be re-expressed in the more familiar form

$$\delta\overline{\overline{E}} = -\theta\,\delta\overline{\overline{H}}-(\overline{\overline{A}}_1\,\delta a_1+\overline{\overline{A}}_2\,\delta a_2+...)+\mu_1\,\delta\overline{\overline{n}}_1+...+\mu_h\,\delta\overline{\overline{n}}_h \tag{140.30}$$

and, compared with our previous fundamental thermodynamic relation (140.11),

$$\delta E = T\,\delta S-(A_1\,\delta a_1+A_2\,\delta a_2+...)+\mu_1\,\delta N_1+...+\mu_h\,\delta N_h. \tag{140.31}$$

In the light of our earlier correlations between thermodynamic and statistical-mechanical quantities under circumstances where changes in the composition of the system were not contemplated, we may now take the thermodynamic quantities—energy E, temperature T, entropy S, potential of the ith component μ_i, number of mols of that component N_i, external force of the jth kind A_j, and corresponding external coordinate a_j—for a system in thermodynamic equilibrium of the so-called 'open' kind where changes in composition are contemplated—as correlated with quantities pertaining to the corresponding grand canonical ensemble by the expressions

$$E \rightleftharpoons \overline{\overline{E}}, \qquad T \rightleftharpoons \frac{\theta}{k}, \qquad S \rightleftharpoons -k\overline{\overline{H}},$$
$$\mu_i \rightleftharpoons N_A\,\mu_i, \qquad N_i \rightleftharpoons \frac{\overline{\overline{n}}_i}{N_A}, \qquad A_j \rightleftharpoons \overline{\overline{A}}_j, \qquad a_j \rightleftharpoons a_j, \tag{140.32}$$

where k is Boltzmann's constant and N_A is Avogadro's number. For future reference it will also be useful for us to note, in accordance with the above and with (140.25), that we can rewrite the correlation of entropy S with statistical-mechanical quantities in the form

$$S \rightleftharpoons -k\frac{\Omega+\mu_1\overline{\overline{n}}_1+...+\mu_h\overline{\overline{n}}_h-\overline{\overline{E}}}{\theta}. \tag{140.33}$$

(*d*) **Conditions for equilibrium when grand canonical ensembles are combined.** We shall now illustrate the appropriate character of the new statistical-mechanical apparatus which has been introduced, by showing that the conditions for statistical equilibrium when two different grand canonical ensembles are placed in connexion with each other would give an explanation of the conditions for thermodynamic equilibrium when the corresponding systems are placed in connexion.

For this purpose let us consider two grand canonical ensembles with the distributions

$$P_{n'_1\ldots n'_h,m'} = e^{\frac{\Omega'+\mu'_1 n'_1+\ldots+\mu'_h n'_h-E_{m'}}{\theta'}} \tag{140.34}$$

and

$$P_{n''_1\ldots n''_h,m''} = e^{\frac{\Omega''+\mu''_1 n''_1+\ldots+\mu''_h n''_h-E_{m''}}{\theta''}}, \tag{140.35}$$

where we use accents to distinguish quantities pertaining respectively to the two distributions. We may regard these two ensembles as representing two separate systems of interest each in a condition of thermodynamic equilibrium. Let us then consider a third ensemble, of higher order in the number of its members, which could be obtained by taking each member of the one ensemble in conjunction with each member of the other. We may regard the new ensemble as representing the two systems of interest taken in conjunction.

Assuming for the moment that the members of the two original ensembles are placed in thermal contact but not in connexion with each other, let us first consider the conditions for *thermal equilibrium*. Each of the two original grand canonical ensembles is itself a collection of petit canonical ensembles, one for each possible composition, normalized to correspond to the representation of that composition in the whole ensemble, and distributed in a manner determined by the parameters θ' and θ'' in the two cases respectively. We can hence apply our previous considerations, § 127, as to thermal flow when the members of canonical ensembles are placed in contact and conclude also for grand canonical ensembles that the condition for no thermal flow in the mean will be expressed by the equality

$$\theta' = \theta''. \tag{140.36}$$

Assuming next that the members of the ensembles which have been combined are allowed to exert mechanical forces on each other, let us consider the conditions for *mechanical equilibrium*. As an example we may assume the introduction of a movable partition which would allow the combined members to exert pressure on each other. As an appropriate expression for an average condition of mechanical equilibrium in the combined ensemble, it will then be reasonable to require an equality

$$\bar{\bar{p}}' = \bar{\bar{p}}'' \tag{140.37}$$

between the mean values of the pressures in the members of the two ensembles, thus assuring that there would be no tendency in the mean for the movable partition to be displaced in either direction. Appropriate consideration could also be given to other kinds of external force, with the general result that we should require the same relation for the mean values of such forces in the ensemble as would be required in a thermodynamic treatment of equilibrium for the definite values which would then be assigned to those forces.

Finally, considering that connexion could be established allowing the passage of matter between the members of the two ensembles, let us study the conditions for *equilibrium towards the transfer of component substances.* To investigate this, let us take the necessary conditions for thermal and mechanical equilibrium as already guaranteed, and, by combining the expressions for probability given by (140.34) and (140.35), write

$$P_{n'_1...n'_h,n''_1...n''_h,m',m''} = e^{\frac{\Omega'+\Omega''+\mu'_1 n'_1+\mu''_1 n''_1+...+\mu'_h n'_h+\mu''_h n''_h-E_{m'}-E_{m''}}{\theta}}, \qquad (140.38)$$

where θ is the common value of that parameter in the two original distributions, as an expression describing the distribution in the combined ensemble when it is set up. Such a distribution would not in general correspond to equilibrium towards the transfer of matter. However, let us now write

$$\begin{aligned} n_1 &= n'_1+n''_1 \\ &\cdot\quad\cdot\quad\cdot\quad\cdot \\ &\cdot\quad\cdot\quad\cdot\quad\cdot \\ n_h &= n'_h+n''_h \end{aligned} \qquad (140.39)$$

as expressions for the total numbers of molecules for the h components in the members of the combined ensemble, put

$$\Omega = \Omega'+\Omega'', \qquad (140.40)$$

and take the possible energies for the interconnected system as approximately given by

$$E_m = E_{m'}+E_{m''} \qquad (140.41)$$

as will be justified for most energy states in typical cases for systems of large volume. And, furthermore, let us *require the equalities*

$$\begin{aligned} \mu_1 &= \mu'_1 = \mu''_1 \\ &\cdot\quad\cdot\quad\cdot\quad\cdot\quad\cdot \\ &\cdot\quad\cdot\quad\cdot\quad\cdot\quad\cdot \\ \mu_h &= \mu'_h = \mu''_h \end{aligned} \qquad (140.42)$$

as holding between the values of these parameters in the two original ensembles. Our expression (140.38) for the initial distribution would then lead—on the gradual establishment of interconnexion (see § 97 (*b*))—to

$$P_{n_1...n_h,m} = e^{\frac{\Omega+\mu_1 n_1+...+\mu_h n_h-E_m}{\theta}}, \qquad (140.43)$$

as a close expression for the distribution in the resulting ensemble. This would be a grand canonical distribution corresponding to equilibrium at temperature $T = \theta/k$, and exhibiting no dependence of probabilities on the separate values of the n_i' and n_i''. Hence we take the equalities (140.42) as the condition for equilibrium with respect to the transfer of matter.

This then gives our statistical-mechanical explanation of the thermodynamic principles governing equilibrium when systems are placed in connexion, since the foregoing requirements (140.36), (140.37), and (140.42) for statistical equilibrium are seen to be natural analogues of our previous thermodynamic requirements (140.6), (140.7), and (140.13) for thermodynamic equilibrium which can be written in the form

$$T' = T'', \qquad p' = p'', \quad \text{and} \quad \mu_i' = \mu_i''. \tag{140.44}$$

To complete our justification for the use of the grand canonical ensemble to represent a thermodynamic system in a condition of equilibrium it would also be necessary to show in typical cases that the fluctuations of composition in the ensemble would be small. This we shall do in the next section in § 141 (*e*), when we are considering fluctuations in general.

(*e*) **Explanation of the Gibbs paradox.** As a further example of the usefulness of the grand canonical ensemble, we may consider the explanation which it provides for the so-called Gibbs paradox as to the entropy of combined systems. When two similar systems are placed in connexion with each other, for example when two samples of the same pure gas at a given temperature and pressure are allowed to mingle by the removal of a separating partition, the entropy of the resulting system is taken as equal to the sum of the entropies of the two original parts. This valid thermodynamic procedure, however, has sometimes seemed paradoxical from the point of view of molecular mechanics, since the entropy would have been regarded as subject to increase by diffusion if the two separate gases had been composed of different kinds of molecules, and it has seemed strange that this increase in entropy should be abolished merely because the two kinds of molecules that diffuse into each other happen to be of the same kind.

The complete statistical-mechanical explanation of the thermodynamic principle, that the entropy of a homogeneous substance is proportional to the amount taken, depends on the recognition of two factors. The first of these factors is the consideration that individual molecules of the same substance when allowed free motion inside a common container are not to be treated as distinguishable one from another in the calculation of entropy. The second of the factors is the consideration that the grand rather than the petit canonical ensemble must be taken as providing the appropriate representation for a system in thermodynamic equilibrium when changes in its composition are contemplated.

In the classical development of statistical mechanics by Gibbs, the first of the above factors was allowed for by the introduction of a special step in which attention was turned away from the *specific phases* of a system, which depend on just which particular molecules are assigned to different regions in the μ-space, to its *generic phases*, which depend only on the numbers of molecules so assigned. In the present quantum mechanical development of statistical mechanics no special step of this kind has to be introduced since the quantum mechanics *per se* already allows for the consideration that individual molecules of the same substance are not in general distinguishable one from another.†

The allowance for this first factor is alone sufficient to lead to a calculated entropy for a homogeneous substance which is nearly but not quite proportional to the amount taken. To illustrate this let us consider the correlation of entropy with statistical-mechanical quantities,

$$S \rightleftharpoons -k\overline{\overline{H}}_{\text{petit}}, \tag{140.45}$$

which we have made at the stage of our development when allowance is made for the indistinguishability of like molecules but when we are treating 'closed' rather than 'open' systems and are not considering the possibility of changes in composition. In accordance with our representation of a 'closed' system at thermodynamic equilibrium by a canonical ensemble, this then gives us, in agreement with (120.5) and (120.6),

$$S = \frac{\overline{\overline{E}}}{T} - \frac{\psi}{T} = \frac{\overline{\overline{E}}}{T} + k\log\sum_m e^{-E_m/kT} \tag{140.46}$$

for the entropy of a system at equilibrium, where $\overline{\overline{E}}$ is the mean energy to be expected for the system and the summation is over all possible energy states m. In the case of a homogeneous substance it is immediately evident that the first of the two terms in this expression is indeed proportional to the amount of the substance taken. With freely moving molecules, however, the second term in (140.46) is not quite proportional to that amount. For example, in the case of a perfect

† An illuminating discussion of the close relation between the introduction of generic phases by Gibbs and the present procedures adopted in the quantum mechanics is given by Epstein in his article, 'Gibbs' Methods in Quantum Statistics', published in *Commentary on the Scientific Writings of J. Willard Gibbs*, vol. ii, Yale University Press, 1936.

monatomic gas, consisting of n atoms of mass m at temperature T and in volume v, we have, in agreement with (135.6),

$$k\log\sum_m e^{-E_m/kT} = nk\left[\log\frac{eg\,v}{h^3 n}(2\pi mkT)^{\frac{3}{2}} - \frac{1}{n}\log\sqrt{(2\pi n)}\right]. \qquad (140.47)$$

This is then seen to be proportional to the amount of gas taken except for a term which becomes negligible as we take large enough amounts of the substance. It may be remarked in a qualitative way that the failure of this procedure to give strict proportionality is due to a failure to take proper account of the possibility for fluctuations in the amount of gas that could be present in the two parts of the whole container when two original samples of the gas are connected. It may also be remarked, however, that the deviation from strict proportionality is too small in typical cases to make any correction for it necessary in practical computations.

Finally, we may now turn to the completely appropriate apparatus for the treatment of 'open' thermodynamic systems, which makes due allowance both for the indistinguishability of like molecules and for the possibility of changes in composition. We then correlate entropy with statistical-mechanical quantities by

$$S \rightleftharpoons -k\overline{\overline{H}}_{\text{grand}}. \qquad (140.48)$$

And, in accordance with our use of the grand canonical ensemble for the representation of equilibrium, we then obtain, in agreement with (140.33),

$$S = \frac{\overline{\overline{E}}}{T} - \frac{\mu_1\overline{\overline{n}}_1 + \ldots + \mu_h\overline{\overline{n}}_h}{T} - \frac{\Omega}{T} \qquad (140.49)$$

for the entropy of a homogeneous substance, where $\overline{\overline{E}}$ is the mean energy to be expected for the substance, and the mean composition to be expected for the system is specified by the mean numbers of molecules $\overline{\overline{n}}_1 \ldots \overline{\overline{n}}_h$ for the different components. This expression then indeed makes the entropy strictly proportional to the amount of substance taken, since we see from the conditions for equilibrium, when portions of the substance are combined, that the quantities $\mu_1 \ldots \mu_h$ would be independent of the amount of substance in agreement with (140.42), and that the quantity Ω would be proportional to the amount taken in agreement with (140.40). We thus obtain a satisfactory statistical-mechanical correlate for the proportionality of the entropy of a homogeneous substance to its amount.

Except for our later consideration of fluctuations in a grand canonical ensemble in § 141 (*e*) of the next section, this must now complete our partial treatment of grand ensembles and their use in the representation of 'open' systems. Once more we emphasize the fundamentality of the classical considerations of Gibbs and their ready applicability for the development of a quantum statistical mechanics.

141. Fluctuations at thermodynamic equilibrium

The general explanation of the principles of thermodynamics in terms of statistical mechanics, which was given in the preceding chapter, and the further applications which have been undertaken in the foregoing parts of the present chapter, have been based on the idea that the thermodynamic behaviour of a system of interest can be correlated with the mean behaviour of the similar systems composing an appropriate representative ensemble, and that the values of thermodynamic quantities applying to the system of interest can be correlated with the mean values of suitable quantities for the members of the ensemble. Thus, for example, the thermodynamic quantities for a 'closed' system in thermodynamic equilibrium, which were selected for special study in the foregoing, the pressure p, energy E, heat capacity C_v, and entropy S, can be regarded from the statistical-mechanical point of view as the mean values of the analogous mechanical quantities in the corresponding canonical ensemble, namely $\overline{\overline{p}} = \partial\overline{\overline{E}}/\partial v$, $\overline{\overline{E}}$, $\overline{\overline{C}} = k\,\partial\overline{\overline{E}}/\partial\theta$, and $-k\overline{\overline{H}}$ respectively. Hence the actual precision of the principles of thermodynamics when applied under typical circumstances must depend on the small fluctuations around the mean which would be exhibited under those conditions by the mechanical analogues of thermodynamic quantities; and a need for amplifying or modifying the methods of thermodynamics would arise whenever the fluctuations become important.

For this reason we shall now give a brief although partial account of the fluctuations to be expected in thermodynamic systems. We shall confine our attention to systems in thermodynamic equilibrium, and hence base our calculations on the canonical or grand canonical distribution as giving the appropriate representation of the condition to be investigated.

By way of summary we shall first present once more the results already calculated for the fluctuations in the numbers of molecules or other component elements in different states in the case of Maxwell-Boltzmann, Einstein-Bose, and Fermi-Dirac systems. These results, however, are of course only indirectly related to fluctuations in the values of the macroscopic variables of thermodynamics. We shall next consider fluctuations in the total energy of a system in thermodynamic equilibrium at a specified temperature. This is of immediate thermodynamic interest since it is only when these fluctuations are small that the usual thermodynamic procedure of simultaneously specifying the temperature and the energy of a system is possible. We shall then consider fluctuations in the external forces exerted by a system in equilibrium. This is also of immediate thermodynamic interest in connexion with the thermodynamic procedure of assigning precise values to such quantities as pressure. We shall then turn to a consideration of the justification for Einstein's somewhat approximate procedure of relating fluctuations in macroscopic variables describing a system to corresponding changes in free energy. The Einstein treatment is of interest because of its ready applicability to problems of observational significance. Finally, we shall consider the fluctuations in composition for an open thermodynamic system that would correspond to its representation by a grand canonical ensemble. This will be of interest in justifying the uses that we have made of the grand canonical ensemble to represent systems having precisely specified compositions from the thermodynamic point of view, and also in providing a method for treating the important observational problem of the fluctuations in density of a fluid.

(*a*) **Fluctuations in the case of Maxwell-Boltzmann, Einstein-Bose, and Fermi-Dirac distributions.** In the case of Maxwell-Boltzmann, Einstein-Bose, and Fermi-Dirac systems we have seen in §§ 113 and 114, for a system composed of n similar weakly interacting elements, that the mean number of elements $\bar{\bar{n}}_k$ in a given energy state k at equilibrium would be given by the respective formulae

$$\bar{\bar{n}}_k = e^{-\alpha-\beta\epsilon_k} \quad \text{(Maxwell-Boltzmann)}, \tag{141.1}$$

$$\bar{\bar{n}}_k = \frac{1}{e^{\alpha+\beta\epsilon_k}-1} \quad \text{(Einstein-Bose)}, \tag{141.2}$$

$$\bar{\bar{n}}_k = \frac{1}{e^{\alpha+\beta\epsilon_k}+1} \quad \text{(Fermi-Dirac)}, \tag{141.3}$$

where α and β are constants whose significance we already understand, and ϵ_k is the energy of the state k. We have also found that the mean square deviations from these numbers would be given by the respective formulae

$$\frac{\overline{\overline{(n_k-\bar{\bar{n}}_k)^2}}}{(\bar{\bar{n}}_k)^2} = \frac{1}{\bar{\bar{n}}_k} \quad \text{(Maxwell-Boltzmann)}, \tag{141.4}$$

$$\frac{\overline{\overline{(n_k-\bar{\bar{n}}_k)^2}}}{(\bar{\bar{n}}_k)^2} = \frac{1}{\bar{\bar{n}}_k}+1 \quad \text{(Einstein-Bose)}, \tag{141.5}$$

$$\frac{\overline{\overline{(n_k-\bar{\bar{n}}_k)^2}}}{(\bar{\bar{n}}_k)^2} = \frac{1}{\bar{\bar{n}}_k}-1 \quad \text{(Fermi-Dirac)}, \tag{141.6}$$

where we restrict ourselves to cases for which $n \gg n_k$, since we have dropped a term $-1/n$ in (141.4), and have used grand canonical ensembles in computing (141.5) and (141.6) instead of petit canonical ensembles corresponding to a definite total number of elements n.

We repeat these formulae here for the sake of inclusion along with other expressions for various kinds of fluctuations. It may be remarked that the methods which we used for their derivation in §§ 113 and 114 could be easily extended (by the performance of further differentiations) to a calculation of fluctuations of order higher than the second; but this would not add much to our immediate physical understanding. It will be noted as the most important characteristic of the mean square fluctuations that they become unimportant in any case for states k that are highly populated.

The expression (141.4) for the fluctuations in a Maxwell-Boltzmann system is of the well-known 'classical' form. It may be used for investigating the fluctuations in the equilibrium distribution of radiation when this is treated as corresponding to the excitation of modes of electromagnetic oscillation in a hollow enclosure.

The expression (141.5) for the fluctuations in an Einstein-Bose system was used by Einstein† for obtaining an alternative calculation of the fluctuations in radiation when treated as consisting of photons. It was emphasized by Einstein in this connexion that the first term in (141.5) could be regarded as corresponding to the fluctuations to be expected in a gas consisting of distinguishable particles and that the second term could be regarded as corresponding to fluctuations arising from the wave-like character of light, thus showing that even in its fluctuation properties light could be regarded as having a dual wave-particle character.

The expression (141.6) for the fluctuations in a Fermi-Dirac system was first obtained by Pauli.‡ It will be noted in this case, as a consequence of the Pauli exclusion principle, that the largest possible mean population of elements in any state k would be unity and that the fluctuation then goes to zero. Thus in the case of conduction electrons, which can be treated as giving a highly degenerate Fermi-Dirac gas even up to ordinary temperatures, we have very small fluctuations in the occupation of those low-lying energy states which are nearly completely filled.

(*b*) **Fluctuations in total energy.** We may now turn to a consideration of the fluctuations in the total energy of a 'closed' system in thermo-

† Einstein, *Berl. Ber.* (1924), p. 261; (1925), p. 3.

‡ Pauli, *Zeits. f. Phys.* **41**, 81 (1927).

dynamic equilibrium at a specified temperature. Such a system would be represented by a canonical ensemble with members distributed over different energies, in correspondence with the circumstance that a system in thermal equilibrium with a large heat bath could assume different values of energy by interchange with that bath.

To obtain a suitable expression for the mean square deviation in energy it is convenient to start with the expression for the mean energy of the members of the canonical ensemble

$$\bar{\bar{E}} = \sum_k e^{\frac{\psi - E_k}{\theta}} E_k, \tag{141.7}$$

and consider the effect of a small variation $\delta\theta$, corresponding to a change to a neighbouring condition of equilibrium of slightly altered temperature. Carrying out such a variation, without change in the volume or other external parameters for the system, we obtain for the variation in mean energy

$$\delta\bar{\bar{E}} = \sum_k e^{\frac{\psi - E_k}{\theta}} \left(\frac{\delta\psi}{\theta} - \frac{\psi - E_k}{\theta^2}\delta\theta\right) E_k = \bar{\bar{E}}\frac{\delta\psi}{\theta} - \frac{\bar{\bar{E}}\psi}{\theta^2}\delta\theta + \frac{\overline{\overline{(E^2)}}}{\theta^2}\delta\theta. \tag{141.8}$$

Furthermore, since the total probability

$$\sum_k e^{\frac{\psi - E_k}{\theta}} = 1 \tag{141.9}$$

for finding the system in one or another state must remain constant, we can also write

$$0 = \sum_k e^{\frac{\psi - E_k}{\theta}} \left(\frac{\delta\psi}{\theta} - \frac{\psi - E_k}{\theta^2}\delta\theta\right) = \frac{\delta\psi}{\theta} - \frac{\psi}{\theta^2}\delta\theta + \frac{\bar{\bar{E}}}{\theta^2}\delta\theta \tag{141.10}$$

as a necessary condition that must be satisfied in making the variation $\delta\theta$. Combining (141.8) with (141.10), we have

$$\delta\bar{\bar{E}} = \frac{\overline{\overline{(E^2)}}}{\theta^2}\delta\theta - \frac{(\bar{\bar{E}})^2}{\theta^2}\delta\theta$$

or

$$\frac{\overline{\overline{(E^2)}} - (\bar{\bar{E}})^2}{(\bar{\bar{E}})^2} = \frac{\theta^2}{(\bar{\bar{E}})^2}\frac{\delta\bar{\bar{E}}}{\delta\theta}. \tag{141.11}$$

Re-expressing the left-hand side by an obvious transformation, and re-expressing the right-hand side by translation into thermodynamic language, we then obtain the desired expression for the mean square fluctuations in energy,

$$\frac{\overline{\overline{(E - \bar{\bar{E}})^2}}}{(\bar{\bar{E}})^2} = \frac{kT^2}{(\bar{\bar{E}})^2} C_v, \tag{141.12}$$

where C_v is the heat capacity of the system, its volume and other external parameters being held constant.

We may first consider the application of this result to the *typical high-temperature situation*, where the energy E and heat capacity C_v of the system would be approximately given by equations of the form

$$E \approx nkT \quad \text{and} \quad C_v \approx nk, \tag{141.13}$$

where n is a number of the order of the number of degrees of freedom of the system. Substituting in (141.12), we then obtain an expression for the fractional mean square deviation in energy,

$$\frac{\overline{\overline{(E-\overline{\overline{E}})^2}}}{(\overline{\overline{E}})^2} \approx \frac{1}{n}, \tag{141.14}$$

which is seen to go to zero as the number of degrees of freedom of the system increases. This is the characteristic kind of result that we can expect in the usual cases where we should wish to apply thermodynamics. It shows for a system of many degrees of freedom, which is at equilibrium at a given temperature, that the probabilities for the different possible values of energy would be highly concentrated in the neighbourhood of the mean energy. It thus provides a justification for the thermodynamic procedure of simultaneously assigning definite values both to the temperature and to the energy of a system under such circumstances. It also provides the grounds for the familiar statement that the principles of thermodynamics may be regarded as becoming exact at the limit of an infinite number of degrees of freedom.

We may next consider the application of our formula for fluctuations in energy to those *special situations* which arise when a system of interest is in a condition to absorb large quantities of energy without appreciable change in temperature. Such a state of affairs would occur in the case of a system containing two different phases—say, for example, solid and liquid—where an absorption of energy would be primarily used for supplying the heat of transition from one phase to the other rather than in producing a rise in temperature. Under such circumstances the heat capacity C_v for the system would become exceedingly large during the transition, and our formula (141.12) would lead to large fluctuations in energy. This finding, which was first appreciated by Gibbs,† again demonstrates the successful character of the statistical-mechanical explanation of thermodynamics since the

† Gibbs, *Elementary Principles in Statistical Mechanics*, Yale University Press, 1902; see in particular the second footnote on p. 75.

large fluctuations in energy now predicted for a canonical ensemble from the statistical viewpoint would occur under those very circumstances where the energy of a system would not be very sharply determined by its temperature from the thermodynamic viewpoint.

Finally, we may consider the application of our formula for energy fluctuations to *quantum mechanical situations*, where the heat capacity of a system is increasing rapidly with temperature owing to the presence of partially excited degrees of freedom. As an illustration we shall take the case of a crystal at very low temperature where the heat capacity is increasing as the cube of the temperature owing to the gradual excitation of its modes of oscillation. In accordance with (137.29) and (137.30), we can then take the energy and heat capacity of a crystal composed of n particles as given by

$$E = \frac{3\pi^4}{5}\frac{nkT^4}{\Theta^3} \quad \text{and} \quad C_v = \frac{12\pi^4}{5}\frac{nkT^3}{\Theta^3}, \tag{141.15}$$

where Θ is a parameter which characterizes the crystal, and we have chosen the energy zero-point so as to make the energy go to zero with the temperature. Substituting into our formula (141.12), we then obtain

$$\frac{\overline{\overline{(E-\overline{\overline{E}})^2}}}{(\overline{\overline{E}})^2} = \frac{20}{3\pi^4}\left(\frac{\Theta}{T}\right)^3\frac{1}{n} \tag{141.16}$$

as an expression for the energy fluctuations at temperature T. Here again we find for any given temperature T that the fluctuations would become negligible as we make the number of particles and hence the number of degrees of freedom greater. Nevertheless, for any given number of particles n, we would always have the possibility of considering temperatures T low enough so that the energy fluctuations would become exceedingly large. Hence at very low temperatures we must regard our usual thermodynamic concepts, which in general make energy and temperature simultaneously determinable quantities, as subject to limitations even for a system of many degrees of freedom and over a wide range of possible conditions.†

In conclusion, it is of course not to be forgotten that the simultaneous assignment of values to the energy and to the temperature of a system becomes in any case impossible when the number of degrees of freedom of the system is very small. This agrees with the large percentage fluctuations in energy which we should expect in the case of a very small system kept in contact with a heat bath.

† Compare Planck, *Theory of Heat*, translated by Brose, Macmillan, 1932, § 136, p. 268.

(*c*) **Fluctuations in an external force.** We may next turn to the fluctuations that would be expected at equilibrium in an external force A exerted by a system corresponding to displacements in some generalized external coordinate a. As a definition (121.6) for the external force A_n when the system is in a given energy state n we have

$$A_n = -\frac{\partial E_n}{\partial a}, \tag{141.17}$$

and hence for the mean force in a canonical ensemble we may write

$$\bar{\bar{A}} = \sum_n e^{\frac{\psi-E_n}{\theta}} A_n = -\sum_n e^{\frac{\psi-E_n}{\theta}} \frac{\partial E_n}{\partial a}. \tag{141.18}$$

To obtain the desired expression for the fluctuations in A, we may start with the necessary expression, holding in the case of a canonical ensemble,

$$\sum_n e^{\frac{\psi-E_n}{\theta}} = 1. \tag{141.19}$$

Regarding the distribution as dependent on the 'temperature' θ and on the external coordinate a, we may then carry out a partial differentiation with respect to the latter quantity. After multiplying by θ, this gives us

$$\sum_n e^{\frac{\psi-E_n}{\theta}} \left(\frac{\partial\psi}{\partial a} - \frac{\partial E_n}{\partial a}\right) = 0. \tag{141.20}$$

Making use of (141.17), this provides in passing the useful relation

$$\frac{\partial\psi}{\partial a} = -\bar{\bar{A}}. \tag{141.21}$$

Differentiating (141.20) a second time, we then obtain

$$\sum_n e^{\frac{\psi-E_n}{\theta}} \left[\frac{1}{\theta}\left(\frac{\partial\psi}{\partial a} - \frac{\partial E_n}{\partial a}\right)^2 + \frac{\partial^2\psi}{\partial a^2} - \frac{\partial^2 E_n}{\partial a^2}\right] = 0,$$

which, with the help of (141.17) and (141.21), can be written in the form

$$\sum_n e^{\frac{\psi-E_n}{\theta}} \left[\frac{1}{\theta}(-\bar{\bar{A}}+A_n)^2 - \frac{\partial\bar{\bar{A}}}{\partial a} + \frac{\partial A_n}{\partial a}\right] = 0. \tag{141.22}$$

This then gives us the desired expression for the mean square deviations in A,

$$\overline{\overline{(A-\bar{\bar{A}})^2}} = \theta\left[\frac{\partial\bar{\bar{A}}}{\partial a} - \overline{\overline{\left(\frac{\partial A}{\partial a}\right)}}\right], \tag{141.23}$$

which can also be rewritten in the form

$$\frac{\overline{\overline{(A-\bar{\bar{A}})^2}}}{(\bar{\bar{A}})^2} = \frac{kT}{(\bar{\bar{A}})^2}\left[\frac{\partial\bar{\bar{A}}}{\partial a} - \overline{\overline{\left(\frac{\partial A}{\partial a}\right)}}\right]. \tag{141.24}$$

The term $\partial\overline{\overline{A}}/\partial a$ in this expression has an immediate macroscopic interpretation as the change in external force with the associated generalized coordinate. The other term $\overline{\overline{(\partial A/\partial a)}}$ has no such immediate interpretation and must be calculated in any specific case from its relation to the mean value of $\partial^2 E_n/\partial a^2$. Hence we cannot make the unqualified assertion that fluctuations in all kinds of external forces will be small.

A specific calculation of the fluctuations in the pressure p of a perfect gas has been made with the help of the above formula by Fowler.† Taking conditions where the calculation could be made classically, and making reasonable assumptions as to the law of force for the molecules at the walls, he was led to fluctuations of the negligible amount

$$\frac{\overline{\overline{(p-\overline{\overline{p}})^2}}}{(\overline{\overline{p}})^2} \approx 4{\cdot}7\times 10^{-12}$$

for a cubic centimetre of gas under standard conditions. This result is of the order $1/n^{\frac{2}{3}}$, where n is the number of molecules of gas.

In closing these remarks on the rigorous calculation of fluctuations in canonical ensembles, it may be mentioned that the methods used above may be readily extended to the calculation of fluctuations in energy $\overline{\overline{(E-\overline{\overline{E}})^n}}$ of any desired order n, and to fluctuations for quantities involving external forces of the types

$$\overline{\overline{(A-\overline{\overline{A}})(E-\overline{\overline{E}})}} \quad \text{and} \quad \overline{\overline{(A_1-\overline{\overline{A}}_1)(A_2-\overline{\overline{A}}_2)}},$$

where A_1 and A_2 are different external forces.‡

(*d*) **Einstein's formula for fluctuations in a macroscopic variable.** In addition to the foregoing rigorous expressions for the treatment of fluctuations in canonical ensembles, a somewhat approximate but important practical method of treating the fluctuations in a macroscopic variable describing the condition of a system has been devised by Einstein.|| We shall continue the present section by showing that Einstein's formula—connecting the probability of a fluctuation with a corresponding increase in free energy—can also be made plausible on the basis of our present correlation of thermodynamic with statistical-

† Fowler, *Statistical Mechanics*, second edition, Cambridge, 1936, chapter xx, p. 756. This chapter also gives a treatment, p. 775, to the fluctuations in the time integral as well as in the instantaneous value of pressure, and may be consulted in general for the specific treatment of many important fluctuation problems.

‡ See Gibbs, *Elementary Principles in Statistical Mechanics*, Yale University Press, 1902, chapter vii.

|| Einstein, *Ann. der Phys.* **33**, 1275 (1910).

mechanical quantities, as well as on the basis of the Boltzmann correlation between entropy and probability that was used by Einstein.

The Einstein formula is to be taken as applying to the fluctuations at equilibrium, around its mean value x_0, in some variable x, which describes the condition of the system of interest from a sufficiently large-scale point of view, so that it would be possible to regard this quantity as macroscopically controllable and appropriate to consider the dependence of the free energy of the system on such a control if it were introduced. According to circumstances, this control might apply to the actual value of the variable or to some limit placed on its values. As an example we could consider the fluctuations, around its mean value v_0 in the volume v of some selected portion of a fluid, taken large enough so that it would be appropriate to consider the different free energies of the system if the volume v of the portion were controlled to different upper limits by the introduction of a suitable adjustable partition.

To treat the equilibrium fluctuations in such a variable x, we may regard the equilibrium condition of the system as represented by a canonical ensemble, and hence may take the probability for any state n of the system as given by an expression of the form

$$P_n = e^{\frac{\psi - E_n}{\theta}}, \tag{141.25}$$

where ψ and θ are parameters and E_n is the energy of the state n. This formula applies to all possible states n of the system and hence to states corresponding to all possible values of the freely fluctuating variable that we are considering. Included among these states n will be those—designated, say, by the symbol $n(x)$—which would be compatible with any definite value x at which we might introduce a control for that variable. For example, returning to our previous illustration, where the fluctuating variable is taken as the volume of a portion of a fluid, there would be included among all possible states n of the fluid those states $n(v)$ which would be compatible with the enclosure of that portion of the fluid inside an inextensible membrane of constant volume which would control the actual volume of the portion to the upper limiting value v. For the total probability of all states $n(x)$ which would be compatible with such a controlled value of the variable x, we can evidently write

$$P(x) = \sum_{n=n(x)} e^{\frac{\psi - E_n}{\theta}} = C \sum_{n=n(x)} e^{-E_n/\theta}, \tag{141.26}$$

where we sum over all those states $n = n(x)$ which are, in fact, compatible with that assigned value, and C is a constant independent of that value.

This expression for the total probability of the states $n(x)$ under consideration may now easily be re-expressed in thermodynamic language by writing

$$\begin{aligned} P(x) &= C\exp\Big(\log \sum_{n=n(x)} e^{-E_n/\theta}\Big) \\ &= C\exp\Big(\frac{1}{kT}\,kT\log \sum_{n=n(x)} e^{-E_n/kT}\Big) \\ &= C\exp\{-A(x)/kT\}, \end{aligned} \tag{141.27}$$

where, in accordance with (122.9), the quantity $A(x)$ is seen to be the free energy that would be ascribed to the system as a whole if the variable under consideration were definitely controlled at the value x instead of being allowed to fluctuate freely. Furthermore, by considering the similar expression which would give the probability $P(x_0)$ for all states $n(x_0)$ compatible with the mean value x_0 for our variable, we can rewrite the above in the convenient form

$$P(x) = P(x_0)\,e^{-\frac{A(x)-A(x_0)}{kT}}, \tag{141.28}$$

where $A(x_0)$ is the free energy that would be ascribed to the system with the variable in question controlled at its mean value x_0.

This formula is not sufficient in general to lead to precise information as to the probabilities for fluctuations of a given extent x, since from a strict point of view $P(x)$ is the total probability for all states $n(x)$ compatible with some physical control which could be introduced on our variable at the point x, and this is not necessarily the same as the probability for the definite value x of that variable. Thus, for example, returning to our previous illustration where the fluctuating variable is the volume v of a portion of the fluid, the symbol $P(v)$ would denote the probability for all states $n(v)$ of the fluid such that the volume of the portion in question would not exceed the value v. Nevertheless, the above formula is often sufficient to give us the qualitative information that the probabilities for fluctuations of x away from its mean value x_0 would be mainly determined by the exponential factor in (141.28), since for any appreciable difference between x and x_0 the increase in free energy $A(x)-A(x_0)$ would be very large compared with kT, and the main effect would in any case be a corresponding exponential decrease in probability as we go to larger fluctuations from the mean. Thus, for example, returning once more to our previous illustration, we realize that the vast majority of states in which the volume of the selected portion of fluid would not be greater than v would be ones in which the volume was nearly equal to v, and hence that the decrease in probability as we go to fluctuations of v farther and farther from v_0 would be essentially determined by the exponential form of the expression given.

In view of the foregoing discussion, it will now often be appropriate, following Einstein, to express the probability dP for finding our fluctuating variable within a definite range x to $x+dx$ in the form

$$dP = f(x)\, e^{-\frac{A(x)-A(x_0)}{kT}}\, dx, \tag{141.29}$$

where $f(x)$ is whatever function of x that may be needed to make the equation valid, and $A(x)$ and $A(x_0)$ have their previous significance. Furthermore, it will often be plausible to assume that $f(x)$ is only a gradually varying function of x and that the main dependence of the probability of x is given by the exponential factor.† Hence, if we now regard the development of $f(x)$ around x_0,

$$f(x) = f(x_0)+\left(\frac{\partial f}{\partial x}\right)_{x=x_0}(x-x_0)+..., \tag{141.30}$$

it will often be allowable to rewrite (141.29) in the approximate form

$$dP = f(x_0)\, e^{-\frac{A(x)-A(x_0)}{kT}}\, dx, \tag{141.31}$$

since this would be a good expression for the probabilities of fluctuations when $x-x_0$ is small, and the probabilities for fluctuations making $x-x_0$ large become in any case so small that it is not then important to have a precisely correct form of expression for them. We are thus led to Einstein's approximate expression for the probabilities of fluctuations in a macroscopic variable x, where $f(x_0)$ is a constant and $A(x)-A(x_0)$ is the increase in free energy for the system as a whole which would correspond to a change in our conceptual macroscopic control on that variable from x_0 to x.

In applying this formula to actual cases it is often convenient to regard $A(x)$ itself as developed around the value $A(x_0)$ in the form

$$A(x)-A(x_0) = \left(\frac{\partial A}{\partial x}\right)_{x=x_0}(x-x_0)+\frac{1}{2!}\left(\frac{\partial^2 A}{\partial x^2}\right)_{x=x_0}(x-x_0)^2+\frac{1}{3!}\left(\frac{\partial^3 A}{\partial x^3}\right)_{x=x_0}(x-x_0)^3+\dots. \tag{141.32}$$

† For example, when x is the volume v of a portion of a fluid, and the external parameter corresponding to x in terms of which A is evaluated may be regarded as the upper limit of this volume, $f(v)$ turns out to be proportional to p/kT, where p is the pressure which the portion of fluid would exert at the volume v. This is indeed a slowly varying function of v.

In substituting this expression into (141.31), we may always take $(\partial A/\partial x)_{x=0} = 0$ and drop the first term, since the free energy of an enclosed system at constant temperature is a minimum at equilibrium, and hence may be taken as a minimum in the present situation at the mean value $x = x_0$ which from a thermodynamic point of view would be regarded as *the* value of x at equilibrium. Furthermore, in substituting this expression it will then in general only be necessary to consider the first of the further terms for which $(\partial^n A/\partial x^n)_{x=x_0}$ does not vanish, since the later terms will be small in the region of small fluctuations which are the important ones to consider.

It will be interesting to investigate the probabilities for fluctuations in the quite common case where the second derivative of the free energy does not vanish. We can then write (141.32) in the form

$$A(x)-A(x_0) \approx \frac{1}{2}\left(\frac{\partial^2 A}{\partial x^2}\right)_{x=x_0}(x-x_0)^2. \tag{141.33}$$

Substituting in (141.31), we have

$$dP = f(x_0)e^{-\frac{1}{2kT}\left(\frac{\partial^2 A}{\partial x^2}\right)_{x=x_0}(x-x_0)^2}\,dx \tag{141.34}$$

for the probability of a given fluctuation in x. With the help of (141.34) we may then compute the mean square fluctuation

$$\overline{\overline{(x-x_0)^2}} = \frac{\int\limits_{-\infty}^{+\infty} e^{-\frac{1}{2kT}\left(\frac{\partial^2 A}{\partial x^2}\right)_{x=x_0}(x-x_0)^2}(x-x_0)^2\,d(x-x_0)}{\int\limits_{-\infty}^{+\infty} e^{-\frac{1}{2kT}\left(\frac{\partial^2 A}{\partial x^2}\right)_{x=x_0}(x-x_0)^2}\,d(x-x_0)}, \tag{141.35}$$

where we integrate over all possible values of $x-x_0$, and do not need to feel concerned by the inaccuracy of our integrands in ranges where $x-x_0$ is large since the contribution to the total integrals will in any case be small in such ranges. With the help of known formulae of integration (Appendix II) we thus obtain

$$\overline{\overline{(x-x_0)^2}} = \frac{kT}{(\partial^2 A/\partial x^2)_{x=x_0}} \tag{141.36}$$

for the mean square fluctuation in the variable x, or, with the help of (141.33),

$$\overline{\overline{(A-A_0)}} = \tfrac{1}{2}kT \tag{141.37}$$

for the mean fluctuation in the corresponding free energy.

We have presented the foregoing account of the Einstein method, in spite of its not very rigorous character, because of its generality and ready applicability in giving an idea of the magnitude of fluctuations that can be expected in practical cases. For example, the above formula at once indicates that we can expect the mean kinetic energy per degree of freedom of any small part of an apparatus to fluctuate by a quantity of the order of $\frac{1}{2}kT$ at thermodynamic equilibrium. The method was actually devised by Einstein in order to discuss the fluctuations in the density of a fluid in the neighbourhood of its critical point which give rise to the phenomenon of critical opalescence. We shall prefer to treat this problem, however, with the help of a discussion of the fluctuations that would be predicted for grand canonical ensembles to which we now turn.

(*e*) **Fluctuations in composition.** In the foregoing parts of this section we have been concerned with fluctuations at thermodynamic equilibrium in quantities pertaining to a 'closed' thermodynamic system. These we have calculated on the assumption that the system of interest could be appropriately represented by a canonical ensemble. We must now consider fluctuations at thermodynamic equilibrium in quantities —in particular in the composition—characterizing an 'open' thermodynamic system which can interchange component substances, with some other system or 'reservoir' with which it is connected in macroscopic equilibrium. These we shall calculate on the assumption that the system of interest, when connected with a sufficiently large reservoir of material can be appropriately represented by a grand canonical ensemble.

To make the desired calculation of the fluctuations to be expected in the amount of any component i contained in a system represented by a grand canonical ensemble, it is convenient to start with the expression (140.24) for the mean number of molecules of that kind in the members of the ensemble

$$\overline{\overline{n}}_i = \sum_{n_1 \dots n_h, m} e^{\frac{\Omega + \mu_1 n_1 + \dots + \mu_h n_h - E_m}{\theta}} n_i, \tag{141.38}$$

and consider the effect of changing to a neighbouring grand canonical distribution by making a small variation $\delta\mu_i$ in the 'potential' for the component i of interest, without changing the remaining 'potentials' $\mu_1 \dots \mu_h$, the 'temperature' θ, or the 'external coordinates' $a_1, a_2, \dots$ such as volume which determine the eigenvalues of energy E_m. Such a change would correspond to a change in the system of interest from one condition of equilibrium to a neighbouring one of altered composition but at the same temperature and with the same values for external coordinates such as volume.

Applying the suggested variation to (141.38), we obtain

$$\delta\overline{\overline{n}}_i = \sum_{n_1\ldots n_h,m} e^{\frac{\Omega+\mu_1 n_1+\ldots+\mu_h n_h-E_m}{\theta}}\left(\frac{n_i}{\theta}\delta\Omega+\frac{n_i^2}{\theta}\delta\mu_i\right)$$

$$= \frac{\overline{\overline{n}}_i}{\theta}\delta\Omega+\frac{\overline{\overline{(n_i^2)}}}{\theta}\delta\mu_i. \tag{141.39}$$

In carrying out such a variation, however, we must retain the validity of the fundamental equation

$$\sum_{n_1\ldots n_h,m} e^{\frac{\Omega+\mu_1 n_1+\ldots+\mu_h n_h-E_m}{\theta}} = 1 \tag{141.40}$$

which makes the total probability of finding a member of the ensemble in one or another possible condition equal to unity. Hence the above variation must be carried out subject to the restriction

$$0 = \sum_{n_1\ldots n_h,m} e^{\frac{\Omega+\mu_1 n_1+\ldots+\mu_h n_h-E_m}{\theta}}\left(\frac{\delta\Omega}{\theta}+\frac{n_i}{\theta}\delta\mu_i\right)$$

$$= \frac{\delta\Omega}{\theta}+\frac{\overline{\overline{n}}_i}{\theta}\delta\mu_i. \tag{141.41}$$

Combining (141.41) with (141.39), we have

$$\delta\overline{\overline{n}}_i = \frac{\overline{\overline{(n_i^2)}}}{\theta}\delta\mu_i-\frac{(\overline{\overline{n}}_i)^2}{\theta}\delta\mu_i$$

or

$$\frac{\overline{\overline{(n_i^2)}}-(\overline{\overline{n}}_i)^2}{(\overline{\overline{n}}_i)^2} = \frac{\theta}{(\overline{\overline{n}}_i)^2}\left(\frac{\delta\overline{\overline{n}}_i}{\delta\mu_i}\right). \tag{141.42}$$

This expression for fluctuations in composition is rigorous and general, in the form as written. For estimating the actual magnitude given by (141.42), however, it is often convenient to evaluate $(\delta\overline{\overline{n}}_i/\delta\mu_i)$ with the help of its thermodynamic analogue $1/(\delta\mu_i/\delta N_i)$. This procedure is in general justified whenever the fluctuations in composition are small enough so that this analogy retains the needed degree of validity, but can lead to incorrect results in the immediate neighbourhood of a point where $(\delta\mu_i/\delta N_i)$ vanishes and its lack of continued proportionality to $(\delta\mu_i/\delta\overline{\overline{n}}_i)$ becomes all-important—e.g. when exceedingly close to the critical point of a fluid.

Making the suggested change to thermodynamic language with the help of (140.32), and re-expressing the left-hand side of (141.42) by an obvious transformation, we then obtain, as a convenient expression for the practical computation of fluctuations in the amount of a component i of an 'open' system at equilibrium,

$$\frac{\overline{(n_i-\overline{\overline{n}}_i)^2}}{(\overline{\overline{n}}_i)^2} = \frac{RT}{\overline{\overline{n}}_i N_i(\delta\mu_i/\delta N_i)}, \tag{141.43}$$

where R is the gas constant per mol, N_i is the number of mols of the component i in the system that would be assigned on the basis of thermodynamic considerations—i.e. $\overline{\overline{n}}_i$ divided by Avogadro's number —and $(\delta\mu_i/\delta N_i)$ is the rate of change in the potential of that substance with its amount, corresponding to a change in composition of the system which leaves the *potentials* of the other components, the temperature, and external parameters such as volume unaltered.

This result now supplies a necessary link in justifying the use that we have made of the grand canonical ensemble to represent an 'open' thermodynamic system in a condition of equilibrium. As already noted in § 140 (*a*), we can expect the quantity $(\delta\mu_i/\delta N_i)$ to be a positive quantity greater than zero, except under special circumstances involving for example the existence of the component i in separate phases. Hence, in typical cases, we see from (141.43) that we can expect the fractional mean square fluctuations in composition to become negligible as we take systems containing a sufficiently large mean number of molecules $\overline{\overline{n}}_i$ of the component under consideration. For example, in case the component i is one of a mixture of perfect gases, we have found (140.20) that we can take

$$\frac{\delta\mu_i}{\delta N_i} = \frac{RT}{N_i}, \tag{141.44}$$

which on substitution in (141.43) gives

$$\frac{\overline{(n_i-\overline{\overline{n}}_i)^2}}{(\overline{\overline{n}}_i)^2} = \frac{1}{\overline{\overline{n}}_i}. \tag{141.45}$$

In accordance with the foregoing, the members of a grand canonical ensemble will in general have compositions highly concentrated about the mean composition in the typical cases where we should wish to use such an ensemble to represent a thermodynamic system, and appropriate treatment can be given to the special circumstances where large fluctuations would be predicted, circumstances which often correspond to cases where there would be indeterminacy even from a thermodynamic point of view. This, then, explains the validity of the thermodynamic procedure of treating 'open' systems as having in general a composition which can be regarded as perfectly definite from a macroscopic point of view.

It is of interest to appreciate that the high concentration around the mean composition in the case of a typical grand canonical distribution makes its properties nearly the same as those of a petit canonical distribution for a composition taken the same as that mean. This, then, makes it unnecessary to give special attention to fluctuations in quantities other than composition, in the case of systems represented by a grand canonical distribution, since our previous study of such fluctuations with the help of petit canonical distributions will give us in general a reliable idea of what is to be expected.

It is also of interest to note that our present quasi-thermodynamic method, of showing that we can in general expect the members of a grand canonical ensemble to be highly concentrated around a mean composition, can now be regarded as providing a new justification for the use that we made of the grand canonical distribution in § 114 as giving a good description of the equilibrium properties of a simple Einstein-Bose or Fermi-Dirac gas composed of a specified number of particles.

Moreover, it will be of interest in this connexion to show that our previous non-thermodynamic discussion of the distribution of the members of a grand canonical ensemble around their most probable particle content, as given in § 114 (*b*), can now itself be translated into more thermodynamic language. For this purpose we have only to note that our previous quantities C and α, used in (114.21) in setting up the grand canonical distribution, and the quantity $\gamma(n)$ defined by (114.24) can be re-expressed in our present terminology for a system of one component by

$$C = e^{\Omega/kT}, \qquad \alpha = -\frac{\mu}{kT}, \qquad \gamma(n) = \frac{\psi(n)}{kT} = \frac{A(n)}{kT}, \tag{141.46}$$

where $A(n)$ is the free energy which would be ascribed to a system of n molecules at temperature T. Comparing with (114.27), we then obtain

$$P(n) = e^{\dfrac{\Omega+\mu\tilde{n}-A(\tilde{n})-\frac{1}{2!}\left(\frac{\partial^2 A}{\partial n^2}\right)_{n=\tilde{n}}(n-\tilde{n})^2-\frac{1}{3!}\left(\frac{\partial^3 A}{\partial n^3}\right)_{n=\tilde{n}}(n-\tilde{n})^3-\frac{1}{4!}\left(\frac{\partial^4 A}{\partial n^4}\right)_{n=\tilde{n}}(n-\tilde{n})^4\ldots}{kT}} \tag{141.47}$$

as an expression for the probability of finding a member of the ensemble containing n molecules, where $\tilde{n}$ is the most probable number of molecules. Furthermore, our previous approximate expression (114.30) for the probability of fluctuations around the most probable number of molecules, when $(\partial^2 A/\partial n^2)_{n=\tilde{n}}$ is not equal to zero and higher derivatives can be neglected, can now be written in the form

$$\frac{\overline{\overline{(n-\tilde{n})^2}}}{(\tilde{n})^2} = \frac{kT}{(\tilde{n})^2(\partial^2 A/\partial n^2)_{n=\tilde{n}}}. \tag{141.48}$$

Equation (141.47) can be used whenever the analogy $A \rightleftharpoons \psi$ is sufficiently valid, and hence sometimes when the analogy $\mu \rightleftharpoons N_A \mu$ breaks down, e.g. at the critical point of a fluid. When $(\partial^2 A/\partial n^2)$ is not too small both analogies hold, and equation (141.48) reduces to (141.43), with $\tilde{n} \approx \overline{\overline{n}} \approx N_A N$ and $(\partial^2 A/\partial n^2) \approx N_A^{-2}(\delta\mu/\delta N)$.

(*f*) **Fluctuations in density of a fluid.** As an important example of the application of fluctuation theory to empirical problems, we may use our general formula (141.43) for mean square fluctuations in composition to investigate the observationally interesting case of fluctuations in the density of a single pure fluid. For this purpose let us consider a small portion of the fluid, located in some specified volume v, and containing in the mean $\overline{\overline{n}}$ molecules, and let us treat this as being an 'open' thermodynamic system which is in equilibrium with the rest of the fluid which serves as a large reservoir for accommodating fluctuations in the amount of fluid in the specified volume v. Representing this small system by a grand canonical distribution, we can then write, in accordance with (141.43),

$$\frac{\overline{\overline{(n-\overline{\overline{n}})^2}}}{(\overline{\overline{n}})^2} = \frac{RT}{\overline{\overline{n}}N(\delta\mu/\delta N)} \tag{141.49}$$

as an expression describing the fluctuations in the number of molecules in the system, where R is the gas constant per mol, N is the number of mols in the volume v from the point of view of thermodynamics—i.e. $\overline{\overline{n}}$ divided by Avogadro's number—and

$$\frac{\delta\mu}{\delta N} = \left(\frac{\partial\mu}{\partial N}\right)_{T,v} \tag{141.50}$$

is the rate at which the Gibbs potential μ for the fluid in the little system would change with the number of mols N in it, holding its volume v and its temperature T constant.

Before making use of the above formula for fluctuations we shall wish to change it into a form similar to that which has actually been employed in the interpretation of observations, and which is based on a knowledge of the equation of state for the fluid rather than of μ, N, v, and T. For this purpose we may make use of the definition of potential

μ in terms of free energy which was given by (140.17) in the preceding section. We can then write

$$\mu = -\int_{\bar{v}_0}^{\bar{v}} p \, d\bar{v} + p\bar{v} \tag{141.51}$$

as an expression for the potential μ of the fluid in our little system, where the first term is the isothermal reversible work per mol necessary to change a portion of the fluid from its standard state of zero free energy where it has the volume $\bar{v}_0$ per mol to the volume $\bar{v}$ per mol which it has in our system, and the second term is the reversible work per mol necessary to introduce a small portion of it into our system against the pressure p which it then has. Substituting

$$\bar{v} = \frac{v}{N}, \tag{141.52}$$

where v is the volume and N the number of mols of fluid for our system, the above can be rewritten in the form

$$\mu = -\int_{\bar{v}_0}^{v/N} p \, d\bar{v} + p\frac{v}{N}, \tag{141.53}$$

which by differentiation then gives

$$\frac{\delta\mu}{\delta N} = \left(\frac{\partial\mu}{\partial N}\right)_{T,v} = p\frac{v}{N^2} - p\frac{v}{N^2} + \frac{v}{N}\left(\frac{\partial p}{\partial N}\right)_{T,v} = \frac{v}{N}\left(\frac{\partial p}{\partial N}\right)_{T,v}. \tag{141.54}$$

Furthermore, we can evidently write

$$N\left(\frac{\partial p}{\partial N}\right)_{T,v} = -v\left(\frac{\partial p}{\partial v}\right)_{T,N} \tag{141.55}$$

since, for any substance describable by an equation of state, the increase in pressure at constant temperature and volume for a given fractional increase in contents must be equal to the increase at constant temperature and contents for the same fractional decrease in volume. Combining with the preceding equation, we then have

$$\frac{\delta\mu}{\delta N} = -\frac{v^2}{N^2}\left(\frac{\partial p}{\partial v}\right)_{T,N}. \tag{141.56}$$

It is also convenient to note that we can evidently write

$$\frac{\overline{\overline{(n-\overline{\overline{n}})^2}}}{(\overline{\overline{n}})^2} = \overline{\overline{\left(\frac{\Delta\rho}{\rho}\right)^2}}, \tag{141.57}$$

where $\Delta\rho$ is a fluctuation in the density of the fluid in our little system around its mean value ρ.

Substituting (141.56) and (141.57) into (141.49), we can then write our expression for the fluctuations in density of a fluid in the familiar forms

$$\overline{\overline{\left(\frac{\Delta\rho}{\rho}\right)^2}} = -\frac{RT}{\dfrac{\bar{\bar{n}}}{N}v^2\left(\dfrac{\partial p}{\partial v}\right)_{T,N}} = -\frac{kT}{v^2\left(\dfrac{\partial p}{\partial v}\right)_{T,N}}. \tag{141.58}$$

We at once note that such fractional fluctuations in density would increase rapidly as we consider smaller and smaller portions of the fluid.

If the fluid in question is a perfect gas with

$$p = \frac{NRT}{v}, \tag{141.59}$$

the above formula leads to the result

$$\overline{\overline{\left(\frac{\Delta\rho}{\rho}\right)^2}} = \frac{1}{\bar{\bar{n}}}, \tag{141.60}$$

where $\bar{\bar{n}}$ is the mean number of molecules of gas in the little volume that has been selected. The percentage fluctuations in density are hence very small unless the portion of gas considered is very small. Nevertheless, such fluctuations play the interesting role of providing small scattering centres for blue light which lead to the blue colour of the sky.

For a fluid in the neighbourhood of its critical point the fluctuations predicted by (141.58) would become very large, since right at the critical point the first and second derivatives of pressure with respect to volume, for a given amount of fluid at the critical temperature, go to zero,

$$\left(\frac{\partial p}{\partial v}\right)_c = \left(\frac{\partial^2 p}{\partial v^2}\right)_c = 0. \tag{141.61}$$

This accounts in a qualitative way for the striking appearance of opalescence in an illuminated fluid as we approach the critical point, the scattering centres now becoming very important and large enough to scatter light of longer wave-length than the blue. This qualitative explanation of the phenomenon of critical opalescence, together with a quantitative discussion of its statistical-mechanical treatment, was first given by Smoluchowski.†

To make a quantitative application of (141.58) to fluctuations in density in the neighbourhood of the critical point, we note, in accordance with (141.61), that the value of $(\partial p/\partial v)$ in that neighbourhood would be closely given by

$$\left(\frac{\partial p}{\partial v}\right)_{T,N} = \left(\frac{\partial^2 p}{\partial T\partial v}\right)_c (T-T_c), \tag{141.62}$$

† Smoluchowski, *Ann. der Phys.* **25**, 205 (1908).

where $(T-T_c)$ is the difference between the actual temperature and the critical temperature. Substituting in (141.58), we obtain

$$\overline{\left(\frac{\Delta\rho}{\rho}\right)^2} = -\frac{kT}{v^2\left(\frac{\partial^2 p}{\partial T \partial v}\right)_c (T-T_c)} \tag{141.63}$$

for the expected fluctuations in density ρ inside a specified volume v containing a portion of a fluid in the neighbourhood of its critical point.†

A quantitative test of this formula was made by Keesom‡ for the substance ethylene by using the scattering of light by the ethylene as providing a method of investigating the fluctuations in its density ρ. In the first place, by comparing the intensities of scattered light of a given wave-length over a range of different temperatures T (11·24–13·53°) in the neighbourhood of the critical temperature (11·18° for the sample of not quite pure ethylene employed), it was possible to show that the value of $(\Delta\rho/\rho)^2$ was indeed quite closely inversely proportional to the distance from the critical temperature $(T-T_c)$. In the second place, by showing that the ratio between the scattering powers for light of wave-lengths in the neighbourhood of the F and of the D Fraunhofer lines decreases as the critical temperature is approached, it was possible to conclude that the fluctuations become important in larger and larger volumes v as $(T-T_c)$ is decreased. Finally, by making a careful determination of the absolute intensity of the scattered light for a particular wave-length (D) and at a particular temperature (11·73°), it was possible to verify the actual magnitude of the fluctuations $(\Delta\rho/\rho)^2$ predicted by the formula (141.63), inserting therein an empirical value for the derivative $(\partial^2 p/\partial T \partial v)_c$.

The empirical check on the theory of fluctuations in the density of a fluid is thus very satisfactory. Other familiar examples, where the

† It will be observed that (141.63) would lead to the prediction of infinite fluctuations at the critical point where $T = T_c$. It must be remembered, however, that the evaluation of $(\delta\mu/\delta\bar{\bar{n}})$ in terms of the thermodynamic $(\delta\mu/\delta N)$—on which the derivation was based—can fail when fluctuations in composition become too large. When (141.63) thus fails, it is possible to use the distribution described by (141.47) for calculating the fluctuations since the thermodynamic analogy $A \rightleftharpoons \psi$ remains valid as we approach the critical point from higher temperatures. At the critical point, noting that the second and third derivatives of A with respect to n will vanish, and using an appropriate equation of state to evaluate the fourth derivative, this then leads to fluctuations $(\Delta\rho/\rho)^2$ of the order $\tilde{n}^{-\frac{1}{2}}$. The failure of $\mu_i \rightleftharpoons N_A\mu_i$ when fluctuations in composition become too large is connected with the necessity of correlating $\bar{\bar{n}}_i$ with N_i in (140.30) and (140.31). The continued validity of $A \rightleftharpoons \psi$ is connected with non-appearance of those quantities in (122.2) and (122.3).

‡ Keesom, *Ann. der Phys.* **35**, 591 (1911).

statistical-mechanical treatment of the fluctuations to be expected in a system at thermodynamic equilibrium has been quantitatively verified, are provided by observations on the distribution of Brownian particles in a gravitational field, by observations on the linear displacements suffered by such particles as time proceeds, and by observations on the torsional oscillations of a small mirror suspended in a high vacuum. The observational existence and empirical measurement of the fluctuations, predicted by statistical mechanics, in quantities to which precise values are assigned from the standpoint of thermodynamics, provides important justification for our belief that statistical mechanics is indeed a more fundamental and more valid science than thermodynamics.

142. Conclusion

This now brings our long book on statistical mechanics to its end by completing our proposed discussion of the statistical-mechanical explanation of the principles of thermodynamics. It is pleasing to have come to this end with the above illustration, which shows so clearly the deeper and more powerful character of the newer science of statistical mechanics.

Much has been omitted from the book in the way of considering important applications of statistical mechanics to problems of physical and chemical interest. In particular we may regret the omission of further and more specific consideration of the applications of statistical mechanics to studies on the actual rates at which physical-chemical systems proceed towards their conditions of equilibrium. This is a field in which much remains to be done and in which the methods of statistical mechanics are specially needed.

It is hoped, however, that our exposition of theory has been sufficiently sound and complete to solve our main task of giving a true insight into the methods that must be employed when we wish to predict the behaviour of a mechanical system on the basis of less knowledge as to its actual state than would in principle be allowable and possible. Such partial knowledge of state is in reality all that we ever do have, and the discipline of statistical mechanics must always remain necessary, whatever changes we may have to make in pure mechanics itself from classical to quantum or to any future form.

APPENDIX I

SYMBOLS FOR QUANTITIES, OPERATORS, AND MATRICES

CONVENTIONS. *Scalar quantities* are denoted by letters in italic or Greek type, e.g. A, a, α. *Vector quantities*, which occur only rarely in the text, are denoted by letters in Clarendon or bold-face type, e.g. **A**, **E**. *Quantum mechanical operators*, which occur frequently in the text, and a few other operators, are also denoted by letters in Clarendon or bold-face type, e.g. **H**, **q**, **p**. *Matrices* are either denoted with the help of double lines, e.g. $\|M\|$, or more frequently by an expression for the elements of the matrix, e.g. M_{ij}.

The *complex conjugate* of a quantity is denoted by an asterisk, e.g. A^*. The *modulus* of a quantity is denoted by single lines, e.g. $|A|$. The *mean value*, or the quantum mechanical expectation value, of a quantity pertaining to a *single system* is denoted by a single bar, e.g. $\bar{A}$. The *mean value* of a quantity for the members of an *ensemble of systems* is denoted by a double bar, e.g. $\bar{\bar{A}}$. The *mode* or most probable value of a quantity is denoted by a tilde, e.g. $\tilde{A}$.

The same letter may be used to denote a *quantum mechanical observable* F, its *eigenvalues* F_k, its *expectation value* $\bar{F}$, the corresponding *operator* **F**, and the *matrix* $\|F\|$ composed of the *elements* F_{ij} which can be obtained with the help of that operator.

Only those symbols are included in the following tables which occur frequently in the text with the significance given. Some of these same symbols appear in a limited context with a different significance.

Scalar Quantities.

$a(k,t)$, $a_k(t)$	generalized probability amplitude for state k at time t.
$a_1, a_2, a_3, \ldots$	external coordinates for a thermodynamic system.
$A_1, A_2, A_3, \ldots$	corresponding external forces.
A	free energy of a system, $E-TS$.
b	critical distance for a molecular collision.
$b(r,t)$, $b_r(t)$	generalized probability amplitude for state r at time t.
c	velocity of light.
c_n	probability coefficient for energy eigenstate n.
$c_k(t)$	probability coefficient for unperturbed energy eigenstate k at time t.
C	constant.
C_v, C_p	heat capacities at constant volume or pressure.

e base of natural logarithms; charge of fundamental particle.
E energy of a system.
E_n energy eigenvalue for state n.
E_k^0 unperturbed energy eigenvalue for state k.
f number of degrees of freedom of a system; fugacity.
$f(q_1 \cdots p_r)$ density of molecular distribution in μ-space.
F thermodynamic potential of a system, $E+pv-TS$.
$F(q,p)$ function of coordinates and momenta of a system.
g quantum weight for a degenerate molecular state.
g_0 quantum weight for a molecule of gas in ground level.
g_c quantum weight for a molecule of crystal in ground level.
g_k quantum weight for a degenerate molecular state k.
g_κ number of molecular quantum states in a group of neighbouring states κ.
G number of classical states—defined by equal volumes in the γ-space—or number of quantum states corresponding to specified condition of a system.
G_0 number of quantum states for a crystal at very low temperature.
G_k quantum weight for a degenerate state k.
G_κ number of quantum states for a system in a group of neighbouring states κ.
h Planck's constant.
H Hamiltonian expression for energy.
H^0 unperturbed Hamiltonian.
H Boltzmann function of the condition of a system.
$\bar{\bar{H}}$ Gibbs function of the condition of a representative ensemble.
i $\sqrt{-1}$.
$i, j, k, l, \ldots, r, s, \ldots$ indices specifying states.
I moment of inertia of diatomic molecule.
J rotational quantum number for a diatomic molecule.
k Boltzmann's constant.
K_p equilibrium constant in terms of partial pressures.
l quantum number for total angular momentum.
L Lagrangian function.

m	mass of particle or molecule; quantum number for a component of angular momentum.
m_0	rest mass.
M_x, M_y, M_z	components of angular momentum for a system.
n	number of molecules or other elements composing a system.
N	number of systems in an ensemble; number of mols in a system.
N_A	Avogadro's number.
p	generalized momentum; pressure.
p_x, p_y, p_z	components of linear momentum for a molecule.
P_x, P_y, P_z	components of linear momentum for a system.
P	probability, coarse-grained density in phase space.
P_k	coarse-grained probability for state k.
P_κ	probability for a group of neighbouring states κ.
q	generalized coordinate.
Q	heat absorbed.
r	radius; number of degrees of freedom for a molecule.
R	gas constant per mol.
s_x, s_y, s_z	spin variables.
S_x, S_y, S_z	components of probability current density.
S	entropy; Hamilton's characteristic function.
t	time.
T	temperature; kinetic energy; time interval.
$u(q_1 \ldots q_f)$	eigensolution not dependent on time.
$u(k, q)$, $u_k(q)$	eigenfunction corresponding to a state k.
$U_s(q_1 \ldots q_n)$, $U_a(q_1 \ldots q_n)$	symmetric and antisymmetric eigenfunctions for a system of n similar molecules.
u_x, u_y, u_z	components of velocity.
v	volume; velocity; vibrational quantum number for a diatomic molecule
$v(r, q)$, $v_r(q)$	eigenfunction corresponding to a state r.
V	potential energy; perturbation energy.
W	work done; Hamilton's principal function.
$W(q, t)$, $W(p, t)$	probability densities in coordinate or in momentum language at time t.
$W(n, t)$	probability for state n at time t.
x	molal fraction.
x, y, z	Cartesian coordinates.
Z	sum-over-states; number.

$Z_{\kappa\nu}$	rate of transition from condition κ to ν.
Z_{kl}^{ji}	rate of classical collisions $\begin{pmatrix} j, i \\ k, l \end{pmatrix}$.
α	parameter in molecular distribution laws.
β	parameter in distribution laws, correlated with $1/kT$.
ϵ	energy of a molecule.
ϵ_0	energy of a molecule in ground-level.
ϵ_k	energy of a molecule in state k.
ϵ_κ	energy of a molecule in condition κ.
θ	distribution parameter in canonical ensembles, correlated with kT.
Θ	characteristic temperature of a crystal.
r, θ, ϕ	polar coordinates.
$R(r), \Theta(\theta), \Phi(\phi)$	eigensolutions dependent on separated variables.
$\kappa, \lambda, \mu, \nu$	indices specifying groups of quantum states.
λ	wave-length.
$\delta\lambda$	solid angle defining a molecular collision.
Λ	heat of vaporization per mol.
μ	reduced mass.
μ_i	distribution parameter for ith component in grand canonical ensemble.
μ_i	Gibbs potential for ith component of a thermodynamic system.
ν	frequency.
π	ratio of circumference to diameter of circle.
ρ	density, fine-grained density in phase space.
σ	symmetry number.
σ_n	density of energy eigensolutions.
$\sigma_x, \sigma_y, \sigma_z$	wave numbers.
τ	time interval used in observation.
$\phi(p, t)$	probability amplitude in momentum language at time t.
$\psi(q, t)$	probability amplitude in coordinate language at time t.
ψ	distribution parameter in canonical ensemble, correlated with free energy A.
$\delta\omega$	infinitesimal range of internal coordinates and external and internal momenta defining a molecular state.
Ω	distribution parameter in grand canonical ensemble.

Operators (quantum mechanical).

F, **G** operators corresponding to observables in general.

H Hamiltonian operator.

$\mathbf{H}^0$ unperturbed Hamiltonian operator.

$\mathbf{M}_x, \mathbf{M}_y, \mathbf{M}_z$ operators corresponding to components of angular momentum.

$\mathbf{M}^2$ operator corresponding to square of total angular momentum.

$\mathbf{P}_x, \mathbf{P}_y, \mathbf{P}_z$ operators corresponding to components of linear momentum.

$\mathbf{p}_k$ operator corresponding to kth generalized momentum.

$\mathbf{q}_k$ operator corresponding to kth generalized coordinate.

s spin operator.

V perturbation operator.

Operators (miscellaneous).

I identity operator.

P operator permuting indices.

S unitary transformation operator to a different mode of quantum mechanical representation.

$\mathbf{U}(t)$ unitary transformation operator from $t = 0$ to $t = t$.

$\sum_i$ summation over indices i.

$\prod_i$ multiplication over indices i.

Matrix Elements.

F_{nm}, G_{nm}, H_{nm}, etc. corresponding to foregoing quantum mechanical operators.

δ_{kl} Kronecker delta $= \begin{cases} 1 \ (k = l) \\ 0 \ (k \neq l). \end{cases}$

I_{kl} elements corresponding to identity operator

$$= \begin{cases} 1 \ (k = l) \\ 0 \ (k \neq l). \end{cases}$$

S_{kr} elements of transformation matrix to a different mode of quantum mechanical representation.

$U_{nk}(t)$ elements of time transformation matrix from $t = 0$ to $t = t$.

APPENDIX II

SOME USEFUL FORMULAE

(*a*) Definite Integrals.†

$$\int_0^\infty e^{-x}x^{n-1}\,dx = \Gamma(n). \tag{1}$$

$$\int_0^\infty e^{-ax}x^{\rho}\,dx = \frac{\Gamma(\rho+1)}{a^{\rho+1}}. \tag{2}$$

$$\int_0^\infty e^{-ax^2}\,dx = \frac{1}{2}\sqrt{\frac{\pi}{a}}. \tag{3}$$

$$\int_0^\infty e^{-ax^2}x^2\,dx = \frac{1}{4}\sqrt{\frac{\pi}{a^3}}. \tag{4}$$

$$\int_0^\infty e^{-ax^2}x^4\,dx = \frac{3}{8}\sqrt{\frac{\pi}{a^5}}. \tag{5}$$

. . . .

. . . .

$$\int_0^\infty e^{-ax^2}x^{2k}\,dx = \frac{1.3\,\ldots\,(2k-1)}{2^{k+1}}\sqrt{\frac{\pi}{a^{2k+1}}}. \tag{6}$$

$$\int_0^\infty e^{-ax^2}x\,dx = \frac{1}{2a}. \tag{7}$$

$$\int_0^\infty e^{-ax^2}x^3\,dx = \frac{1}{2a^2}. \tag{8}$$

$$\int_0^\infty e^{-ax^2}x^5\,dx = \frac{1}{a^3}. \tag{9}$$

. . . .

. . . .

$$\int_0^\infty e^{-ax^2}x^{2k+1}\,dx = \frac{k!}{2a^{k+1}}. \tag{10}$$

† See, for example, B. O. Peirce, *Short Table of Integrals*, Ginn and Co., Boston.

(b) DIRICHLET INTEGRAL.†

General Formula.

$$I = \int \dots \iint x_1^{i_1-1} x_2^{i_2-1} \dots x_n^{i_n-1}\, dx_1\, dx_2 \dots dx_n$$

$$= \frac{a_1^{i_1} a_2^{i_2} \dots a_n^{i_n}}{p_1 p_2 \dots p_n} \frac{\Gamma\left(\frac{i_1}{p_1}\right)\Gamma\left(\frac{i_2}{p_2}\right)\dots\Gamma\left(\frac{i_n}{p_n}\right)}{\Gamma\left(\frac{i_1}{p_1}+\frac{i_2}{p_2}+\dots+\frac{i_n}{p_n}+1\right)} c^{\frac{i_1}{p_1}+\frac{i_2}{p_2}+\dots+\frac{i_n}{p_n}}, \tag{11}$$

when integrated over all *positive* values of x subject to the restriction

$$\left(\frac{x_1}{a_1}\right)^{p_1} + \left(\frac{x_2}{a_2}\right)^{p_2} + \dots + \left(\frac{x_n}{a_n}\right)^{p_n} \not> c, \tag{12}$$

with $i_1, i_2, \dots, i_n$; $a_1, a_2, \dots, a_n$; $p_1, p_2, \dots, p_n$; and c greater than zero.

Volume of Ellipsoid.

$$V = \iiint dxdydz = \tfrac{4}{3}\pi abcr^3, \tag{13}$$

when integrated over *all* values of x, y, z inside the ellipsoid

$$\left(\frac{x}{a}\right)^2 + \left(\frac{y}{b}\right)^2 + \left(\frac{z}{c}\right)^2 = r^2. \tag{14}$$

(c) FOURIER INTEGRAL.‡

For any continuous function $f(x)$ for which the integral $\int_{-\infty}^{+\infty} |f(x)|\, dx$ exists, we can write

$$f(x) = \frac{1}{\sqrt{(2\pi)}} \int_{-\infty}^{+\infty} g(y) e^{-ixy}\, dy, \tag{15}$$

with

$$g(y) = \frac{1}{\sqrt{(2\pi)}} \int_{-\infty}^{+\infty} f(x) e^{ixy}\, dx. \tag{16}$$

Similarly for a continuous function of several variables $f(x_1 \dots x_n)$ for which the integral $\int_{-\infty}^{+\infty} \dots \int_{-\infty}^{+\infty} |f(x_1 \dots x_n)|\, dx_1 \dots dx_n$ exists we can write

$$f(x_1 \dots x_n) = \frac{1}{(2\pi)^{n/2}} \int_{-\infty}^{+\infty} \dots \int_{-\infty}^{+\infty} g(y_1 \dots y_n) e^{-i(x_1 y_1 + \dots + x_n y_n)}\, dy_1 \dots dy_n, \tag{17}$$

with

$$g(y_1 \dots y_n) = \frac{1}{(2\pi)^{n/2}} \int_{-\infty}^{+\infty} \dots \int_{-\infty}^{+\infty} f(x_1 \dots x_n) e^{i(x_1 y_1 + \dots + x_n y_n)}\, dx_1 \dots dx_n. \tag{18}$$

† See, for example, Edwards, *Integral Calculus*, Macmillan, London, 1922, vol. ii, chapter xxv.

‡ See, for example, Courant and Hilbert, *Methoden der mathematischen Physik*, Springer, Berlin, 1924, chapter ii, § 6.

SUBJECT INDEX

NAME INDEX

A CATALOG OF SELECTED

DOVER BOOKS

IN SCIENCE AND MATHEMATICS

Astronomy

BURNHAM'S CELESTIAL HANDBOOK, Robert Burnham, Jr. Thorough guide to the stars beyond our solar system. Exhaustive treatment. Alphabetical by constellation: Andromeda to Cetus in Vol. 1; Chamaeleon to Orion in Vol. 2; and Pavo to Vulpecula in Vol. 3. Hundreds of illustrations. Index in Vol. 3. 2,000pp. 6¹/₈ x 9¹/₄.
Vol. I: 0-486-23567-X
Vol. II: 0-486-23568-8
Vol. III: 0-486-23673-0

EXPLORING THE MOON THROUGH BINOCULARS AND SMALL TELESCOPES, Ernest H. Cherrington, Jr. Informative, profusely illustrated guide to locating and identifying craters, rills, seas, mountains, other lunar features. Newly revised and updated with special section of new photos. Over 100 photos and diagrams. 240pp. 8¹/₄ x 11. 0-486-24491-1

THE EXTRATERRESTRIAL LIFE DEBATE, 1750–1900, Michael J. Crowe. First detailed, scholarly study in English of the many ideas that developed from 1750 to 1900 regarding the existence of intelligent extraterrestrial life. Examines ideas of Kant, Herschel, Voltaire, Percival Lowell, many other scientists and thinkers. 16 illustrations. 704pp. 5³/₈ x 8¹/₂. 0-486-40675-X

THEORIES OF THE WORLD FROM ANTIQUITY TO THE COPERNICAN REVOLUTION, Michael J. Crowe. Newly revised edition of an accessible, enlightening book re-creates the change from an earth-centered to a sun-centered conception of the solar system. 242pp. 5³/₈ x 8¹/₂. 0-486-41444-2

ARISTARCHUS OF SAMOS: The Ancient Copernicus, Sir Thomas Heath. Heath's history of astronomy ranges from Homer and Hesiod to Aristarchus and includes quotes from numerous thinkers, compilers, and scholasticists from Thales and Anaximander through Pythagoras, Plato, Aristotle, and Heraclides. 34 figures. 448pp. 5³/₈ x 8¹/₂.
0-486-43886-4

A COMPLETE MANUAL OF AMATEUR ASTRONOMY: TOOLS AND TECHNIQUES FOR ASTRONOMICAL OBSERVATIONS, P. Clay Sherrod with Thomas L. Koed. Concise, highly readable book discusses: selecting, setting up and maintaining a telescope; amateur studies of the sun; lunar topography and occultations; observations of Mars, Jupiter, Saturn, the minor planets and the stars; an introduction to photoelectric photometry; more. 1981 ed. 124 figures. 25 halftones. 37 tables. 335pp. 6¹/₂ x 9¹/₄. 0-486-42820-8

AMATEUR ASTRONOMER'S HANDBOOK, J. B. Sidgwick. Timeless, comprehensive coverage of telescopes, mirrors, lenses, mountings, telescope drives, micrometers, spectroscopes, more. 189 illustrations. 576pp. 5⁵/₈ x 8¹/₄. (Available in U.S. only.)
0-486-24034-7

STAR LORE: Myths, Legends, and Facts, William Tyler Olcott. Captivating retellings of the origins and histories of ancient star groups include Pegasus, Ursa Major, Pleiades, signs of the zodiac, and other constellations. "Classic."—Sky & Telescope. 58 illustrations. 544pp. 5³/₈ x 8¹/₂. 0-486-43581-4

Chemistry

THE SCEPTICAL CHYMIST: THE CLASSIC 1661 TEXT, Robert Boyle. Boyle defines the term "element," asserting that all natural phenomena can be explained by the motion and organization of primary particles. 1911 ed. viii+232pp. $5^3/_8$ x $8^1/_2$. 0-486-42825-7

RADIOACTIVE SUBSTANCES, Marie Curie. Here is the celebrated scientist's doctoral thesis, the prelude to her receipt of the 1903 Nobel Prize. Curie discusses establishing atomic character of radioactivity found in compounds of uranium and thorium; extraction from pitchblende of polonium and radium; isolation of pure radium chloride; determination of atomic weight of radium; plus electric, photographic, luminous, heat, color effects of radioactivity. ii+94pp. $5^3/_8$ x $8^1/_2$. 0-486-42550-9

CHEMICAL MAGIC, Leonard A. Ford. Second Edition, Revised by E. Winston Grundmeier. Over 100 unusual stunts demonstrating cold fire, dust explosions, much more. Text explains scientific principles and stresses safety precautions. 128pp. $5^3/_8$ x $8^1/_2$. 0-486-67628-5

MOLECULAR THEORY OF CAPILLARITY, J. S. Rowlinson and B. Widom. History of surface phenomena offers critical and detailed examination and assessment of modern theories, focusing on statistical mechanics and application of results in mean-field approximation to model systems. 1989 edition. 352pp. $5^3/_8$ x $8^1/_2$. 0-486-42544-4

CHEMICAL AND CATALYTIC REACTION ENGINEERING, James J. Carberry. Designed to offer background for managing chemical reactions, this text examines behavior of chemical reactions and reactors; fluid-fluid and fluid-solid reaction systems; heterogeneous catalysis and catalytic kinetics; more. 1976 edition. 672pp. $6^1/_8$ x $9^1/_4$. 0-486-41736-0 $31.95

ELEMENTS OF CHEMISTRY, Antoine Lavoisier. Monumental classic by founder of modern chemistry in remarkable reprint of rare 1790 Kerr translation. A must for every student of chemistry or the history of science. 539pp. $5^3/_8$ x $8^1/_2$. 0-486-64624-6

MOLECULES AND RADIATION: An Introduction to Modern Molecular Spectroscopy. Second Edition, Jeffrey I. Steinfeld. This unified treatment introduces upper-level undergraduates and graduate students to the concepts and the methods of molecular spectroscopy and applications to quantum electronics, lasers, and related optical phenomena. 1985 edition. 512pp. $5^3/_8$ x $8^1/_2$. 0-486-44152-0

A SHORT HISTORY OF CHEMISTRY, J. R. Partington. Classic exposition explores origins of chemistry, alchemy, early medical chemistry, nature of atmosphere, theory of valency, laws and structure of atomic theory, much more. 428pp. $5^3/_8$ x $8^1/_2$. (Available in U.S. only.) 0-486-65977-1

GENERAL CHEMISTRY, Linus Pauling. Revised 3rd edition of classic first-year text by Nobel laureate. Atomic and molecular structure, quantum mechanics, statistical mechanics, thermodynamics correlated with descriptive chemistry. Problems. 992pp. $5^3/_8$ x $8^1/_2$. 0-486-65622-5

ELECTRON CORRELATION IN MOLECULES, S. Wilson. This text addresses one of theoretical chemistry's central problems. Topics include molecular electronic structure, independent electron models, electron correlation, the linked diagram theorem, and related topics. 1984 edition. 304pp. $5^3/_8$ x $8^1/_2$. 0-486-45879-2

Engineering

DE RE METALLICA, Georgius Agricola. The famous Hoover translation of greatest treatise on technological chemistry, engineering, geology, mining of early modern times (1556). All 289 original woodcuts. 638pp. $6^3/_4$ x 11. 0-486-60006-8

FUNDAMENTALS OF ASTRODYNAMICS, Roger Bate et al. Modern approach developed by U.S. Air Force Academy. Designed as a first course. Problems, exercises. Numerous illustrations. 455pp. $5^3/_8$ x $8^1/_2$. 0-486-60061-0

DYNAMICS OF FLUIDS IN POROUS MEDIA, Jacob Bear. For advanced students of ground water hydrology, soil mechanics and physics, drainage and irrigation engineering and more. 335 illustrations. Exercises, with answers. 784pp. $6^1/_8$ x $9^1/_4$. 0-486-65675-6

THEORY OF VISCOELASTICITY (SECOND EDITION), Richard M. Christensen. Complete consistent description of the linear theory of the viscoelastic behavior of materials. Problem-solving techniques discussed. 1982 edition. 29 figures. xiv+364pp. $6^1/_8$ x $9^1/_4$. 0-486-42880-X

MECHANICS, J. P. Den Hartog. A classic introductory text or refresher. Hundreds of applications and design problems illuminate fundamentals of trusses, loaded beams and cables, etc. 334 answered problems. 462pp. $5^3/_8$ x $8^1/_2$. 0-486-60754-2

MECHANICAL VIBRATIONS, J. P. Den Hartog. Classic textbook offers lucid explanations and illustrative models, applying theories of vibrations to a variety of practical industrial engineering problems. Numerous figures. 233 problems, solutions. Appendix. Index. Preface. 436pp. $5^3/_8$ x $8^1/_2$. 0-486-64785-4

STRENGTH OF MATERIALS, J. P. Den Hartog. Full, clear treatment of basic material (tension, torsion, bending, etc.) plus advanced material on engineering methods, applications. 350 answered problems. 323pp. $5^3/_8$ x $8^1/_2$. 0-486-60755-0

A HISTORY OF MECHANICS, René Dugas. Monumental study of mechanical principles from antiquity to quantum mechanics. Contributions of ancient Greeks, Galileo, Leonardo, Kepler, Lagrange, many others. 671pp. $5^3/_8$ x $8^1/_2$. 0-486-65632-2

STABILITY THEORY AND ITS APPLICATIONS TO STRUCTURAL MECHANICS, Clive L. Dym. Self-contained text focuses on Koiter postbuckling analyses, with mathematical notions of stability of motion. Basing minimum energy principles for static stability upon dynamic concepts of stability of motion, it develops asymptotic buckling and postbuckling analyses from potential energy considerations, with applications to columns, plates, and arches. 1974 ed. 208pp. $5^3/_8$ x $8^1/_2$. 0-486-42541-X

BASIC ELECTRICITY, U.S. Bureau of Naval Personnel. Originally a training course; best nontechnical coverage. Topics include batteries, circuits, conductors, AC and DC, inductance and capacitance, generators, motors, transformers, amplifiers, etc. Many questions with answers. 349 illustrations. 1969 edition. 448pp. $6^1/_2$ x $9^1/_4$. 0-486-20973-3

ROCKETS, Robert Goddard. Two of the most significant publications in the history of rocketry and jet propulsion: "A Method of Reaching Extreme Altitudes" (1919) and "Liquid Propellant Rocket Development" (1936). 128pp. 5³/₈ x 8¹/₂. 0-486-42537-1

STATISTICAL MECHANICS: PRINCIPLES AND APPLICATIONS, Terrell L. Hill. Standard text covers fundamentals of statistical mechanics, applications to fluctuation theory, imperfect gases, distribution functions, more. 448pp. 5³/₈ x 8¹/₂. 0-486-65390-0

ENGINEERING AND TECHNOLOGY 1650–1750: ILLUSTRATIONS AND TEXTS FROM ORIGINAL SOURCES, Martin Jensen. Highly readable text with more than 200 contemporary drawings and detailed engravings of engineering projects dealing with surveying, leveling, materials, hand tools, lifting equipment, transport and erection, piling, bailing, water supply, hydraulic engineering, and more. Among the specific projects outlined-transporting a 50-ton stone to the Louvre, erecting an obelisk, building timber locks, and dredging canals. 207pp. 8³/₈ x 11¹/₄. 0-486-42232-1

THE VARIATIONAL PRINCIPLES OF MECHANICS, Cornelius Lanczos. Graduate level coverage of calculus of variations, equations of motion, relativistic mechanics, more. First inexpensive paperbound edition of classic treatise. Index. Bibliography. 418pp. 5³/₈ x 8¹/₂. 0-486-65067-7

PROTECTION OF ELECTRONIC CIRCUITS FROM OVERVOLTAGES, Ronald B. Standler. Five-part treatment presents practical rules and strategies for circuits designed to protect electronic systems from damage by transient overvoltages. 1989 ed. xxiv+434pp. 6¹/₈ x 9¹/₄. 0-486-42552-5

ROTARY WING AERODYNAMICS, W. Z. Stepniewski. Clear, concise text covers aerodynamic phenomena of the rotor and offers guidelines for helicopter performance evaluation. Originally prepared for NASA. 537 figures. 640pp. 6¹/₈ x 9¹/₄. 0-486-64647-5

INTRODUCTION TO SPACE DYNAMICS, William Tyrrell Thomson. Comprehensive, classic introduction to space-flight engineering for advanced undergraduate and graduate students. Includes vector algebra, kinematics, transformation of coordinates. Bibliography. Index. 352pp. 5³/₈ x 8¹/₂. 0-486-65113-4

HISTORY OF STRENGTH OF MATERIALS, Stephen P. Timoshenko. Excellent historical survey of the strength of materials with many references to the theories of elasticity and structure. 245 figures. 452pp. 5³/₈ x 8¹/₂. 0-486-61187-6

ANALYTICAL FRACTURE MECHANICS, David J. Unger. Self-contained text supplements standard fracture mechanics texts by focusing on analytical methods for determining crack-tip stress and strain fields. 336pp. 6¹/₈ x 9¹/₄. 0-486-41737-9

STATISTICAL MECHANICS OF ELASTICITY, J. H. Weiner. Advanced, self-contained treatment illustrates general principles and elastic behavior of solids. Part 1, based on classical mechanics, studies thermoelastic behavior of crystalline and polymeric solids. Part 2, based on quantum mechanics, focuses on interatomic force laws, behavior of solids, and thermally activated processes. For students of physics and chemistry and for polymer physicists. 1983 ed. 96 figures. 496pp. 5³/₈ x 8¹/₂. 0-486-42260-7

Mathematics

FUNCTIONAL ANALYSIS (Second Corrected Edition), George Bachman and Lawrence Narici. Excellent treatment of subject geared toward students with background in linear algebra, advanced calculus, physics and engineering. Text covers introduction to inner-product spaces, normed, metric spaces, and topological spaces; complete orthonormal sets, the Hahn-Banach Theorem and its consequences, and many other related subjects. 1966 ed. 544pp. $6^1/_8$ x $9^1/_4$. 0-486-40251-7

DIFFERENTIAL MANIFOLDS, Antoni A. Kosinski. Introductory text for advanced undergraduates and graduate students presents systematic study of the topological structure of smooth manifolds, starting with elements of theory and concluding with method of surgery. 1993 edition. 288pp. $5^3/_8$ x $8^1/_2$. 0-486-46244-7

VECTOR AND TENSOR ANALYSIS WITH APPLICATIONS, A. I. Borisenko and I. E. Tarapov. Concise introduction. Worked-out problems, solutions, exercises. 257pp. $5^3/_8$ x $8^1/_4$. 0-486-63833-2

AN INTRODUCTION TO ORDINARY DIFFERENTIAL EQUATIONS, Earl A. Coddington. A thorough and systematic first course in elementary differential equations for undergraduates in mathematics and science, with many exercises and problems (with answers). Index. 304pp. $5^3/_8$ x $8^1/_2$. 0-486-65942-9

FOURIER SERIES AND ORTHOGONAL FUNCTIONS, Harry F. Davis. An incisive text combining theory and practical example to introduce Fourier series, orthogonal functions and applications of the Fourier method to boundary-value problems. 570 exercises. Answers and notes. 416pp. $5^3/_8$ x $8^1/_2$. 0-486-65973-9

COMPUTABILITY AND UNSOLVABILITY, Martin Davis. Classic graduate-level introduction to theory of computability, usually referred to as theory of recurrent functions. New preface and appendix. 288pp. $5^3/_8$ x $8^1/_2$. 0-486-61471-9

AN INTRODUCTION TO MATHEMATICAL ANALYSIS, Robert A. Rankin. Dealing chiefly with functions of a single real variable, this text by a distinguished educator introduces limits, continuity, differentiability, integration, convergence of infinite series, double series, and infinite products. 1963 edition. 624pp. $5^3/_8$ x $8^1/_2$. 0-486-46251-X

METHODS OF NUMERICAL INTEGRATION (SECOND EDITION), Philip J. Davis and Philip Rabinowitz. Requiring only a background in calculus, this text covers approximate integration over finite and infinite intervals, error analysis, approximate integration in two or more dimensions, and automatic integration. 1984 edition. 624pp. $5^3/_8$ x $8^1/_2$. 0-486-45339-1

INTRODUCTION TO LINEAR ALGEBRA AND DIFFERENTIAL EQUATIONS, John W. Dettman. Excellent text covers complex numbers, determinants, orthonormal bases, Laplace transforms, much more. Exercises with solutions. Undergraduate level. 416pp. $5^3/_8$ x $8^1/_2$. 0-486-65191-6

RIEMANN'S ZETA FUNCTION, H. M. Edwards. Superb, high-level study of landmark 1859 publication entitled "On the Number of Primes Less Than a Given Magnitude" traces developments in mathematical theory that it inspired. xiv+315pp. $5^3/_8$ x $8^1/_2$. 0-486-41740-9

CALCULUS OF VARIATIONS WITH APPLICATIONS, George M. Ewing. Applications-oriented introduction to variational theory develops insight and promotes understanding of specialized books, research papers. Suitable for advanced undergraduate/graduate students as primary, supplementary text. 352pp. $5^3/_8$ x $8^1/_2$. 0-486-64856-7

MATHEMATICIAN'S DELIGHT, W. W. Sawyer. "Recommended with confidence" by *The Times Literary Supplement,* this lively survey was written by a renowned teacher. It starts with arithmetic and algebra, gradually proceeding to trigonometry and calculus. 1943 edition. 240pp. $5^3/_8$ x $8^1/_2$. 0-486-46240-4

ADVANCED EUCLIDEAN GEOMETRY, Roger A. Johnson. This classic text explores the geometry of the triangle and the circle, concentrating on extensions of Euclidean theory, and examining in detail many relatively recent theorems. 1929 edition. 336pp. $5^3/_8$ x $8^1/_2$. 0-486-46237-4

COUNTEREXAMPLES IN ANALYSIS, Bernard R. Gelbaum and John M. H. Olmsted. These counterexamples deal mostly with the part of analysis known as "real variables." The first half covers the real number system, and the second half encompasses higher dimensions. 1962 edition. xxiv+198pp. $5^3/_8$ x $8^1/_2$. 0-486-42875-3

CATASTROPHE THEORY FOR SCIENTISTS AND ENGINEERS, Robert Gilmore. Advanced-level treatment describes mathematics of theory grounded in the work of Poincaré, R. Thom, other mathematicians. Also important applications to problems in mathematics, physics, chemistry and engineering. 1981 edition. References. 28 tables. 397 black-and-white illustrations. xvii + 666pp. $6^1/_8$ x $9^1/_4$. 0-486-67539-4

COMPLEX VARIABLES: Second Edition, Robert B. Ash and W. P. Novinger. Suitable for advanced undergraduates and graduate students, this newly revised treatment covers Cauchy theorem and its applications, analytic functions, and the prime number theorem. Numerous problems and solutions. 2004 edition. 224pp. $6^1/_2$ x $9^1/_4$. 0-486-46250-1

NUMERICAL METHODS FOR SCIENTISTS AND ENGINEERS, Richard Hamming. Classic text stresses frequency approach in coverage of algorithms, polynomial approximation, Fourier approximation, exponential approximation, other topics. Revised and enlarged 2nd edition. 721pp. $5^3/_8$ x $8^1/_2$. 0-486-65241-6

INTRODUCTION TO NUMERICAL ANALYSIS (2nd Edition), F. B. Hildebrand. Classic, fundamental treatment covers computation, approximation, interpolation, numerical differentiation and integration, other topics. 150 new problems. 669pp. $5^3/_8$ x $8^1/_2$. 0-486-65363-3

MARKOV PROCESSES AND POTENTIAL THEORY, Robert M. Blumental and Ronald K. Getoor. This graduate-level text explores the relationship between Markov processes and potential theory in terms of excessive functions, multiplicative functionals and subprocesses, additive functionals and their potentials, and dual processes. 1968 edition. 320pp. $5^3/_8$ x $8^1/_2$. 0-486-46263-3

ABSTRACT SETS AND FINITE ORDINALS: An Introduction to the Study of Set Theory, G. B. Keene. This text unites logical and philosophical aspects of set theory in a manner intelligible to mathematicians without training in formal logic and to logicians without a mathematical background. 1961 edition. 112pp. $5^3/_8$ x $8^1/_2$. 0-486-46249-8

INTRODUCTORY REAL ANALYSIS, A.N. Kolmogorov, S. V. Fomin. Translated by Richard A. Silverman. Self-contained, evenly paced introduction to real and functional analysis. Some 350 problems. 403pp. $5^3/_8$ x $8^1/_2$. 0-486-61226-0

APPLIED ANALYSIS, Cornelius Lanczos. Classic work on analysis and design of finite processes for approximating solution of analytical problems. Algebraic equations, matrices, harmonic analysis, quadrature methods, much more. 559pp. $5^3/_8$ x $8^1/_2$. 0-486-65656-X

AN INTRODUCTION TO ALGEBRAIC STRUCTURES, Joseph Landin. Superb self-contained text covers "abstract algebra": sets and numbers, theory of groups, theory of rings, much more. Numerous well-chosen examples, exercises. 247pp. $5^3/_8$ x $8^1/_2$. 0-486-65940-2

QUALITATIVE THEORY OF DIFFERENTIAL EQUATIONS, V. V. Nemytskii and V.V. Stepanov. Classic graduate-level text by two prominent Soviet mathematicians covers classical differential equations as well as topological dynamics and ergodic theory. Bibliographies. 523pp. $5^3/_8$ x $8^1/_2$. 0-486-65954-2

THEORY OF MATRICES, Sam Perlis. Outstanding text covering rank, nonsingularity and inverses in connection with the development of canonical matrices under the relation of equivalence, and without the intervention of determinants. Includes exercises. 237pp. $5^3/_8$ x $8^1/_2$. 0-486-66810-X

INTRODUCTION TO ANALYSIS, Maxwell Rosenlicht. Unusually clear, accessible coverage of set theory, real number system, metric spaces, continuous functions, Riemann integration, multiple integrals, more. Wide range of problems. Undergraduate level. Bibliography. 254pp. $5^3/_8$ x $8^1/_2$. 0-486-65038-3

MODERN NONLINEAR EQUATIONS, Thomas L. Saaty. Emphasizes practical solution of problems; covers seven types of equations. ". . . a welcome contribution to the existing literature. . . ."—*Math Reviews*. 490pp. $5^3/_8$ x $8^1/_2$. 0-486-64232-1

MATRICES AND LINEAR ALGEBRA, Hans Schneider and George Phillip Barker. Basic textbook covers theory of matrices and its applications to systems of linear equations and related topics such as determinants, eigenvalues and differential equations. Numerous exercises. 432pp. $5^3/_8$ x $8^1/_2$. 0-486-66014-1

LINEAR ALGEBRA, Georgi E. Shilov. Determinants, linear spaces, matrix algebras, similar topics. For advanced undergraduates, graduates. Silverman translation. 387pp. $5^3/_8$ x $8^1/_2$. 0-486-63518-X

MATHEMATICAL METHODS OF GAME AND ECONOMIC THEORY: Revised Edition, Jean-Pierre Aubin. This text begins with optimization theory and convex analysis, followed by topics in game theory and mathematical economics, and concluding with an introduction to nonlinear analysis and control theory. 1982 edition. 656pp. $6^1/_8$ x $9^1/_4$. 0-486-46265-X

SET THEORY AND LOGIC, Robert R. Stoll. Lucid introduction to unified theory of mathematical concepts. Set theory and logic seen as tools for conceptual understanding of real number system. 496pp. $5^3/_8$ x $8^1/_4$. 0-486-63829-4

TENSOR CALCULUS, J.L. Synge and A. Schild. Widely used introductory text covers spaces and tensors, basic operations in Riemannian space, non-Riemannian spaces, etc. 324pp. $5^3/_8$ x $8^1/_4$. 0-486-63612-7

ORDINARY DIFFERENTIAL EQUATIONS, Morris Tenenbaum and Harry Pollard. Exhaustive survey of ordinary differential equations for undergraduates in mathematics, engineering, science. Thorough analysis of theorems. Diagrams. Bibliography. Index. 818pp. $5^3/_8$ x $8^1/_2$. 0-486-64940-7

INTEGRAL EQUATIONS, F. G. Tricomi. Authoritative, well-written treatment of extremely useful mathematical tool with wide applications. Volterra Equations, Fredholm Equations, much more. Advanced undergraduate to graduate level. Exercises. Bibliography. 238pp. $5^3/_8$ x $8^1/_2$. 0-486-64828-1

FOURIER SERIES, Georgi P. Tolstov. Translated by Richard A. Silverman. A valuable addition to the literature on the subject, moving clearly from subject to subject and theorem to theorem. 107 problems, answers. 336pp. $5^3/_8$ x $8^1/_2$. 0-486-63317-9

INTRODUCTION TO MATHEMATICAL THINKING, Friedrich Waismann. Examinations of arithmetic, geometry, and theory of integers; rational and natural numbers; complete induction; limit and point of accumulation; remarkable curves; complex and hypercomplex numbers, more. 1959 ed. 27 figures. xii+260pp. $5^3/_8$ x $8^1/_2$. 0-486-42804-8

THE RADON TRANSFORM AND SOME OF ITS APPLICATIONS, Stanley R. Deans. Of value to mathematicians, physicists, and engineers, this excellent introduction covers both theory and applications, including a rich array of examples and literature. Revised and updated by the author. 1993 edition. 304pp. $6^1/_8$ x $9^1/_4$. 0-486-46241-2

CALCULUS OF VARIATIONS, Robert Weinstock. Basic introduction covering isoperimetric problems, theory of elasticity, quantum mechanics, electrostatics, etc. Exercises throughout. 326pp. $5^3/_8$ x $8^1/_2$. 0-486-63069-2

THE CONTINUUM: A CRITICAL EXAMINATION OF THE FOUNDATION OF ANALYSIS, Hermann Weyl. Classic of 20th-century foundational research deals with the conceptual problem posed by the continuum. 156pp. $5^3/_8$ x $8^1/_2$. 0-486-67982-9

CHALLENGING MATHEMATICAL PROBLEMS WITH ELEMENTARY SOLUTIONS, A. M. Yaglom and I. M. Yaglom. Over 170 challenging problems on probability theory, combinatorial analysis, points and lines, topology, convex polygons, many other topics. Solutions. Total of 445pp. $5^3/_8$ x $8^1/_2$. Two-vol. set.
Vol. I: 0-486-65536-9 Vol. II: 0-486-65537-7

INTRODUCTION TO PARTIAL DIFFERENTIAL EQUATIONS WITH APPLICATIONS, E. C. Zachmanoglou and Dale W. Thoe. Essentials of partial differential equations applied to common problems in engineering and the physical sciences. Problems and answers. 416pp. $5^3/_8$ x $8^1/_2$. 0-486-65251-3

STOCHASTIC PROCESSES AND FILTERING THEORY, Andrew H. Jazwinski. This unified treatment presents material previously available only in journals, and in terms accessible to engineering students. Although theory is emphasized, it discusses numerous practical applications as well. 1970 edition. 400pp. $5^3/_8$ x $8^1/_2$. 0-486-46274-9

Math—Decision Theory, Statistics, Probability

INTRODUCTION TO PROBABILITY, John E. Freund. Featured topics include permutations and factorials, probabilities and odds, frequency interpretation, mathematical expectation, decision-making, postulates of probability, rule of elimination, much more. Exercises with some solutions. Summary. 1973 edition. 247pp. 5³/₈ x 8¹/₂.
0-486-67549-1

STATISTICAL AND INDUCTIVE PROBABILITIES, Hugues Leblanc. This treatment addresses a decades-old dispute among probability theorists, asserting that both statistical and inductive probabilities may be treated as sentence-theoretic measurements, and that the latter qualify as estimates of the former. 1962 edition. 160pp. 5³/₈ x 8¹/₂.
0-486-44980-7

APPLIED MULTIVARIATE ANALYSIS: Using Bayesian and Frequentist Methods of Inference, Second Edition, S. James Press. This two-part treatment deals with foundations as well as models and applications. Topics include continuous multivariate distributions; regression and analysis of variance; factor analysis and latent structure analysis; and structuring multivariate populations. 1982 edition. 692pp. 5³/₈ x 8¹/₂. 0-486-44236-5

LINEAR PROGRAMMING AND ECONOMIC ANALYSIS, Robert Dorfman, Paul A. Samuelson and Robert M. Solow. First comprehensive treatment of linear programming in standard economic analysis. Game theory, modern welfare economics, Leontief input-output, more. 525pp. 5³/₈ x 8¹/₂. 0-486-65491-5

PROBABILITY: AN INTRODUCTION, Samuel Goldberg. Excellent basic text covers set theory, probability theory for finite sample spaces, binomial theorem, much more. 360 problems. Bibliographies. 322pp. 5³/₈ x 8¹/₂. 0-486-65252-1

GAMES AND DECISIONS: INTRODUCTION AND CRITICAL SURVEY, R. Duncan Luce and Howard Raiffa. Superb nontechnical introduction to game theory, primarily applied to social sciences. Utility theory, zero-sum games, n-person games, decision-making, much more. Bibliography. 509pp. 5³/₈ x 8¹/₂. 0-486-65943-7

INTRODUCTION TO THE THEORY OF GAMES, J. C. C. McKinsey. This comprehensive overview of the mathematical theory of games illustrates applications to situations involving conflicts of interest, including economic, social, political, and military contexts. Appropriate for advanced undergraduate and graduate courses; advanced calculus a prerequisite. 1952 ed. x+372pp. 5³/₈ x 8¹/₂. 0-486-42811-7

FIFTY CHALLENGING PROBLEMS IN PROBABILITY WITH SOLUTIONS, Frederick Mosteller. Remarkable puzzlers, graded in difficulty, illustrate elementary and advanced aspects of probability. Detailed solutions. 88pp. 5³/₈ x 8¹/₂. 0-486-65355-2

PROBABILITY THEORY: A CONCISE COURSE, Y. A. Rozanov. Highly readable, self-contained introduction covers combination of events, dependent events, Bernoulli trials, etc. 148pp. 5³/₈ x 8¹/₄. 0-486-63544-9

THE STATISTICAL ANALYSIS OF EXPERIMENTAL DATA, John Mandel. First half of book presents fundamental mathematical definitions, concepts and facts while remaining half deals with statistics primarily as an interpretive tool. Well-written text, numerous worked examples with step-by-step presentation. Includes 116 tables. 448pp. 5³/₈ x 8¹/₂. 0-486-64666-1

Math—Geometry and Topology

ELEMENTARY CONCEPTS OF TOPOLOGY, Paul Alexandroff. Elegant, intuitive approach to topology from set-theoretic topology to Betti groups; how concepts of topology are useful in math and physics. 25 figures. 57pp. $5^3/_8$ x $8^1/_2$. 0-486-60747-X

A LONG WAY FROM EUCLID, Constance Reid. Lively guide by a prominent historian focuses on the role of Euclid's Elements in subsequent mathematical developments. Elementary algebra and plane geometry are sole prerequisites. 80 drawings. 1963 edition. 304pp. $5^3/_8$ x $8^1/_2$. 0-486-43613-6

EXPERIMENTS IN TOPOLOGY, Stephen Barr. Classic, lively explanation of one of the byways of mathematics. Klein bottles, Moebius strips, projective planes, map coloring, problem of the Koenigsberg bridges, much more, described with clarity and wit. 43 figures. 210pp. $5^3/_8$ x $8^1/_2$. 0-486-25933-1

THE GEOMETRY OF RENÉ DESCARTES, René Descartes. The great work founded analytical geometry. Original French text, Descartes's own diagrams, together with definitive Smith-Latham translation. 244pp. $5^3/_8$ x $8^1/_2$. 0-486-60068-8

EUCLIDEAN GEOMETRY AND TRANSFORMATIONS, Clayton W. Dodge. This introduction to Euclidean geometry emphasizes transformations, particularly isometries and similarities. Suitable for undergraduate courses, it includes numerous examples, many with detailed answers. 1972 ed. viii+296pp. $6^1/_8$ x $9^1/_4$. 0-486-43476-1

EXCURSIONS IN GEOMETRY, C. Stanley Ogilvy. A straightedge, compass, and a little thought are all that's needed to discover the intellectual excitement of geometry. Harmonic division and Apollonian circles, inversive geometry, hexlet, Golden Section, more. 132 illustrations. 192pp. $5^3/_8$ x $8^1/_2$. 0-486-26530-7

THE THIRTEEN BOOKS OF EUCLID'S ELEMENTS, translated with introduction and commentary by Sir Thomas L. Heath. Definitive edition. Textual and linguistic notes, mathematical analysis. 2,500 years of critical commentary. Unabridged. 1,414pp. $5^3/_8$ x $8^1/_2$. Three-vol. set.

Vol. I: 0-486-60088-2 Vol. II: 0-486-60089-0 Vol. III: 0-486-60090-4

SPACE AND GEOMETRY: IN THE LIGHT OF PHYSIOLOGICAL, PSYCHOLOGICAL AND PHYSICAL INQUIRY, Ernst Mach. Three essays by an eminent philosopher and scientist explore the nature, origin, and development of our concepts of space, with a distinctness and precision suitable for undergraduate students and other readers. 1906 ed. vi+148pp. $5^3/_8$ x $8^1/_2$. 0-486-43909-7

GEOMETRY OF COMPLEX NUMBERS, Hans Schwerdtfeger. Illuminating, widely praised book on analytic geometry of circles, the Moebius transformation, and two-dimensional non-Euclidean geometries. 200pp. $5^5/_8$ x $8^1/_4$. 0-486-63830-8

DIFFERENTIAL GEOMETRY, Heinrich W. Guggenheimer. Local differential geometry as an application of advanced calculus and linear algebra. Curvature, transformation groups, surfaces, more. Exercises. 62 figures. 378pp. $5^3/_8$ x $8^1/_2$. 0-486-63433-7

History of Math

THE WORKS OF ARCHIMEDES, Archimedes (T. L. Heath, ed.). Topics include the famous problems of the ratio of the areas of a cylinder and an inscribed sphere; the measurement of a circle; the properties of conoids, spheroids, and spirals; and the quadrature of the parabola. Informative introduction. clxxxvi+326pp. 5³/₈ x 8¹/₂. 0-486-42084-1

A SHORT ACCOUNT OF THE HISTORY OF MATHEMATICS, W. W. Rouse Ball. One of clearest, most authoritative surveys from the Egyptians and Phoenicians through 19th-century figures such as Grassman, Galois, Riemann. Fourth edition. 522pp. 5³/₈ x 8¹/₂. 0-486-20630-0

THE HISTORY OF THE CALCULUS AND ITS CONCEPTUAL DEVELOPMENT, Carl B. Boyer. Origins in antiquity, medieval contributions, work of Newton, Leibniz, rigorous formulation. Treatment is verbal. 346pp. 5³/₈ x 8¹/₂. 0-486-60509-4

THE HISTORICAL ROOTS OF ELEMENTARY MATHEMATICS, Lucas N. H. Bunt, Phillip S. Jones, and Jack D. Bedient. Fundamental underpinnings of modern arithmetic, algebra, geometry and number systems derived from ancient civilizations. 320pp. 5³/₈ x 8¹/₂. 0-486-25563-8

THE HISTORY OF THE CALCULUS AND ITS CONCEPTUAL DEVELOPMENT, Carl B. Boyer. Fluent description of the development of both the integral and differential calculus—its early beginnings in antiquity, medieval contributions, and a consideration of Newton and Leibniz. 368pp. 5³/₈ x 8¹/₂. 0-486-60509-4

GAMES, GODS & GAMBLING: A HISTORY OF PROBABILITY AND STATISTICAL IDEAS, F. N. David. Episodes from the lives of Galileo, Fermat, Pascal, and others illustrate this fascinating account of the roots of mathematics. Features thought-provoking references to classics, archaeology, biography, poetry. 1962 edition. 304pp. 5³/₈ x 8¹/₂. (Available in U.S. only.) 0-486-40023-9

OF MEN AND NUMBERS: THE STORY OF THE GREAT MATHEMATICIANS, Jane Muir. Fascinating accounts of the lives and accomplishments of history's greatest mathematical minds—Pythagoras, Descartes, Euler, Pascal, Cantor, many more. Anecdotal, illuminating. 30 diagrams. Bibliography. 256pp. 5³/₈ x 8¹/₂. 0-486-28973-7

HISTORY OF MATHEMATICS, David E. Smith. Nontechnical survey from ancient Greece and Orient to late 19th century; evolution of arithmetic, geometry, trigonometry, calculating devices, algebra, the calculus. 362 illustrations. 1,355pp. 5³/₈ x 8¹/₂. Two-vol. set. Vol. I: 0-486-20429-4 Vol. II: 0-486-20430-8

A CONCISE HISTORY OF MATHEMATICS, Dirk J. Struik. The best brief history of mathematics. Stresses origins and covers every major figure from ancient Near East to 19th century. 41 illustrations. 195pp. 5³/₈ x 8¹/₂. 0-486-60255-9

Physics

OPTICAL RESONANCE AND TWO-LEVEL ATOMS, L. Allen and J. H. Eberly. Clear, comprehensive introduction to basic principles behind all quantum optical resonance phenomena. 53 illustrations. Preface. Index. 256pp. 5³/₈ x 8¹/₂. 0-486-65533-4

QUANTUM THEORY, David Bohm. This advanced undergraduate-level text presents the quantum theory in terms of qualitative and imaginative concepts, followed by specific applications worked out in mathematical detail. Preface. Index. 655pp. 5³/₈ x 8¹/₂. 0-486-65969-0

ATOMIC PHYSICS (8th EDITION), Max Born. Nobel laureate's lucid treatment of kinetic theory of gases, elementary particles, nuclear atom, wave-corpuscles, atomic structure and spectral lines, much more. Over 40 appendices, bibliography. 495pp. 5³/₈ x 8¹/₂. 0-486-65984-4

A SOPHISTICATE'S PRIMER OF RELATIVITY, P. W. Bridgman. Geared toward readers already acquainted with special relativity, this book transcends the view of theory as a working tool to answer natural questions: What is a frame of reference? What is a "law of nature"? What is the role of the "observer"? Extensive treatment, written in terms accessible to those without a scientific background. 1983 ed. xlviii+172pp. 5³/₈ x 8¹/₂. 0-486-42549-5

AN INTRODUCTION TO HAMILTONIAN OPTICS, H. A. Buchdahl. Detailed account of the Hamiltonian treatment of aberration theory in geometrical optics. Many classes of optical systems defined in terms of the symmetries they possess. Problems with detailed solutions. 1970 edition. xv + 360pp. 5³/₈ x 8¹/₂. 0-486-67597-1

PRIMER OF QUANTUM MECHANICS, Marvin Chester. Introductory text examines the classical quantum bead on a track: its state and representations; operator eigenvalues; harmonic oscillator and bound bead in a symmetric force field; and bead in a spherical shell. Other topics include spin, matrices, and the structure of quantum mechanics; the simplest atom; indistinguishable particles; and stationary-state perturbation theory. 1992 ed. xiv+314pp. 6¹/₈ x 9¹/₄. 0-486-42878-8

LECTURES ON QUANTUM MECHANICS, Paul A. M. Dirac. Four concise, brilliant lectures on mathematical methods in quantum mechanics from Nobel Prize-winning quantum pioneer build on idea of visualizing quantum theory through the use of classical mechanics. 96pp. 5³/₈ x 8¹/₂. 0-486-41713-1

THIRTY YEARS THAT SHOOK PHYSICS: THE STORY OF QUANTUM THEORY, George Gamow. Lucid, accessible introduction to influential theory of energy and matter. Careful explanations of Dirac's anti-particles, Bohr's model of the atom, much more. 12 plates. Numerous drawings. 240pp. 5³/₈ x 8¹/₂. 0-486-24895-X

ELECTRONIC STRUCTURE AND THE PROPERTIES OF SOLIDS: THE PHYSICS OF THE CHEMICAL BOND, Walter A. Harrison. Innovative text offers basic understanding of the electronic structure of covalent and ionic solids, simple metals, transition metals and their compounds. Problems. 1980 edition. 582pp. 6¹/₈ x 9¹/₄. 0-486-66021-4

HYDRODYNAMIC AND HYDROMAGNETIC STABILITY, S. Chandrasekhar. Lucid examination of the Rayleigh-Benard problem; clear coverage of the theory of instabilities causing convection. 704pp. 5³/₈ x 8¹/₄. 0-486-64071-X

INVESTIGATIONS ON THE THEORY OF THE BROWNIAN MOVEMENT, Albert Einstein. Five papers (1905–8) investigating dynamics of Brownian motion and evolving elementary theory. Notes by R. Fürth. 122pp. 5³/₈ x 8¹/₂. 0-486-60304-0

THE PHYSICS OF WAVES, William C. Elmore and Mark A. Heald. Unique overview of classical wave theory. Acoustics, optics, electromagnetic radiation, more. Ideal as classroom text or for self-study. Problems. 477pp. 5³/₈ x 8¹/₂. 0-486-64926-1

GRAVITY, George Gamow. Distinguished physicist and teacher takes reader-friendly look at three scientists whose work unlocked many of the mysteries behind the laws of physics: Galileo, Newton, and Einstein. Most of the book focuses on Newton's ideas, with a concluding chapter on post-Einsteinian speculations concerning the relationship between gravity and other physical phenomena. 160pp. 5³/₈ x 8¹/₂. 0-486-42563-0

PHYSICAL PRINCIPLES OF THE QUANTUM THEORY, Werner Heisenberg. Nobel Laureate discusses quantum theory, uncertainty, wave mechanics, work of Dirac, Schroedinger, Compton, Wilson, Einstein, etc. 184pp. 5³/₈ x 8¹/₂. 0-486-60113-7

ATOMIC SPECTRA AND ATOMIC STRUCTURE, Gerhard Herzberg. One of best introductions; especially for specialist in other fields. Treatment is physical rather than mathematical. 80 illustrations. 257pp. 5³/₈ x 8¹/₂. 0-486-60115-3

AN INTRODUCTION TO STATISTICAL THERMODYNAMICS, Terrell L. Hill. Excellent basic text offers wide-ranging coverage of quantum statistical mechanics, systems of interacting molecules, quantum statistics, more. 523pp. 5³/₈ x 8¹/₂. 0-486-65242-4

THEORETICAL PHYSICS, Georg Joos, with Ira M. Freeman. Classic overview covers essential math, mechanics, electromagnetic theory, thermodynamics, quantum mechanics, nuclear physics, other topics. First paperback edition. xxiii + 885pp. 5³/₈ x 8¹/₂. 0-486-65227-0

PROBLEMS AND SOLUTIONS IN QUANTUM CHEMISTRY AND PHYSICS, Charles S. Johnson, Jr. and Lee G. Pedersen. Unusually varied problems, detailed solutions in coverage of quantum mechanics, wave mechanics, angular momentum, molecular spectroscopy, more. 280 problems plus 139 supplementary exercises. 430pp. 6¹/₂ x 9¹/₄. 0-486-65236-X

THEORETICAL SOLID STATE PHYSICS, Vol. 1: Perfect Lattices in Equilibrium; Vol. II: Non-Equilibrium and Disorder, William Jones and Norman H. March. Monumental reference work covers fundamental theory of equilibrium properties of perfect crystalline solids, non-equilibrium properties, defects and disordered systems. Appendices. Problems. Preface. Diagrams. Index. Bibliography. Total of 1,301pp. 5³/₈ x 8¹/₂. Two volumes. Vol. I: 0-486-65015-4 Vol. II: 0-486-65016-2

WHAT IS RELATIVITY? L. D. Landau and G. B. Rumer. Written by a Nobel Prize physicist and his distinguished colleague, this compelling book explains the special theory of relativity to readers with no scientific background, using such familiar objects as trains, rulers, and clocks. 1960 ed. vi+72pp. 5³/₈ x 8¹/₂. 0-486-42806-0